Handbook of Distance Learning for Real-Time and Asynchronous Information Technology Education

Solomon Negash
Kennesaw State University, USA

Michael E. Whitman
Kennesaw State University, USA

Amy B. Woszczynski
Kennesaw State University, USA

Ken Hoganson
Kennesaw State University, USA

Herbert Mattord
Kennesaw State University, USA

Information Science REFERENCE

INFORMATION SCIENCE REFERENCE

Hershey · New York

Acquisitions Editor:	Kristin Klinger
Development Editor:	Kristin Roth
Senior Managing Editor:	Jennifer Neidig
Managing Editor:	Jamie Snavely
Assistant Managing Editor:	Carole Coulson
Copy Editor:	Ashlee Kunkel
Typesetter:	Michael Brehm
Cover Design:	Lisa Tosheff
Printed at:	Yurchak Printing Inc.

Published in the United States of America by
Information Science Reference (an imprint of IGI Global)
701 E. Chocolate Avenue, Suite 200
Hershey PA 17033
Tel: 717-533-8845
Fax: 717-533-8661
E-mail: cust@igi-global.com
Web site: http://www.igi-global.com

and in the United Kingdom by
Information Science Reference (an imprint of IGI Global)
3 Henrietta Street
Covent Garden
London WC2E 8LU
Tel: 44 20 7240 0856
Fax: 44 20 7379 0609
Web site: http://www.eurospanbookstore.com

Library of Congress Cataloging-in-Publication Data

Handbook of distance learning for real-time and asynchronous information technology education / Solomon Negash ... [et al.], editors.
p. cm.
Includes bibliographical references and index.
Summary: "This book looks at solutions that provide the best fits of distance learning technologies for the teacher and learner presented by sharing teacher experiences in information technology education"--Provided by publisher.
ISBN 978-1-59904-964-9 (hardcover : alk. paper) -- ISBN 978-1-59904-965-6 (ebook : alk. paper)
1. Distance education--Computer-assisted instruction. 2. Information technology. I. Negash, Solomon, 1960-
LC5803.C65H36 2008
371.3'58--dc22

2008007838

British Cataloguing in Publication Data
A Cataloguing in Publication record for this book is available from the British Library.

Table of Contents

Section II
Effectiveness and Motivation

Section III
Interaction and Collaboration

Detailed Table of Contents

This chapter identifies six e-learning classifications to understand the different forms of e-learning and demonstrates the differences and similarities of the classifications with classroom examples, including a pilot empirical study. It argues that understanding the different e-learning classifications is a prerequisite to understanding the effectiveness of specific e-learning formats. In order to understand effectiveness, or lack thereof of an e-learning environment, more precise terminology which describes the format of delivery is needed. To address this issue, this chapter provides six e-learning classifications.

This chapter focuses on the designing and development of blended learning environment for adult education, and especially the education of teachers. The author argues that the best combination of advanced learning technologies of synchronous and asynchronous learning is conducive to the formation of new learning environments, which, under certain pedagogical conditions, will adequately meet the special needs of adult students. Particular emphasis is given to the designing and development of a pedagogical blended learning model, based on the principles of transformation adult theory and constructivism. A case study of a blended environment of teachers' training is presented.

The development of learning communities has become an acknowledged goal of educators at all levels. As education continues to move into online environments, virtual learning communities develop for several reasons, including social networking, small group task completions, and authentic discussions for topics of mutual professional interest. The sense of presence and copresence with others is also found to be significant in developing Internet-based learning communities. This chapter illustrates the experiences with current learning communities that form in a 3D immersive world designed for education.

The continued and increasing use of online training raises the question of whether the most effective training methods applied in live instruction will carry over to different online environments in the long run. Behavior modeling (BM) approach—teaching through demonstration—has been proven as the most effective approach in a face-to-face (F2F) environment. This chapter compares F2F, online synchronous, and online asynchronous classes in a quasi-experiment using the BM approach. The results were compared to see which produced the best performance, as measured by knowledge near-transfer and knowledge far-transfer effectiveness. Overall satisfaction with training was also measured.

This chapter proposes a framework that links student performance and satisfaction to the learning environment and course delivery and empirically evaluates the framework. The results show that a well-designed distance education course can lead to a high level of student satisfaction, but classroom-based students

chapter the nature and strength of such relationship is of more interest. This work discusses ramifications of scale-related economies for public policy. The chapter will also recommends how to achieve minimum efficient scale (MES) size so that scale-related economies are achieved.

Foreword

As the world during the late 1980s and early 1990 stood poised on the brink of the Information Age, speculation ran rampant about the impact that the new and emerging information and communication technologies would have on business, on government, on social relationships, on defense policy, and yes, on education as well.[1] Optimists argued that because of the new and emerging information and communication technologies, humankind was on the verge of entering a new golden age in which constraints imposed by time, distance, and location would be overcome and fall by the wayside. Conversely, pessimists asserted that at best, the world would continue on as before, and that at worst, new and emerging information technologies would help the rich become richer and make the poor poorer, would make bad information indistinguishable from good information, and spawn new generations of humans so dependent on the new technologies that they could accomplish little on their own.[2]

We are now some two decades into the Information Age, and reality has proven more complex than either the optimists or the pessimists predicted.

This is nowhere more true than in higher education, where optimistic early assumptions that new information and communication technologies would make classrooms irrelevant, drive the cost of higher education down, and enable faculty to teach greater numbers of students more effectively proved unfounded, and where pessimistic earlier assumptions that higher education would continue on as in earlier eras proved wrong.

Rather, the Information Age has brought a much more complex higher education environment. Traditional classrooms remain but are increasingly becoming "bricks and clicks" wired classrooms. Many campuses are now partially or fully enclosed in wireless clouds that enable students to access the Internet from within the cloud. And hundreds of thousands, even millions, of students never set foot within a classroom. Some faculty have extensively incorporated the new technologies into their teaching and learned new teaching methodologies. Others have utilized the new technologies and methodologies more cautiously. Still others remain wedded to traditional ways of teaching.

As for students, distance learning technologies based on the new and emerging information technologies have proven to be a godsend to many. For other students, the new and emerging technologies are a helpful addition to traditional ways of learning. And in still other instances, Information Age technologies have been irrelevant or even detrimental to the educational process.

The purpose of this book and the authors who have contributed to it is to present a broad sampling of the efforts that college and university faculty members have initiated to take advantage of the capabilities that Information Age technologies provide to higher education, to assess what has worked and what has not worked, and to better fit the needs of students and faculty to the educational process. For anyone interested in how the Information Age has impacted higher education, this book is valuable reading.

Daniel S. Papp, PhD
President, Kennesaw State University

REFERENCES

Alberts, D. S., & Papp, D. S. (Eds.). (1997). *Information age anthology: Volume 1.* Washington, D.C.: National Defense University.

ENDNOTES

1 Many technologies led to the rise of the Information Age, but eight stand out. They are: (1) advanced semiconductors, (2) advanced computers, (3) fiber optics, (4) cellular technology, (5) satellite technology, (6) advanced networking, (7) improved human-computer interaction, and (8) digital transmission and digital compression.

2 For discussions of the impact of the new and emerging information and communication technologies on a broad array of human activities, refer to Alberts and Papp (1997).

Daniel S. Papp *is president of Kennesaw State University. Prior to being named president by the Board of Regents, Papp served as senior vice chancellor for academics and fiscal affairs of the university system of Georgia. He has directed educational programs for Yamacraw, Georgia's initiative to become the global leader in broadband technologies and components. Papp has also served as interim president of Southern Polytechnic State University and executive assistant to the president at Georgia Tech. His academic specialties include international security policy, U.S. and Russian foreign and defense policies, and international system change. He is the author or editor of 10 books on these topics, including the biography of former U.S. Secretary of State Dean Rusk. He also has published more than 60 journal articles and chapters in edited books.*

Foreword

Distance learning means different things to different people. For some, distance learning is in sharp contrast to the traditional face-to-face classroom, integrating little more than interactive video between geographically separated campuses of training locations. To others, distance learning is an entirely new medium for instruction; it is a new instructional strategy distinct from the typical "bricks and mortar" classroom setting where students and professors interact over Internet-delivered video and audio conferencing, share collaborative projects among students, or participate in synchronous or asynchronous instruction opportunities.

Regardless of your individual bent toward this newest instructional delivery vehicle, distance learning has matured as a viable, effective, and efficient training medium for a number of reasons. The geometric rise in the amount and quality of information available to individuals continues to explode. The global community has evolved to the point where rapid change is the rule, not the exception. Professional and educational training opportunities have broadened opportunities for advancement even for those located in remote or dispersed locations. In any environment where people need improved access to information, need to share resources, or where learners, teachers, administrators, and subject matter specialists must travel to remote locations in order to communicate with one another, distance learning is preordained for consideration.

Whether its implementation is a success or a failure (and, in either case, what makes for that distinction) is the fodder for researchers and investigators like Solomon Negash and his team of editors and contributing authors, many of whom I have had the pleasure of involving in other projects related to teaching and learning with technology. Several of the contributors have provided their expertise in publications of my own, such as the *International Journal of Information Communication and Technology Education* (*IJICTE*) and *Online and Distance Learning* reference source.

The *Handbook of Distance Learning for Real-Time and Asynchronous Information Technology Education* offers a rich resource that combines the pedagogical foundations for teaching online with practical considerations that promote successful learning. Of particular note is the dual classification format used in the text to create an atmosphere focusing on the importance of the individual while simultaneously suggesting ways to overcome learning barriers via collaboration. Synchronous and asynchronous tools are the crux of effective online learning, yet few publications infuse pedagogy and best practice into a common core of tools for effective implementation of technology for teaching at a distance. This text does exactly that and, as such, has assured itself a place in the ready-reference library of online educators.

Too, the *Handbook* addresses critical areas of research and practice related to adult learners, collaborative technologies, teaching and learning, and best practice. The editorial team has discovered contributors steeped in investigation and implementation who make their stories a must-read for educational technologists and distance educators alike. Divided into learning environments, effectiveness and motivation, collaboration and interaction, teaching in the classroom, and adoption and economic analysis, the text provides a broad brush scrutiny of 17 of the most up-to-the-minute topics in this rapidly changing medium.

The *Handbook of Distance Learning for Real-Time and Asynchronous Information Technology Education* is destined to take its rightful place with other similar contributions to the advancement of online and distance education.

Lawrence A. Tomei,
Robert Morris University

Lawrence A. Tomei *is the associate vice president for academic affairs and associate professor of education, Robert Morris University. He earned a BSBA from the University of Akron (1972) and entered the U.S. Air Force, serving until his retirement as a Lieutenant Colonel in 1994. Dr. Tomei completed his MPA and MEd at the University of Oklahoma (1975, 1978) and EdD from USC (1983). His articles and books on instructional technology include Online and Distance Learning (2008), Integrating ICT Into the Classroom (2007), Taxonomy for the Technology Domain (2005), Challenges of Teaching with Technology Across the Curriculum (2003), Technology Facade (2002), Teaching Digitally: Integrating Technology Into the Classroom (2001), Professional Portfolios for Teachers (1999), and Technology Literacy Applications in Learning Environments (Chapter 1, Defining Instructional Technology Literacy) (2004).*

Preface

OVERVIEW

Distance learning (DL) has been defined in many ways, for this book we adopted the following: Distance learning results from a technological separation of teacher and learner which frees the necessity of traveling to a fixed place in order to be trained (Keegan, 1995; Valentine, 2002). This definition includes asynchronous learning with no fixed time and place and synchronous learning with fixed time but not fixed place.

Distance learning delivery mechanisms have progressed from correspondence in the 1850s (Morabito, 1997; Valentine, 2002), to telecourse in the 1950s and 1960s (Freed, 1999a), to open universities in the 1970s (Nasseh, 1997), to online distance learning in the 1980s (Morabito, 1997), and to Internet-based distance learning in the 1990s (Morabito, 1997). Along with this progress, online DL technologies and the associated cost have transformed from answering machines that recorded students' messages for telecourse instructors in the 1970s, where it cost $900 per answering machine (Freed, 1999b), to Internet-based applications that were unthinkable three decades ago (Alavi, Marakasand, & Yoo, 2002; Dagada & Jakovljevic, 2004; DeNeui & Dodge, 2006).

While DL and the associated technologies progressed, a chasm between teacher and learner seem to grow between the "digital natives" of today's learners and their teachers who are considered as "digital immigrants" (VanSlyke, 2003; Hsu, 2007; Prensky, 2001; Ferris & Wilder, 2006). This book shares experiences of teachers and how they incorporated DL technologies in the classroom.

THE CHALLENGE

Teachers have incorporated DL technologies in varying forms; some are shown in this book. While many success stories exist, there are several studies that present shortcoming of DL education. Piccoli, Ahmad, and Ives (2001) found that DL learners are less satisfied when the subject mater is unfamiliar (complex), like databases; dropout rates for online courses were found to be higher than courses offered in traditional classrooms (Levy, 2005; Simpson, 2004; Terry, 2001).

The challenge for the teacher is to identify what works and what does not.

THE SOLUTION: CONTRIBUTION OF THIS BOOK

Finding a solution that best fits the needs of the teacher and learner requires sustained research that uncovers the effectiveness of DL technologies in the learning experience (Alavi & Leidner, 2001; Hodges,

2005). This book contributes towards this solution by sharing teachers' experiences in information technology (IT) education.

In IT, unlike many other fields, the need to support the unique perspective of technologically advanced students and deliver technology-rich content presents unique challenges. In the early days of distance learning, a video taped lecture may have sufficed for the bulk of the content delivery. Today's IT students need the ability to interact with their instructor in near-real time, interact with their peers and project team members, and access and manipulate technology tools in the pursuit of their educational objectives. In other fields, like the humanities and liberal arts, the vast majority of the content is delivered by the instructor and textbook, supported by outside materials. In the IT fields (specifically including information systems and computer science), virtually all of the curriculum include the need to explore IT in the content, requiring the instructor and student to have integrated interaction with the technology.

Fundamental pedagogical changes are taking place as faculty begins to experiment with the use of technologies to support the delivery of curriculum to learners unable to participate in traditional classroom instruction. The vast majority of faculty members begin with a clean slate, experimenting using available technologies, without the benefit of the lessons learned from other faculty members who have faced the same challenges. The purpose of this book is to disseminate the challenges, successes, and failures of colleagues in their search for innovative and effective distance learning education.

ORGANIZATION OF THE BOOK

The book is organized into five sections with 18 chapters: Section I: Learning Environments consists of the first four chapters; Section II: Effectiveness and Motivation consists of Chapters V through VII; Section III: Interaction and Collaboration consists of Chapters VIII through XI; Section IV: Course Design and Classroom Teaching consists of Chapters XII through XV; and Section V: Economic Analysis and Adoption Consists of Chapters XVI thorough XVIII. A brief description of each of the chapters follows.

Chapter I proposes six DL classifications and demonstrates the differences and similarities of the classifications with classroom examples, including a pilot empirical study from the author's experience. It argues that understanding the different e-learning classifications is a prerequisite to understanding the effectiveness of specific e-learning formats. How does the reader distinguish e-learning success and/or failure if the format used is not understood? For example, a learning format with a Web site link to download lecture notes is different from one that uses interactive communication between learner and instructor and the later is different from one that uses "live" audio and video. In order to understand effectiveness, or lack thereof of an e-learning environment, more precise terminology which describes the format of delivery is needed. E-learning classifications can aid researchers in identifying learning effectiveness for specific formats and how it alters student learning experience.

Chapter II focuses on the design and development of blended learning environments for adult education, and especially the education of teachers. The author argues that the best combination of advanced learning technologies of synchronous and asynchronous learning is conducive to the formation of new learning environments. The chapter also presents a blended environment case study of teachers' training.

Chapter III illustrates the findings and experiences of various communities of learners formed within a 3D immersive Internet-based virtual world developed for graduate education. This award winning 3D learning community describes how students and instructors collaborate across time and distance. Students, faculty, and guests, graphically represented by avatars, move through the 3D world spaces interacting with each other and with artifacts within the worlds. These artifacts may be linked to different resources, Web pages, and tools necessary to provide content and support for various kinds of synchronous and asynchronous interactions. The authors show how small and large group shared workspace tools enable interactive conversations in text chats, threaded discussion boards, audio chats, group sharing of documents, and Web pages.

Chapter IV presents a quasi experiment to compare behavior modeling (teaching through demonstration), proven as the most effective training method for live instruction, in three environments: face-to-face, online synchronous, and online asynchronous. Overall satisfaction and performance as measured by knowledge near-transfer and knowledge far-transfer effectiveness is evaluated. The authors conclude by stating that when conducting software training, it may be almost as effective to use online training (synchronous or asynchronous) as it is to use a more costly face-to-face training in the long term. In the short term the face-to-face knowledge transfer model still seems to be the most effective approach to improve knowledge transfer in the short term.

Chapter V proposes a framework that links student performance and satisfaction to the learning environment and course delivery. The study empirically evaluates the proposed framework using the traditional classroom setting and distance education setting. The authors conclude that a well-designed distance education course can lead to a high level of student satisfaction, but classroom-based students can achieve even higher satisfaction if they also are given access to learning material on the Internet.

Chapter VI introduces how to differentiate instruction in an online environment. The study reviews the literature on differentiation and its connection and impact to online learning and discusses the principles that guide differentiated instruction. The authors posit that the "one size fits all" approach is not realistic for either face-to-face or online setting and provide online learning environment strategies that respond to the diverse needs of learners.

Chapter VII explores student motivation to engage in origination and distant site in an IP-based teleconferencing. The study posits that understanding student motivation for participating in IP teleconferencing as part of a class lecture will inform teachers on how to incorporate it in the curriculum. The authors examine three studies on student motivation to understand the benefits of teleconference-based DL.

Chapter VIII presents six requirements for next generation groupware systems to improve team cooperation and awareness in DL settings. The requirements are grouping, communication and discussion, specialization, collaboration by sharing tasks and resources, coordination of actions, and conflict resolution. The authors use two case studies to illustrate how the five requirements can be realized; they elaborate on how an ideal collaborative education tool can be used to construct a shared mental model among students in a team to improve their effectiveness.

Chapter IX reports survey findings on the impact of chat on facilitating participation in collaborative group learning processes and enhancing understanding of course content from a sociocultural constructivist perspective. The study used a qualitative case study of a distant course exemplifying the innovative instructional application of online synchronous (chat) interaction in virtual tutorials. The results reveal factors that affected both student perception and use of participation opportunities in chat tutorials, and understanding of course content. The authors conclude by recommending that the design of learning environments should encompass physical and virtual instructional contexts to avoid reliance on any one mode which could needlessly limit the range of interactions permitted in distance educational programs.

Chapter X investigates the factors that encourage student interaction and collaboration in both process- and product-oriented computer mediated communication tasks in a Web-based course that adopts interactive learning tasks as its core learning activities. The authors analyzed a postcourse survey questionnaire from three online classes and posit that some of the important factors that influence participation and contribute to sustained online interaction and collaboration are the structure of the online discussion, group size, group cohesion, strictly enforced deadlines, direct link of interactive learning activities to the assessment, and the differences in process- and product-driven interactive learning tasks.

Chapter XI proposes a four step model of greeting, message, reminder, and conclusion (GMRC) to gain a closer relationship between teachers and students in a DL environment. The authors posit that when using the GMRC approach, teachers can relate their concerns with each DL learner's specific questions and needs. The authors provide examples to support their proposed model.

Chapter XII presents a framework for developing Web courses, demonstrates the design and application of an online course, and discusses the experimental results for the selected course. The study compares speed of loading, file size, security, and flexibility of different development tools based on analytical discussions and experimental results; a sample course implementation that integrates the proposed principles and selected tools is presented. The authors conclude by presenting design rules of thumb for online Web courses.

Chapter XIII provides the lessons learned from teaching information security in a DL setting. The case study identified successful DL techniques and technologies for teaching information security. The authors found that lecture recording and virtual private network (VPN) technologies were relevant for teaching online information security courses. The later, VPN technology, was used to support hands-on laboratory exercises virtually.

Chapter XIV examines the challenges and opportunities of teaching computer programming in management information systems (MIS) curriculum in general and teaching computer programming instructions for MIS curriculum in particular. The study describes a hybrid computer programming course for MIS curriculum that embraces an assignment-centric design, self-paced assignment delivery, low involvement multimedia tracing instructional objectives, and online synchronous and asynchronous communication. The authors employed survey methodology to evaluate the course and observed two opportunities that impact MIS research and practice: the integration of ICT for instructional purposes, and the development, use, and validation of instruments designed to monitor our courses.

Chapter XV provides a primer on establishing relationships with high schools to deliver college-level IT curriculum in an asynchronous learning environment. The study describes the curriculum, provides details of the asynchronous online learning environment used in the program, and discusses the challenges and key lessons learned. The authors posit that the college environment, in which professors have local autonomy over curriculum delivery and instruction, differs from a public high school environment where curriculum has rigid standards that must be achieved, along with guidelines on methods of delivery. The authors state that forming a politically savvy team aware of how to navigate the high school environment is a must for ensuring success.

Chapter XVI presents an in-depth study of the factors influencing asynchronous distance learning courses purchase decision. The study identifies motivators and inhibitors of distance course adoption among consumers, focusing on the impact of relations with the medium, service considerations, and perceived purchase risk. The empirical study results show that perceived course utility, lack of mistrust in the organizing institution (service considerations), and satisfaction with the use of Internet when doing this type of training (relations with the medium) determine the asynchronous distance learning course purchase intention. The authors conclude by providing a set of recommendations to positively influence the purchase decision of asynchronous DL courses.

Chapter XVII analyzes e-learning from an industry perspective by evaluating the use of ICT technologies for university teaching. A scenario framework developed for the study of ICT impact on knowledge industries is applied to an e-learning case study. The study outlines a scenario framework for analyzing ICT impact on knowledge services, discusses different types of e-learning from the authors' experiences, and provides an analysis of the market for e-learning. The authors posit that the most important lesson from the experiences is that although a substantial part of the learning can be done by use of ICT, it is essential for students to meet occasionally; once personal contact among students and fellow teachers is established, interactive learning by use of online communication can be performed much more efficiently.

Chapter XVIII evaluates the relationship between the size of student enrollment in distance learning education and unit operational costs. Per conventional wisdom, the authors posit that the larger the size of the DL educational facility in terms of student enrollments, the lower the unit capital and unit operating costs; empirical evidence in the correlation between enrollments and average total costs is unmistakable, if not significant. The study looks at the nature and strength of these relationships. The authors conclude by suggesting minimum efficient scale (MES) to achieve economies of scale.

CONCLUSION

This book shares lessons learned from hands-on experience in teaching in synchronous and asynchronous DL. The book discusses DL issues ranging from learning environments to course design and technologies used in the classroom. The first section, learning environment, identifies different formats, presents the design of blended learning environment, and discusses the experience of 3D learning communities and a longitudinal experiment comparing face-to-face, synchronous, and asynchronous learning environments.

The second section, effectiveness and motivation, presents a framework for designing an effective DL course, shares lessons learned on how to differentiate DL courses to meet learners needs, and discusses student motivation to participate in teleconferencing. The third section, interaction and collaboration, presents suggestions on how to improve team collaborations in DL courses, a discussion on lessons learned from virtual tutorial moderated by synchronous chat, and recommendations on factors that promote online discussion and collaborations. The last section, economic analysis and adoption, presents the motivation for purchase decisions of DL courses, discusses the impact of DL technologies on knowledge industries, and compares the nature and strength of relationship between DL enrollment and operational costs.

REFERENCES

Alavi, M., & Leidner, D. E. (2001). Research commentary: Technology mediated learning-a call for greater depth and breadth of research. *Information Systems Research, 12*(1), 1-10.

Alavi, M., Marakasand, G. M., & Yoo, Y. (2002). A comparative study of distributed learning environments on learning outcomes. *Information Systems Research, 13*(4), 404-415.

Dagada, R., & Jakovljevic, M. (2004). Where have all the trainers gone? E-learning strategies and tools in the corporate training environment. In *Proceedings of the 2004 Annual Research Conference of the South African Institute of Computer Scientists and Information Technologists on IT Research in Developing Countries* (pp. 194-203). Stellenbosch, Western Cape, South Africa.

DeNeui, D. L., & Dodge, T. L. (2006). Asynchronous learning networks and student outcomes: The utility of online learning components in hybrid courses. *Journal of Instructional Psychology, 33*(4), 256-259.

Freed, K. (1999a). *A history of distance learning: The rise of the telecourse, part 1 of 3.* Retrieved July 22, 2007, from http://www.media-visions.com/ed-distlrn1.html

Freed, K. (1999b). *A history of distance learning: The rise of the telecourse, part 3 of 3*. Retrieved July 22, 2007, from http://www.media-visions.com/ed-distlrn1.html

Hodges, C. B. (2005). Self-regulation in Web-based courses: A review and the need for research. *The Quarterly Review of Distance Education*, *6*(4), 375-383.

Hsu, J. (2007). Innovative technologies for education and learning: Education and knowledge-oriented applications of blogs, wikis, podcasts, and more. *International Journal of Information and Communication Technology Education, 3*(3), 70-89.

Keegan, D. (1995). *Distance education technology for the new millennium: Compressed video teaching* (Eric Document Reproduction Service No. ED 389 931). ZIFF Papiere. Hagen, Germany: Institute for Research into Distance Education..

Levy, Y. (2005). Comparing dropout and persistence in e-learning courses. *Computers & Education, 48*(2), 185-204.

Morabito, M. G. (1997). Online distance education: Historical perspective and practical application. *Dissertation.com*. ISBN: 1 58112 057 5. Retrieved July 22, 2007, from http://www.bookpump.com/dps/pdf-b/1120575b.pdf

Nasseh, B. (1997). *A brief history of distance learning*. Retrieved July 22, 2007, from http://www.seniornet.org/edu/art/history.html

Piccoli, G., Ahmad, R., & Ives, B. (2001). Web-based virtual learning environments: A research framework and a preliminary assessment of effectiveness in basic IT skills training. *MIS Quarterly, 25*(4), 401-426.

Prensky, M. (2001a). Digital natives, digital immigrants. *On the Horizon, 9*(5), 1-6. Retrieved July 22, 2007, from http://www.marcprensky.com/writing/Prensky%20-%20Digital%20Natives,%20Digital%20Immigrants%20-%20Part1.pdf

Prensky, M. (2001b). Digital natives, digital immigrants, part II: Do they really think differently? *9*(6), 1-6. Retrieved July 22, 2007, from http://www.marcprensky.com/writing/Prensky%20-%20Digital%20Natives,%20Digital%20Immigrants%20-%20Part2.pdf

Simpson, O. (2004). The impact on retention of interventions to support distance learning students. *Open Learning, 19*(1), 79-95.

Terry, N. (2001). Assessing enrollment and attrition rates for the online MBA. *THE Journal, 28*(7), 64-68.

Valentine, D. (2002). Distance learning: Promises, problems, and possibilities. *Online Journal of Distance Learning Administration, 5*(3). Retrieved July 22, 2007, from http://www.westga.edu/~distance/ojdla/fall53/valentine53.html

VanSlyke, T. (2003). Digital natives, digital immigrants: Some thoughts from the generation gap. *The technology resource archives, University of North Carolina*. Retrieved July 22, 2007, from http://technologysource.org/article/digital_natives_digital_immigrants/

Solomon Negash
Kennesaw State University

Section I
Learning Environments

Chapter I
E-Learning Classifications: Differences and Similarities

Solomon Negash
Kennesaw State University, USA

Marlene V. Wilcox
Bradley University, USA

ABSTRACT

This chapter identifies six e-learning classifications to understand the different forms of e-learning and demonstrates the differences and similarities of the classifications with classroom examples, including a pilot empirical study from the authors' experience. It argues that understanding the different e-learning classifications is a prerequisite to understanding the effectiveness of specific e-learning formats. How does the reader distinguish e-learning success and/or failure if the format used is not understood? For example, a learning format with a Web site link to download lecture notes is different from one that uses interactive communication between learner and instructor and the latter is different from one that uses "live" audio and video. In order to understand effectiveness, or lack thereof of an e-learning environment, more precise terminology which describes the format of delivery is needed. To address this issue, this chapter provides the following six e-learning classifications: e-learning with physical presence and without e-communication (face-to-face), e-learning without presence and without e-communication (self-learning), e-learning without presence and with e-communication (asynchronous), e-learning with virtual presence and with e-communication (synchronous), e-learning with occasional presence and with e-communication (blended/hybrid-asynchronous), and e-learning with presence and with e-communication (blended/hybrid-synchronous). E-learning classifications can aid researchers in identifying learning effectiveness for specific formats and how it alters the student learning experience.

INTRODUCTION

Technology is transforming the delivery of education in unthinkable ways (DeNeui & Dodge, 2006). The impact and influence of technology can be seen rippling through academe and industry as more and more institutions of higher education and corporations offer, or plan to offer, Web-based courses (Alavi, Marakasand, & Yoo, 2002; Dagada & Jakovljevic, 2004).

There is a call for studies that enable researchers to gain a deeper understanding into the effectiveness of the use of technologies for e-learning (Alavi & Leidner, 2001; Alavi et al., 2002). Such studies need to be qualified by differentiating among e-learning formats.

Brown and Liedholm (2002) compared the outcomes of three different formats for a course in the principles of microeconomics (face-to-face, hybrid, and virtual) and found that the students in the virtual course did not perform as well as the students in the face-to-face classroom settings and that differences between students in the face-to-face and hybrid sections vs. those in the virtual section were shown to increase with the complexity of the subject matter. Piccoli, Ahmadand, and Ives (2001) found that the level of student satisfaction in e-learning environments for difficult (or unfamiliar) topics like Microsoft Access dropped when compared to familiar topics like Microsoft Word and Microsoft Excel. Brown and Liedholm (2002) found that students in virtual classes performed worse on exams than those in face-to-face classes where the exam questions required more complex applications of basic concepts. Brown and Liedholm (2002) conclude that ultimately there is some form of penalty for selecting a course that is completely online. These studies, while important, do not distinguish among the different e-learning formats used to conduct the courses; they are based on the premise that the e-learning formats are the same.

Studies on success and failure of e-learning presuppose that all online learning deliveries are the same, but there are differences. Those who cite the failure of e-learning formats often cite lack of support for students, lack of instructor availability, lack of content richness, and lack of performance assessment. Of course, it all depends on the course content being offered; but it also depends on the course delivery format. For example, an online class where the learner is provided only a Web site link to download the lecture notes is different from one where the learner has interactive communication with the instructor. The latter is also different from an e-learning class that provides the learner with "live" audio and video vs. one that does not.

In order to understand the effectiveness, or lack thereof, of an e-learning environment, more precise terminology which describes the format of delivery is needed, since all online instruction delivery formats are not equal; different content require different delivery formats. Technology advances have provided many tools for e-learning but without a clear understanding of the format of delivery it is difficult to assess the overall effectiveness of the environment. The question arises as to what classification can be used to understand the different e-learning formats. To help address this issue, this chapter provides an e-learning classification and demonstrates with a classroom example from the authors' experience.

There are seven sections in this chapter. First, we identify six classifications and describe them briefly. We then describe learning management systems (LMS) and give some examples. In the third section, we discuss e-learning environments and six dimensions that distinguish e-learning environments from face-to-face classrooms. The fourth section provides an example of each classification, followed by a pilot empirical study and a framework for e-learning environment effectiveness in section five. Sections six and seven provide a discussion and the conclusion.

E-LEARNING CLASSIFICATIONS

Falch (2004) proposes four types of e-learning classifications: e-learning without presence and without communication, e-learning without presence but with communication, e-learning combined with occasional presence, and e-learning used as a tool in classroom teaching.

Following Falch's (2004) presence/communication classification, we have redefined the terms "presence" and "communication" and expanded the classifications to six in order to make a distinction between physical presence and virtual presence. The six classifications are outlined in Table 1.

In order to understand the differences between classifications it is important to differentiate between content delivery and content access. In this classification we consider presence available as "Yes" only if the instructor and learner are simultaneously available during content delivery, either physically or virtually. We classify e-communication available as "Yes" only if e-communication exists between instructor and learner at the time of instruction delivery or e-communication is the primary communication medium for completing the course.

Brief descriptions of the six e-learning classifications are provided in this section; more details and examples are given in later sections. The descriptions are as follows:

Type I: E-Learning with Physical Presence and Without E-Communication (Face-to-Face)

This is the traditional face-to-face classroom setting. The traditional face-to-face classroom is classified as e-learning because of the prevalence of e-learning tools used to support instruction delivery in classrooms today. In this format both the instructor and learner are physically present in the classroom at the time of content delivery, therefore presence is available. An example of Type I e-learning is a traditional class that utilizes PowerPoint slides, video clips, and multimedia to deliver content. Many face-to-face classrooms also take advantage of e-learning technologies outside the classroom, for example, when there is interaction between the learner and instructor and among learners using discussion boards and also e-mail. In addition, lecture notes and PowerPoint slides may be posted online for students to access and assignment schedules may be set up online. It

Table 1. E-Learning classifications

Classification	Presence*	eCommunication**	Alias
Type I	Yes	No	Face-to-Face
Type II	No	No	Self-Learning
Type III	No	Yes	Asynchronous
Type IV	Yes	Yes	Synchronous
Type V	Occasional	Yes	Blended/Hybrid-asynchronous
Type VI	Yes	Yes	Blended/Hybrid-synchronous

** Presence is defined as real-time presence where both instructor and learner are present at the time of content delivery; it includes physical and virtual presence*
*** E-communication refers to whether the content delivery includes electronic communication or not.*

should be noted that in a traditional face-to-face classroom, e-learning tools do not have to be used for instruction; however, it is common today for many e-learning tools to be used for content delivery. The primary communication between learner and instructor takes place in the classroom or is handled through office visits or phone calls; e-communication is therefore classified as "No," or not available.

Type II: E-Learning without Presence and Without E-Communication (Self-Learning)

This type of e-learning is a self-learning approach. Learners receive the content media and learn on their own. There is no presence, neither physical nor virtual in this format. There is also no communication, e-communication, or otherwise between the learner and the instructor. With this e-learning format, the learner typically receives prerecorded content or accesses archived recordings. Communication between the learner and instructor (or the group that distributes the content) is limited to support or to other noncontent issues like replacing damaged media or receiving supplemental material. Type II e-learning is content delivered on a specific subject or application using recorded media like a CD ROM or DVD.

Type III: E-Learning without Presence and with E-Communication (Asynchronous)

In this format the instructor and learner do not meet during content delivery and there is no presence, neither physical nor virtual; presence is therefore classified as "No" or not available. With this format, the instructor prerecords the content (content delivery) and the learner accesses content (content access) at a later time (i.e., content delivery and content access happen independently so there is a time delay between content delivery and access). In this environment, the instructor and learner communicate frequently using a number of e-learning technologies. A Type III e-learning format is the typical format most people think of when they think about "online learning." Even though the instructor and learner do not meet at the time of content delivery, there is, however, rich interaction using e-learning technologies like threaded discussion boards and e-mail and instructors may post lecture notes for online access and schedule assignments online. E-communication is not available at the time of content delivery, however, e-communication is the primary mode of communication for the asynchronous format; e-communication is therefore categorized as "Yes," or available.

Type IV: E-Learning with Virtual Presence and with E-Communication (Synchronous)

This is synchronous e-learning, also referred to as "real-time." In synchronous e-learning the instructor and learner do not meet physically, however, they always meet virtually during content delivery, therefore, presence is classified as available, or "Yes." In this format e-communication is used extensively and the virtual class is mediated by e-learning technologies; e-communication is therefore classified as available, or "Yes." The technologies used in a Type IV e-learning environment include all of the technologies used in asynchronous e-learning in addition to synchronous technologies such as "live" audio, "live" video, chat, and instant messaging.

Type V: E-Learning with Occasional Presence and With E-Communication (Blended/Hybrid-Asynchronous)

This is a blended or hybrid e-learning format with occasional presence. In this format content is delivered through occasional physical meetings (face-to-face classroom, possibly once a month) between the instructor and learner and via e-learning technologies for the remainder of the time. This arrangement is a combination of face-to-face and asynchronous e-learning. In this format e-communication is used extensively just like the asynchronous format; therefore e-communication is classified as available, or "Yes." Presence, on the other hand, is occasional; there is physical presence during the face-to-face portion and no physical or virtual presence during the asynchronous portion, therefore presence is categorized as "occasional."

Type VI: E-Learning with Presence and with E-Communication (Blended/Hybrid-Synchronous)

This is a blended or hybrid e-learning format with presence at all times. In this format e-communication is used extensively just like with a synchronous format; e-communication is therefore classified as available, or "Yes." In this environment, presence alternates between physical and virtual. Some class sessions are conducted with physical presence (i.e., in a traditional face-to-face classroom setting) and the remaining class sessions are conducted with virtual presence (i.e., synchronously). With this format the learner and instructor meet at the same time, sometimes physically and other times virtually; nevertheless, presence exists at all times. In this format, presence is therefore classified as "Yes," or available. An example of Type VI e-learning is where the instructor and learner use the classroom for part of the time and for the other part they use live audio/video for their virtual meetings. In both cases, meetings take place with both participants available at the same time, which is a combination of face-to-face and synchronous e-learning.

LEARNING MANAGEMENT SYSTEM (LMS)

Learning management systems (LMS) facilitate the planning, management, and delivery of content for e-learning; it is therefore important to mention them here briefly. LMSs can maintain a list of student enrollment in a course, manage course access with logins, lecture files and lecture notes, support quizzes and assessments, schedule assignments, support e-mail communication, manage discussion forums, facilitate project teams, and support chat. These systems support many-to-many communication among learners and between learners and instructors.

A search for "learning management system" on Wikipedia (http://wikipedia.org) results in a listing of 35 commercial and 12 open source LMS products. See Table 2 for a partial listing.

Some LMSs include technologies for creating content, such as assignments and quizzes, and provide support for instant messaging, "live" audio, "live" video, and white boards. These types of LMSs can host asynchronous e-learning and some are even capable of hosting synchronous e-learning.

E-Learning System: An Example

There are many e-learning systems capable of supporting all six e-learning classifications. Cogburn and Hurup (2006) conducted a lab performance test at Syracuse University to compare nine types of Web conferencing software capable of supporting postsecondary teaching. A summary of their study, listed alphabetically by product, is provided in Table 3. We encourage the reader to look at their study for further details.

In order to help illustrate the six e-learning classifications, we describe our experience with one of the nine e-learning systems, Marratech[1] (http://www.marratech.com), along with one LMS system, WebCT-Vista[2] (http://webct.com). While we have experience with other e-learning

Table 2. Sample learning management systems*

Product	URL	Availability
ANGEL Learning	http://angellearning.com/	Commercial
Apex Learning	http://www.apexlearning.com/	Commercial
Blackboard	http://www.blackboard.com/us/index.Bb	Commercial
Bodington	http://bodington.org/	Open source
Claroline	http://www.claroline.net/	Open source
Desire2Learn	http://www.desire2learn.com/	Commercial
eCollege	http://www.ecollege.com/indexflash.learn	Commercial
iCohere	http://www.icohere.com	Commercial
.LRN	http://dotlrn.org/users/	Open source
Moodle	http://moodle.org/	Open source
OLAT	http://www.olat.org/public/index.html	Open source
Open Campus	http://campus.dokeos.com/index.php	Open source
Reliant	http://reliantlive.com/index.htm	Commercial
Sakai	http://www.sakaiproject.org/	Open source
SimplyDigi	http://www.simplydigi.com/Welcome.aspx	Commercial
Scholar360	http://www.scholar360.com/	Commercial
WebCT	http://www.webct.com/	Commercial

**Selected based on their Web site's indication of higher education solutions for clients*

Table 3. Synchronous e-learning systems

Product	Report Card*	Installation**	Cross Platform***
Adobe Breeze	B+	In-house	Yes
Elluminate Live	A-	In-house	Yes
e/pop Web Conferencing	B-	In-house	No
Genesys Meeting Center	C+	Hosted	No
Marratech	B	In-house	Yes
Microsoft Office Live Meeting	C+	Hosted	No
Raindance Meeting Edition	C+	Hosted	No
Saba Centra Live	B+	In-house	No
WebEx Meeting Center	A-	Hosted	No

** Overall grade assigned by the reviewers*
*** Installation indicates whether the application was installed at the lab or hosted by the vendor*
**** Cross platform is defined as running on all three major operating systems: Windows, Macintosh, and Linux*

systems including Elluminate Live, Horizon Wimba, eCollege, e/pop, and Blackboard, our experience with Marratech includes nine semester courses conducted over a 1 year period. We have also used WebCT-Vista since its debut in 2006, and WebCT for several years prior to that. In this section we used a combination of Marratech and WebCT-Vista to illustrate our experience in the six e-learning classifications.

Type I: E-Learning with Physical Presence and without E-Communication (Face-to-Face)

A traditional classroom supported by WebCT-Vista. We have taught many traditional face-to-face classes augmented by WebCT-Vista's LMS. We posted lecture notes (PowerPoint slides) and assignments on WebCT-Vista and enforced assignment due dates through WebCT-Vista. Discussion board and e-mail communication between students and instructor and among students was facilitated using WebCT-Vista. Student access to the course Web site (hosted within WebCT-Vista) was managed through a login in WebCT-Vista. The student roster was populated by the registrar and only students who registered for the course had access to the course content. As instructors we added teaching assistants and guest speakers as needed. During the course instruction, we were physically present in the classroom and although our primary communication took place in the classroom, e-communications were used to augment the course.

Type II: E-Learning without Presence and without E-Communication (Self-Learning)

For a data warehousing and business intelligence class we posted a prerecording of a SQL server installation for our students; students downloaded the archived instructions and learned the application on their own. We also provided instruction on downloading, installing, and using the Marratech system. Students once again learned the process on their own. In both instances, with the exception of a couple of students, the students learned the content on their own without presence, of the instructor, that is, with "No" e-communication. Other examples occurred where the learner purchases instructional CD to learn different application software independently.

Type III: E-Learning without Presence and with E-Communication (Asynchronous)

Prerecorded Marratech sessions with WebCT-Vista support. While some of our colleagues used this format for an entire semester, our experience is limited to a few sessions. We recorded lectures in advance with full video and audio. The recorded sessions were placed within WebCT-Vista where students were able to download and access the instruction material at their own pace. All WebCT-Vista features described in Type I above were applied here. We found the asynchronous approach very convenient during instructor absence (i.e., during travel to conferences or emergencies). We did not meet with the students during the asynchronous sessions but we had extensive e-communication through WebCT-Vista.

Type IV: E-Learning with Virtual Presence and with E-Communication (Synchronous)

"Live" Marratech sessions supplemented with WebCT-Vista. We conducted several classes in this format. One course was conducted entirely with a synchronous format without any physical contact with students. In a typical session as instructors, we entered a virtual room, uploaded the PowerPoint slides, and turned on audio and video. In the virtual room, we appeared as a talking-head, in a 20 inch x 18 inch (50 cm x 45 cm) window. A thumbnail with a picture and username was also shown in the display window. In this setting, we also had synchronous chat with our students; the system time stamped the messages and included the sender username. All WebCT-Vista features described in Type I above were applied here. We used the whiteboard area to display PowerPoint slides and to present the lecture to students who were present via audio/video connection from their home. Students who had full-duplex audio were able to ask questions or make comments at any time. Students were given "presenter" privileges when they lead discussions or presented a project. The "live" audio/video link allowed us to be virtually present at all times. We also used e-communication during content delivery and content access.

Type V: E-Learning with Occasional Presence and with E-Communication (Blended/Hybrid-Asynchronous)

Face-to-face classroom combined with prerecorded Marratech sessions supplemented by WebCT-Vista. When conference travels or emergencies arose, we prerecorded the class lecture using the Marratech system and uploaded the recorded session to WebCT-Vista. We have also used this option when we wanted to target the face-to-face classroom for discussion and collaborations; in these cases we posted the prerecorded content in advance. Students were able to learn the material at their own pace and come to class for the discussion and collaboration. All WebCT-Vista features described in Type I above were applied here. We met with students during the face-to-face sessions but not during the asynchronous sessions; presence was therefore occasional. We used WebCT-Vista for communication with students and to enable students to interact with each other. E-communication in these instances was therefore "Yes."

Type VI: E-Learning with Presence and with E-Communication (Blended/Hybrid-Synchronous)

We combined physical presence (face-to-face) and virtual synchronous presence (Marratech) along with e-communication support from WebCT-Vista. Some of our classes were scheduled with the options to attend classes online. The face-to-face sessions were always in progress in these classes but students were given the option to attend 50% of the classes online. In these class sessions, when students joined the online session they joined the "live" class in progress with the instructor and those students who had chosen to attend in the face-to-face format. The majority of the students who did not utilize the online option and instead attended all class session in the face-to-face format indicated that they did not make use of the online option because they were already on campus, had scheduled classes back-to-back, and did not have time to go home to participate in the online class. Students who choose to take advantage of the online option had the opportunity to ask questions and participate in the class discussion during the "live" session. Unlike in the asynchronous mode, the synchronous hybrid/blended mode had participants' presence inside and outside of the classroom during instruction. The WebCT-Vista features described in Type I above were applied here and e-communication was supported by WebCT-Vista.

A summary of the examples of e-learning systems is outlined in Table 4.

The Marratech interface used in the courses discussed in the examples is depicted in Figure 1.

The Marratech user interface shows a large whiteboard on the left; this is where we displayed the PowerPoint slides. On the right hand side there are three stacked panes with a talking head, a list of participants, and a chat window.

Table 4. Summary of e-learning systems

Classification	Presence	E-communication
Type I: face-to-face	Physical	post lecture notes schedule assignments discussion and e-mail outside classroom
Type II: self-learner	None	None
Type III: asynchronous	None	includes all listed for Type I audio/video lecture recordings
Type IV: synchronous	Virtual	includes all listed for Type III "live" audio "live" video synchronous chat
Type V: blended/hybrid-asynchronous	Physical	includes all listed for Type III
Type VI: blended/hybrid-synchronous	Physical and Virtual	includes all listed for Type IV

Figure 1. Marratech user interface

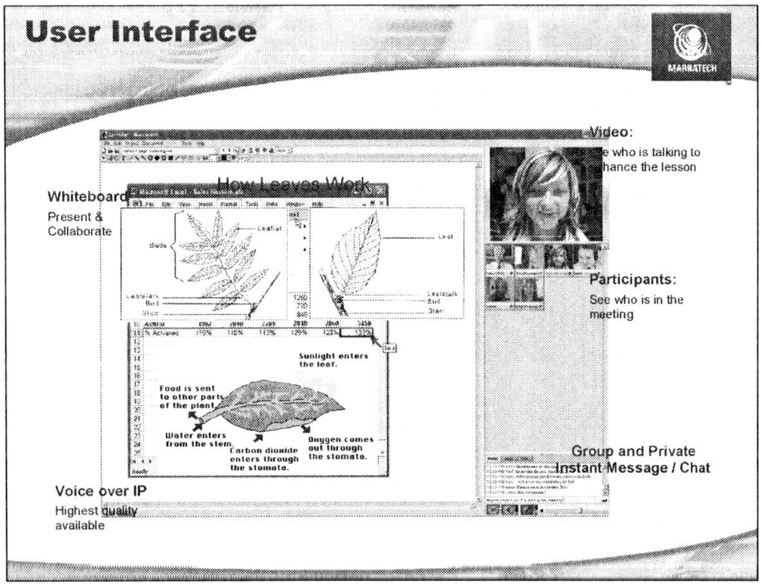

E-LEARNING ENVIRONMENTS

E-learning is the general term used for computer-enhanced learning. It differs from distance learning because in e-learning, a computer is a prerequisite. Distance learning, however, may use computers but is not required. Advances in information technology (IT) continually expand the capabilities of e-learning (Seng & Al-Hawamdeh, 2001). Cogburn and Hurup (2006) identified 15 must-haves for Web conferencing. VoIP, video, participant roles, interactive capabilities for participants, diverse session content options, live application sharing, recording and archiving capabilities, break-out rooms, bandwidth management, accessibility, security, integration, session management, customization and support, cross-platform functionality, and compliance with the

Table 5. E-learning technologies and features

Accessibility for disabled	E-mail	Screen casts
Application sharing	Educational animation	Security
Archiving	Electronic voting	Session management
Audio	Games	Simulations
Bandwidth management	Hypermedia	Text chat
Blogs	Instant messaging	VoIP
Break-out rooms	Interactive participants	Video
Computer aided assessment	Learning management systems	Webinars
Content access options	MP3 players	White board
Cross-platform functionality	Palm pilots	Wikis
Customization and support	Assigned Participant roles	
Discussion board	Podcasts	

Americans with Disabilities Act. We have used a number of these features in our classrooms and they have enhanced the students learning experience. Table 5 provides a partial listing of technologies that can be employed in e-learning.

Content delivery in e-learning utilizes many of these technologies. The extent to which these technologies are used varies from instructor to instructor as well as from learner to learner.

Piccoli et al. (2001) use the term virtual learning environments (VLEs) to describe e-learning environments and they defined them as "computer-based environments that are relatively open systems which allow interactions and encounters with other participants and providing access to a wide range of resources" (Piccoli et al., 2001, p. 402; Wilson, 1996).

E-learning environments can be characterized by six dimensions which distinguish them from traditional classrooms and computer aided instruction. These dimensions are time, place, space, technology, interaction, and control (Piccoli et al., 2001). We adopted the basic definitions from Piccoli et al. (2001) and expanded them to differentiate between synchronous and asynchronous communication. The six dimensions are further discussed below:

Time is defined as "the timing of instruction" (Piccoli et al., 2001, p. 404). In an asynchronous e-learning environment the learner decides the timing of instruction access. "When instruction is delivered asynchronously in [an e-Learning environment], participants retain control over when they engage in the learning experience. Learners determine the time and pace of instruction" (Piccoli et al., 2001, p. 404), the time constraints for learners in asynchronous e-learning environments are therefore removed (Piccoli et al., 2001). In synchronous e-learning environments two time modalities exist: time of instruction delivery and time of accessing archived sessions. At the time of instruction delivery the learner has to be present, albeit virtually. In a synchronous format learners do not have control over when they can engage in the learning experience and time constraints for the learner are the same as in a face-to-face delivery, where learners have to meet with the instructor and other learners at a specified class time. When accessing archived sessions, the learner decides when to access instruction; in this case the time constraint is removed. This is similar to an asynchronous e-learning environment.

Time flexibility and learner control are found to be benefits of e-learning environments (Piccoli et al., 2001), however, synchronous e-learning environments fix the delivery time, eliminating this advantage. In asynchronous e-learning environments, the learner has a greater degree of control during the time of instruction access. Learner control in synchronous e-learning environments, however, takes on a different form. In synchronous e-learning environments, the responsibility for learner control is retained by the instructor and the burden of time management is removed from the learner. In synchronous e-learning environments the familiar face-to-face classroom environment is maintained.

Place is defined as "the physical location of instruction" (Piccoli et al., 2001, p. 404). In an asynchronous e-learning environment there is no formal class meeting and learners can access instruction from "anywhere" (e.g., home or work). In synchronous e-learning environments learners can also access instruction from "anywhere." However, because synchronous e-learning environments have a formal class meeting, learners must coordinate their time with the scheduled class session.

Space is defined as "the collection of material and resources available to the learner" (Piccoli et al., 2001, p. 404). "While it is possible to expand the traditional model of classroom-based instruction to include the variety of resources available in [e-Learning environments], generally these materials remain only a secondary resource in instructor-led classroom education" (Piccoli et al., 2001, p. 404). In asynchronous e-learning environments timing for instruction access is independent of instruction delivery; therefore the learner controls the pace of learning. Because learners control the pace of learning they can access a wide array of resources as often as desired. The same is true when accessing archived sessions for synchronous environments. In a synchronous classroom, however, because learners have to be present at the time of content delivery the array of resources available to the learner is limited by the instructor's presence. Instructor control of content in the synchronous mode is managed by the instructor despite the fact that the student is in a different location. In the Marratech e-learning system, described earlier, as the instructor changed to a new page the learner was redirected to the same page as the instructor.

Technology is defined as "the collection of tools used to deliver learning material and to facilitate many-to-many communication among participants" (Piccoli et al., 2001, p. 404). "In [asynchronous e-Learning environment] technology is used to deliver learning material and to facilitate many-to-many communication among distributed participants" (Piccoli et al., 2001, p. 404). Many technologies including text, hypertext, graphics, streaming audio, streaming video, computer animation and simulation, embedded tests, dynamic content, e-mail, and online threaded discussion boards are used in asynchronous e-learning environments. Synchronous e-learning environments use live audio, live video, synchronous chat, and desktop videoconferencing in addition to the technologies used in asynchronous e-learning environments.

Interaction is defined as "the degree of contact and educational content exchange among learners and between learners and instructors" (Piccoli et al., 2001, p. 404). "[Asynchronous e-Learning environments] rely on information and communication technologies to create the venue for knowledge transfer and to monitor the progress of learning. [E-Learning environments] are open systems that allow for communication and interaction among participants" (Piccoli et al., 2001, p. 404). In an asynchronous format, interaction with the instructor and among learners can take place at the time of content access; however, content delivery is a one way communication from instructor to learner. In synchronous e-learning environments, on the other hand, learners can interact with the instructor and among learners at the time of instruction delivery. Interaction in synchronous e-learning environments for access to instruction material (archived sessions) is the same as in asynchronous classrooms. Synchronous e-learning environments such as Marratech provide private interaction between learner and instructor and among learners during content delivery.

Control is defined as "the extent to which the learner can control the instructional presentation" (Piccoli et al., 2001, p. 404). "A certain degree of learner control can be built into traditional classroom instruction, but [asynchronous e-Learning environments] have the potential to provide far greater personalization of instruction and a much higher degree of learner control than traditional classroom education. Traditional learning environments do allow students, when outside of the classroom, to control the pace and sequence of material, and the time and place of their study. Asynchronous e-Learning environments], however, provide this flexibility during instruction as well." In an asynchronous e-learning environment a learner can control the pace and sequence of content access (Piccoli et al., 2001), however, asynchronous learners do not have control over the delivery of content. Archived sessions of synchronous classrooms provide the same level of control as asynchronous environments. Learner control in a synchronous e-learning environment is limited during instruction delivery since it is controlled by the instructor. For example, when using Marratech, learners are able to move around the instruction material presented to them during an online class session at a pace and sequence they chose, but they are redirected to the instructor-led page each time the instructor changes the page. In an archived session however, participants have control over the pace and sequence just like in the asynchronous classrooms.

PILOT STUDY-TYPE VI: HYBRID/BLENDED SYNCHRONOUS E-LEARNING

Piccoli et al. (2001) propose a framework to test the effectiveness of e-learning environments. Their framework, shown in Figure 2, depict dimensions and antecedents of e-learning environments.

The design dimensions in the framework include learning models, technology, learner control, content, and interaction. The human dimensions include learners (students) and instructors. Effectiveness is measured by performance, self-efficacy, and satisfaction.

A pilot study using the constructs in this framework was conducted to compare a Type VI: blended/hybrid-synchronous e-learning environment to a Type I: traditional face-to-face classroom.

Examples of blended/hybrid-synchronous e-learning are not easily attainable, therefore we included an empirical pilot study comparing a blended/hybrid-synchronous e-learning to a traditional face-to-face classroom.

In synchronous e-learning environments learners use networked resources and a computer based interface to access the learning material and to communicate with classmates and instructors (Piccoli et al., 2001). We therefore hypothesize:

H1: *Students in synchronous hybrid e-Learning environments will report higher levels of computer self-efficacy than their counterparts in traditional learning environments.*

The general student population is used to the traditional learning environment (face-to-face classroom instruction) (Simon, Grover, Teng, & Whitcomb, 1996). Some studies have found satisfaction in traditional environments to be higher than e-learning environments (Maki, Maki, Patterson, & Whittaker, 2000). Therefore we hypothesize:

H2: *Students in traditional learning environments will report higher levels of satisfaction than students in virtual learning environments.*

The university setting, course description, learning environment, and results of the pilot study are discussed below.

Figure 2. Dimensions and antecedents of e-learning environment effectiveness (adopted from Piccoli et al., 2001)

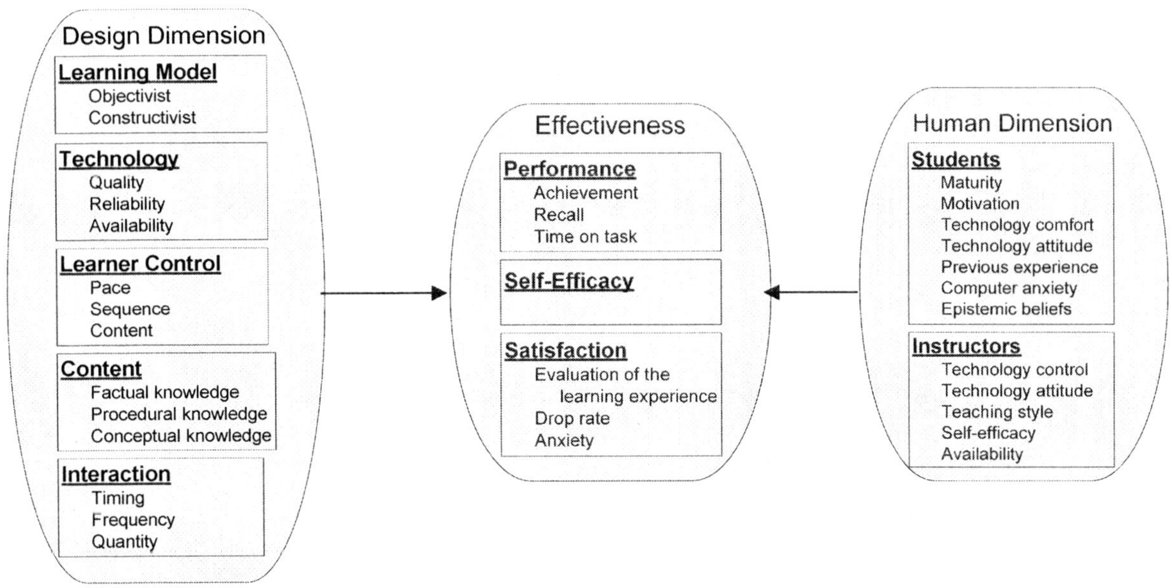

The University Setting and the Courses

The setting for the study was a large, public 4-year AACSB-accredited University with an enrollment of over 20,000 students. Three courses were examined in the study: a systems analysis and design (undergraduate) course, a project management (graduate) course, and an IT resource management (undergraduate) course.

The systems analysis and design course is a required course for all information systems and computer science students, and a prerequisite for all upper division core courses. A term project was used to practice the course content and students had to work in groups to complete the project. As part of the project, students were required to select an organization for their project, identify requirements, and develop a proposed information system. The modeling language used was unified modeling language (UML). Four major outputs were expected from the term projects: an activity diagram, class diagram, sequence diagram, and method specifications. A take-home midterm and final exam were administered for the course. The exams consisted of a case study which required the students to create the four major outputs specified above.

The IT resource management course is a capstone course for undergraduate information systems (IS) majors. This course is taken after students have completed 90 semester credit hours and is typically taken by senior students. The aim of the course is to bring together the concepts from the core course requirements in the IS program. Students were evaluated through their case study analyses, oral presentations, and term research papers.

The project management course is a core requirement of the Masters degree in the IS program. In this course, students are assigned individual projects. No exams are administered for the course. Instead student performance is assessed based on six assignments and a simulation project. Students are required to submit a write-up of their assignments in addition to making class presentations. The simulation project ran for six weeks.

The Learning Environment

The Marratech and WebCT-Vista technologies described above were used for the project management and systems analysis and design classes. For the third course, IT resources management, WebCT-Vista and Camtasia Studio[3] were the technologies used.

The recordings for the systems analysis and design and project management classes were completed in the classroom; sessions were recorded at the same time the face-to-face lectures were delivered. Students in these classes were given the e-learning option for half of the scheduled classes. With the e-learning option, students connected to the "live" classroom from locations other than the classroom, that is, from home. Some students selected the e-learning option—attending half of the classes outside of the classroom—while others attended all classes in a face-to-face environment.

Results

Students from all three courses participated in the survey online. A total of 63 students completed the survey with 30% (19) graduate and 70% (44) undergraduate. The distribution of the participant age ranges is shown in Table 6.

Table 6. Subject participation by age

Age Range (Years)	No. of Students	Percentage
19-23	4	6
24-29	22	35
30-35	16	25
36-40	8	13
41-45	0	0
46-49	5	8
>50	3	5
No Response	5	8

The gender mix of survey participants was 70% (44) male and 21% (13) female, 10% (6) did not provide a response to this question. All the graduate students were enrolled in the Masters of IS program. Graduate students accounted for 30% (19) of the total survey participants. Undergraduate students accounted for 62% (39). Over two-thirds (70%) of the undergraduate students were IS majors and the balance were computer science (CS) majors. They were comprised of 43% (27) seniors, 30% (19) juniors, 16% (10) sophomores. Eight percent (5) of the participants did not respond to this question.

All respondents indicated that they had computer and Internet access from home. Computer experience for participants was reported as 73% professional users, 17% frequent users, and 2% reported being somewhat experienced; 3 respondents did not answer this question. Eighty-nine percent of respondents said they enjoyed working with computers while only 2% indicated that they felt threatened by computers.

On a scale of 1 to 10, with 10 being the highest, a large number of respondents rated themselves high for self-efficacy (over 70% of the participants).

Satisfaction with the class experience was measured on a 5-point Likert scale with 5 being very satisfying. Over 90% of the respondents from each of the courses reported their satisfaction as either a 4 or 5.

DISCUSSION

In this section we discuss the pilot study results, differences in asynchronous and synchronous e-learning environment, hybrid-learning, limitations and future study.

Pilot Study Results

For the purpose of this study students were classified as traditional classroom learners or hybrid/blended-synchronous e-learning learners. The traditional classroom students were those students that attended all classes in a face-to-face format. Hybrid/blended-synchronous e-learning students were those students who attended some of the classes in the synchronous hybrid e-learning format. In the pilot study 18 respondents (29%) indicated that they used the synchronous hybrid e-learning format and 44 respondents (70%) reported using the traditional classroom format. One student did not respond to the question.

Each respondent was asked a set of 10 questions on self-efficacy. The questions are listed in Table 7.

A T-test was used to determine if significant differences exist between e-learners and traditional classroom learners. The results are shown in Table 8.

Self-efficacy Questions 1, 3, 4, 5, 6, 7, 8, and 10 resulted in slightly higher means for those in the

Table 7. Self-efficacy questions

	I could complete the job using the software package...
1	...if there was no one around to tell me what to do as I go
2	...if I had never used a package like it before.
3	...if I had only the software manuals for reference.
4	...if I had seen someone else using it before trying it myself.
5	...if I could call someone for help if I got stuck.
6	...if someone else had helped me get started.
7	...if I had a lot of time to complete the job for which the software was provided.
8	...if I had just the built-in help facility for assistance.
9	...if someone showed me how to do it first.
10	...if I had used similar packages before this one to do the same job.

Table 8. Self-efficacy responses for research groups (Traditional Class format = 39 cases; e-Learning = 18 cases)

Self-Efficacy Question	Mean e-Learning	Mean Traditional classroom	T-Test	
			t	Sig.
1	7.22	7.23	.013	.990
2	6.44	6.59	.218	.828
3	8.28	7.31	-1.488	.142
4	8.39	7.82	-1.092	.280
5	8.72	7.69	-1.548	.127
6	8.89	8.33	-1.032	.307
7	7.56	8.51	1.375	.175
8	7.83	7.38	-.668	.507
9	8.44	8.92	.805	.424
10	8.67	8.56	-.147	.883

e-learning group, while Questions 2 and 9 were slightly higher for the traditional classroom group. Self-efficacy ratings between the two groups were not found to be significantly different.

The first hypothesis (H1) stated that students who tend to choose the e-learning environment would have a higher level of computer self-efficacy. This hypothesis was not supported by the data, which indicates that the two groups had similar levels of self-efficacy. Further analysis of the data indicated that factors other than self-efficacy determined the students desire to participate in the synchronous hybrid e-learning. In the two classes where synchronous hybrid e-learning was offered almost all respondents (94%) stated they were already on campus for another class just before/after this one and did not have time to drive home for the online class and therefore chose to attend the face-to-face format.

Satisfaction responses for the research groups are shown in Table 9. The two research groups of synchronous hybrid e-learning and traditional face-to-face classroom did not show differences in satisfaction.

The second hypothesis (H2) stated that students in the traditional classroom setting would report higher levels of satisfaction when the subject level is complex. This hypothesis was not supported by the data. The Chi-Square test indicated that the two groups were not significantly different (χ^2=2.714, p=.438).

When asked whether they would take another e-learning class, 91% of the respondents indicated they would by selecting a 4 or 5 on a 5-point Likert scale. Eighty-seven percent of the respondents said they did not regret enrolling in this online class, and 83% said they would recommend this online class format to their friends.

Asynchronous and Synchronous Differences

Four of the six classifications (Type III, IV, V, and VI) involve some form of e-communication. The key differentiator for e-communication among the four classifications is the mode of communication (i.e., asynchronous or synchronous).

Asynchronous communication is communication that is "time-delayed or time-deferred computer mediated mode of delivery" (Seng & Al-Hawamdeh, 2001, p. 238). In an asynchronous environment, the sender and receiver do not have to be present at the same time for communications to occur. Examples of the mode of delivery in asynchronous communication are e-mail and threaded discussion boards (Seng & Al-Hawamdeh, 2001). Synchronous (real-time) communication on the other hand is communication that takes place concurrently. In a synchronous environment, the sender and receiver have to be present at the same time in order for communication to take place (e.g., video-conferencing) (Seng & Al-Hawamdeh, 2001).

Piccoli et al. (2001) identify five student challenges when using an asynchronous e-learning environment:

Table 9. Satisfaction responses for research groups

Satisfaction with the class	Synchronous hybrid e-Learning		Traditional classroom	
1=Very Dissatisfying	0	0.0%	0	0.0%
2=Somewhat Dissatisfying	0	0.0%	1	2.6%
3=Undecided	1	6.0%	1	3.0%
4=Somewhat Satisfying	4	22.0%	16	41.0%
5=Very Satisfying	13	72.0%	21	53.8%
Total	18		39	

1. **Difficulty managing the high degree of control:** In traditional classrooms the instructor provides direction and structure. Asynchronous e-learning environments avail a high degree of control for participants, however, participants are challenged when managing the high degree of control in the absence of instructor direction and structure. In a synchronous e-learning environment on the other hand, students are able to participate in a familiar strategy that consists of instructor direction and structure.
2. **Overburdened by the shift of responsibility and control:** In an asynchronous e-learning environment the instructor is not present at the time of instruction access. When a concept is not clear, learners are unable to ask the instructor questions in real-time. In a synchronous e-learning environment however, the instructor is present at the time of instruction delivery and learners can ask questions in real-time.
3. **Feeling isolated:** Asynchronous learners access instruction material independent of the instructor and classmates and learners do not engage in real-time interaction and therefore may feel isolated. Synchronous learners, in contrast, can see the instructor and fellow students at the time of instruction delivery by connecting via a webcam to the synchronous e-learning environment, thereby reducing feelings of isolation. A participant in our pilot study commented by saying, "With the live video and audio connections I feel like I am in the [traditional] classroom."
4. **Experiencing anxiety:** Participants who experience anxiety at the time of instruction delivery in synchronous e-learning environments are able to get immediate assistance through audio and video communication. This feature, however, is not available to participants in asynchronous environment where there would be a time delay in getting access to help.
5. **Difficulty in time management:** Asynchronous learners have to manage their instruction access time, primarily because there is no fixed-time for instruction access. The flexible "anytime" access in an asynchronous environment creates time management challenges, whereas synchronous (or face-to-face) environments have fixed-time where learners sign up with prior knowledge about the time constraints, thus there are no new time management challenges.

In our experience with synchronous e-learning environment courses, there was no indication that our students encountered these challenges.

Synchronous virtual learning environments are not without issues, they pose their own learner challenges too. Some of these challenges are:

1. **Technology investment:** Learners in both asynchronous and synchronous e-learning environments must have access to computers and all learners in our pilot study indicated that they had computer access at home. In our classes learners were required to use a headset (a basic headset costs about $10.00). Our department purchased basic headsets (bulk rate of $6.00 per headset) and provided them to students who wished to use them as loaners. In our classes, video from the instructor was always available, but students had the option to install a Webcam on their end (a basic Webcam costs about $40.00) to view the class in session. Investment for a university includes individuals with expertise to support real time communication and investment in a high speed Internet connection (Seng & Al-Hawamdeh, 2001).
2. **Technology glitches:** Instruction delivery, online real-time, may be interrupted due to technology glitches including LMS system errors and system connectivity errors, such as video frames freezing while being transmitted over the Internet and audio breaking

up and becoming distorted (Seng & Al-Hawamdeh, 2001). These challenges are unique to the individual setup. In an asynchronous format the student is responsible for dealing with the issues. In a synchronous format, because the instructor is leading a "live" class session the instructor is responsible for delivering instruction content and is also responsible for the delivery medium. When audio reception for one learner malfunctions it distracts, if not interrupts, other learners as well. In our situation, when a lack of bandwidth delayed communication, we asked learners to disable their video. On some occasions the instructor video was also disabled to overcome bandwidth shortages. In some instances, we encountered echo problems with audio; this often happened when one or more participants were not using headsets. In these cases we shifted to a walkie-talkie mode where everyone turned off their audio except the person actively speaking.

3. **Virtual presence:** Participants have to be in a location where they have access to a computer and a high speed Internet connection in order to participate in the synchronous classrooms. This may pose a challenge for students who are taking other classes on campus or are away from their equipment (Negash & Wilcox, 2007).
4. **Technical expertise:** Participants, both instructors and students, need to be comfortable with computers, they may also need to have some level of technical know-how in order to conduct or participate in a synchronous e-learning environment.

Hybrid/Blended E-Learning

Blended and hybrid e-learning formats are used interchangeably in this chapter. In blended/hybrid e-learning instruction, delivery combines presence and no presence and the type of presence can be physical or virtual. The blended/hybrid format can be one of the three combination formats:

- Asynchronous e-learning (no presence) with face-to-face (physical presence)
- Asynchronous e-learning (no presence) with synchronous (virtual presence)
- Synchronous (virtual presence) with face-to-face (physical presence).

The amount of face-to-face time in a blended/hybrid format varies greatly from institution to institution. Some institutions conduct the first and last class sessions of a semester course with presence and conduct the balance without presence. Other institutions hold 25% of the classes with presence and the other 75% without, while others conduct 50% of the class with presence and the balance without. Still there are some institutions that conduct their courses with 100% presence through a combination of physical and virtual presence. While there are no standards prescribing the proportion of presence/no-presence or physical-presence/virtual-presence in blended/hybrid e-learning environments, it would be useful to develop a standard that serves all stakeholders including instructors, learners, and institutions (Ranganathan, Negash, & Wilcox, 2007).

To date the research discussion on e-learning, for the most part, has been with respect to asynchronous format. In the debate over the value of asynchronous vs. traditional face-to-face courses, the promise of the hybrid model for e-learning has largely gone unnoticed and is just now starting to garner some attention from academics across a variety of disciplines (Brunner 2006).

There is no standard definition of a "hybrid course" (Brunner, 2006; DeNeui & Dodge, 2006). The definition adopted here is one "in which a significant portion of the learning activities have been moved online, and time traditionally spent in the classroom is reduced but not eliminated" (Garnham & Kaleta, 2002, p. 1). Also somewhat unclear, is how much time in a hybrid course is actual face-to-face and how much online (Brunner, 2006; Ranganathan et al., 2007).

Hybrid/blended models have started to gain attention because they offer the opportunity to apply the best features of online education and those of the traditional classroom to active independent learning (Garnham & Kaleta, 2002, p. 1). Brunner (2006) outlines the strengths of a hybrid model as: "1) student performance and retention increase, 2) time and flexibility for students is greater, 3) colors on the teaching palette multiply, 4) depth of community enhances the learning environment, 5) the breadth of 'interaction' is enlarged, 6) it allows for a gradual transition from face-to-face to online learning, 6) expectations are higher" (pp. 230-233).

Web, Gill, and Poe (2005) found that students' online discussions may enhance learning in case methods when taught using a hybrid approach. In a study conducted to compare traditional and technology-assisted instruction methods in eight sections of a business communications class, where live vs. hybrid formats were compared, an improvement in writing skills was found in students who participated in the hybrid course, particularly for those whom English is a second language (Sauers & Walker, 2004). McCray (2000) found courses which combine online learning with the traditional classroom can help students to become more engaged in rich classroom interactions by appealing to different learning styles through variety in content delivery.

In theological education, Delamarter and Brunner (2005) investigated hybrid models and found that student satisfaction and learning outcomes were higher for the students in hybrid courses than those in both asynchronous and face-to-face classroom settings.

Limitations

The sample size for the pilot study discussed above consisted of 63 responses, which is relatively small; therefore the collection of additional data to further validate the findings is a natural extension of the study. The results of this pilot study may also be limited to the specific courses examined and for this reason the study may not be generalizable to other courses, universities, or environments.

FUTURE RESEARCH DIRECTION

Additional research needs to be undertaken to gain a clearer understanding of the different e-learning classifications and a broader study to understand the effectiveness of each classification. Further, a comparison between the classifications is needed.

In the study, we found similar self-efficacy levels between users of e-learning format and face-to-face format; this may be because all the students in the courses either majored in computer science or information systems and this was a group that had more exposure to computers. A broader study evaluating students from noncomputer majors would be useful to provide further understanding of whether a higher level of technical skill is required for individuals participating in e-learning formats.

In the pilot study we found similar levels of satisfaction between students who chose the face-to-face and e-learning formats. This too presents an area that needs further examination in a broader study.

CONCLUSION

There is a tendency to characterize all e-learning modalities as if they are identical. In reality, learning outcomes, workload, success/failure rates, and pedagogical needs differ among the different modalities. Recognizing these differences is an important first step in understanding the effectiveness of e-learning environments. We conclude by highlighting some general issues with e-learning:

- **Instructor workload:** Teaching in an e-learning environment requires training and experience. Training and experience developed in a traditional face-to-face format do not easily translate into e-learning success. A professor with expertise in a subject area can probably walk into a traditional classroom and teach the course content with little preparation. The same expert professor, however, may not be able to achieve instant success in the e-learning environment just because the professor has been successful in a face-to-face setting. Teaching in an e-learning environment takes a considerable amount of time and effort. In the authors' experience, an e-learning format requires as much as 2-3 times more preparation time than a similar face-to-face class given the same level of content expertise, not to mention the increased level of interaction and communication with students.
- **Student workload:** Workload issues are not limited to instructors as e-learning formats also increase student workload. Students often assume that e-learning classes require less time and are therefore easier than traditional face-to-face classes. On the contrary, e-learning formats require more time. E-learning formats inherently shift some of the learning responsibility to the student, and as a result, student workload in this environment increases when compared to the workload in a traditional class.
- **Student expectations:** Student expectations for faculty availability are different for e-learning than in a traditional classroom. When students come to the face-to-face class they expect to see and talk to the professor. In e-learning environments, students often expect to get immediate response from the instructor anytime they log into the e-learning medium, regardless of time of day. For example, when there are due dates for assignments or exams, the instructor may not check e-mail from students just prior to the due date and as a result last minute student questions may not get addressed. This can potentially create disputes around whether or not the student could have earned a higher grade had the student's question been answered prior to the due date. To avoid such problems the instructor has to setup a standard response policy in the course syllabus. For example, the instructor can state that electronic communications must allow for at least a 24 hour response time during weekdays and 48 hour response time during weekends.

Many of the difficulties reported by students when using asynchronous e-learning environment were not found in our synchronous e-learning experience. These difficulties included difficulty managing the high degree of control, overburdened by the shift of responsibility and control, feelings of isolation, experiencing anxiety, and difficulty in time management.

This chapter has proposed and discussed six e-learning formats: e-learning with physical presence and without e-communication, e-learning without presence and without e-communication, e-learning without presence and with e-communication, e-learning with virtual presence and with e-communication, e-learning with occasional presence and with e-communication, and e-learning with presence and with e-communication. An empirical study comparing a blended/hybrid-synchronous e-learning format was also presented. Based on our study, we believe e-learning classifications have different levels of effectiveness. The differences may be the result of differences between the asynchronous and synchronous formats. Synchronous formats may have the potential to provide solutions for some of the challenges faced in an asynchronous format.

We encourage researchers to further study the proposed classifications, identify differences and similarities of the classifications, and evaluate the learning effectiveness of each classification.

REFERENCES

Alavi, M., & Leidner, D. E. (2001). Research commentary: Technology mediated learning-a call for greater depth and breadth of research. *Information Systems Research, 12*(1), 1-10.

Alavi, M., Marakasand, G. M., & Yoo, Y. (2002). A comparative study of distributed learning environments on learning outcomes. *Information Systems Research, 13*(4), 404-415.

Brunner, D. L. (2006). The potential of the hybrid course vis-a-vis online and traditional courses. *Teaching Theology and Religion, 9*(4), 229-235.

Cogburn, D. L., & Hurup, D. (2006, April 13). The world is our campus: Synchronous collaboration software lets universities unite colleagues, students, and researchers from all over the globe. *Network computing for IT by IT* (pp. 57-68). Retrieved August 6, 2006, from http://www.networkcomputing.com/showArticle.jhtml?articleID=184428959&pgno=1

Dagada, R., & Jakovljevic, M. (2004). Where have all the trainers gone? E-learning strategies and tools in the corporate training environment. In *Proceedings of the 2004 Annual Research Conference of the South African Institute of Computer Scientists and Information Technologists on IT Research in Developing Countries* (pp. 194-203). Stellenbosch, Western Cape, South Africa.

DeNeui, D. L., & Dodge, T. L. (2006). Asynchronous learning networks and student outcomes: The utility of online learning components in hybrid courses. *Journal of Instructional Psychology, 33*(4), 256-259.

Garnham, C., & Kaleta, R. (2002, March 20). Introduction to hybrid courses. *Teaching with Technology Today, 8*(6), 1-3. Retrieved May 5, 2007, from http://www.uwsa.edu/ttt/articles/garnham.htm

Hodges, C. B. (2005). Self-regulation in Web-based courses: A review and the need for research. *The Quarterly Review of Distance Education, 6*(4), 375-383.

Maki, R. H., Maki, W. S., Patterson, M., & Whittaker, P. D. (2000). Evaluation of a Web-based introductory psychology course: I. Learning and satisfaction in on-line versus lecture courses. *Behavior Research Methods, Instruments, and Computers, 32*(2), 230-239.

Negash, S., & Wilcox, M. V. (2007). Synchronous hybrid e-learning: Teaching complex information systems classes online. In *Proceedings of the 18th Annual International Information Resources Management Association Conference*, Vancouver, British Columbia, Canada.

Piccoli, G., Ahmadand, R., & Ives, B. (2001). Web-based virtual learning environments: A research framework and a preliminary assessment of effectiveness in basic IT skills training. *MIS Quarterly, 25*(4), 401-426.

Ranganathan, S., Negash, S., & Wilcox, M. V. (2007). Hybrid learning: Balancing face-to-face and online class sessions. In *Proceedings of the Tenth Annual Conference of the. Southern Association for Information Systems,* Jacksonville, FL.

Sauers, D., & Walker, R. C. (2004). A comparison of traditional and technology-assisted instructional methods in the business communication classroom. *Business Communication Quarterly, 67*(4), 430-442.

Seng, L. C., & Al-Hawamdeh, S. (2001). New mode of course delivery for virtual classroom. *Aslib Proceedings, 53*(6), 238-242.

Simon, S. J., Grover, V., Teng, J. T., & Whitcomb, K. (1996). The relationship of information systems training methods and cognitive ability to end-user satisfaction, comprehension, and skill transfer: A longitudinal field study. *Information Systems Research, 7*(4), 466-490.

ADDITIONAL READING

Brown, B. W. (2002). Can Web Courses Replace the Classroom in Principles of Microeconomics. *The American Economic Review*, pp. 444-448.

Delamarter, S. and Brunner, D.L. (2005). Theological Education and Hybrid Models of Distance Learning. *Theological Education, 40*(2).

Dodero, J. M., Fernandez, C., & Sanz, D. (2003). An experience on students' participation in blended vs. online styles of learning. *ACM SIGCSE Bulletin, 35*(4), 39-42.

Falch, M. (2004). *A Study on Practical Experiences with using E-learning Methodologies and Cooperative Transnational Development Methodology.* CTI Working Paper, no. 97, 2004, Center for Tele-Information, Technical University of Denmark. Retrieved February 17, 2007 from http://www.cict.dtu.dk/upload/centre/cict/publications/working%20papers/ctiwp97.pdf

Fox, M. (2002). Keeping the blended promise. *E-Learning, 3*(3), 26-29.

Hale, R. L., & Heiphetz, A. (2006). Rewind the teacher: A case for technology. *Distance Learning, 3*(4), 72-76.

Hardesty, B. S. (2007, January 5). E-learning: Successes and failures. *Chronicle of Higher Education, 53,* B20.

Hiltz, S. R., & Turoff, M. (2005). Education goes digital: The evolution of online learning and the revolution in higher education. *Communications of the ACM, 48*(10), 59-64.

Kazmer, M., & Haythornthwaite, C. (2005). Multiple perspectives on online learning. *ACM SIGGROUP Bulletin, 25*(1), 7-11.

Knutsen, D., Knutsen, E., & Slazinski, E. (2003). *Employing new advances in IP videoconferencing to enhance teaching and learning through the use of a hybrid distance learning course.* Paper presented at the Conference On Information Technology Education, Lafayette, IN.

Lorenzetti, J. P. (2004). For quality and cost effectiveness, build a hybrid program. *Distance Education Report, 8*(21), 1-2.

Mansour, B. E., & Mupinga, D. M. (2007). Student's positive and negative experiences in hybrid and online classes. *College Student Journal, 41*(1), 242-249.

McCloud, R. (2004). Does an online course work in computer science? *Journal of Computing Sciences in Colleges, 19*(5), 260-269.

McCray, G. E. (2000). The hybrid course: Merging on-line instruction and the traditional classroom Information. *Technology and Management,* 1, 307-327.

Mortera-Gutierrez, F. (2006). Faculty best practices using blended learning in e-learning and face-to-face instruction. *International Journal on E-Learning, 5*(3), 313-337.

Nelson, M., Bhagyavati, Miles, G., Settle, A., Shaffer, D., Watts, J., et al. (2005). Online teaching practices (both best and worst). *Journal of Computing Sciences in Colleges, 21*(2), 223-230.

Olapiriyakul, K., & Scher, J. M. (2006). A guide to establishing hybrid learning courses: Employing information technology to create a new learning experience, and a case study. *The Internet and Higher Education, 9*(4), 287.

Reay, J. (2003). Blended learning: A fusion for the future. *Knowledge Management Review, 4*(3), 2.

Smith, P. W., & Lyons, K. A. (2004). E-learning basics: Essay. User experience in the first ARISE distributed classroom. *eLearn, 2004*(3), 2.

Sprague, D., Maddux, C., Ferdig, R., & Albion, P. (2007). Online education: Issues and research questions. *Journal of Technology and Teacher Education, 15*(2), 157-166.

Tabor, S. W. (2007). Narrowing the distance. *Quarterly Review of Distance Education, 8*(1), 47-57.

Vaughan, N. (2007). Perspectives on blended learning in higher education. *International Journal on E-Learning, 6*(1), 81-94.

Vrasidas, C. (2004). *Engineering e-learning systems (ELS): Issues of pedaogy and design in e-learning systems.* Paper presented at the Symposium on Applied Computing Nicosia, Cyprus

Webb, H. W., Gill, G., & Poe, G. (2005). Teaching with the case method online: Pure versus hybrid approaches. *Decision Sciences Journal of Innovative Education*, *3*(2), 223-250.

Wilson, B. G. (1996). *Constructivist learning environments: Case studies in instructional design.* Englewood Cliffs, NJ: Educational Technology Publications.

Young, J. R. (2002, March 22). 'Hybrid' teaching seeks to end the divide between traditional and online instruction. *Chronicle of Higher Education, 48,* A33.

ENDNOTES

1. Marratech was recently acquired by Google (http://google.com)
2. WebCT recently merged with Blackboard (http://blackboard.com)
3. Camtasia Studio is a product specially designed for recording and publishing video presentations .

Chapter II
Blending Interactive Videoconferencing and Asynchronous Learning in Adult Education: Towards a Constructivism Pedagogical Approach-A Case Study at the University of Crete (E.DIA.M.ME.)

Panagiotes S. Anastasiades
University of Crete, Greece

ABSTRACT

This chapter focuses on the designing and development of blended learning environments for adult education, and especially the education of teachers. The author argues that the best combination of advanced learning technologies of synchronous and asynchronous learning is conducive to the formation of new learning environments, which, under certain pedagogical conditions, will adequately meet the special needs of adult students. Particular emphasis is given to the designing and development of a pedagogical blended learning model based on the principles of transformation adult theory and constructivism. This model implements advanced learning technologies in a pedagogical context, aiming at the formation of a collaborative blended learning environment, which will encourage critical thinking and reflection, providing the necessary conditions for a polymorphic distant education for teachers. Finally, we present a case study of a blended environment of teachers' training designed by the Center of Intercultural and Migration Studies (E.DIA.M.ME.) at the University of Crete.

INTRODUCTION

Education is now playing the most important role in the Information Era (Anasstasiades, 2002a; Raptis & Rapti, 2004). In the modern world of globalisation and the Internet, knowledge constitutes the main productive element of economy (Tapscott, 1999). Although the pace of change in the working and social environment may influence evolution and renewal of all human activities positively, it does depreciate knowledge and skills to an unprecedented extent and contributes to the establishment of digital divide (Anastasiades, 2005a; Norris & Pippa, 2001; United Nations, 2003). The influence of age on the use of information and communication technologies (ICT) (Eastman & Iyer, 2004), the updating and enhancing knowledge, and skills of citizens in the emerging information society designate lifelong learning as an essential condition for their harmonious and productive integration into the new social and working conditions (Anastasiades, 2005b).

ICT form a new teaching and learning environment in all levels of education, mainly adult education. Educational institutions throughout the world are designing and applying distant teaching environments, which take into consideration the specific needs of adult students and hope to provide flexibility regarding the location, time, and pace of learning (Anastasiades, & Spantidakis, 2006).

For many years, distance education (DE) was regarded as a technological and organisational entity according to prevailing technology-centered perceptions (Bates, 1995). This view led to the downgrade of the pedagogical aspect of learning and teaching (Massialas, 1989; Paulsen, 2003).

This chapter demonstrates the view that new technologies should not be considered as a neutral teaching medium (Lionarakis, 2006) but, instead, be implemented under pedagogical conditions aiming at the development of critical thinking (Brusilovsky, 1999; De Bra, Eklund, Kobsa, Brusilovsky, & Hall, 2000; Kemmis, 1985; Kostoula & Makrakis, 2006; Mezirow 1981) through their creative integration into the social and cultural context (Carr & Kemmis, 2002).

The contents of the chapter are as follows:

In the second section we provide the theoretical frame of e-learning and describe the technologies of synchronous and asynchronous transmission and the *blended* learning environments. Emphasis is given to the pedagogical conditions of the designing Web-based learning environments.

The third section outlines the basic principles, methodology, and characteristics of the proposed pedagogical model, which is based on the rudiments of adult education emphasizing the transforming learning, the constructivism theory, and the fundamental principles of DE by American Distance Education Consortium (ADEC).

In the fourth section we analyse key issues of designing and developing an asynchronous learning environment, which are the basic characteristics, the functions, and designing models of an asynchronous learning environment, focusing on the designing principles and the phases of development.

The main issue of the fifth section is the designing and developing of synchronous learning environments. We describe the characteristics and the methodology of designing collaborative environments emphasizing interactive videoconferencing and live transmission of lectures via Internet.

Finally, in the sixth section we present a case study on the designing and developing of a blended Web-based learning environment, which has been applied for 4 years by the EDIAMME of the University of Crete and aims at the training of teachers around the world.

NEW TECHNOLOGIES AND DISTANCE EDUCATION

The dynamic appearance of digital technology in the recent years, the advanced potential of telecommunications, and the explosion of the Internet and technological systems have formed a completely different environment in the daily and working life of millions of people around the world.

Education strives to keep pace with the needs of an era which dictates open, flexible, student-centered systems without the necessary presence of students in a classroom. Distance education is challenged to tackle effectively the emerging new needs.

Theoretical Frame of Distance Education

It is a fact that there have been many definitions of distance learning (DL). Most of them include the separation of the teacher from the student, the impact of an educational organisation, the use of media to form a communication system with the aim of reinforcing interaction, and so forth.

Desmond Keegan (1996) suggests one of the most complete definitions of DL, which includes five fundamental principles:

1. The permanent separation of the teacher from the student during the whole learning process.
2. The impact of an educational institution on the planning of pacing, the production of the teaching package, and the provision of academic and learning support.
3. The implementation of technology and media (i.e., printed material, video, radio frequencies, or personal computer) for the transmission of the content and the provision of interaction.
4. The provision of two-way interaction and communication.
5. The permanent absence of a student group, so that students learn more as individuals rather than as a group (Keegan, 1996, p. 50). Keegan supports that the fifth principle should be reconsidered, since the appearance of student groups is feasible through the implementation of technology (Keegan, 1996, pp. 46-47).

In DL, the term 'distance' refers to the possibility of studying within a frame of physical-geographical distance between the tutor and the student but without space or time limitations, which differentiates DL from face-to-face education. Physical distance is no longer a hindrance since the variety of media and their appropriate implementation contribute to an education of high quality (Lionarakis, 1999).

For Wedemeyer (1981), the most important element of DL was student's autonomy. He defined ten characteristic traits of an educational system which emphasise student independence and the use of technology to promote it. DL aims at the motivation of the students to learn on their own and act towards self-directed learning (Lionarakis, 2001).

For some researchers, DL comprises aberrance from face-to-face education. Holmberg (1986) claims that it constitutes a separate form of education. Keegan (1988) claims that DL is a separate field, parallel and complementary to conventional education. However, Shale (1988) thinks that the content of face-to-face educational process is the same as that of DL process. Cropley and Kahl (1983) compare and contrast DL with face-to-face education in terms of psychological aspect and concluded that neither of these sets of principles comprises a separate form.

The advances of OCTs, the globalisation of economy, and the new learning theories necessitate the reconsideration and further development of the traditional approaches to DL policies.

The impact of new technologies and their use in education leads Desmond Keegan (1995) to realise that the Internet connection between the tutor and the students at different locations can create a virtual classroom. Michael Moore and Greg Kearsley (1996) considered DL to be an educational process which is performed at different locations and needs the implementation of special techniques for lesson planning, the use of technological media and communication systems, and a specific organisational and administrative support. New technologies redefine our perception and definition of DL (Hanson, Maushak, Schlosser, Anderson, Sorensen, & Simonson, 1997). Mike Simmonson provides one of the latest definitions, according to which DL is a typical educational system where students are located in different areas and are linked to the tutor and to each other through interactive means of media telecommunication systems (Simonson, Smaldino, Albright, & Zvacek, 2000; Simonson, 2002).

Advanced Learning Technologies

The learning and instruction process are changing with the use of the personal computer as a knowledge tool (Raptis & Rapti, 2004). We are led towards a student-centered environment, which overcomes the barrier of distance, and the different location of people no longer considerably affects the communication between them.

ICT play a significant role in the modern social, financial, and educational reality. Networks, videoconferencing, and a wealth of technological and communications applications are the components of the global village (Rheingold, 2000). An important condition for the success of the introduction of ICT in the reform of the educational process and learning culture (Raptis & Rapti, 2004) is the achievement of teaching approaches which will serve the needs of the new teaching and learning environment, combined with the constant training and encouragement of the most important part in the educational process, which is the teacher (Vosniadou & Kollias, 2001).

Educational technologies, educational multimedia, and distance learning are the forerunners of a new era in education (Harley, 2001). Technological tools, as it is, are the first step in the transition of the contemporary traditional classroom towards the new model of virtual classroom (Norton, 2001) and hybrid school (Anastasiades, 2004; Rosbottom, 2001). At the same time, it is compulsory to create a pedagogical model as the theoretical base which will define the frame to integrate the new educational technologies.

Teaching technology refers to the theory and implementation of planning, developing, using, running, and evaluating the process and learning materials which are conducive to learning (Seels & Richey, 1994, p. 1).

The term educational technology does not simply refer to the material and technological media (e.g., software, PC) but to a systematic approach aiming at the improvement of human learning by emphasizing the needs of the tutor and the teaching process.

The term learning technologies refers to the variety of technologies which facilitate the learning and teaching process focusing on the student and the learning process.

ICT are the result of the collaboration of informatics and communications engineering. Computer networks and hypermedia systems comprise the new environment of advanced learning technologies (ALT) underlining the development of learning environments on the Internet. They aim at the enhancement of interaction between students, tutors, and learning material and tools, thus providing new prospects for DE (Wegner, 2001; Grigoriadou, Papanikolaou, Cotronis, Velentzas, & Filokyprou, 1999).

ALT alters the concept of space and time in the educational process and expands distance leaning. In face-to-face tuition the simultaneous presence of tutors and students in the classroom is one of the most important elements unchanged throughout time.

Advanced Learning Technologies and Asynchronous Learning Environments

The learning process depends directly on the participation of tutors and students within a community (Bruner, 1990; Vygotsky, 1978). Students create their own learning conditions through social interaction, exchanging ideas with their peers or others who share common interests (Bruner, 1990; Solomon, 1987; Tobin, 1990; Vygotsky, 1978).

A learning environment is a complex system comprised of associated factors which influence interactive learning with and irrespective of individual and cultural differences (Salomon, 1995). Some of these factors in a learning environment are the location, the prearranged behaviors, the expectations and understanding, the set context for the achievement of clear goals guided by an individual with responsibility and authority, and the technological support.

The main aim of a complete Web-based learning environment is to guarantee the essential conditions (i.e., pedagogical, administrative and organisational) which will fulfill distance learning through the use of advanced learning Internet technologies (Anastasiades, 2003).

The first phase of implementing a Web-based learning environment focused on:

1. The learning management systems, which contributed to the automatisation of administrative and organisational procedures, such as registration of students, curriculum, organisation of learning activities, learning resources management, monitoring performance of students, performance reports, and so forth.
2. The content management systems (CMS), which emphasized the production, management, research, and distribution of the learning material.
3. The virtual learning environments (VLE), which constitute an information environment especially designed for the encouragement of interaction among the participants (tutors and students). Students not only have an active role but also are fundamental contributors to the configuration of the virtual environment, with the aim of rendering it a forum which will promote collaborative learning through a variety of learning activities (Dillenburg, Scheider, & Syntena, 2002; Dillenburg, 1999).

VLE support the communication and collaboration between a student and a tutor (Dori, Barak, & Adir, 2003; Light, Nesbitt, Light, & White, 2000), the communication among students (Guzdial & Turns, 2000), and the formation of a collaborative environment for the tutors (Nachmias & Mioduser, 2000; Sheremetov & Arenas, 2002).

Only a few years ago there were attempts of complete systems—educational platforms, such as the WebCT (Clark, 2002), the Blackboard (Yi & Hwang, 2003), and the Stellar (Stellar, 2003)—which provide a user-friendly environment for the application of DE through the Internet.

In distance learning the simultaneous presence of students and tutors is not necessary (*asynchronous DE*) as they can choose the location, the time, and, in most cases, the pace they will participate in a 'self-directed learning' (Lionarakis, 1998).

Asynchronous Education includes:

- **Self-teaching,** where the main tool of learning is the educational material (e.g., books, CBT, Internet, etc.) and the student can decide on the pace of learning. Such an example is foreign languages learning through multimedia, books, cassettes, and so forth.
- **Semiautonomous learning,** in which the student can study the learning material and communicate at prearranged meetings with the assigned tutor by means of face-face meetings, e-mail, chat rooms, forums, and so forth.
- **Collaborative learning,** in which the communication between the tutor and the students is asynchronous; the students study individually following an arranged schedule of assignments.

The Web-based learning environment provides considerable advantages to the students participating in distance learning programs but also to those who study in the conventional way and wish to enhance the benefits of this process (Cornford & Pollock, 2003; Dearing, 1997; Forsyth, 2001; Maier & Warren, 2000; Ryan, Scott, Freeman, & Patel, 2000).

We should consider that the implementation of a Web-based learning environment varies from one institution to another, as the features of the Web-based learning environment are defined by the aims of the institution and not vice versa. For example, an institution may use a Web-based learning environment to support and improve its conventional courses. On the other hand, the use of a Web-based learning environment may completely replace the traditional system, dependent on the physical presence of tutors and students. In this case, the administrative, communicative, and learning activities of the institution are conducted through the Internet.

Synchronous Learning Technologies and Distance Learning: Interactive Videoconferencing

Advanced technologies of synchronous transmission define a new perspective in education, mainly DE, as under pedagogical conditions they can create a dynamic environment of collaboration and two-way interaction (Anastasiades, 2003a).

In the future, videoconferencing will play a significant role in the field of DE (Abbott, et al., 1993; Chen & Willits, 1998; Fillion, Limayem, & Bouchard, 1999; Martin, 2005) and adult education (Anastasiades, 2000; Dallat et al., 1992).

Interactive videoconferencing (IVC) technology allows students at two or more distant locations to create a collaborative environment at the same time (Gibson & Cohen, 2003; Suthers, 2001). The communication may include data and graphics exchange (Brown, 2001; Finn, Sellen, & Wilbur, 1997) and data sharing (Gürer, Kozma, & Millán, 1999).

Typically, a computer mediated conference (CMC) is based on text but increasingly it includes drawings, photographs, and other images (e.g., emoticons). Such examples are e-mails, chat rooms, discussion boards, text messaging, instant messaging, shared databases, or application-specific groupware.

Videoconferencing can contribute to the creation of an interactive environment of collaborative distance learning effectively under specific pedagogical and technological conditions. Within this context, teachers and students at two or more locations will be able to come into contact, communicate, and collaborate in real time through sound, live image, and data (Damanakis & Anastasiades, 2005).

By applying (IVC) we can create a learning environment which features interaction and flexibility, fosters collaboration, uses a variety of media, allows access to multiple information sources, and demands an affordable cost (Sullivan, Jolly, Foster, & Tompkins, 1994).

For the IVC to be effective, it should meet certain requirements such as the time-consuming teaching preparation, training of the lecturers, use of technology, and so forth (Lawyer-Brook, 1991). Particular difficulties are connected to the application of videoconferencing between remote locations, the technical or organisational restrictions at some classes, and the management of different environments at schools (Barker, 1991).

International literature provides numerous references to the educational usage of videoconferencing, mainly in higher/tertiary education, dealing with teaching (Coventry, 2000; Mitchell et al., 1993; Pitcher, Davidson, & Goldfinch, 2000;bReed & Woodruff, 1995; Unruh, 2000), broadband issues (Smyth, 2005; Hearnshaw, 2000), and cost (Twigg, 2002). A number of studies suggest new teaching methods aiming at the enhancement of interaction and there are guides advising effective videoconferencing (Digital Bridges: K-12 videoconferencing; Hayden, 1999; Robinson, 1997). However, there is not adequate research focusing on the designing and application of a holistic pedagogical model for the utilisation of videoconferencing in education.

Blended Learning Environments

Synchronous and asynchronous tele-education are not competitive forms but they can—and sometimes have to—be combined to create a *blended* collaborative learning environment (Anastasiades, 2005c). The pedagogical approach, on which we base the designing of a blended collaborative learning environment, is a key issue analysed further on.

Pedagogical Conditions for the Development of Distance Learning Courses

Distance learning takes advantage of the constant advances of multimedia and Internet technology and emerges as an exciting and promising field (since students and tutors expect much from it), offering new prospects to learning and access to information and knowledge (Makrakis, 2000). A great many scholars argue that the era of an open, flexible, student-centered, interactive learning of high quality, free of spatial and time restrictions is forthcoming.

However, we should have moderate expectations regarding the impact of ALT on the every day practice of distance learning (Simonson et al., 2000).

Significant reservations refer to:

1. The effect of digital divide. Not all individuals have the basic skills of Internet use, which deters the wide access of the population to the digital era of DE (Anastasiades, 2005b). We should also consider that the telecommunications capacity of an average user cannot support all the applications provided by ALT, while the access cost is relatively high in many countries (Richter, 1999).
2. The risk of using ICT as a method and not as a tool, which will result in the distortion of the pedagogical principles of distance learning (Lionarakis, 1999). Furthermore, we should take into consideration that nowadays, in the Internet era, DL may be regarded as an easy or inexpensive way to respond to the increasing demand for educational opportunities. The main approach worldwide and especially in the USA regards DL more as a technological and organisational entity, without focusing on its pedagogical aspect, the qualitative production of the educational package, and the procedure supporting the student to discover learning. It is then crucial to underline that learning cannot be seen as a product, which is transmitted through the teaching process from a source to another, or from one field to another or even from an empirical-philosophical domain to another (Lionarakis, 2003).
3. The risk of underestimating the quality of the provided education due to programs of questionable quality which are designed for profit (Connick, 1999). Illustrative example of this is the phrase 'digital mills of degrees,' coined by David F. Noble, Prof. of History at York University, Toronto. James Dunderstadt (1997), honorary President of the University of Michigan, refers to a ubiquitous university, where digital networks can minimise the restrictions of time and location, or even of reality itself. Digital technology will enable individuals to engage in learning at any place, at any time.
4. The possibility that the prolonged daily activity on the PC lead to isolation and deprivation of social interaction, with the following negative social and psychological consequences (Kokkos & Lionarakis, 1999).

Figure 1. The blended learning model

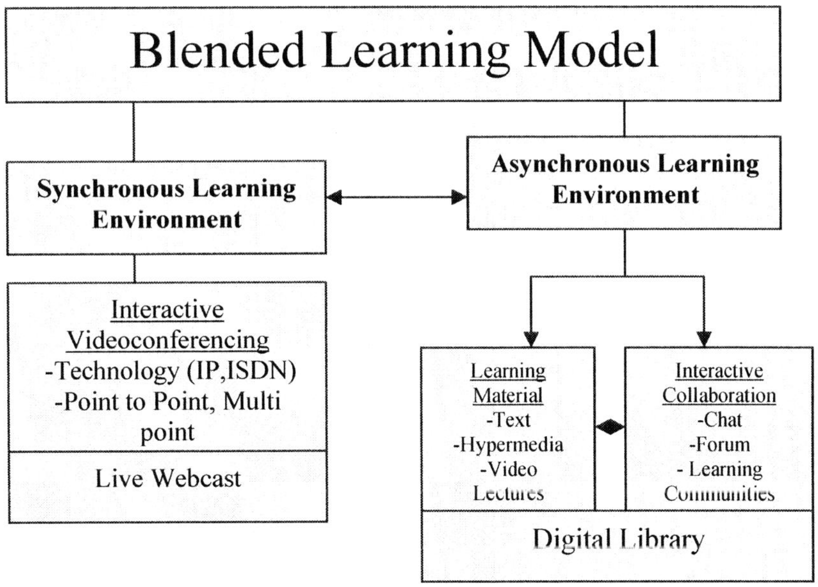

BLENDED LEANING MODEL: THE PEDAGOGICAL APPROACH

General Description

The proposed pedagogical approach suggests the functional combination (blended learning model) of advanced learning technologies of synchronous (videoconferences) and asynchronous (Web-based learning platform) learning (see Figure 1) in order to provide an interactive learning environment

Going from face-to-face teaching to the new blended learning environment for adults is not an easy process, as it requires optimal combination of learning theories, principles of distance learning, principles of adult theory, interactive media, and instructional methods and techniques. .

Adult learners include working adults with family responsibilities, older workers who may not feel confident about returning to school, and people who are currently in the workforce and who need to upgrade skills and knowledge (McIntyre, 1997). So we have to meet the needs of adult learners in developing blended learning courses, and find out the optimal combination of learning theories that match with our learning goals.

The proposed pedagogical approach is based on three pillars:

1. Adult theory.
2. Learning theory
3. Distance learning basic assumptions and principles

The Adult Learning Theory

The adult education (AE) literature generally supports the idea that teaching adults should be approached in a different way than teaching children.

Figure 2. The pedagogical approach: Steps and methodology

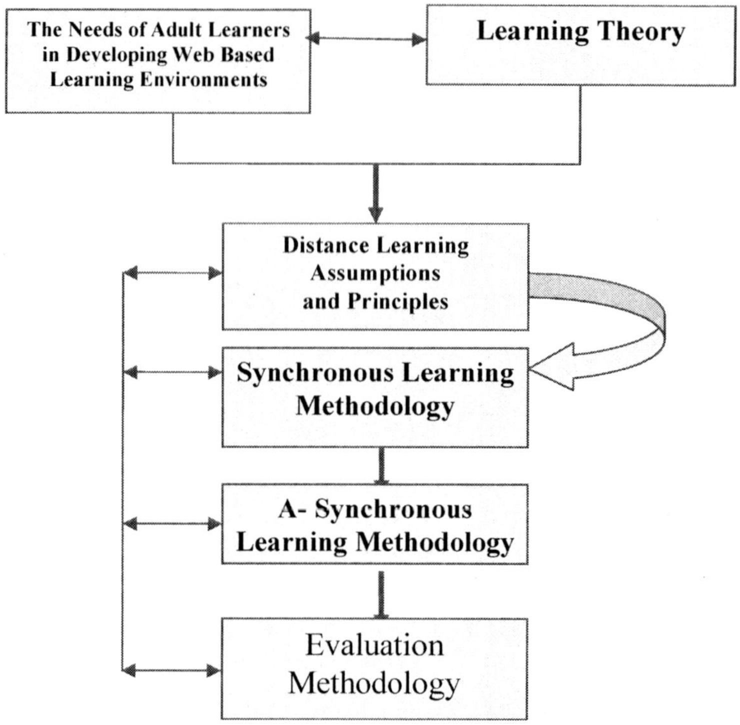

Rogers (1969) advocates an unstructured method of teaching where the teacher's role is that of a facilitator and the student is allowed the freedom to pursue self-discovered learning activities. A summary of Rogers' ideas about what he terms 'experiential learning.'

Knowles (1970) introduces the concept of andragogy as 'the art and science of helping adults learn.' He contrasts andragogy to the more traditional pedagogy, which he argues is not always appropriate for teaching adults on the basis of four crucial assumptions about the characteristics of adult learners that are different from the assumptions about child learners on which traditional pedagogy is based.

Hiltz (1994) reports that the virtual classroom environment resulted in better mastery of course materials, greater student satisfaction, and a higher level of student-reported learning than traditional classroom experiences. According to Stilborne and Williams, (1996), distance adult learners will learn only what they feel they need to learn, learn by comparing past experience with new experience, need immediate feedback concerning their progress, want their learning to be practical, try to avoid failure (dispositional barrier), and finally, do not all learn the same way (personal learning styles). Research on learning processes in face-to-face groups indicates that development of social climate is important in order to make students feel like insiders in the learning environment, thus contributing to students' motivation, involvement, and contentment (Chan & Rapman, 1999).

During the last years, research in the field of adult education has systematically investigated the way in which adult students perceive and interpret reality according to their own needs and experiences (Kokkos, 2007). Studying the ways of adult learning, Rogers (2003) emphasises not the cognitive procedure as much, but the relationship between teacher and students and argues that the goal of the trainer in adult education should be to enhance the ability of the participants to define themselves and nurture active citizens. Emancipation in adult education has been an imperative issue for many distinguished scholars. Jarvis argues that emancipation is a fundamental goal in AE (Jarvis, 2004; Kokkos, 2007). Freire also refers to the ultimate value of emancipation of adult students, taking into account his social and cultural experience from the Third World countries (Freire, 1984).

Thus, the focal point of contemporary scholars is the needs of adult students, which more often than not are expressed in a distorted way and lead to phobic behaviors toward new knowledge (Illeris, 2002, 2003). An issue of utmost importance is to realise that adult students should be supported by trainers so as to understand the causes creating their needs within their individual social and cultural identity. This critical understanding aims at their release from beliefs which worsen their introversion and the assumption of an active personal and social role based on the current trends (Kokkos, 2007).

But how can we transform the sentiments of adult students critically in order to lead them to emancipation? The answer to this key issue can be given by the '**transforming learning theory**,' introduced by Mezirow (1980, 1981, 1985, 1990, 1991) and further analysed by τον Brookfield (1985, 1986, 1987, 1990, 1991, 1992, 1995, 1996, 2001). Mezirow's thoughts are based on the critical theory of the School of Frankfurt.

Mezirow (1981) discusses his theory that critical reflection and awareness of 'why we attach the meaning we do to reality' may be two of the most significant distinguishing characteristics of adult learning. According to Mezirow, the role of the educator is to help the learner focus on and examine the assumptions that underlie their beliefs, feelings, and actions, to assess the consequences of these assumptions, identify and explore alternative sets of assumptions, and test the validity of assumptions through effective participation in reflective dialog. Transformative learning involves becoming more reflective and critical, being more open to the perspectives of others, being less defensive, and more accepting of new ideas. An elaboration of the idea of critical reflection in an attempt to develop a theoretical foundation for explaining how transformations occur in adult learning. Brookfield (1994) addresses four major research areas in adult learning, including self-directed learning and critical reflection (or transformational learning), in order to explore the claim that adult learning is a discretely separate domain that has little connection to learning in childhood or adolescence.

The Learning Theory

According to our approach, adults need to construct their own understanding of each concept so that the primary role of teaching is not to lecture, explain, or otherwise attempt to 'transfer' knowledge, but to create situations for students that will foster their making of the necessary mental constructions (Schiller & Mitchell, 1993; Scnurr & Smith, 1995). According to Garrison (2006), we have to establish a climate that will create a community of inquiry and a critical reflection and discourse that will support systematic inquiry. So according to our approach **constructivism** is the optimal learning theory in order to achieve our learning goals (Anastasiades, 2005c, 2006a).

Distance Learning Basic Assumptions and Principles

According to the proposed theoretical background we adopt the ADEC (1999) guiding principles for distance teaching and learning (Anastasiades, 2006a).

The current methodology accepts the following principles:

Figure 3. The proposed blended learning pedagogical approach for adults

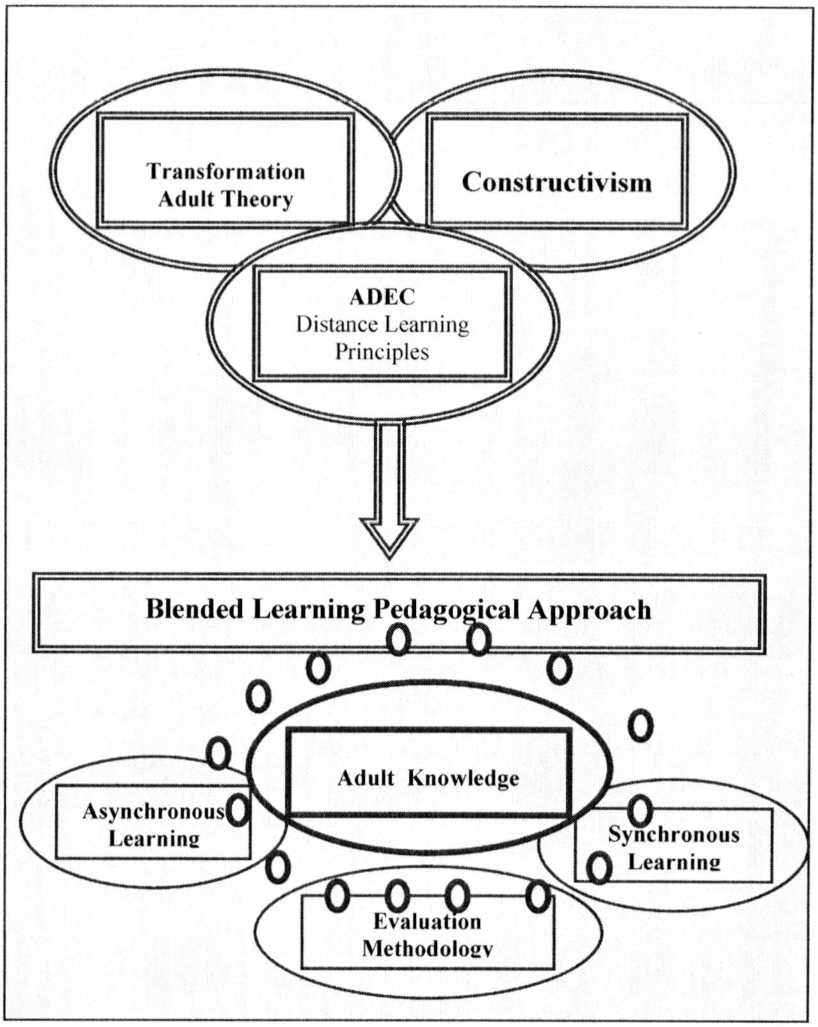

- The learning experience must have a clear purpose with tightly focused outcomes and objectives.
- The learner is actively engaged.
- The learning environment makes appropriate use of a variety of media.
- Learning environments must include problem-based as well as knowledge-based learning.
- Learning experiences should support interaction and the development of communities of interest.

DESIGNING AN ASYNCHRONOUS LEARNING ENVIRONMENT

Components

According to Claus (2003), the development of a Web-based learning environment is modeled by four main factors:

- The content
- The format
- The technological infrastructure
- The pedagogical perspective

Figure 4. Asynchronous learning environment: Subsystems

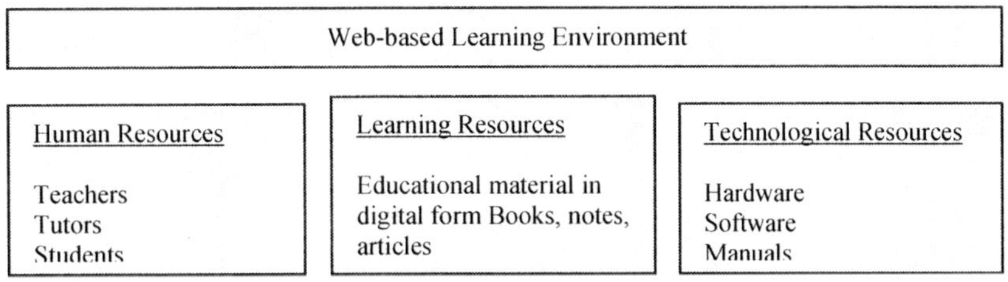

Through a systemic approach, a Web-based learning environment comprises of three subsystems:

1. The subsystem of pedagogical organisation of learning, which regards the human resources (i.e., teachers, tutors, and students) as members of a collaborative learning community. Most scholars recommend constructive learning, which is quite difficult to apply (Mikropoulos, 2000).
2. The subsystem of technological organisation, namely the tools, patterns, and methods of designing and building of the learning environment according to the needs of the learning community.
3. The subsystem of social organisation, which promotes the culture of collaborative learning in the context of a student-centered teaching approach.

The components of a Web-based learning environment are the human, learning, and technological resources.

In its modern form a Web-based learning environment is a user-friendly environment, which is at the service of its human resources (teachers, tutors, students and system managers) with the aim of supporting:

- The designing and organisation of courses
- The development of learning content
- The provision of the necessary resources in digital form (i.e., notes, articles, visual and audio material, books, etc.).
- The formation of an environment of interactive communication (i.e., chat, forum, e-mail, etc.)
- The completion of organisational and administrative procedures (i.e., registration, assessment, performance reports ,etc.)
- The organisation of a wide variety of learning activities

Fundamental Functions and Features

For DL to meet the needs of the support of collaborative learning and active participation of the students, it must follow some basic principles, which, according to Broady (1996), are: learning goals and content presentation, interactions, assessment and measurement, instructional media and tools, and learner support and services.

The fundamental functions that comprise a complete Web-based learning environment are the following:

Designing Patterns of a Web-Based Learning Environment

The designing of learning environments should take into consideration various psychological, philosophical, and technological stances (Hannafin & Land, 1997). The traditional models of teaching development of conventional learning environments (Bobbitt, 1918; Dick & Carey, 1996;

Table 1. Web-based learning environment: The fundamental

Functions	Services available	Benefits for students	Benefits for teachers and tutors
1. Management	Access policy online diary	• Access to course • Course schedule • Submission of assignments • Grades • Update	• Access to secrecy area • Class management • Submission of grades • Monitoring students' performance • Diary: reports
2. Production	• Tools: designing methods of teaching material • Designing patterns • Reusable learning objects	• Preparation of assignments in digital form	• Production of teaching material individually or in collaboration with others
3. Storage classification	• Selection of database • Reusable files • Metadata	**Storage:** Classification of assignments	**Storage:** Classification of teaching material and learning activities
4. Structure of supporting functions	• Frequently asked questions • Glossary/ terminology • External links • Search engines • Access to useful tools • Annotations reference material	**Support to:** • Administrative and learning procedures • Search engines • Study methods	**Support to:** • Administrative and learning procedures • Search engines • Study methods
5. Provision of content	• Multi-user interface • Navigator • Client software • Plug in • Transmission of content in: • Printed form • CD	**Internet access to:** • Teaching material • Additional notes • Alternative sources • Slides • Lectures	Ability to post and update teaching material on platform
6. Communication	• E-mail • Notice boards, chat rooms , forums • Frequently asked questions, etc. • Online users awareness • Data sharing: Application sharing • Videoconferencing	• Communication with teacher/other students • Exchange of information • Joint/collaborative preparation of assignments (group-work) • Social interaction	• Communication/collaboration with students and teachers • Development of teaching activities and research
7. Assessment	**Evaluation of system performance:** • Formative • Summative	**Assessment of:** • The teacher • The material • User-friendliness, etc.	Assessment of system
8. Support to procedures of students' assessment	Assessment of students' performance	Self-assessment	• Assessment criteria • Authoring questioning tolls • Monitoring of performance • Reports
9. Management of learning needs	• Diagnosis of students' learning needs • Diagnosis of students' learning styles	Supports the individual choice of learning style	Grouping of students sharing common goals, learning needs and styles, etc.

Gagne, Briggs, & Wager, 1994; Smith & Regan, 1993) are linear or structured and based on traditional principles of objectivism-positivism.

Following the contemporary learning theories, there seems to be a transition from a teacher-centred to a student-centred approach, from teaching to learning, from individual learning, to a learning derived from a collaborative environment, within the context of a learning community. Such considerations originate from the field of cognitive theories of social interaction (i.e., distributed cognition theory [Pea, 1995, & Salomon, 1995] and activity theory [Engestrom, 1987]).

In the recent years there have been proposals recommending the designing of a Web-based learning environment (Graham, McNeil, & Pettiford, 2000; Horton, 2000; Jolliffe, Ritter, & Stevens, 2001; Karpov & Haywood, 1998). Most of them converge as to the point that the important elements of designing a Web-based learning environment are the learning goals, the learning activities, the role of tutors and students, the connection between learning goals and teaching material, the assessment, and the social content of learning.

Roberts (1995) suggests a user-friendly model called 'a template for converting classroom courses to distributed, asynchronous courses' (http://www.unc.edu/cit/iat-archive/publications/roberts/template.html), which emphasises the definition of learning goals, the adaptation of the teaching material to the defined learning goals, the choice and best combination of the appropriate learning theories, the choice of technological means, and, finally, the formation of a collaborative environment.

In a systemic approach towards the designing of a Web-based learning environment (Cobb, 1994; Jonassen, 1992; Philips, 1995) the model of problem solving is adopted and includes four phases: analysis, designing, development-implementation, and assessment-revision.

Since the early 90s (Salomon, 1992; Kagan, 1994) emphasis has been given on the designing of collaborative Web-based environments. These attempts focus on the fact that the student should be capable of solving problems, collaborating with others, being responsible for their pacing, and being rewarded for achieving their goals within the group (Reiser, 2001). However, several issues are raised which are connected to the development (Jonassen, 1997; Van Berlo, 2000) and the quality of such systems regarding the achievement of the set learning goals (Hakkinen, Järvelä, & Byman, 2001).

The formation of collaborative Web-based learning environments offers a great many significant advantages (Connell, 1994). However, its success depends to a great extent on the designing of an environment which aims at the encouragement and support of the active participation of students (Mason & Bacsich, 1998).

The most recent research (Strijbos, Martens, & Jochems, 2004) recommends the following six steps in the formation of a collaborative Web-based learning environment: defining the learning goals, selecting the expected interaction, selecting the responsibilities of the human resources, deciding whether it will be a structured collaborative environment or not, and finally, defining the technological means to support the application.

Fundamentals of Designing a Web-Based DE Environment

The designing of a Web-based learning environment should be based on pedagogical principles, the understanding of learning material and the set learning goals, and definitely, the awareness of the way advanced Internet technologies can contribute to attaining these goals (Colis & Moonen, 2001).

The designing of a Web-based learning environment requires special attention, as we need to adhere to DE principles when we develop the teaching material, plan our teaching strategy, and decide on the combination of the technological means to support the attainment of the set learning goals (Anastasiades, 2002). In the proposed designing we follow the basic principles which should determine a learning environment according to the methodology of the American Distance Education Consortium (http://www.adec.edu).

- A distance learning environment should be regulated by clear learning goals and focus on predefined expected outcomes, consid-

Figure 5. Basic principles of American Distance Education Consortium (http://www.adec.edu/)

Defining goals Student-centred—Flexible	**Active participation of students** *Learning by doing, case-based learning*
Supporting collaborative learning Development of Communities	**Best combination of technological means**
ADEC Principles	

ering the special characteristics and needs of students within the context of an open, flexible, and student-centred approach

- The student should be encouraged to actively participate throughout the learning course, associating learning by doing, learning by reflection, case-learning study, and learning by exploring. Relating learning goals with real life learning experiences is a major priority.
- The learning environment should combine the use of technological means in order to attain the set learning goals considering the different learning styles of students. The selection of the technological means depends on the nature of content, the access to technology of the learning group, and the general educational philosophy of the teaching staff.
- The learning environment should encourage interaction among the human resources by ensuring the appropriate conditions and actively supporting the development of communities which share common interests with the aim of achieving collaborative learning.

Strong emphasis is placed on interaction, as it is regarded as one of the fundamental factors in achieving the learning objectives. The proposed interactive environment implements the theory of three types of interaction (Moore, 1989) and that of Paulsen's methodology (1977) and is illustrated in Table 2.

Phases of Development of a Web-Based Learning Environment

The designing of a Web-based learning environment proposed in this chapter applies the methodology of dividing the process into phases and implements the relevant approaches.

This model recommends four phases of development: analysis, designing, application, and assessment.

Each phase comprises of specific actions and demands the most of the human, learning, and technological resources of the system.

Synchronous Learning Environment: Designing Aspects

The main objective of videoconferencing is not to replace face-to-face conventional teach-

Table 2. Learning events based on interaction

Method / Interaction	Learner—Content	Learner—Teacher	Learner—Learner
One: alone (e.g. www)	Web pages with Graphics, Audio, Video, Quizzes, Interactive Checks on Progress		
One to One (e.g. e-mail)		E-mail, chat, online diary, Tutor-Marked Assignments	E-mail, chat (social and/or academic)
One to Many (e.g. Bulletin board)		E-mail, mailing list, group chat, discussion board	E-mail, mailing list, group chat, discussion board
Many to Many (e.g. Conferencing)		Group chat, discussion board	Group chat, discussion board, Group projects, Peer-based evaluation

Table 3. Phases of development of a Web-based learning environment

Phases	Description	Expected outcomes
1st Phase: Analysis		
Action 1: Analysis of needs justification of necessity of the Web-based course	• Which learning needs are covered? • Which are the learning groups we aim at? • Is the proposed plan feasible in teaching, technological and financial terms? • Will there be certification and in which form? • Will the students be charged?	**Report:** • Outline • Schedule of actions • Budget
Action 2: Analysis of the minimum necessary characteristics of the learning group	• Basic PC and Internet skills • Does the attendance require specific technological tools? How will the student be provided with them? • Which is the students' capacity for communication? • Will there be support for students not having the appropriate equipment of technology and communication?	**Report:** • Defining the minimum requirements of technological means and communication • Supporting system for students (loans, collaboration with organisations, etc.) • Pilot plan for students
Action 3: Analysis of the institution characteristics	• How will the teaching staff be trained in the new system? • Which are the necessary characteristics of the human and technological resources? • Which will the form of support of the new system be (internal development, purchase of services)?	**Report:** • Standardisation of the required human and technological resources • Support planning and alternative application plans • Statute
2nd Phase: Designing		
Action 1: Pedagogical planning	• **Defining the model:** synchronous, asynchronous, combined, hybrid? • **Learning theories:** we usually combine strategies from the most popular learning theories (behaviourism, cognitive) according to the learning goals and the students' profile	**Recommendation:** • Guidelines to pedagogical designing
Action 2: Designing the development of the teaching material	• Which will the methodology be? • Who will be involved, in what way and capacity? • Internal development or outsourcing? • monitoring tools	1. Manual for designing teaching material 2. Procedure plan 3. Curriculum
Action 3: Designing asynchronous transmission technologies	• What will the pedagogical characteristics of the asynchronous platform be? • Which basic functions will it support? • Open source or market research? • Purchase of equipment or access to outbound sources? • What are the necessary human and technological resources to develop, maintain and update it? • Internal development or outsourcing?	• Development plan of asynchronous transmission technologies
Action 4: Designing asynchronous transmission technologies	• What will the characteristics of videoconferencing be? • Which basic functions will it support? • Purchase of equipment or resort to other solutions? • What human and technological resources are required to operate the system?	1. Development plan of the videoconferencing system 2. Recommendation for the classroom
Action 5: Assessment plan	• What will be assessed? • How? • When? • By whom?	**Assessment methodology:** • of asynchronous learning • of synchronous education

continued on following page

Table 3. continued

Phases	Description	Expected outcomes
3rd Phase: Application		
Action 1: Designing the platform	• Application of the pedagogical planning • Purchase, installation and operation of equipment • Completion of applications • Grouping of material and software • Piloting • Training	1. Operation of platform 2. Teaching staff training on the usage and the pedagogical strategies
Action 2: Developing the teaching material	1. Compilation of the basic course manual applying DE method (author-assistant author) 2. Feedback from the DE methodology expert 3. Converting the material into HTML 4. Feedback (author, tutor, program co-ordinator) 5. Adaptation of material 6. Posting the material on the Internet (platform manager) 7. Feedback (platform manager, author, tutor, etc.)	1. Posting the material on the platform 2. Pilot operation of the platform 3. Assessment 4. Revision
Action 3: Applying synchronous education	• Layout of classroom • Purchase, installation and operation of technological equipment • Supplementary equipment • Training	1. Preparation of teleconferencing classroom 2. Training in the pedagogical application of synchronous transmission 3. Pilot operation 4. Assessment 5. Revision
Action 4: Defining the statute of human resources	• Statute of teaching staff, administrative staff and students	1. teaching staff guide 2. administrative staff guide 3. student guide
Action 5: Information, promotion	• Information of the potential	• Promotion
Action 6: Maintenance, update		• Maintenance guidelines
4th phase: Assessment		
	• Learning effectiveness • User-friendliness • Usability • Resolvability/preservability	**Formative and summative assessment of:** • Asynchronous platform • Synchronous services

ing but to come into supplementary operation (Anastasiades, 2006b; Hanor & Hayden, 2003). Berge και Mrozowski (2001), who studied educational videoconferences from 1990 since 1999, concluded that the most important issue for the successful outcome of a videoconference was not only the technology provided but also the educational methodology that was constructed and followed by the educators. IVC have to create an environment in which social dialogue, discourse and interaction, problem-based learning, negotiation of meaning, and construction of knowledge must be the basic goals of the whole process, so constructivism is the appropriate pedagogy for videoconferencing (Jonassen, Howland, Moore, & Marra, 2003; Anastasiades 2006b).

The proposed IVC pedagogical model requires a specific pedagogic approach (Anastasiades, 2003c, 2006a) designed to achieve the best possible results which include:

1. The development of the IVC pyramid
2. The delineation of a communication model
3. The designing of classroom architecture
4. The selection of a technological infrastructure
5. The design of an organisational and supporting model
6. The evaluation methodology

Figure 6. The IVC pyramid for adult learning

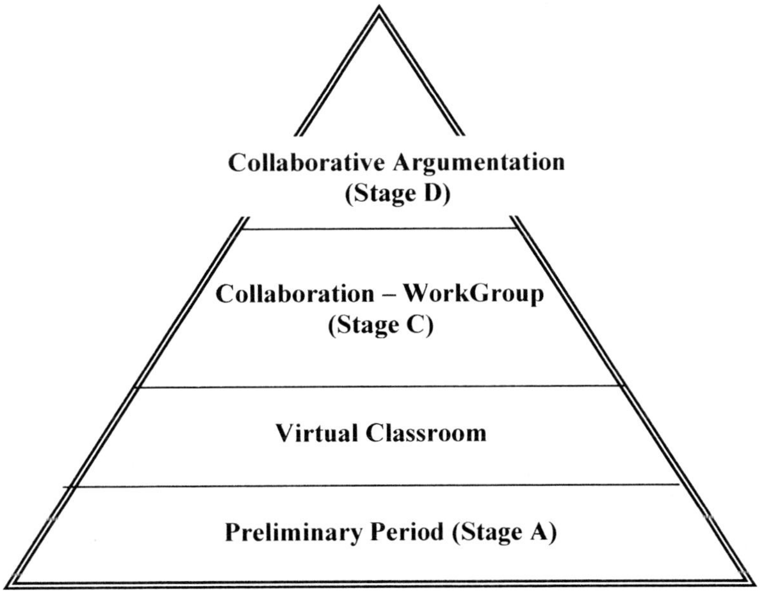

The IVC Pyramid

The adult students will get in contact with the new teaching model progressively, in order to become a part of the new learning environment as smoothly as possible.

This will be achieved through the implementation of four stages

Preliminary Period (Stage A)

Based on the above methodology, the steps of the Phase A are described as follows:

A1 Infrastructures: Provision of all the necessary technology (i.e., VC software, hardware, communicational status, networking of classrooms, and technical staff) in order to support the instructor on technical matters.

A2 Familiarisation of the instructors with the new reality:

- Seminar providing basic knowledge on VC in teaching and learning
- Familiarisation of instructors with the basic traits of the new learning environment (i.e., new roles, pedagogic concerns etc.).
- Knowing how to use the required technological VC tools and applications.
- Psychopedagogic approach of the new environment.
- Techniques of encouraging and motivating students.

A3 Preparation of students:

- Introductory briefings on distance learning via IVC.
- It is very important to choose and prepare a student as a class motivator. The student will have the responsibility to motivate the other students to get involved into the learning process and to support the administration background (mention and fix small technical problems, etc.). The motivator will be the interface between teacher and students in distance mode.

A4 Introductory VC: The introductory VC course will be organised by the instructor and will be attended by both the students of the instructor's class and the students form remote classrooms. At a scheduled time the two classes will be ready for action. In this phase we need to familiarise as smoothly as they can the instructor, the motivators, and the students with the concept of VC.

A5 Evaluation:

- **Students' evaluation** (evaluation of knowledge, experience from the teaching, etc.) by the teachers.
- **Teachers' evaluation** (noting down problems, ways of solving them, alternate approaches, etc.) by the teachers themselves in cooperation with other teachers, specialists and so forth.
- **Evaluation of the VC** (satisfaction, interactivity, familiarising, etc.)

Creation of the Virtual Classroom (Stage B)

The creation of an interactive learning environment is attempted in this phase, where the whole VC course period will be covered by lecturing, questions, and interactive dialogues in order to negotiate the adult learners needs, find out the prior knowledge, and to start building up the knowledge construction. The instructor must have the ability to manage the whole virtual classroom and the motivators to be active to the other classrooms. This phase is aiming at the creation of the necessary conditions in order to unfold all the activities that take place in a conventional classroom. Teachers, motivators and students familiarise themselves with the idea of the virtual classroom.

Evaluation:

- **Students' evaluation** (evaluation of knowledge, experience from the teaching, etc.) by the teachers.
- **Instructor' evaluation** (noting down problems, ways of solving them, alternate approaches, etc.) by the teachers themselves in cooperation with other teachers, specialists, and so forth.

Collaboration by Distance: Development of Joint Activities Between Remote Classrooms. (Phase C)

In this phase we have an attempt to create an open collaborative environment between the students of the remote classrooms, by creating work groups that will be in collaboration from a distance, in order to carry through a joint activity. During this phase, the students are the leading actors in the new collaborative environment, while the teacher and the motivators play a rather supervising-guiding role, interfering whenever they find it necessary.

In this stage we try to engage teachers and learners in collaborative learning activities according to research and practice in computer supported collaborative learning (Dillenbourg, Baker, Blaye, & O'Malley, 1996; O'Malley, 1995). According to the proposed methodology participants engage collaboratively in knowledge construction and negotiate with one another to reach a common shared understanding about a particular topic in order to achieve a joint project (Anastasiades 2003a; Dillenbourg & Traum, 1999)

Evaluation:

- **Self-evaluation of the students** (evaluation of the procedure by the students themselves).
- **Students' evaluation by the teachers** (noting down problems, ways of solving them, alternate approaches, etc.) by the students themselves in cooperation with other students, specialists, and so forth.

Collaborative Argumentations (Stage D)

Each group of the local and distant sites presents their collaborative projects to the virtual classroom and the facilitator organises an interactive collaborative argumentation. Learners have to explain their position, focusing their cognitive activities on the problem so that different perspectives are essential to discuss them collaboratively (Fischer, Kollar, Mandl, & Haake, in press)

Evaluation:

- **Self-evaluation of the students** (evaluation of the procedure by the students themselves).
- **Students' evaluation by the teachers** (noting down problems, ways of solving them, alternate approaches, etc.) by the students themselves in cooperation with other students, specialists, and so forth.

The Communication Model

This methodology is combined with the application of the three models method of University of Maryland, University College, (UMUC), and particularly model A (IDE, 1996) available online at http://www.umuc.edu/IDE/modeldata.html, that is a virtual classroom composed of groups of students in two or more distant locations. The collaboration of teachers and motivators is an innovative proposal, which support effectively the whole process.

The Classroom Architecture Model

The aim of a distance learning methodology is to develop an interactive learning virtual space in which learning communities with common interests can collaborate 'face-to-face' each other. This is easy to manage, if we have small courses (8-10 students). But what we have to do if we a need to manage a medium or big audience? The methodology chosen is that of separating the audience into active and passive.

The active audience consists of 6-8 students who sit in a triangle on the top of which we find the teacher and the blackboard. The students of the active and passive audience alternate during the lessons, so that all the students of each class have an active role experience.

The conceptual convergence of model A (IDE, 1996) with the proposed classroom architecture model forms a transitory approach of original methodology which from now on will be referred

Figure 7. The communication model (point to point IVC)

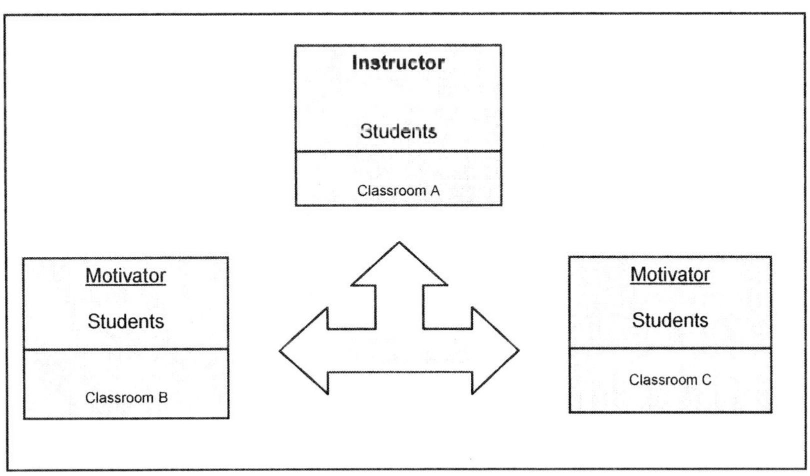

Figure 8. The classroom architecture model

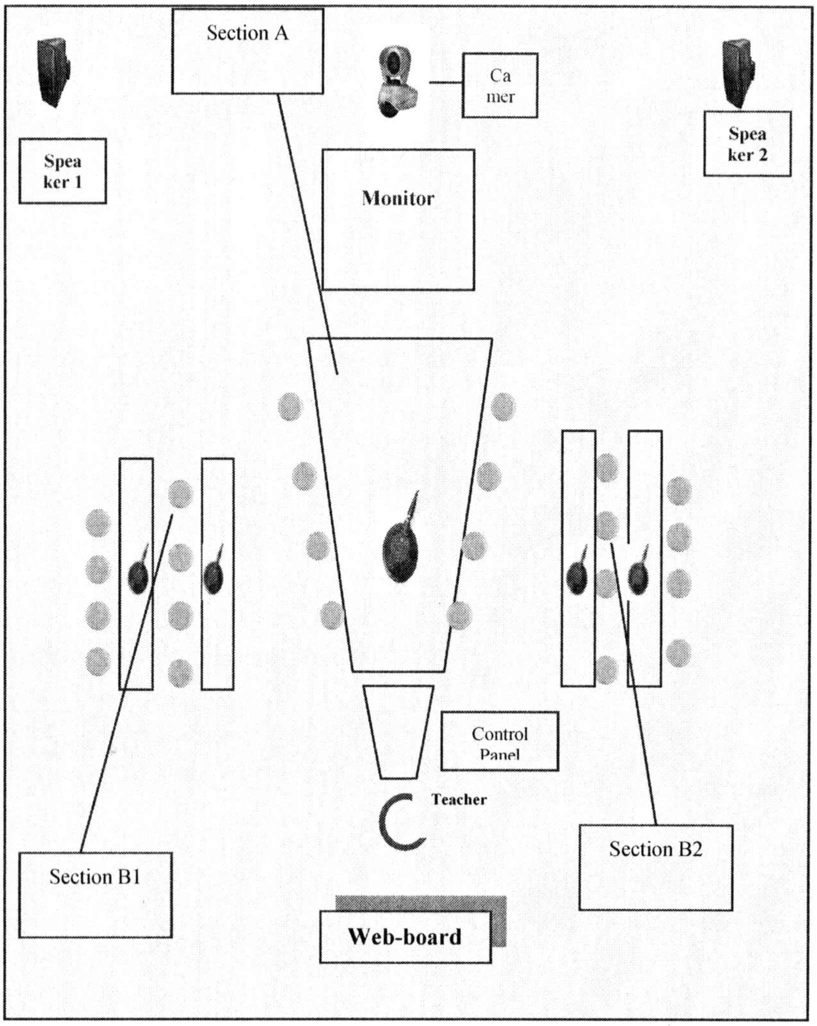

to as transition methodology. Transition methodology ensures equal participation of all the students of the class in the new learning environment, without upsetting the relation of the students to the existing structure of the traditional school schedule.

The Technological Infrastructure

The main components of the proposed synchronous learning environment are as follows:

1. Videoconferencing system H.320/H.323 and MCU
2. Interactive whiteboard facilities
3. Streaming media encoder/server
4. Unidirectional condenser boundary microphone

The Organisation and Support Model

The organisational and support model constituted three committees. The monitoring committee, which is composed of certain representatives of institutions, has the general monitoring responsibility of the project. The research committee has the planning and implementation responsibility of

Figure 9. The communication model (Multipoint IVC)

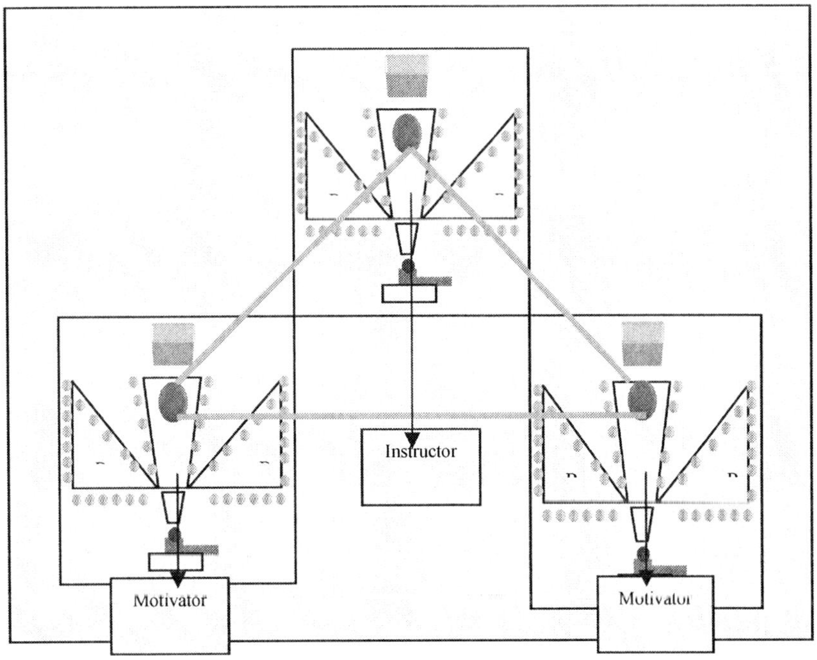

the project. Finally, the organisational committee has a general supporting role.

A CASE STUDY AT THE UNIVERSITY OF CRETE

General Description

The whole project, PAIDEIA OMOGENWN (http://ediamme.edc.uoc.gr/diaspora/), aims to continue, develop, and promote Greek language and culture, to primary and secondary students of Greek origin, who live and study abroad, as well as non-Greek speaking students who want to learn the Greek language and become participants of the Greek culture (Damanakis, 1987). The implementation of the specific program started in June 1997 and continued until December 2004. It was funded by the Greek Ministry of Education (25%) and the European Union (75%). The Ministry of Education in Greece, with its various departments, supervises the program, while its implementation has been assigned to the University of Crete and more specifically to the Center of Intercultural and Migration Studies (E.DIA.M.ME.) at the Department of Education of the university with Director Professor Michali Damanaki.

One of the most important topics of the project concerns the implementation of a complete e-learning environment adopted and developed for the training for the continuous training of teachers who teach Greek as a second and foreign language through e-learning.

Based on the initial plan, learning materials were designed and developed for three courses:

1. Socialisation and Education in the Diaspora (Prof. Damanakis)
2. History and Culture in the Modern Greek (Prof. Xourdakis)
3. Topics in Modern Greek Literature (Instructor Mitrofanis)

This material was designed to work in an environment with both synchronous and asynchronous e-learning, whose goal is to create an 'learning community' on the Web between instructors of Greek descent working all over the world and teaching the Greek language. Within the framework of this 'community,' all instructors of Greek descent will have the opportunity to communicate with their colleagues, as well as with special scientists (Damanakis & Anastasiades, 2005). They will be able to discuss educational questions and issues of daily concern and also get trained on specialised subject matters that will help them in their job Also, all the learning materials and articles that have been published within the program can be found online, for the support of e-learning (synchronous and asynchronous), but also for those who wish to get generally informed (http://www.uoc.gr/diaspora).

The Target Group

Based on the suggested methodology (Anastasiades, 2005c; 2006a), Greek teachers abroad were separated in three target groups, depending on their ability to make use of computers and the Internet.

The first group is made up of teachers who lack basic ICT skills. The second group is made up of teachers who do possess qualifications in information technology but the telecommunications services of the country they live in do not allow for access to the Internet. The third group is made up of teachers who both possess basic ICT skills and have access to the Internet.

Our effort concerns the implementation of a complete Web-based learning environment, focusing on the third group and particularly on countries such as the Australia, USA, Canada, Germany, Sweden, Great Britain, France, and other European countries. (see Table 4)

Table 4. Teachers applications/country (September-November 2006)—EDIANNE edited by Petraki

Country	Teachers' Applications
Great Britain	13
Egypt	20
Australia	52
Belgium	30
France	1
Germany	202
Georgia	9
Switzerland	6
USA	17
Kazakhstan	9
Canada	7
Kirgistan	1
Congo	4
Libya	2
Luxemburg	3
South Africa	17
Holland	11
Uzbekistan	3
Saudi Arabia	5
Sweden	16
Syria	2

The Blended Learning Environment: Form Theory to Practice

The proposed methodology (Anastasiades, 2005c; 2006a) developed aims to create a blended learning environment, which:

- Will support the study of learning material through (the method of) distance learning;
- Will provide the trainee with evaluation and self-assessment methods;
- Will facilitate the distribution of necessary information to system users;
- Will encourage the development of an interactive environment through the provision

of the suitable communication tools; and

- Will functionally support the administrative part of the learning process.

Special focus will be given on the creation of an open collaborative learning environment where the instructor and the trainees shall communicate in both synchronous and asynchronous manner; those trained will study in their own time, although subject to a predetermined timetable for the delivery of projects.

At this point it should be noted that the methodology for the evaluation system of the asynchronous platform is currently being developed.

The Asynchronous Learning Environment

The asynchronous learning environment developed the e-class platform, which is based on the philosophy of freeware software developed by the group of asynchronous learning team of Greek Academic Internet GUnet (Anastasiades, 2005c). The platform of asynchronous learning is designed in accordance with the characteristics of the learning environment (see http://elearn.edc.uoc.gr/).

The introductory interface of the platform contains information with regard to the educational programs, the teaching method, the cost, the terms and conditions for studying, the studies' certification, and so forth. It further provides a friendly and functional environment through which the regions corresponding to the system's human resources are activated, that is, the professor, the assistant, the student or trainee, the visitor, and finally the administrator.

The current learning environment provides users with the following additional possibilities:

1. Configuration of Web pages for registration and attendance of courses for students. Registration will be made through clearly determined steps that each student must follow (i.e., expression of interest, completion of questionnaire, collection of personal information, registration and mission of code to student, etc.).
2. Configuration of Web pages for the posting and management of courses for teachers. Specifically for the idea of courses there will be models of texts to be used by the professors-assistants in writing the content, which will then be used in the asynchronous environment under minimal possible intervention by the user.
3. Configuration of Web sites for taking the necessary action undertaken by the assistants of each course (tutors).
4. Figuration of learning material in categories and in chronological order (deliveries, exercises, questions, work, etc.)

Picture 1. The asynchronous learning platform

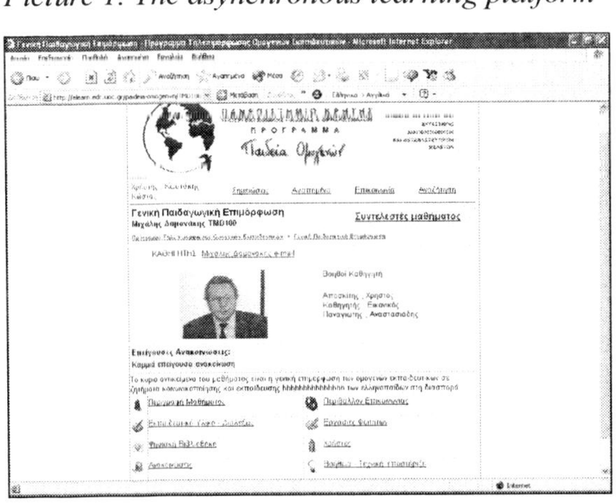

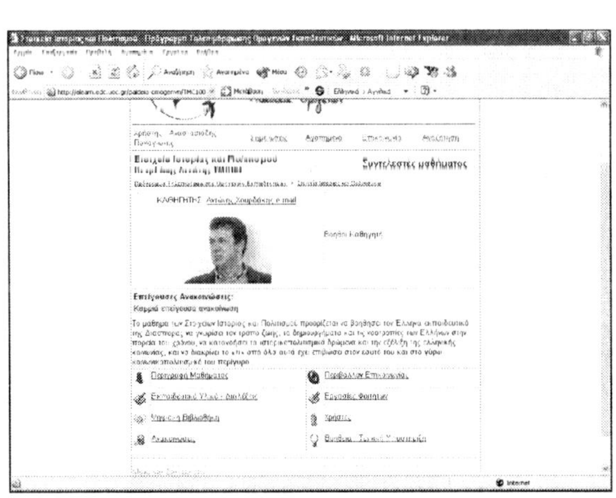

Figure 10. The asynchronous learning environment (http://elearn.edc.uoc.gr/)

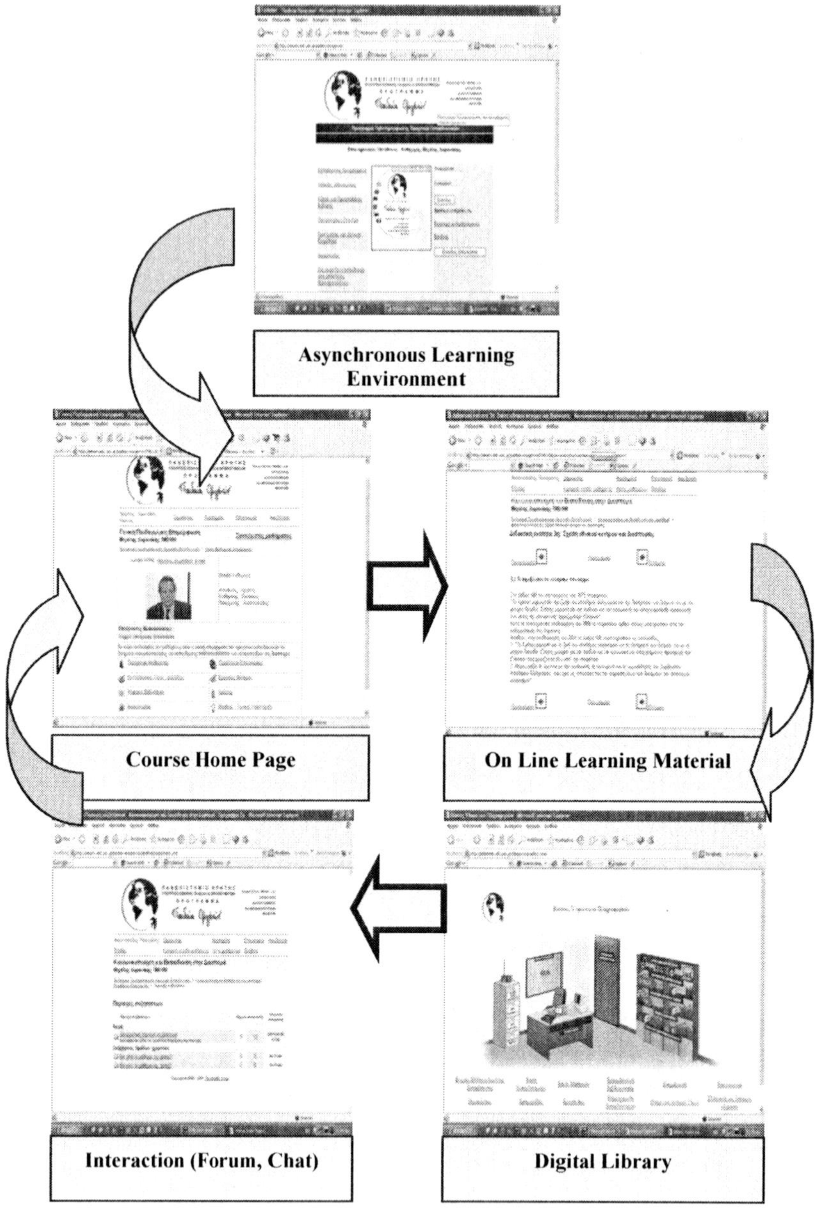

5. Development of chat rooms to facilitate the direct communication between students, assistants, and professors
6. Configuration of Web sites for help and frequently asked questions in the e- learning environment.
7. Incorporation of a search engine to assist in the location of Web sites, the list of courses depending on the content and/or keywords.
8. Incorporation of an evaluation system of the educational process by students, in various stages of the process (initial, middle, final stage).
9. Automation of production of copies in CD-ROM or in printed form, based on the content that has already been posted on Web sites in the asynchronous platform.

The Synchronous Learning Environment

The designing of a synchronous learning environment should take into consideration the needs of the students, especially when students are located in different countries and cities and are diversified in terms of technology, organisation, and administration of learning activities and geographical scatter (Anastasiades, 2006a). The problem further deteriorates in the areas where there is no access to videoconferencing rooms due to lack of infrastructure, the prohibitive cost, or lack of trained staff.

In our effort to meet the requirements of a complex environment we designed the applied synchronous learning based on:

1. The layout of a technologically advanced videoconferencing room in the University of Crete, which can link up to eight remote locations simultaneously.
2. The educational application of videoconferencing not only in well-equipped rooms but also at VC designated areas at schools.
3. The live broadcast of lectures via Internet in real time.

Figure 11. The basic characteristics of the synchronous learning environment

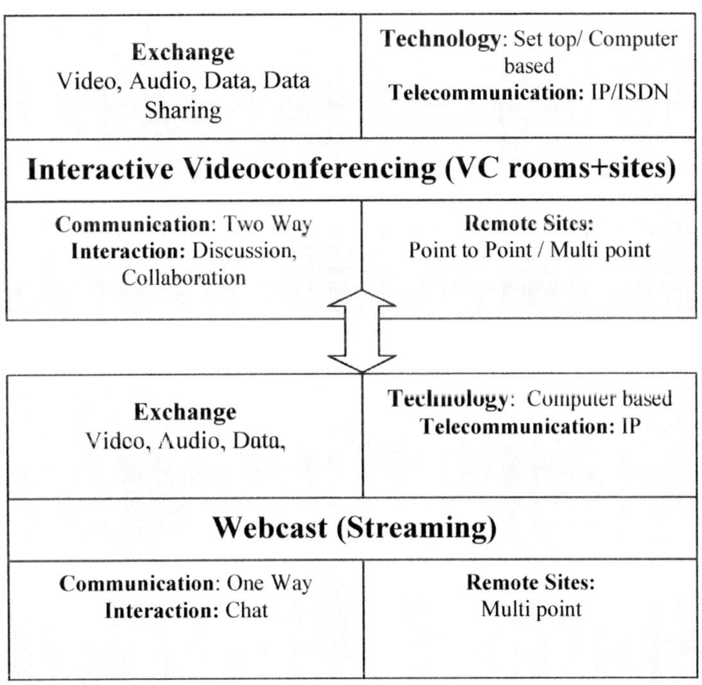

IVC Room in the University of Crete

Our IVC room has a VCON HD5000 Videoconference System.

Transmission speed: H.323: 64Kbps-4Mbps, H.320: 64Kbps-384Kbps
Video standards: H.261, H.263, H.264 (up to 1M)

Picture 2. The IVC room in the University of Crete

Picture 4. IVC sites in Düsseldorf , Bielefeld (Germany)

CONCLUSION

The new advanced learning technologies of synchronous and asynchronous transmission allow for the implementation of a new learning environment, which is flexible in terms of location, time, and pace of learning. In this chapter we present the principles and methodology of a pedagogical blended learning model, based on the adult education principles (emphasising on

Picture 3. IVC room in Melburne (Australia)

Picture 5. IVC Rooms in Melbourne, Sydney, Adelaida (Australia) and IVC sites in Düsseldorf, Bielefeld (Germany) A Multipoint IVC

Picture 6. Web cast: Live streaming (http://Webcast.ucnet.uoc.gr/)

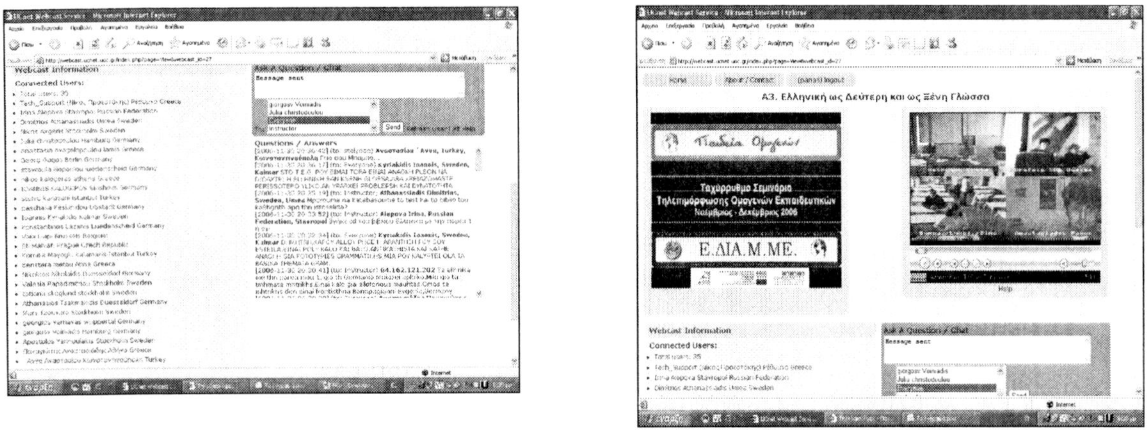

Picture 7. Dec 2005 Prof Damanakis (Greece) Audience in Melbourne (Australia) Course: General Pedagogy

Picture 8. Dec 2005 Prof Xourdakis (Greece) Audience in Melbourne (Australia) Course: History and Culture

Picture 9. Dec 2005 Giannis Mitrofanis (Greece) Audience in Melbourne (Australia) Course: Modern Greek Literature

Picture 10. March 2006: Prof. Katsimali (Greece) Audience in: Melbourne, Sydney & Adelaide (Australia) Course: Language

Picture 11. March 2006: Prof Damanakis Audience in: Melbourne & Adelaide (Australia) and Bilefeld & Düsseldorf (Germany)

Picture 12. March 2006: Prof Xatzidaki Audience in: Bilefeld & Düsseldorf (Germany) Course: Language

Picture 13. December 2006: Prof Xatzidaki, Prof Katsimali (Course: Language) IVC Audience in: Bilefeld & Düsseldorf (Germany) Live Webcast Audience in 7 countries: Germany, Sweden, Turkey, Netherlands, Belgium, Czech Republic, Georgia , Egypt, United Kingdom, Russian Federation, Luxembourg, Denmark

Picture 14. December 2006: Prof Papadogiannakis & Prof Nikoloudaki (Course: Language) IVC Audience in: Bilefeld & Düsseldorf (Germany) Web cast in 7 countries: Germany, Sweden, Turkey, Netherlands, Belgium, Czech Republic, Georgia , Egypt, United Kingdom, Russian Federation, Luxembourg, Denmark

Picture 15. IVC Audience in: Bilefeld & Düsseldorf (Germany). Live Webcast Audience in 7 countries: Germany, Sweden, Turkey, Netherlands, Belgium, Czech Republic, Georgia , Egypt, United Kingdom, Russian Federation, Luxembourg, Denmark

December 2006: Prof Xourdakis & Prof Karagiorgos (Course: History and Culture)

December 2006: Prof Spadidakis (Course: Educational Maultimedia)

the transforming learning theory), the constructivism theories, and the DE principles of ADEC. Applying our pedagogical approach, we defined the guidelines and phases of development of an asynchronous learning environment as well as the steps and methodology of a synchronous learning environment, focusing on interactive videoconferencing.

The proposed model is the cornerstone of the designing and application of a significant DE program for the training of teachers throughout the world, which is implemented by EDIAMME of the University of Crete and is under constant assessment and adaptation.

FUTURE RESEARCH DIRECTIONS

Adult education is one of the most important priorities of contemporary information society. Millions of adults all over the world look for training programs in order to improve their knowledge and skills in their professional, social, and personal arena.

In the next years, blended learning will become one of the most important educational processes of adult learning all over the world, as it is open and flexible and can be transferred to the learner's educational environment, pace, and time.

The introduction of ICT into education significantly changes its structure providing adults with new learning environments. On the other side, most theories on adult education have been planned and implemented into face-to-face teaching environments, resulting in many implementation problems due to their unconsidered design and adaptation in these blended-based learning environments.

Planning, development, and implementation of blended learning environments for adult learners demand a preparation of holistic pedagogical approaches with respect to their individual characteristics.

Many questions arise which have not yet been answered. Some of these questions are:

- Are conventional-based instruction principles adequate for these blended learning environments? Does the use of ICTs add new facts and what are these?
- Which are the adults' characteristics that require new educational experiences through the use of Internet? Which is the best way for training adults in blended learning environments? Are there any peculiarities and what are these?
- Which are the most suitable learning theories which can support the learning goals effectively in blended learning environments? Is there a need for changes and adjustments and what these may be?

Further research must focus on blended learning pedagogy for adult learners.

REFERENCES

Abbott, L. et al. (1993). *Videoconferencing and distance learning. Faculty of education and department of adult and continuing education.* Ireland: University of Ulster.

ADEC Guiding Principles for Distance Learning and Teaching. (1999). *American Distance Education Consortium.* Retrieved February 21, 2008, from http://www.adec.edu/admin/papers/distanceteaching_ principles.html

Anastasiades, P. (2002a). The theory of information reversal. *Computers and Society, ACM Special Interest Group on Computers and Society (ACM SIGCAS), 32*(2), 10-16.

Anastasiades, P. (2002b). *Towards the global information society: The enactment of a regulatory framework as a factor of transparency and social cohesion* (LNCS 2510, pp. 527-535). Berlin/Heidelberg: Springer-Verlag.

Anastasiades, P. (2003a). Distance learning in elementary schools in Cyprus: The evaluation methodology and results. *Computers & Education, 40*(1), 17-40. Elsevier Science.

Anastasiades, P. (2003b, March 24-29). Synchronous distance learning courses: Evaluation methodology. In *Proceedings of the International Conference SITE 200: Society for Information Technology and teacher Education*, Nashville/New Mexico.

Anastasiades, P. (2003c, April 1-5). Building up a pedagogical model for synchronous distance learning courses (selected paper). In *Proceedings of the 13th International Conference on College Teaching and Learning*, Florida.

Anastasiades, P. (2004, June 27-30). *From the conventional classroom to the new hybrid learning environment: Steps and methodology.* Paper presented at the Eleventh Literacy and Education Research Network Conference on Learning, Cojímar Pedagogical Convention Centre. Havana, Cuba: Common Ground.

Anastasiades, P. (2005a, November 11-13). Lifelong e-learning and digital divide. In *Proceedings of the Third International Conference on Open and Distance Learning.* Patra, Greece: Greek Open University.

Anastasiades, P. (2005b, July 14-17). Web based education and digital divide towards the European lifelong learning society. In *Proceedings of the 3rd International Conference on Education and Information Systems: Technologies and Applications*, Orlando, FL.

Anastasiades, P. (2005c, June 27-July 2). Synchronous vs asynchronous learning? Principles, methodology and implementation policy of a blended learning environment for lifelong learning, at the University of Crete. In *Proceedings of the ED-MEDIA 2005 World Conference on Educational Multimedia, Hypermedia and Telecommunications, (AACE),* Montreal. Association for the Advancement of Computing in Education.

Anastasiades, P. (2005d, October 24-28). Lifelong learning and Greek diaspora towards the global information society: Suggested e-learning monitoring methodology (e-LMM). In *Proceedings of the E-Learn 2005--World Conference on E-Learning in Corporate, Government, Healthcare, and Higher Education,* Vancouver, Canada.

Anastasiades, P. (2006a, July 5-7). Interactive videoconferencing in lifelong learning: Methodology and implementation policy at the University of Crete (E.DIA.M.ME). In *Proceedings of the Diverse 2006, 6th International Conference on Video and Videoconferencing in Education*. Scotland, Glasgow: Caledonian University.

Anastasiades, P. (2006b, July 5-7). Interactive videoconferencing in K- 9 education: "ODUSSEAS 2000-2004" a case study in elementary schools in Greece and Cyprus. In *Proceedings of the Diverse 2006 1 6th International Conference on Video and Videoconferencing in Education.* Scotland, Glasgow: Caledonian University.

Anastasiades, P., & Spantidakis, J. (2006, June 26-30). Advanced learning technologies and the new hybrid learning environment towards the knowledge society: Steps and implementation policy. In *Proceedings of the ED-MEDIA World Conference on Educational Multimedia, Hypermedia and Telecommunications (ED MEDIA 2006).* Orlando, FL: Association for the Advancement of Computing in Education (AACE).

Barrows, H. S. (1986). A taxonomy of problem-based learning methods. *Medical Education, 20,* 481-486.

Bastiaens, T., & Martens, R. (2000). Conditions for Web-based learning with real events. In B. Abbey (Ed.), *Instructional and cognitive impacts of Web-based education* (pp. 1-32). London: Idea Group.

Bates, A. W. (1995). *Technology: Open learning and distance education.* New York: Routledge.

Broady, E. (1996). Learner attitudes towards self-direction. E. Broady & M. M. Kenning (Eds.), *Promoting learner autonomy in university language teaching* (pp. 215-236). London: CILT.

Brookfield, S. D. (1990). *The skillful teacher.* San Francisco: Jossey-Bass.

Brookfield, S. D. (Ed). (1985). The continuing educator and self-directed learning in the community. In self-directed learning: From theory to practice. *New Directions for Continuing Education* (25). San Francisco: Jossey-Bass.

Brookfield, S. D. (1986). *Understanding and facilitating adult learning.* San Francisco: Jossey-Bass.

Brookfield, S. D. (1987). *Developing critical thinkers.* San Francisco: Jossey-Bass

Brookfield, S. D. (1991). The development of critical reflection in adulthood. *New Education, 13*(1), 39-48.

Brookfield, S. D. (1992). Developing criteria for formal theory building in adult education. *Adult Ed. Q. 42*(2), 79-93.

Brookfield, S. D. (1994). Adult learning: An overview. In A. Tuinjman (Ed.), *International encyclopedia of education.* Oxford: Pergamon Press.

Brookfield, S. D. (1995). *Becoming a critically reflective teacher.* San Francisco: Jossey-Bass.

Brookfield, S. D. (1996). Adult learning an overview. In A. Tuijnman (Ed.), *International encyclopedia of adult education and training* (pp.175, 380). Oxford: Pergamon.

Brookfield, S. D. (2001). Repositioning ideology critique in a critical theory of adult learning. *Adult Education Quarterly, 52*(1), 7-22.

Brown, S. (2001). *Views on videoconferencing higher education and research opportunities in the UK (HERO).* Retrieved February 21, 2008, from http://www.hero.ac.uk/inside_he/archive/views_on_videoconferencin883.cfm?

Bruner, J. S. (1990). *Acts of meaning.* Cambridge: Harvard University Press.

Brusilovsky, P. (1996). Methods and techniques of adaptive hypermedia. *User Modeling and User-Adapted Interaction, 6*(2/3), 87-129.

Brusilovsky, P. (1999). Adaptive and intelligent technologies for Web-based education. *Kunstliche Intelligenz: Special Issue on Intelligent Systems and Teleteaching, 4,* 19-25.

Brusilovsky, P. (2001). Adaptive hypermedia. *User Modeling and User-Adapted Interaction, 11*(1/2), 111-127.

Carr, W., & Kemmis, S. (2002). *For a critical educational theory. Education, knowledge and action-research.* Athens: Kodikas.

Chan, T., & Rapman, J. (1999). Flow in Web based instructional activity: An exploratory research project. *International Journal of Educational Communications, 5*(3), 225-237.

Chen, Y., & Willits, F. (1998). A path analysis of the concepts in Moore's theory of transactional distance in a videoconferencing environment. *Journal of Distance Education, 13*(2), 51-65.

Clark, J. (2002). A product review of WebCT. *Internet and Higher Education, 5,* 79-82.

Cobb, P. (1994). Where is the mind? Constructivist and sociocultural perspectives on mathematical development. *Educational Researcher, 23*(7), 13-20.

Colis & Moonen (2001). *Flexible learning in a digital world: Experiences and expectations.* London: Kogan Page.

Connell, R. W. (1994). Poverty and education. *Harvard Educational Review, 64*, 125-149.

Connick, P. (1999). *The distance learner guide*. Upper Saddle River, NJ: Prentice

Cornford, H. J., & Pollock, N. (2003). *Putting the university online: Information, technology and organisational change*. Buckingham, UK: The Society for Research into Higher Education and Open University Press.

Coventry, L. (2000). *Videoconferencing in higher education*. Heriot Watt University, Edinburg: Institute for Computer Based Learning. Retrieved February 21, 2008, from http://www.man.ac.uk/MVC/SIMA/video3/contents.html

Cropley, A. J., & Kahl, T. N. (1983). Distance education and distance learning: Some psychological considerations. *Distance Education, 4*(1), 27-39.

Damanakis & Anastasiades (2005). Life long & distance learning and the diaspora: Implementing a virtual learning environment at the University of Crete. *Themes in Education: Special Issue Information & Communication Technologies in Diaspora, 6*(1), 83-96.

Dearing, R. (1997). Higher education in the learning society. *Report of the Committee of Inquiry into Higher Education*. Retrieved September 15, 2005, form http://www.ncl.ac.uk/ncihe/index.htm

De Bra, P., Eklund, J., Kobsa, A., Brusilovsky, P., & Hall, W. (1999). *Adaptive hypermedia: Purpose, methods and techniques*. Paper presented at the 10th ACM Conference on Hypertext and Hypermedia (pp. 199-200).

Dick, W., & Carey, L. (1996). *The systematic design of instruction*. New York: Harper Collins College Publishers.

Digital Bridges: K-12 videoconferencing, Note: another very good resource. Its "Teacher's Guide to Videoconferencing" is valuable and comprehensive, with emphasis on phased approach to planning, delivery and evaluation. A list of reference literature could be updated (the latest referenced articles were from 1999). http://www.netc.org/digitalbridges/vc/index.html

Dillenbourg, P. (1999). What do you mean by "collaborative learning"? In P. Dillenbourg (Ed.), *Collaborative learning: Cognitive and computational approaches* (pp. 1-19). Amsterdam: Pergamon.

Dillenbourg, P., Baker, M., Blaye, A., & O'Malley, C. (1996). The evolution of research on collaborative learning. In E. Spada & P. Reiman (Eds.), *Learning human and machine* (pp. 189-211). Oxford: Elsevier.

Dillenbourg, P., & Traum, D. (1999). Does a shared screen make a shared solution? In C. Hoadly & J. Roschelle (Eds.), *Proceedings of the Third Conference on Computer Supported Collaborative Learning* (pp. 127-135). California: Stanford University.

Dori, Y. J., Barak, M., & Adir, N. (2003). A Web-based chemistry course as a means to foster freshmen learning. *Journal of Chemical Education, 80*(9), 1084-1092.

Dunderstadt, J. (1997). The future of the university in an age of knowledge. *Journal of Asynchronous Learning Networks, 1*(1), 78-88.

Eastman, J., & Iyer, R. (2004). The elderly's uses and attitudes towards the Internet. *Journal of Consumer Marketing, 21*(3), 208-220.

Engestrom, Y. (1987). *Learning by expanding: An activity theoretical approach to developmental research*. Helsinki, Finland: Orienta Konsultit Oy.

Fillion, G., Limayem, M., & Bouchard (1999). Videoconferencing in distance education: a study of student perceptions in the lecture context. *Innovations in Education and Training International, 36*, 302.

Finn, K. E., Sellen, A. J., & Wilbur, S. B. (Eds.). (1997). *Video-mediated communication.* Mahwah, NJ: Lawrence Erlbaum Associates.

Fischer, F., Kollar, I., Mandl, H., & Haake, J. M. (Eds.). (in press). Scripting computer-supported collaborative learning. *Cognitive, computational and educational perspectives.* New York: Springer.

Forsyth, I. (2001). *Teaching & learning materials & the Internet.* London: Kogan Page.

Freire, P. (1984). *The importance of reading and the liberation process.* Mexico: Siglo XXI.

Gagné, R., Briggs, L., & Wager, L. (1994). *Principles of instructional design.* Fort Worth, TX: HBJ College Publishers.

Garrison, D. R. (2006). Online collaboration principles. *Journal of Asynchronous Learning Networks, 10*(1), 25-34.

Gibson, C., & Cohen, S. (Eds.). (2003). *Virtual teams that work: Creating conditions for effective virtual teams.* San Francisco: Jossey-Bass/Wiley.

Graham, D., McNeil, J., & Pettiford, L. (2000). *Untangled Web: Developing teaching on the Internet.* Harlow, UK: Pearson Education.

Grigoriadou, M., Papanikolaou, K., Cotronis, Y., Velentzas, C., & Filokyprou, G. (1999). Designing and implementing a Web-based course. In *Proceedings of International Conference of Computer Based Learning in Science*, Enschede, Netherlands.

Gürer, D., Kozma, R., & Millán, E. (1999). Impact of shared applications and implications for the design of collaborative learning environments. In S. Lajoie (Ed.), *Proceedings for the ninth world conference on artificial intelligence in education* (AI-ED99) (pp. 439-445). Amsterdam: IOS Press.

Guzdial, M., & Turns, J. (2000). Effective discussion through a computer-mediated anchored forum. *Journal of the Learning Sciences, 9,* 437-469.

Häkkinen, P., Järvelä, S., & Byman, A. (2001). Sharing and making perspectives in Web-based conferencing. In P. Dillenbourg, A. Eurelings, & K., Hakkarainen (Ed.), *European perspectives on computer-supported collaborative learning* (pp. 285-292). In Proceedings of the 1st European Conference on Computer-Supported Collaborative Learning, Maastricht University, Maastricht.

Hannafin, M., & Land, S. (1997). The foundations and assumptions of technology enhanced student-centered learning environments. *Instructional Science, 25,* 167-202.

Hanson, D., Maushak, N., Schlosser, C., Anderson, M., Sorensen, C., & Simonson, M. (1997). *Distance education: Review of the literature* (2nd ed.). Washington, D.C./Ames, IA: Association for Educational Communications and Technology and Research Institute or Studies in Education.

Harley, D. (2001). Higher education in the digital age: Planning for an uncertain future. *Syllabus: New Dimensions in Education Technology, 15*(2), 10-12.

Hayden, K. L. (1999). *Videoconferencing in K-12 education: A Delphi study of characteristics and critical strategies to support constructivist learning experiences.* Retrieved February 21, 2008, from http://hale.pepperdine.edu/~kahayden/dissertation.html

Hearnshaw, D. (1999). Capitalising on the strengths and availability of desktop videoconference. *Active Learning, 9,* 52-59.

Hiltz, S. R. (1994). *The virtual classroom: Learning without limits via computer networks.* Norwood, NJ: Ablex.

Holmberg, B. (1986). *Growth and structure of distance education.* London: Croom Helm.

Horton, S. (2000). *Web teaching guide: A practical approach to creating course Web sites.* London: Yale University Press.

IDE. (1996). Three models of distance education. *University of Maryland University College.* Retrieved February 21, 2008, from http://www.umuc.edu/IDE/modeldata.html

Illeris, K. (2002). *The three dimensions of learning: Contemporary learning theory in the tension field between the cognitive, the emotional, and the social.* UK: Leicester.

Illeris, K. (2003) Towards a contemporary and comprehensive theory of learning. *International Journal of Lifelong Education, 22*(4), 396-406.

Jarvis, P. (2004). *Adult and continuing education. Theory and practice.* Athens: Metaixmio.

Jolliffe, A., Ritter, J., & Stevens, D. (2001). *The online handbook: Developing and using Web based learning.* London: Kogan Page.

Jonassen, D. H. (1992). Objectivism versus constructivism: Do we need a new philosophical paradigm? *Educational Technology Research and Development, 39*(3), 5-14.

Jonassen, D. (1997). Instructional design models for well-structured and ill-structured problem-solving learning outcomes. *Educational Technology Research & Development, 45*, 65-94.

Jonassen, D., Howland, J., Moore, J., & Marra, R. (2003). *Learning to solve problems with technology, a constructivist perspective.* Upper Saddle River, NJ: Merrill Prentice Hall.

Kagan, S. (1994). *Cooperative learning.* San Juan: Capistrano.

Karpov, Y. V., &. Haywood, H. C. (1998, January) Two ways to elaborate Vygotsky's concept of mediation. Implications for instruction. *American Psychologis,* 27-36.

Keegan, D. (1986). On defining distance education. In D. Sewart, D. Keegan, & B. Holm-berg (Eds.), *Distance education: International perspectives* (pp. 6-33). London: Routledge.

Keegan, D. (1995). *Distance education technology for the new millennium: Compressed video teaching* (ERIC Document Reproduction Service No. ED 389 931).

Keegan, D. (1996). *The foundations of distance education.* London: Croom Helm.

Kemmis, S. (1985). Action research and the politics of reflection. In D. Boude et al. (Eds.), *Reflection turning experience into learning.* London: Kogan Page.

Knowles, M. S. (1970). *The modern practice of adult education: Andragogy versus pedagogy.* New York: Association Press.

Kokkos, A. (2007). Special characteristics and aims of adult education. *Adult Education,* **9**, 4-19.

Kokkos, A., & Lionarakis, A. (1998). *Open and distance learning, relations teachers and students* (Vol. B). Hellenic Open University Press.

Kostoula, N., & Makrakis (2006). *Intercultural and education.* Athens.

Lawyer-Brook, D. (1991). *The use of distance education to deliver, chapter I services.* Denver: Region 5 R-TAC.

Light, P., Nesbitt, E., Light, V., & White, S. (2000). Variety is the spice of life: Student use of CMC in the context of campus based study. *Computers & Education, 34*, 257-267.

Lionarakis, A. (1998). Polymorfic education: A pedagogical framework for open and distance learning in universities in a digital era. *Transformation, Innovation and Tradition – Roles and Perespectives of Open and Distance Learning, 2*, 499-504. Italy: University of Bologna.

Lionarakis, A. (2001). Qualitative approaches of the design and production ofdistance, polymorphic educational material. In Makrakis (ed), *Proceedings of the Pan-Hellenic Conference with International Participation for New Technologies in Education and in Distance Education* (Vol. B). Atrapos Editions.

Lionarakis, A. (2003, June 15-18). A *preliminary framework for a theory of open and distance learning: The evolution of its complexity.* Paper presented at the 12th Conference of the European Distance Education Network, Rhodes.

Lionarakis, A. (2006). *Open and distance learning, theory and practice.* Athens: Probobos.

Maier, P., & Warren, A. (2000). *Integrating technology in learning and teaching.* London: Kogan Page.

Martin, M. (2000). *Videoconferencing in teaching and learning: Case studies and guidelines.* Western Education and Library Board.

Mason, R., & Bacsich, P. (1998). Embedding computer conferencing into university teaching. *Computers in Education, 30*(3/4), 249-258.

Massialas, B. (1989). *Claims and counterclaims of the educational uses of microcomputers.* Paper presented at the Third International Pedagogical Conference, Athens, Greece.

Mezirow, J. (1980). How critical reflection triggers transformative learning. In J. Mezirow (Ed.), *Fostering critical reflection in adulthood: A guide to transformative and emancipatory learning.* San Francisco: Jossey-Bass.

Mezirow, J. (1981). A critical theory of adult learning and education. *Adult Education, 32*(1), 3-24.

Mezirow, J. (1985). A critical theory of self-directed learning. In self-directed learning: From theory to practice. *New Directions for Continuing Education*, (25). San Francisco: Jossey-Bass.

Mezirow, J. (1991). *Transformative dimensions of adult learning.* San Francisco: Jossey-Bass.

Mikropoulos, T. A. (2000). Design, development and evaluation of advanced learning environments. An overall approach, HERMES. *Advanced systems for teaching and learning over the world wide Web* (pp. B42-B52), Samos.

Mitchell, J. et al. (1993). *Video-conferencing in higher education in Australia.* Canberra: AGPS.

Moore, M. G. (1989). Editorial: Three types of interaction. *American Journal of Distance Education, 3*(2), 1-16.

Moore, M. G., & Kearsley, G. (1996). *Distance education: A systems view.* Wadsworth Pub Co.

Nachmias, R. & Mioduser, D. (2000). Web-supported emergent-collaboration in higher education courses. *Educational Technology & Society, 3*(3), 94-104.

Norris & Pippa. (2001). *Digital divide: Civic engagement, information poverty and the Internet worldwide.* Cambridge: University Press.

Norton, B. (2001). Non-participation, imagined communities and the language classroom. In P. Breen (Ed.), *Learner contributions to language learning: New directions in research* (pp. 159-171). Essex: Pearson Education Limited.

O'Malley, C. (1995a). *Computer supported collaborative learning.*, Berlin: Springer Verlag.

O'Malley, C. E. (Ed.). (1995b). *Computer supported collaborative learning.* Springer-Verlag.

Paulsen, M. F. (1997). *Teaching methods and techniques for computer mediated communication.* Paper presented at the 18th ICDE World Conference. Pensylvania: State University.

Paulsen, M. F. (2003). *Online education and learning management systems. Global e-learning in a Scandinavian perspective.* Oslo: NKI Forlaget.

Pea, R. (1995). *Practices of distributed intelligence and designs for education, distributed cognitions: Psychological and educational considerations* (pp. 47-87). Cambridge, England: Cambridge University Press,.

Phillips, D. C. (1995). The good, the bad, and the ugly: The many faces of constructivism. *Educational Researcher, 24*(7), 5-12.

Pitcher, N., Davidson, K., & Goldfinch, J. (2000). Videoconferencing in higher education. *Innovations in Education and Training International, 37*, 199.

Raptis & Rapti (2004). *Teaching and learning towards the information age*. Athens.

Reed, J., & Woodruff, M. (1995). An introduction to using videoconferencing technology for teaching. The Distance Educator Newsletter, Fall, Scott, T D and Pitcher, N (1997) SUMSMAN: Scottish, from: http://www.kn.pacbell.com/wired/vidconf/Using.html

Reiser, R. (2001). A history of instructional design and technology. Part 2: A history of instructional design. *Educational Technology Research & Development, 49*, 57-67.

Rheingold, H. (2000). *The virtual community: Homesteading on the electronic frontier* (2nd ed.). Cambridge, MA: MIT Press.

Richter, M. (1999). Overseas Internet users protest high cost of access. *The New York Times*, p. G3.

Roberts, L. H. (1995). *A template for converting classroom courses to distributed, asynchronous courses*. ITS, Center for Instructional Technology, The University of North Carolina at Chapel Hill. Retrieved February 21, 2008, from http://www.unc.edu/cit/iat-archive/publications/roberts/template.html

Robinson, A. (1997). Adult learning by videoconferencing. In J. Field (Ed.), *Electronic pathways. Adult learning and the new communication technologies*. Leicester: National Organisation for Adult Learning (NIACE).

Rogers, C. R. (1969). *Freedom to learn*. Columbus, OH: Merrill.

Rogers, A. (2003). *What is the difference? A new critique of adult learning and teaching*. Leicester: NIACE.

Rosbottom, J. (2001). Hybrid learning: A safe route into Web-based open and distance learning for the Computer Science teacher. *ACM Association for Computing Machinery Sigcse Bulletin, 33*(3), 89-92.

Ryan, S., Scott, B., Freeman, H., & Patel, D. (2000). *The virtual university: The Internet and resource-based learning*. London: Kogan Page.

Salomon, G. (1992a). New challenges for educational research: Studying the individual within learning environments. *Scandinavian Journal of Educational Research, 36*(3), 167-182.

Salomon, G. (1992b). What does the design of effective CSCL require and how do we study its effects? *SIGCUE Outlook, Special Issue on CSCL, 21*(3), 62-68.

Salomon, G. (1995). *Distributed cognitions: Psychological and educational considerations*. In B. B. Seels & R. C. Richey (Eds.), *Instructional technology: The definitions and domains of the filed*. Washington D.C.: Association for Educational Communications and Technology.

Schiller, J., & Mitchell, J. (1993). Interacting at a distance. Staff and student perceptions of teaching and learning via video conferencing. *Australian Journal of Educational Technology, 9*(1), 41-58.

Scnurr, C., & Smith, C. (1995). *Video conferencing in education: Meeting teachers and learners support and training needs a report to the advisory group on computer graphics* (SIMA Report Series ISSN 1356-5370). Retrieved February 21, 2008, from http://www.agocg.ac.uk/mmedia.htm

Shale, D. (1988). Toward a reconceptualization of distance education. *The American Journal of Distance Education, 2*(3), 25-35.

Sheremetov, L., & Arenas, A. G. (2002). *EVA: An interactive Web-based collaborative learning environment Computers & Education, 39*, 161-182.

Simonson, M. (2002). *Teaching and learning at a distance: Foundations of distance education.* Prentice Hall.

Simonson, M., Smaldino, S., Albright, M., & Zvacek, S. (2000). *Teaching and learning at a distance: Foundations of distance learning.* Upper Saddle River, New Jersey: Merrill.

Smith, P. L., & Ragan, T. J. (1993). *Instructional design.* Upper Saddle River, NJ: Prentice-Hall.

Smyth, R. (2005). Exploring the usefulness of broadband videoconferencing for student-centred distance learning in tertiary science. In C. McLoughlin & A. Taji (Eds.), *Student centred teaching in science.* New York: Howarth.

Solomon (1987). Social influences on construction of pupil's understanding of science. *Studies in Science Education, 14*, 63-82.

Stellar. (2003). *Course management system.* Retrieved October 25, 2005, from http://stellar.mit.edu/

Stilborne, L., & Lindy, W. (1996). *Meeting the needs of adult learners in developing courses for the Internet.* Retrieved November 2007, from http://www.isoc.org/isoc/whatis/conferences/inet/96/proceedings/c4/c4_2.htm

Strijbos, J. W., Martens, R. L., & Jochems, W. M. G. (2004). Designing for interaction: Six steps to designing computer-supported group-based learning. *Computers & Education, 42*(4), 403-424. Elsevier.

Sullivan, M., Jolly, D., Foster, D., & Tompkins, R. (1994). *Local heroes: Bringing te le communications to rural, small schools.* Austin, TX: Southwest Educational Development Laboratory.

Suthers, D. (2001, January 3-6). Collaborative representations: Supporting face to face and online knowledge-building discourse. In *Proceedings of the 34th Hawai`i International Conference on the System Sciences (HICSS-34)*, Maui, Hawaii [CD-ROM]. Institute of Electrical and Electronics Engineers.

Tapscott, D. (1995). *Digital economy. Promise and peril in the age of networked intelligence.* McGraw-Hill.

Tobin, K. (1990). Social constructivist perspectives on the reform of science education. *Australian Science Teachers Journal, 36*(4), 29-35.

Twigg, C. (2002). Improving quality and reducing costs: Designs for effective learning using information technology. *The Observatory on Borderless Higher Education.* Retrieved March, 3, 2003, from http://www.obhe.ac.uk/products/reports/

United Nations. (2003). *Address by UN Secretary-General to the World Summit on the Information Society, Geneva, Switzerland.* Retrieved April 5, 2005, from www.itu.int/wsis/geneva/coverage/statements/opening/annan.html

Unruh, D. L. (2000). Desktop videoconferencing the promise and problems of delivery of Web-based training. *Internet and Higher Education, 3*, 183-99.

Van Berlo, M. P. W. (2000). Empirical validation of team training ID-guidelines. In *Proceedings of the 44th Annual Meeting of the Human Factors and Ergonomics Society*, San Diego, (Vol. 2, pp. 394-397).

Vosniadou, S. & Kollias, V. (2001). Information and communications technology and the problem of teacher training: Myths, dreams and the harsh reality. *Themes in Education, 2*(4), 341-365.

Vygotsky, L. S. (1978a). *Mind in society.* Cambridge, MA: Harvard University Press.

Vygotsky, L. (1978b). Interaction between learning and development. In E. Souberman (Ed.), *Mind in society.* Cambridge, MA: Harvard University Press.

Wedemeyer, C. (1981). *Learning at the backdoor.* Madison, WI: University of Wisconsin Press.

Wegner, E. (2001). *Supporting communities of practice. A research and consulting report.*

Yi, M. Y., & Hwang (2003). Predicting the use of Web-based information systems: Self-efficiency, enjoyment, learning goal orientation, and the technology acceptance model. *International Journal of Human–Computer Studies, 59*(4), 431-449.

ADDITIONAL READING

Anastasiades, P (2003, January). Distance learning in elementary schools in Cyprus: The evaluation methodology and results. *Computers & Education, 40*(1), 17-40. Elsevier Science.

Anastasiades, P. (2005, June 27-July 2). Synchronous vs. asynchronous learning? Principles, methodology and implementation policy of a blended learning environment for lifelong learning, at the University of Crete. In *Proceedings of the ED-MEDIA 2005 World Conference on Educational Multimedia, Hypermedia and Telecommunications, (AACE),* Montreal, Canada. Association for the Advancement of Computing in Education.

Anastasiades, P. (2006a, July 5-7). Interactive videoconferencing in lifelong learning: Methodology and implementation policy at the University of Crete (E.DIA.M.ME). In *Proceedings of the Diverse 2006, 6th International Conference on Video and Videoconferencing in Education,* Glasgow, Caledonian University, Scotland.

Anastasiades, P. (2006b, July 5-7). Interactive videoconferencing in K- 9 education: "ODUSSEAS 2000-2004" a case study in elementary schools in Greece and Cyprus. In *Proceedings of the Diverse 2006 16th International Conference on Video and Videoconferencing in Education, Glasgow Caledonian University, Scotland, Glasgow.*

Bates, A. W. (1995). *Technology: Open learning and distance education.* New York: Routledge.

Brookfield, S. D. (2001). Repositioning ideology critique in a critical theory of adult learning. *Adult Education Quarterly, 52*(1), 7-22.

Brown, S. (2001). *Views on videoconferencing higher education and research opportunities in the UK (HERO), March issues.* Retrieved February 27, 2008, from http://www.hero.ac.uk/inside_he/archive/views_on_videoconferencin883.cfm

Brusilovsky, P. (2001). Adaptive hypermedia. *User Modeling and User-Adapted Interaction, 11*(1/2), 111-127.

Cornford, J., & Pollock, N. (2003). *Putting the university online: Information, technology and organisational change.* Buckingham, UK: The Society for Research into Higher Education and Open University Press.

Damanakis & Anastasiades (2005). Life long & distance learning and the Diaspora: Implementing a virtual learning environment at the University of Crete. *Themes in Education, Special Issue, Information & Communication Technologies in Diaspora, 6*(1), 83-96.

Dillenbourg, P. (1999). What do you mean by "collaborative learning"? In P. Dillenbourg (Ed.), *Collaborative learning: Cognitive and computational approaches* (pp. 1-19). Amsterdam: Pergamon.

Garrison, D. R. (2006). Online collaboration principles. *Journal of Asynchronous Learning Networks, 10*(1), 25-34.

Grigoriadou, M., Papanikolaou, K., Cotronis, Y., Velentzas, C., & Filokyprou, G. (1999). Designing and implementing a Web-based course. In *Proceedings of International Conference of Computer Based Learning in Science*, Enschede, Netherlands.

Hayden, K. L. (1999). *Videoconferencing in K-12 education: A Delphi study of characteristics and critical strategies to support constructivist learning experiences.* Retrieved February 26, 2008, from http://hale.pepperdine.edu/~kahayden/dissertation.html

Holmberg, B. (1995). *The sphere of distance-education theory revisited* (ERIC Documentation Reproduction Service No. ED 386 578.).

IDE. (1996). Three models of distance education. *University of Maryland University College.* Retrieved February 26, 2008, from http://www.umuc.edu/IDE/modeldata.html

Illeris, K. (2003). Towards a contemporary and comprehensive theory of learning. *International Journal of Lifelong Education, 22*(4), 396-406.

Jarvis, P. (2004). Adult and continuing education. *Theory and practice.* Athens: Metaixmio.

Jonassen, D., Howland, J. Moore, J., & Marra, R. (2003). *Learning to solve problems with technology, a constructivist perspective.* Upper Saddle River, NJ: Merrill Prentice Hall

Keegan, D. (1996). *The foundations of distance education.* London: Croom Helm.

Kemmis, S. (1985). Action research and the politics of reflection. In D. Boude, et al (Eds.), *Reflection turning experience into learning.* London: Kogan Page.

Kokkos, A. (2007). Special characteristics and aims of adult education. *Adult Education, 9*, 4-19.

Lionarakis, A. (2003, June 15-18). *A preliminary framework for a theory of open and distance learning: The evolution of its complexity.* Paper presented at the 12th Conference of the European Distance Education Network.

Lionarakis, A. (2006). *Open and distance learning, theory and practice.* Athens: Probobos.

Mezirow, J. (1981). A critical theory of adult learning and education. *Adult Education, 32*(1), 3-24.

Mezirow, J. (1985). A critical theory of self-directed learning. In S. Brookfield (Ed.), *Self-directed learning: From theory to practice. New directions for continuing education* (No. 25). San Francisco: Jossey-Bass.

Mezirow, J. (1991). *Transformative dimensions of adult learning.* San Francisco: Jossey-Bass.

Mikropoulos, T. A. (2000). Design, development and evaluation of advanced learning environments. An overall approach. *Advanced Systems for Teaching and Learning over the World Wide Web*, B42-B52.

Moore, M. G., & Kearsley, G. (1996). *Distance education: A systems view.* Wadsworth Pub Co.

Nachmias, R., & Mioduser, D. (2000). A. Oren and J. Ram, Web-supported emergent-collaboration in higher education courses. *Educational Technology & Society, 3*(3), 94-104.

Paulsen, M. F. (2003). Online education and learning management systems. *Global e-learning in a Scandinavian perspective.* Oslo: NKI Forlaget.

Picciano, A. (2001) *Distance learning.* Ohio: Merrill Prentice Hall.

Rheingold, H. (2000). *The virtual community: Homesteading on the electronic frontier* (2nd ed.). Cambridge, MA: MIT Press.

Rogers, A. (2003). *What is the difference? A new critique of adult learning and teaching.* Leicester: NIACE.

Simonson, M. (2002). *Teaching and learning at a distance: Foundations of distance education.* Prentice Hall.

Vosniadou, S., & Kollias, V. (2001). Information and communications technology and the problem of teacher training: Myths, dreams and the harsh reality. *Themes in Education, 2*(4), 341- 365.

Chapter III
Teaching IT Through Learning Communities in a 3D Immersive World:
The Evolution of Online Instruction

Richard E. Riedl
Appalachian State University, USA

Regis M. Gilman
Appalachian State University, USA

John H. Tashner
Appalachian State University, USA

Stephen C. Bronack
Appalachian State University, USA

Amy Cheney
Appalachian State University, USA

Robert Sanders
Appalachian State University, USA

Roma Angel
Appalachian State University, USA

ABSTRACT

The development of learning communities has become an acknowledged goal of educators at all levels. As education continues to move into online environments, virtual learning communities develop for several reasons, including social networking, small group task completions, and authentic discussions for topics of mutual professional interest. The sense of presence and copresence with others is also found to be significant in developing Internet-based learning communities. This chapter illustrates the experiences with current learning communities that form in a 3D immersive world designed for education. Faculty at Appalachian State University (ASU) have developed and taught the graduate instructional technology program in an award-winning 3D world setting for several years. Additional ASU faculty and program areas are currently transitioning into this environment. Further, colleagues from major universities in other countries are using this environment for their students to work and to collaborate across time and distance. Telecommunications technologies in education (exposing the graduate students to the breadth of IT experiences and knowledge required), hypermedia, and advanced Web design are examples of IT-related courses offered in the graduate program. The results of these experiences highlight the efficacy of this tool toward the formation of authentic communities within 3D Internet-based worlds as online distance education environments continue to evolve.

INTRODUCTION

New technologies for collaboration have generated increasing interest in the formation of various kinds of online learning communities for distance education. A wide range of distributed learning communities are currently involved in training, education, gaming, social networking, and other emerging online endeavors. These distributed learning communities are available in different forms and demonstrate underlying frameworks that include collaborative text-based environments, Web-based text and graphical multiuser domains, and the more sophisticated CAVEs (projection-based automatic virtual environments). Each of the above presents its own unique technologies and possibilities for online distributed collaboration and learning. Each presents opportunities for group interactions in different ways that bring a sense of community to the task. This chapter will focus on the findings and experiences of various communities of learners formed within a 3D immersive Internet-based virtual world developed for graduate education.

Descriptions of a 3D Internet-based learning environment—called Appalachian Educational Technology Zone (AET Zone)—used by the instructional technology program in the Department of Leadership and Educational Studies at Appalachian State University have been noted in other research (e.g., Bronack, Riedl, & Tashner, in press; Riedl, Bronack, & Tashner, 2005; Tashner, Bronack, & Riedl, 2005). An Active Worlds universe server (http://www.activeworlds.com/) serves as the current platform for AET Zone, and provides a means to build virtual worlds for students, instructors, and other invited guests to meet and to work together in ways not found in other learning environments currently available. AET Zone may be characterized by significant components of space, movement, physical presence and copresence, conversational tools with small and large group shared workspaces, and metaphors and artifacts that assist with collaboration and learning online in unique and powerful ways. Students, faculty, and guests, graphically represented by avatars, move through the 3D world spaces interacting with each other and with artifacts within the worlds. These artifacts may be linked to different resources, Web pages, and tools necessary to provide content and support for various kinds of synchronous and asynchronous interactions. Small and large group shared workspace tools enable interactive conversations in text chats, threaded discussion boards, and audio chats. Group sharing of documents, Web pages, and other types of application software also are available within the virtual world.

Typical students in this graduate program are mid-career K-12 classroom teachers who want to learn more in-depth ways to integrate technology into their curriculum, or who want to become instructional technology specialists in their schools or chief technology officers (CTO) at the district level. Many of the students in the program teach within a 100-mile radius of the institution. However, recent initiatives have expanded opportunities to enroll K-12 teachers in a totally online experience. For example, several Mexican teachers from the D'Amicis School in Puebla, Mexico, and faculty and students in Griffith University in Brisbane, Australia, are working within AET Zone. Without the ability to depend on face-to-face contact, these international collaborations are challenging us to rethink the way we develop and enhance the sense of community in distance educational settings.

The instructional technology program at Appalachian State University uses a cohort model, where students enroll and move though the program together through a specific sequence of courses. Students and faculty currently meet face-to-face regularly at the beginning of the program, with reduced numbers and frequency of meetings as the members of a cohort become more comfortable working within the virtual world and gain understanding of course structures and expectations. While the virtual world is used for each class, the number of face-to-face meetings rapidly decreases after the first several courses to only an orientation class at the beginning and a final class session for student presentations at the end. A handful of courses during the final phase of the program are conducted completely within the virtual world, with no concurrent face-to-face meetings.

A set of four cohorts, consisting of 80 students who had experienced at least 2 years in the program, were asked several questions concerning ways they would describe their experiences as learners in this immersive 3D world. An informal qualitative analysis was conducted for the common themes expressed through the aggregated responses. These are presented and discussed below.

BASIC TENET

Conceptual Framework

A *conceptual framework* (Reich College of Education, 2005), based upon social constructivism (Vygotsky, 1978), was developed by the College of Education and provides a clear foundation that guides teaching and learning within AET Zone. These basic concepts are:

- Learning occurs through participation in a community of practice
- Knowledge is socially constructed and learning is social in nature in a community of practice
- Learners proceed through stages of development from novice to expert under the guidance of more experienced and knowledgeable mentors in the community of practice
- An identifiable knowledge base that is both general in nature and also specific to specialties emerges from the community of practice
- All professional educators develop a set of dispositions reflecting attitudes, beliefs, and values common to the community of practice

AET Zone reflects these assumptions about teaching and learning, and provides a powerful space through which effective learning communities are formed and nurtured. Students know and can see when their colleagues are logged into the world. They can approach other students and talk to them about life, work, or the latest news. Through these interactions, both planned and serendipitous, students begin to create knowledge together. They talk about the work they are doing in class, they share ideas, processes, and resources with one another, and they contribute to the base of knowledge that exists in their field. Throughout this process, they move from novice to expert, both in terms of knowledge and skills, but also in terms of their abilities to work collaboratively within a virtual learning environment using tools previously unknown to them. Their beliefs about teaching and learning are challenged, refined, and shaped by the process of learning together in an authentic social world of dialogue and discovery (Sanders & McKeown, 2007).

Differences Between Conventional Classrooms, Traditional Distance Education and Emerging Environments

Table 1 describes the differences between conventional classrooms, traditional forms of distance education, and emerging educational environments such as AET Zone. These characteristics are based on observations of what occurs in each environment. One key factor is the continuity and persistence of the AET Zone setting in which students and faculty run into each other during all times of the day and night regardless of physical location. While one can argue that similar persistence and continuity can and does

Table 1. Analysis of the principles of the RCOE conceptual framework

	Conventional Instruction[1]	Current Distance Education[2]	AET Zone[3]
Knowledge is socially constructed and learning is social in nature	**Usually only within the context of each individual class**	**Rarely and if so within the context of an individual class**	Within the entire virtual world community
Learning occurs through participation in a community of practice	**Usually only within the context of each individual class**	**Rarely and if so within the context of an individual class**	Regularly throughout the entire virtual world community
The development of educators proceeds through stages from novice to expert under the guidance of more experienced and knowledgeable mentors in the community of practice	**Rarely; contact with mentors usually limited to the course instructor**	**Rarely; contact with mentors usually limited to the course instructor**	Exposure to and interaction with a wide range of mentors throughout the virtual world community
An identifiable knowledge base emerges out of the community of practice that is both general for all educators and specific to specialties and content areas	**Limited by lack of exposure to the broader community of practice**	**Limited by lack of exposure to the broader community of practice**	Regular contact with the broader community of practice develops a full and shared knowledge base
All professional educators develop a set of dispositions reflecting attitudes, beliefs, and values common to the community of practice	Limited by lack of exposure to the broader community of practice	Limited by lack of exposure to the broader community of practice	Regular contact with the broader community of practice leads to sharing of beliefs and values leading to dispositions that are part of that community of practice

occur on traditional campuses, it should be noted that there is a distinct *discontinuity* between the confines of the classroom setting and the rest of the campus setting. In AET Zone, the learning environment and the social environment are one and the same. Thus, the community of practice is more explicit and becomes a more obvious factor in the experiences of students and faculty.

LEARNING COMMUNITIES

Learning communities have been characterized in many ways, and some division exists in current literature on the actual meaning of learning communities. "Communities of learners," according to some, are groups formed to increase their understandings or knowledge base in specific areas. Jonnasen (1997) cites the following necessary components for a learning community: active, constructive, collaborative, intentional, complex, contextual, conversational, and reflective. Others use the term "community of practice" which seems to indicate communities of similar practitioners who are currently exploring various aspects of their endeavors together. Wenger (1998) states that communities of practice include: "a *joint enterprise* as understood and continually renegotiated by its members..., *mutual engagement* that bind members together into a social entity.... and the *shared repertoire* of communal resources (routines, sensibilities, artifacts, vocabulary, styles, etc.) that members have developed over time." Others use the terms "learning communities" and "communities of practice" interchangeably.

Developing Online Communities

In either case, the literature suggests several main themes that emerge as useful guides for developing online virtual communities. An overview from a recent conference on building learning communities states that such communities:

Foster peer-to-peer collaboration, communication, interaction, resource sharing, negotiation and social construction of meaning, and expressions of support of encouragement among students. A blended or online learning community must have its own meeting or gathering space, as well as a defined set of members' roles and norms for resolving disputes. ("Academic Impressions," 2006)

A key element in the development of the community in AET Zone is that faculty members who teach in this environment stop thinking of students in one section of a class as "their" students but instead they interact with all students across sections and across classes. The "flattening" of their thinking is trickling down to students as well. Students are meeting each other online, learning what they have in common and how they differ, and then forming effective online partnerships and communities around real-world projects and activities (Sanders, Bronack, Cheney, Tashner, Reidl, & Gilman, 2007). Students just beginning in programs are interacting with students who are nearing graduation. Students in school administration, library science, higher education, and reading programs are interacting with each other and with instructional technology majors. Virtual worlds such as AET Zone are moving distance education efforts toward realizing the full potential of what distance learning might become. The virtual world serves as a catalyst for a learning community that reaches far beyond what normal classroom settings have been able to accomplish. Zhao and Kuh (2004) support this goal, asserting, "Learning communities are associated with enhanced academic performance, integration of academic and social experiences, gains in multiple areas of skill, competence, and knowledge, and overall satisfaction with the college experience" (p. 130).

The communities forming between and among students are beginning to resemble what Wilson and Ryder (2006) describe as "dynamic learning communities." Such communities are defined as "groups of people who form a learning community generally characterized by the following: distributed control; commitment to the generation and sharing of new knowledge; flexible and negotiated learning activities; autonomous community members; high levels of dialogue, interaction, and collaboration; a shared goal, problem, or project that brings a common focus and incentive to work together." These dynamic communities of learners are the ultimate goal in the process of applying social constructivist theory in the design and development of tools and spaces to support effective Internet-based communities for learning.

Common Themes in Learning Communities

Several common themes consistently emerge from these descriptions of learning communities. Communication, collaboration, and support are central to their development and maintenance. Other factors include shared resources and authentic reasons to join together. Recently emerging research and the emergence of 3D Internet-based environments for teaching and learning suggest the importance of the sense of presence and copresence in the development and evolution of online communities (Schroeder, Steed, Axelsson, Heldal, Abelin, Widestrom, et al., 2001). Using such characteristics as both a vision and a guide, the instructional technology graduate program has been studying ways to develop an environment that continues to foster and to support a wide variety of learning communities that may be identified with these characteristics.

Development and support of communities within 3D immersive worlds used for learning require consideration of how students will move through the course environments in collaborative ways, how to provide means to enhance the communication between students, guests, and instructors, and how to ensure participants will interact with the various resources in the environment that contribute to building meaningful communities of learners.

Figure 1. A community of learners collaborating in AET Zone

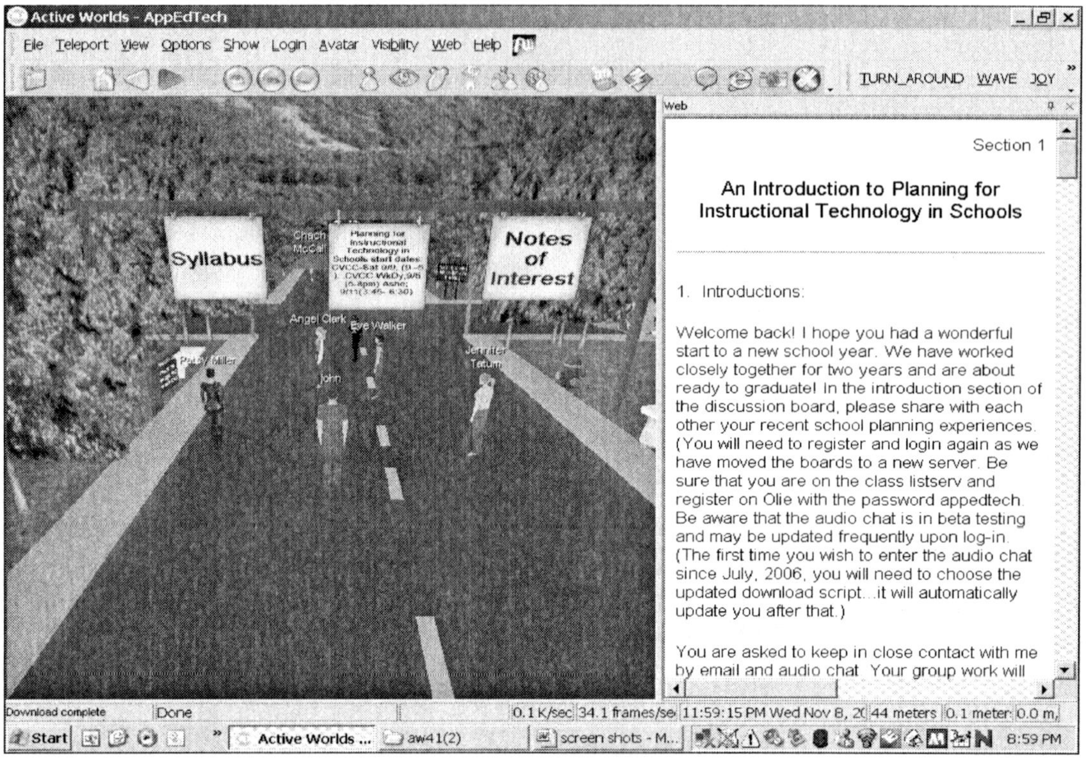

Collaboration

Participants in courses and other activities within AET Zone express a strong sense of collaboration by those engaged in learning within the virtual world. This collaboration exists between students in a specific cohort as well as between students from different cohorts. In fact, students from one section of a course often collaborate on specific tasks with students from other cohorts enrolled in sections of the same course, thereby increasing their collaborative resources exponentially.

Additionally, students cite many instances of working with other students from different program areas who were also taking different courses within the virtual world. It was indicated that students felt a strong collaboration with instructors, who served as knowledge guides rather than sole sources of expertise, as well. Additionally, students know that the course resources (including fellow students and faculty) will remain available to them through the AET Zone following completion of the course, and for graduates, even after completion of the degree program. They are free to visit other courses, to access various resources, and to engage students in other courses as resources in the learning process.

COMMUNICATION

Learning is a social process which reveals a conflict between what is already known and what is being observed (Brooks & Brooks, 1993). To resolve this conflict, an effective learning process requires interaction between learners and content, between learners and their peers, and between learners and those more expert than they (Levin & Ben-Jacob, 1998). Tools for communication, topics about which to communicate, and an authentic need to communicate are requisite factors for effective communication to be sustained within learning communities or communities of practice.

Figure 2.

Synchronous Communication

In 3D immersive worlds, several kinds of communication tools are found to be necessary to support ongoing tasks and community building. Synchronous tools such as text-based and audio chat capabilities are critical parts of the infrastructure necessary for creating learning communities. Such tools provide a means of working together at the same time in ways not otherwise possible. According to a recent analysis (Tashner et al., 2005), participants are able to develop and work together on authentic projects and topics because of the communication tools provided.

Asynchronous Communication

Asynchronous tools, however, also are important to the participants as ways of sharing ideas, research, and practice over time. For instance, a well-defined threaded discussion board provides opportunities for participants to share ideas, opinions, practices, and research. This communication tool also provides for the element of reflection that is not immediately available in synchronous environments. It is noted that the blend of these two communication tools within virtual worlds such as AET Zone enable a greater opportunity for interactions between and among participants.

Both formal and informal communication occurs in AET Zone and throughout the IT courses. Analysis further suggests that the informal communication is a powerful contributor to effective learning within 3D immersive worlds. Informal communication may spring up casually as faculty and students move around together in the world. Just as students on campus are thought to learn a great deal of content outside of structured classroom environments, so too, informal discussions in 3D immersive worlds may provide similar results. For example, students may join an audio chat room while simultaneously walking through the virtual world exploring together the artifacts that are present. Participants also may explore other topics of mutual interest that may or may not be part of their formal curriculum or agenda, but may still be tangentially relevant. This is an essential element of collaboration, communication, and community building.

SENSE OF PRESENCE AND COPRESENCE

Much contemporary Web-based instruction is characterized by "essentialists" view of teaching and learning. That is, certain "essential" things are to be learned as set forth by the instructor. Information flows in one direction, from the instructor and auxiliary materials to the student. Interactions that occur in such environments generally are limited to those between a single student and the student's instructor or in limited cases, between students enrolled in the same class. The student then shows the instructor by summative assessments that learning has occurred. When done online such environments lack many of the interactions and social aspects of learning that characterize communications within 3D immersive worlds. Emerging constructivist paradigms as noted above can be used as guiding principles in designing environments in which students engage in discussions with others across sections of the same class, different classes, and even different programs to deal with and to solve problems of interest from different perspectives. Such interactions include different forms of student to student, groups of students, instructors, and other experts interacting in various configurations to develop perspectives, to solve tasks, or to explore issues of mutual interest.

As 3D multiplayer games emerged in the late 1990s, researchers became interested in exploring these types of richer participant interactions taking place within gaming environments. Research suggests that social networks are powerful components of online multiplayer games (Jakobsson & Taylor, 2003). Drawing from ethnographical and constructivist approaches, Manninen (2001) offers a taxonomy to conceptualize these forms of interactions based on components such as language-based communications, avatar appearance, body language (subconscious), and physical contact. Research has also focused on roles that presence and copresence may play in enhancing participant interactions within virtual worlds (Schroeder, 2002). While the term "virtual" has recently been applied to many different types of technologies and mediated environments, Schroeder's definition of "virtual reality" focuses on the common elements linking these technologies and environments together, specifically, "a computer-generated display that allows or compels the user (or users) to have a feeling of being present in an environment other than the one they are actually in and to interact with that environment" (p. 2).

Presence

Schroeder (2002) argues that shared virtual environments "combine a high degree of presence with a high degree of co-presence because the sense of being in another place and of being there with another person reinforces each other" (p. 5). Furthermore, "presence and co-presence will be affected by the extent of experience with the medium" (Schroeder, 2006, p. 439). The more familiar and comfortable users are with the medium and the social norms of the virtual environment, the more their sense of presence and copresence will be heightened. However, regardless of the users' competence and proficiency working in a virtual environment, two users' "connected presence" in that environment will have an impact on the overall experience for both users, simply as a result of being in the environment together (in a similar way to how ethnographers have noted that the act of observing influences that which is being observed).

As an immersive 3D environment, AET Zone allows participants to "see" each other via representative avatars. Each participant moves his or her avatar through the virtual world using a keyboard or a mouse. As one moves, one's perspective changes; thus what the environment looks like changes. This change in perspective as one moves creates a sense of "presence." A participant has the perception of being somewhere else. In addition, as one observes others in the environment, one has a feeling of being somewhere else with someone else or "copresence." These concepts lead one to experience a connected presence or mutual awareness of others. As the mutual awareness increases, so does the desire for and feeling of heightened engagement in the world and in the activities conducted within the world. Emerging from these feelings is a strong theme of the importance of both presence and copresence in developing learning communities. Students report that the feeling of isolation and working alone diminished as they become accustomed to working in the environment. This was of particular import to both retention and to the individual successes of students toward their educational goals (Tashner et al., 2005).

Interestingly, presence is evidenced in several ways. The foremost is sensing that you are actually somewhere different than your physical location. As you move through the world and you sense the movement, your perspective changes and you become "there" as well as "here." Some students will desire to change their personas on a frequent basis by changing their avatar. When asked why, they state that they were "feeling different.[sic]" On the other hand, some students do not see "themselves" in the same way.

Copresence

Copresence is characterized as being "there" with "someone else," though the "someone else" is represented by an avatar. We have noticed that adults take into the 3D world some parts of their personalities and cultural more that they exhibit in the outside world. For instance, if one avatar gets too close to another, the second one will move in order to preserve "personal space." Novice students must learn to minimize windows so that they can "see" when others are trying to communicate with them. Some become disheartened when they speak to another avatar and the "other" ignores them. Yet, we have also seen a reluctance to meet "others" outside their class, to converse with "strangers" in the 3D world. Such issues are worked through by assignments to meet others, explore courses together with "persons" you do not know, and many other techniques as needed. However, these examples demonstrate the importance of understanding the concepts of presence and copresence in immersive worlds.

Role of Presence and Copresence in Online Communities

The sense of presence and copresence are critical factors in creating and maintaining deeply engaging online communities. As participants gain more of a sense of being somewhere and with somebody else, communication and collaboration are dramatically enhanced. According to Ahuna (2006), when constructs such as communication and collaboration combine to support the formation of community, "a semantic world of sharing knowledge, solving problems, working as a team, playing, building, quarreling, cooperating, planning and forming relationships develop." The following screen shot illustrates an overview of the Network Basics building. Students move through the building, walking across the various components, clicking on components to access descriptions and resource information. For a slightly different perspective, students may choose to float above the floor. Various tools to enhance the cognitive awareness and understanding of the concepts and constructs are available to the learners. Group interaction is encouraged as an important piece of the learning process, in developing the learning communities, increasing collaboration, and to increase levels of content understanding. The combination of communication and small group shared collaboration tools with a sense of presence and copresence provides opportunities for developing authentic learning environments for Internet-based learning that goes far beyond attempts to replicate traditional classroom instruction using typical Web-based applications.

Our experiences with graduate students in the AET Zone suggest that many forms of communities evolve as needed. Some will develop for specific tasks and time periods and then dissolve. These include task oriented communities, for example, where students will form groups to read and to discuss specific books and to inform other larger groups of what they are learning in various discussion formats. Hence, a group of four students may find themselves in discussions on ideas with eighty other students. Another example might be a task involving the development and implementation of certain projects that include ideas, knowledge and resources shared among a larger group who have similar interests. Others will remain intact for longer periods of time. For instance, different forms of social groups also have been noted in AET Zone that are more persistent. One group who met online each week to work on assignments decided to meet together at a different time for dinner. They cooked the same dinner, drank the same wine and met, not face-to-face, but connected inside the 3D world in an audio chat room to enjoy each other's company for a while.

Certainly, our experiences in thinking about the roles of presence and copresence in AET Zone help us understand the importance of these sensory inputs in Internet-based instruction. However, we are deeply aware that we are dealing with very complex variables. We are exploring new questions that emerge from our observations. How might we develop a deeper sense of belongingness to these communities? Are there pedagogical ways to provide social networking within a series of courses or is it even desirable?

Instead of information flow in one direction only from a source to a receiver, many other possibilities emerge. The result is a vibrant, active, participatory, and engaging environment developed for community members to build new knowledge based upon the foundation presented by the group.

Figure 3.

METAPHORICAL GRAPHICAL USER INTERFACES

One striking feature of AET Zone is its extensive use of metaphors in the design of the graphical user interface. As students move through the world, they find themselves in plazas, gardens, frontiers, and suburbia. Every space in the 3D world is built upon a metaphor or a series of metaphors to provide students with access to content, context, and tools for navigation. We have been very deliberate in our selection of metaphors in our designs and believe that thoughtful and reflective choices about the metaphors to use are important to the success our students have working within the virtual world.

Cates (1994) cites Lakoff and Johnson in defining a metaphor as "understanding and experiencing one kind of thing in terms of another." One thing, often familiar, is a figurative representation of the other, often abstract or unfamiliar. According to Nicholson and Sarker (2002), Aristotle understood the value of a metaphor when he said, "Ordinary words convey only what we know already; it is from metaphor that we can best get hold of something fresh."

Some suggest that simply a virtual representation of a physical space or artifact is not metaphorical, but rather, the virtual representation must be different in its representation (e.g., Cates, 1996). According to Cates (1996), a graphical user interface (GUI) that is metaphorical must be based on either an explicit or implicit metaphor, but it makes little difference as to whether the metaphor is obvious to the user or not. The important aspect is that the metaphor works to provide some insight into or aid in understanding of that idea, concept, or thing it represents.

According to Black and later expanded upon by Cates (1994), there are two types of metaphors: *underlying* or *primary* and *auxiliary* or *secondary*. An underlying metaphor is the main metaphor used. For example, in one of the courses taught in the instructional technology program, the underlying metaphor of the Wild West was used as the main metaphor throughout the course space within the virtual world. An auxiliary metaphor is one that is consistent with the underlying metaphor and is used to support or enhance this main metaphor. In the case of the aforementioned course, examples of auxiliary metaphors might include a "saloon" for meeting and conversing, a "general store" for finding useful content, and a "haystack" that links to useful search engines.

Complimentary Metaphors

Complimentary metaphors are those that enhance the online teaching and learning environment. These are complementarily aligned with one another to assist learners in developing a "conceptual framework of understanding through which the learner can further enhance prior knowledge and conceptualize a higher level of understanding towards the knowledge being obtained" (Henry & Crawford, 2001, p. 3). Henry and Crawford further suggest that through the utilization of these metaphorical graphical user interfaces (MGUI), "a sense of community is presented to the learner, and in turn, a collaborative e-learning environment is well on its way towards realization."(p. 4) This community emerges out of an immersive environment in which students "collaborate on projects, work in teams, and create material and artifacts together… Students assume a variety of roles…and students must negotiate as they will have to negotiate in the adult world" (Marshall, 2000, p. 5). For this to occur, auxiliary metaphors selected must be complementary to the underlying metaphor employed.

The effective use of metaphors in an online learning environment can be valuable in offering students a model to assist in understanding more abstract concepts in more familiar, concrete terms and can help students understand a concept and content more quickly than without the use of the metaphor by helping students learn and understand how things should work (Bishop & Cates, 1996;

Cates, 1994). Bishop and Cates (1996) note existing literature that supports the position that content can be better learned through the interaction with metaphorical graphical user interfaces by providing both "superficial and deep similarities between familiar and novel situations." Ultimately, it is the finding of these and other studies that the use of metaphors helps students build knowledge, develop higher level thinking skills, build community, and gain a more universal understanding of the subject matter being taught (Bishop & Cates, 1996; Henry & Crawford, 2001).

The goal in the use of underlying metaphors is to enhance the students' learning experience by providing a device that allows each to interact with the instruction and the content in ways more familiar, and, as a result, more accessible, to them. Well-crafted auxiliary metaphors complement the underlying metaphor and the overall learning experience.

Confounding Metaphors

It should be noted, however, that the ineffective use of metaphors can have a deleterious effect on even the most well-designed and well-intended learning environments. Some have suggested that gratuitous or fantastical metaphors can be in conflict with the tenets that ground effective meaning-making (e.g., Nicholson & Sarker, 2002). While this assertion is in stark contrast with the value of metaphor discussed above, it does provide an important reminder that problems can arise in the use of metaphors, especially those that are auxiliary to the underlying metaphor.

Form should follow function, and the selection of underlying and auxiliary metaphors (form) should enhance and complement the teaching and learning tools and activities (function) embedded in a virtual world. One challenge in the inclusion of metaphors is the overdependence on their use within the interface design (Cates, 1996; Nelson, 1990). There are times when the poor choice of metaphors overshadows the instructional design of the content and the virtual world in which the content is presented. When this happens, students must reconstruct what they think they know and understand about the content and virtual world with which they are working. Again, according to Cates (1994, p. 103), "When users are faced with such an auxiliary metaphor [confounding] they are required to reconstruct the environment radically, envisioning a book [for example] that is unlike any that the user has ever seen. Users seem unlikely to make such radical reconstructions.... When users come to this conclusion, the benefits of the underlying metaphor are greatly reduced." It is even possible that students may reject and disengage from the virtual world completely if the cognitive dissonance created by the confounding metaphor is too great.

Multiple studies warn that metaphors used incorrectly or out of context can make it difficult for learners to engage effectively within environments such as virtual worlds (e.g., Rosendahl-Kreitman, 1990; Semper, 1990; Vertelney, Arent, & Lieberman, 1990). Misalignment or inappropriate linkage between metaphors and expectations may result in a debilitating tension for learners (Burge & Carter, 1997). Barrie (1996) explains this tension as created out of a "pause in the cadence of the composition...producing a reaction of tension and anticipation" and notes an emotional reaction occurs when this pause or interruption occurs, often resulting in frustration or even feelings of incompetence. Rohrer (1995) offers a different twist, suggesting that there is also a tension taking place between the literal and figurative, or "magical," qualities of the metaphors being used, and this tension extends to the a tension between the user and the computer itself, which is viewed as an "other—a sentient being with a consciousness of its own (and usually a malevolent consciousness at that)."

When it comes to metaphors, it seems, numbers count. Incorporating too few or too many metaphors can pose problems for users as well. Learners may not have enough to make sense of the interface nor to understand the content to be learned if there are too few metaphors employed. On the other hand, too many metaphors can be overwhelming to a learner and lead to cognitive overload (Cates, 1994). Regardless, any use of metaphor has the potential of requiring learners to translate not only the content and instruction being delivered into more familiar and understandable terms, but also to force them to work through another cognitive layer posed by the use of the metaphor. When the layers overload, the use of metaphors may hinder—rather than help—learners make sense of the information at hand (Lohr & Heng-Yu, 2003). As previously mentioned, the poor use of metaphors can ultimately cause learners to abandon the metaphor altogether (Rohrer, 1995).

What is the lesson for designers of user interfaces for virtual worlds, then? The lesson is clear. To design an effective virtual world for learning, it is essential to use complementary metaphorical strategies that foster the development of community and to avoid becoming enamored with the metaphors themselves.

SUPPORT

An additional theme that has emerged from our work with virtual worlds for learning is that of "support." Support is expressed in many forms but in this case, the concept is of peer and instructor support. It is often expressed as assistance that is usually available whenever one requested it. Whether the online library resources, an individual course, or even professional assistance is needed, there is an instructor or peer ready to offer support. In social constructivist learning communities, such as AET Zone, participants move along a developmental continuum from novice to expert. Indeed, in each course and throughout the program, students represent various aspects of this continuum at various points in each individual's personal development. As each becomes more aware of others through planned and serendipitous interactions, and as each becomes increasingly comfortable with others, their collective working relationships weave a complex support network for and by all participants. Bender (2003) suggests that a feeling of belonging within a chosen community of practice is requisite for effective learning. Both feeling supported and feeling supportive play an integral role in this important construct of belonging.

LEADERSHIP

It is in these same 3D communities that participants find themselves alternately leading and being led and where some participants unexpectedly find themselves becoming leaders and identifying with leadership roles. The leadership theme emerges as personal as well as organizational leadership.

Working collaboratively and communicating together in learning communities enhances the leadership skills and comfort levels of participants, with self-reported transfer to their own teaching and learning environments. Students who spend time in AET Zone, for example, report a heightened sense of awareness of their own expertise arising from interactions and participation in the various communities in which they work and learn. They express an increase in personal and professional self-confidence, which they indicate is transferred into their professional practices.

In addition to leadership dynamics that emerge naturally from participation in 3D immersive communities of learners, deliberate leadership-focused prompts, such as case studies, provide problem-based learning situations in cross-disciplinary contexts that foster deeper levels of development. In AET Zone, for example, participants are immersed in authentic circumstances requiring the development of leadership "voice" in the "safety" of the virtual community. Participants are given opportunities to "try out" various responses to situations in an effort to solve problems within organizational communities. In the virtual environment mistakes can be made, consequences examined, and corrections tried without fear of real consequence or penalty. This type of natural yet safe learning is necessary for developing better leaders. Thus, experimenting with shared leadership skills can become a natural consequence of learning within the 3D community and, as well, can be a result of responding to deliberately conceived situations requiring leadership thought and decisions (Angel, Sanders, & Tashner, 2005; Sanders & Angel, 2005). In short, the 3D environment is a rich context for learning personal leadership skills and, as well, for applying those skills in real work situations outside this environment.

FUTURE RESEARCH DIRECTIONS

The convergence of sophisticated gaming platforms, communications technologies, social networking trends, and educational needs provide rich opportunities for future research. In this time of global transition, we are changing paradigms of what it means to teach and to learn. Rather than trying to address old problems with new questions, we must begin to ask new questions about new problems.

Online educational environments started by attempting to recreate the four-wall classroom that had been successful in the past. Should online learning continue to try to be the same as its face-to-face counterpart? Can it be unique in its approach, using different methods and tools for teaching and learning? How might a newer generation of online learning platforms containing more immersive and engaging environments add value to learning? Furthermore, should we be asking additional questions about how specific technologies allow us to expand beyond the four walls of a traditional classroom and transcend borders, cultures, and perspectives to create active participatory groups of learners? The development of online pedagogies to create the teaching and learning models needed for a 21st century education is a field ripe for research.

Especially important in future research may be the applications of social constructivism pedagogies to online environments. Social constructivism is fundamentally about the social construction of knowledge through participation in communities of practice. Through interaction and communication, collaboration and mentoring, learners become a part of and contribute to this community of practice. Researchers have just begun to explore the effects of various kinds of online collaboration and communications between students, instructors, and colleagues in developing these communities of practice. Questions subsequently begin to emerge regarding the value that might be added by the use of tools and processes whose purpose is to enhance synchronous and asynchronous collaborations and communications in the context of social constructivist learning environments.

Specific research questions to be asked include: What constitutes online learning communities and how might they be developed and expanded? To what extent do learning communities enhance online learning? What is the added value of the participants' sense of presence and copresence in online environments? What tools are needed to assist them in high level functioning? Additionally, one might ask: What kinds of skills and attitudes are needed by educational leaders to move and to support students and teachers as their organizations move into 21st century learning environments? How might current educational leaders develop such needed skills and attitudes?

Finally, the need for new assessment methods and tools is critical if we are serious about teaching toward higher levels of "critical thinking" and performance. The current testing movement in the United States is geared toward measuring lower level cognitive skills, thus creating a mismatch between what is measured and what is stated as goals for 21st century success. Teaching and learning in online environments, especially those that are built upon a foundation of social constructivism, will require new assessment tools and measures in order to know and understand the learning that takes place in these 21st century learning communities. Basic research must consider what an educated person looks like in the 21st century. What kinds of educational experiences are needed to develop 21st century individuals? In a world where the same knowledge base is accessible by everyone, what does "knowing," mean? What skills and knowledge are needed by individuals to be deemed "educated"? The answers to these questions will assist researchers in the development of valid and reliable assessment tools that are more consistent with a social constructivist approach to online teaching and learning.

CONCLUSION

Those who have learned and taught within AET Zone report a variety of positive experiences, advantages, and learning outcomes from their work in this environment and through the powerful learning communities that develop within. While participating in this social constructivist, immersive environment, students enrolled in these courses use a variety of tools and metaphors for communication and collaboration, fostering various forms of learning communities. These include the many virtual communities developing within AET Zone for social networking, small group task completions, and authentic discussions on topics of mutual professional interest. Feedback from multiple cohorts across time and distance suggests the strong sense of presence and copresence felt while in AET Zone is a critical factor that fosters the development of useful learning communities which, in turn, facilitate practical, useful learning.

The shared experience of AET Zone participants has a number of other outcomes as well. Students share a variety of resources both during and after their period of formal participation. They also report that the environment provides support for learning, both from instructors and peers. As a result, students' sense of leadership and vision is heightened.

Virtual worlds such as AET Zone are unique environments for teaching and learning. The tools, support, and constructivist pedagogies embedded within AET Zone lend themselves readily to the creation of learning communities. Clearly, there is a need for well-designed research studies to develop a body of literature that will guide educators as they continue to move into emerging online environments for teaching and learning.

REFERENCES

Academic impressions building learning communities: Strategies for collaborative learning on-line conference overview. (2006). Retrieved October 12, 2006, from https://www.academicimpressions.com/conferences/1006-collaborative-communities.php

Ahuna, C. (2006). *Online game communities are social in nature.* Retrieved February 12, 2006, from http://switch.sjsu.edu/v7n1/articles/cindy02.html

Angel, R. B., Sanders, R. L., & Tashner, J. H. (2005, March). *Constructing learning communities through Web-based environments: Problem based learning in cross-disciplinary social constructivist frameworks.* Phoenix, AZ: Society for Information Technology and Teacher Education.

Barrie, T. (1996). *Spiritual path, sacred place: Myth, ritual, and meaning in architecture.* Boston: Shambhala.

Bender, T. (2003). *Discussion-based online teaching to enhance student learning; Theory, practice, and assessment.* Sterling, VA: Stylus Publishing.

Bishop, M. J., & Cates, W. M. (1996). A door is a big wooden thing with a knob: Getting a handle on metaphorical interface design. In *Proceedings of Selected Research and Development Presentations at the 1996 National Convention of the Association for Educational Communications and Technology* (pp. 80-88). Indianapolis, IN (ERIC Document Reproduction Service No. ED397779).

Brooks, J. G., & Brooks, M. G. (1993). *The case for constructivist classrooms.* Alexandria, VA: ASCD Publications.

Bronack, S., Riedl, R., & Tashner, J. (in press). Learning in the zone: A social constructivist framework for distance education in a 3D virtual world. *Journal Interactive Learning Environments.*

Burge, E. J. & Carter, N. M. (1997). *It's building, but is it designing? Constructing Internet-based learning environments.* Paper presented at the World Conference of the International Council for Distance Education, University Park, PA (ERIC Document Reproduction Service No. ED412333).

Cates, W. M. (1994). Designing hypermedia is hell: Metaphor's role in instructional design. In *Proceedings of Selected Research and Development Presentations at the 1994 National Convention of the Association for Educational Communications and Technology* (pp. 95-108). Indianapolis, IN (ERIC Document Reproduction Service No. ED373706).

Cates, W. M. (1996). Towards a taxonomy of metaphorical graphical user interfaces: Demands and implementations. In *Proceedings of Selected Research and Development Presentations at the 1996 National Convention of the Association for Educational Communications and Technology* (pp. 101-110). Indianapolis, IN (ERIC Document Reproduction Service No. ED397781).

Henry, A., & Crawford, C. M. (2001). Creating a collaborative Web-based environment through the inclusion of metaphorically enhanced graphics. In *Proceedings of WebNet 2001: World Conference on the World Wide Web and Internet* (pp. 1-8). Orlando, FL (ERIC Document Reproduction Service No. ED462914).

Jakobsson, M., & Taylor, T. L. (2003). The Sopranos meets EverQuest: Social networking in massively multiplayer online games. *Melbourne DAC.* Retrieved February 9, 2007, from http://hypertext.rmit.edu.au/dac/papers/Jakobsson.pdf

Jonassen, D. (1997, Spring). *INSYS 527 designing constructivist learning environments.* Retrieved October 12, 2006, from http://www.coe.missouri.edu/~jonassen/INSYS527.html

Levin, D. S., & Ben-Jacob, M. G. (1998, November). *Using collaboration in support of distance learning.* Paper presented at the WebNet 98 World Conference of the WWW, Internet and Intranet Proceedings, Orlando, FL (ERIC Document Reproduction Service No. ED 427 716).

Lohr, L. L., & Heng-Yu, K. (2003). Development of a Web-based template for active learning. *Quarterly Review of Distance Education, 4*(3), 213-227.

Manninen, T. (2001). Rich interactions in the context of networked virtual environments: Experiences gained from the multi-player games domain. In A. Blandford, J. Vandersoickt, & P. Gray (Eds.), *Joint Proceedings of HCI 2001 and IHM 2002 Conference* (pp. 383-398). Springer-Verlag.

Marshall, G. (2000). Models, metaphors and measures: Issues in distance learning. *Educational Media International, 37*(1), 2-8.

Nelson, T. (1990). The right way to think about software design. In B. Laurel (Ed.), *The art of human computer interface design* (pp. 235-243). Reading, MA: Addison-Wesley.

Nicholson, J., & Sarker, S. (2002). Unearthing hidden assumptions regarding on-line education: The use of myths and metaphors. In *Proceedings of the International Academy for Information Management (IAIM) Annual Conference: International Conference on Informatics Education Research (ICIER)* (pp. 298-306). Barcelona, Spain (ERIC Document Reproduction Service No. ED481748).

Reich College of Education – Appalachian State University, Boone, NC. (2005). *Conceptual framework.* Retrieved November 6, 2006, from http://ced.appstate.edu/about/conceptualframework.aspx

Riedl, R., Bronack, S., & Tashner, J. (2005, January). *3D Web-based worlds for instruction.* Phoenix: The Society for Information and Teacher Education.

Rohrer, T. (1995). *Metaphors we compute by: Bringing magic into interface design.* Retrieved August 4, 2004, from University of Oregon, Department of Philosophy Web site: http://philosophy.uoregon.edu/metaphor/gui4web.htm

Rosendahl-Kreitman, K. (1990). The challenge of interface design: Creating quality experience for the user. *Multimedia interface design.* Santa Clara, CA: Multimedia Computing Corporation.

Sanders, R. L., & Angel, R. B. (2005, August). *Shared decision-making: Case study analysis to promote cross-program dialogue between administrators and media coordinators.* Paper presented at the International Conference on Computers and Advanced Technology in Education, Oranjestad, Aruba.

Sanders, R. L., Bronack, S., Cheney, A., Tashner, J., Reidl, R., & Gilman, R. (2007, February). *Education in the zone: Dynamic learning communities in a 3D virtual world.* Paper presented at the IADIS International Conference of Web Based Communities 2007, Salamanca, Spain.

Sanders, R. L., & McKeown, L. (2007, January). *Promoting reflection through action learning in a 3D virtual world.* Paper presented at the Association of Library and Information Science Educators Annual Conference, Seattle, WA.

Schroeder, R. (Ed). (2002). *The social life of avatars: Presence and interaction in shared virtual environments* (pp. 1-18). Great Britain: Springer-Verlag/London Limited.

Schroeder, R. (2002). Social interaction in virtual environments: Key issues, common themes, and a framework for research. In R. Schroeder (Ed.), *The social life of avatars: Presence and interaction in shared virtual environments.* London: Springer.

Schroeder, R. (2006). Being there together and the future of connected presence. *Presence: Teleoperatores and Virtual Environments, 15*(4), 438-454.

Schroeder, R., Steed, A., Axelsson, A., Heldal, I., Abelin, A., Widestrom, J., et al. (2001). Collaborating in networked immersive spaces: As good as being there together? *Computers and Graphics, 25*, 781-788.

Semper, R. (1990). Hypercard and education: Reflections on the hyperboom. In S. Ambron & K. Hooper (Eds.), *Learning with interactive multimedia: Developing and using multimedia tools in education* (pp. 52-67). Redmond, WA: Microsoft Corporation.

Tashner, J., Bronack, S., & Riedl, R., (2005, March). *Virtual worlds: Further development of Web-based teaching.* Paper presented at the Hawaii International Conference on Education, Honolulu.

Vertelney, L., Arent, M., & Lieberman, H. (1990). Two disciplines in search of an interface: Reflections on the design process. In B. Laurel (Ed.), *The art of human computer interface design* (pp. 45-55). Reading, MA: Addison-Wesley.

Vygotsky, L. S. (1978). *Mind in society: The development of higher psychological processes.* Cambridge, MA: Harvard University Press.

Wenger, E. (1998, June). Communities of practice: Learning as a social system. *Systems thinker.* Retrieved October 16, 2006, from http://www.co-i-l.com/coil/knowledge-garden/cop/lss.shtml

Wilson, B., & Ryder, M. (2006). *Dynamic learning communities: An alternative to designed instructional systems.* Retrieved October 6, 2006, from http://carbon.cudenver.edu/~mryder/dlc.html

Zhao, C. M., & Kuh, G. D. (2004). Adding value: Learning communities and student engagement. *Research in Higher Education, 45*(2), 115-138.

ADDITIONAL READING

Palloff, R. M., & Pratt, K. (2001). *Lessons from the cyberspace classroom: The realities of online teaching.* San Francisco: Jossey-Bass.

Palloff, R. M., & Pratt, K. (2003). *The virtual student: A profile and guide to working with online learners.* San Francisco: Jossey- Bass.

Palloff, R. M., & Pratt, K. (2004). *Collaborating online: Learning together in community.* San Francisco: Jossey-Bass.

Palloff, R. M., & Pratt, K. (2007). Building online learning communities: Effective strategies for the virtual classroom. *Building learning communities in cyberspace* (2nd ed.). San Francisco: Jossey-Bass.

ENDNOTES

1 Typical context is one teacher with many students meeting in a classroom for a finite amount of time and in a class that is not necessarily connected with other classes or other experiences.

2 Typical context is one teacher with many students who are in many different locations and in a class that is not necessarily connected with other classes or experiences.

3 Students and instructors of many classes intermingle at many different times and locations... Alumni and other experts are available throughout the virtual world and at many different times.

Chapter IV
Online Synchronous vs. Asynchronous Software Training Through the Behavioral Modeling Approach: A Longitudinal Field Experiment

Charlie C. Chen
Appalachian State University, USA

R. S. Shaw
Tamkang University, Taiwan

ABSTRACT

The continued and increasing use of online training raises the question of whether the most effective training methods applied in live instruction will carry over to different online environments in the long run. Behavior Modeling (BM) approach teaching through demonstration—has been proven as the most effective approach in a face-to-face (F2F) environment. A quasi-experiment was conducted with 96 undergraduate students who were taking a Microsoft SQL Server 2000 course in a university in Taiwan. The BM approach was employed in three learning environments—F2F, online synchronous and online asynchronous classes. The results were compared to see which produced the best performance, as measured by knowledge near-transfer and knowledge far-transfer effectiveness. Overall satisfaction with training was also measured. The results of the experiment indicate that during a long duration of training no significant difference in learning outcomes could be detected across the three learning environments.

CONCEPTUAL FOUNDATIONS

The Internet's proliferation creates a wealth of opportunities to deploy alternative online learning environments to facilitate many users in their learning processes. The information technology (IT) skills training market represented 76% of the entire online learning market in year 2000, according to a Jupiter Research report (CyberAtlas, 2003). The worldwide corporate online learning market may grow to $24 billion ($18 billion in the U.S.) by 2006 with a compound annual growth rate of 35.6% (IDC, 2002). The burgeoning online learning/training market, and the increasing training budgets of businesses and schools has provided these key users of online training and marketing tools with practical reasons, as well as compelling research motives, to investigate the effectiveness of training and education in different online formats.

Online learning differs primarily from the traditional face-to-face (F2F) learning in that it is a user-centered, rather than instructor-centered, learning mode. Other benefits of substituting online learning for F2F learning include (1) self-paced instruction; (2) the ability to incorporate text, graphics, audio and video into the training; (3) opportunity for high levels of interactivity; (4) a written record of discussions and instructions; (5) low-cost operation; and (6) access to a worldwide audiences (Aniebonam, 2000). In addition, online learning can remove a certain degree of space and time limitations, speed up the learning process for motivated learners, lower economic costs of attending F2F classes and have higher information accessibility and availability.

Although IT has changed the training and educational approaches and environments, the ultimate goal of learning has not changed, that is, to transfer knowledge to students and allow them to apply the acquired knowledge in real situations. In the field of IT, the success of software training can be assessed with a trainee's IT skills of, and knowledge of the use of, particular software to solve problems. Surprisingly, after attending a training session, very few students know how to properly apply the acquired knowledge and skills to real situations. This raises an important issue, that is, how to improve knowledge transfer capability of learners in different online learning environments.

The importance of knowledge transfer is self-evident. However, the knowledge transfer process does not occur naturally. There is a need to assist learners in transferring their acquired knowledge into future applications. One effective approach to assisting the learning transfer process is "behavior modeling" (BM). This approach teaches learners through demonstration and hands-on experience. Simon, Grover, Teng, and Whitcomb (1996) and Compeau and Higgins (1995) found that in the field of information technology, BM is the most effective approach compared to the other two knowledge transfer approaches: exploration—teaching through practice on relevant example, and instruction—teaching software characteristics.

Distance education is defined as "teaching through the use of telecommunications technologies to transmit and receive various materials through voice, video and data" (Bielefield & Cheeseman, 1997, p. 141). In the same token, Leidner and Jarvenpaa (1995) define distance learning as "the transmission of a course from one location to another" (p. 274). These definitions provide an analogy to distance learning in the field of information technology or online software training. Online software training can be the transmission of instructional IT programming or contents to geographically dispersed individuals or groups.

There are two general modes of online learning: synchronous and asynchronous modes. Each mode can be marshaled with IT tools to deliver software training. Case in point, audio and video conferences are two types of online synchronous training mediums. Online asynchronous training mediums range from Web pages, file download, e-mail, e-mail list, newsgroup, forum, chat, response pad, whiteboard and to screen sharing. Built on his personal distance training and education experiences since 197 -+ 1, Horton (2000) suggests that online synchronous and asynchronous learning and training be designed for different purposes. Incorporating synchronous learning demands the control of schedule, time, people, class size, video and audio equipment and place. These factors constrain the possibility of reaching large numbers of students at any given time and in any given place.

However, the BM approach for trainees can be a problem in online asynchronous and synchronous training. For instance, any demonstration presented by a live instructor would need to be replaced with a scripted or videotaped demonstration in asynchronous mode, and live transmission or Webcam in synchronous mode. This raises several important questions. Can the scripted, videotaped, live transmission or Webcam approach still be as effective as the traditional classroom? How receptive are students to different online learning environments with differential degrees of student-centered interaction compared to an instructor-centered F2F environment? Most importantly, it is an unknown but interesting question to ask whether knowledge can be effectively transferred in different online environments. This research is to address these important issues faced by any instructor who intends to apply the BM approach in either online synchronous and asynchronous environments.

BEHAVIORAL MODELING AND KNOWLEDGE TRANSFER

Social learning theory is the basis of the behavior modeling approach. Therefore, it is important to assess the applicability of the theory and approach in the online learning environment. Learning outcomes can be measured by different types of knowledge transfer and end-user satisfaction.

Behavior Modeling in Online Environments

Bandura (1977) proposed the Social Learning Theory to explain the interactive learning process between individuals and their social environment. He asserted a series of social learning needs take place to direct an individual from biological and self-centered response to social and group behaviors. Since the social learning process takes place within a society, individuals learn to establish their behavior models by observing and imitating other individuals' behaviors or through the enforcement of the media and environment. Online learning in different environments needs to be delivered via different media. Different online learning environments, therefore, may have different degrees of enforcement to learners' individual behaviors.

Learning by modeling or observing people's behaviors may be more effective than learning by trial-and-error because the former approach can avoid unnecessary mistakes and harm. Modeling an instructor's behaviors empowers students to (1) learn new behavior from the instructor, (2) self-evaluate their behaviors against the instructor's and (3) enforce students' current behavior.

Learning by modeling takes place in four sequential steps: (1) attention, (2) retention, (3) motor reproduction and (4) motivation and reinforcement (Bandura, 1977). Enforcement forces, such as the duration of training, praise, motivation and attention of others, allows learning to move along these four steps against counter forces. Enforcement forces, such as retention enhancement and practice, can contribute to better cognitive learning (Yi & Davis, 2001).

Lewin (1951) argued that the effectiveness of Behavior Modeling is a function of people interacting within an environment. The BM approach is different from learning by adaptation. The former approach teaches through demonstration, while the latter approach influences the behaviors of learners by reward and punishment (Skinner,

1938). The BM approach was first applied in the training of interpersonal communication and management skills (Decker & Nathan, 1985). Gist, Schwoerer and Rosen (1989) further applied the training to the context of information technology.

BM may be readily employed in face-to-face instruction, but cannot be easily simulated in online asynchronous instruction, which lacks the interactive immediacy necessary for optimally effective instructor demonstration and correction. The richness of information media in online synchronous instruction is another constraint and may also have less enforcement force than F2F instruction to the learning outcomes. For example, in a live training class, the instructor is able to demonstrate a software process and immediately ask the students to repeat the activity under the instructor's close supervision. However, in an online asynchronous situation where there is no live instructor, the demonstration loses the benefit of that immediate feedback. In the same token, in an online synchronous situation bandwidth constraints and compromised reciprocity may undermine the enforcement force of the demonstration. In both online environments, enforcement forces can be further compromised with the missing of "learning by doing," another key element of F2F BM training (McGehee & Tullar, 1978).

Therefore, there is a strong possibility that the BM approach cannot be fully replicated in either the online synchronous or asynchronous situation and will not be as effective a method in online training as in the traditional environment.

Knowledge Transfer

Knowledge transfer is the application of acquired skills and knowledge into different situations. Unless the transferring process occurs, learning has little value. The applied situations could be similar or novel to the learning situation. Depending on the situation, knowledge transfer can take place in different formats. In general, there are four different types of knowledge transfer.

Positive Transfer vs. Negative Transfer

Positive transfer of learning means that learning in one situation stimulates and helps learning in another situation. Negative transfer of learning hinders the application of learning in one situation to other situations. Positive learning experience can be enhanced via analogy, informed instruction (Paris, Cross & Lipson, 1984), tutorial (Morris, Shaw & Perney, 1990) and so forth. Learning effectiveness can be improved by triggering positive learning and mitigating negative learning experience.

Near Transfer vs. Far Transfer

Salomon and Perkins (1988) argued that transfer of learning could have a differential degree of transfer. The effectiveness of near-transfer learning depends on the learner's ability to solve problems similar to those encountered in the learning context. For instance, learning how to add two digit numbers allows learners to add three digit numbers. Near-transfer learning occurs in two similar situations and at a lower level. Therefore, the level of learning is more easily acquired and applied. In contrast, applying the acquired skills and knowledge in two dissimilar and sometimes novel situations is much harder to achieve. For instance, a table tennis player can apply skills of playing pinball to playing tennis. Although both sports look similar on the surface, the techniques to control pinballs and tennis balls are very different. The learning transfer is much harder to be acquired and retained. Therefore, the transfer is defined as far-transfer learning. Near-transfer and far-transfer of knowledge seem to be the most widely used measures of learning outcomes in the field of information technology since learners must utilize the knowledge learned in a computing environment.

Specific Transfer and General Transfer

Depending on learning content, there are two different learning transfers: specific transfer and general transfer (Bruner, 1996). The former refers to the extension and association of habit and skills. The latter refers to the transfer of principles and attitudes that can be used to deepen the understanding of basic concepts.

Lateral Transfer and Vertical Transfer

Gagne (1992) asserted that the transfer of learning includes lateral and vertical transfers. Lateral learning is to apply one domain of knowledge to another domain. Lateral learning does not follow step-by-step instruction and is considered as provocative learning. Vertical learning means that a higher level of learning needs to be created by integrating acquired skills, and experiences with new situations. Vertical transfer of learning is analytical and sequential.

Other Knowledge Transfer Theories

Theories related to knowledge transfer are not limited to the above mentioned ones. For instance, the theory of identical elements asserts that the more identical elements different learning contains, the more efficient the transfer of learning (Thorndike, 1949). Baldwin and Ford (1988) proposed a general training theory to classify three categories of factors affecting transfer of training: (1) training inputs, (2) training outputs and (3) conditions of transfer. The situated learning theory argues that individuals are affected by learning environment when trying to solve practical problems. Therefore, the interaction between learners and the environment is an important factor that needs to be taken into account when measuring the transfer of learning. Finally, the theory of formal discipline argues that knowledge transfer skills can be acquired by training learner's sensuality, such as thinking, judgment, classification, imagination, creation and so forth.

The objective of this study was to investigate the impacts of the learning environment in online and offline formats on the transfer of learning. The situational changes rationalize the adoption of situated learning theory. To accomplish this objective, we sought to train end-user to learn how to use Microsoft SQL server 2000 software. Therefore, we adopted the near-transfer and far-transfer measures of learning outcomes for our information technology related experiment.

HYPOTHESES

Hypotheses are formulated to investigate whether the BM approach is as effective in online synchronous and asynchronous environments as in the traditional face-to-face environment. We measured learning outcomes by trainees' performances in near-transfer and far-transfer tasks, as well as overall satisfaction levels. The study also considered the importance of time variant. Hence, training and performance measurement were conducted over five weeks.

Knowledge Near-Transfer (KNT) Tasks

H1: End-users trained using F2F behavior modeling perform near transfer information system tasks better than those trained in asynchronous behavior modeling.

H2: End-users trained in F2F behavior modeling perform near-transfer information system tasks better than those trained in synchronous behavior modeling.

H3: End-users trained in synchronous behavior modeling perform near-transfer information system tasks better than those trained in asynchronous behavior modeling.

Knowledge Far-Transfer (KFT) Tasks

H4: End-users trained in F2F behavior modeling perform far-transfer information system tasks better than those trained in asynchronous behavior modeling.

H5: End-users trained in F2F behavior modeling perform far-transfer information system tasks better than those trained in synchronous behavior modeling.

H6: End-users trained in synchronous behavior modeling perform far-transfer information system tasks better than those trained in asynchronous behavior modeling.

Overall Satisfaction

H7: End-users trained in synchronous behavior modeling have a higher overall satisfaction level than those trained in asynchronous behavior modeling.

RESEARCH DESIGN

This study applied Simon, Grover, Teng and Whitcomb's (1996) well-constructed software training theory to experimentally test behavior modeling training in three learning environments — F2F, online asynchronous and online synchronous environments. In doing so, it should be possible to detect the effects of the single independent variable (training environment) on training outcomes. The experiment was conducted in a field setting that enabled the study to garner greater external validity than would be the case with a laboratory experiment. A field experiment methodology has the merits of "testing theory" and "obtaining answers to practical questions" (Kerlinger & Lee, 2000). The exploratory nature of the study requires that variables (e.g., training environments and subject areas of study) under investigation be manipulated.

Subjects Control

The setting for the field experiment was the Tamkang University in Taiwan. The experiment was prompted by the need of 96 college sophomores, who are Management Information Systems (MIS) majors, to learn a Microsoft SQL Server 2000 software program in a database processing course. The schedule agreed on with the faculty at Tamkang University was to run the experiment for an hour training each week for four weeks. The author's graduate assistant Ms. Lin helped administer the experiment to collect the data. The subject pool had a mean age of 22 years. Subjects who participated in the structured experiment had little database-related experience. Their intellectual levels are relatively the same because subjects scored the same range of scores in a national entrance exam. The national entrance exam system has been adopted for more than 40 years in Taiwan and is considered a relatively reliable test. Subjects' individual backgrounds should not have influence on learning outcomes.

For the purposes of this study, subjects were chosen if they lacked a theoretical and procedural understanding of the particular subject area being tested. Participants were given a pretraining questionnaire that includes important study units on Microsoft SQL Server 2000. Two experts of the domain administered the Delphi study to finalize the study units and questionnaires. This is to improve the content validity. The subjects voluntarily answered whether they knew those study units and answered their database-related experiences. Based on their answers, a correlation test of database and usage experience of the target system showed no significant differences among three experimental groups. Subjects of the study may be considered representative of novice end-users. Many studies (Ahrens & Sankar 1993; Santhanam & Sein, 1994) support using students as experimental subjects to represent the general populations. Hence, all subjects' questionnaires were used for further data analysis. This segmentation was used to mitigate the effects of computer literacy and experience on the findings, thereby improving the internal validity of the study.

Training Treatments

Face-to-face BM (FBM) is instructor-centered training while online Asynchronous BM (ABM) and Synchronous BM (SBM) are learner-centered training. Course materials used in online learning environments were created to properly reflect the key elements of a behavior modeling approach. AniCam simulation software was used to record the demonstration of instruction. Hyperlink structure was used to help users assimilate nonlateral conceptual, and procedural knowledge.

Feedback activities of behavior modeling approach in online asynchronous environment are supported with e-mail and hyperlinks. SBM differs from ABM in providing feedback functions via real-time discussion forums. Training materials integrate key elements of behavior modeling approach: (1) control of three different learning environments, (2) demonstration of the instructor, (3) continuous feedback (verbal feedback in F2F and online synchronous environments; e-mail feedback in the online asynchronous environment). Three training environments were designed to maximize the effect of size on their differences (Figure 1).

Training Procedures

The experimental study lasted for four weeks. There was a 50 minute training session each week for each class. Figure 2 shows the experimental procedures used at each time period. The X's, Y's and Z's represent online asynchronous BM training, online synchronous BM training and F2F BM training methods, respectively. The subscripts next to each alphabet indicate the ith observation or training session, respectively. Before executing experimental treatments (the pretest period O1), the instructor asked the subjects to complete a short questionnaire soliciting demographic information, database software-related experience and attitudes towards learning in the subject's assigned online learning environment (Pretest). Approximately one-third of the subjects pooled received the same experimental treatment for four straight weeks (Week1 to Week4). The assigning process was random on the class basis. Randomizing the execution of O4 and O5 in Week2 and Week3 for Group A and Group B can help avoid possible confounding results from the interactive effects of the pretest of O1 and O3. This randomization process can further ensure that difference in learning outcomes of O6 is not possibly due to the sensitization of the participants after the pretest and the interaction of their sensitization, O4 and O5 (Kerlinger & Lee, 2000).

Before or after each training session, subjects were asked to complete database design tasks using the MS SQL commands to assess their prior knowledge in the trained subjects and immediate learning outcomes that involve both near-transfer and far-transfer knowledge. On week five, students were evaluated again for their attitude changes towards the e-learning sessions and performance in near and far-transfer tasks (Post-test). The final exam concludes the five-week training sessions.

Training materials were designed to integrate key elements of the three training environments, as illustrated in Figure 3. Course materials used in the online asynchronous training session were stored on the school's server for students to learn at their own pace after each training session was completed. At the end of the experiment, students were asked about their affect for their learning environments.

Outcomes Measurement

Regardless of the teaching environment, computer training is intended to instill in users a level of competency in using the system and to improve their satisfaction with the system. A user's competency in using a system is contingent upon the

Figure 1. Differences of behavior modeling approach in three learning modes

Online Learning Environments		Off-line Learning Environment
Asynchronous BM (ABM)	**Synchronous BM (SBM)**	**Face-to-Face BM (FBM)**
• Scripted demonstration of step-by-step instructions • Deductive/inductive complementary learning • Trainees choose one of two relevant examples to practice • Without online reference sources • Trainee control	• Webcam-delivered demonstration of step-by-step instructions • Deductive/inductive complementary learning • Instructor chooses examples that are relevant to trainees' majors • Without online reference sources • Trainer/trainee partially control	• Demonstration of a live instructor to learn step-by-step • Deductive/inductive complementary learning • Live instructor chooses examples that are relevant to trainees' majors • Without online reference sources • Trainer control

Figure 2. Experimental procedures

GROUP	Pretest	Week1	Week2	Week3	Week4	Post-test
Group A	O1	O2 X1 O3	X2	O4 X3 O5	X4	O6
Group B	O1	O2 Y1 O3	O4 Y2 O5	Y3	Y4	O6
Group C	O1	O2 Z1 O3	O4 Z2 O5	Z3	Z4	O6

Oi = Questionnaire and Tests
Xi = ABM (Online Asynchronous BM Training)
Yi = SBM (Online Synchronous BM Training)
Zi = FBM (F2F BM Training)

user's knowledge absorption capacity. Ramsden (1988) finds that effective teaching needs to align students with situations where they are encouraged to think deeper and more holistically. Kirkpatrick (1967) also suggests that learning effectiveness needs to be evaluated by students' reactions, learning and knowledge transfer. The levels of knowledge absorbed by students, Bayman and Mayer (1988) suggest, may include syntactic, semantic, schematic and strategic knowledge. Mennecke, Crossland and Killingsworth (2000) believe that experts of one particular knowledge domain possess more strategic and semantic knowledge than novices. Knowledge levels, as Simon, Grover, Teng and Whitcomb (1996) suggest, can be categorized as near-transfer, far-transfer or problem solving. Near-transfer knowledge is necessary for being able to understand software commands and procedures. This type of knowledge is important for a trainee to be able to use software in a step-by-step fashion. Far-transfer knowledge seeks to ensure that a trainee has the ability to combine two or more near-transfer tasks to solve more complicated problems.

Both the use of software and information systems and the satisfaction levels of using them are useful surrogates to properly measure the effectiveness of an information system (Ives, Olson, & Baroudi, 1983). The end-user satisfaction level has been widely adopted as an important factor contributing to the success of end-user software training. Since the study was to replicate Simon, Grover, Teng and Whitcomb's (1996) research in a dissimilar environment, near-knowledge and far-knowledge transfer, and end-user overall satisfaction levels were adopted in this study to

Figure 3. Delivery mechanisms of behavior modeling approaches

	FBM (F2F Behavior Modeling)	ABM (Asynchronous Behavior Modeling)	SBM (Synchronous Behavior Modeling)
Course Materials	Instructor demonstrates the use of software along with PowerPoint slides Covered three study subjects within forty five minutes each week	Course materials covered by FBM was pre-recorded and stored in a server. No instructor was present to assist the learning process of students. Students learned at their own path and completed their study within forty five minutes.	Instructor was present, but broadcasted steaming video from a broadcast room. Instructor conducted the real-time discussion with students on a BBS station.
Information Systems Tools	Instructor, PowerPoint, and PC	AniCam, PowerPoint and Acrobat Reader	AniCam, PowerPoint, Stream Author v.2.5 (Authoring Tool) and Acrobat Reader
Target System	SQL Server2000 Personal Edition	SQL Server2000 Personal Edition	SQL Server2000 Personal Edition
Pretest Questionnaire	Learning Experience and Style Questionnaires	Learning Experience and Style Questionnaires	Learning Experience and Style Questionnaires
The First Week	First Training Session First Learning Outcomes Test	First Training Session First Learning Outcomes Test	First Training Session First Learning Outcomes Test
The Second Week	Second Training Session Second Learning Outcomes Test	Second Training Session	Second Training Session Second Learning Outcomes Test
The Third Week	Third Training Session	Third Training Session Second Learning Outcomes Test	Third Training Session
The Fourth Week	Comprehensive Test (Third Learning Outcomes Test)	Comprehensive Test (Third Learning Outcomes Test)	Comprehensive Test (Third Learning Outcomes Test)
Post-test Questionnaire	Measure End-User Satisfaction	Measure End-User Satisfaction	Measure End-User Satisfaction

measure training outcomes. Cronbach's alpha reliability for Simon et al.'s (1996) instrument to measure satisfaction is r = 0.98. Users need to use the Likert scale from one to five to answer 12 test items related to their satisfaction with the use of online system.

DATA ANALYSIS

Table 1 shows the means and standard deviations for the scores at each treatment period. Table 2 shows F and P values of the dependent variables (near-transfer and far-transfer task performances, and overall satisfaction) across treatment groups and in different times. Pretest scores (Q1, Q2 and Q4) in varying weeks were used to tell apart students with prior experiences and knowledge on the studied topics. After learning in a weekly session, students participated in a post-test. Their scores (Q3 and Q5) were used for KNT effectiveness comparison across training sessions. Scores of Q6 are KFT effectiveness and end-user satisfaction levels. A cursory examination of means (Table 1) indicates that no patterns can be identified for near-transfer performance from time Week1 to Week5. Subjects in ABM performed better than those in FBM, followed by SBM at Week1 while at Week2 and Week3 the order was changed to FBM>ABM>SBM and SBM>FBM>ABM, re-

Table 1. Descriptive statistics - means (Standard Deviations)

	ABM (N=40)	SBM (N=26)	FBM (N=30)	Overall (N=96)
KNT (Week 1)	27.63 (5.77)	25.38 (7.06)	26.00 (7.24)	26.51 (6.61)
KNT (Week 2)	28.50 (3.43)	28.46 (3.09)	29.83 (0.91)	28.91 (2.83)
KNT (Week 3)	56.05 (14.06)	64.27 (17.52)	62.27 (12.71)	60.22 (14.98)
KNT (Week 5)	71.70 (16.21)	70.50 (19.78)	67.00 (17.27)	69.91 (17.49)
KFT (Post-test)	9.63 (1.33)	9.42 (1.63)	8.83 (2.15)	9.32 (1.72)
OS (Post-test)	38.10 (6.87)	39.38 (9.21)		38.61 (7.83)

Table 2. Performance on differentlLearning outcomes over five weeks

	F	p-value	Power
KNT (Week 1)	1.035	0.359	0.337
KNT (Week 2)	2.415	0.095*	0.605
KNT (Week 3)	2.891	0.061*	0.677
KNT (Post-test)	0.634	0.532	0.246
KFT (Post-test)	1.913	0.153	0.517
OS (Post-test)	0.420	0.519	0.169

Table 3. Results for training methods

Variable	Hypothesis	Result in Correct Direction?	Significant p-value? (n.s.—not significant)
Week1	H1: FBM > ABM	F	n.s. (p=0.312)
	H2: FBM > SBM	T	n.s. (p=0.729)
	H3: SBM > ABM	F	n.s. (p=0.182)
Week2	H1: FBM > ABM	T	p=0.051
	H2: FBM > SBM	T	p = 0.069
	H3: SBM > ABM	T	n.s. (p=0.956)
Week3	H1: FBM > ABM	T	p=0.083
	H2: FBM > SBM	F	n.s. (p=0.612)
	H3: SBM > ABM	T	p = 0.029
Week4	H1: FBM > ABM	F	n.s. (p=2.71)
	H2: FBM > SBM	F	n.s. (p=0.459)
	H3: SBM > ABM	F	n.s. (p=0.787)
Post Test (KFT)	H4: FBM > ABM	F	p=0.057
	H5: FBM > SBM	F	n.s. (p=0.2)
	H6: SBM > ABM	F	n.s. (p=0.639)
Post Test (OS)	H7: SBM > ABM	F	n.s. (p=0.579)

spectively. The findings are not in agreement with a consistent pattern as predicted by Hypotheses H1 and H2. For KFT tasks, subjects in ABM performed better than those in SBM, followed by FBM. This is the reversed order of a pattern as predicted by Hypotheses H3 and H4. The measurement of overall satisfaction level somewhat follows the predicted patterns of Hypotheses H5 and H6.

We took a closer look at the mean difference at the significance level of 0.05. The study used one-way ANCOVA to analyze the effects of behavior modeling approach on learning outcomes over different time. Levene's Test (1960) was used to examine the variance homogeneity of three groups. Its F-statistics showed that KNT was 3.04 (p=0.053) at Week1, 13.01 (p=0.000) at Week2, 1.71 (p=0.187) at Week3, and 0.47 (p=0.627) at the Post-test. In contrast, the F-statistics of Levene's Test for KFT and OS were 7.64 (p=0.001) and 1.75 (p=0.191) at the Post-test. With the exception of KNT at Week2 and KNF at the Post-test, all dependent variables met the $p > 0.05$ criterion for assuring homogeneity of variances. The heteroscedasticity of variances for these two exceptions suggested that the statistical test results may not be valid. As such, the following discussion will ignore these two variances and focus on KNT and OS. For other effects that show significance, the study adopts the Scheffe post-test to analyze data. In addition, Pearson Correlation Analysis was used to assess the carry-over effects of different training sessions.

ANCOVA was performed using the general linear model approach; the results are presented in Table 3. It shows that the treatment effects are significant for KNT (Week2) and KNT (Week3) with F-statistics of 2.415 (p=0.095) and 2.891 (p=0.061), confirming a univariate treatment effect of learning environments on the dependent variable: KNT. However, the treatment effects are not salient for other dependent variables: KFT and OS. These lacks of effect may have been due to small effect sizes.

Least-Squares Deconvolution (LSD) was used to test cross-correlations for KNT (Week2) and KNT (Week3). LSD is a cross-correlation technique for computing average profiles. LSD is very similar to most other cross-correlation techniques, though slightly more sophisticated in the sense that it cleans the crosscorrelation profile from the autocorrelation profile (Donati, 2003). For KNT (Week 2), the LSD results indicate that subjects in FBM perform better than those in ABM (p=0.051) and SBM (p=0.069). This supported the Hypotheses 1 and 2. However, Hypothesis 3 cannot be supported because the mean difference between ABM and SBM is not significant. For KNT (Week3), the LSD results indicate that (1) subjects in FBM performed better than those in ABM (p=0.083), and (2) subjects in SBM performed better than those in ABM (p=0.029). Hypotheses 4 and 6 are supported. Worthy to be noted is that H4 is upheld but in the reversed direction. This indicates that ABM is a more effective method than FBM at improving knowledge far transfer.

Four out of nine hypotheses in total are supported. Although not all hypothesized relationships are fully supported, the results obtained are interesting. The most intriguing result is that although there is statistically-justified reason for preferring FBM to ABM or SBM or software training, the pattern of results is not persistent in the long run. FBM resulted in better outcomes than ABM and SBM at Week2, and than ABM at Week3 for KNT. Although it never does so at a statistically significant level, subjects in ABM performed better than those in SBM, followed by those in FBM for KNT (Post-test) and KFT (Post-test). One interpretation of this is that either ABM or SBM training is no worse than FBM training across all dependent variables. The pattern of results for FBM suggests that trainers might choose ABM or SBM, which should to be a less costly alternative to FBM, without making any significant sacrifices in either learning or trainee reaction outcomes.

Another result of interest is that, with respect to the three online asynchronous training methods, the pattern of results suggests that FBM might be the best for KNT in the short term. Of the nine hypotheses concerning relationships between these methods, four are in the expected direction, and significantly so. This indicates that use of ABM or SBM may be a better—and certainly no worse—software training strategy in the long term.

IMPLICATIONS FOR RESEARCH

This article studied the impact of training duration on performance and trainee reactions. Trainees were exposed to the same training methods with different degrees of social presence for different durations. These findings indicated that training duration and social presence have little impacts on learning outcomes. Despite this, the findings here raise additional questions for research.

It may be more important to investigate the impacts of information richness (Fulk, 1993) features of online training media on training outcomes. Future studies might vary the social presence features of training media or their combination with social presence features (e.g., with instructor's feedbacks versus discussion boards, e-mail response or playback features). Information richness may be a more influential factor affecting the performance of training approaches.

It may also be useful to replicate the experimental equivalence of FBM, ABM and SBM methods of software training with different software and subjects. Since in the long term different treatments have similar impacts on learning outcomes, it may be practical to demonstrate the cost-based advantage of ABM over SBM, and SBM over FBM for software training in practical settings.

Another way to improve the reliability of the study is to manipulate some useful blocking variables. A series of comparative studies can be conducted to assess the impact of individualism as a cultural characteristic, computer self-efficacy, task complexity (simple tasks vs. fuzzy tasks), professional backgrounds and the ratio of the training duration to the quantity of information to be processed, among others.

Learning style may be an important factor to consider in the online learning environment. According to social learning theory, learners interact with the learning environment to change their behavior. Learning style is situational and can vary with different learning environments. Therefore, it is possible that the combination of training methods, learning style and social presence information richness (SPIR) attributes may jointly determine learning outcomes. This is not the case for BM approach in F2F environment. The self-paced online learning environment may alter the assertion. Hence, it may be necessary to conduct longitudinal studies of the influence of learning style on learning performance and trainee reaction.

IMPLICATIONS FOR PRACTICE

The largest implication for practice is that ABM and SBM may provide cost-effective substitutes for FBM without significant reductions in training outcomes in the long term. While it may still be true that FBM is still the most effective approach to improve KNT in the short term, ABM and SBM have similar leverage in KFT in the short term and KNT in the long term. Regardless of training environments, trainees have same satisfaction levels in the near- and long-term. These findings strongly indicate that the cost issue is more important than learning effectiveness. When given the options to decide which BM approach to take in the long term, nonperformance issues (teacher and facility availability, trainee's preferences, location and convenience issues) have to be first taken into account.

CONCLUSION

The success of an online training strategy depends on its effectiveness in improving learning outcomes. This study, built on well-accepted frameworks for training research (Bostrom, Olfman & Sein, 1990; Simon & Werner 1996), examines the relative effectiveness of the behavior modeling approach in online synchronous, online asynchronous and face-to-face environments. The results from this experiment provide an empirical basis for the development of an online behavior modeling strategy: (1) FBM is more effective than ABM and SBM for knowledge transfer in the short term (KNT), and (2) ABM and SBM are as effective as FBM for knowledge transfer and overall satisfaction in the long term (KFT).

What is learned from this study can be summarized as follows: When conducting software training, it may be almost as effective to use online training (synchronous or asynchronous) as it is to use a more costly face-to-face training in the long term. In the short term face-to-face knowledge transfer model still seems to be the most effective approach to improve knowledge transfer in the short term.

The limitation of this experimental study is that it was conducted with a homogeneous group with Taiwanese cultural and educational backgrounds. Therefore, this study may be constrained with the generalizability of its findings to different cultural contexts.

Hofstede (1997) stated that the domains of education, management and organization have nurtured the values context that differs from one country to another. Cultural influences have been discerned in the study of Internet usage (Lederer, Maupin, Sena & Zhuang, 2000; Moon & Kim, 2001; Straub, 1997) and Web site design (Chu, 1999; Svastisinha, 1999). Users from different cultures have different perceptions about the usefulness and ease of use regarding different information systems (Straub, 1994). E-learning systems may differ based on the cultural backgrounds of the learners to improve their satisfaction levels and cognitive gains. Benefits of the congruence may include the improvement of (1) global e-learning adoption rate and (2) learning outcomes (attitude and cognitive gains). From the perspective of research design (Kerlinger & Lee, 2000), a cross-cultural study to replicate the study with American or European subjects may further validate and extend the generalizability of the findings.

The study has accomplished its major goal; it provides evidence as to the relative effectiveness of the behavior modeling approach in different learning environments for software training. This research somewhat improves the generalizability of theories on the behavior modeling approach in different learning environments.

REFERENCES

Ahrens, J. D., & Sankar, C. S. (1993). Tailoring database training for end users. *MIS Quarterly, 17*(4), 419-439.

Aniebonam, M. C. (2000, October). Effective distance learning methods as a curriculum delivery tool in diverse university environments: The case of traditional vs. historically black colleges and universities. *Communications of the Association for Information Systems, 4*(8), 1-35.

Baldwin, T.T., & Ford, J.K. (1988). Transfer of training: A review and directions for future research. *Personnel Psychology, 41*, 63-105.

Bandura, A. (1977). *Social learning theory*. Morristown, NJ: General Learning Press.

Bayman, P., & Mayer, R. E. (1988). Using conceptual models to teach BASIC computer programming. *Journal of Educational Psychology, 80*(3), 291-298.

Bielefield, A., & Cheeseman, L. (1997). *Technology and copyright law*. New York: Neal-Schuman Publishers, Inc.

Bostrom, R. P., Olfman, L., & Sein, M. K. (1990). The importance of learning style in end-user training. *MIS Quarterly, 14*(1), 101-109.

Bruner, J. (1996). *Toward a theory of instruction.* New York: Norton.

Chu, G.-L. (1999). *The relationships between cultural differences among American and Chinese university students and the design of personal pages on the World Wide Web.* Unpublished doctoral dissertation, University of Georgia.

Compeau, D. R., & Higgins, C. A. (1995). Application of social cognitive theory to training for computer skills. *Information Systems Research, 6*(2), 118-143.

CyberAtlas. (2003). E-Learning market expanding beyond IT training. Jupiter Research. Retrieve July 28, 2006, from *http://cyberatlas.internet.com/markets/education/article/0,,5951_914901,00.html*

Decker, P. J., & Nathan, B. R. (1985). *Behavior modeling training.* New York: Praeger.

Donati, J. (2003). Least-squares deconvolution of Stellar Spectra. Retrieved July 28, 2006, from http://webast.ast.obs-mip.fr/people/donati/multi.html

Fulk, J. (1993). Social construction of communication technology. *Academy of Management Journal, 36*, 921-950.

Gagne, R. M. (1992). *Principles of instructional design.* New York: Holt, Rinehart and Winston, Inc.

Gist, M. E., Schwoerer, C., & Rosen, B. (1989). Effects of alternative training methods on self-efficacy and performance in computer software training. *Journal of Applied Psychology, 74*, 884-891.

Hofstede, G. (1997) *Cultures and organizations: Software of the mind.* New York: McGraw-Hill.

Horton, W. (2000). *Designing Web-based training: How to teach anyone anything anywhere anytime.* New York: John Wiley & Sons.

IDC. (2002, September 30). While corporate training markets will not live up to earlier forecasts, IDC suggests reasons for optimism, particularly e-learning. Retrieved July 28, 2006, from *http://www.idc.com/getdoc.jhtml?containerId=pr2002_09_17_150550*

Ives, B., Olson, M., & Baroudi, S. (1983). The measurement of user information satisfaction. *Communications of the ACM, 26*, 785-793.

Kerlinger, F. N., & Lee, H. B. (2000). *Foundations of behavioral research.* New York: Harcourt Brace College Publishers.

Kirpatrick, D. L. (Ed.). (1967). *Evaluation of training: Training and development handbook.* New York: McGraw-Hill.

Lederer, A. L., Maupin, D. J., Maupin, M. P., Sena, M.P. & Zhuang, Y. (2000). The technology acceptance model and the World Wide Web. *Decision Support Systems, 29*, 269-282.

Leidner, D. E., & Jarvenpaa, S. L. (1995). The use of information technology to enhance management school education: A theoretical view. *MIS Quarterly, 19*, 265-291.

Levene, H. (1960). In I. Olkin et al. (Eds.) *Contributions to probability and statistics: Essays in honor of Harold Hotelling.* (pp. 278-292). Stanford University Press.

Lewin, K. (1951). *Field theory in social science: Selected theoretical papers.* New York: Harper and Row.

McGehee, W., & Tullar, W. (1978). A note on evaluating behavior modification and behavior modeling as industrial training techniques. *Personal Psychology, 31*, 477-484.

Mennecke, B. E., Crossland, M. D., & Killingsworth, B. L. (2000). Is a map more than a picture? The role of SDSS technology, subject characteristics, and problem complexity on map reading and problem solving. *MIS Quarterly, 24*(4), 601-627.

Moon, J., & Kim, Y. (2001). Extending the TAM for a World Wide Web context. *Information & Management, 38*, 217-230.

Morris, D., Shaw, B., & Perney, J. (1990). Helping low readers in grades 2 and 3: An after-school volunteer tutoring program. *The Elementary School Journal, 91*, 133-150.

Paris, S. G., Cross, D. R., & Lipson, M. Y. (1984). Informed strategies for learning: A program to improve children's reading awareness and comprehension. *Journal of Educational Psychology, 7*, 1239-1252.

Ramsden, P. (Ed.). (1988). Context and strategy: Situational influences on learning. In *Learning strategies and learning styles*. New York: Plenum Press.

Salomon, G., & Perkins, D. N. (1988). Teaching for transfer. *Educational Leadership, 46*(1), 22-35.

Santhanam, R., & Sein, M. K. (1994). Improving end-user proficiency: Effects of conceptual training and nature of interaction. *Information Systems Research, 5*(4), 378-399.

Simon, S. J., Grover, V., Teng, J. T. C., & Whitcomb, K. (1996). The relationship of information system training methods and cognitive ability to end-user satisfaction, comprehension, and skill transfer: A longitudinal field study. *Information Systems Research, 7*(4), 466-490.

Simon, S. J., & Werner, J. M. (1996). Computer training through behavior modeling, self-paced, and instructional approaches: A field experiment. *Journal of Applied Psychology, 81*(6), 648-659.

Skinner, B. F. (1938). *The behavior of organisms: An experimental analysis*. New York: Appleton-Century Company, Incorporated.

Straub, D. W. (1994). The effect of culture on IT diffusion: E-mail and FAX in Japan and the U.S. *Information Systems Research, 5*(1), 23-47.

Straub, D., Keil, M., & Brenner, W. (1997). Testing the technology acceptance model across cultures: A three country study. *Information and Management, 33*, 1-11.

Svastisinha, R. W. (1999). *Wahhn: Web-based design. Wind and human comfort for Thailand*. Unpublished doctoral dissertation, University of Southern California.

Thorndike, R. L. (1949). *Personnel selection: Test and measurement techniques*. New York: John Wiley & Sons.

Yi, M. Y., & Davis, F. D. (2001). Improving computer training effectiveness for decision technologies: Behavior modeling and retention enhancement. *Decision Sciences, 32*(3), 521-544.

This work was previously published in Journal of Distance Education Technologies, 4(4), edited by T. Shih, pp. 88-102, copyright 2006 by IGI Publishing, formerly known as Idea Group Publishing (an imprint of IGI Global).

Section II
Effectiveness and Motivation

Chapter V
A Framework for Distance Education Effectiveness:
An Illustration Using a Business Statistics Course

Murali Shanker
Kent State University, USA

Michael Y. Hu
Kent State University, USA

ABSTRACT

Distance education is now an integral part of offering courses in many institutions. With increasing access to the Internet, the importance of distance education will only grow. But, to date, the specific benefits that distance education brings to student learning objectives remain unclear. We first propose a framework that links student performance and satisfaction to the learning environment and course delivery. Next, we empirically evaluate our framework using data from a Business Statistics course that we offer in the traditional classroom setting and as a distance-education course. Our results show that a well-designed distance education course can lead to a high level of student satisfaction, but classroom-based students can achieve even higher satisfaction, if they also are given access to learning material on the Internet. This indicates that material for an effective distance-education course also can be used to supplement in-class teaching in order to increase satisfaction with student learning objectives.

INTRODUCTION

Distance education has created a substantial impact on students, faculty, and institutions. Distance education classes now are routinely available to many students. In a survey conducted by the National Center for Education Statistics, the percentage of two- and four-year degree-granting institutions offering distance education classes increased by 11% from 1995 to 1997. The number of courses being offered nearly doubled in the same time period (Sikora & Carrol, 2002). The effect of distance education also has been significant for faculty. In a study conducted by Lewis et al. (1999), nearly 6% of all faculty members in Title IV degree-granting institutions was involved in distance education classes, and about 9% offered courses using non-face-to-face mediums (Lewis et al., 1999). Studies also indicate that distance education faculty members bear a higher burden of teaching. Bradburn and Zimbler (2002) found that, on average, faculty members teaching distance education classes had more sections and more course preparations than faculty members who only taught face-to-face.

Institutions are also at a crossroads. While the trend to offer more distance education classes is clear, with increasing competition for limited resources, many institutions face questions concerning lack of fit with mission, program development costs, and technological infrastructure, among others (Bradburn & Zimbler, 2002). These questions need to be answered if distance education is to fulfill its potential.

Cost aside, it is clear that students, faculty, and institutions benefit from distance education. But, currently, the benefits of distance education are neither clearly defined nor can they be easily measured. A brief tally from 1992 to 2002 indicates that there were 22 papers finding significant positive effects and 26 not finding significant benefits in using distance education (Russel, 2003a, 2003b). While these studies varied in subject and in the choice of performance metrics, it is still too early to conclude what specific benefits students and institutions can reap from distance education. Importantly, the role distance that education plays in the overall attainment of student learning objectives remains unanswered.

Research efforts continuously have been extended to explain the effectiveness of distance education, and typically, these comparisons are made with traditional classroom education. But in order to clearly evaluate the effects of distance education, factors like student learning styles, delivery of content, course characteristics, and technology also need to be considered. Then, with increasing research, a clearer picture will emerge on factors that lead to a successful implementation of distance education. This study hopes to add to this body of research. We first propose a framework that links student performance and satisfaction with the learning environment and course delivery. Then, we empirically examine our framework and provide more evidence to the growing body of research on distance-education effectiveness. As part of our empirical data, we also show how a Business Statistics course can be offered over the Web.

The ubiquity of the Internet certainly has been a key factor in the rise of distance education. Web-based classes especially occupy a special niche, as their growth has been a result of this spread of the Internet. In this article, we review cases of instruction for two groups of students, those enrolled in a Web-based class vs. those receiving traditional classroom instructions. We propose a framework for studying distance education. We argue that the education environment, whether it is Web-based or classroom-based, will govern how a course is to be designed, and that course design is a critical factor in the overall determinant of student satisfaction level. The primary intent in proposing such a framework is to force us to have a deeper thinking about the overall problem setting. That is, we need to first identify the key structural components leading to satisfaction and how those components are interconnected. Then, after empirical findings are gathered from students, we can be in a better position to pinpoint the potential factors of student satisfaction.

The rest of the article is organized as follows. The next section discusses our framework, linking the learning environment and course design to student satisfaction. This is followed by a description and design of the undergraduate course that we use to study and illustrate our findings. The undergraduate course, Business Statistics, displays many characteristics in order to be successfully administered as a Web class. In addition to discussing course structure in this section, we also present the tools and techniques specifically developed for the Web-based class. Then, we present our results, followed by the Conclusion section.

A FRAMEWORK

From the learning environment to course design and delivery, Web-based and classroom-based educations provide faculty and students with different challenges. Figure 1 describes a framework that relates these challenges to student satisfaction in our study. To be effective, an educator first must have a clear understanding of the differences between Web-based and classroom-based learning environments. Courses then must be selected and designed to suit the learning environment. Student satisfaction then is largely the result of the implementation of the course design. This, in turn, leads to a better understanding of the learning environments and hopefully an improvement in course offerings.

As shown in Figure 1, the learning environment is influenced by several factors. Face-to-face interaction is a predominant part of classroom education but plays a minimal role in Web classes. This face-to-face interaction provides an environment where the delivery of instruction via audio-visual means is instantaneous and synchronized with interactions between students and faculty. This allows the instructor to know whether the intended message is communicated clearly to the students. The message-response-feedback is usually iterative and complete, and any breakdown in communication can be corrected immediately. Similarly, the pace and scope of course material coverage can be adjusted accordingly. Furthermore, factors like facial expression and body language all help to bring about more effective communication between instructor and student. Thus, courses that require constant interaction and effective two-way communication, like case-based classes, are suited ideally for classroom education.

Communication plays an important role in the learning environment. Both learning environments can use synchronous and asynchronous communication tools like chat, peer-to-peer, e-mail, videoconferencing, and electronic blackboards. While certain distance-learning classes, like virtual classrooms using VTEL (VTEL), can duplicate the synchronous face-to-face communication of classroom environments, for most Web classes similar to that illustrated in this article, communication is usually one-way, and any two-way communication is likely to be asynchronous. Thus, courses that require a constant flow of exchange of ideas and discussion are likely to be more difficult to implement in Web-based education. The lack of instant feedback as there is in classroom instruction means that instructors need to plan in detail ahead of time how course materials should be covered. Furthermore, many institutions allow students flexibility in the duration required to complete Web classes. This requires significant up-front work from the instructor, as all course content, testing, and assessment modules have to

Figure 1. A framework

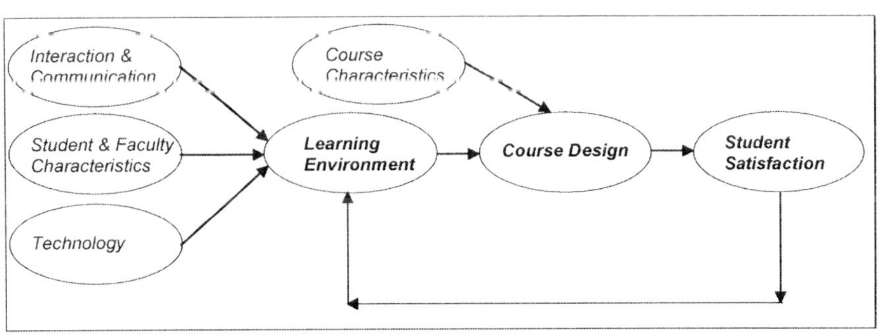

be available at the beginning of the term. Thus, once the course starts, it becomes difficult to make changes to any of the modules. As such, Web-based course content and delivery tend to be static for the term but offer uniform course coverage across sections. Classroom instruction, on the other hand, is usually more dynamic and has greater variability of course coverage, as the instructor can adjust delivery and content during the term to shifting student needs. Therefore, designing an effective Web-based course requires significant design effort to accommodate different student learning styles and abilities.

The up-front work required for a Web class and the flexibility in the duration allowed for students to take a Web class provide some additional advantages. Students now can review course content at any time. This allows students to become active participants in their learning. While cooperative learning (Millis & Cottell, 1997) usually is stressed in traditional classes in order to increase student participation, to be successful in Web classes, active student learning becomes a prerequisite. To facilitate this, learning tools, including course navigation, must be well-designed in Web classes.

Technology plays a greater role in Web-based education. Instructors and students need to be comfortable with technology in order to fully utilize the Web environment. While technology is used in classroom education, lack of technological competence there usually can be compensated for by face-to-face interaction. No such solution exists for Web classes. As such, faculty and students who are uncomfortable with technology are likely to be intimidated by Web classes.

Recent research also indicates that student personality traits affect performance in Web-based classes (Schniederjans & Kim, 2005). While classroom education by its more dynamic nature and greater interactivity can compensate for such traits, it is difficult to do so in Web classes. Thus, the selections of students, in addition to faculty, become important considerations in offering Web classes.

Clearly, the learning environments influence the success of courses. But, in order for a Web class to be successful, it is equally important to

Table 1. Differences in learning environments

Dimension	Web-based	Classroom-based
Interaction and communication		
Type	Virtual, 1-way	Virtual, Direct, 2-way
Mode	Audio, Visual	Audio, Visual, Direct
Timing	Asynchronous	Synchronous
Technology	Required	Optional
Course Design and Characteristics		
Structure	Static	Variable and dynamic
Content repeatability	May be reviewed repeatedly	Class times are predetermined
Content variability	Consistent and identical for all classes	Varies from class to class
Assessments	Restricted. Suitable for questions that are easy to generate and grade	Flexible
Navigation	Flexible	Predefined
Student-faculty contact	Irregular	Regular

consider course characteristics. Courses that can be adapted easily to the Web learning environment are likely to be well received as a Web class. For example, because of the need for constant two-way interaction, case-based courses generally are not well suited for the Web environment. But, courses where concepts and examples can be constructed easily and presented using software tools may be better suited for Web-based education. Such courses allow students to learn through interactivity and repeatability at their own pace, thus satisfying diverse student learning capabilities. Table 1 summarizes our observations of the two learning environments.

In order for a Web-based class to be successful, it should exploit the characteristics of the learning environment. The Business Statistics course that we discuss in the next section has many characteristics that make it suitable to be offered as a Web class. This course also was offered in the classroom, thus allowing us to compare the satisfaction between the two classes.

COURSE DESIGN: BUSINESS STATISTICS

The business statistics course considered in this study is an introductory course open to all majors but required for business majors. This course covers basic concepts and applications, with emphasis on intuitive statistical thinking. Topics include descriptive statistics, observational studies and experiments, sampling distributions, hypothesis testing and confidence intervals, and regression analysis.

Students have the option of taking this course in a classroom setting or as a Web-based course. Every semester, multiple classroom sections are offered, but the Web-based section is offered only once a year. Average enrollment for each classroom section is around 150, and for the Web-based section, around 56. Classroom sections meet twice a week with the instructor for 75-minute sessions each time. There is no face-to-face interaction between the instructor and the Web-based students. Communication between the instructor and the classroom students is predominantly face-to-face and through e-mail. Communication between the instructor and Web-based students is through instant messaging, e-mail, and electronic bulletin boards. Both groups of students were welcome to see the instructor for additional help.

The course material was divided into 10 chapters. In addition to the textbook, multimedia content was created for this course. This content, available on CD or on the Internet, contained animated presentations of all topics, interactive exercises, practice problems, class notes to print, copies of old exams, and the syllabus. The only requirements to access this multimedia content were a Web browser with Flash (Macromedia Flash) and Java (Sun Java) plugin enabled, and access to the free Adobe Acrobat Reader (Adobe Acrobat) for printing the class notes. All students had equal access to all course materials.

In addition to common course materials, both classroom and Web-based students were assessed similarly. Students were required to take eight quizzes and six examinations, which were administered through WebCT (WebCT). Each quiz had 15 questions and took approximately 40 minutes. Examinations had 25 questions and were 75 minutes long on average. Question types for both quizzes and examinations included multiple choice, calculated, and short answer. All questions were drawn from a central database of questions. There was one difference between how the testing was administered between classroom and Web sections. For classroom-based sections, the quizzes and examinations only could be taken during specific time periods. Quizzes for a topic usually were administered after the topic was covered in class. As Web-based students could cover topics at their own pace, no restrictions were placed on when they could take the tests. All quizzes and examinations were available on the first day of the semester for these students. They could take the quizzes and examinations in any order; the only requirement was that all testing had to be completed before the end of the semester. While the technology existed to restrict students with certain IP addresses, it was impractical to do so for Web-based students. In the end, no restriction was placed on the location from where students could take their tests. All tests were open book, and the final grading scale was the same for all students. In any given year, a single instructor was responsible for all sections of this course. The results in this article come from sections taught by the same instructor.

Although classroom and Web sections used the same material and were tested similarly, the manner in which the classes progressed differed. Cooperative learning was encouraged for classroom students. Class notes provided the outline of the day's lecture. The instructor would give a brief lecture explaining the concepts. This was followed by examples. Data for examples usually were drawn from the class itself, so students were involved in the data collection process. Students then were given additional problems that they solved in groups. Sometimes, group activities took the entire class. In such cases, the instructor functioned more as a facilitator than as a lecturer in a typical classroom setting. As such, the classroom setting provided students with an interactive learning environment in which they could explore both the theoretical and practical aspects of statistical thinking.

Animated presentations were created to capture much of this interactive learning atmosphere of the classroom environment and to transfer them to the virtual classroom. Thus, animation was used to depict the concepts graphically, and voice over was used to explain what was being shown. As it was impractical to collect data in real time, predefined examples with data collected from previous classes were used to illustrate concepts. Interactive exercises were created to mimic the group activities that students do in a classroom. For example, Figure 2 shows a simulation experiment to illustrate probabilities. In class, students would use a random number generator to do this experiment, with one student generating random numbers, while the other performed the experiment. For the animated presentations, the random number generator was built into the system, so a single student could perform the simulation. Additional exercises were created to allow Web students to explore the topics further. Figure 3 shows an example that relates p-values, Type I, and Type II errors. Students could interactively change the decision point to see what happens to the errors. During the course of listening and seeing these presentations, students had access to all navigation buttons that one typically finds on a DVD player. They could Stop, Play, Fast Forward, Rewind, or move to the next topic at any time. Figure 4 shows a typical Flash presentation with navigational controls. The presentations also automatically paused at predefined points and presented students with practice questions. Thus, these animated presentations were meant to serve as a substitute for in-class lectures and interaction between instructor and student. Figure 5 shows the front page to access all Web-based materials.

As discussed previously, significant effort was spent on the design of the statistics course in order for it to be offered as a Web-based course. Being predominantly quantitative, students' understandings of the material were explored mainly through numerical examples and problems. Examples to illustrate concepts can be created with Web-friendly programming tools like Java or Flash. Assessment also can be done easily, as a question on a single concept can be administered to many students just by changing the numerical values of the problem. As such, while each student can be tested on the same concept, they receive different questions.

Our research objectives are multifold. In the previous two sections, we examined the differences between the Web-based and classroom-based learning environments and discussed characteristics that we feel are essential to consider, if

Figure 2. An animated simulation experiment

There are three doors. Behind one door is a car. Behind each of the other two doors is a donkey. As a contestant, you are asked to select a door, with the idea that you will receive the prize that is behind that door. The game host knows what is behind each door. After you select a door, the host opens one of the remaining doors that has a donkey behind it. Note that no matter which door you select, at least one of the remaining doors has a donkey behind it for the host to open. The host then gives you the following two options:

1. Stay with the door you originally selected and receive the prize behind it.
2. Switch to the other remaining closed door and receive the prize behind it.

What is the probability of winning the car if you stay? What is the probability of winning the car if you switch? Will switching increase your chance of winning the car?

Try the simulation below to see which is the winning strategy.

How the simulation works:

1. Click on one of the doors. The computer will then open one of the remaining doors with a donkey behind it.
2. If your strategy is to remain with the original choice, click on that door again. That is, for example, if you clicked on door 2 first, click on door 2 again. If your strategy is to switch, click on the other door which is closed (not door 2).
3. The computer will keep a tally of each choice.

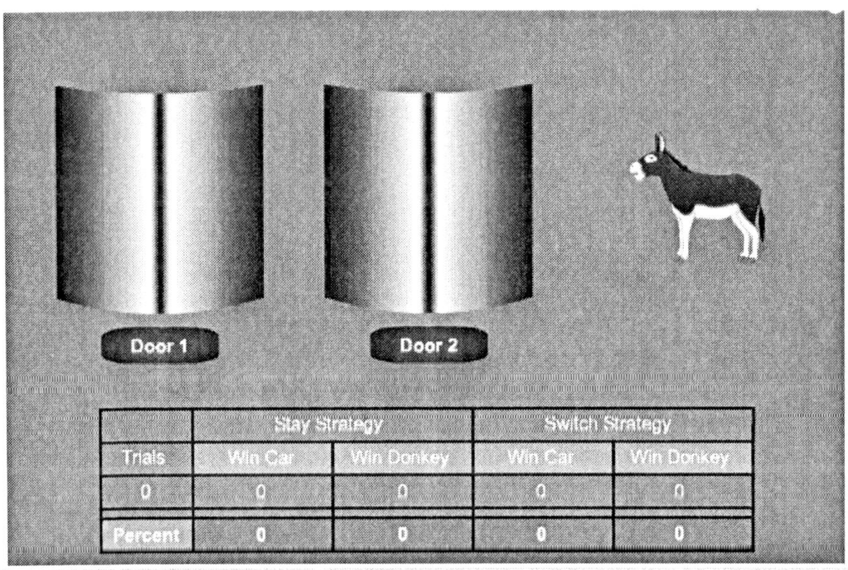

Figure 3. Relating p-values to type I and II errors

Decision Making and P-Values for a Right-Tailed Hypothesis

Below is an interactive exercise that you can use to calculate *Type I* and *II* errors, and *p-values*. Just to jog your memory, here are some definitions:

- *P-values*: The chance of observing the sample result or something more extreme under the null hypothesis
- *Type I error* (a): The chance of rejecting the null, when in fact it is true.
- *Type II error* (b): The chance of concluding the null hypothesis when the alternative is true.

The problem is based on the following hypothesis:

H_0: Bag A
H_1: Bag B

Remember that larger values support Bag B, the alternative hypothesis. Therefore, the direction of extreme is to the **right**. This is an example of a *right tailed hypothesis*.

- So, when calculating the Type I error, we will be calculating the area to the right.
- As the *p*-value calculations are also based on the null hypothesis and direction of extreme, that area is also calculated to the right.
- As the *p*-value represents the chance of observing the sample result under the null, the smaller the p-value, the more it supports the alternative hypothesis.
- If the p-value <= a, we reject the null hypothesis in favour of the alternative.

How to use the interactive exercise:

- The red line represents your decision rule. All values to the right of the red line represent the areas where you would reject the null hypothesis in favour of the alternative hypotheis. That is, you would choose Bag B instead of Bag A. Therefore, when calculating the Type I error, we are calculating this area under the null hypothesis (the number of values under Bag A that is to the right of this line). As you move the line, you will see the change in the Type I error.
- Once a decision rule is made (i.e., you have moved your red line), the area to the left represents the region where you would conclude Bag A. As the Type II error is when you conclude Bag A under the alternative, you can calculate this from the graph. Just count the number of values under Bag B to the left of this line.
- Finally, *p*-value is the area to the right of the selected sample point under the null. Move your cursor so that it points to a value under the null hypothesis, and you will see the *p*-value. Note that whenever *p*-value > a, your conclusion is not to reject the null hypothesis. Try it.

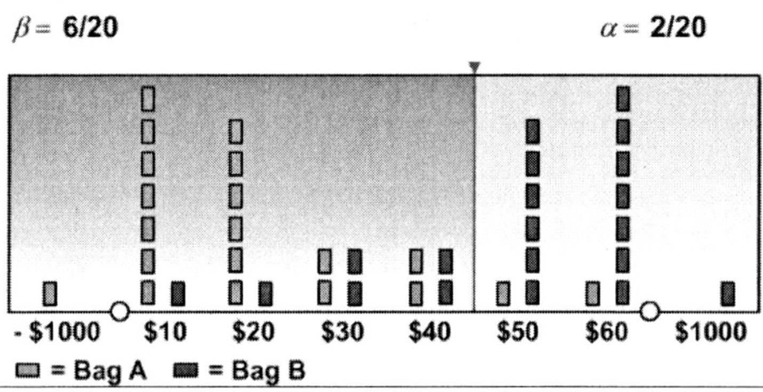

Figure 4. Navigation controls in interactive exercises

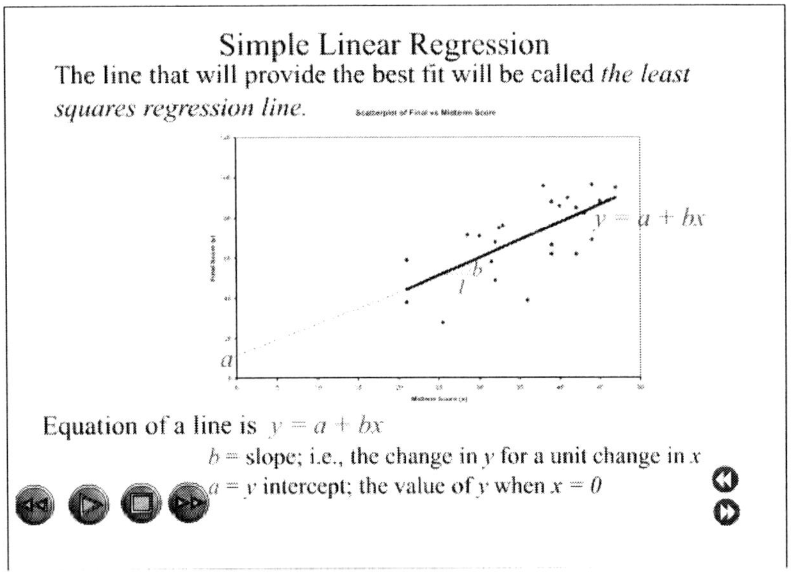

Web-based courses are to be well received. The design and suitability of the Statistics course for Web students also were discussed. In the following sections, we empirically evaluate our framework and observations by answering two questions: (1) what is the satisfaction of students taking the Web based class? and (2) how does Web-based student satisfaction compare with those taking the traditional classroom sections?

The next section presents our results. We first discuss student characteristics in our empirical study.

RESULTS

Student Characteristics

A total of eight sections of classroom courses were offered over a period of two years. As the effectiveness of the Web-based courses was still being tested, only one Web course was offered each year over the two-year period. During these years, there were no policy changes that would have affected the characteristics for either the classroom or the Web students. All course materials and sections were developed and taught by the same instructor, thus removing the instructor as a source of variation between the two courses. Students in both courses were exposed to identical course content. As such, for the purposes of this study, all classroom students will be considered as one group and the Web-based students as the second group.

A total of 113 students participated in the Web-based class and 1,027 in the classroom setting. At the beginning of the each semester, a questionnaire survey was administered to assess the demographic profile of the students taking the Web-based and traditional classroom courses. The questionnaire contained questions relating to age, gender, distance from home to campus, average number of work hours per week, and the average number of hours spent on their computer per week. The last question pertained to a measure of proficiency level in the use of computers. It

Figure 5. Front page to access the course

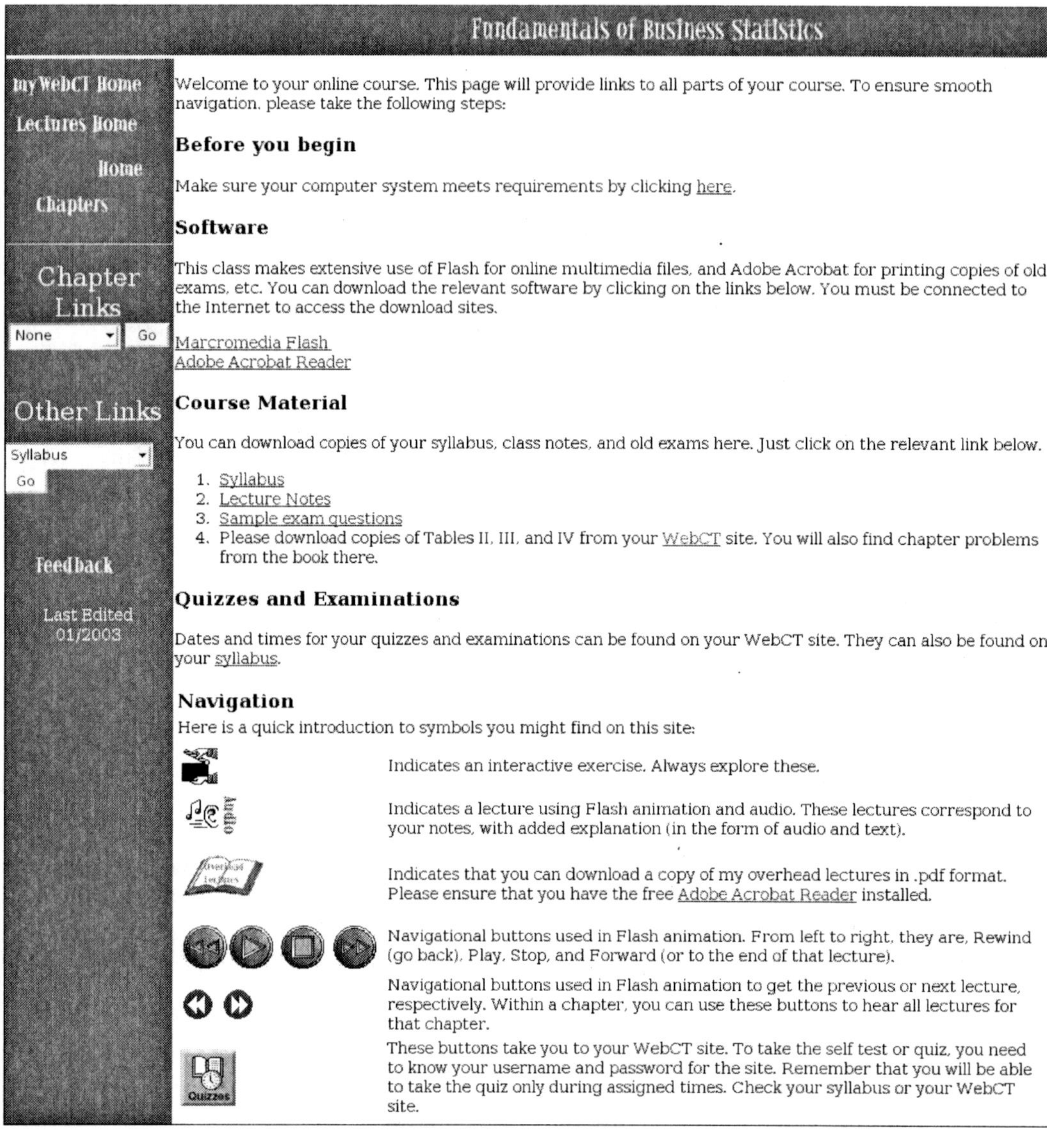

was expected that students taking the Web-based course were more proficient than those taking the traditional lecture courses.

Results in Table 2 show that the average age of students taking the Web-based course was 22.71 years vs. 21.41 for the other group. There was a larger percentage (56.96%) of males taking the traditional class than the Web-based course (49.40%). At the same time, 32.13% of the students taking the Web-based course lived more than 20 miles from campus as compared to only 25.73% for the other group. Students in the Web-based course spent more time on their work than the other students: 21.10 hours each week vs. 17.23 hours. As for computer proficiency level, 64.7% in the Web-based course spent more than 10 hours each week on their computers at home compared to 42.19% in the other group of students.

These results agree with results in the literature about the profile of students taking distance-learn-

Table 2. Beginning of semester survey results

Demographics	Web-based	Classroom-based
Age	$\overline{X} = 22.21$, $n = 83$	$\overline{X} = 21.41$, $n = 841$
Gender	% Male = 49.40	% Male = 56.96
Distance from University		
< 20 miles:	57 (67.87%)	632 (74.27%)
≥ 20 miles:	27 (32.13%)	219 (25.73%)
Work hours per week	$\overline{X} = 21.10$	$\overline{X} = 17.23$
Course primarily taken at:		
Home:	55 (65.48%)	81 (9.53%)
Place of employment:	4 (4.76%)	2 (0.24%)
The main university campus:	22 (26.19%)	760 (89.41%)
A distance learning site:	2 (2.38%)	0 (0.00%)
A remote campus:	0 (0.00%)	4 (0.47%)
Other:	1 (1.19%)	3 (0.35%)
Hours spent on computer:		
≥ 10	55 (64.70%)	362 (42.19%)
< 10	27 (31.77%)	464 (54.08%)
0	3 (3.53%)	32 (3.73%)

ing classes. Many of them take it because it is convenient, and usually, they are more comfortable with technology than traditional students. The relatively higher age and greater proportion of female students in the Web sections again support the contention that flexibility and convenience in taking the class overrides the disadvantages of not having face-to-face student-faculty interaction.

The next section empirically examines the satisfaction of Web-based students. We then examine the differences in satisfaction between Web-based and classroom-based sections.

Student Satisfaction of Web-Based Instruction

As discussed in the previous sections and shown in Table 1, in order for a Web-based course to be well received, it has to satisfy student expectations along multiple dimensions. Course delivery refers to the experience of the student with respect to the quality of the delivery of the course content. For even a well-designed course, technical problems are likely to detract from the educational experience and provide poor student satisfaction. Since the Web-based course is provided completely over the Internet, the quality and speed of connection, therefore, is paramount. While most students on campus have access to broadband connections, 32% of all Web-based students live more than 20 miles from campus. Many of these students still log on to the campus network using dial-up connections. To ensure that all students receive a good quality of delivery, the interactive exercises and related course content was optimized for dial-up connections. Using video sparingly also minimized transmission overhead. Audio was converted to mp3 files, and Flash modules were optimized for 56K modems. In addition, each chapter was broken up into several Flash modules with an average file size of less than 100K.

In order to determine student experience with course delivery and experience, a second survey was administered at the end of the semester (EOS). Every attempt was made to ensure the anonymity of the students. As such, the two sets of survey responses, one at the beginning of the semester and the other at the end of the semester, cannot be paired at the student level.

Two sets of items were identified and selected from the formal battery of items used by researchers in distance learning. The first set of four items related to the quality of delivery and the second set of six to content. Both sets entail a four-point Likert scale varying from Strongly Disagree to Strongly Agree. Table 3A shows the results of the responses of Web-based students to these questions in the EOS survey. It is clear that the vast majority of Web-based students found the delivery of course content satisfactory. Students also were generally satisfied with the speed of access to the network and with getting help when needed.

Quality of course delivery is only one of the characteristics for successful distance education. Course design is also a key to success. Without the teacher-student, face-to-face interaction, tools must be provided to simulate and to test students' critical thinking abilities. As part of the EOS survey, students were asked to rate the instructor's ability to provide such an environment along several dimensions.

A large percentage of students agreed that the instructor was successful in providing an environment that stimulated independent thinking (88%) and that the ideas in the course were summarized effectively (85%). Furthermore, 81% of the students reported that they learned a great deal from this course (Table 3B). In distance learning, communication of results also plays an

Table 3. Web students' responses to quality of delivery and course content (Note: Number of students (row percentage in brackets))

Question	Strongly Disagree	Disagree	Agree	Strongly Agree
A: Quality of Delivery				
I spend too much time accessing the institution's network	18 (29)	35 (56)	8 (13)	1 (2)
The use of WebCT for online examinations worked as it should	0 (0)	5 (8)	29 (48)	27 (44)
The use of Multimedia Lectures and Interactive Exercises worked as they should	3 (5)	6 (10)	32 (52)	21 (34)
It is easy to contact the site administrator when I have a problem	3 (5)	5 (8)	34 (55)	20 (32)
B: Satisfaction with Course Content				
The course was well organized		8 (17)	19 (41)	20 (43)
The instructor gave clear explanations		4 (8)	21 (45)	15 (32)
I learned a great deal from this instructor	1 (2)	8 (18)	26 (55)	12 (26)
Students were kept informed of their progress			16 (34)	31 (66)
The instructor stimulated independent thinking		6 (12)	27 (58)	14 (30)
The instructor synthesized, integrated, or summarized ideas effectively	1 (2)	6 (13)	29 (62)	11 (23)

important role. Nearly all students were satisfied with being informed about their progress. Clearly, by the dimensions measured here, most students were satisfied with the delivery and content of the Web-based course.

Web vs. Classroom Student Satisfaction

Five additional items were recorded relating to various aspects of a course. These items were anchored with a four-point Likert scale. One additional item addressing the overall satisfaction levels also was included in the EOS survey. For the Web students, these six items were included in the EOS survey. Separate surveys containing only these six items were administered at the end of the semester to classroom students.

A majority of students felt positively about the course they were taking (Table 4). Table 4 also shows that a higher percentage of the students taking the traditional courses expressed stronger agreement (Table 4A). These findings also are consistent with the overall satisfaction level (Table 4B). Students enrolled in the traditional courses were more satisfied with their experience in the course than those enrolled in the Web-based course.

Table 4. Course comparison (Note: Number of students (row percentage in brackets))

Question	Section	Strongly Disagree	Disagree	Agree	Strongly Agree
A: Experience					
I am more comfortable participating in discussions in this course than in other courses	**Web**	6 (12.50)	20 (41.67)	16 (33.33)	6 (12.50)
$\chi^2 = 51.63, p = 0.0001$	**Class**	72 (8.35)	285 (33.06)	436 (50.58)	69 (8.00)
I feel comfortable telling the instructor of this course when I disagree with something he/she said	**Web**	3 (6.38)	16 (34.04)	23 (48.94)	5 (10.64)
$\chi^2 = 50.83, p = 0.0001$	**Class**	53 (6.21)	221 (25.88)	506 (59.25)	74 (8.67)
I am better able to understand the ideas and concepts taught in this course	**Web**	4 (5.88)	22 (32.35)	36 (52.94)	6 (8.82)
$\chi^2 = 17.72, p = 0.0018$	**Class**	28 (3.11)	138 (15.33)	571 (63.44)	163 (18.11)
I am better able to visualize the ideas and concepts taught in this course	**Web**	3 (4.48)	24 (35.82)	35 (52.24)	5 (7.46)
$\chi^2 = 17.51, p = 0.0015$	**Class**	27 (3.00)	154 (17.13)	554 (61.62)	164 (18.24)
Because of the way this course uses electronic communication, I spend more time studying	**Web**	3 (4.48)	25 (37.31)	29 (43.28)	10 (14.93)
$\chi^2 = 6.16, p = 0.1876$	**Class**	34 (0.00)	238 (27.77)	513 (59.86)	106 (12.37)

		Very Dissatisfied	Dissatisfied	Satisfied	Very Satisfied
B: Overall Satisfaction					
Overall, I have been	**Web**	3 (4.41)	10 (14.71)	28 (41.18)	27 (39.71)
$\chi^2 = 7.81, p = 0.05$	**Class**	34 (3.74)	68 (7.49)	511 (56.28)	295 (32.49)

Satisfaction and experience of both groups of students is important from all perspectives. In order for institutions to provide comparable learning experiences on the Web, it is necessary to understand and to implement good practices for distance education. At the same time, it is important to see if tools and techniques geared toward distance learning also could be used successfully in a more efficient manner in a traditional classroom setting.

CONCLUSION AND DISCUSSION

Distance education is here to stay. It will take on a greater role in the delivery of higher education as colleges look for ways to serve as many students as possible in light of scarcity of resources. As information technology becomes a way of life, both students and faculty will become more attuned to this new environment. But this proficiency in the new environment is still tempered with the understanding that distance education will not completely replace traditional classroom instruction. To what extent distance education can and should be used and how it can be used to supplement classroom education are the basic intents of this study.

Most studies in this area directly compare classroom and Web-based learning in terms of their effectiveness. As stated previously, performance outcomes mostly are a function of the learning environments and course design. Without laying out the course structure in each of the learning environments and course design, it would be difficult for one to establish any cause-and-effect relationships. Furthermore, special care needs to be exercised in selecting courses and faculty as potential candidates for Web-based education. Recent results also indicate that Web-based education may not benefit all students and that student personality traits have a significant impact on achievements scores in Web classes (Schniederjans & Kim, 2005). In contrast, currently, most students often follow a self-reflective procedure by way of deciding whether to sign up for Web-based or classroom courses.

This study first proposes a framework linking the learning environments with course design and performance. Then, student performance, as measured by their satisfaction, can be traced back to the learning environment and course design.

On the whole, students taking the Web-based business Statistics course devoted more time to their work, lived farther away from campus, and were more computer literate. Given these characteristics, students found the delivery and course design of the Web course satisfactory.

Comparing the Web-based course students with the traditional classroom students, it is somewhat surprising to note that traditional students were even more satisfied with the course offerings. This higher level of satisfaction most likely can be attributed to face-to-face interaction in the classroom. This environment possibly motivates students to be more involved and engaged in their learning.

In order to be successful in the Web-based environment, a student has to exercise a high degree of self-discipline. Simultaneity of stimulus and response play the role in holding students' attentions in the classroom. It is clear that a well-designed Web course can provide a satisfactory learning environment for students. For the particular course that we consider in our study, augmenting a traditional classroom setting with Web-enhanced lectures provided an even greater satisfaction. Clearly, this is an impetus to consider how Web-based tools could be used to improve current classroom education.

REFERENCES

Adobe Acrobat. http://www.adobe.com

Bradburn, E.M., & Zimbler, L. (2002). *Distance education instruction by postsecondary faculty and staff: Fall 1998*. National Center for Education Statistics. Retrieved from http://nces.ed.gov/pubs2002/2002155.pdf

Lewis, L., Snow, K., Farris, E., Levin, D., & Greene, B. (1999). *Distance education at postsecondary education institutions: 1997-98*. National Center for Education Statistics. Retrieved from http://nces.ed.gov/pubs2000/2000013.pdf

Macromedia Flash. http://www.macromedia.com

Millis, B.J., & Cottell Jr., P.G. (1997). *Cooperative learning for higher education faculty*. Oryx Press.

Russel, T. (2003a). *The no significant difference phenomenon*. TeleEducation New Brunswick. Retrieved from http://teleeducation.nb.ca/nosignificant difference/

Russel, T. (2003b). *The significant difference phenomenon*. TeleEducation New Brunswick. Retrieved from http://teleeducation.nb.ca/significant difference/

Schniederjans, M.J., & Kim, E.B. (2005). Relationship of student undergraduate achievement and personality characteristics in a total Web-based environment: An empirical study. *Decision Sciences Journal of Innovative Education, 3*(2), 205-221.

Sikora, A., & Carrol, D. (2002). *A profile of participation in distance education: 1999-2000*. National Center for Education Statistics. Retrieved from http://nces.ed.gov/pubs2003/2003154.pdf

Sun Java. http://java.sun.com

VTEL. http://www.vtel.com

WebCT. http://www.webct.com

This work was previously published in International Journal of Web-Based Learning and Teaching Technologies, Vol. 1, Issue 2, edited by L. Esnault, pp. 1-17, copyright 2006 by IGI Publishing, formerly known as Idea Group Publishing (an imprint of IGI Global).

Chapter VI
Differentiating Instruction to Meet the Needs of Online Learners

Silvia Braidic
California University of Pennsylvania, USA

ABSTRACT

This chapter introduces how to differentiate instruction in an online environment. Fostering successful online learning communities to meet the diverse needs of students is a challenging task. Since the "one size fits all" approach is not realistic in a face-to-face or online setting, it is essential as an instructor to take time to understand differentiation and to work in creating an online learning environment that responds to the diverse needs of learners. It is our responsibility to ensure that the teaching and learning that takes place online is not only accessible, but of quality. The author hopes that developing an understanding of differentiation and specific instructional strategies to differentiate online will inform the learner of ways to maximize learning by addressing the diverse needs of students.

INTRODUCTION

Teaching is complex. It involves careful preparation and the planning of objectives and learning experiences. Effective educators set high expectations for all students and select strategies to propel student learning. As an educator, it is essential to create a sense of community in which students feel significant and respected. We realize that not all students are alike. A central focus of the educator is to maximize the capacity of each student. When teaching in a face-to-face classroom at the university level, there will invariably exist a diverse group of students with various levels of readiness, interests, and learning profiles. Students must be thought of as individuals in order to help differentiate the classroom, thereby bringing more meaning to their learning. The same is true in an online classroom. Although networked learning offers us new opportunities to build collaboration and creativity into the teaching and learning process, these innovations also pose numerous challenges

(Zhu, Payette, & DeZure, 2003). This chapter attempts to address the following questions by reviewing the literature on differentiation and its connection to and impact on online learning. It offers ideas for differentiating in an online environment. The conclusion section discusses some implications to online learning and offers recommendations for future research.

- What is differentiated instruction?
- What are the principles that guide differentiated instruction?
- What is the impact to online learning?
- Why differentiate? Its basis in theory and research.
- How can you differentiate?
- Conclusion
- Future Research Directions

WHAT IS DIFFERENTIATED INSTRUCTION?

Not all students are alike. This is true with students in a face-to-face or online setting. Based on this knowledge, differentiated instruction applies an approach to teaching and learning that gives students multiple options for taking in information and making sense of ideas. Differentiation has come to mean consistently using a variety of instructional approaches to modify content, process, and/or products in response to learning readiness, interest, and learning profile of academically diverse students (Tomlinson, 1999).

BACKGROUND: WHAT ARE THE PRINCIPLES THAT GUIDE DIFFERENTIATION?

Tomlinson (1999) presents a few key ideas about differentiation as it relates to the traditional face-to-face classroom. Consider these principles in relation to your online classroom.

Principle 1: The Teacher Focuses on the Essentials

In a differentiated classroom, the teacher carefully fashions instruction around the essential concepts, principles, and skills of each subject (Tomlinson, 1999). The standards tell us what our students need to know and be able to do in the K-12 setting. The same is true at the university level when preparing students for a particular field. Most professions have national and/or state organizations that publish a clearly defined set of standards. These standards guide our programs. Whether you are teaching face-to-face or online, the standards are the same.

Principle 2: The Teacher Attends to Student Differences

Students differ in terms of readiness, interest, and learning profile. As instructors, we need to take time to understand, appreciate, and build upon student differences. The pedagogical theory that guides differentiation is constructivism, which is the belief that learning happens when the learner makes meaning out of information (Benjamin, 2005). Just as we have a variety of learners in the face-to-face setting, the same is true in an online environment. Online education must capitalize on student's unique approaches to learning, says Nishikant Sonwalkar (2003). In order to do so, we must design and implement programs of study and courses that differentiate to meet the needs of the students we serve. The research indicates that individuals learn in accordance with their readiness to do so. Jenson (2000) indicates that moderate challenge is critical. When a task is not challenging enough, students become bored. Yet if a task is too challenging, students become anxious. Also, when interest is tapped, learning is more likely to be rewarding and students become more autonomous learners. When designing a learning environment, helping students to discover and pursue their passions can maximize their engagement with learning, their productivity, and their individual talents. Finally, individuals vary in preference for conditions of learning and that consideration of the multiple intelligences (Gardner, 1991) is also important. As an instructor, it is essential to create an interesting and engaging online environment that can help settle and focus the learner just as you would want to do in a face-to-face classroom setting.

Principle 3: Assessment and Instruction are Inseparable

Assessment and instruction go hand-in-hand. They are both ongoing. Just as in a face-to-face class, diagnostic, formative and summative assessments must be considered by instructors. In creating a differentiated online classroom, instructors must ask themselves if students have the opportunity for ongoing feedback, if they have time to revisit ideas and concepts to connect or extend them, and if students have time to reflect and set goals. In an online environment, ongoing assessment plays a crucial role.

Principle 4: The Teacher Modifies Content, Process, and Products

In a face-to-face setting, instructors take time to assess on a daily or weekly basis. By doing so, the instructor can modify content, process, or product at any point. In considering online courses, the key is to offer variety. Content, process, products, readiness, interest, and learning style are all considered in the design and implementation of the online course. This process is accomplished through discussion formats, chats, and individual and group work.

Principle 5: All Students Participate in Respectful Work

Instructors take time to understand their students as learners. They take time to try and meet students where they are in order to move forward. With the instructors' understanding of their diverse needs, students in an online class can engage in work that challenges them, but does not overwhelm them. Differentiated instruction adopts the concept of "readiness." Difficulty of skills taught should be slightly in advance of the student's current level of mastery. This is grounded in the work of Lev Vygotsky (1978) and the zone of proximal development, which is the range at which learning takes place. The classroom research by Fisher et al. (1980) strongly supports this concept. Researchers found that in classrooms where individuals were performing at a level of about 80% accuracy, students learned more and felt better about themselves and the subject area under study (Fisher, 1980 in Tomlinson, 2000). Brain research appears to affirm this as well (Jensen, 1998) explaining that learning occurs when the learner experiences neither boredom nor anxiety, when the learner is neither over challenged nor under challenged.

Principle 6: The Teacher and Students Collaborate in Learning

"Teachers are the chief architects of learning, but students should assist with the design and building" (Tomlinson, 1999, p. 12). In an online environment, you as the instructor are the architect for the course. It is your responsibility to establish the course goals and objectives and to vary the instructional approaches based on the purpose and diverse needs of the students. But, students need to contribute to and take responsibility for their own learning as well. Students in an online class will come from various backgrounds, have different interests, and the like. It is important to consider what they know, what they want to know, and even how they may go about learning the content and skills related to the course. Together, you can work to set goals, monitor progress, and collaborate in learning. By offering variety in the learning and opportunities for student choice, collaboration takes place.

Principle 7: The Teacher Balances Group and Individual Norms

In a differentiated class, instructors understand both group and individual norms. The instructor does not focus on making everyone the same. Instead, the instructor meets the student where they are, and works to create the conditions for that student to become the best that the student can be.

Principle 8: The Teacher and Students Work Together Flexibly

To address the diverse needs of students in a class, instructors must be flexible, working with students in respect to content, process, and products based on readiness, interest, and learning profile. Consider whole class, small group, and individual learning activities throughout the design and implementation of the course. Also, take time to consider the use of various materials, both teacher- and student-selected.

ISSUES, CONTROVERSIES, PROBLEMS: WHAT IS THE IMPACT TO ONLINE LEARNING?

Differentiated instruction is a process of teaching and learning for students of differing abilities in the same class. The intent of differentiating instruction is to maximize each student's growth and individual success by meeting each student where the student is and assisting in the learning process. The model of differentiated instruction requires teachers to be flexible in their approach to teaching and learning.

A major challenge confronting universities is a focus on how to change the fundamental structure of teaching and learning through the use of emerging technologies. As instructors at the university level, we will need to think about how to better respond to the growing demand for lifelong learning, and how to satisfy the increasingly diverse needs of individual students (Hartmon, 2004). The challenge is determining how technology and pedagogy fit together in the online environment. The thought of what online courses and programs can produce is captivating. Online, we as instructors will be able to offer a very rich learning environment worldwide to learners who may not otherwise have access to that kind of education. The charge for online education is how can we most effectively meet the needs of the diverse population we will serve? Can the established theories of learning grounded in the research show their effectiveness online? Ragan, the Director of Instructional Design and Development at Penn State's World Campus, indicates that skills for online effectiveness fall into two broad categories: the design skills needed for authoring a course and the teaching skills necessary for delivering the course online. Clearly, the most important role of the online instructor is to model effective teaching and accept "the responsibility of keeping discussions tracked, contributing special knowledge and insights, weaving together various discussion threads and course components, and maintaining group harmony" (Rohfeld & Hiemstra, 1995, p. 91). Differentiating instruction is good teaching and it is here to stay.

WHY DIFFERENTIATE? ITS BASIS IN THEORY AND RESEARCH

As an instructor, it is important for you to investigate instructional approaches that are sound for the online environment. Also, knowledge of the biological and cognitive factors that influence learning provides a foundation for understanding how students learn. In order for learning to occur, whether in person or online, we must meet the needs of students, giving them a safe and supportive environment in which to learn and grow.

What Do We Know from Brain Research?

During the last two decades, research in the neurosciences has revealed new understanding about how the brain grows, develops, and learns. This information has important implications for what educators do in classrooms. As educators, we need to ensure that our classes are not focused solely on dispensing knowledge, but more on developing individuals who will know what knowledge and skills are important for their continued success in the complex world of the 21st century. (Sousa, 2003) With our knowledge of educational practices, we must determine if and how brain research informs that practice. Given our vast background of knowledge about teaching and learning, educators are in the best position to know how the research does—or does not—supplement, explain, or validate current practices.

We need to differentiate instruction because we cannot do otherwise. We know too much about student variance to pretend that it does not exist or that it is unimportant. We know too much about the art of teaching to assume it can happen effectively in template fashion. (Tomlinson, 1999, p. 31)

Differentiation is rooted in educational theories. Brain research offers that individuals learn in accordance with their readiness to do so. "Tasks must be at the proper level of difficulty to be and to remain motivated: tasks that are too easy become boring; tasks that are too difficult cause frustration" (National Research Council, 1999).

As an instructor, it is essential to create an interesting and engaging online environment that can help settle and focus the learner just as you would want to do in a face-to-face class setting. In order to do so, let us consider creating a brain-based classroom that supports differentiated instruction. Creating an environment in which students feel comfortable with the instructor and other students, as well as with various instructional approaches will help to establish an engaging class setting. Gregory and Chapman (2002) suggest that a brain-based classroom should focus on a variety of factors, two of them being brain organization and building safe environments. Just as you take time to determine the classroom set up in a face-to-face setting, it is just as important to take the time to do so in an online environment. In this way, the online classroom will be organized and safe in contributing to students' willingness to learn and take risks. Consider the following suggestions when organizing your course shell:

Course Syllabus

The course syllabus is typically the first formal document that students in your class will view. A syllabus serves as a guide for the student by providing them with information such as the following: course description, objectives, course materials, outline of course content, teaching methodology, grading policy, course assessments, academic integrity statement, netiquette, and library resources.

Professor Introduction

The first day of contact with students is important whether it be in a face-to-face or online setting. Taking time to introduce yourself to the class through a video and/or audio introduction, in addition to a bio placed in your syllabus, can set the tone for the semester.

Office

Just as in the traditional sense, an "office" can also be created by the instructor online. There, students may post questions which relate to the course content. Instructors can also set scheduled office hours whereby students are assured that the instructor will be available for consultation during that particular time.

Class Lounge

The course lounge provides students with an opportunity to introduce and acquaint themselves with one another. The course lounge is an excellent way to begin the semester; it can serve as a place where students share information about themselves and where they can casually meet as a class throughout the semester.

Online Student Journal

Online journals provide students with the opportunity to reflect and to set goals which relate to the course content and real experiences. Journals can be created in various formats, each relating to questions to which students must respond. Instructors may ask students to consider a K-W-L format. In this way, students can take a moment to reflect upon course-related content which they already KNOW, and that which they WANT to know. They can also take the time to revisit and share what they have LEARNED. A natural progression to this, especially at the end of a semester, is to take time to set goals for future work as it relates to ongoing learning in these areas.

Units

Units have established goals and objectives and align with the overall course goals. A visually pleasing introduction is provided which captivates the learner and draws the learner's attention to the unit focus. Throughout the course of a unit, students may engage in chats, threaded or audio discussions, live classroom, or a variety of other learning experiences.

Chat Rooms

Establishing chat rooms to be utilized throughout the course of the semester (both teacher- and student-selected) will help create a sense of community in the class. The chat allows the users to interact with each other via a text-based format (Blackboard, 2005).

Discussion Boards/Voice Boards

The discussion/voice board is a communication medium for posting and responding to text or audio messages. Conversations are grouped as threads which contain a main posting and all related replies to that posting (Blackboard, 2005). Discussion/voice boards provide students with an opportunity to participate in large- or small-group activities.

Live Classrooms

Users can ask questions, draw on the whiteboard, and participate in breakout sessions from the virtual classroom (Blackboard, 2005). In the live or virtual classroom, instructors can allow students to raise their hand to ask questions, poll, share PowerPoint presentations, or even engage students to take the lead in a breakout session.

What Do We Know About Sensory Approaches to Learning? Auditory-Visual-Kinesthetic

In 1987, Rita and Ken Dunn proposed a model in which learning styles were classified as auditory, visual, or kinesthetic. Visual learners learn best through their sense of sight when cues are provided in written or pictorial form. Auditory learners learn best through hearing. In other words, some students will respond best to spoken cues and others to auditory ones. Still others learn best by doing or experiencing, often referred to as kinesthetic. By thinking about your online course and creating different activities for each of these different sensory approaches to learning, it would be easy to address some of the diversity among students. Online, you can combine a synchronous Web presentation with a conference call discussion or online threaded discussion either through text or voice. In considering the sensory approach to learning, ask yourself the following:

Have I planned to accommodate the senses in my online environment through (Dunn & Dunn, 1987):

- Activities that involve spoken and heard material: voice boards for discussion, voice-mails, chats, recordings, interviews, and live classroom? (auditory learners)
- Activities that include information that can be seen or read: graphic organizers, summaries in each unit, use of color, note-taking, pictures, diagrams, illustrations, photos, video clips, and streaming video? (visual learners)
- Activities that allow students to handle and manipulate materials: writing, drawing, equipment, and tools? (tactile learners) Activities that allow students to do and move and become physically involved: field work and projects? (kinesthetic learners)

What Do We Know About Learning Styles?

Silver, Strong, and Perini (2000) base their ideas of learning styles on the work of Carl Jung (1933) who conceptualized four dimensions of personality: thinking, sensing, feeling, and intuition. They also use the work of Isabel Myers (1985) who adapted Jung's ideas to develop the famous Myers-Briggs type indicator (MBTI). This model differs from sensory approach to learning in that it focuses on personality theory and not a sensory-channel model. The individual components of their categories are (Langa & Yost, 2007):

Sensing-Thinking Learners (ST) or Mastery Style

Silver, Hanson, Strong and Schwartz (2003) indicate that sensing-thinking learners can be characterized as realistic, practical, and matter-of-fact. This type of learner is efficient and results oriented. This type prefers action to words and involvement to theory, and has a high energy level for doing things that are pragmatic, logical, and useful. They rely on thinking to make decisions, and are concerned about logical consequences more than personal feelings. These learners perceive the world in terms of thing tangible to the senses, rather than abstract or symbolic ideas, theories, or models. They are objective, efficient, and goal-oriented. For the instructor, it is essential to present information and provide practice opportunities for students to exercise their new learning. Students who prefer this style or possess this strength learn best through procedures. They like to perform calculations and computations. They also enjoy learning through observation, memorization, practicing, and sequencing. This needs to be done in order to remember important skills and information.

Intuitive-Thinking Learners (NT) or Understanding Style

Silver et al. (2003) indicate that intuitive-thinking learners can be characterized as theoretical, intellectual, and knowledge-oriented. They are logical probers who want to understand complex problems. They like to be challenged intellectually and to think things through for themselves. Facile with language, they are able to speak, debate, and write extensively on a subject they have studied. These learners are always asking why and looking for logical relationships. They are interested in abstract ideas, possibilities, and the meanings of things beyond what is concrete. Students who prefer this style learn best conceptually. They use higher-level thinking skills to compare and contrast, analyze and summarize, establish cause and effect, and support or refute ideas. As an instructor, you will need to present data for the students to process. Also, you will need to probe students' explanations in order to help them develop reasoning skills and an understanding of concepts.

Intuitive-Feeling Learners (NF) or Self-Expressive Style

Silver et al. (2003) indicate that intuitive-feeling learners are characterized as curious, insightful, imaginative, and creative. These unconventional students prefer to follow their own path to learning. Tending to be nonconformists, they dislike rules and routines. They are learners who need self-expression and who excel when allowed to use original ideas and solutions while problem solving. Students with this style preference describe learning that produces original work using creative application and synthesis of old skills and information. These students will like to use information in new ways. As the instructor, you will need to present students with challenges and problems to solve. You must require students to reorganize their thinking.

Sensing-Feeling Learners (SF) or Interpersonal Style

Silver et al. (2003) indicate that sensing-feeling learners are characterized as sociable, friendly, and interpersonally-oriented. These are emotionally involved students who are interested in learning about situations concerning living things rather than cold, hard facts. Ever helpful, they care deeply about people and need to interact with others while learning by sharing ideas. These students excel in a cooperative learning environment. As an instructor, you will need to provide opportunities for cooperative learning, real-life contexts, and connections to everyday life. These students learn best contextually.

Because learners have different learning styles or a combination of styles, online educators should design activities that address their modes of learning in order to provide significant experiences for each class participating. In considering the learning styles approach, ask yourself the following:

Have I provided learning opportunities related to the four learning styles in my online environment?

- To acquire knowledge and skills through drill, memorization, repetition, practice and application (i.e., drill/repetition activities, demonstrations, projects, objective tests, and checklists).
- To acquire knowledge and skills through personal sharing of feelings and judgments, individual and social awareness, and collaborative group work (i.e., independent work, essays, debate, arguments, and open-ended questions).
- To think, reason, and defend conclusions through observing and describing data, comparing and contrasting, and identifying patterns and concepts (i.e., open ended discussion, projects, and portfolio).
- To acquire knowledge and skills through creative and divergent thinking, visualization and imagination, problem-solving, and metaphorical thinking (i.e., group project, cooperative learning, personal sharing/journaling, and surveying).

As an instructor, think about how this learning style model can be applied in your online setting. Whether you are developing activities, developing assessment tools, or planning a lesson/unit for your online course, elements that will address the learning styles of the students can be incorporated.

What Do We Know About Multiple Intelligences?

In 1983, Howard Gardner of Harvard University introduced his theory of multiple intelligences in his book *Frames of Mind*. In the book, Gardner suggests that intelligence is not merely a single, discrete number (IQ) that is determined by the answers to a series of items on a test, measuring primarily an individual's verbal and mathematical abilities. Rather, he proposes that humans possess many intelligences and that the mind's problem-solving capacities are multifaceted. "The concept of style designates a general approach that an individual can apply equally to every conceivable content. In contrast, an intelligence is a capacity, with its component processes, that is geared to a specific content in the world (such as musical sounds or spatial patterns)" (Armstrong, 2000, p. 10).

Howard Gardner claims that all human beings have multiple intelligences. These multiple intelligences can be nurtured and strengthened, or ignored and weakened. He believes each individual has eight intelligences:

Verbal-Linguistic Intelligence

Well-developed verbal skills and sensitivity to the sounds, meanings and rhythms of words. As an instructor, consider the following: essays, audio recordings, reports, interviews, research project, quizzes/tests, journals, discussions, observations/findings, oral report, voice board, and written assignments.

Mathematical-Logical Intelligence

The ability to think conceptually and abstractly, and capacity to discern logical or numerical patterns. As an instructor, consider the following: research, problem-solving, outlines, predictions, calculations, statistics/data, analyzing a situation, classifying/ranking/comparing, interpretation of evidence/data, use of statistics, graphic organizers, and timelines.

Musical Intelligence

The ability to produce and appreciate rhythm, pitch and timber. As an instructor consider the following: music, poetry, jingles, background sounds and noises, compositions, and recordings.

Visual-Spatial Intelligence

The capacity to think in images and pictures, to visualize accurately and abstractly. As an instructor, consider the following: color/lines/shapes, creative design, visualizations, graphic organizers, visuals, art media, poster, charts, brochures, pictures, illustrations, cartoons, illustrations of events, and diagrams.

Bodily-Kinesthetic Intelligence

The ability to control one's body movements and to handle objects skillfully. As an instructor, consider the following: inventions, participation in the field, hands-on experiences, simulations, role-play, field trip, and demonstrations.

Interpersonal Intelligence

The capacity to detect and respond appropriately to the moods, motivations, and desires of others. As an instructor, consider the following: communication with others via e-mail, discussions, chats, cooperative learning, role-play, tutoring sessions, jigsaw, and interviews.

Intrapersonal Intelligence

The capacity to be self-aware and in tune with inner feelings, values, beliefs, and thinking processes. As an instructor, consider the following: one-on-one conferencing with a classmate or instructor, journal entries, surveys, inventories, exams, self-studies, contracts, personal choices, independent work, portfolio, and personal reflections.

Naturalist Intelligence

The ability to recognize and categorize plants, animals, and other objects in nature. As an instructor, consider the following: classifications, problem solving in environmental situations, research, real-life situations, nature sounds, and pictures.

The implications of MI theory extend far beyond classroom instruction. At heart, the theory of multiple intelligences calls for nothing short of a fundamental change in the way schools are structured. Consider the impact of MI theory to online instruction. It delivers to educators everywhere the strong message that students who show up for school (online or in person) at the beginning of each day have the right to be provided with experiences that activate and develop all of their intelligences.

Now that we have had an opportunity to understand the basis of theory and research as it relates to the need for differentiation, let us consider how to differentiate.

SOLUTIONS AND RECOMMENDATIONS: HOW DO YOU DIFFERENTIATE?

When differentiating instruction, three questions that instructors must ask themselves are:

1. What do I want my students to know, understand and be able to do? (content)
2. What will I do instructionally to get my students to learn? (process)
3. How will my students show what they have learned? (product)

As instructors, not only can we differentiate content, process, or product, but we may do so according to our students' readiness, interest, and learning profile. Readiness refers to the skill level and background knowledge that students bring with them to the class. Interest refers to the student's preference within the curricular area. Finally, learning profile includes visual, auditory, or kinesthetic learners as well as other factors such as grouping arrangements. As you explore how to differentiate content, process, and product, keep these areas in mind.

How Do You Differentiate Content?

Content is what students are to learn. The standards set the context for the content that students will learn in terms of what they should know and be able to do. In an online environment, content is differentiated when you preassess students' knowledge and skills, then match the learners with appropriate learning activities according to their readiness. Content is also differentiated when you give students choices about topics to explore in greater depth. Finally, content can be differentiated when you provide students with basic and advanced resources that match their current levels of understanding.

Although content, process, and product are intertwined, think of content in the following context:

- Tomlinson (1999) defined content as "what the teacher wants the students to learn and the materials or mechanisms through which that is accomplished. It is the subject matter or unit being taught" (p. 11).
- Content is what we teach, and what we want students to learn. It can also be thought of as input (Tomlinson, 1995).
- Content is what a student should come to know (facts), understand (concepts and principles), and be able to do (skills) as a result of a given segment of study (a lesson, a learning experience, a unit). Content is input. It encompasses the means by which students will become acquainted with information (through textbooks, supplemental readings, videos, field trips, speakers, demonstrations, lectures, or computer programs). (Northey, 2003, p. 43)

It is vital to be clear about what is essential in content. Clarity about what really matters in the disciplines enables us to teach for understanding. Having a goal does matter, since we cannot teach (and students cannot learn) everything. We ought to therefore take care to teach that which is most durable and useful. Curricular goals are the springboard from which differentiation ought to begin. If, as a teacher, one is foggy about precisely what students should know, understand, and be able to do as the result of a course, unit, or lesson, instruction may be differentiated, but it is likely to generate multiple versions of fog. Furthermore, if the instructor is uncertain of the precise outcomes for a unit (and how a particular lesson or product serves those outcomes), the instructor will be unable to preassess students' proximity to those outcomes effectively, and thus be uncertain of how to craft the start of the learning journey for students whose proficiencies vary. In an effectively differentiated classroom, the same powerful understanding-based goals will nearly always "belong" to everyone. An instructor will begin by preassessing learners' proficiency with those goals. With that information in hand, the instructor can assist some students in developing precursor proficiencies necessary for continued growth, and other students in extending their competencies related to the goals. Moreover, the instructor has a road map for the learning journey that directs ongoing assessment and adjustment of teaching and learning plans throughout the unit, just as it directs construction of the unit.

There are a variety of ways to differentiate content according to a student's readiness, interest, and learning profile. As you think about differentiating content, process, and product, keep in mind that they are interrelated. We only look at them here separately to make it more manageable.

When differentiating content, take time to first preassess learners' proficiency with the goals. Then, develop alternatives in the delivery method and content. Because students typically vary in their prior knowledge and skill levels, responsive teachers target their instruction to address significant gaps. Some suggestions, based on Tomlinson's (1999) work for differentiating content are included in the list below:

- Multiple texts and supplementary print resources
- Varied computer programs
- Varied audio-visuals
- Varied support mechanisms (can include audio tapes, computers, study partners, reading buddies, mentors, etc.)
- Note taking organizers
- Varied time allotments
- Contracts
- Compacting
- Complex instruction
- Group investigation

How Do You Differentiate Products?

Products are the vehicle through which students demonstrate what they have learned. A good product causes students to rethink what they have learned, apply what they can do, extend their understanding and skill, and become involved in both critical and creative thought. Anyone concerned about teaching and learning is automatically interested in assessment. Assessment provides us with evidence to help answer important questions: "Did the student learn it?"; "To what extent does the student understand?"; and "How might I adjust my teaching to be more effective for learners with varying needs?" Taking time to first determine acceptable evidence and then plan teaching and learning activities is important. By considering in advance the assessment evidence needed to validate that the desired results have been achieved, teaching becomes more purposeful and focused. Also, with clarity about what constitutes evidence that students have achieved desired results, teachers have a consistent framework within which they can make modifications for their students' readiness levels, interests, and learning preferences. (Tomlinson & McTighe, 2006)

Tomlinson (1999) defines products as "vehicles through which students demonstrate and extend what they have learned" (p. 11). Products should help students—individually or in groups—rethink, use, and extend what they have learned over a long period of time (i.e., a unit, a semester, or even a year). Products are important not only because they represent students' extensive understandings and applications, but also because they are the element of curriculum that students can most directly "own." For that reason, well-designed product assignments can be highly motivating because they will bear their creator's thumbprint. (Tomlinson, 1995)

Tomlinson and McTighe (2006) present three assessment principles in their book *Integrating Differentiated Instruction and Understanding by Design*. They are as follows.

Assessment Principle #1: Consider Photo Albums Vs. Snapshots

In other words, reliable assessment demands multiple sources of evidence. In a classroom, a variety of assessments may be used to gather evidence of learning (McTighe & Wiggins, 2004), including:

- Selected-response format (e.g., multiple choice, true-false) quizzes and tests
- Written or oral responses to academic prompts (short answer format)
- Performance assessment tasks, yielding:
    - Extended written products (e.g., essays, lab reports)
    - Visual products (e.g., PowerPoint shows, murals)
    - Oral performances (e.g., oral reports, foreign-language dialogue)
    - Demonstrations (e.g., skill performances in P.E.)
- Long-term, "authentic" projects
- Portfolios
- Reflective journals or learning logs
- Informal, ongoing observations of students (e.g., teacher note taking, probing questions, exit cards)
- Formal observations of students using observable indicators or criterion list
- Student self-assessments
- Peer reviews and peer response groups

Assessment Principle #2: Match the Measures with the Goals

Assessments must provide an appropriate measure of a given goal. Consider three types of educational goals:

- **Declarative knowledge:** What students should know and understand
- **procedural knowledge:** What students should be able to do
- **Dispositions:** What attitudes or habits of mind students should display (Marzano, 1992)

Assessment Principle #3: Form Follows Function

The way in which we design and use classroom assessment should be directly influenced by the answers to four questions: What are we assessing? Why are we assessing? For whom are the results intended? How will the results be used?

Classroom assessments serve different purposes:

- Diagnostic (or preassessments)
- Formative
- Summative

In an online environment it is important to consider how you will utilize each of these in a variety of ways. Consider how multiple intelligences, learning styles, or sensory channels may be utilized for approaches to diagnostic, formative, and summative assessment.

There are a variety of ways to differentiate products according to a student's readiness, interest, and learning profile. As you think about differentiating content, process, and product, keep in mind that they are interrelated. The list below, adapted from Tomlinson (1999) suggests some ways that you can differentiate products.

- Tiered product assignments
- Independent study
- Multiple intelligence-based products
- Complex instruction
- Group investigation
- Range of media or formats to express students' knowledge, understanding, and skill
- Visual, auditory, and kinesthetic product options

Products can take many forms. In fact, it is the flexibility of products that make them so potentially powerful in classrooms sensitive to learner variance. If, as a student, I can show the teacher that I have come to know, understand, and do the nonnegotiables of the unit, how I do so may be open. Tests are certainly one form of product. Nonetheless, when tests are the only form of student product, many students find that their ability to show what they know is restricted. With tests, it is important to remember that the goal should not be regurgitation of information, but rather, demonstration of the capacity to use knowledge and skills appropriately. It is also important to remember that tests should enable rather than impede a student's ability to show how much the student has learned. Thus, some students may need to provide an audio response to a test, or may need additional time. The table above illustrates just a few ways in which you can differentiate products in response to student readiness, interest, and learning profile.

How Do You Differentiate Process?

Process is the "how" of teaching. Process refers to the activities that are designed to help students think about and make sense of the key principles and information of the content they are learning. In an online environment, process plays a critical role in differentiating to meet the needs of the students.

Tomlinson (1999) defined process as "the activities designed to ensure that students use key skills to make sense out of essential ideas and information. It is the method in which students acquire the skills" (p. 11). The line between process and content is a blurred one. "When students encounter new ideas or information, they need time to run the input through their own filters of meaning. As they try to analyze, apply, question, or solve a problem using the material, they have to make sense of it before it becomes 'theirs.' This processing or sense-making is an essential component of instruction because without it, students either lose the ideas or confuse them" (Tomlinson, 1995, p. 53). Think of process as how students gain an understanding of the main idea(s) of the unit (i.e., the activities used).

Any effective activity is essentially a sense-making process, designed to help a student progress from a current point of understanding. Students process and make sense of ideas and information most easily when their classroom activities:

- Have a clear purpose
- Focus on a few key ideas
- Guide them in understanding the ideas and the relationships among them
- Offer opportunities to explore ideas through varied modes/intelligences (e.g., visual, kinesthetic, auditory, spatial, and musical)
- Help them relate new information to previous understandings
- Match their level of readiness (Tomlinson 1995)

There are a variety of ways to differentiate process according to a student's readiness, interest, and learning profile. As you think about differentiating content, process, and product, keep in mind that they are interrelated. Teachers must select instructional strategies that support responsive teaching, that is, strategies that lend themselves to addressing readiness, interest, and learning profile. Having access to a variety of approaches to teaching and learning gives teachers the agility to reach out to all students and give them time to process the information. It will nearly always be the case that some students prefer certain instructional approaches over others.

The list below, adapted from Tomlinson (1999), suggests some ways that you can differentiate process:

- Varied questioning
- Tiered activities
- Multiple intelligences assignments/activities
- Graphic organizers
- Simulations or real world scenarios
- Learning logs
- Flexible grouping
- Independent projects/study/field work
- Choice boards
- Journals
- Role-playing
- Agendas
- Task cards
- Tic-Tac-Toe
- Discussions/Chats
- Varying amount of support

Instructors should consider a diverse set of instructional strategies when teaching online so that they can most effectively meet the needs of the students they serve. A few instructional strategies for online teaching are elaborated on below.

Questioning

Questioning provides an opportunity to engage in ongoing assessment throughout the semester; it allows an instructor to ask a variety of questions in terms of complexity, so that one can address the diverse needs in the class and meet students where they are. Questioning can occur in threaded discussions, via e-mails (text and voice), assignments, projects, and real world problem-based scenarios. Questioning is at the heart of classroom practice. In fact, research in classroom behavior indicates that cueing and questioning might account for as much as 80% of what occurs in a given classroom on a given day (Marzano, Pickering, & Pollock, 2001). Also, providing students with an opportunity to not only answer questions, but to ask them as well, helps learners to become self-reflective and goal-oriented.

Discussion

The discussion method is the most popular pedagogical technique used in the online classroom. It is important to understand how to design and maintain an online discussion. When a variety of higher-order questions are used to initiate discussion, and probing follow-up questions are employed, the discussion method can provide a forum to enhance constructive thinking. Learners can be exposed to multiple perspectives and view issues from the perspective of others. Students may be forced to examine the assumptions which underlie their values, beliefs, and actions (Brookfield, 1991). Unstructured problems and the complex and ambiguous nature of many topics can be examined. In a constructivist learning environment, the instructor always needs to keep in mind that when facilitating online discussion, asking the right questions is almost always more important than giving the right answers.

Grouping Strategies

Cooperative learning falls under a general category of "grouping strategies." According to Johnson and Johnson (1999), there are five defining elements of cooperative learning:

- **"Positive Interdependence":** A sense of "sink or swim together"
- **Face-to-face (or computer-to-computer in an online environment) promotive interaction:** Helping each other learn, applauding success and efforts.
- **Individual and group accountability:** Each of us has to contribute to the group achieving its goals
- **Interpersonal and small group skills:** Communication, trust, leadership, decision making, and conflict resolution
- **Group processing:** Reflecting on how well the team is functioning and how to function even better. (Marzano et al. 2001)

Presentations

This approach has been utilized in the traditional and online classroom. In an online class, taking time to create a narrated PowerPoint provides students with a visual and auditory presentation. It gives the instructor an opportunity to highlight main ideas and stress important points.

Real World Scenarios

Real life scenarios are an essential learning tool. Throughout a course, taking time to pose real world scenarios as they relate to the course content allows students the opportunity to blend theory with practice and also bring their personal experiences into play.

Field Work

Field work provides students with an opportunity to integrate theory with practice in a meaningful way in a real world setting. It also provides students with an opportunity to engage in diverse field settings and work on assignments that offer some student choice.

Chats

Online conversations take place in real time in chat rooms. The chat allows the users to interact with each other via a text-based chat (Blackboard, 2005). When a user posts a message to a chat room, every other user who is viewing the chat room sees the message and can respond immediately. Participating in a chat room is like participating in a face-to-face group discussion.

CONCLUSION

When thinking about how to differentiate in an online environment, always start with your instructional goals and outcomes. What is it that you want your students to learn? What are the knowledge, skills, and dispositions? How is this related to standards? Once you have established goals and outcomes, you must determine acceptable evidence of student learning. Then, decisions for differentiation should be based on the focus of instruction. Consider and determine whether you are differentiating content, process, product, or all three. Also, determine whether the focus of the differentiation is readiness, interest, and/or learning profile. The principles of differentiation should be kept in mind throughout the process.

The information in this chapter has provided the reader with details related to the theory and research that supports differentiation and how this may look in an online setting. Fostering successful online learning communities to meet the diverse needs of university or K-12 students is a challenging task. Since the "one size fits all" approach is not realistic in a face-to-face or online setting, it is essential as an instructor to take time to understand differentiation and work to create an online learning environment that responds to the diverse needs of learners. Implications for standards for effective online teaching are important and need to be part of this process. Educational leaders must consider how to make differentiated instruction an integral part of the online environment. This requires staff development and ongoing support. As educators, it is our responsibility to ensure that the teaching and learning processes which take place online are as empowering and comprehensive as they are accessible (Zhu et al. 2003).

FUTURE RESEARCH DIRECTIONS

As many universities and K-12 schools move toward serving a broad range of students in an online setting, it is important to assist instructors in developing classrooms responsive to the needs of academically diverse learners they will serve. Understanding what can facilitate appropriately differentiated instruction in an online setting is essential for instructors so that they can create learning communities to address the diverse needs of learners. If a university or K-12 school system is to establish online classrooms in which instructors can effectively address needs of academically diverse learners, intensive and sustained staff development will be required. A focus on standards for quality online teaching will be of the essence. One of the National Education Technology Plan action goals for improving the use of educational technology is to "support e-learning" and one of the strategies within this goal is to "enable every teacher to participate in e-learning training" (U.S. Department of Education, 2005, pp. 41-42). Useful insights which merit further study include:

1. What are the characteristics of successful K-12 and university online teachers?
2. Instructors transitioning from a face-to-face classroom setting to online teaching do not automatically know how to address academic diversity in this setting and therefore are in need of appropriate staff development. What are the most effective training, mentoring, and support systems for online teachers?
3. Without clear standards for quality online teaching embraced by an educational system, instructors may believe that making occasional minor modifications from the face-to-face setting to an online setting is adequate. Should online professional development be required for the preparation and credentialing of online teachers?

REFERENCES

Armstrong, T. (2000). *Multiple intelligences in the classroom*. Alexandria, VA: Association for Supervision and Curriculum Development.

Benjamin, A. (2005). *Differentiated instruction using technology*. Larchmont, NY: Eye on Education.

Blackboard academic suite – instructor manual. (2005). Blackboard Inc.

Brookfield, S. D. (1991). Discussion. In M. W. Galbraith (Ed.), *Adult learning methods* (pp. 187-204). Malabar, FL: Krieger Publishing Company.

Dunn, K., & Dunn, R. (1987). Dispelling outmoded beliefs about student learning. *Educational Leadership, 44,* 6.

Fisher, C., Berliner, D., Filby, N., Marliave, R., Cahen, L., & Dishaw, M. (1980). Teaching behaviors, academic learning time, and student achievement: An overview. In C. Denham & A. Lieberman (Eds.), *Time to learn* (pp. 7-32). Washington, DC: National Institutes of Education.

Gardner, H. (1991). *The unschooled mind: How children think and how schools should teach*. New York: Basic Books.

Gregory, G., & Chapman, C. (2002). *Differentiated instructional strategies – one size doesn't fit all*. Thousand Oaks, CA: Corwin Press.

Hartman, J. (2004). *The horizontal university: E-learning as a catalyst for organizational transformation.*

Jensen, E. (1998). *Teaching with the brain in mind.* Alexandria, VA: ASCD.

Jensen, E. (2000). *Different brains, different learners.* San Diego: The Brain Store. Retrieved February 9, 2008, from www.thebrainstore.com

Johnson, T. & Johnson, R. (1999). *Learning together and alone: Cooperative, competitive and individualistic learning.* Boston: Allyn and Bacon.

Jung, Carl (1923). *Psychological types* (H.G. Baynes, Trans.). New York: Harcourt, Brace & Co.

Langa, M. & Yost, J. (2007). *Curriculum mapping for differentiated instruction.* Corwin Press: Sage Publications. Thousand Oaks, CA.

Marzano, R. (1992). A different kind of classroom: Teaching with dimensions of learning. Alexandria, VA: Association for Supervision and Curriculum Development.

Marzano, R., Pickering, D., & Pollock, J. (2001). *Classroom instruction that works: Research-based strategies for increasing student achievement.* Alexandria, VA: Association for Supervision and Curriculum Development.

McTighe, J. & Wiggins, G. (2004). *Understanding by design professional development workbook.* Alexandria, VA: Association for Supervision and Curriculum Development.

Myers, I. (1985). *Manual: The Myer-Briggs Type Indicator.* Palo Alto, Ca: Consulting Psychologist Press.

National Research Council (1999). *How people learn: Brain, mind, experience, and school.* Washington, DC: National Academy Press.

Northey, S. (2005). *Handbook on differentiated instruction for middle and high schools.* Larchmont, NY: Eye on Education.

Rohfeld, R. W., & Hiemstra, R. (1995). Moderating discussions in the electronic classroom. In Z. Berge & M. Collins (Eds.), *Computer mediated communication and the online classroom (Vol. 3): Distance learning* (pp. 91-104). Cresskill, NJ: Hampton Press.

Silver, H., Hanson, J., Strong, R., & Schwartz, P. (2003*) Teaching styles and strategies.* Ho-Ho-Kus, NJ: Thoughtful Education Press.

Silver, H.F., Strong, R.W., & Perini, M.J. (2000). *So each may learn: Integrating learning styles and multiple intelligences.* Alexandria, VA: Association for Supervision and Curriculum Development.

Sonwalkar, N. (2003). Logging in with Nishikant Sonwalkar: Online education must capitalize on students' unique approaches to learning. *Chronicle of Higher Education – Distance Education.*

Sousa, D. (2003). How the gifted brain learns. Thousand Oaks, CA: Corwin Press Incorporated.

Tomlinson, C. A. (1995). *How to differentiate instruction in mixed ability classrooms.* Alexandria, VA: ASCD.

Tomlinson, C. A. (1999). *The differentiated classroom: Responding to the needs of all learners.* Alexandria, VA: ASCD.

Tomlinson, C. A. (2000). *Leadership for differentiating schools and classrooms.* Alexandria, VA: ASCD.

Tomlinson, C. A., & McTighe, J. (2006). *Integrating differentiated insiruction and understanding by design.* Alexandria, VA: ASCD.

U.S. Department of Education. (2005). Toward a new golden age in American education: How the Internet, the law and today's students are revolutionizing expectations. *National Education Technology Plan 2004*. Retrieved February 9, 2008, from www.ed.gov/about/offices/list/os/technology/plan/2004/index.html

Winograd, D. *A medium for collaborative learning.* Retrieved February 9, 2008, from www.Emoderators.com

Zhu, E., Payette, P., & DeZure, D. (2003). *An introduction to teaching online* (CRLT Occasional Papers No. 18). University of Michigan.

Additional Reading

ALN Web Center. (2000). *Web center: Learning networks effectiveness research.* Retrieved February 9, 2008, from http://www.alnresearch.org/index.jsp

Blomeyer, R. L. (2006). Professional development for effective teaching and online learning. *Virtual school report, Connections Academy.* Retrieved February 9, 2008, from www.connectionsacademy.com/pdfs/VirtualNewsSpring2006.pdf

Bourne, J., & Moore, J. (2004). *Elements of quality online education* (Vol. 5).

Braidic, S. (2007, October-December). I.Q. – I question: Teacher and student questioning in an online environment. *International Journal of Information and Communication Technology Education (IJICTE), 3*(4).

Carbonara, D. (Ed.). (2005). *Technology literacy applications in learning environments.* Hershey, PA: IGI Global, Inc.

Cavanaugh, C., Gillian, K., Kromey, J., Hess, M., & Blomeyer, R. (October, 2004). The effects of distance education on K-12 student outcomes: A meta-analysis. *Learning point associates.* Retrieved February 9, 2008 from www.ncrel.org/tech/distance/k12distance.pdf

Clark, J., & DiMartino, J. (April, 2004). A personal prescription for engagement. *Principal Leadership, 4*(8), 19-23.

Davidson, K., & Decker, T. (2006). *Bloom's and beyond.* Marion, IL: Pieces of Learning Publishing.

Dede, C., Korte, S., Nelson, R., Valdez, G., & Ward, D. (September, 2005). *Transforming learning for the 21st century: An economic imperative.* Naperville, IL: Learning Point Associates.

Garrison, D., & Anderson, T. (2003). *E-learning in the 21st century.* London: Routledge.

Gold, S. (May, 2001). A constructivist approach to online training for online teachers. *Journal of Asynchronous Learning Networks, 5*(1).

Lorenzo, G., & Moore, J. (Ed.). (2002). *The Sloan consortium report to the nation: Five pillars of quality online education.* Sloan-C.

Moore, J. (Ed.). (2005). *The Sloan consortium quality framework and the five pillars.* Sloan-C.

Muelinburg, L., & Berge, Z. (2000). *The moderators' homepage: A framework for designing questions for online learning.* Retrieved February 9, 2008, from http://www.emoderators.com/moderators/muilenburg.html

O'Neil, H. (February 2003). What works in distance learning? *University of Southern California/CRESST.* Retrieved February 9, 2008, from www.adlnet.gov/downloads/124.cfm

Rockwell, S. K., Schauer, J., Fritz, S. M., & Marx, D. B. (2000, Summer). Faculty education, assistance, and support needed to deliver education via distance. *Online Journal of Distance Education Administration, 3*(2).

Smith, R., Clark, T., & Blomeyer, R. (November, 2005). *A synthesis of new research on K-12 online learning.* Naperville, IL: Learning Point Associates. Retrieved February 9, 2008, from www.ncrel.org/tech/synthesis/synthesis.pdf

Southern Region Education Board. (2003, April). *Essential principles of high-quality online teaching.* Atlanta: Author. Retrieved February 9, 2008, from www.sreb.org

Southern Region Education Board. (2005, July). *Technical guidelines for digital learning content.* Atlanta: Author. Retrieved February 9, 2008, from www.sreb.org

Southern Region Education Board. (2006a, August). *Standards for quality online teaching.* Atlanta: Author. Retrieved February 9, 2008, from www.sreb.org

Southern Region Education Board. (2006b, November). *Standards for quality online courses.* Atlanta: Author. Retrieved February 9, 2008, from www.sreb.org

Sprague, D., & Dede, C. (1999, September). Constructivism in the classroom: If I teach this way, am I doing my job? *Learning and Leading with Technology, 27*(1), 6-9, 16-17.

U.S. Department of Education. (2005). Toward a new golden age in American education: How the Internet, the law and today's students are revolutionizing expectations. *National Education Technology Plan 2004.* Retrieved February 9, 2008, from www.ed.gov/about/offices/list/os/technology/plan/2004/index.html

U.S. Department of Education, Office of Postsecondary Education. (2006, March). *Evidence of quality in distance education programs drawn from interviews with the accreditation community.* Retrieved February 9, 2008 from www.itcnetwork.org/Accreditation-EvidenceofQualityinDEPrograms.pdf

United States Distance Learning Association. (2006). Distance learning for educators, trainers, and leaders. *United States Distance Learning Association Journal 3*(1).

What works in distance learning? (2003). Retrieved February 9, 2008, from http://fusion.jointadlcolab.org/wwindl/

Chapter VII
Exploring Student Motivations for IP Teleconferencing in Distance Education

Thomas F. Stafford
University of Memphis, USA

Keith Lindsey
Trinity University, USA

ABSTRACT

This chapter explores the various motivations students have for engaging in both origination site and distant site teleconferenced sections of an information systems course, enabled by Internet protocol (IP)-based teleconferencing. While in the past many distance learning courses have been asynchronous Web-based offerings, technology and cost advantages now available through IP teleconferencing provide for synchronous course offerings that can serve several physical locations at the same time while retaining the converged media advantages of Internet delivery. To better understand how this new capability can be incorporated into future curricula, it is important to understand student motivations for participating in IP teleconferencing as part of a lecture section for a class delivered across geographically dispersed collegiate campuses. Theoretical perspectives of student motivations for engaging in distance education are examined, and the results of three specific studies of student motivations for IP teleconferencing and multimedia-enhanced instruction are examined and discussed.

OVERVIEW

Distance education (DE) is a popular delivery modality in view of the cost effectiveness and operational efficiencies it brings to course delivery (Allen, Mabry, Mattrey, Bourhis, Titsworth, & Burrell, 2004). This is one reason that prompts administrators to learn more about the economic efficiency of various DE alternatives, such as teleconferencing, computer-mediated delivery, and hybrid mixtures of models (Chang, 2004). The use of Web-based technologies to both supplement and replace traditional lecture courses has become the popular solution from the administrative perspective (Berger & Topol, 2001; Casini & Vincino, 2003) and some say that Web-based asynchronous instruction is now the leading DE delivery mode (Chang, 2004). Even so, it appears that asynchronous learning approaches that include no lecture sessions are not as successful as initially expected (Ginsberg & Foster, 1998; Hara, 1998; Johnson, 2000; Wilkes, Simon, & Brooks, 2006). The question is: What are the key benefits of teleconference-based distance learning as opposed to the strictly computer-based asynchronous approaches?

To help answer this question, three specific studies concerning teleconference-based DE are reported and reviewed here to clarify the nature of student motivations and responses to teleconferencing and technology-mediated support in DE courses, providing an empirical basis for discussing critical considerations in the choice to expand live instruction course delivery with teleconference extension and with Web-based course supplements. Each of these studies was conducted independently, but the same introductory information systems course was used for all three studies.

One empirical perspective develops the concept of media uses and gratifications for distance education, particularly demonstrating the role of social motivations for engaging in teleconferenced DE courses. Another perspective examines converged Web and teleconference technologies in the development of multisection distance education course offerings. This draws upon work of Newcomer and Stafford (2001) on dual-classroom pedagogy, as adapted in Stafford and Simon's (2002) innovative instruction study, and is developed here as a case-based demonstration of the "high-technology adjunct" approach to DE. The empirical discussion concludes with the report of a study that examines differential motivations of teleconferencing students at geographically dispersed sites, using Internet protocol (IP) teleconferencing. As these three exploratory studies are compared, some common understandings begin to appear. The chapter is organized in the following manner. First, theoretical perspectives concerning student motivations in DE are discussed. Then the results of three exploratory studies are presented. The first study deals with social gratifications for teleconferenced courses, the second study discusses a new technology-enabled educational opportunity called "the high-tech adjünct," and the final study examines student preferences for distant and local teleconference sections. Following those three studies, a summary and conclusions are provided.

THEORETICAL PERSPECTIVES

Asynchronous Web-based delivery of lectures is popular with administrators due to cost considerations, but for students, the lack of live interaction with instructors in strictly asynchronous courses is challenging when frequent clarifications or elaborations on course material are required, as frequently may be experienced in technical courses (Flowers, Pascarella, & Pierson, 2000; Gloster & Doss, 2000). Even so, there is a case to be made for promoting the use of teleconferencing technology in the classroom, given the beneficial influences of hands-on experience in preparing students for future high-tech careers (Alavi, 1994). Yet, when considering all technologically-mediated approaches to course delivery, students might not accept Web-mediated asynchronous courses as comparable substitutes to live instruction, and one typically sees higher dropout rates and lower retention rates in asynchronous Web-based classes than in traditional lecture-format classes (Brewer, 2004).

Live interaction instruction makes important contributions to the education process (Abler & Wells, 2005). These benefits come in the form of social presence in technology-mediated courses, reducing alienation, and providing participants with the sense and benefits of a traditional classroom or seminar room (He, Zhang, & Cheng,

2004). In other words, the presence of live instruction, even if it is technologically-delivered, as in the case of a teleconference of a live course, provides an increased social quality of the course interaction which can help students overcome frustrations they may feel when not able to directly interact with instructors and classmates, as would be the case in most asynchronous courses (Hara, 1998).

Internet technology serves increasingly important support roles in DE, but computers will never totally substitute for the learning experience students receive from an instructor (Stafford, 2005), nor are Web-based offerings ever completely satisfactory to students (Hara, 1998). Videoconferencing of classes can overcome the clear social limitations of computer instructed course offerings (Abler & Wells, 2005), since part of the live interaction experience involves important information from social cues (Stiefelhagen, Chen, & Yang, 2005). IP teleconferencing provides for the student-teacher interaction as well as the enhanced learning outcomes available with hyperlinked multimedia.

This chapter covers motivational aspects of student involvement in technologically-mediated education, specifically examining teleconferenced DE. Three aspects of student involvement and technological support for course delivery are discussed, each of which possesses differing theoretical underpinnings:

1. In the first section, basic Internet-based motivations related to motivations for engaging in IP teleconferencing as a course delivery mode are examined. Here, media-use theory is adapted and developed into a framework that we refer to as the uses and gratifications perspective.
2. In the second section, the consideration of Web-mediated course support as a technological adjunct is introduced, applying the principles of media richness theory to the educational setting to enhance students' achievement of the educational task.
3. And in the third section, the principles of transactional distance theory are applied to find distinctions between student motivations for teleconference origin course sections and distant receiving sections.

SOCIAL GRATIFICATIONS FOR TELECONFERENCED COURSES

The use of technology in education has not always been as successful as envisioned (Ginsberg & Foster, 1998; Johnson, 2000), nor has it been as widely adopted for classroom use as initially expected (Miller, Martineau, & Clark, 2000). Part of the problem lies in an incomplete understanding of the value that technology students find in the information technology used in conjunction with training and education (Stafford, 2005). Teleconference-based DE technologies have not always met with great approval from students, but Web-based technologies used to supplement standard teleconferencing techniques can increase student satisfaction (Berger & Topol, 2001; Casini & Vincino, 2003). The problem with understanding how best to integrate Internet support for DE offerings is compounded by the general lack of research on the nature of student motivations to utilize classroom technology (Stöttinger & Schlegelmilch, 2002).

This section of the chapter examines student motivations for technology use associated with educational teleconferencing by applying the uses and gratifications perspective (U&G) from media-use theory. U&G is a research tradition from mass communications used to understand motivations for the use of emerging media in the early days of radio and television (Herzog, 1944; Katz, 1950; Klapper, 1963), and more recently as regards the Internet (Eighmey & McCord, 1998; Stafford, 2000, 2005). In an extension of the U&G approach, here we examine the use of information resources in the classroom from a media-centric Internet model related to IP teleconferencing combined with Web site support.

Technological Innovation in Education

Despite a rapid diffusion of technology in education, its use is not guaranteed to produce greater instructional quality if not properly understood or deployed (Alavi & Leidner, 2001; Johnson, 2000). We are specifically interested in increasing our knowledge of how technology supports large and dispersed courses. It is a routine matter to provide a DE course by teleconference (Alavi, Wheeler, & Valacich, 1995; Alavi, Yoo, & Vogel, 1997; Rovai, 2001); it may even be desirable to do so in order to boost exposure of undergraduate students to key instructional personnel, since "duplicating" a large, local live class section with a teleconference that links a second large section in a distant location has a specific impact on terminally degreed faculty coverage ratios, which is often an issue for accreditation (Stafford, 2005).

Although the use of computer-mediated course information has become a common practice in many areas of education, this practice is a virtual necessity in large-enrollment courses (Karakaya, Ainscough, & Chopoorian, 2001). Introductory information systems courses typically cover a broad base of topic-area knowledge in large-enrollment formats, as compared to the more specialized upper-division classes, and are good contexts in which to leverage technology to expand coverage and increase student satisfaction and learning. As compared to traditional physical distribution of lecture support materials in regular classes, the wide reach and timeliness of Web-based delivery of instructional materials is compelling when dealing with a geographically dispersed group of on- and off-campus students in the multisection high enrollment format of an introductory information systems course.

Students today have less time for school, and the conflicts of career, family, and personal lives require them to find ways to do more with less available time, specifically in regards to attending lectures, performing class work, and completing course assignments (Stafford, 2005). The use of information technology can provide time-starved students with quick content access and time shifted information delivery (Eastman & Swift, 2001; Rehg, 1999), which can help overcome the time strictures of modern careers and lifestyles.

Interactive technology as part of a course enhances the learning process (Huang & Lu, 2003; Uiterwijk, Seoane, Mitchell, & Welch, 1998), but it has been shown that students tend to welcome the opportunity to work with most any form of educational technology *aside* from teleconferencing (Hamer, 2001). Yet, teleconference-based DE is a leading technology (Evans, 2001), and despite student reluctance the integration of the technology into courses can actually convert DE programs into desirable and beneficial offerings for students (Berger & Topol, 2001; Casini & Vincino, 2003; Flowers et al., 2000). Plus, completely asynchronous Web-based delivery of courses, though widely practiced by many universities and colleges, has met with limited success (Ginsberg & Foster, 1998; Johnson, 2000), apparently because students are poorly motivated to enroll and complete courses where there is no interaction with the instructor, and where course content requires frequent clarification or elaboration (Flowers et al., 2000; Gloster & Doss, 2000).

Uses and Gratifications for the Internet

The U&G theoretical perspective evolved in the communications theory literature as a method for profiling audience motivations for use of radio and early television media (Katz, 1950; Klapper, 1963), as well as more recent television innovations such as cable television, video recorders, and television remote controls (Cutler & Danowski, 1980). It is now routinely used to understand the role of user motivations for Internet use (Stafford, Stafford, & Schkade, 2004) as well as student motivations for educational teleconferencing (Stafford, 2005).

U&G focuses on motivated use of a medium. This is a "how and why" approach to understanding usage motivations, since "gratifications" are typically defined as some aspect of satisfaction reported by users, related to the actual *use* of the medium (Herzog, 1944). The approach of the U&G perspective is what people do *with* a medium (Klapper, 1963), given that people are considered to be essentially motivated as opposed to random in their media use (Katz, 1950). Hence, an individual's media choices are motivated by particular self-defined uses and goals (Lin, 1977) and these are described in the U&G approach.

Previous U&G research demonstrates a general dichotomy of user motivations, balancing the preference for media content vs. the enjoyment of the media usage experience (Cutler & Danowski, 1980; Levy & Windahl, 1984). Results from U&G research on the traditional media suggest that people are motivated by two broad dimensions which are characterized as *content gratifications* and *process gratifications* (Cutler & Danowski, 1980), with content gratifications referencing the messages carried by the medium and process gratifications concerning the act of use itself. By analogy, Internet users may be motivated by enjoyment of the usage process, characterized by random browsing and site navigation (Hoffman & Novak, 1996), and users of specific Web sites might be motivated by specific site-related informational content, such as news and weather, or travel information (Stafford et al., 2004).

Studies of Internet U&G indicate a third major source of motivations in this new medium, as compared to prior traditional media research: the social interactions that the Internet fosters and the gratifications users derive from the interactive and communicative aspects of the medium (Stafford et al., 2004; Stafford, 2005).

Uses and Gratifications for Student Internet Use

Social motivations for Internet use in the classroom can be important, since one of the most fundamental services that classroom technology can provide for students is communicative (Evans, 2001). Content gratifications for Internet use can also be important, since most technology-supported learning is based on some form of information content (Stöttinger & Schlegelmilch, 2002). The question is: To what extent are social and content gratifications for educational teleconferencing in play, contrasted to the more basic gratification of usage processes? Three Internet U&G scales, recently validated for use in assessing Internet user motivations (Stafford et al., 2004), are applied here to investigate the context of user motivations for IP teleconferences in the classroom.

Method

The data collection process involved administration of a questionnaire to 85 students enrolled in two sections of an introductory information systems course at a major southern university. Students participated in exchange for extra credit points in the course.

Measures

The scales used, illustrated in the Appendix, were a previously developed set of content-validated indicators of three primary Internet U&G: process, content, and social (Stafford et al, 2004). Internal consistency figures for the scales were excellent, with the process factor producing a coefficient alpha of .82, the content factor producing .85, and the social factor returning an alpha of .8. Since the previously developed scales were originally developed to assess user gratifications for Internet use, the new context explored here—that of the DE classroom—suggests the approach of an exploratory analysis to examine the factor structure that would arise in the classroom teleconference scenario.

Data Collection

Two sections of an introductory information systems course were surveyed. Each of the two sections included both local and distant rooms. The main room for the course was the origination room for a DE teleconference link, and the second dedicated room comprised of a remote facility that received the teleconference transmission. Sixty-two students were surveyed in the origination room and 23 were surveyed in the remote room, and a course Web site delivered informational content to both groups of students in the form of lecture note postings, readings, and study notes. Sixty percent of the students were male, 40% were female; 40% were in the 18-24 age group, 48.2% were 25-34, 10.6% were 35-44, and 1.2% were 45-54. Analysis was performed using SPSS, applying factor analysis with principle components extraction and varimax rotation. Principle components extraction was used to obtain an initial factor analysis in this exploratory research and varimax rotation was used to simplify the interpretation of the results. Four factors were retained, using the eigenvalue greater than 1 rule. These factors are displayed in Table 1.

Results

The four factors retained accounted for 70% of the variance, and the initial factor, characterized by content gratifications, accounted for 30.7% of variance. This factor was characterized by strong loadings from variables such as education, information, knowledge, and learning. This motivational dimension directly references the informational content that users seek, which is indicative of the knowledge seeking function of a DE classroom. The professor utilizes Web-based support resources to support the geographically dispersed class, and has noticed that many students appear to have a strong motivation for seeking the informational content in teleconferenced classes as evidenced by use of the associated Web-based support sites.

The second factor produced 18.5% of the variance, and was characterized by strong loadings from variables such as chatting, friends, interaction, and people. This represents the influence of

Table 1. Uses and gratifications factor loadings (item assignments in bold type)

	Content Variance: 30.8 α = .87	**Social** Variance: 18.4 α = .80	**Process 1: Search** Variance: 11.9 α = .77	**Process 2: Surf** Variance: 8.9 α = .72
Chatting	-.111	**.761**	.013	.245
Friends	.06	**.913**	.063	-.025
Interaction	.003	**.873**	.114	.015
People	.212	**.885**	.004	.05
Resources	.456	.007	**.530**	-.105
Search Engines	.03	.002	**.921**	.105
Searching	.089	.009	**.894**	.245
Surfing	.07	.196	.307	**.774**
Technology	.324	.023	.099	**.644**
Web sites	.335	.056	.016	**.839**
Education	**.831**	.157	.10	.140
Information	**.750**	.05	.243	.08
Knowledge	**.862**	.016	-.028	.227
Learning	**.861**	-.04	.012	.230
Research	.04	.025	.015	.03

social gratifications. It has been demonstrated that students in DE courses are likely to feel apart from and potentially alienated from the rest of course members due to the mediated interface through which the course is delivered (Berger & Topol, 2001; Gloster & Doss, 2000); the presence of this social dimension of motivation in the teleconferenced course speaks to the need students might have to stay in touch with the professor and other students, and not be "out of sight" in a DE scenario. Social motivations are related to the interactive and distinctly interpersonal social environment of live-interaction classes (Evans, 1986).

The third and fourth factors appeared to be different variations on the usage process gratification often found in U&G work. They were characterized by strong loadings from variables that have previously loaded together on a single Internet process-related factor, but in this analysis the general process factor actually diverged into two distinct different process-based factors. Of the two process-related factors, factor three, representing nearly 12% of variance, was characterized by strong loadings from variables such as resources, search engines, and searching. These are process-oriented variables that appear to characterize the process of searching, specifically, and would relate to the DE course support Web page. In the fourth factor, the variables are characteristic of the playful process of Internet use, or browsing, also related to the course support site.

Discussion

This survey assessed U&G as motivations for participation in a teleconference-based DE course in IT. The key implication of the motivational dimensions found in analysis is that Internet technology serves an important support role in DE, though the social benefits of teleconferencing are readily seen in the appearance of the second, social, factor. While students appear to have specific needs for Web-mediated course content in DE courses, their needs for social interaction are also clearly seen. As has been shown in other research, the computer, itself, cannot fully substitute for the interactive and social learning experience students receive from an instructor (Flowers et al., 2000), though computers do provide very convenient enhancements and supplements to instructor activities.

Hence, a recommended approach to teleconferenced DE is that courses ought to combine both teleconference and Web-based instructional resources in order to maximize the combined effect of instructor guidance and Web-based content delivery. The content factor is information-oriented, and its strong variance component (about 31%) shows that the content that instructors supply through a course support Web site is an important part of what motivates students in a DE class. But the social factor has important implications as well. Alavi et al.'s work (1995, 1997) suggests that fostering collaborative interactions among students in separate sections is a synergistic learning technique, and the use of teleconference can contribute to this process as can course-related Web resources. This combination of teleconferencing lectures and supporting Web sites is something we characterize as the high-tech adjunct, and this is discussed in more detail in the next section of this chapter.

THE HIGH TECHNOLOGY ADJUNCT

Despite the demonstrable social usage gratifications for student involvement in course teleconferences, as compared to the asynchronous and impersonal approach to Web-based learning, students on the distant end of an educational teleconference can feel more removed and isolated from class than do the students on the local end of the teleconference, who are experiencing the lecture from the room in which the lecture originates (Stafford, 2005). As this isolation can lead to greater ambiguity and uncertainty toward the educational task, it is appropriate to look to media richness theory (Daft & Lengel, 1984) for cues about ways to enhance the social presence and more effectively change the level of students' understanding. Despite the evidence that the teleconferences, themselves, provide social support in DE settings, there are also indications that integrating rich media Web-based course-support sites into the teleconferenced class context can ameliorate higher levels of isolation among the distant-side participants in teleconferenced courses (Stafford, 2005). Every bit, it appears, helps and this suggests the utility of a hybrid model comprised of multimedia Web sites combined with teleconference-based instruction (Allen et al., 2004). We call this approach "the high-tech adjunct."

The Problem of Supporting Large and Diverse Enrollments

Although the use and distribution of computer generated lecture material has become a common practice in many areas of education, this practice is a virtual necessity in large-enrollment courses (Karakaya et al., 2001), simply because the Web is the only delivery mechanism that brings wide reach and timeliness of availability to a diverse group of on- and off-campus students. In large sections, the volume of assignments generated for grading can also be a logistical challenge.

Use of Internet technology to display, complete, and submit homework assignments has a dual synergy in management information systems (MIS) classes, since students use the very technology for their assignments that they are learning about in class and will later use in industry (Allen, Wedman, & Folk, 2001; Alavi et al., 1995, 1997). Hence, online multimedia technology is both the delivery agent and *also the supplemental agent* in assuring the successful execution of pedagogical and contact plans (Lincoln, 2001).

Students welcome and even seek time-saving approaches to education (Stafford, 2005). These can include nonsynchronous delivery (Eastman & Swift, 2001; Rehg, 1999) and synchronous-interactive technologies (Uiterwijk et al., 1998). To students, the key benefits of converged Internet support in the DE class are more accessibility, heightened communication, and better access to class resources (Seepanski & von Wahlde, 1998); increased learning motivation arises from such student-centered environments (Miller, Martineau, & Clark, 2000).

From the delivery side, the key problems to be solved in an introductory MIS class are the size of the classes and the difficulties in ensuring consistency of delivery for topical material. When students have similar educational experiences, delivery of an assured body of knowledge can be represented to accrediting agencies, and courses for functional area majors can be based on an assumption of commonality of background knowledge on the part of the student. It is beneficial that the technology which is used to deliver large-enrollment DE courses also can serve to motivate students enrolled in such programs, if used synergistically. That approach is demonstrated here in the context of a case study of a large multisection introduction to MIS course at a major Southern university. We consider the convergence of Web-based asynchronous support with teleconferenced synchronous delivery modes in the form of what we characterize as the high-tech adjunct to such large enrollment courses.

Method

The approach used to demonstrate the high-tech adjunct is a case analysis of the large enrollment introduction class to MIS in a major southern university, the same course that was studied in the U&G research. The approach is interpretive and the observations made are meant to illustrate the theoretical concept, under development, of the high-tech adjunct to the DE classroom. The study spanned an academic year of course coverage for an all-majors required introductory MIS course in the College of Business.

Level of Students

Most of the students exposed to our course are sophomores who are completing the course as a prerequisite to other courses they require for their specific academic majors in the College of Business. The course is also taken by students in majors other than business in order to meet the university's computer literacy mandate. While it cannot be denied that all sorts of students, both graduate and undergraduate, can benefit from innovative instructional approaches that integrate technology into the curriculum content and delivery, the specific impact of the course discussed here is at the undergraduate level.

Number of Students

Enrollments for business courses typically fluctuate with market conditions closely related to the national economy. While the economic conditions in the United States have been generally good, the period of time associated with the introduction of the uniform, technologically-integrated curriculum that is discussed here roughly corresponds with the recent academic marketplace for the MIS undergraduate major.

Enrollments for the design year under discussion, considering all sections offered in Fall, Spring, and Summer semesters, were 984 students. Age and demographic characteristics are generally characteristic of undergraduates everywhere, that is, somewhat skewed toward the 18-24 age group, ethnically mixed, but largely Caucasian, and evenly split between men and women. Our city is the home of several large corporations, and offers plentiful part-time and full-time employment opportunities for undergraduate students, so many are involved with work at local firms.

Major Educational Objectives

The high-tech adjunct is the concept of utilizing the Internet and associated telecommunications as the venue for the construction of virtual partnerships with publishing house and corporate media support resources, such that course support sites at a university can provide seamless linkages with content provider sites. Text materials, readings, assignments, even student submission of homework and supplemental professorial support in learning can all be technologically mediated though technological course adjuncts such as are described here.

It is known that the in-class lecture is still a potent vehicle for learning purposes. This is particularly the case when the instructor is a subject matter expert, and leverages that expertise by describing personal experiences or sharing pertinent anecdotal information (Miller et al., 2000). Yet, beyond that context, students tend to differ greatly and differences are most pronounced in the large section courses, where the evidence tends to suggest that technological approaches to education work well when they are student-centered (Young, 2001). This means that technology must serve the needs of students in learning, and in the large course contexts, where technology mediated learning might provide incrementally more advantages and improvements in the learning experience for the broad majority of students (Karakaya et al., 2001), the opportunity to reach greater numbers of students with the efficacy of technological innovations is potent.

Indications are that computers and information technology make the greatest contribution to the process of learning when they are *integrated* into instructional approaches, as opposed to being designated as the conduit of instruction (e.g., Flowers et al., 2000). To that end, these are the educational objectives of the integrative instruction technique practiced in this large-enrollment computer literacy course:

- The full utilization of professorial World Wide Web sites as a central delivery point for course ancillary materials, such as lecture files, copies of course assignments, and even distribution of course syllabi.

- To utilize information technology in the classroom on a daily basis, including
  - Presentation of lecture materials
  - Provision of technology demonstrations in support of lecture topics
  - Demonstration of technological skills to be utilized in class assignments
  - Demonstration of Web-linked resources provided by virtual partners to the course, specifically,
  - The course text publisher
  - The online resources of a major business news organization

The outcome of the provision of course content in the manner suggested is the realization of increased technological literacy levels among undergraduate students at the College of Business. An ancillary outcome and pragmatic objective is also the increased recruitment of undeclared and undecided majors to MIS, achieved through the delivery of a compelling and motivating educational experience in the required introductory course taken by nonmajors.

Unique Features of the Approach

The instructional Web site is a central aspect of this approach to course delivery, but teleconferencing is also used to provide lecture section access to the diverse student groups. Both in-person live sections and teleconferenced live sections of the class are available for enrollment, as the student desires. Yet, there are actually three unique constituencies in provision of technology-mediated education, of which only one is the student. The university *and* the publisher of course materials also share common interests and can be brought together through the course support page to meet the increasing demands of content production and course delivery (Muniz, Billingsley, & Brill, 2002). That is the approach practiced here, where technologically-facilitated partnerships for common cause content production and educational delivery are developed between the text publisher, the instructors, and the virtual partnerships discussed here under the rubric of "high-tech adjuncts." These represent an important aspect of the approach used in the course sequence for the provision of live class lectures on and off campus integrated with technologically-mediated learning and telecommunication aids.

In this particular case, three levels of Web-based course support were integrated. There was a master site for the multisection course, forming a repository from which homework assignments and assigned supplementary reading material could be distributed, but there were also individual instructor course pages cross-linked with the course master site to more uniquely serve the needs of individual course sections. Finally, each site—both the course master site and individual instructor sites—were cross-linked with the course text publisher's site, where study materials, interactive practice tests, media linkages, and homework assignments were delivered.

Impact of the Approach

Distance education courses offered by teleconference can boost exposure of undergraduate students to key instructional personnel, since "duplicating" a large, local live class section with a teleconference-linked, second large section in a distant location has a specific impact on terminally degreed faculty coverage, which is often an issue for business accreditation.

What goes beyond the ordinary is the way in which the sections are integrated with each other and with the instructor *and* with the learning materials at hand. Access to digitized course content (Seepanski & von Wahlde, 1998), and the general digitization of the course resource base (Hill & Hannafin, 2001) are critical aspects of the modern DE approach. In this case, where a DE section of the course is combined under the direction of a single professor with several other large local sections of the high-enrollment course, it is the instructor's course Web site that serves as the nexus for publisher, instructor, and student access *and* contact in the course.

Designed to facilitate single-point access to the full range of digital resources, the instructor's site is extensively hot-linked to the sites of the course partners, where extracurricular readings and online homework assignments are provided, in addition to pedagogical tools including chapter summaries and practice tests. The instructor site also serves as the origination site for copies of lecture presentations, downloadable descriptions for all course assignments, and demonstrations of technologies to be used in the assigned projects.

Since several large-enrollment sections are offered by a single instructor, asynchronous telecommunications is often useful in the delivery of the course, so homework that is assigned by Web link and completed online at a business media partner site is also delivered by e-mail to a central grading address. Students are able to use e-mail to make inquires of the instructor about course matters, but are also encouraged to download (through a site-provided link) and make use of instant messaging software, since instructors typically interact with students by providing support and answering questions most times they happen to log on to the Internet.

While the use of instant messaging is best thought of as a convenience for students, who appreciate the ability to quickly ask questions about class matters and receive online chats about technical tutorials and live-time remote software demonstrations, it is the student on the distant end of these multiple section courses that best benefits from the combination of synchronous and asynchronous telecommunications utilities. Receiving lectures by teleconference, these distant-end students also like e-mail and instant messaging for increasing their social presence in the course activity. These students typically do not drive to the main campus for in-person meetings with the instructor, nor do they have the opportunity to interact in person during class presentations, other than across the mediated teleconference link, so additional and supplemental synchronous and asynchronous communications are useful.

The converged mix of teleconference and Web-based course resources was also used to serve the special needs of a deaf student enrolled in the course. This student was provided with American Sign Language interpreters to translate the lecture, but frequently used instant messenger (specifically, America Online's AIM client for instant messaging services) and e-mail to interact directly with the instructor, because interpreters were not provided for out-of-class discussions. These were carried on by IM and e-mail, which were provided as part of Web-mediated course support.

Course Content

Course content was provided in the form of an introductory text in information systems published by a prominent academic business press publishing house. Content for lectures was delivered in the form of Microsoft PowerPoint ™ slides, but unlike the typical situation in which professors provide students with local PowerPoint™ handouts or local downloads, in this case the slides, study guides, and self-tests used for the course were linked in through the publisher via a link on the instructor's course support instructor site.

The publisher content also provided linkages and integrated content from the online edition of *The Wall Street Journal,* which was used for purposes of stimulating class discussion, for extracurricular reading, and as frequent content focus points for online homework assignments. Though this content was provided through the text publisher, the instructor provided hotlinks to the *Journal* companion site resources from his course support site, so that students had the convenience of a single point of contact for all Web resources across the virtual partnerships. This unique use of news media linkages resulted in an online "reading room" area where students were assigned specific technology articles to read, and then required to complete an interactive homework assignment on the article topic to be forwarded by e-mail to the instructor for grading.

Course Organization

The organization of the multisection courses is unremarkable in its simplicity. Since the instructor Web site serves from the very first day of the course as the primary conduit of class information outside of live lecture (no paper syllabi are distributed; the Web site provides the only student access to course descriptions and schedules), it is a common matter to direct students to the Web site for ongoing scheduling information regarding test times, assignment deadlines, and lecture schedules.

Any number of media outlets might support any number of publishers of textbooks and course materials, both of which, individually, could be generally combined in a form best characterized as custom electronic course packs to provide content support for the instructor.

Presentation of Course Material

In the technology-mediated classroom, the instructor evolves from an information deliverer to a learning environment creator, and a facilitator in a problem solving process (Allen et al., 2001). Even though the prominent delivery mode of material in the course is the physical lecture format, in both live and teleconferenced synchronous versions, there are numerous asynchronous "adjuncts" to the lecture process in the form of ancillary readings and resources provided by the virtual partners, as well as both synchronous and asynchronous off-hour telecommunications links with the instructor, using e-mail and instant messaging. It is a common event to supplement discussion and demonstrations of computer applications in class with later, off-hour interactions via instant messenger where the professor guides students on a step-by-step basis through some particular computer application, each operating from a home personal computer with the application activated for examination and a common point of reference.

Hence, it can be said that the presentation of material in lecture, even though it is delivered to one of two simultaneous sections by teleconference, is rather traditional. It is the use of telecommunications to enhance the after-class informal interactions that distinguishes the particular combination of methods in the approach of the high-tech adjunct. Often, examples and tutorials found on the Web site can provide quick answers to student questions about how to use a specific application discussed in class; when this does not suffice, students will frequently bypass the traditional office hour in-person visit to the professor for guidance in favor of an Web-mediated chat to learn how to work with a specific technology. This has the dual synergy of using technology to *learn about* technology, which has been shown to be particularly useful for building confidence and job skills in students (Alavi et al., 1995, 1997).

Effectiveness of the Approach

It is clear that students learn more about technology when they use technology as part of the learning experience. Whether this high-tech adjunct approach—utilizing extensive Web site linkages to off-campus virtual partners, and Internet chat utilities to supplement the delivery of course content to off-campus distant classroom sites—is more or less effective than standard "chalk-and-talk" approaches remains to be empirically demonstrated. Anecdotally, technology integration in the classroom works (e.g., Young, 2001); however, leading edge educators are so busy developing and delivering content that their literary documentation processes have tended to lag their practice, since the development of Web-enabled courses is substantially more time consuming than normal course preparation (Miller et al., 2000).

Discussion

Our introductory MIS concepts course is technologically integrated with Web resources of a university, a publication house, and a media outlet. The professor weaves together the resources provided by the virtually-linked partners through a central course Web site in order to provide students with seamless access to a wide variety of materials supporting and supplementing the course and its text. Part of the supplementation involves interactive access to the media outlet resources as graded exercises which not only provide students with a compelling and intuitive technical interface for doing homework, but also stimulates greater use, interest, and experience with a key technological resources linked to career success. Enhancing instructor contact with off-hour synchronous and asynchronous communications over the Internet ensures that students will enjoy the maximum learning benefit from the course at any time they choose to apply themselves to its content, either in or out of class.

PREFERENCES FOR DISTANT AND LOCAL TELECONFERENCE SECTIONS

The live interaction experiences students receive from an instructor provide important social cues that contribute to understanding, which cannot be fully duplicated through computer-based learning (Stafford, 2005; Stiefelhagen et al., 2005, Hara, 1998). A critically important goal, then, of technology-mediated DE systems is to increase the amount of social presence in order to provide participants with the sense and benefits of a traditional classroom or seminar room (He et al., 2004), thus overcoming the frustrations that students tend to feel when they are not able to directly interact with instructors and classmates (Hara, 1998).

There are customer satisfaction issues with live teleconferenced delivery. Not all sections of DE teleconferenced classes are perceived equally. Students on the remote end often tend to feel isolated from the main origination section, despite the ameliorating social presence effects of two-way teleconferenced interactions (Lemak, Shin, Reed, & Montgomery, 2005; Stafford, 2005). Student ratings of teacher effectiveness have also been seen to suffer on the distant end of a teleconference (Lemak et al., 2005).

Lemak and colleagues (Lemak et al., 2005; Lemak, Reed, & Montgomery, 2003) characterize the challenge of student motivations for course participation and evaluations of course delivery as one of "transactional distance," which is an interesting expansion of the social presence literature in DE research. The sense of this approach is that students tend to feel apart from a DE-delivered course if dialogue (two-way live interaction with the instructor) is low, and if structure (construed as formulaic provision of material not customized to the current course) is high. In view of the transaction distance perspective, a cut-and-dried arrangement of "canned" course materials delivered by an instructor who does not seek response and interaction from students will possess a very high level of transaction distance and result in poor student motivation. Technically speaking, even the live local section of a DE course can suffer from transaction distance issues if the preparation is by rote and the instructor does not take care to interact with the students. Even so, we expect that the greatest transaction distance challenge will be on the distant end of a live DE delivery.

Demographic Differences

In view of changing demographic trends among students, there will likely be differential responses from students in terms of their preferences for technology-mediated DE courses. Evidence from the field shows that younger students are more comfortable with the technology involved in DE courses than older students (Parnell & Carraher, 2003), so there may be a tendency for younger students to be more satisfied with the distant end of a teleconferenced course. However, the increased numbers of older students in the modern student body—students who are more likely to hold jobs, attend part-time, and need flexible course solutions—implies that there will be increasing numbers of students who can be expected to be *less* comfortable with DE technology while at the same time requiring more flexibility in course offerings and delivery modalities (Kirschner, 2005).

In spite of their expected lower degrees of technical proficiency, older students should then prefer the flexibility provided by DE course offerings even if they are not as comfortable with the technology (Parnell & Carraher, 2003). The increasing scheduling challenges faced by the increasingly older student demographic (Brewer, 2004; Kirschner, 2005) means that older students are more likely to appreciate DE courses, as compared to younger students.

Social Differences

In videoconference formats, social cues are strongly represented, in comparison to strictly computer-mediated instruction formats (Alavi et al., 1997; Brewer, 2004). It is likely that highly technical courses will require more social interaction (Abler & Wells, 2005), since it has been shown that increased social presence aids in the successful delivery of complex and highly technical course content (He et al., 2004; Stafford, 2005). Hence, we can expect that students with high social motivations for DE course participation will have significantly better perceptions of the course and course delivery technology than students in the local section of an introductory technology course.

Theoretical Expectations

In sum, the literature available on student responses to teleconferenced DE courses suggests four specific outcomes that might be predicted:

1. There will be a preference among students for the local section of teleconferenced DE course sections on information technology. Teleconference origin sections are generally richer and more interactive than distant receiving sections.
2. Younger students are likely to be more technically oriented, and this should result in greater appreciation of DE technology. It would be reasonable to expect younger students to have positive perceptions of teleconferenced courses.
3. Older students have more complicated lives than younger students. Teleconferenced courses should be perceived as more useful to this group of students.
4. Students with high social orientation will likely respond better to teleconferenced course deliveries than students with low social orientation.

These expectations were examined in a format that consisted of two sections of introductory information systems classes. One section was located at a private university in the Southwest, the other a major Southern university. Instructors in each course, at the geographically separated sites, transmitted teleconferenced lectures from one class to the other in order to assess differential student responses to the technology.

Method

The student volunteers in both courses were awarded bonus points for their participation. Data for initial analysis was collected at a time when the Southern campus was operating as the origination section and the Southwestern campus was serving as the distant site. A total of 63 students participated across both sections, with 48 at the distant site and 15 at the origination site, including 21 females and 42 males. Among the two broad age groupings of college students discussed by Kirschner (2005), 51 were in the 18-24 age grouping and 12 were 25-34. The distribution of ages was analyzed and is reported in Table 2. One note of concern in this study is the large number of traditional college-age students in the distant section, and the possibility that data provided by this large subgroup may overwhelm the remaining data. To prevent that effect, the hypotheses not specifically related to age were analyzed by comparing only the responses within each age group. Future research should carefully control the size of each population in order to create more generalizable results.

Data was analyzed in SPSS, using analysis of variance techniques. Results indicated that the theoretical expectations of reactions to local and distant sections of a teleconferenced course were generally confirmed.

Results

A number of the measures could be employed to describe differential gratifications for a teleconferenced DE course. Expressed satisfaction with the teleconference is an overall indication of student reactions to the teleconferencing technology. An expressed belief that the learning goals of the course had been met is also indicative of course satisfaction, and the perceptions of usefulness for various sections of the teleconferenced course

Table 2. Age distribution between DE sections (count)

DE Section	18-24	25-34
Local Section	7	8
Distant Section	44	4

could also indicate a relative level of appreciation for each particularly delivery mode and course section. Students were given a 47 item questionnaire that had them rate various measures relating to use, usefulness, and satisfaction with regard to technology in general and teleconferencing in particular.

The first expectation examined here is that students will generally be more satisfied with the local section of a teleconference course. Though significant differences were expected between the distant and local sections on the measure of satisfaction with the teleconferenced course (agree/disagree, 7 point Likert format), an even more interesting result was noted as a result of further analysis due to the potential interaction of the age variable. To avoid this interaction, the data for each age group was analyzed separately using analysis of variance. As shown in Table 3, students in the traditional college age group (18-24) had a preference for the local section, but students in the nontraditional group (25-34) showed no differences across both sections.

Means analysis indicated higher average agreement with the satisfaction measure among the 18-24 year old students in the local section than in the distant section, but no significant difference among the 25-34 year old students.

The second expectation investigated is that younger students in both sections would consider themselves more technologically oriented, but as shown in Table 4, this proved not to be the case. In fact, the 25-34 age group in both sections of the DE course reported significantly stronger levels of self-perceived technical competency than did the 18-24 year-olds from both sections.

We expected that older students would find teleconferenced courses more useful, and that more social students would also find teleconference courses more useful. These approaches explore antecedents of course appreciation as well as the social presence advantages of teleconferencing. Commonly-used measures of perceived usefulness are widely available to investigate

Table 3. Course satisfaction between DE sections (mean/sd)

DE Section	Satisfaction with Teleconference	Learning Goals Met
Local Section	5.88 (1.219)	5.29 (1.047)
Distant Section	2.42 (1.541)	2.31 (1.518)
	$F_{1,61} = 76.795$, $p < .0001$	$F_{1,61} = 51.451$, $p < .0001$

Table 4. Technical mastery by age (mean/sd)

Age Grouping	Technical Mastery
18-24	10.059 (3.414)
25-34	15.000 (2.216)
	$F_{1,61} = 22.719$, $p < .0001$ Overall mean = 11.000 (3.755)

usefulness perceptions of a course, and Stafford's (2005) social gratification scales have also been used to demonstrate social gratifications for the DE course. Using the social gratification scale, comprised of four-7 point Likert format scales anchored by "chatting," "friends," "interaction," and "people," a sample mean of 12.9 ($\sigma = 7.5$) was used to create a dichotomous mean-split classification for purposes of testing.

As shown in Table 5, older students tend to express more satisfaction with a teleconferenced course, in general. They found the course more useful, were more satisfied with the course, and were more assured that their learning goals had been met in the course than younger students.

The social orientation of students was calculated based on the way students reported that they used the internet to interact with others, and based on this calculation, students were assigned into either the high or low social orientation group. As shown in Table 6, also as expected,

more social students were more satisfied with the course, were more assured their learning goals had been met, and found the course generally more useful. Teleconference technology provides greater social presence, and this is inherently more appealing to the more socially oriented student.

Discussion

There are several forces currently shaping higher education. Advanced technology has enabled IP teleconferencing, which provides more options for high-quality delivery of instruction to more students in more locations, all at reduced costs. At the same time, the ranks of nontraditional students are increasing and the traditional student population is in decline, resulting in a much more diverse student body to be served with the available technological advances. Understanding the differential responses and motivations of these differing constituencies is an important challenge in modern DE, and the results discussed here are one step in that direction.

This research lends support to theoretical expectations concerning technology-mediated DE systems. First, it demonstrated that traditional college age students prefer the local section of a teleconferenced DE course, while nontraditional students do not share this preference. By contrast, older nontraditional students rate teleconferenced DE courses higher for satisfaction and practical usefulness dimensions than do younger students. Lastly, the social presence advantages expected from teleconferencing approaches to DE appeal quite a bit more to socially-oriented students than they do to students with a low social orientation, as measured by teleconference satisfaction measures.

What was not expected, but may provide interesting implications for course design in periods of diverse enrollment, is the finding that older, nontraditional students (who could reasonably be expected to derive relatively greater benefits from teleconferenced courses) are also probably

Table 5. Course appreciation by age (mean/sd)

Age Grouping	Perceived Usefulness	Learning Goals Met	Satisfaction with Teleconference
18-24	14.608 (4.355)	2.47 (1.617)	2.73 (1.801)
25-34	17.750 (5.446)	5.42 (1.084)	5.75 (1.712)
	$F_{1,61} = 4.591$, $p = .036$ Overall mean = 15.2 (4.7)	$F_{1,61} = 35.810$, $p < .0001$ Overall mean = 3.0 (1.9)	$F_{1,61} = 27.883$, $p < .0001$ Overall mean = 3.3 (2.1)

Table 6. Course satisfaction between DE sections (mean/sd)

Social Orientation	Perceived Usefulness	Learning Goals Met	Satisfaction with Teleconference
Low	12.921 (3.773)	1.92 (1.148)	2.11 (1.203)
High	18.680 (3.783)	4.72 (1.595)	5.12 (1.965)
	$F_{1,61} = 35.057$, $p < .0001$ Overall mean = 15.2 (4.7)	$F_{1,61} = 65.628$, $p < .0001$ Overall mean = 3.0 (1.9)	$F_{1,61} = 57.175$, $p < .0001$ Overall mean = 3.3 (2.1)

more comfortable with the use of teleconferencing technology, based on their higher reported degrees of self-perceived technical competency. This is a tentative finding, and, due to the small sample size in this study, should be further evaluated. This may suggest a preliminary specification for course design that promotes the enrollment of traditional students and students with low social orientation on the origination section of teleconferenced courses, while also suggesting that nontraditional students are the better targets to whom remote locations of a teleconference should be promoted.

Granted, this evidence also demonstrates a general preference by all students for the local section of a teleconferenced course. Yet, it appears that nontraditional students may be in a better position to make use of and, in fact, appreciate the delivery of lectures to remote sites via teleconferencing. Even so, it is useful to know which student groups are more or less likely to respond positively to modern DE offerings.

FUTURE DIRECTIONS

An unexpected, but serendipitous, finding of this series of studies was that the group that could likely derive the greatest benefits from Internet-enabled DE (i.e., older, nontraditional students) also may be more comfortable with the use of that technology, based on their higher reported degrees of self-perceived technical competency. This is an outcome that deserves further exploration, particularly in light of the midcareer and transitional segments of the baby-boom generation in education.

An implication of this research for further exploration involves the arrangement whereby courses containing significant technological content could be held in a traditional lecture or seminar on a main campus, with traditional students and students with low social orientation on the transmitting end, and with distant classrooms supplemented by Internet-enabled teleconferencing at one or more remote locations which are convenient for nontraditional students.

This chapter also provided results of an investigation of student gratifications for teleconferencing in the classroom. One opportunity for further study would be to research the student preferences between the remote section of an Internet-enabled teleconference and the more standard computer-mediated course. In so doing, the costs of the two methods would be more similar, and thus these could reasonably be expected to be alternatives that an administrator might consider. Perhaps more importantly, the variances attributable to technology would be more equally distributed between these two alternatives, and a true estimation of the value of interactivity, as provided by the Internet, might be understood.

CONCLUSION

Teleconferencing for DE is both popular and efficacious. Although students are not as interested in the modality as universities are, in view of potent operational cost benefits to its use, there are clear motivational benefits to using live teleconferencing to enhance DE offerings and to boost student motivations to engage in DE courses. Of late, a new generation of teleconferencing technology is coming to market and finding leading edge use in the corporate boardroom (Dunlap, 2007), and both higher resolution video and more interactive services can be expected to be come a standard part of teleconferencing set-ups in the near future. This can only have beneficial outcomes for the continued use of teleconferencing as a DE alternative in large enrollment courses, as discussed here.

Teleconferencing has already been integrated with other Web-based course support technologies and the approach of the high-tech adjunct, but continued investigation of the synergies between live transmission of lectures to remote points coupled with Web-based course support should lead to ongoing increases in efficiency and effectiveness for DE offerings. As the nature of work changes and as demographic changes in the population make themselves felt at the higher education level, universities and colleges will continue to find teleconferencing technologies a flexible and capable modality for course delivery in new and innovative formats designed to meet the new and more demanding needs of modern students. Though students will likely continue to have preferences for live in-person instructor-led course sessions from a quality perspective, it can be seen that substituting teleconferenced contact with instructors continues to provide important social support and orientation benefits to students who are unable or unwilling to meet live class session in person.

As students continue to gain technical skills, it is likely that their reaction to and appreciation of technologically-mediated instruction modalities such as teleconferencing will grow. As this transpires, educators will find additional opportunities to expand course coverage and delivery options through the use of teleconferencing technologies.

REFERENCES

Abler, R. T., & Wells, I. G. (2005). Distributed engineering education: Evolution of the telecollaboration stations for individualized distance learning. *IEEE Transactions on Education, 48*(3), 490-496.

Abraham, T. (2002). Evaluating the virtual management information systems classroom. *Journal of Information Systems Education, 13*(2), 125-133.

Alavi, M. (1994). Computer-mediated collaborative learning: An empirical evaluation. *MIS Quarterly, 18*, 159-174.

Alavi, M., & Leidner, D. E. (2001). Research commentary: Technology-mediated learning – a call for greater depth and breadth of research. *Information Systems Research, 12*, 1-10.

Alavi, M., Marakas, G. M., & Yoo, Y. (2002). A comparative study of distributed learning environments on learning outcomes. *Information Systems Research, 13*(4), 404-417.

Alavi, M., Wheeler, B. C., & Valacich, J. S. (1995). Using IT to reengineer business education: An exploratory investigation of collaborative telelearning. *MIS Quarterly, 19*, 293-312.

Alavi, M., Yoo, Y., & Vogel, D. R. (1997). Using information technology to add value to management education. *Academy of Management Journal, 40*(6), 1310-1333.

Allen, M., Mabry, E., Mattrey, M., Bourhis, J., Titsworth, S., & Burrell, N. (2004). Evaluating the effectiveness of distance learning: A comparison using meta-analysis. *Journal of Communication, 54*(3), 403-420.

Allen, G. K., Wedman, J. F., & Folk, L. C. (2001). Looking beyond the valley: A five year case study of course innovation. *Innovative Higher Education, 26*(2), 103-119.

Berger, K. A., & Topol, M. T. (2001). Technology to enhance learning: Use of a Web site platform in traditional classes and distance learning. *Marketing Education Review, 11*(3), 15-26.

Brewer, P. D. (2004). An examination of alternative instructional methods. *Delta Pi Epsilon Journal, 46*(2), 92-104.

Casini, M., & Vincino, A. (2003). The automatic control telelab: A user-friendly interface for distance learning. *IEEE Transactions on Education, 46*, 252-257.

Chang, S. (2004). High tech vs. high touch in distance education. *International Journal of Distance Education Technologies, 2*(2), i-iii.

Cutler, N. E., & Danowski, J. A. (1980). Process gratification in aging cohorts. *Journalism Quarterly, 57*, 269-77.

Daft, R. L., & Lengel, R. H. (1984). Information richness: A new approach to managerial behavior and organizational design. In L. L. Cummings & B. M. Staw (Eds.), *Research in organizational behavior 6* (pp. 191-233). Homewood, IL: JAI Press.

Dunlap, T. (2007). *IBM, Cisco team up on chat, video platform.* Retrieved March 2007, from http://www.intranetjournal.com/articles/200703/ij_03_07_07a.html

Eastman, J. K., & Swift, C. O. (2001). New horizons in distance education: The online learner-centered marketing class. *Journal of Marketing Education, 23*(1), 25-34.

Eighmey, J., & McCord, L. (1998). Adding value in the information age: Uses and gratifications of sites on the World Wide Web. *Journal of Business Research, 41,* 187-194.

Evans, J. R. (1986). Creative thinking and innovative education in the decision sciences. *Decision Sciences, 17*(2), 250-163.

Evans, J. R. (2001). The emerging role of the Internet in marketing education: From traditional teaching to technology-based education. *Marketing Education Review, 11*, 1-14.

Flowers, L., Pascarella, E. T., & Pierson, C. T. (2000). Information technology use and cognitive outcomes in the first year of college. *The Journal of Higher Education, 71*(6), 637-667.

Ginsberg, R. B., & Foster, K. R. (1998). The wired classroom. *IEEE Spectrum, 35*(8), 44-51.

Gloster, C., Jr., & Doss, C. (2000). A distance education course in computer engineering at NC State University. *Computers in Education Journal, 10*, 22-26.

Hamer, L. O. (2001). Distance learning technologies as facilitators of learning and learning-related student activities. *Marketing Education Review, 11*, 55-67.

Hara, N. (1998, October 14-17). Students' perspectives in a Web-based distance education course. In *Proceedings of the Mid-Western Educational Research Association*. Retrieved February, 2007, from http://php.ucs.indiana.edu/~nhara/paper/mwera98.htm

He, A., Zhang, G., & Cheng, Z. (2004). A design of real-time and interactive distance education. *International Journal of Distance Education Technologies*, *2*(2), 1-12.

Herzog, H. (1944). What do we really know about day-time serial listeners? In P. Lazarsfeld & F. Stanton (Eds.), *Radio research 1942-1943*. New York: Duel, Sloan and Pearce.

Hill, J. R., & Hannafin, M. J. (2001). Teaching and learning in digital environments: The resurgence of resource-based learning. *Educational Technology Research and Development*, *49*(3), 37-52.

Hoffman, D. L., & Novak, T. P. (1996). Marketing in hypermedia computer-mediated environments: Conceptual foundations. *Journal of Marketing*, *60*, 50-68.

Huang, H., & Lu, C. (2003). Java-based distance learning environment for electronic instruments. *IEEE Transactions on Education*, *46*(1), 88-94.

Johnson, J. D. (2000). Levels of success in implementing information technologies. *Innovative Higher Education*, *25*(1), 59-75.

Karakaya, F., Ainscough, T. L., & Chopoorian, J. (2001). The effects of class size and learning style on student performance in a multimedia-based marketing course. *Journal of Marketing Education*, *23*(2), 84-90.

Katz, E. (1950). Mass communication research and the study of popular culture: An editorial note on a possible future for this journal. *Studies in Public Communication*, *2*, 1-6.

Kirschner, A. (2005). Alma mater in the time of TiVo. *Chronicle of Higher Education*, *52*. Retrieved March, 2007, from http://chronicle.com/weekly/v52/i16/16b00601.htm

Klapper, J. T. (1963). Mass communication research: An old road resurveyed. *Public Opinion Quarterly*, *27*, 515-527.

Lemak, D. L., Montgomery, J. C., & Reed, R. (2003). Instructor effectiveness in distance education: The case of technology and transactional distance. In *Proceedings of the 2003 Academy of Management Conference*.

Lemak, D. L., Shin, S. J., Reed, R., & Montgomery, J. C. (2005). Technology, transactional distance and instructor effectiveness: An empirical investigation. *Academy of Management Learning & Education*, *4*(2), 150.

Levy, M. R., & Windahl, S. (1984). Audience activity and gratifications: A conceptual clarification and exploration. *Communication Research*, *1*, 51-78.

Lin, N. (1977). Communication effects: Review and commentary. In B. Ruben (Ed.), *Communication yearbook 1*. New Brunswick, NJ: Transaction Books.

Lincoln, D. J. (2001). Marketing educator Internet adoption in 1998 versus 2000: Significant progress and remaining obstacles. *Journal of Marketing Education*, *23*(2), 103-116.

Miller, J. W., Martineau, L. P., & Clark, R. C. (2000). Technology infusion and higher education: Changing teaching and learning. *Innovative Higher Education*, *24*, 227-241.

Muniz, A. M., Jr., Billingsley, W., & Brill, T. (2002). The GoReader launch: Developing marketing strategy for an innovative education technology. *Journal of Interactive Marketing*, *16*(1), 67-88.

Newcomer, J., & Stafford, T. F. (2001). Teaching simultaneously in dual classrooms: Tips for when you're asked to use videoconferencing as an instructional tool. In *Proceedings of the Southwest Academy of Management*, SWFAD, Houston.

Parnell, J. A., & Carraher, S. (2003). The management education by Internet readiness (MEBIR) scale: Developing a scale to assess personal readiness for Internet-mediated management education. *Journal of Management Education*, *27*(4), 431-446.

Rehg, J. A. (1999). Developing Web-based courses using an online development guide and templates. *Computers in Education Journal*, *9*, 51-55.

Rovai, A. (2001). Building classroom community at a distance: A case study. *Educational Technology Research and Development*, *49*, 1042-1629.

Seepanski, J. M., & von Wahlde, B. (1998). Mega-system collaboration: Cross-continent consortial cooperation. *Information Technology and Libraries*, *17*(1), 30-35.

Stafford, T. F. (2000). Internet as metamedium: Emerging uses of the World Wide Web – a tutorial. In H.M. Chung (Ed.), *Proceedings of the Association for Information Systems Conference, AIS*, Long Beach, CA.

Stafford, T. F. (2005). Understanding motivations for Internet use in distance education. *IEEE Transactions on Education*, *48*(2), 301-306.

Stafford, T. F., & Simon, J. C. (2002). High-tech adjuncts: Using technology-mediated virtual partnerships to facilitate the delivery of information systems course content. In *Proceedings of the 2002 Decision Sciences Institute Conference, DSI*, Atlanta.

Stafford, T. F., Stafford, M. R., & Schkade, L. L. (2004). Determining uses and gratifications for the Internet. *Decision Sciences*, *35*(2), 259-288.

Stiefelhagen, R., Chen, X., & Yang, J. (2005). Capturing interactions in meetings with omnidirectional cameras. *International Journal of Distance Education Technologies*, *3*(3), 34-47.

Stöttinger, B., & Schlegelmilch, B. B. (2002). Information and communication technologies in tertiary education: A 'customer' perspective. *Marketing Education Review*, *12*, 63-72.

Uiterwijk, J., Seoane, D., Mitchell, L., & Welch, J. (1998). The virtual classroom. *InfoWorld*, *20*(47), 64-70.

Wilkes, R. B., Simon, J. C., & Brooks, L. D. (2006). A comparison of faculty and undergraduate students' perceptions of online degree programs. *Journal of Information Systems Education*, *17*(2), 131-140.

Young, M. R. (2001). Windowed, wired and webbed: Now what? *Journal of Marketing Education*, *23*(1), 45-54.

ADDITIONAL READING

The work of Alavi is an excellent starting point for researchers interested in the foundations of behavioral research on the use of teleconferenced learning in education settings. Suggested readings from this stream of distance education research include:

Alavi, M. (1994). Computer-mediated collaborative learning: An empirical evaluation. *MIS Quarterly*, *18*, 159-174.

Alavi, M., & Leidner, D. E. (2001). Research commentary: Technology-mediated learning – a call for greater depth and breadth of research. *Information Systems Research*, *12*, 1-10.

Alavi, M., Marakas, G. M., & Yoo, Y. (2002). A comparative study of distributed learning environments on learning outcomes. *Information Systems Research*, *13*(4), 404-417.

Alavi, M., Wheeler, B. C., & Valacich, J. S. (1995). Using IT to reengineer business education: An exploratory investigation of collaborative telelearning. *MIS Quarterly*, *19*, 293-312.

Alavi, M., Yoo, Y., & Vogel, D. R. (1997). Using information technology to add value to management education. *Academy of Management Journal, 40*(6), 1310-1333.

For those interested in technology usage motivations related to student use of distance learning technology, see Stafford's uses and gratifications perspective on distance education technology use is useful reading:

Stafford, T. F. (2005). Understanding motivations for Internet use in distance education. *IEEE Transactions on Education, 48*(2), 301-306.

Stafford, T. F., Stafford, M. R., & Schkade, L. L. (2004). Determining uses and gratifications for the Internet. *Decision Sciences, 35*(2), 259-288.

For those interested in assessment of distance education technologies and practices, the work of Abraham, as well as that of Alavi et al., are instructive for evaluative purposes:

Abraham, T. (2002). Evaluating the virtual management information systems classroom. *Journal of Information Systems Education, 13*(2), 125-133.

Alavi, M., Marakas, G. M., & Yoo, Y. (2002). A comparative study of distributed learning environments on learning outcomes. *Information Systems Research, 13*(4), 404-417.

APPENDIX: INTERNET USES AND GRATIFICATIONS SCALES

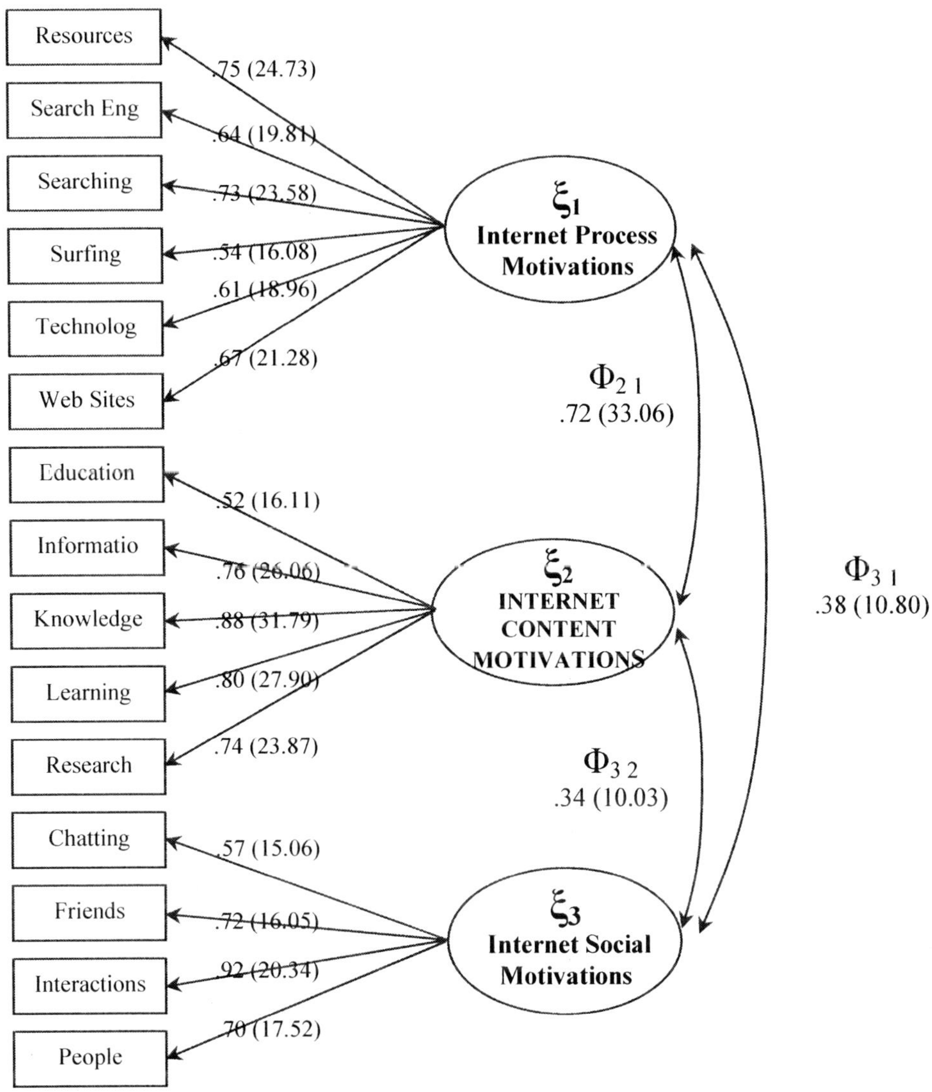

$\chi^2_{(79)}$ = 242.82 (p = .000)
GFI = .97, AGFI = .95
RMSR = .11, SRMSR = .043
NFI = .96, CFI = .97

Section III
Interaction and Collaboration

Chapter VIII
Collaborative Technology: Improving Team Cooperation and Awareness in Distance Learning for IT Education

Levent Yilmaz
Auburn University, USA

ABSTRACT

This chapter presents a set of requirements for next generation groupware systems to improve team cooperation and awareness in distance learning settings. The premise of the chapter is based on the observation that in distance learning online asynchronous (e.g., e-mail, conference tools) or synchronous (e.g., chat) mechanisms are used to facilitate collaboration and coordination to complete necessary tasks. However, students are neither trained in basic principles regarding how effective cooperation takes place, nor means for their realization. Basic methods of cooperation are delineated along with a set of requirements based on a critical analysis of the elements of cooperation and team awareness. The means for realizing these elements are also discussed to present strategies to develop the proposed elements. Two scenarios are examined to demonstrate the utility of collaboration to provide deep integration of communication and task accomplishment within a unified coherent framework.

INTRODUCTION

Information technology (IT) organizations increasingly rely on teams to address a variety of complex and challenging tasks (Salas & Fiore, 2004). Large complex software intensive system design, development, and management require considerable effort in collaboration and coordination among peers. Hence, teams have become an integral and essential component in every IT organization (Yilmaz & Phillips, 2006). Organizations believe that teams, effective teamwork, and engineers with proper skills to function in IT team projects can provide a competitive edge (Ellis, Gibbs, & Rein, 1991). Providing IT students the necessary educational tools and their principled use have the potential to improve the students' cooperation and cognition skills, which are critical for students to succeed in today's global economy (Carmel, 1999). Awareness of this trend influenced instructors to incorporate group projects and research assignments into curriculum.

In distance learning, team projects are assigned to a group of students who are geographically dispersed. Hence, online asynchronous (e.g., e-mail, conference tools) or synchronous (e.g., chat) mechanisms are used to facilitate collaboration and coordination to complete necessary tasks. However, students are neither trained in basic principles regarding how effective cooperation takes place, nor means for their realization. Furthermore, existing educational collaborative technologies (i.e., groupware) provide a classical organizational view of communication and lack implicit coordination facilities that support shared mental model construction (team cognition). Awareness of other group members is a critical building block in the construction of team cognition, and consequently *computational support for awareness* in collaborative education tools is crucial for supporting team cognition (Endsley, 1995) in distributed student groups.

This chapter presents a framework and a strategy to help mitigate these shortcomings in existing distance learning groupware for IT education. First, the characteristics of team work in distance learning are elaborated to set the stage for discussing requirements for next generation collaborative education tools. Specifically, we compare how such teams are different than conventional teams in IT education. The significance of team cognition in conjunction with cooperation is emphasized in the distance learning context. Then, we discuss why cooperation in distance learning is difficult to do right. In particular, the problems that pull apart a student team in a virtual education environment are highlighted. In particular, dispersion, breakdown of traditional coordination and control mechanisms, team cohesion, and the substance (richness) of communication are presented as challenges (Carmel, 1999). A number of mitigation strategies, one of which is advanced collaborative instruction technologies, are proposed to counter the challenges. The chapter will then focus on a proposed framework for advancing the state of the art in collaborative instruction technology. The framework is based on fundamental methods of cooperation. The methods and underlying principles are then used to propose strategies to augment additional collaborative technologies to improve team cognition and cooperation skills of students. We will consider both generic collaborative technologies (e.g., e-mail, audio-conferencing, video-conferencing, and groupware platforms) and collaborative technology to support task specific solution for IT education.

BACKGROUND

Project awareness (Gutwin & Greenberg, 2004) is something people take for granted in everyday world. This is mainly due the fact that acquiring such awareness information is natural and simple; as such, it is rarely considered as an intentional activity. As a consequence, it is often overlooked in the design of educational tools and collaboration support frameworks such as groupware systems. The problem is that maintaining that awareness has proved difficult in current distributed educational groupware systems in which interaction mechanisms are poor and information resources are not designed to promote awareness.

Group Project Awareness

There are three main reasons why most educational groupware does not support project awareness. First, the input and output mechanisms are not capable of handling perceptual information available to face-to-face settings. Second, the amount of information generated by a user is much less in a virtual setting than a physical workspace. Third, the education groupware systems do not present even the limited awareness information to the user.

Table 1. Traditional vs. virtual teams

Traditional Student Team	in Distance Education
Colocated members	Distributed members
Face-to-face interaction	Electronic communication
Mostly informal communication	Continuous structured communication
Information distribution	Information access (pull)
Sharing completed work	Continuous sharing of incomplete work
Transparent process	Computer-visible process
Culture learned through osmosis	Culture learned through electronic communication and produced artifacts

Awareness is often defined as the knowledge created through the interaction of an agent (i.e., user) with its environment (Endsley, 1995). The four basic characteristics of awareness, defined by Adams, Tenney, and Pew (1995), Endsley (1995), and Norman (1993), are as follows:

- Awareness is the knowledge of the state of the environment.
- Since the state of the environment changes over time, awareness must be kept up to date.
- Awareness is maintained by interacting with the environment.
- Awareness is not the purpose; rather it is a means to achieve the purpose (e.g., the task at hand).

Several types of awareness have been investigated in the literature. Conversational awareness (Clark, 1996), casual awareness of others in work groups (Borning & Travers, 1991), and situational awareness (McNeese, Salas, & Endsley, 2001) are among the major types reported in the literature. Situation assessment and situation awareness studies in cognitive psychology provide a wealth of information and results that can be leveraged to improve the state of the technology in educational groupware. Workspace awareness involves perception and understanding of others' interaction with the shared project workspace. Hence, workspace should focus on understanding of people in the project workspace, as opposed to workspace itself. Furthermore, workspace awareness is based on the observations of the events within the workspace. The physical structure of the workspace and the nature of the artifacts influence team cognition and provide a baseline for the external representation of the team's joint activity and its external memory. However, a number of characteristics pertaining to distance learning in IT education complicates gathering information that is necessary to construct workspace awareness.

Challenges and Issues in Distance Learning for IT Education

Virtual student groups in distance education are viewed by many as being team-based because activities are necessarily defined by the project and associated tasks. The key difference between traditional student teams and virtual teams is that members are no longer colocated. Some may be working out of their home. Table 1 depicts major differences between traditional and virtual teams.

The above differences impose various forces on the effectiveness of team work. In particular, geographic dispersion, coordination breakdown, communication substance, and team cohesion are effected as a result of distance learning. Each one of these forces will be analyzed to better understand their impact and sound solutions will be suggested to counter and mitigate them.

- **Dispersion:** The discussion of dispersion vs. colocation highlights what we know intuitively, that is, it is harder to manage from a distance. Shorter project timelines via the ability to give feedback quickly and shorter communication lines are well-known (Rafii, 1995).
- **Breakdown of Traditional Control and Coordination:** The overhead of control and coordination experienced by students is significant. Coordination is the act of integrating each task to a coherent unit so that it contributes to the overall objective of the project (Ferber, 1994). Distributed/dispersed teams create further burdens; primarily the informal ones due to interdependency among tasks and difficulty in effective decision making in online environments.
- **Communication Richness:** The substance of communication refers to the richness of interaction. Rich communication entails two-way interaction involving more than one sensory channel (Trevino, Daft, & Lengel, 1987). Team members need rich interaction styles to collaborate and convey information accurately and timely manner.
- **Team Cohesiveness:** A "real" team has a collective responsibility for the project, shares responsibility for managing the tasks, has a common goal, and works together on tasks that are interdependent. Cohesion is one of the differences between a successful and unsuccessful team. Distance is impediment to building relationships of trust, and it may take considerable time in distance learning settings.

Technologies and Methodologies for Effective Team-based Distance Learning

Various promising technologies and methodologies counter the challenges discussed above. Table 2 depicts the main categories of such technologies.

While collaborative technology has been promoted as a significant enabler that facilitates team cooperation, its focus is mainly on mechanisms to improve collaboration and explicit coordination of activities. Yet, it is widely acknowledged that sustaining effective operation of teams relies on establishing shared mental models via implicit coordination strategies (Espinosa, Lerch, & Kraut, 2004). While the links between theoretical approaches form cognitive science and education groupware are not lacking in theory, far fewer

Table 2. Effective methodologies/technologies for improving team effectiveness

Methodology/Technology	Synopsis
Collaborative technology	Collaborative technology supports two key communication objectives: fostering informal communication between peers in a team and bringing new forms of formal communication via groupware software that facilitates location transparency to bring distant team members closer. The effects of distance, coordination breakdown, and communication loss are counterbalanced with proper collaborative technologies.
Telecommunications infrastructure	Telecommunication infrastructure is the foundation for enabling reliable, high bandwidth network. Virtual private networks are increasingly common channel connecting professional teams and can be used in educational domains.
Instructional techniques	Instructional techniques pertain to allocation of roles in teams and paying attention to human factors issues such as rewarding, recognizing effective teamwork, and managing conflict between members of a team.
Team building methodologies	To facilitate effective education in a team context, the formation of teams should take into account the impact of team workload, team size, team composition, team structure, and team cohesion.

methodologies are used in existing groupware systems to foster their development.

Collaborative technology in educational groupware systems that support distance education often (1) serves as a team memory and knowledge center, (2) provides a basis to inform all members regarding tasks, status information, people, and other dynamic project information, (3) reduces duplication of effort, (4) supports coordination of activities as well as the workflow, and (5) helps maintain quality of artifacts produced by students. However, education groupware systems provide limited support for group workspace awareness. Configuration management of artifacts, project status information, notification services, activity scheduling and tasking, and team memory and knowledge center features are among the needed functions to minimally support workspace awareness in collaborative distance education settings. Collaborative situations constrain collaboration to the environment in which interaction takes place among people, the type of systems they use to support distributed collaboration, and the tasks that people do.

- **Environment—Shared workspace:** The environment is the shared project workspace through which users exchange and share artifacts related to their activities.
- **Systems—Distributed groupware:** Distributed groupware allows members of a team to work from different locations at the same or at different times depending on whether the system is real-time or asynchronous.
- **Tasks:** The main tasks in shared workspace are the creation, manipulation, and navigation through artifacts and execution of activities that pertain to these tasks.
- **Groups—Small and focused groups:** The activities in these workspaces are carried out by groups of two to five students, who engage in collaborative tasks. The students often shift back and forth between individual and shared activities during a work session.

REQUIREMENTS FOR NEXT GENERATION EDUCATIONAL GROUPWARE: ISSUES AND PROBLEMS

Student group project management, one of the most cooperation-intensive activities in group projects, presents significant difficulties when project members are distributed. Because of lack of social contact, geographically distributed students without appropriate tool support may have trouble in attaining a consistent and coherent understanding of the status of the project. The position advocated in this chapter is that alleviation of these difficulties requires integrating proper educational groupware systems that support cooperation in students' project workspace. The question is why the work environments in typical educational groupware systems do not support group projects effectively and efficiently. Furthermore, what functionalities might a cooperation component of a groupware system support to correct these deficiencies? To better respond to this question we need to elaborate on the characteristics and elements of cooperation.

Basic Methods of Cooperation

The concept of interaction is central to the issue of cooperation. Interaction occurs when two or more students are brought into a dynamic relationship through a set of reciprocal actions. Depending on the compatibility or incompatibility of *goals*, the availability of *resources* and *skills*, interaction situations can be classified as *independence, cooperation*, or *antagonism*. The type of interaction of interest for the purpose of group project management is *cooperation*. To develop groupware systems, it is essential to understand the means for realization of the conditions for cooperation.

The first method, called *grouping*, consists very simply of arranging team members to obtain a more or less homogeneous unit in space or a communication network. The second method, called *communication,* entails having a system that facilitates exchanging messages between members of a team. *Specialization* is the process through which team members become more and more adapted to their tasks. Implicit coordination via team cognition facilities are needed to take advantage of the specialization of team members. *Collaboration by sharing tasks and resources* involves techniques to distribute resources, information, and tasks among students based on their roles and specialization. *Coordination of actions* refers to coordinating the activities of individual team members and integrates each task to a coherent unit so that it contributes to the overall objective. Finally, *conflict resolution via arbitration and negotiation* pertains to resolving disagreements and preventing the performance of the team from deteriorating.

Grouping

Arranging students into small groups to form either a homogeneous unit or a social network allowing them to behave as if they are physically side by side is a first step in enabling cooperative behavior. One can consider a group as a distributed organism. There are specialists in finding the right resources, and specialists in using the required tools in an efficient and effective manner.

Communication and Discussion

Many educational IT projects in distance learning environments share project artifacts via e-mailed text documents. Students may communicate further by phone, e-mail, video conference, or chat to resolve ambiguities in the document and refine it. The lack of sophisticated integrative tools that integrate project management and communication leads to frequent context switching between task accomplishment and interactions. Fragmentation of content across several media due to multiple channels of communication inhibits establishing a common shared model of the project. As a result, students need more intuitive and seamless integration of informal collaboration facilities with the project management and awareness space. That is, educational groupware systems should provide facilities for not only documenting the artifacts developed during the group project but for also holding rich contextual discussions around the project needs. Navigation from a particular discussion to a requirement and design artifact or the other way around would enable flexible and easy dissemination of ongoing discussions.

Specialization

Specialization is the process through which students play roles and adapt to specific tasks during the project. It is often difficult to include students in a group that can specialize in all tasks. Carrying out a task with a reasonable level of performance implies possession of specific structural and behavioral characteristics, which do not allow other tasks to be carried out efficiently. Also, there are many roads to learning. That is why group project designs often require establishing role definitions for students at the very beginning. Students bring different talents and styles of learning. Students rich in hands-on experience and problem solving activities such as programming and design may not do so well with problem analysis activities. Students need the opportunity to show their talents and learn in ways that work for them.

Collaboration by Sharing Tasks and Resources

Collaboration consists of several students working together on a project, a common task. We consider collaboration mechanisms as being all those facilities that enable students playing specific roles to distribute tasks, information, and resources to solve the problems pertaining to the project under consideration. In an educational groupware system, collaboration requires workspace aware-

Table 3. Elements of project workspace awareness

Category	Element	Issues of Interest
Who	Presence	Is anyone in the workspace?
	Identity	Who is participating?
	Task assignment	Who is doing what?
What	Action	What are they doing?
	Goal	What intention is the action part of?
	Artifact	What object are they working on?
Where	Location	Where are they working?
	Extent	What artifacts can they access?

Table 4. Elements of social network

	Students	Knowledge	Resource	Task	Team
	Social interaction network	Knowledge acquisition network	Capacity network	Task allocation network	Assignment network
Students	who knows who	who knows what	who has what	who is assigned to what	who is assigned to what team
Knowledge		what knowledge is needed to derive X	what knowledge is needed to use Y	what knowledge is needed for Z	what knowledge is located where
Resource			what resources can be used with resource Y	what resources are needed for Z	what resources are located where
Task				what task precedes task Z	what tasks are performed where
Team					which teams work with which teams

ness. The elements of workspace awareness are shown in Table 3.

The awareness of presence and identity is simply the knowledge that there are others involved in the project, and who they are. Task assignment pertains to mapping tasks to individuals in a group.

The second category in Table 3 depicts the issues pertaining to awareness about what actions are being performed by whom and for what objective. Finally, the third category deals with the location information as well as the knowledge about the extent of access of individuals in the project. On the other hand, social organization perspective imposed on the interactions among teams and engineers depends on the management style, culture, norms, values, and configuration of the social networks presented in Table 1.

Coordination of Actions

Managing the activities of a number of students requires carrying out supplementary tasks, which are not necessarily productive. However, these tasks aim to ensure that the productive actions can be accomplished in a consistent and coherent manner to fulfill the requirements imposed by the overall process. The action coordination phase involves the definition of the order of actions to be carried out.

Figure 1. Required cooperation activities of the groupware system

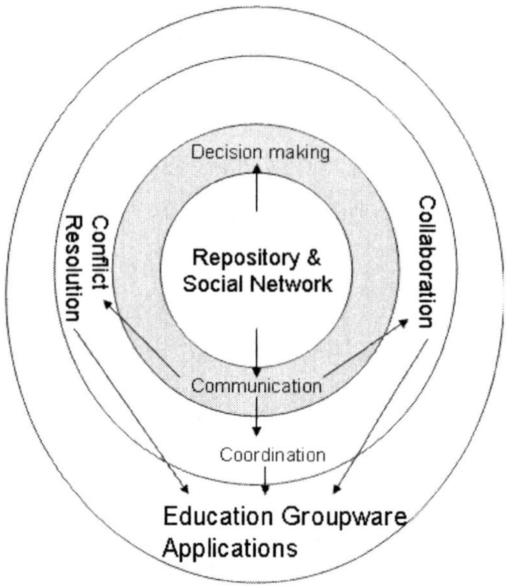

Conflict Resolution

Collaboration among students in producing the desired and necessary artifacts of the group project may result in conflicts due to limitations of existing resources as well as goal incompatibilities. Arbitration and negotiation are two of the means used in resolving conflicts, to stop disagreements between students from turning into open struggles, and to prevent the performance of the team as a whole from deteriorating. Arbitration is based on the definition of rules of behavior, which act as constraints on the group of students in engaging and furthering the progress of the project. Their effect is to limit conflicts by avoiding situations that are conducive to goal and action incompatibilities. The negotiation strategy, on the other hand, is based on the premise that agents can resolve their conflicts by seeking a bilateral agreement through a negotiation process.

The functions and methods of cooperation are interconnected. Figure 1 illustrates the elements of the cooperation activities. At the center of the system, the social network and the related data that capture resource, student, team, task, and knowledge pertaining to artifacts are stored. Communication and decision making subsystems operate over the core knowledge-base and social network to facilitate collaboration, coordination, and conflict resolution functions defined at the outer layer. Increasing the efficiency and effectiveness of cooperation and workspace awareness involves improving the performance levels of students. This requires the application of methods such as grouping and differentiation of roles. However, these methods themselves raise issues such as distribution of tasks, increased level of conflicts due to access to shared resources, and lack of coordination. The methods used in communication, task allocation, coordination of actions, and conflict resolution aim to address these issues and to improve awareness and hence performance levels of the group. However, these techniques need to be structured within a groupware system that takes the way in which students and heir roles are positioned in relation to each other and how they can effectively work together.

SUPPORTING AWARENESS IN DISTRIBUTED EDUCATIONAL GROUPWARE: TWO CASE STUDIES

In IT education, despite the criticality of fully understanding a problem prior to solution generation phase, research has shown (e.g., Moreland & Levine, 1992) that teams show little inclination to engage in the problem formulation aspect of the problem solving task. Using the basic methods of cooperation discussed above, in this section, we will elaborate on how an ideal collaborative education tool can be used to construct a shared mental model among students in a team to improve their effectiveness.

Figure 2. Interaction of students playing roles to coordinate their actions

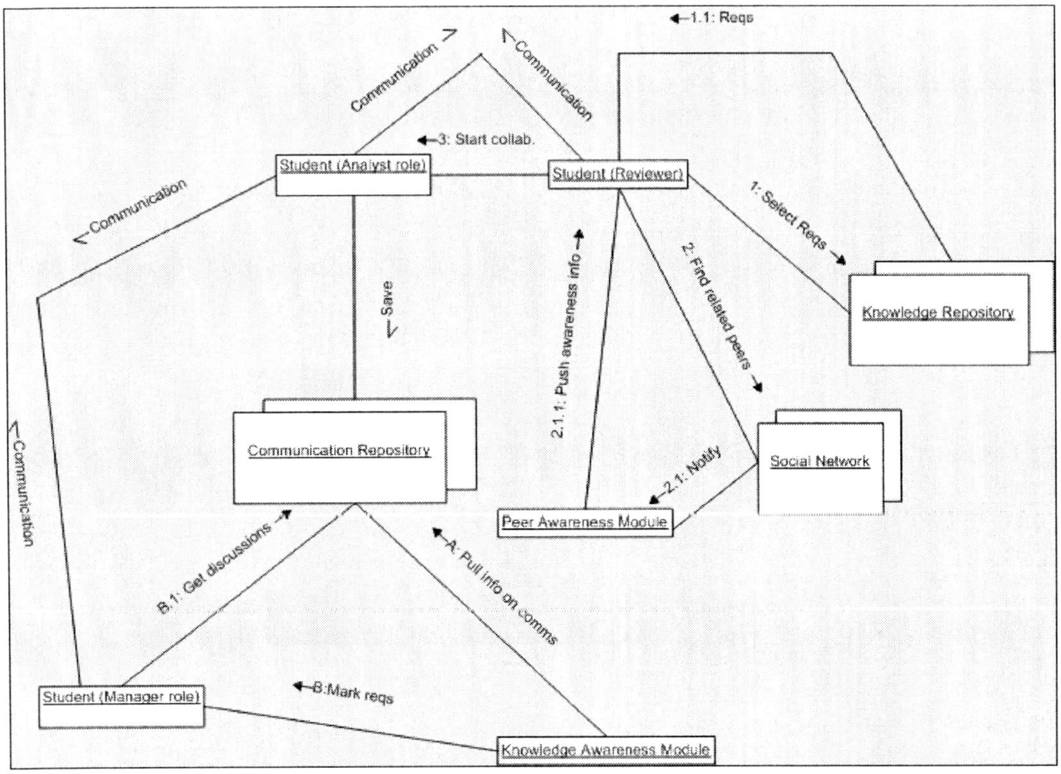

Managing Communication and Discussion

When students collaborate in a project workspace, they shift seamlessly back and forth between individual and shared work (Dourish & Belotti, 1992; Gaver, 1991). The degree to which people are working together is called *coupling*. The reasons why the members of a team move from loose to tight coupling is they need to discuss or decide on the contents of a project deliverable, to plan the next activity in the process, or that their current task needs engagement of another student.

Consider for instance a scenario where a student who acts as a reviewer needs to identify the right stakeholders (team members) to discuss the project requirements. In this scenario, the student looks up the online status of the student's peers in the group using the workspace awareness service and initiates communication depending on the availability of the peer student. To denote the significance of context explicitly and to facilitate ease of navigation, the student who initiates the communication embeds links to the requirements in the student's messages. The communication history is logged to the repository and marked to facilitate awareness so that other team members can track ongoing discussions.

Figure 2 is a collaboration diagram that depicts the above scenario in terms of the interaction among students playing roles to coordinate their actions. The scenario presented in Figure 2 implies a number of tool considerations, some of which are presented in the previous section. The interaction starts when a student with the reviewer role consults with the knowledge repository to identify the requirements or analysis artifacts that are being developed during the problem formulation phase. The student consults with the shared knowledge repository to discover and locate requirements of interest. The student then uses the social network component, the services of which are listed in Table 3. The network notifies the peer awareness module so that it can push the necessary information to the reviewer so that reviewer can contact peers. Following the identification of peers, the student contacts the student, who is discovered to be responsible with the requirement of interest to start collaboration. Further asynchronous communications between these two students ensue after the start of the collaboration, during which the connection is established. To facilitate engagement and participation of others in the ongoing discussion, the student with the analyst role saves the communication logs to the communication repository.

In case the manager of the project desires to view ongoing discussions, the manager needs to access communication information. To facilitate this, the knowledge awareness module frequently pulls information on communications and marks the requirements over which the discussion is being conducted. The student, who plays the manager role, contacts the communication repository to retrieve communication logs of interest based on the information the student receives from the knowledge awareness module.

Managing Changes

Student projects and the associated learning processes are often incremental and iterative. Students add new detail and refine and revise their models as they gather new information and improve their insight about the problem. As students cooperate over the artifacts that reflect

Figure 3. Interaction of students managing change as learning takes place

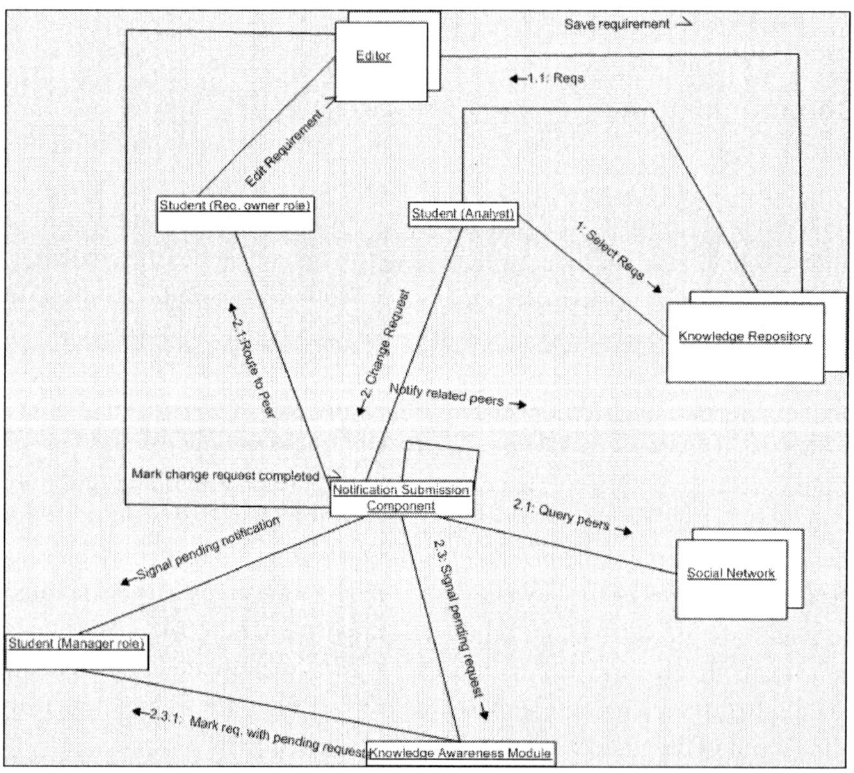

their understanding of the problem and application domain, they often detect gaps and ambiguities. Unlike in colocated face-to-face student a project, relying on informal communication to propagate the changes in a distributed setting is extremely difficult, if not unrealistic. Group project managers need to make sure that the right information pertaining to a change reaches the right people in the group and that proper actions are taken before the delivery deadline imposed by the assignment. This suggests that change management needs to be a critical component of project workspace awareness subsystems of the educational groupware systems. Facilities for editing changes are necessary but not sufficient to attain comprehensive workspace awareness. An awareness system should proactively notify students, who are involved in or affected by the change. Figure 3 presents the interaction scenario depicting such a case.

In this scenario, a change request submitted by a student playing the role of an analyst is routed using the notification submission component to a student, who is responsible for editing the analysis model associated with the change request. The interaction scenario records the context of change by linking the edit to the change request and the request as completed. Upon receiving the message indicating the completion of the change request, the notification component notifies the students whose tasks are related to the completed change. These students receive change notifica tion requests.

Another challenge in distance learning within the context of the above discussion is that knowledge is often fragmented, making reuse a difficult task. Especially when one or more students drop the course and hence the group project, the team members may not have easy access to knowledge produced by the previous team members. A collaboration mechanism embedded to an education groupware system provides a project workspace awareness framework to navigate the data efficiently and locate the relevant information.

FUTURE RESEARCH DIRECTIONS

The requirements presented in this chapter provide a basis for future research directions to improve the state of the art in educational groupware. Specifically, the need for group project awareness calls for developing and integrating frameworks that improve students' understanding of the project status. The social networking concepts are promoted in this chapter to promote team cohesiveness and effective cooperation. It is argued that social interaction, knowledge acquisition, capacity, assignment, and task allocation networks need to be at the core of educational groupware systems. The cooperation mechanisms such as collaboration, coordination, and conflict resolution are expected to operate over this core knowledgebase. As depicted in the case studies section, the notion of change management is expected to be critical. Shared workspace design for the creation, manipulation, and navigation through artifacts and execution of activities that pertain to learning tasks will provide significant research challenges. Human-computer interaction research in conjunction with social network design concepts will enable the development of next generation educational groupware systems. Group project awareness will require situation awareness facilities that rely on perception and understanding capabilities that enhance a student's comprehension and expectation of the status of the global view of the project state space.

CONCLUSION

To address the challenges of distance learning in IT education, engineers who build next generation educational groupware system solutions must make cooperation and project workspace awareness a centerpiece of the tool architecture. This chapter outlined a set of requirements based on a critical analysis of the elements of cooperation and team awareness. The means for realizing these elements are also discussed to present strategies to develop the proposed elements. Two scenarios are examined to demonstrate the utility of collaboration to provide deep integration of communication and task accomplishment within a unified coherent framework.

Specifically, we argued for tool features that facilitate:

- Informal collaboration services to manage ad hoc interaction during problem formulation and analysis as well as problem solving;
- Structured cooperation services to manage change during the learning process by reducing student interaction;
- Incorporation of workspace awareness functionality to provide visual cues and notification about pending requests; and
- Knowledge management techniques to make sense of the content in the case when team cohesion is not well-formed or sustained throughout the duration of the group project.

It is expected that solutions that promote the above features would be the adaptation of the concept of collaborative development environments. However, most such tools do not provide the rest of the requirements such as conflict resolution, coordination of actions, and support for social networks. By including such features, we believe next generation education groupware support systems will create a seamless environment that eliminates or automates most of the mundane tasks and provide mechanisms that encourage creativity through high-bandwidth communication among students.

REFERENCES

Adams, M., Tenney, Y., & Pew, R. (1995). Situation awareness and the cognitive management of complex systems. *Human Factors, 37*, 85-104.

Borning, A., & Travers, M. (1991). Two approaches to causal interaction over computer and video networks. In *Proceedings of the Conference on Human Factors in Computing Systems* (pp. 13-19). New York: ACM.

Carmel, E. (1999). *Global software development*. Prentice Hall.

Clark, H. (1996). *Using language*. Cambridge, England: Cambridge University Press.

Dourish, P., & Belotti, V. (1992). Awareness and coordination in shared workspaces. In *Proceedings of the Conference on Computer-Supported Cooperative Work* (pp. 107-114). New York: ACM.

Ellis, C., Gibbs, S., & Rein, G. (1991). Groupware: Some issues and experiences. *Communications of the ACM, 34*(1), 38-58.

Endsley, M. (1995). Toward a theory of situation awareness in dynamic systems. *Human Factors, 37*(1), 32-64.

Espinosa, A. J., Lerch, J. F., & Kraut, E. R. (2004). Explicit versus implicit coordination mechanisms and task dependencies: One size does not fit all. In E. Salas & S. M. Fiore (Eds.), *Team cognition: Understanding the factors that drive performance and process* (pp. 107-129). Washington, D.C.: American Psychological Association.

Ferber, M. (1994). *Multi-agent systems: An introduction to distributed artificial intelligence*. Addison-Wesley.

Gaver, W. (1991). Sound support for collaboration. In L. Bannon, M. Robinson, & K. Schmidt (Eds.), *Proceedings of the Second European Conference on Computer-Supported Cooperative Work* (pp. 293-308). Amsterdam: Klewer.

Gutwin, C., & Greenberg, S. (2004). The importance of awareness for team cognition in distributed collaboration. In E. Salas & S. M. Fiore (Eds.), *Team cognition: Understanding the factors that drive performance and process* (pp. 177-201). Washington, D.C.: American Psychological Association.

McNeese, M., Salas, E., & Endsley, M. (2001). *New trends in cooperative activities: Understanding system dynamic in complex environments.* San Diego: Human factors and Ergonomics Society Press.

Moreland, R. L., & Levine, M. J. (1992). Problem identification in groups. *Group processes and productivity* (pp. 17-47).

Norman, D. (1993). *Things that make us smart.* Reading, MA: Addison-Wesley.

Rafii, F. (1995). How important is physical co-location to product development success. *Business Horizons, 38*(1) 78-84.

Salas, E., & Fiore, M. S. (2004). Why team cognition: An overview. *Team cognition: Understanding the factors that drive process and performance* (pp. 3-8). Washington, D.C.: American Psychological Association.

Trevino, L. K., Daft, H. R., & Lengel, L. R. (1987). Media symbolism, media richness, and media choice in organizations. *Communication Research, 14*(5), 87-96.

Yilmaz, L., & Phillips, J. (2006). Organization-theoretic perspective for simulation modeling of agile software processes. In *Proceedings of the SPW/ProSim2006 Workshop,* Shanghai, China, (LNCS 3966, pp. 234-241). Springer-Verlag.

ADDITIONAL READING

Fisher, G. (1995). Distributed cognition, learning webs, and domain-oriented design environment. In *Proceedings of the Conference on Computer-supported Collaborative Learning* (pp. 125-129).

Goodsell, A., Maher, M., Tinto, V., Smith, B. L., & MacGregor, J. (Eds.). (1990). *Collaborative learning: A sourcebook for higher education.* University Park, PA: National Center on Postsecondary Teaching, Learning, and Secondary Assessment.

McLellan, H. (1993). Evaluation in a situated learning environment. *Educational Technology, 33*(3), 39-45.

National Research Council. (2002). *Enhancing undergraduate education with information technology: A workshop summary.* Center for Education, Division of Behavioral and Social Sciences and Education. Washington, D.C.: National Academy Press.

Chapter IX
Chatting to Learn: A Case Study on Student Experiences of Online Moderated Synchronous Discussions in Virtual Tutorials

Lim Hwee Ling
The Petroleum Institute, UAE

Fay Sudweeks
Murdoch University, Australia

ABSTRACT

As most research on educational computer-mediated communication (CMC) interaction has focused on the asynchronous mode, less is known about the impact of the synchronous CMC mode on online learning processes. This chapter presents a qualitative case study of a distant course exemplifying the innovative instructional application of online synchronous (chat) interaction in virtual tutorials. While chat interaction has primarily been researched for its effectiveness in supporting social-emotional aspects of learning, this chapter reports survey findings on its impact on facilitating participation in collaborative group learning processes and enhancing understanding of course content from a sociocultural constructivist perspective. The results reveal factors that affected both student perception and use of participation opportunities in chat tutorials, and understanding of course content. The findings present implications for the pedagogical design of online synchronous collaborative-constructivist learning activities that enhance understanding of course content through dialogic participation in the learning process.

INTRODUCTION

In distance education, online interaction between learning parties is largely facilitated by computer-mediated communication (CMC) technologies. Most research on educational CMC interaction has focused on the asynchronous mode which is widely held to offer learners greater convenience as well as extended time for participation and reflection. However, less is known about the impact of the synchronous CMC mode on the online learning process which stem largely from the underutilization of the real-time mode in the design of most distance courses. This chapter presents a qualitative case study of an online undergraduate course that exemplifies the innovative instructional application of online synchronous (chat) interaction in virtual tutorials. While chat interaction has primarily been researched for its effectiveness in supporting social-emotional aspects of learning, this chapter reports survey findings covering its impact on facilitating participation in collaborative group learning processes and enhancing understanding of course content from a sociocultural constructivist perspective. The implications of the findings are discussed and recommendations are made regarding the pedagogical design of online synchronous collaborative-constructivist learning activities. Finally, several possible areas for future research are suggested.

BACKGROUND

Interaction and the Online Learning Process

From a sociocultural constructivist perspective of learning (Vygotsky, 1962), dialogic interactions between members of a learning community are crucial for supporting meaning negotiation that leads to knowledge construction. In online educational contexts, as students and tutors share individual understandings of concepts, intellectual growth is supported by the availability of *scaffolding* or guidance from the learning parties with interaction mediated by language and various CMC technologies such as e-mail, discussion forums, and chat rooms.

Synchronous and asynchronous CMC technologies offer different capabilities for facilitating interaction in online learning environments (Ngwenya, Annand, & Wang, 2004). The *asynchronous* CMC mode supports delayed-time dialogue with interactions largely manifested as text-based contributions which could be composed, sent, and accessed without time and proximity constraints. However, the *synchronous* CMC mode requires communicating parties to be "present" at the same time for the dialogue to occur through services and applications such as voice over IP, desktop video conferencing, and Internet relay chat. Online synchronous (chat) interactions are mainly manifested as textual messages, composed and sent by parties who are simultaneously logged in chat rooms. Rather than having the facility to order messages in topical or temporal order, as in the case of asynchronous discussion threads, chat messages appear chronologically on screen with preceding exchanges scrolling up and then off each party's computer screen at a speed corresponding to the pace of the overall conversation (Werry, 1996), offering a potentially permanent record of the proceedings, which is generally not retrievable unless deliberately saved by the user.

Research on Quality of Online Educational Interaction

In higher education, the quality of online asynchronous interaction has been extensively examined from a constructivist approach for indications of sustained reflection associated with knowledge building (Garrison, Anderson, & Archer, 2001). The asynchronous mode is assumed to support extended reflection (Harasim, Hiltz, Teles, & Turoff, 1995) and provide the time needed for learners to move beyond information sharing to reach higher level *integration* and *resolution* phases of the critical thinking process where shared information is synthesized and new knowledge created (Garrison, Anderson, & Archer, 2000). A number of studies have analyzed the quality of online asynchronous discussions for the presence of cognitive and/or social-emotional dimensions considered necessary to develop student critical thinking and collaborative skills (e.g., Booth & Hulten, 2004; De Laat & Lally, 2004; Garrison,

2003; Garrison et al., 2001; Hara, Bonk, & Angeli, 2000; McLoughlin & Luca, 1999; Meyer, 2004).

In contrast, there is sparser research on the quality of online synchronous interaction in higher education. Researchers have observed that chat has only recently been used for instructional purposes (Murphy & Collins, 1997). This could be due to perceptions such as "promoting active asynchronous discussion is the best way to support interactivity in the online course" (Palloff & Pratt, 2003, pp. 24-25) and that chat is useful primarily for building social relations in distant learning groups (Lapadat, 2002). Additionally, the synchronicity and conversational characteristics (Kortti, 1999) of chat interaction have led to unfavourable comparisons with asynchronous CMC on aspects of time constraint for extended reflection on learning, the availability of participation opportunities due to competition for the "speaking" floor (Meyer, 2003), and additional skills (i.e., typing, language fluency) required of tutors and learners for managing or coping with chat interaction and its discourse (Dykes & Schwier, 2003; Warschauer, 1996).

However, other studies have contended that the sense of immediacy afforded by real-time interaction reduces *transactional distance* (Moore & Kearsley, 1996) between distant learners and enhances social-emotional aspects of collaborative learning and work group processes (Chou, 2002; Mercer, 2003; Schwier & Balbar, 2002; Sudweeks & Simoff, 2000). The capability of the synchronous mode to "contract" time could make it particularly appropriate for instructional activities that require interactivity, spontaneity, and fast decision making (Murphy & Collins, 1997). Additionally, the conversational characteristics of chat discourse reflect face-to-face classroom exchanges that are familiar to learners and faculty, hence facilitating the transfer of formal patterns of behaviour acquired in physical classrooms to virtual learning environments (Crook & Light, 2002). Furthermore, the largely text-based chat medium is assumed to filter out visual and social cues (Kiesler, Siegel, & McGuire, 1984), encourage greater self-disclosure that builds ties which bind online communities (Haythornthwaite, Kazmer, Robins, & Shoemaker, 2000), and enable learners to have (or perceive to have) equal opportunities for contributing to discussions.

Online Learning Experiences and Participation in Educational Interaction

Studies on student perceptions of distance learning experiences have generally yielded mixed findings. Current course management systems, supported by better synchronous and asynchronous technologies, are held to offer high quality interaction and enable a wide range of teaching approaches to enhance learning. The networked learning model for higher education proposed by Harasim et al. (1995) would move students from physical learning situations to globally connected learning communities, offer interactive instructional activities, support opportunities for communication between all parties in the learning process, and ultimately lead to "improvements in cognition and social interaction" (p. 273). On the contrary, Hara and Kling's (1999) study on student experiences with a Web-based course revealed frustrations over the nature of online asynchronous interactions (lack of timely feedback and visual cues), management of communication (unclear task instructions), and technical problems that could impede learning and have significant impact when students eventually give up on the formal content of the course.

However, a number of studies reported learner satisfaction with factors associated with CMC supported interaction such as convenience and availability of scaffolding or guidance from instructors/peers (McLoughlin & Luca, 1999; Thomas, Jones, Packham, & Miller, 2004). Other studies found evidence of pedagogical benefits in terms of CMC-facilitated collaborative knowledge construction and critical thinking development in online learning groups (Armitt, Slack, Green, & Beer, 2002; Newman, Johnson, Cochrane, & Webb, 1997).

Regarding online synchronous learning experiences, several studies suggested that student perceptions could be affected by the extent to which the chat learning activities are integrated into the course design, namely, framed within formal instructional objectives, schedules, and assessment (Cox, Carr, & Hall, 2004; Pilkington, Bennett, & Vaughan, 2000; Spencer & Hiltz, 2003). Given the sociocultural constructivist view that learning is constituted in the interaction, a particularly crucial aspect of student experiences of knowledge-building processes would be the availability of opportunities to participate in learning conversations.

The literature highlights several main factors, summarized below, that could affect student perceptions of participation opportunities in educational chat interaction:

- The *text-based chat CMC medium*, which displays rapid speed of discussion (Dykes & Schwier, 2003) and multiple concurrent discussion threads that could impact on interactional coherence and discussion focus in the absence of visual turn-taking cues (Herring, 1999; Pilkington & Walker, 2004);
- The *activity characteristics*, which include mandated participation in assessed instructional activities (Sudweeks & Simoff, 2000), tutor facilitation style (Cox et al., 2004; Kneser, Pilkington, & Treasure-Jones, 2001), and student moderation style (Chou, 2002); and
- The *participant characteristics*, which encompass English language proficiency (Warschauer, 1996), prior experience with the chat medium and its linguistic conventions (Murphy & Collins, 1997), and gender (Chou, 2002).

Essentially, studies on synchronous CMC interaction have largely focused on its effectiveness in enhancing social-emotional aspects of collaborative learning and work group processes while its role in supporting knowledge construction or greater understanding of course content through dialogic participation remains unclear. Such a situation highlights the need to further current understanding on the impact of chat interaction in facilitating online learning processes for a more pedagogically effective integration of the synchronous CMC technology into course designs, as well as to justify current and future provisions of such services. The next section describes a hybrid undergraduate course which exemplifies the innovative instructional application of chat interaction in collaborative group learning and formed the case context for this study.

THE CASE

The case is an undergraduate unit of study (*organizational informatics*) offered by the School of Information Technology at Murdoch University (Perth, Western Australia). This section describes the pedagogical framework of the unit, its virtual learning environment, the case participants, and conduct of the online tutorial instructional event.

About Organizational Informatics

The unit of study was originally a postgraduate course available from Sydney University in 1998. In 1999, it was modified and trialled as a third-year undergraduate unit at Murdoch University. Currently, the *organizational informatics* (OI) unit, which focuses on computer-mediated work processes, is available in the second semester (13 weeks) of each academic year to third-year Murdoch students.

The OI unit aims to develop skills associated with "organizational aspects of the design and development of information systems" (Sudweeks, 2004, p. 90), including skills in critical assessment and management of issues related to knowledge building organizations by facilitating knowledge construction through reflection. The unit adopts a hybrid/blended course delivery design that offers face-to-face lectures and online synchronous (chat) tutorials to *internal* and *external* students who, respectively, undergo the course on campus and via a distance learning mode.

The two main learning activities in the OI unit are a collaborative group project and chat tutorial discussions. Earlier studies by the coauthor examined the *collaborative group project* in terms of the following areas: student satisfaction with the collaborative work process (Sudweeks, 2003a)and patterns in group communication, group dynamics, and student perceptions of online learning in general and the group project task in particular (Sudweeks, 2003b). Of greater relevance to this chapter is the *chat tutorial activity* which was utilized as a case in several studies. For instance, Sudweeks (2004) examined changes in computer-mediated group processes over time, focusing on developmental and leadership characteristics of asynchronous and synchronous computer-mediated groups, of which the chat tutorials in the unit constituted the case for the synchronous computer-mediated group. Sudweeks and Simoff (2000) studied the chat tutorial activity for its effect on student motivation and participation, while Sudweeks and Simoff (2005) examined emergent leaders in collaborative virtual groups.

In 2005, the unit assessment components (Table 1) included a group project involving the collaborative planning and presentation of a proposal for a major event, and reflective journals that incorporated critiques on set-readings and reflections on tutorial discussions. As this chapter focuses on interaction situated in the chat tutorials, three areas of assessment, namely, *reflective journals*, *tutorial presentations*, and *discussion participation*, that complement and support the tutorial activity are described below.

Reflective journals are student critiques of set-readings that are expected to include "reactions to the articles for each topic, and how they relate to the lectures, other topics and other material" (Sudweeks, 2005, p. 4). The main pedagogical objective of this assessment/learning task is to enable students to experience "critically reviewing and recording ... thoughts about the readings for the unit, as well as from a variety of other sources" (Sudweeks, 2005, p. 4). Hence, in each journal (about 500 words in length), the student is expected to review the reading and pose at least one question related to the issue(s) in the reading for further discussion during the chat tutorial. Students are required to submit a journal each week to the tutorial group's private bulletin board prior to the tutorial session to enable group members to read each other's critiques and the scheduled student presenter to collate questions and/or issues to raise during the discussion.

In the 13-week semester, compulsory one-hour chat tutorials are held weekly (Weeks 2-13) with the final session in Week 13 reserved for online presentations of the group projects. The tutorials are conducted in a seminar style, moderated by one or two student presenters in WebCT chat rooms and facilitated by the tutor. *Tutorial presentations* by scheduled student presenters are assessed ac-

Table 1. Organizational informatics assessment components (Sudweeks, 2005)

Assessment Components	Component weight
1. Research essay (individual)	(15%)
2. Proposal for a major event (group)	(15%)
3. Reflective journals (individual)	(20%)
4. Tutorial presentation (individual)	(10%)
5. Discussion participation (individual)	(5%)
6. Examination (individual)	(35%)

cording to the following criteria: provision of "a clear [brief] summary, identification of key issues, knowledge of the topic, expressions of opinions on the topic(s), efforts to stimulate discussion, and management of the group discussion" (Sudweeks, 2005, p. 5).

To ensure active involvement during tutorials, *discussion participation* is assessed by the tutor and peers based on the level and quality of participation, participant effort, and sense of responsibility. Students are required to submit a peer assessment form to the tutor via e-mail at the end of the semester.

Essentially, the online synchronous interaction involving critical discussion during chat tutorials is framed by formal learning objectives, schedules, and assessment. Hence, the OI unit constitutes a single, particularly *information rich* case (Patton, 2002; Yin, 1994) from which one could potentially learn most (Stake, 1995) regarding the impact of chat interaction in facilitating online learning processes.

The Virtual Learning Environment

The main learning resources for the OI unit are a print *resource materials* reader (336 pages) and electronic resources (including electronic copies of all articles from the resource materials reader as well as links to relevant Web sites) available from the unit home page (Figure 1) which is hosted on WebCT. WebCT is a commercial learning management system adopted by Murdoch University as its university-wide virtual learning environment (VLE).

Online learning resources for the unit were initially organized into three categories: materials for *learning tasks*, *learning resources*, and *learning supports* (Sudweeks, 2003a). According to Sudweeks (2003b), due to the need to "encourage more social cooperative learning" (p. 175), a new collaborative online group project (which involves the development of a proposal for a major event) was introduced in 2002 which prompted modifications to the VLE design to reflect the additional learner support necessary for facilitating online communication and group work. The structure of the VLE was therefore extended to four categories: resources for *communication*, *resources*, *learner support*, and *assessment* (Figure 2).

Since then, the unit coordinator has further refined the range of learning resources available from the unit Web site. A possible interpretation of the VLE structure in 2005 is presented in Figure

Figure 1. 2005 Organizational Informatics home page

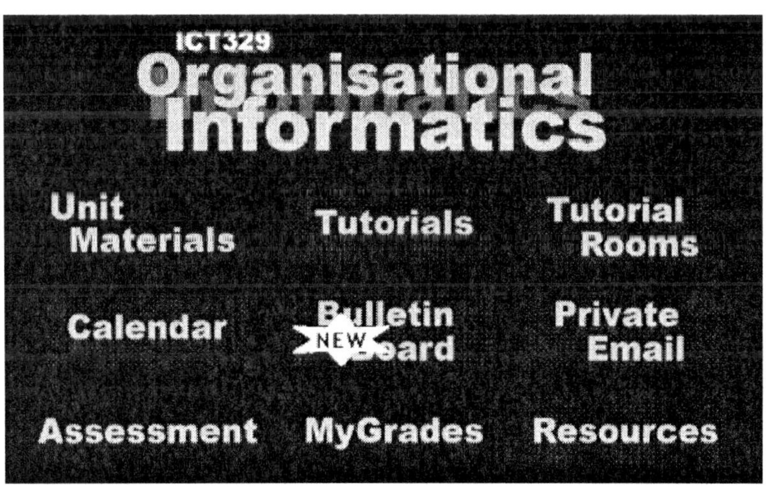

Figure 2. Extended organizational informatics VLE (Sudweeks, 2003b, p. 176)

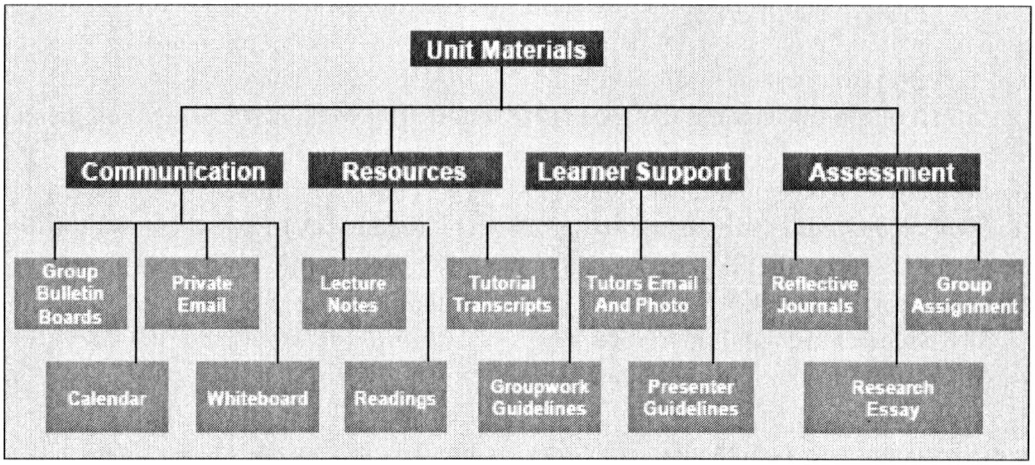

Figure 3. 2005 representation of organizational informatics VLE (adapted from Sudweeks, 2003b, p. 176)

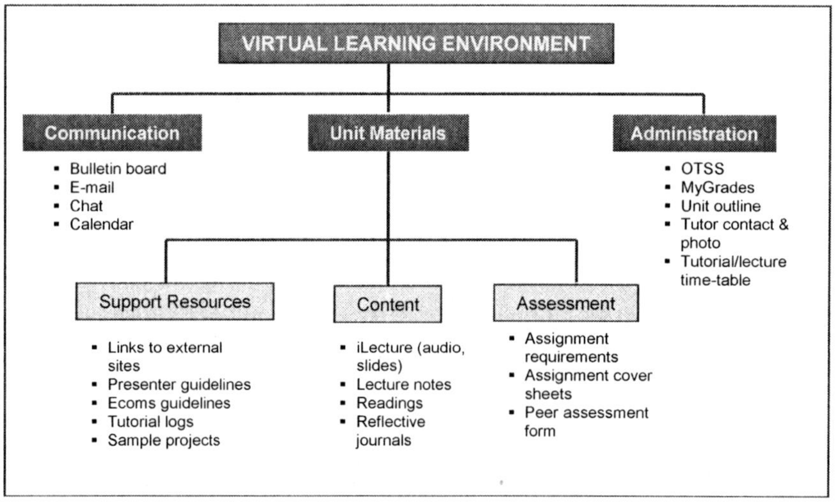

3. It should be noted that the VLE elements are not assigned to mutually exclusive categories and that in actual practice, some elements perform overlapping functions. For instance, the *calendar* could be a communication tool for conveying noteworthy events and an administration tool for organizing public and/or private diary entries. Similarly, *the tutor contact details/photo* could function as an administration element or a supporting resource element for establishing social presence of the online instructor.

From this perspective, the VLE for the OI unit is organized into three main components: *communication*, *unit materials*, and *administration*. The *communication* component includes synchronous and asynchronous communication tools such as WebCT chat (Figure 4), bulletin boards, private e-mail, and a common calendar.

Figure 4. WebCT chat facility

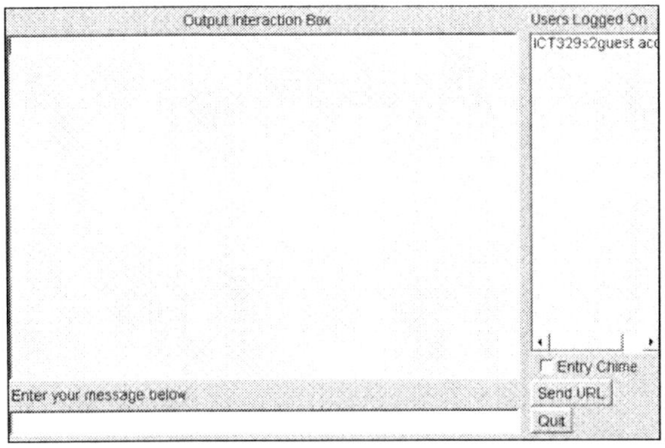

The *administrative* component supports course organization services such as self-enrolment in tutorial groups through the *online tutorial signup system* (OTSS), the distribution of grades, access to lecture/tutorial schedules, and other unit administrative documents.

The *unit materials* component is retained as "the hub of the site" (Sudweeks, 2003b, p. 174) and had expanded significantly since its representation in Figure 2. The component consists three subcategories of learning materials: *content materials*, *support resources*, and *assessment resources*. Content materials and support resources provide access to main and secondary instructional materials such as iLecture (streamed audio files), lecture notes, and links to external sites. The assessment subcategory provides access to assignment resources such as project requirements and peer assessment forms.

The OI Unit Pedagogical Framework

The pedagogical framework of the OI unit is based on the social constructivist view of learning (Vygotsky, 1962) as "a cycle of interpretation, evaluation and reflection of content evolving into individual and shared knowledge" (Sudweeks & Simoff, 2000, Section 3). In congruence with the unit's constructivist basis, instructional strategies emphasize "collaboration, personal autonomy, generativity, reflectivity, active engagement, personal relevance, and pluralism" (Sudweeks, 2004, p. 83). Hence, main learning activities, namely, the collaborative group project and chat tutorial discussions, are designed to facilitate students' construction of knowledge through participation and reflection.

Reflecting the networked learning model (Harasim et al., 1995) that also underlies the OI instructional design, there is significant use of the VLE as "a digital educational environment" (Sudweeks, 2004, p. 92) where students could access an extensive range of resources for their educational needs and the management of learning processes. The VLE also provides online spaces where communities of learners could gather in synchronous and asynchronous environments such as chat rooms and bulletin boards, hence reducing the transactional distance (Moore & Kearsley, 1996) usually perceived by students in distance courses. Moreover, there is extensive use of CMC to not only support interaction during chat tutorials and the group work processes for the collaborative team project, but also to facilitate unit administration or assessment, such as electronic submission of coursework to the tutor via e-mail or posting of journals to the bulletin board.

The Online Synchronous Tutorial

In 2005, there were four tutorial groups with 9 to 15 students in each group. All groups underwent equivalent learning activities and two of the four available tutorial groups (i.e., G1 and G4, in Table 2) were selected for a comparative study covering the impact of chat interaction on their collaborative learning processes.

The chat tutorials are designed to introduce students, in an active and experiential way, to the theory and practice of computer-mediated work processes which are directly relevant to the course topics (Table 3). The weekly one-hour tutorials are conducted in a seminar style, with a tutor-facilitator and one or two student presenters moderating the discussion in WebCT chat rooms.

Table 2. Characteristics of tutorial groups 1 and 4

Characteristics	Group 1	Group 4
Group tutor	Rachel^ (Part-time staff)	Fay^ (Full-time staff)
Group size	15 students, 1 tutor	9 students, 1 tutor, Lim^ (researcher)
Enrolment status	13 internal, 2 external students*	4 internal, 5 external students*
Nationality	Majority of international students, minority of Australian students	Majority of Australian students, minority of international students
English language proficiency	Majority of ESL/EFL speakers, minority of native English speakers	All native English speakers
Gender	3 female and 12 male students	1 female and 8 male students

^ *Other than the authors (Lim, Fay), all names are pseudonyms to protect the privacy of the participants.*
* *Internal and external students, respectively, undergo the course on-campus and through distance learning mode.*

Table 3. OI unit content topics (from 2005 Resources Materials reader)

Organizational Informatics Content Topics	
• Computer mediated communication • Organizational design and group processes • Organizational culture • Virtual organizations and communities • Work in the information age • Globalization	• Computer•mediated collaborative work • Organizational decision support systems • Systems theory • Managing information and information technology

The presenter role is rotated among all the students in each tutorial group.

In more detail, for tutorial sessions with *two presenters* (Figure 5), each presenter moderates a half-hour discussion slot based on the critique of one reading and adopts the participant role when not presenting. Before the tutorial, each presenter prepares brief critiques on at least two of the week's readings. One critique is posted on the group's bulletin board and the other is presented during the tutorial. In addition, each presenter prepares questions and collates questions from other students in the group (drawn from journals submitted in the group's bulletin board) for highlighting issues related to the reading and stimulating the discussion.

For tutorial sessions with *one presenter* (Figure 6), the sole presenter also prepares brief critiques on at least two of the week's readings before the tutorial and discusses both critiques during the tutorial. The sole presenter moderates the discussion for the entire session based on critiques of two readings.

During the tutorial, the presenter starts the discussion by highlighting main issues in the selected reading based on the presenter's critical evaluation of the article. The presenter is expected to moderate the discussion by "posing pertinent questions that bring out the main issues of the articles, stimulating discussions and encouraging participation by all members" (Sudweeks, 2003c, section 3). The tutor is present as a facilitator throughout the session and evaluates the presenter's performance as well as the extent of participation by other students in the discussion. The other students are expected to participate

Figure 5. Tutorial session with two presenters

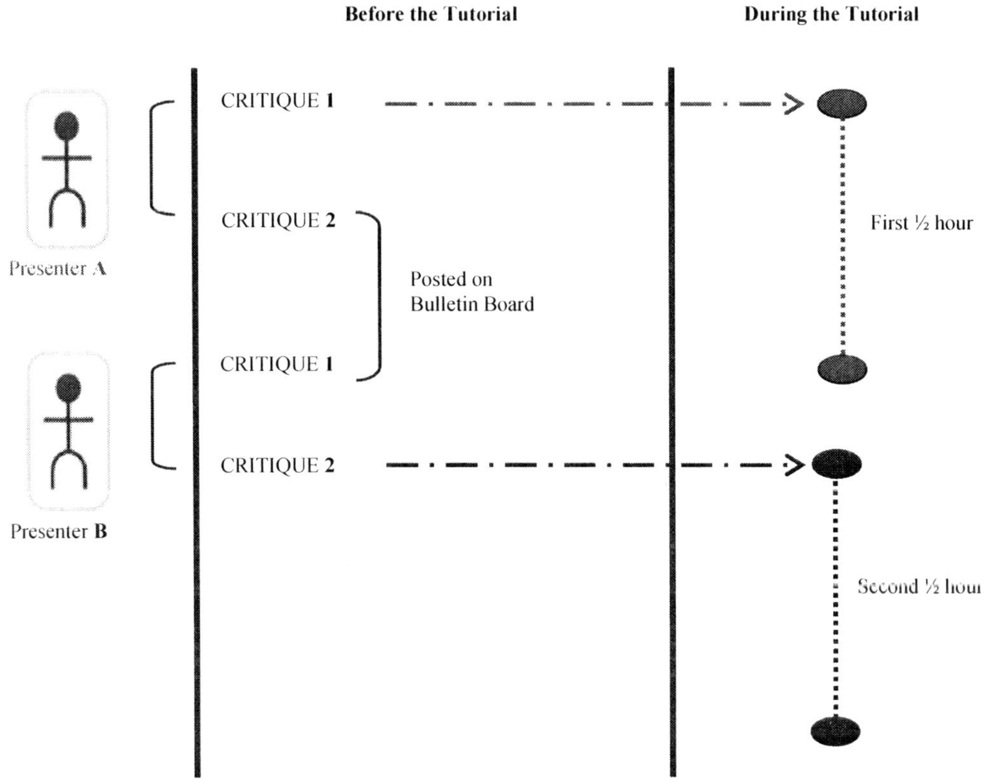

Figure 6. Tutorial session with one presenter

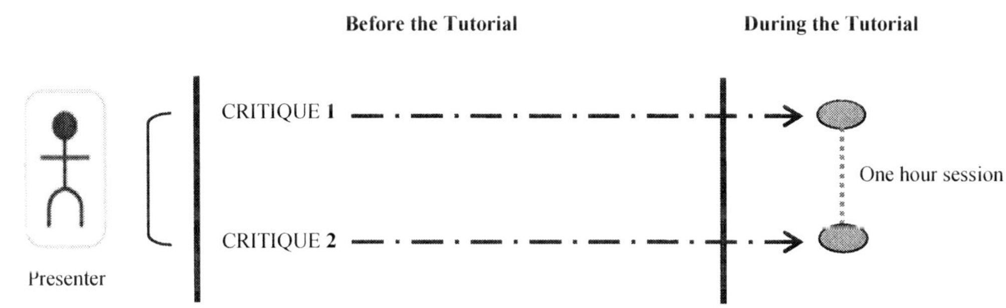

actively during discussions and evaluate the presenter as part of *peer assessment of participation* with the aid of archived discussion logs. In the peer assessment form, students are required to evaluate each other's level and quality of participation, effort, and sense of responsibility displayed in discussions (excluding academic and language abilities) on a seven-point rating scale from 0 to 5.

Preparation for tutorial activity is supported by online resources that include the following: *reflective journal* which states the requirements for the critique; *ecoms guidelines* which highlights CMC conventions and netiquette; and *guidelines for tutorial presenters* which states the responsibilities of the presenter and provides presenters with strategies for managing discussions and enhancing interaction, as well as technical instructions on procedures for communicating textual information via the synchronous CMC medium.

Essentially, the constructivist pedagogical framework of the OI unit is reflected in the tutorial activity which involves critical review of readings, dialogic exchange of multiple perspectives, and student reflection on learning with the aid of archived tutorial logs. Additionally, the tutorials also function as supportive virtual learning environments which reflect the *community of inquiry* (COI) model (Garrison et al., 2000) conceived as comprising three mutually interacting and reinforcing elements of cognitive, social, and teaching presences supported in online instructional environments by CMC technologies.

The presence and interactions between these three elements in the COI model are considered "crucial prerequisites for a successful higher education experience" (Garrison et al., 2000, p. 2). The *cognitive* presence reflects the intellectual climate of the learning environment with the instructional objectives justifying its existence to the participants. The perception of an open or unthreatening *social* climate facilitates the knowledge sharing process necessary to sustain cognitive presence while the *teaching* presence structures and mediates all the components (Anderson, Rourke, Garrison, & Archer, 2001; Garrison, 2003). As student presenters moderate by drawing less confident members into discussions, supporting views of others, and keeping discussions relevant under the guidance of the tutor-facilitator, they would be involved in establishing teaching presence in the online learning environment. Moreover, as student participants share individual knowledge and negotiate new understandings during dialogic interaction, they would essentially be engaged in providing social and cognitive support to each other.

RESULTS AND DISCUSSION

This section reports and discuses a subset of findings, drawn from a wider study (Lim, 2006), focusing specifically on the impact of chat interaction during the virtual tutorials on facilitating participation in the collaborative learning process and enhancing understanding of course content. At the end of the semester in November 2005, a Web survey was administered to 23 student respondents from both tutorial groups with return rates of 93% (G1) and 89% (G4). While the whole survey by Lim (2006) covered different aspects of the online learning experience, this chapter presents a subset of findings on student perceptions of (a) availability and exercise of participation opportunities and (b) factors that motivated/inhibited participation and affected understanding of course content during tutorial discussions. These aspects of the online collaborative learning process are assumed to be empirically observable through examining participant self-reflections on learning experiences in chat tutorials.

A self-administered, nonanonymous Web questionnaire, comprising closed and open-ended questions, was created with *Remark Web Survey* (Principia Products, 2005) software which also supports data retrieval and processing. Responses to closed questions were precoded by the survey software, hence minimal data processing was necessary before the application of descriptive statistical analysis. Data from open-ended questions were postcoded using categories that emerged from interpretive content analysis of the responses. The units of analysis for the survey data are the tutorial group and individual participants. Quotes from the survey responses are used here in tandem with extracts from transcripts of chat tutorial discussions to elaborate on some of the survey results, thus providing "rich" descriptions that add to the credibility of findings by qualitative research standards (Denzin & Lincoln, 2000).

Table 4. Groups 1 and 4: Presence and use of participation opportunities

		SA*	A*	D*	SD*	UJ*
I had plenty of opportunities to participate in the discussion	G1	3 (23.1%)	8 (61.5%)	2 (15.4%)	0 (0.0%)	0 (0.0%)
	G4	3 (37.5%)	5 (62.5%)	0 (0.0%)	0 (0.0%)	0 (0.0%)
I was able to make best use of the opportunities available for participation	G1	4 (30.8%)	7 (53.8%)	1 (7.7%)	0 (0.0%)	1 (7.7%)
	G4	0 (0.0%)	7 (87.5%)	1 (12.5%)	0 (0.0%)	0 (0.0%)

**SA = strongly agree; A = agree; D = disagree; SD = strongly disagree; UJ = unable to judge*

Perception of Participation Opportunities

Results in Table 4 show that participation opportunities in discussions were perceived to be *present* and *exercised* by most respondents, with greater agreement found in G4. Since there were contrary experiences reported in both groups, possible factors affecting participation were further explored. The results are presented and discussed below.

Factors that Motivated and Inhibited Participation

Respondents were asked five sets of questions covering a range of factors motivating and inhibiting participation. Sets 1 to 4 were closed questions that examined factors located from the literature: *roles*, *facilitation style*, *assessment*, and *turn-taking behaviour*. Set 5 comprised open-ended questions that captured other factors stated by respondents as affecting participation during discussions. Even though both groups underwent equivalent learning activities, given the different group profile (Table 2), it was not unexpected that certain factors were found to motivate participation within one group more than another.

Essentially, responses to the five sets of questions showed that participation in G4 was largely encouraged by the following factors:

- **The presenter role,** in which all aspects of online communication and management of discussion were regarded as effective;
- **The tutor facilitation style,** which supported the presenter in the management and stimulation of discussion;
- **Tutor assessment of participation,** which encouraged more activity; and
- **Turn-taking behaviour,** which indicated greater tendencies towards making early and additional contributions to discussions.

However, participation in G1 was mainly motivated by

- **The presenter facilitation style,** which stimulated participation and ensured relevance of discussion; and
- **Tutor and peer assessment** of participation.

In other words, while G1 participation was largely motivated by *peer*-related factors (facilitation, assessment), G4 participation was mainly encouraged by *tutor*-related factors (facilitation, assessment) with the greater ease reported in the presenter role attributable to the level of tutor support received by G4 respondents in the online communication and management of discussions. Lim (2006) found different extents of learning support to be provided by the two tutors. Overall, Rachel (G1) was minimally involved in guiding the learning process, whereas Fay (G4) displayed greater efforts to scaffold interactions by clarifying content issues, sharing information, and managing discussions. The more intense involvement by the G4 tutor could be due to Fay's additional role as unit coordinator with the accompanying implication that she had a higher stake in ensuring the success of the learning process.

Regarding *turn-taking behaviour*, while G4 respondents were less likely to refrain from making early and additional contributions to discussions, G1 reported a greater tendency to avoid making additional contributions when others had expressed similar ideas, preferring to let discussions develop before joining in. Although such turn-taking behaviours by G1 conform to the rules of "orderly talk" (Sacks, Schegloff, & Jefferson, 1974) that add to discourse coherence, the avoidance of opportunities to participate implies a reduced involvement in the learning process, which could undermine the unit's pedagogical assumption that active participation in the dialogic sharing of individual understandings supports knowledge building.

Common Factors Affecting Participation and Understanding of Content

Given the sociocultural constructivist view adopted in this study, that learning is constituted in the interaction, factors common to both groups that affect participation and understanding of course content during the chat tutorials are therefore of particular interest. A deeper awareness of their combinatory effect could serve to guide the pedagogical design of collaborative-constructivist group learning activities that considers the impact of the CMC mode on facilitating learning conversations from which participants could *appropriate* (Rogoff, 1990) the resulting shared understandings.

Respondents were asked the following set of open-ended questions in the survey:

Q.6: Were there other factors that encouraged or motivated you to contribute to tutorial discussions in this unit?

Q.7: Were there other factors that discouraged or inhibited you from contributing to tutorial discussions in this unit?

Q.11: What were the 1 or 2 specific things in the online tutorials that affected your understanding of the course topics?

The common factors that emerged from responses to these questions were the synchronous CMC medium, the presenter, and quality of online interaction. Findings on these factors, which positively and negatively affected (I) *participation opportunities* and (II) *understanding of course content*, are discussed below.

(I) Impact of Factors on Participation Opportunities

Regarding the impact on availability and use of participation opportunities, the *synchronous CMC medium* was found to encourage expression of views and provide a novel learning experience that generated greater collaborative efforts. However, it also presented difficulties for complete expression of thought attributed to the rapid speed and reduced nonverbal cues characteristic of the text-based chat medium.

The main factor i think that because it was not face-to-face i felt abit more at ease at putting forward my opinions. The tutorial being online really did help. Gave me more confidence. [Scott]

At times I found that I had a lot of things to say, but by the time I had thought of how to word my comments appropriately and typed them, the discussion had moved on. This is similar to what would happen in face-to-face communications, but seemed to either occur more often, or become more noticeable when it happened. [Jack]

The presenters' different abilities in facilitating, stimulating participation and ensuring relevance of discussion were found to both motivate and inhibit participation. While participation was encouraged when "tutorial presenters throw questions" [Diane], difficulties were experienced when "the presenter asks questions which are totally unrelevant to the topic" [Wendy].

Although the *quality of online interaction* was reported to motivate contribution to discussion when reflecting the presence and acceptance of different perspectives, participation was inhibited when there was dominance of discussion by certain parties that compounded the difficulties of turn-allocation and ensuring the visibility of own contributions in an online environment.

Well I guess what encouraged me... was that everyone in the tutorial group was open and accepting of other ideas and feelings. They were all willing to listen. [Robin]

Sometimes I feel that by contributing during a persons presentation of the tutorial, that it will either be overseen, or disrupt the flow of the presentation. [Colin]

(II) Impact of Factors on Understanding of Course Content

Concerning the impact of these three factors on understanding of course content, some respondents stated that the *synchronous CMC medium* had a positive impact on their understanding of course content by reducing inhibitions leading to greater willingness to discuss issues and exchange ideas.

Everyone could discuss issues without being shy. Hence a lot of ideas could be exchanged. [Diane]

Just recently there was a tutorial where many of the participants didn't understand the topic very well. after several explanations from both the presenter and the supervisor, I think everyone, including myself, understood the topic better. In a classroom, this may not have been as easy, as the presenter may not have been so forward in their 'teachings'. [Jack]

However, other respondents maintained that the chat medium led to superficial discussions and added to difficulties in comprehending messages attributed to the speed and reduced nonverbal cues characteristic of the text-based medium.

... lack of elaborate discussion and ability to express physical and facial communication. [James]

... harder to understand how someone expresses words in text ... [Ian]

While a respondent noted that the *presenter's moderation skill* ("[t]he way the topics were explained by the people presenting" [Eric]), enhanced understanding of course content and difficult concepts, another respondent stated that understanding of the topics was affected when "the presenter is focusing on a topic too specific within the readings" [Wendy] thus failing to develop discussion threads beyond the immediate issues in the set-readings.

The *quality of online interaction* was held to have enhanced learning when it enabled:

- sharing of real-life examples and work experiences;
- exchange of different perspectives or interpretations of the set-readings; and
- active engagement reflected by the presence of questions and responses that clarified meanings of concepts or issues.

Differing interpretations of the weekly readings, and also the work experiences and perspectives tutorial members brough to the discussion. [Pete]

People's opinions on the related readings. As we did critiques we gave our point of view on the readinds, then in the tutorials, you got to see what other people thought and at times it went against what the readings were about. [Scott]

That we as a group discussed the readings themes, points etc... I sometimes found I didn't understand some things... but was able to after the chat tutorial ... [Robin]

Essentially, survey responses on the quality of online interaction indicate student appreciation of the different perspectives shared as part of the learning process. There was also awareness of the significance of active engagement as the presence of questions and responses, which led to self-reflection or reconsideration of individual understandings during the construction of learning conversations. These student self-reflections on the impact of these factors on understanding of course content were corroborated by exchanges from the transcripts of chat tutorial discussions shown below. In Example 1, during a discussion on using soft system methodology to improve information organization in the workplace, Evan shared a case drawn from his work experience where the lack of a proper documentation system led to adverse financial results in a company.

In Example 2, the topic of national culture as defined by Hofstede's model initially generated debate on its applicability to Internet culture. The main discussion thread was then extended by the different interpretations exchanged by Jason, Derek, and Sam on adaptation strategies of business organizations and the societies in which they are located.

Example 3 illustrates active engagement by participants with the extended exchange of questions and responses that clarified meaning. In a discussion on group decision support systems (GDSS), questions were posed by Robin, Lim, and Pete for clarifications on the definition of a GDSS. The extended responses from the online

Example 1. Sharing of work experiences in abridged exchange

Evan>>	you would surprise the number of big projects I have had to fix up after people have just thought they would give it a go
Fay>>	can you give us an example evan?
Evan>>	Cant mention names but a large confectionary company recently upgraded their infrastructure with no project plan and
Evan>>	the result was have to restore the Windows Infrastructure and start from scratch, end up costing them about $20K
Evan>>	more than it should have
Robin>>	wow... just shows you how much having a project plan can be on a big project

Example 2. Different interpretations of readings in abridged exchange

Diane>>	Internet culture itself differs in different orgs
Wendy>>	actually i wud c Internet as having a very general culture :S
Jason>>	difference is a part of live..whether it be in culture or character so an organisation has to embrace that learn on working with it....
Alvin>>	yeah, i agree
Derek>>	But to flip that, societies that refuse to adapt their culture to that of the multinational organisations can often find themselves passed over by the organisations
Sam>>	ya but normaly the company will adapt to the culture of the country.....or else the have no business
Rachel>>	good point sam

Example 3. Active engagement with questions and responses that clarified meaning in abridged exchange

Pete>>	Question from Jack: Is there any Practical DGSS, either real or conceptual, which would actually do what it would be required to do: support the group decision making process? This ties back to Hwee's question - has anyone used a GDSS?
Evan>>	Not in a formal way
Fay>>	i've used a system that had a model similar to a nominal group technique
Robin>>	could you give an example of when you used it fay
Fay>>	we separated into co-located groups and each group brainstorm ideas on a feasibility study of the division of arts
Fay>>	and then we all looked at the ideas and evaluated them
Fay>>	the advantage being that everything was then recorded
Lim>>	so is GDSS a decision making methodology or is it a software system? I'm confused
Robin>>	yes so am i
Eric>>	From the example it looks like it can be both
Fay>>	both lim
Pete>>	So its a methodology which can have varying levels of software support?
Fay>>	here's what i said before - basically a gdss comprises groupware + dss capabilities + telecommunications
Lim>>	but that definition emphasizes the technical features
Fay>>	but it is also a decision methodology usually of brainstorming, analysis and evaluation
Pete>>	I think the Bannon article emphasises the CMC but not the DSS
Lim>>	ok, now its clearer
Robin>>	yes i can understand it easier now

Example 4. Absence of clarification on meanings in abridged exchange

Alan>>	So how do these differ from soft systems methodology?
Rachel>>	anyone?
Diane>>	in soft systems.....our PW affects our ideas....and our ideas affect our PW? 2 way?
Tony>>	what differs from what alan

tutor (Fay) and contributions from other students (Eric, Evan) helped to enhance understanding of the concept.

However, it is acknowledged that the sheer quantity of information shared could prove daunting for cognitive processing during the rapid chat discussions. One respondent said, "misinterpretation and understanding the interpretation differently from the topic" [Tony] could occur during discussions. Hence, the presence of diverse and/or contradictory messages may not necessarily further understanding when they are not clarified or followed up during the discussion (Example 4).

Overall, the results established that chat interaction facilitated participation in collaborative group learning process as most respondents reported the availability and use of opportunities to contribute to tutorial discussions. Possible factors affecting participation were further explored and roles, facilitation style, assessment, and turn-taking behaviour were expectedly found to motivate participation within one group more than another given the different group profiles. Of greater interest was the impact of factors that are common to both groups. The common factors of the synchronous CMC medium, moderation skill of presenters, and quality of online interaction were found to have both positively and negatively affected participation and understanding of course content.

CONCLUSION AND RECOMMENDATIONS

In conclusion, this chapter presented a qualitative case-based study examining real-time instruction in higher education. Specifically, this chapter introduced a distance undergraduate course which exemplifies the rare yet innovative instructional application of moderated online synchronous interaction in virtual tutorials. Findings were presented and discussed regarding student experiences of chat interaction in virtual tutorials, focusing on the impact of the real-time CMC medium on participation and understanding of course content. Given the sociocultural constructivist view on learning, that interaction supports meaning negotiation that builds new knowledge, the availability of opportunities to participate is therefore considered essential to the learning process. Findings of different perceptions of the availability and exercise of participation opportunities during chat tutorials prompted further analyses which identified factors that affected participation in both tutorial groups. In addition, student perceptions of the extent to which the chat tutorial experience enhanced their understanding of content were found to be mixed. Three main factors common to both groups—*the synchronous CMC medium*; *the presenter*; and *quality of online interaction*—were found to both positively and negatively affect participation in discussions and understanding of course content.

The constructivist assumptions of this study locate it at the paradigmatic level within the qualitative research framework. Hence, the research process reflects an interpretive approach involving the study of phenomena in their natural settings in order to illuminate and gain greater understanding of the online learning processes of a single informative case. Such knowledge gained from the interpretive analysis of participant self-reports corroborated by the chat transcript data are not claimed to be generalizable to wider populations. However, implications drawn from the findings regarding the pedagogical design of online synchronous collaborative learning activities may be extrapolated, in the form of recommendations, to similar contexts "in the sense of pointing out lessons learned and potential applications to future efforts" (Patton, 2002, p. 584).

From the research reported in this chapter, there are specific recommendations for the pedagogical design of online collaborative learning activities. Since the three common factors transcend differences in groups and do not exclusively exert a positive or negative impact, it is recommended that the combinatory effect of these factors be considered in designing effective online collaborative-constructivist group learning activities that encourage participation and minimize potential sources of frustration over the nature of chat interaction that may impede learning. More broadly, it is recommended that the design of learning environments should encompass physical and virtual instructional contexts, as in the case of the OI unit, to avoid reliance on any one mode which could needlessly limit the range of interactions permitted in distance educational programs. The hybrid course delivery design adopted by the OI unit enables educational interaction to be experienced via face-to-face lectures and online instructional contexts (chat tutorial room, bulletin board) facilitated by synchronous and asynchronous CMC technologies. The totality of the OI unit learning environment therefore supports participation in the sharing of individual understandings through a range of communication channels and contribution by learners at various levels of intensity.

These recommendations will be of interest to researchers concerned with the use of technology for online learning, higher education professionals responsible for the design and delivery of distance learning programmes, as well as promoters of educational technology who may benefit from a greater understanding of the role of synchronous CMC medium in supporting the learning process.

FUTURE RESEARCH DIRECTIONS

In its areas of inquiry, this study is essentially cross-disciplinary since it involves education, information and communication technology

(ICT), and educational technology, hence presenting several potential areas for future research in these fields.

The single-case study approach adopted by this study enabled an in-depth investigation of one particularly informative case (the OI unit) and a comparison of the impact of chat interaction on the online learning process of two tutorial groups (i.e., G1 and G4) within the case. Although unique cases are, by definition, not easily available, there is scope for further research. Future studies could adopt a methodological design that encompasses all the tutorial groups available in the OI unit. Alternatively, the OI unit could be investigated in comparison to other units offering similar, albeit not identical, CMC facilitated learning contexts and experiences.

Given the hybrid or blended course delivery design of the OI unit, one tutorial group could be examined in greater depth in terms of the relationship between learning processes that are supported by the entire range of face-to-face, online asynchronous and synchronous instructional environments afforded by the OI unit. Additionally, the students could be surveyed at different intervals of the course, rather than once at the end of the semester, to investigate finer changes in their perceptions of learning experiences over an extended period of time. Such research efforts could yield valuable insights on the appropriate incorporation of the various CMC technologies in supporting online educational processes. Moreover, the findings could provide timely feedback to online tutors regarding the effective management of instructional events.

Finally, this study has mainly presented findings from the analysis of survey data on student perceptions of online learning experiences. While self-reports of experiences offer one perspective on the phenomena, further insight could be gained from the analyst's interpretation of interactions from the transcripts of chat tutorial discussions. Further research effort in analyzing the synchronous computer-mediated discourse present in the archived discussion logs could enable triangulation of methods and data that provides a more holistic and richer account of the construction of learning conversations.

REFERENCES

Anderson, T., Rourke, L., Garrison, D., & Archer, W. (2001). Assessing teaching presence in a computer conferencing context. *Journal of Asynchronous Learning Networks, 5*(2), 1-17.

Armitt, G., Slack, F., Green, S., & Beer, M. (2002, January). *The development of deep learning during a synchronous collaborative on-line course.* Paper presented at the CSCL 2002, Boulder, Colorado.

Booth, S., & Hulten, M. (2004). Opening dimensions of variation: An empirical study of learning in a Web-based discussion. In P. Goodyear, S. Banks, V. Hodgson, & D. McConnell (Eds.), *Advances in research on networked learning* (Vol. 4, pp. 153-174). MA: Kluwer Academic Publishers.

Chou, C. (2002). *A comparative content analysis of student interaction in synchronous and asynchronous learning networks.* Paper presented at the 35th Annual Hawaii International Conference on System Sciences, Hawaii.

Cox, G., Carr, T., & Hall, M. (2004). Evaluating the use of synchronous communication in two blended courses. *Journal of Computer Assisted Learning, 20*, 183-193.

Crook, C., & Light, P. (2002). Virtual society and the cultural practice of study. In S. Woolgar (Ed.), *Virtual society? Technology, cyberbole, reality.* Oxford: Oxford University Press.

De Laat, M., & Lally, V. (2004). Complexity, theory and praxis: Researching collaborative learning and tutoring processes in networked learning community. In P. Goodyear, S. Banks, V. Hodgson, & D. McConnell (Eds.), *Advances in research on networked learning* (Vol. 4, pp. 11-42). MA: Kluwer Academic Publishers.

Denzin, N., & Lincoln, Y. (2000). Introduction: The discipline and practice of qualitative research. In N. Denzin & Y. Lincoln (Eds.), *Handbook of qualitative research* (2nd ed., pp. 1-28). London: Sage Publications.

Dykes, M., & Schwier, R. (2003, Spring). Content and community redux: Instructor and student Interpretations of online communication in a graduate seminar. *Canadian Journal of Learning and Technology, 29*(2).

Garrison, D. (2003). Cognitive presence for effective asynchronous online learning: The role of reflective inquiry, self-direction and metacognition. In J. Bourne & J. Moore (Eds.), *Elements of quality online education: Practice and direction* (Vol. 4). Needham, MA: The Sloan Consortium.

Garrison, D., Anderson, T., & Archer, W. (2000). Critical inquiry in a text-based environment: Computer conferencing in higher education. *Internet and Higher Education, 11*(2), 1-14.

Garrison, D., Anderson, T., & Archer, W. (2001). Critical thinking, cognitive presence, and computer conferencing in distance education. *American Journal of Distance Education, 15*(1), 7-23.

Hara, N., Bonk, C. J., & Angeli, C. (2000). Content analysis of online discussions in an applied educational psychology course. *Instructional Science, 28*, 115-152.

Hara, N., & Kling, R. (1999). Students' frustrations with a Web-based distance education course. *First Monday, 4*(12).

Harasim, L., Hiltz, S. R., Teles, L., & Turoff, M. (1995). Network learning: A paradigm for the twenty-first century. *Learning networks: A field guide to teaching and learning online* (pp. 271-278). Cambridge. MA: MIT Press.

Haythornthwaite, C., Kazmer, M., Robins, J., & Shoemaker, S. (2000). Community development among distance learners: Temporal and technological dimensions. *Journal of Computer Mediated Communication, 6*(1).

Herring, S. (1999). Interactional coherence in CMC. *Journal of Computer Mediated Communication, 4*(4).

Kiesler, S., Siegel, J., & McGuire, T. (1984). Social psychological aspects of computer-mediated communication. *American Psychologist, 39*(10), 1123-1134.

Kneser, C., Pilkington, R., & Treasure-Jones, T. (2001). The tutor's role: An investigation of the power of exchange structure analysis to identify different roles in CMC seminars. *International Journal of Artificial Intelligence in Education, 12*, 63-84.

Kortti, H. (1999). *On some similarities between discourse in the IRC and the conventions of spoken English.* Retrieved 9 November, 2004, from http://www.student.oulu.fi/~hkortti/proseminar-final.html

Lapadat, J. (2002). Written interaction: A key component in online learning. *Journal of Computer-Mediated Communication, 7*(4).

Lim, H. L. (2006). *Constructing learning conversations: A study of the discourse and learner experiences of online synchronous discussions.* Unpublished doctoral thesis, Murdoch University, Perth, Australia.

McLoughlin, C., & Luca, J. (1999). *Lonely outpourings or reasoned dialogue? An analysis of text-based conferencing as a tool to support learning.* Paper presented at the ASCILITE 99, Brisbane, Australia.

Mercer, D. (2003). *Using synchronous communication for online social constructivist learning.* Paper presented at the 2003 CADE-ACED Conference, St Johns, Newfoundland.

Meyer, K. (2003). Face-to-face versus threaded discussions: The role of time and higher-order thinking. *Journal of Asynchronous Learning Networks, 7*(3), 55-65.

Meyer, K. (2004). Evaluating online discussions: Four different frames of analysis. *Journal of Asynchronous Learning Networks, 8*(2), 101-114.

Moore, M., & Kearsley, G. (1996). *Distance education: A systems view.* CA: Wadsworth Publishing Company.

Murphy, K., & Collins, M. (1997). Communication conventions in instructional electronic chats. *First Monday, 2*(11).

Newman, D., Johnson, C., Webb, B., & Cochrane, C. (1997). Evaluating the quality of learning in computer supported co-operative learning. *Journal of the American Society for Information Science, 48*(6), 484-495.

Ngwenya, J., Annand, D., & Wang, E. (2004). Supporting asynchronous discussions among online learners. In T. Anderson & F. Elloumi (Eds.), *Theory and practice of online learning* (pp. 319-347). Canada: Athabasca University.

Palloff, R., & Pratt, K. (2003). *The virtual student: A profile and guide to working with online learners.* San Francisco: Jossey-Bass.

Patton, M. Q. (2002). *Qualitative research and evaluation methods* (3rd ed.). Thousand Oaks, CA: Sage.

Pilkington, R., Bennett, C., & Vaughan, S. (2000). An evaluation of computer mediated communication to support group discussion in continuing education. *Educational Technology and Society, 3*(3), 349-359.

Pilkington, R., & Walker, S. (2004). Facilitating debate in networked learning: Reflecting on online synchronous discussion in higher education. In P. Goodyear, S. Banks, V. Hodgson, & D. McConnell (Eds.), *Advances in research on networked learning* (Vol. 4, pp. 67-90). MA: Kluwer Academic Publishers.

Principia Products. (2005). *Remark Web Survey®* (Version 2) [Computer software]. Author.

Rogoff, B. (1990). *Apprenticeship in thinking.* New York: Oxford University Press.

Sacks, H., Schegloff, E., & Jefferson, G. (1974). A simplest systematics for the organization of turn-taking for conversation. *Language, 50*(4), 696-735.

Schwier, R., & Balbar, S. (2002, Spring). The interplay of content and community in synchronous and asynchronous communication: Virtual communication in a graduate seminar. *Canadian Journal of Learning and Technology, 28*(2).

Spencer, D., & Hiltz, S. (2003). *A field study of use of synchronous chat in online courses.* Paper presented at the 36th Annual Hawaii International Conference in System Sciences (HICSS 03), Big Island, Hawaii.

Stake, R. (1995). *The art of case study research.* Thousand Oaks, CA: Sage Publications.

Sudweeks, F. (2003a). *Promoting cooperation and collaboration in a Web-based learning environment.* Paper presented at the 2003 Informing Science and Information Technology Education Conference, Informing Science Institute, Santa Rosa, CA.

Sudweeks, F. (2003b). Connecting students with group work. In C. Constantinou & Z. Zacharia (Eds.), *Computer-based learning in science* (Vol. 1, pp. 173-183). Nicosia, Cyprus: University of Cyprus.

Sudweeks, F. (2003c). *The reflective learner: A framework for reflective e-learning.* Paper presented at the ICIER03, Seattle, WA.

Sudweeks, F. (2004). *Development and leadership in computer-mediated collaborative groups.* Unpublished doctoral thesis, Murdoch University, Perth, Australia.

Sudweeks, F. (2005). *Unit outline*. Murdoch University, School of Information Technology.

Sudweeks, F., & Simoff, S. (2000). *Participation and reflection in virtual workshops.* Paper presented at the 3rd Western Australian Workshop on Information Systems Research, Perth, Australia.

Sudweeks, F., & Simoff, S. (2005). *Leading conversations: Communication behaviour of emergent leaders in virtual teams.* Paper presented at the 38th Hawaii International Conference on System Sciences (HICSS05), Hawaii.

Thomas, B., Jones, P., Packham, G., & Miller, C. (2004, April 5-7). *Student perceptions of effective e-moderation: A qualitative investigation of E-college Wales.* Paper presented at the Networked Learning Conference 2004, Lancaster University, England.

Vygotsky, L. (1962). *Thought and language.* Cambridge, MA: MIT Press.

Warschauer, M. (1996). Comparing face-to-face and electronic discussion in the second language classroom. *CALICO Journal, 13*(2-3), 7-26.

Werry, C. (1996). Linguistic and interactional features of Internet relay chat. In S. Herring (Ed.), *Computer-mediated communication* (pp. 47-64). Philadelphia: John Benjamins Publishing Company.

Yin, R. (1994). *Case study research: Design and methods* (Vol. 5). Thousand Oaks, CA: Sage Publications.

ADDITIONAL READING

Anderson, T. (2003). Getting the mix right again: An updated and theoretical rationale for interaction. *International Review of Research in Open and Distance Learning, 4*(2).

Bonk, C. J., Daytner, K., Daytner, G., Dennen, V., & Malikowski, S. (2001). Using Web-based cases to enhance, extend, and transform pre-service teacher training: Two years in review. In C. D. Maddux & D. LaMont Johnson (Eds.), *The Web in higher education: Assessing the impact and fulfilling the potential* (pp. 189-211). New York: The Haworth Press, Inc.

Bonk, C., & Reynolds, T. (1997). Learner-centered Web instruction for higher-order thinking, teamwork and apprenticeship. In B. Khan (Ed.), *Web-based instruction* (pp. 167-178). Englewood Cliffs, NJ: Educational Technology Publications.

Carr, T., Cox, G., Eden, A., & Loopuyt, M. (2002). *An analysis of face to face and online learning conversations in three mixed mode courses.* Paper presented at the Multimedia Educational Group (MEG) Colloquium October 2002, Sport Science Institute of South Africa.

Chickering, A., & Gamson, A. (1987). Seven principles for good practice in undergraduate education. *AAHE Bulletin, 39*(7), 3-7.

Cobb, P. (1994). Where is the mind? Constructivist and sociocultural perspectives on mathematical development. *Educational Researcher, 23*(7), 13-20.

Cooney, D. (1998). Sharing aspects within ASPECTS: Real-time collaboration in the high school English classroom. In C. J. Bonk & K. S. King (Eds.), *Electronic collaborators: Learner-centered technologies for literacy, apprenticeship, and discourse* (pp. 263-287). NJ: Lawrence Erlbaum Associates.

Couper, M., Traugott, M., & Lamias, M. (2001, Summer). Web survey design and administration. *Public Opinion Quarterly, 65*(2), 230-253.

December, J. (1993, July 8). *Characteristics of oral culture in discourse on the Net.* Paper presented at the 12th Annual Penn State Conference on Rhetoric and Composition, University Park, Pennsylvania.

Duemer, L., Fontenot, D., Gumfory, K., & Kallus, M. (2002). The use of online synchronous discussion groups to enhance community formation and professional identity development. *The Journal of Interactive Online Learning, 1*(2).

Duffy, T., & Cunningham, D. J. (1996). Constructivism: Implications for the design and delivery of instruction. In D. Jonassen (Ed.), *Handbook of research for educational communications and technology* (pp. 170-198). New York: Simon & Schuster Macmillan.

Ertmer, P., & Newby, T. (1993). Behaviourism, cognitivism, constructivism. Comparing critical features. *Performance Improvement Quarterly, 6*(4), 50-70.

Fricker, R., Jr., & Rand, M. (2002). Advantages and disadvantages of Internet research surveys: Evidence from the literature. *Field Methods, 14*(4), 347-367.

Goh, S.-C., & Tobin, K. (1999). Student and teacher perspectives in computer-mediated learning environments in teacher education. *Learning Environments Research, 2*, 169-190.

Hancock, J., & Dunham, P. (2001). Language use in computer-mediated communication: The role of coordination devices. *Discourse Processes, 31*(1), 91-110.

Harasim, L., Calvert, T., & Groeneboer, C. (1997). Virtual-U: A Web-based system to support collaborative learning. In B. Khan (Ed.), *Web-based instruction* (pp. 149-158). Englewood Cliffs, NJ: Educational Technology Publications.

Herring, S. (2003). Computer-mediated discourse. In D. Schiffrin, D. Tannen, & H. Hamilton (Eds.), *The handbook of discourse analysis* (pp. 612-634). Oxford: Blackwell.

Johnson, D., & Johnson, R. (1996). Cooperation and the use of technology. In D. Jonassen (Ed.), *Handbook of research for educational communications and technology* (pp. 1017-1044). New York: Simon & Schuster Macmillan.

Jonassen, D., Davidson, M., Collins, M., Campbell, J., & Haag, B. (1995). Constructivism and computer-mediated communication in distance education. *The American Journal of Distance Education, 9*(2), 7-26.

Kanuka, H., & Anderson, T. (1998). Online social interchange, discord and knowledge construction. *Journal of Distance Education, 13*(1), 57-74.

Kanuka, H., & Garrison, D. (2004). Cognitive presence in online learning. *Journal of Computing in Higher Education, 15*(2), 1-18.

Kumar, A., Kumar, P., & Basu, S. C. (2002). Student perceptions of virtual education: An exploratory study. In M. Khosrow-Pour (Ed.), *Web-based instructional learning* (pp. 132-141). London: IRM Press.

McIsaac, M., & Gunawardena, C. (1996). Distance education. In D. Jonassen (Ed.), *Handbook of research for educational communications and technology* (pp. 403-437). New York: Simon & Schuster Macmillan.

McKlin, T., Harmon, S., Evans, W., & Jones, M. (2002). *Cognitive presence in Web-based learning, a content analysis of students' online discussion.* Paper presented at the ITFORUM 2002.

Pawan, P., Paulus, T., Yalcin, S., & Chang, C. (2003). Online learning: patterns of engagement and interaction among in-service teachers. *Language Learning and Technology, 7*(3), 119-140.

Chapter X
What Factors Promote Sustained Online Discussions and Collaborative Learning in a Web-Based Course?

Xinchun Wang
California State University–Fresno, USA

ABSTRACT

Although the pedagogical advantages of online interactive learning are well known, much needs to be done in instructional design of applicable collaborative learning tasks that motivate sustained student participation and interaction. This study investigates the factors that encourage student interaction and collaboration in both process and product oriented computer mediated communication (CMC) tasks in a Web-based course that adopts interactive learning tasks as its core learning activities. The analysis of a post course survey questionnaire collected from three online classes suggest that among others, the structure of the online discussion, group size and group cohesion, strictly enforced deadlines, direct link of interactive learning activities to the assessment, and the differences in process and product driven interactive learning tasks are some of the important factors that influence participation and contribute to sustained online interaction and collaboration.

INTRODUCTION

Theoretical Framework

The pedagogical advantages of student interaction in collaborative construction of knowledge are grounded in the social constructivist perspective of learning. From the social constructivist perspective, all learning is inherently social in nature. Vygotsky's theory of the Zone of Proximal Development posits that learners benefit most from social interactions concerning tasks they cannot do alone but can do in collaboration with more knowledgeable or more experienced peers (Kern, 1995). Knowledge is discovered and constructed through negotiation, or collective sense making. Pedagogically sound tasks in an online learning environment should, therefore, reflect social learning and collaborative construction of knowledge.

In designing and implementing online collaborative learning tasks, educators also draw heavily from Bakhtin's social theories to support their models of social interaction in collaborative construction of meaning in an online learning environment (Duin & Hansen, 1994; Wang & Teles, 1998; Wu, 2003). A speaker gives voice to a thought, an utterance, this utterance, though representing the ideas of an individual, reflects a social environment that is shared. The listener interprets the utterances in a purposeful, conscious act, in terms of his or her own concept of the social context, in terms of what the words mean to him or her individually. Therefore, speech and writing are dialogical in that the meaning of an utterance is created by both the speaker/writer and listener/reader through social interaction (Duin & Hansen, 1994). Pedagogically sound online learning tasks should therefore facilitate such online interactive learning for knowledge construction.

Interactive Learning and Online Collaboration

From a student's perspective, online interaction in learning takes place at two different levels: interaction with content and interaction with instructors and between peers (Gao & Lehman, 2003). There is evidence that pedagogically well-designed interactive learning tasks actually increase rather than decrease student access to instructors; increase interactions between instructors and among students; and increase students involvement of course content as well (Lavooy & Newlin, 2003; Mouza, Kaplan, & Espinet; 2000; Wu, 2003). Interactive learning tasks also promote greater equality of participation (Mouza, Kaplan & Espinet, 2000), more extensive opinion giving and exchanges (Summer & Hostetler, 2002), empower shy students to participate, and promote more student-centered learning (Kern, 1995; Wang & Teles, 1998)

At the level of interaction with content, students benefit more from producing explanations than receiving explanations. Such proactive learning engages students in a higher level of thinking than the reactive type of learning (Gao & Lehman, 2003; Wu, 2003). To promote such proactive learning, online course instructors need to integrate more active learning tasks that require more production than reception of explanations. Therefore, tasks that require written explanations should be considered over multiple choice type of reading comprehension in interpreting learning materials. Computer Meditated Communication in both synchronous and asynchronous discussion forums is inherently supportive of tasks for exchange of such written explanations. Furthermore, the systems can also archive written explanations posted in online forums and can be easily accessed and retrieved for references.

Although CMC supports interaction and collaborative learning, it also has inherent shortcomings. Disadvantages include the time it takes to exchange messages and the increased difficulties in expressing ideas clearly in a context reduced learning environment and the difficulty in coordinating and clarifying ideas (Sumner & Hostetler, 2002). The increased time it takes to reach consensus and decisions (Kuhl, 2002; Sumner & Hostetler, 2002) and to produce a final product (Macdonald, 2003). Given all these difficulties students need to overcome in order to collaborate effectively in interactive learning environment, online instructors need to address these obstacles with careful instructional design and provide support for collaborative learning with appropriate interactive learning tasks.

Factors That Influence Student Online Interaction and Collaboration

Research has also shown that computer mediated communicative tasks require more active role of students than traditional instruction in the face-to-face environment does (Wang & Teles, 1998). Students need to be willing to send a formal written question rather than have a casual conversation with peers or with the instructor in order to have their questions answered (Kuhl, 2002). To communicate effectively with peers and the instructor, students need to create the context through written messages, which requires the writing skills to identify their problems and express them precisely in order to have the questions answered. Team work and negotiation for meaning are necessary skills in CMC that cannot be assumed. Students need to learn to be familiar with the discourse of the discipline and academic genre for an online synchronous and asynchronous forum (Kuhl, 2002; Macdonald, 2003).

In addition to negotiation skills online, previous research has identified a number of other factors that influence student participation and interaction in a Web-based learning environment. Among others, the assessment of collaborative learning tasks plays a crucial role in ensuring student participation (Kear, 2004; Kear & Heap, 1999; Macdonald, 2003). In general, assessed collaborative learning tasks attract student participation at the cost of unassessed tasks. Furthermore, grade for discussion was also positively related to students' perceived learning (Jiang & Ting, 2000).

The structure of discussion in CMC is found to be another important factor in ensuring the amount of participation and level of interaction and collaboration among the peers. Such structure includes the size of the discussion groups, the nature and types of discussion topics (Williams & Pury, 2002), and whether the collaboration emphasizes the process of learning or the end product of such collaboration, or both (Kear, 2004; Kear & Heap, 1999; Macdonald, 2003). Online collaboration can be either process or product oriented. Forum discussions regarding course contents or related issues are commonly process oriented as the sharing of ideas help learners understand the issues without necessarily leading to a final product. Students are assessed individually based on their participation and quality of their contributions. Alternatively, online interaction and collaboration may lead to a final product such as an essay, a project, or a Web page, and so forth. There can be two assessment elements to such tasks, a common grade for the group for the overall quality of the collaborative product and individual grades for the contribution of each individual to the collaborative endeavor (Kear, 2004; Kear & Heap, 1999; Macdonald, 2003).

Finally, like any other form of learning, learning collaboratively in an online course is also characterized by individual differences. Collaboration as a process of participating to the knowledge communities is not an equal process to all the members of the community (Leinonen, Järvelä & Lipponen, 2003).

To summarize, online negotiation skills, the direct link between collaborative tasks and assessment, the structure of online discussions such as the nature and types of discussion topics, the size of the group, and the differences between process and product oriented collaborative tasks are some of the factors that influence student participation, interaction, and collaboration.

It is important to note that some of the above findings are based on experiments that are not a part of an online course (Gao & Lehman, 2003). Others have based their studies on courses that integrate some collaborative tasks in mainly student-instructor/tutor interaction type courses (Kear, 2004; Kear & Heap, 1999; Leinonen, Järvelä, & Lipponen, 2003; Macdonald, 2003; Williams & Pury, 2002). Web-based courses that employ collaborative learning tasks that form the essential course syllabus are less studied. While the advantages of student interaction and collaborative learning in Web-based learning environment has long been recognized, what remains to be identified are what instructional design of course tasks and activities that promote sustained and consistent student interaction and collaboration for knowledge construction.

Moreover, there is also evidence that online interactive learning and collaboration are not always sustainable and students' participation in CMC collaborative tasks may wane after the assessed tasks that require the postings are completed (Macdonald, 2003). In a recent survey on college student's attitudes toward participation in electronic discussions, Williams and Pury (2002, p. 1) found that "contrary to much literature on electronic collaboration suggesting students enjoy online collaboration, our students did not enjoy online discussion regardless of whether the discussion was optional or mandatory." Much needs to be done to explore factors that promote sustained student interest in online interactive learning and collaboration.

The Study

Through a post course survey, this study investigates the factors that promote sustained student participation in computer-mediated discussions as the core interactive learning tasks in a small group setting in an upper division undergraduate course that was offered entirely online. It also examines students' attitudes toward process and product oriented interactive and collaborative learning. The research questions are:

1. What factors encourage sustained participation, interaction, and collaboration in asynchronous discussion forums in a Web-based course?
2. What interactive learning tasks are sustainable and what are not?
3. Are there any differences in student attitudes toward process and product oriented online collaborative learning tasks? If so, what are the factors that influence students' different perspectives toward such tasks?
4. What pedagogical implications do the findings have?

COURSE INFORMATION AND DATA COLLECTION

Course Information

The course under study was an upper division general education course in Bilingualism and Bilingual Education delivered entirely on Blackboard in Spring and Fall 2004 at a state university in California. A total of 60 students, 22 in the Spring semester class, and 20 and 18 students in the two Fall semester classes completed the course. All were local students who took the course online because the same course offered face to face conflicted with their schedules. Some students lived over an hour of driving distance from campus (not uncommon in Central California) and chose to take the online course to avoid commute. According to student self-report, all had taken at least one Web-enhanced course and were familiar with the Blackboard interface, although most of these courses used Blackboard for downloading course materials and lecture notes rather than integrating interactive learning activities. About 20% of the students reported they had taken at least one Web-based course. It was not clear how many of them experienced interactive learning online.

Collaborative Tasks and Their Assessment

Forum discussions on course readings and related issues formed the core interactive learning activities that were 45% of the course grade. These were process oriented interactive learning tasks for which individual grades were assigned for each student based on their quantity and the quality of postings in the forums. Small groups of 4-6 people were formed at the beginning of the semester for the weekly asynchronous group forums. During the 16 week semester, a total of 18 discussion forums were completed in each online group. For each forum, the instructor assigned a reading chapter along with comprehension questions and discussion topics to help the students to grasp the contents. Students divided the reading questions among themselves in their groups and posted the answers to each question for the first round of postings. They were also required to make comments on at least one peer's answers in the second round of postings to carry on the discussions.

To ensure participation, strict deadlines for each round of postings were enforced and each student's answers to the questions and comment messages were assessed by the instructor who assigned up to 3% of the course grade for participation of each discussion forum. After each forum was completed, the moderator of each group (in each group, students rotated as moderators) was required to summarize the discussions and post the summary messages in a class forum that was accessible to all groups. These general class forums were intended to provide the students an opportunity to learn what was going on in other group forums that they did not have access to. This way, they did not need to read the numerous messages of 3-4 other groups but could still learn the gist of other group discussions. Although the summary messages were required, they were not graded. However, the summary of group discussions in a whole class discussion forum was eliminated in the Fall semester classes because it was not popular based on the input from the Spring semester class post course survey.

For the entire semester, the mean postings of each student in group forums ranged from 62-77 messages. On average, each student posted 3.5-4.3 in each of the 18 discussion forums. Although there was some variation in number of messages posted across groups and classes, most students did more than the minimum requirement of posting two rounds of messages in each discussion forum. Messages posted in the course related forums outside the group discussion forums were not included in the calculation because they were either inquiries or socialization in nature. Moderators' postings of summary messages in the class discussions forum were not included either because these postings were not enforced in the two Fall classes.

The other major collaborative task was a product oriented group project that constituted 12% of the course grade for which all the students in the same group received a common grade based on the level of collaboration and the quality of the final written report. There was no individual assessment component for the group project. The interdependent grading (a common grade for all members of a group only) was aimed at promoting more collaboration among the peers to produce a true collaborative product with individual contributions. The group project was closely related to one of the course themes on types of bilingual education programs. Each student was required to visit a local school to interview a bilingual teacher to gain firsthand information about bilingual education programs implemented in Central California. Students then shared and synthesized the interview data to produce a group report. They were not required to meet face-to-face for the group project but exchanged information in an online forum that was mostly procedural to plan, negotiate, to reach agreement and to produce the final product. The process of planning and producing the project required negotiation, cooperation, and collaboration among peers to actually arrive at consensus to produce a report. Though not graded, the progress of each group in the online forums was closely monitored by the instructor. The deadline for submitting the group project was strictly imposed to ensure the completion of the work.

Other course activities included two individual written assignments (8%) and three online exams (35%) that assessed the learning outcomes of the course readings and group discussions. Table 1 summarizes the course activities and grading.

Table 1. Course activities and grading

Activities	Grading	Description
Weekly group forums	45%	Structured discussions on course readings
Weekly class forums	0%	Required postings of moderator's summaries from each weekly group forum (Spring Semester class only)
Group project	12%	Final product graded interdependently (same grade for each member of the group)
Individual assignments	8%	No interaction among students required
Three exams	35%	Online exams on course contents to assess outcome of learning

Data Collection: Post Course Survey Data

At the end of the semester, an online survey was administered in each class to collect information about students' learning experience and their attitudes toward the course, in particular, their experience with online collaboration in both the weekly conference discussions and the group project. The survey questionnaire, which consisted of 17 multiple choice questions and 4 open-ended questions (see Appendix) was uploaded to the survey area of the course on Blackboard. Students were able to access and complete the survey questionnaire anonymously during the week after the final exam. Blackboard automatically calculated the results of the multiple choice questions in percentage. The transcripts of the survey responses for all three classes were printed out for analysis. 16 of the 22 Spring semester students and 37 of the 38 Fall semester students completed the survey questionnaire. Therefore, the analysis of the survey data was based on the 53 completed questionnaires.

RESULTS

Students' Attitudes towards Online Discussions and Collaborative Learning

Table 2 presents student responses to the question "what are your thoughts about the structure of the course?" Overall, 92.5 % of the students preferred the collaborative learning in the form of small group discussions to the weekly online quizzes (7.5%) if given the choices. Additionally, the first open-ended question asked the students to describe their experience with the forum discussions. Among the 47 students who answered this question, only 1 student expressed negative experience with the discussion forums. Three students commented that their experience was mixed. The majority, 43 students (91.5 %), expressed their experience with this form of learning ranged from positive to extremely positive.

What factors encouraged students to participate in this form of active and interactive learning throughout the semester? Did the students really

Table 2. Students' responses to "what are your thoughts about the structure of the course?" (N = 53)

Choices	% Reponses	Chi²
I like the way the course is structured in terms of forum discussions because we learn from each other.	92.5%	41.679*
I prefer weekly quizzes based on the readings rather than answering questions and joining the group discussions.	7.5%	

**Unless otherwise specified, the P values of the Chi² is <0.0001 in this study.*

Table 3. Students' views about group discussions (N = 53)

Survey Questions	% Responses				
	Strongly agree	Agree	Disagree	Strongly disagree	Chi²
My answers to the questions and comments on peers' messages help me to understand the readings better.	30%	62%	8%	0%	49.717
My peers' answers/comments helped me understand the readings better.	32%	57%	11%	0%	39.453
I learned more from online discussions than I would have learned from lectures.	25%	47%	25%	2%	21.792
The online discussion is helpful because we collaborate more and learn from each other more.	38%	55%	6%	2%	41.415
The group cohesion and mutual trust is an important factor in our group.	53%	36%	11%	0%	36.132

think they learned from building on each other's insights? What were the effects of such learning as reflected by students' responses in the survey data? The survey questionnaire addressed these issues in a number of questions. Table 3 summarizes students' responses to the effectiveness of group discussions.

Chi Square analyses of students' responses to the questions in Table 3 along the scale of strongly agree to strongly disagree were all significant beyond 0.0001 level. (Unless otherwise stated, Chi Square analyses reported in this study were significant beyond the level of 0.0001.) About 90% of the students agreed or strongly agreed that answering questions and participating in discussions helped them understand the readings better and that online discussion was helpful because they collaborated more and learned more from each other. Additionally, 72% of the students responded that they learned more from online discussions than they would have learned from the lectures. Furthermore, 89% of the students responded, saying group cohesion and mutual trust was an important factor in their group.

Factors that Affect Level of Participation and Sustained Interaction

Assessment

Table 4 summarizes students' responses to the level of participation in their group discussions if the postings were not required and graded. Overall, 51% of the students responded they would post some but not as many messages, 21% said they would post very few, and 8% responded they would not post any messages at all! Only 21% responded they would post the same number of messages. One might argue that the survey data may not reflect the real level of participation in discussions if the postings were not required or assessed because all the postings in this course were actually required and assessed. Therefore, a firm claim of the effect of assessment on forum contributions must be tested with a treatment group whose postings in forums were assessed and compared with a control group whose postings in forums were optional and unassessed. Neverthe-

Table 4. Students' responses to "would you post the same number of messages as you actually did over the semester if these postings were optional, not required or graded?"

Choices	% Responses	Chi²
Yes, I will post the same number of messages	21%	21.491
I will post some messages but not as many	51%	
I will post very few messages	21%	
I will not post any messages	8%	

less, students' responses to this survey question still reflect the "if not" situation because they had just completed the weekly postings for the entire semester and such learning experience would certainly affect their responses. Therefore, the "if not assessed" situation was contrasted against the real situation of "assessed" postings.

There was further evidence that tasks that were not directly linked to assessment did not attract as much attention and were difficult to sustain. For example, in the first offering of the course in the Spring semester, the moderators posted summary messages of each weekly group forum in a class forum by the deadline as required. However, these messages seldom attracted voluntary comments. Table 5 summarizes the Spring semester students' responses to a survey question on the whole class forums. The results showed that while 38% of the students acknowledged that it was an important way to learn the ideas of the other groups, which may indicate these students had read summary messages from other groups, 44% of the students reported they seldom read these messages. 19% of the students responded that the class forum should be eliminated.

As the whole class forum was not popular with the majority of the students in the first offering of the course, this task was eliminated in the Fall semester classes. To investigate whether students missed the level of input from other groups, the post course survey asked Fall semester students questions about their thoughts on the input from other groups. Table 6 summarizes the responses from the two Fall semester classes.

Seventy-eight percent of the students felt that participating in their own group discussion was sufficient to learn the course contents and it would have taken too much time to read and respond to the summary messages from other groups. 14% responded that every group should have summarized their forum discussions each week and posted it to a general forum so that interested students could comment on the discussions in other groups. Very few students, 8% in all, wanted other group members to read their postings or missed the discussions in other groups. Therefore, with or without the summary postings in the whole class forums, the survey data suggest that the majority of students showed the same lack of interest in participating whole class discussions that were not graded.

Table 5. Spring semester students' responses to the tasks of "group summaries in the main message board" (N = 16)

Choices	% Responses	Chi²
Is relevant and is an important way to learn the ideas of other groups	38%	10.12*
I seldom read these summaries	44%	
Can be eliminated because I have never read the group summaries	19%	

$*p < 0.0063$

Table 6. Fall semester students' attitudes toward other group discussions (N = 37)

Choices	% Responses	Chi²
I wanted other group members to read our group discussions and I also missed the discussions in other groups.	8%	90.392
Every group should have summarized their forum discussions each week and post it to a general forum so that interested students could comment on the discussions in other groups.	14%	
Participating my own group discussion is sufficient for me to understand the course contents. It would take too much time to read and respond to summary messages from other groups.	78%	

Deadlines

Table 7 presents students' responses to the importance of deadlines in the weekly postings of group forums.

Overall, 93% of the students agreed or strongly agreed that the imposed deadlines for postings had an important impact on their participation in collaborative learning and in getting the tasks done in a timely fashion.

Group Formation

Table 8 summarizes students' responses to the question on group formation.

62% of the students responded that they preferred to work with the same people for their group discussions and 30% expressed that they did not have any preferences. Only 8% responded that they wanted to work with different people because they felt that they would also learn from other students they never interacted with in this course.

Table 7. Students' attitudes toward deadlines in group discussions (N = 53)

Statement	% Responses				
	Strongly agree	Agree	Disagree	Strongly disagree	Chi²
The deadlines for the readings and postings in each forum are very important because they help to complete the readings and the course	51%	42%	7%	0%	39.755

Table 8. Fall Semester students' responses to "what is your view about group formation?" (N = 37)

Choices	% Responses	Chi²
I want to work with the same group members the way it is now because we know each other better.	62%	44.244
I want to work with different people in a group every few weeks because we will learn from other students we never meet.	8%	
It will not make a difference to me working with the same people or different people in a group.	30%	

Chi Square analysis yielded highly significant differences between the responses.

It appears that the group as a community for online learning established deep roots in this course. Recall that the class level discussion forums in the form of summaries from each group forum was eliminated. Except for some course related general forums in which questions regarding course activities were exchanged, students generally did not have access to the majority of the fellow students in their class. It would not have been surprising if students had expressed their desires to learn the discussions in other groups through some form of exchanges on a class level, or, through reshuffling groups. Yet, the survey responses suggest that at least two thirds of the students did not express the need to work outside their fixed groups.

It is important to note that the survey data reflected the student views towards their working groups that were fixed for the entire semester. If they actually had the chance to work in different groups in this online course, they might have different views. To explore the advantages and disadvantages of fixed or dynamic small groups in a Web-based course that uses weekly forum discussions, both group types need to be included in the data in future studies.

Process vs. Product Oriented Collaboration

Table 9 presents students' responses to a question that allowed for multiple choices about the group project.

Two of the choices provided in the answers were aimed at assessing whether the assignment itself was important for the course in the students' eyes because the importance of the group project may affect their overall performance. As seen in Table 9, 70% of the students from all three classes responded positively about the importance of this group project and agreed they learned a lot through doing it. However, 30% of the students felt that it could be an individual project focus-

Table 9. The group project about bilingual programs in our local schools (N = 53)

Multiple choices (choose all that apply)	% Responses
Is a good assignment and I learned a lot through doing the project.	70%
Makes the course readings more meaningful and more relevant to me.	68%
Is a good assignment but takes too much time to complete.	17%
Could be an individual assignment focusing on one school rather than a group project that involves more collaboration.	30%
Is not very important for this course.	4%

Table 10. Fall semester students' response to the group project (N = 37)

Choices	% Responses	Chi²
I prefer individual work leading to a project of my own even though I only have information about one school.	32%	1.162*
I prefer to collaborate with peers the way it is now because it is not a problem with me to collaborate.	24%	
I prefer to collaborate with others for a group projected but I do not like to depend on other people's schedule because some just do not get their work done on time.	24%	
Even though it is hard to collaborative for the group project, it is still worth doing it because we learn more about our bilingual programs in different schools through doing it together.	22%	

**P = 0.072*

ing on one school rather than a group project that involved more collaboration. The Fall semester postcourse survey asked an additional question about the group project and the responses are summarized in Table 10.

While 24% of the students preferred to work with peers because they had no problems to collaborate, exactly another 24% of them did not like to depend on other peoples' schedules because some just did not get the work done on time. Similarly, although 22% of the students felt it worthwhile to collaborate for the group project despite the fact that it was difficult, 32% preferred individual work leading to a project of their own even though they would not accomplish as much. Chi Square analysis failed to yield significant differences between student responses to this question. Compared to 92.5% positive responses toward collaboration in forum discussions, students' attitudes toward online collaboration in producing the group project were mixed.

Such differences were also reflected in some student comments on the group project in the open-end questions. As the open-ended questions did not address the group project directly, only 11 students expressed their views about the group project in their responses to the question about their likes and dislikes about the course (Questions 18b), and the question about any changes they wanted to recommend to improve the course (question 18d). Of these 11 students, one commented that she liked the group project the most about this course. However, 10 expressed their dislikes or frustrations about the group project. One student wrote "I think it's too inconvenient to try and get a group project together online. I also don't like having someone's performance affect my grade. I would rather do the project on my own." It appears that the end product type of collaborative tasks demands more consensus-building collaboration. When students were timed for such intensive interaction and collaboration, they became less enthusiastic about it.

DISCUSSION

A number of issues can be identified in answering the four research questions raised earlier. The following two sections discuss these questions along with the research findings.

Factors that Promote Sustained Online Small Group Discussions

Survey data suggest that a number of factors contributed to the sustained small group discussions in this course. Among others, the structure of discussions with carefully prepared discussion questions, small groups with fixed group members for interactive learning activities, the direct link between participation and assessment, and the strictly imposed deadlines for each forum were the main factors that contributed to the sustained the interactive learning in this Web course.

Previous studies suggest that topics that are not relevant to the course contents or not related to students' life experience do not attract participation and are not sustainable in online discussions (Williams & Pury, 2002). One of the factors that might have contributed to the sustained online small group discussions in this study was that students not only always had "something to say" in each forum but knew exactly what specific questions they were expected to answer in advance. These written exercises required in the first round of postings kept each individual student accountable for knowing the contents through reading. Therefore, students' interaction with the course readings, the first level of interaction with the material, was enhanced by producing written answers to be commented by peers in the group forums. Predetermined specific comprehension questions and thought provoking topic questions for each reading assignment helped students to focus on the learning contents and provided continuous discussion topics for the weekly group forums. Such proactive learning not only engaged students in a higher level of thinking (Gao &

Lehman, 2003; Wu, 2004) than the reactive type of learning but also kept the students accountable for participating in the weekly forums. The enthusiasm in group discussions never waned forum after forum because each forum focused on a new reading chapter.

Furthermore, the comment messages required students to exchange information by building on each other's ideas to negotiate for meaning and to collaboratively construction knowledge. Such interaction between peers and between students and instructors provided another level of interaction for learning. Students' positive experience with the semester long forum discussions was related to the benefits of proactive learning and learning from each other for knowledge construction. While the advantages of online interactive learning have long been proved in previous studies (Kern, 1995; Lavooy & Newlin, 2003; Mouza, Kaplan & Espinet; 2000; Summer & Hostetler, 2002; Wang & Teles, 1998; Wu, 2003), this study provided new data for the use of small group discussion as the core interactive learning tasks through the application of carefully prepared discussion questions that elicits proactive learning and through peer interaction and collaboration. When online collaborative learning tasks become main course pedagogy, such interactive learning is likely to be more sustainable and effective.

Previous studies have also indicated that collaboration as a process of participating to the knowledge communities is not an equal process to all the members of the community (Leinonen, Järvelä & Lipponen, 2003) and the size of online learning community affects the level of comforts which influences the level of participation (Williams & Pury, 2002). The current finings suggest that a group of 4-6 members can be an efficient and active learning community in which the members tend to generate sufficient responses from each other. On the other hand, the number of messages produced by each member was manageable and easy to keep track of. However, caution must be taken on this finding as the current study did not experience with other group sizes. Future studies need to test different group sizes with different learning tasks.

The survey data also indicated that students believed that group cohesion and mutual trust was the main factor of their groups. Furthermore, the majority of students said they preferred working with the same members of the group for the entire semester rather than rotating the peers. Obviously, it takes time to establish such mutual trust, even in a small group of 4-6 members. Therefore, it is very likely that the group cohesion and mutual trust comes from the semester long interaction, cooperation, and collaboration online. A small number of students expressed the desire to work with different peers and some did not show preference in working with the same peers or not. Future studies need to investigate the benefits and disadvantages of dynamic group formations in which students are given the chance to work with different online peers during the semester.

Survey data also indicated that assessment played a crucial role in motivating the students to participate in the semester-long group discussions week after week. Although over 90% of the students claimed that they learned more from reading and commenting on peers' messages, many admitted they would not have posted as many messages if the postings had not been required and assessed. The data support the previous research findings that the assessment of collaborative learning tasks plays a crucial role in ensuring student participation. Macdonald (2003) reported that students actively contributed to the discussions when the tasks were assessed but participation of discussions waned when the postings became optional. Grade for discussion was also positively related to students' perceived learning (Jiang & Ting 2000). Apparently, any optional interactive learning tasks would not have sustained for the entire semester.

Current data also suggest that required postings that were not directly linked to assessment did not attract equal amount of attention as the graded postings. This was clearly demonstrated in the lack of interest in participating the forum discussions at the whole class level in the Spring class. Williams and Pury (2002, p. 1) reported that "contrary to much literature on electronic collaboration suggesting students enjoy online collaboration, our students did not enjoy online discussion regardless of whether the discussion was optional or mandatory." It was not clear whether their "mandatory" participation of discussions was enforced by direct assessment of the actual postings in the forum discussions. This study provided further evidence that direct assessment of student interactive learning in CMC promotes sustained participation and interaction and also affects the level of participation and interaction.

Another important factor that appeared to have contributed to the completion of each discussion forum on time for the entire semester was the strictly imposed deadlines for each round of posting in each discussion forum. Student responses to survey questions suggest that required postings alone were not sufficient for guaranteed participation and interaction within a time frame. Strict deadlines seemed to be the best solution to complete the weekly forums on schedule. Therefore, the importance of imposing deadlines cannot be overemphasized for even directly assessed interactive learning activities.

Process and Product Oriented Interactive Learning

Very few studies have dealt with the differences between process and product oriented interactive learning tasks and how these differences influence peer interaction and collaboration (Kear, 2004; Kear & Heap, 1999; Macdonald, 2003). This Web-based course applied both process and product orientated interactive learning tasks that required different types and levels of interaction and collaboration. As discussed earlier, in the weekly group forums, the debate and exchange of ideas focused on the process of learning that did not lead to a final product. In contrast, the group project was a product driven collaborative task in that the interaction and collaboration among the peers through sharing and exchange of ideas and negotiation must help to reach certain consensus to produce a group report. Survey data suggest that students were more enthusiastic about process oriented group discussions than the group project even though 70% of the students agreed that the group project was a good assignment and they learned a lot through doing it.

Among others, the main reasons for students' frustration about the group project were the difficulties in reaching agreement according to a time frame, especially in the online environment. The differences in working pace and conflicts of schedules, and, perhaps more importantly, differences in level of devotion to the collaborative task in online environment made it more difficult for the peers to reach consensus in the process of doing the group project. The early birds who preferred to start and complete their parts of the work in a timely fashion conflicted with those who procrastinated in getting the work done. As peers in the same group would receive a common grade only for their project, there was pressure for them to compromise to reach agreements in completing the project.

Although the common grade can be used as a useful instructional strategy to implement end product driven collaborative tasks to encourage collaboration, the frustration and stress caused by the schedule conflicts and different levels of devotion toward such collaboration calls for more careful instructional design of such tasks. Perhaps some form of individual grading in addition to the interdependent grading are necessary to measure each individual student's efforts and contribution. In fact, Kear and Heap (1999) reported that students expressed a preference for a higher individual grade component when both common and individual grades were assigned for their group project. It is important to balance the level of collaboration among the students and the individual flexibility of online learning. Future studies need to address the pedagogical design of end product driven collaborative tasks in Web-based courses.

CONCLUSIONS AND RECOMMENDATIONS

This study identified some important factors that promote sustained online small group discussions as main interactive learning tasks in a Web-based course. Among other things, the structure of the online discussion, group size and group cohesion, strictly enforced deadlines, direct link of the interactive learning tasks to the assessment, and strictly imposed deadlines are some of the important factors that influence participation and motivate sustained online interaction and collaboration. The differences in process and product driven interactive learning tasks also have a different impact on student online collaboration. In general, students were more enthusiastic about process oriented than product driven collaborative tasks.

Finally, as the current data are based on one Web-based course that was mainly a reading course, the findings may not be generalized into a broad scope. Because of this limitation, the current findings may not be directly applicable to other courses that have a different online pedagogical approach. Yet, a few recommendations may be made for designing and implementing similar interactive learning activities to promote sustained and effective online collaboration.

- Although a very good tool for promoting interactive learning and collaboration, online discussion is not always sustainable if not well planned and structured. It is recommended that instructors carefully design each forum discussion with direct involvement of course contents with predetermined specific questions to engage students in a high level of thinking through providing written answers to the topics for which peer critiques are required.
- To continue to motivate the students, link the assessment with all interactive learning tasks utilizing specific grading scales.
- Impose strict deadlines for each round of postings in each discussion forum.
- Form small groups of 4-6 as learning communities for discussions so the peers will have sufficient input from each other yet still find it easy to keep track of all the postings in each new thread.
- Use process oriented interactive learning tasks to facilitate continuous online interaction and collaboration and yet still give each student sufficient amount of freedom in completing the assessed learning tasks.
- When design product oriented interactive learning tasks, much care needs to be taken in order to prepare the students to reach consensus. Give sufficient time for completing such learning assignment. Incorporate both common and individual grades in grading a group project.

ACKNOWLEDGMENT

The author thanks Sarah Maddison, Terese Thonus, and Ondine Gage-Serio for their insightful comments on earlier versions of the paper. The author also appreciates many helpful comments from the Associate Editor. Thanks are also due to Dawn Truelsen for her assistance in online course design using Blackboard.

REFERENCES

Duin, H., & Hansen, C. (1994). Reading and writing on computer networks as social construction and social interaction. In C. Selfe & S. Hilligoss

(Eds.), *Literacy and computers: The complications of teaching and learning with technology* (pp. 89-112). New York: The Modern Language Association.

Gao, T., & Lehman, J. D. (2003). The effects of different levels of interaction on the achievement and motivational perceptions of college students in a Web-based learning environment. *Journal of Interactive Learning Research, 14*(4), 367-387.

Jiang, M., & Ting, E. (2000) A study of factors influencing students' perceived learning in a Web-based course environment. *International Journal of Educational Telecommunications, 6*(4), 317-338.

Kear, K. (2004). Peer learning using asynchronous discussion systems in distance education. *Open Learning, 19*(2), 151-164.

Kear, K., & Heap, N. (1999). Technology-supported group work in distance learning. *Active Learning, 10,* 21-26.

Kern, R. (1995). Restructuring classroom interaction with networked computers: Effects on quantity and characteristics of language production. *The Modern Language Journal, 79*, 457- 476.

Kuhl, D. (2002). *Investigating online learning communities.* U.S. Department of Education Office of Educational Research and Improvement (OERI).

Lavooy, M. J., & Newlin, M. H. (2003). Computer mediated communication: Online instruction and interactivity. *Journal of Interactive Learning Research, 14*(2), 157-165.

Leinonen, P., Järvelä, S., & Lipponen, L. (2003). Individual students' interpretations of their contribution to the computer-mediated discussions. *Journal of Interactive Learning Research, 14*(1), 99-122.

Macdonald, J. (2003). Assessing online collaborative learning: Process and product. *Computers and Education, 40*, 377-391.

Mouza, C., Kaplan, D., & Espinet, I. (2000). *A Web-based model for online collaboration between distance learning and campus students* (IR020521). Office of Educational Research and Improvement. U.S. Department of Education.

Sumner, M., & Hostetler, D. (2002). A comparative study of computer conferencing and face-to-face communications in systems design. *Journal of Interactive Learning Research, 13*(3), 277-291.

Wang, X., & Teles, L. (1998) Online collaboration and the role of the instructor in two university credit courses. In. T. W. Chan, A. Collins, & J. Lin (Eds.), *Global Education on the Net, Proceedings of the Sixth International Conference on Computers in Education* (Vol. 1, pp. 154-161). Beijing/Heidelberg: China High Education Press and Springer-Verlag.

Williams, S., & Pury, C. (2002). Student attitudes toward participation in electronic discussions. *International Journal of Educational Technology, 3*(1), 1-15.

Wu, A. (2003). Supporting electronic discourse: Principles of design from a social constructivist perspective. *Journal of Interactive Learning Research, 14*(2), 167-184.

APPENDIX: SURVEY QUESTIONNAIRE

1. Is this your first Web-based (entirely online) course?
   a. ___Yes.
   b. ___No, I already took one entirely online course before this one.
   c. ___No, I took two or more other entirely online courses before this one.
   d. ___I took one or more Web-enhanced course (partially online) before this Web-based (entirely online course).
   e. ___No, I have never taken any Web-based nor Web-enhanced course.

2. This reading course is structured on group discussions with individual and group assignments. What are your thoughts about the structure of the course?
   a. ___I like the way the course is structured in terms of forum discussions because we learn from each other.
   b. ___I prefer weekly quizzes based on the readings rather than answering questions and joining the group discussions.

3. Will you post the same number of messages as you actually did over the semester if these postings were optional, not required and graded?
   a. ___Yes, I will post the same number of messages.
   b. ___I will post some messages but not as many.
   c. ___I will post very few messages.
   d. ___I will not post any messages.

4. Please circle one answer for each of the following:

   a. In our group forums, my answers to the questions and comments on peers' messages help me to understand the contents/readings of the course better.

   strongly agree agree disagree strongly disagree

   b. My peers' answers/comments helped me to understand the readings better.

   strongly agree agree disagree strongly disagree

   c. I learned more through online discussions than I would have learned from the lectures.

   strongly agree agree disagree strongly disagree

d. The online discussion is helpful because we collaborate more with each other and support each other.

strongly agree agree disagree strongly disagree

e. The group cohesion and mutual trust is an important factor in our group forums.

strongly agree agree disagree strongly disagree

f. I prefer individual work to group work and would have done better if I did not have to collaborate with my peers in my group for **discussions**.

strongly agree agree disagree strongly disagree

g. I prefer individual work to group work and would have done better if I did not have to collaborate with my peers in the group for the **final project**.

strongly agree agree disagree strongly disagree

h. The deadlines for the readings and postings in each forum are very important because they help me to complete the readings and the course.

strongly agree agree disagree strongly disagree

i. The overall course contents are interesting and I have learned a lot about bilingualism and bilingual education from taking this course.

strongly agree agree disagree strongly disagree

5. Choose one of the following:

a. ___ I wanted other group members to read our group discussions and I also missed the discussions in other groups.

b. ___ Every group should have summarized their forum discussions each week and post it to a general forum so that interested students could comment on the discussions in other groups.

c. ___ Reading and responding to peers' messages in our own group discussions is sufficient for me to understand the course contents. It would take too much time to read and respond to summary messages from other groups.

6. What is your view about group formation?
   a. ___I want to work with the same group members the way it is now because we know each other better.
   b. ___I want to work with different people in a group every few weeks because we will learn from other students we never meet.
   c. ___It will not make a difference to me working with the same people or different people in a group.

7. The pace of the course, including readings and postings
   a. ___Is neither too fast nor too slow for me.
   b. ___Is too fast for me because I always try to catch up with the readings.
   c. ___Is too slow for me and we could have read more chapters.
   d. ___Should be OK for a course like this but I found it too fast for me because I work many hours a week and have limited time for course work.

8. Course documents:
   a. ___I printed out all the lecture notes and review guides (or some of them) because they are helpful.
   b. ___I read the lecture notes and the review guides online but did not print them all.
   c. ___I never printed out nor read the lecture notes and the review guides because they are not essential for me.

9. The videos on reserve in the music library are used in all other face to face sessions of the same course. I found these videos
   a. ___worth seeing because they are informative and very relevant to the course content.
   b. ___relevant to the course content, but it is hard for me to make special trips to the university to watch them all.
   c. ___are not relevant to the course content and can be omitted.

10. You took all the three exams online in this semester. Do you think the online exam should be kept the way they are now, or do you prefer to take these exams in a classroom on a certain date?
    a. ___I prefer online exams the way they are now.
    b. ___I prefer to come to a classroom to write the exams.
    c. ___I have no preference.

11. Exam format:
    a. ___ I prefer multiple choice exams.
    b. ___ I prefer essay question type of exams.
    c. ___ It does not make a difference for me.

12. The group project about bilingual programs in our local schools (circle all the answers that apply to you)
    a. ___Is a good assignment and I learned a lot through doing the project.
    b. ___Makes the course readings more meaningful and more relevant to me.
    c. ___Is a good assignment but takes too much time to complete.
    d. ___Could be an individual assignment focusing on one school rather than a group project that involves more collaboration.
    e. ___Is not very important for this course.

13. For the group project:
    a. ___I prefer individual work leading to a project of my own even though I only have information about one school.
    b. ___I prefer to collaborate with peers the way it is now because it is not a problem with me to collaborate.
    c. ___I prefer to collaborate with others for a group projected but I do not like to depend on other people's schedule because some just do not get their work done on time.
    d. ___Even though it is hard to collaborative for the group project, it is still worth doing it because we learn more about our bilingual programs in different schools through doing it together.

14. Overall, my experience with this Web-based course
    a. ___Is very positive.
    b. ___Is positive.
    c. ___Is negative.
    d. ___Is very negative.

15. Experience with the Blackboard and the online forums: circle all apply to you.
    a. ___I found it challenging at the beginning but quickly picked up and like it now.
    b. ___The interface is straightforward and easy to learn, although I was not very experienced with any online courses.
    c. ___It was never a problem for me because I am good at technology.
    d. ___It was a plus because I learned the technology as well as the course contents.

16. If I have the choice in future,
    a. ___I will take a similar Web-based course.
    b. ___I will not choose to take a similar Web-based course.
    c. ___It will not make a difference, Web-based or face-to-face version.

17. Would you recommend a friend to take this Web-based course?
    a. ___Yes
    b. ___No
    c. ___Not sure

18. Please take some time to answer the following questions:
    a. Please describe your experience with the forum discussion part of the course. (positive, negative, expectation, effect on learning, etc. anything you think is relevant)
    b. What do you like the most, or dislike the most about this course?
    c. In your opinion, what are the most important elements for a Web-based course like this to be successful?
    d. To improve the course for future students, what changes do you recommend?

This work was previously published in International Journal of Web-Based Learning and Teaching Technologies, Vol. 2, Issue 1, edited by L. Esnault, pp. 17-38, copyright 2007 by IGI Publishing, formerly known as Idea Group Publishing (an imprint of IGI Global).

Chapter XI
Achieving a Working Balance Between Technology and Personal Contact Within a Classroom Environment

Stephen Springer
Texas State University, USA

ABSTRACT

This chapter addresses the author's model to assist faculty members in gaining a closer relationship with distance learning students. The model that will be discussed consists of a greeting, message, reminder, and conclusion (GMRC). The GMRC will provide concrete recommendations designed to lead the faculty through the four steps. Using these steps in writing and responding to electronic messages demonstrates to the distance learning student that in fact the faculty member is concerned with each learner and the learner's specific questions and needs. It is a practical application of human relations theory and is based on ideas generated by counseling theory. In addition, the chapter will take the reader through issues and examples that will arise during the duration of discussions and exchange of information using electronic messages. It is the intent of the author to provide not simply a theoretical model, but a model that can be learned and applied immediately upon completion of reviewing the article.

INTRODUCTION

It is clear that both the younger adult students as well as the more seasoned adults are seeking efficiency and independence in learning. Williams (2006) documents that utilizing the Web for instruction is being used by institutions and public schools all over the world. Computer techniques, chemistry, business, and other content areas are being taught through online courses (Williams, 2006). The adult learning market is seeking ways to more efficiently learn course requirements without 45 contact hours in a crowded or inaccessible classroom. Furthermore, the adult is often rushed and harried in life and cannot consistently attend class on any campus. The efficiency in learning, as well as the quest for independence, has also marked a new era of problems involving standards and contact with the instructors. In addition to taking classes, advising and mentoring remain important. The advisor remains important to graduate students and is a key in the student's success (Polson, 2003). Balancing learning and advising through the Internet is a common issue that has been studied and documented. However, there is much to be learned and discussed. Satisfaction of students with e-learning remains an element of discussion, as does retention and motivation. This all impacts the instructor and the student relationship. Mason and Renne (2006) reinforced the importance of the tutor's and teacher's skills in dealing with the students in their recent publication. Relationships are important even within a distance learning environment.

Independence in itself is certainly no vice, and promotes stretching the borders of individual learning upon demand. The power to learn upon demand is reaching the many sectors of even traditional universities. As noted, distance learning is pervasive in all learning groups. Singh (2004) makes it clear that if the distance learning can be well managed than it can be a positive learning experience or perhaps even better than the traditional classroom learning. At the same time, questions still remain unsolved and unanswered in regard to the very bedrock of learning through electronic means. Distance learning instructors need different skills than the traditional instructors (Bower & Hardy, 2004). What skills do they need and what models can be provided for new instructors who communicate with students through e-mail? Unless academic-based recommendations are provided, it is possible that the same questions could continue to haunt the academic environment far into the 21st century.

Beyond independence and intrinsically involved in this academic mix is the issue of speed and progress in obtaining the credential. Learners see some of the traditional methods of the university as no longer relevant. However, the ability to access information and advisement at the touch of a button should not replace the advisor's responsibility. Technology can certainly enhance student advisement but it does not replace the advisor (Wagner, 2001).

Attendance in some institutions has become an issue because some students may not want to experience a dry lecture or hear something they could obtain from an Internet posting. In essence, a person has a significant amount of power to learn and may choose to learn what are perceived educational needs. If a lecture is not relevant or is something that an individual already has mastered, it is difficult for the lecturer to obtain and retain the attention of the learner. Therefore, some learners see it as waste of time for them to attend the "live" lectures. Education has attempted to monitor the learner's own progress and sometimes there has been a bogging down or a series of barriers that the students have faced. Perhaps as education has attempted to rush and meet the needs of the student's distance learning aspirations there has been a loss of the personal touch. Payne (2005) documented a study of multiple universities and concluded that some of those participating in the study did not like losing contact with advisors. However, Payne indicated that they did adjust and made it work. This teacher/advisor/student relationship has provided countless elements of support through the centuries. The personal and professional skills for online instructors still should involve the skills listed by Brewer, Dejonge, and Stout (2001) that include choosing words carefully, encouraging e-mail, and scheduling office hours. Certainly another issue is in regard to the students' reasons for using e-mail. Faculty may see students as using it to enhance learning, and on the other hand, faculty may see the students as using electronic excuses for their lack of performance (Duran, Kelly, & Keaten, 2005). Perhaps the truth may lie between the extremes and the type of individuals using the e-mail may also provide some clues in this discussion. Finch, Keaten, and Kelly (2004) note that reticent students will talk to teachers before or after class; however, they prefer asynchronous channels of communication. This information may also provide another incentive for the instructor to establish a stronger professional relationship with the instructor's students through electronic methodology.

The world continues to debate and discuss these same issues. Nanyang Technological University in Singapore has as one of their goals "'to humanize' e-learning and develop quality and interactive and engaging content" (Shih & Hung, 2007, p. 224). To some, the relationship between the student and faculty members is a precursor to learning. Others would argue that "content" is the important element and there is a lack of need for interaction with the faculty member. This remains a strong challenge for distance learning. If only content is important, then the best practices including faculty contact and student contact, discussed by Testa, would no longer be applicable and something would be missing (Cole, 2000). Once the individual has graduated and attempts to move into the market place, the lack of social skills has serious implications for our society.

The remaining section of this chapter will focus on the difficult task of identifying a balance between the benefits of distance learning and faculty being able to provide personal and individual attention for their students.

THE DILEMMA OF BALANCING DISTANCE LEARNING AND PERSONAL CONTACT

The dilemma of the balance between distance learning and personal attention is debated between faculty, students administration, and informed observers. How can we maintain such a balance? In addition, we also are dealing with maintaining academic integrity and standards as well as still providing for the unique learner in the 21st century. This difficult element of distance learning standards and accreditation adds one more difficult area that the faculty must address. There are several assumptions by the author that should be recognized before progressing to the model provided in this chapter.

First, it is assumed that interaction within an educational environment facilitates and fosters learning. The discussion of educational topics within a controlled environment theoretically allows the learner to hear and respond to controversial subject matter and respond to diverse opinions. In our pluralistic society this is not only important but is imperative to be able to move forward in this century. Without discourse with others of conflicting views, the logjams in such places as congress will reach a point of absolutely no progress. Therefore, the case can be supported that listening, critiquing, and responding to diverse viewpoints is not only healthy but practical and stimulating for our society.

Second, it is assumed that most individuals inherently desire social interaction yet find it difficult due to time and space demands to pursue relationships or discourse. It is easier to send a curt e-mail than it is to demonstrate concern for the individual through personal or phone contact.

A third issue that must be discussed when dealing with these issues is the question of what is the university experience. This academic discourse between the learner and the professor is a treasured experience and one not to be taken lightly. In addition, the total experience involves a composite of new learning, reflections, interactions, challenges, and trials. The difficult question faced by the academic community is that we may now be heading for a disjointed academic environment where distance learners receive only limited feedback and interaction with others. They may not be given the opportunity to be enriched by contact and open discussion. This limitation on the actual thought and reasoning may become a difficult barrier for the distance learner of today to face tomorrow. Are the distance learners receiving the same educational exposure as the more traditional students?

Next we must recognize the fourth area in our dilemma. It must be fully understood that the distance learning revolution will not be reversed nor will traditional lecturing be restored to its 20th century glory. Time is marching on and this topic of discussion and debate will disappear from the radar in a few years. Therefore, the considerations for learning in the university must be tempered with the needs and desires of the students. Although some would argue that the students should not set the stage for the class nor the education system, they will continue to evaluate the program with their feet; they will gravitate toward greener learning pastures that offer a distance learning option.

The final assumption that cannot be overlooked forms the essence of this chapter. There is a method to balance the needs of the learning community with the needs of the 21st century student. Educators cannot despair due to the changes in their students or their means of learning. Can the "importance of community" be established as noted by Bowles (2004) in collaborative e-learning? Each generation will bring new challenges and at the same time there will be new means to cope with the problems that exist. The remaining sections of this chapter are designed to address the changes and discuss the greeting, message, reminder, and conclusion (GMRC) model that will be useful to the faculty member in interactions with students.

INTRODUCTION OF THE GREETING, MESSAGE, REMINDER AND CONCLUSION MODEL (GMRC)

The model, developed by the author, consists of greeting, message, reminder, and conclusion. It involves providing faculty members suggestions and methodology to build trust and conversation within the e-mail message. The author is a licensed professional counselor and the model is based on several of his eclectic beliefs utilized in his private counseling practice for over 20 years. These principles are: there must be respect for the dignity of all persons; individual's desire response and feedback; individuals can solve their own problems; and without a clear direction people tend to fail. Although these principles may appear to be simplistic, these form the foundation for both the author's counseling methodology and the GMRC model. What is the model and what does it mean to the professor, counselor and administrator?

The GMRC model is actually four parts to a message that is sent to a student who is communicating with the instructor via the Internet or other electronic message boards. The four parts to each message are carefully woven into the response to an inquiry by a student or a message that the instructor desires to convey to the students. In essence, the four parts of the model provide a template to follow in electronic communication (greeting, message, reminder and conclusion). Although these steps are not difficult to understand, the application of the model becomes tenuous.

This consistent application of the model is one of the difficulties that a convert to the theory will face. Although most individuals will agree that civility, respect, and specific information are valuable in all messages, it may be difficult to apply these principles in a consistent manner. The next sections of this chapter will analyze the steps and discuss how to incorporate these into the daily routine of dealing with e-mail.

Greeting

The first step in the message to any student, whether it is a large number of students or simply one-on-one e-mail message, is to provide a responsible and responsive greeting. In reality this type of greeting demonstrates respect to the person or persons addressed and assure them that this e-mail is both important and timely in regard to their needs.

Before reviewing responsible and responsive e-mail greetings it is important to understand respect in the greeting. Individuals who have "titles" should be addressed by that title. An example for this is the number of soldiers returning from active duty. They may be in a situation that they are still on duty and have served their country well even with resulting injuries. Addressing them as LT or SGT can demonstrate respect from the instructor. In addition there are other means by which respect can come across to the students.

Older adult students may be called by their first names by professors. On the other hand professors may require or expect that Billy who is 60 years old or Mary who is 48 call them by their professional titles. Although there is nothing new about this issue, it becomes very evident in a situation where written communication is the primary source of communication. As the professor continues to call students by first names in the communication, there may be the feeling that indeed the student warrants a "Mr." or a "Ms." in the e-mail. It is something to contemplate in the area of respect.

Certainly one area often debated and which reaches limited resolution is the informal greeting used by some such as "hey…wanted to let you guys know…" These types of greetings are debated in the sense that some propose these help the student to understand the professor is with the times. The opposite view is that sometimes extremely informal greetings can confuse the reader. The reader may feel that the class is simply a "buddy" situation and the professor is a "personal friend." Perhaps many of those reading this chapter and who have been professors may see that we may strive to be in the middle. We want to be up on the times and show the students our interest in their lives. However, we want them to know that we are ethical and will uphold our institution's standards of teaching.

Now that the discussion has centered on issues in regard to greetings, there is need to examine the responsible and the responsive greeting. Although these two words appear the same, they are indeed different and are both important to the faculty member who sets the tone for the class that is being taught online or the advisement that is being provided. Initially it is important to closely examine the responsible greeting. This greeting assumes that the faculty member is fully responsible for word choice both in the greeting and in the entire body of the e-mail communication.

Responsible greetings are sometimes overlooked in the sense that the faculty member may be in a hurry to respond to the student and can easily overlook the greeting. The greeting provides three distinct impressions to the reader:

1. The writer cares or does not care about the student as a person
2. The writer does remember the student
3. The writer does respect the student's opinions and academic work

The faculty member may indicate in the greeting comments about such areas as the student's illness, problems the student has had recently, or some e-mail that was notable. This sets the tone that the faculty member does indeed separate the student from the other students in the sense of the student's individual needs. This can be done in as little as one or two short sentences. Comments to the entire group of students in one e-mail can also be made using this concept. They can all benefit from a greeting that acknowledges their class as being unique from the professor's other teaching assignments.

The professor could begin the e-mail by mentioning such things as, "Sorry that you missed last Tuesday's e-mail assignment …I hope you are doing better." In addition, such comments as, "Glad to receive your e-mail, I enjoyed reading your last critique of the situation in Europe," can be very encouraging. Finally, it may be useful to acknowledge the students problems with the last e-mail and then proceed with the updated version. An example may be something to this effect, "Mr. Smith, glad to hear the assignment is coming along well. As you know I was disappointed with the last assignment and am looking forward to seeing your latest one."

Certainly responsive greetings are difficult to initiate if the faculty member has not been on the system in several days or has allowed too much time to elapse. This type of greeting implies that it has not been long since the faculty member and the student have communicated. It reinforces the point that the student has responsibility to maintain timely communication. Sometimes it has been clear that the student has been away or has had some issues that have severely impeded the response or initiation of electronic communication.

It is agreed that some of the issues provided by students may be spurious or inconsequential. However, it is true that life does come at you fast and there will be issues that are affecting the outcome of a course. Therefore, responsive greetings are timely in content. It profits little to respond to a week old e-mail about a person's cold or minor health issues when likely it has subsided. In addition, responsive greetings acknowledge the time line. Sometimes this helps the "tardy" student to realize where the student is in the course.

Faculty may use the responsive approach in almost every manner, particularly in reigning in students who may hide behind electronics as an excuse or crutch for slow work performance. Alden (1998) points out in his book on training in Web based instruction that we should ask specific questions about the "lurkers" situation and progress. One issue that sometimes frustrates faculty members who are trying to maintain close communication with the students is that some students use various ISPs and then close the accounts. What can happen is that the students e-mail the professor but they seldom check their college or university e-mail account and may not forward their academic account to their personal e-mail. The faculty member continues to reply to the student's college account and the account begins to fill and the box is overloaded with unread mail.

In the case above, both the student and the professor have a tremendous amount of frustration. The faculty member continues to send the material or responses and the student has not acknowledged receipt. This may continue until the faculty member or the student realizes that the method for communication in the class or in academic advising is the college or university mail system.

How long could such a comedy of errors continue in an academic environment? It could last way into the semester. Some classes, taught by distance faculty, find that students continue to bypass the university e-mail and struggle because they are not receiving updates. They never set up their university e-mail system! Although these are extreme situations, it is recognized that Murphy's Law continues to work overtime in the electronic world.

Whatever the situation, the greeting in an e-mail continues to be extremely important. Not only does it set the tone for the message, but it also provides the opportunity for the faculty and students to remind one another of their shared environment and bond with each other through electronics.

Message

The second part of the GMRC is in reality the "body" or the message that needs to be conveyed to the student. Once the "greeting" has been made in the e-mail it is important to move to the essence of the message. This is the second step in the GMRC model and it is a particularly important part of the communication. Simply sending messages to send messages is not what the faculty member has time to do. So, there must be a clear reason why the e-mail has been sent.

Although it is not in the context of this chapter to discuss the title of the e-mail, it is important that the title line be clearly something to remove it from the appearance that the message is spam or is not important to the course. Therefore, the title must be related to the message. Good strong and eye-catching titles are imperative such as "history examination review," "mid-term math information," or "spring advisement" may be very useful to the student.

Titles that do not convey the content of the e-mail or which do not separate it from the normal inbox materials may go unread and the professor's message will become lost in cyberspace. Titles such as "hello class," "hi team," or "another e-mail" may not always spark an urgent response. Students receive as much or more spam and e-mails than do faculty members so it is imperative that the student knows the e-mail is important and timely.

A final point regarding the title as related to the "M" or message in the GMRC model is the potential use of a time sensitive date in the title of the message. If the faculty member is discussing an upcoming examination it can be useful to indicate "mid-term examination for March 28th." Although not always necessary to include a time in our title, it is our desire to help students excel in learning and such a reminder may spur a student to update the student's calendar. This will help the student to be more attentive to upcoming academic benchmarks in the course.

Moving back to the discussion of the "M," there are several issues that must be addressed. First, the message must be written with positive guidelines and expectations as opposed to writing e-mail with what not to do or negative expectations. Teaching a course is also modeling behavior that is expected in the academic environment. By providing clear expectations in the "message" part of the communication, the professor provides the guidelines and information necessary for the student to respond and fully understand the implications of the message in the professor's course. Ambiguous messages or messages that are too lengthy may be too much for the reader to absorb.

At times a faculty member may be asked by potential students to tell them about the college major or discipline that the faculty member represents. In general, the best response will be to give them some basic guidelines in the first message. The major details of the degree program might not be addressed in the first e-mail. Instead, providing a link to a Web site and directions from there to obtain the information can be useful. Too much information in the original e-mail takes the chance of misrepresenting already clear information in the online course or the Web site information. It remains more positive to use the already "clear" explanations in existence and refer the student to those along with a brief explanation. This does help avoid duplication and lengthy e-mails. It is best to simply show the way and not clog up the system with more redundant explanations.

Attempting to avoid negative information in providing direction in the "message" is something that is noble in its intent. This can help to shape the relationship between the student and the faculty member. Indicating "for full credit consideration the assignments must be submitted by …" is much more tolerable than "don't forget to do the assignment on time" or " too many have been late and I can't accept that anymore." Although professors are "human" in the sense of showing anger or frustration, they set the standard. Part of that standard is to teach students civility and respect, which remains an element of discussion within the academic environment. The professor's ability to continue modeling positive communication is a gallant attempt to address society's problems one at a time.

How much information is too much information in the message? Normally e-mails are not intended to replace other information sources such as information in the course management system being used. However, in teaching a class the individualization of the work for each student must be to some extent truly individualized with e-mail communication. There are three guidelines to consider in dealing with the information level in an e-mail response:

1. Is the information located somewhere else electronically where a student may be directed easily?
2. Is the information something that has been repeated before, indicating that a richer means of commutation needs to be utilized, such as a phone call?
3. Is the information provided in the e-mail essential or could be it be shortened and the same information be conveyed?

Finally, in dealing with the message content, can the writer as the professor understand the message? Is the message so disjointed that it makes little or no sense when read for understanding? As established in communication studies, there are obligations both for the sender and receiver. The final check prior to the e-mail being sent should be for the clarity of the content. It is possible using the GMRC model that all sections are well written and complete with the exception that the message section is disjointed! It may not be obvious to the writer because the instructor knows what he or she wants to say. However, it may not be understandable to the student!

Reminder

Once the message has been provided, there will generally be something the professor may want to provide that particular student in regard to the class. Although some could argue this part of the model might be redundant or unnecessary, it could also be that it provides one last opportunity for the student to receive some positive strokes or some information that the student could have overlooked. The reminder needs not be something that is extremely weighty or for that matter immediately germane to the topic discussed. It should be something that is a helpful reminder and demonstrates individuality of the e-mail.

In classes taught by the author or in response to undergraduate students, they are often reminded of the cooperative education project course they will need to enroll in during their last semester. Although not critical to remind them in the e-mail, the weaker students will be greatly helped with this. At times they are overwhelmed with information and their planning skills are lacking. A friendly statement or "reminder" in the e-mail can be very helpful. Perhaps that one statement will help to put a person on track and promote that person's collegiate education more than even suspected.

A simple statement could be, "I noted that you were fairly far in your degree program and wanted to mention that you might benefit by checking the information on our Web site in regard to our internship. It will be coming up soon." This statement format is one the author has used to encourage students to prepare for other courses while completing the current courses. In our degree program the internship is an important part of the program, therefore this simple statement can really benefit our students in looking ahead.

Other short statements in the reminder section of the GMRC can be useful. Some professors may want to mention their examination date in a homework assignment. Again, this can be something individual for those students who are having a difficult time or it can be a blanket statement that is designed to cover the entire class. Some of the potential uses of the reminder are:

1. Examinations
2. Projects
3. Discussion room responses required
4. Change in professors response days for e-mail or phone calls
5. Indication of new class beginning next semester

Although most of these may be on the class syllabus, a reminder may be very helpful for those who have schedules that are completely filled.

At this juncture it is important to address the criticism for "babying" the students in using the reminder method. It is true that much of the "reminder" may be material that is not essential, but positive to provide. Students who have difficulty may have their lives changed simply because one professor did provide more information than was necessary in a course. This willingness to travel the "extra mile" is time consuming on an individual basis. However, students may wake up and provide a positive response after receiving such information. In the author's experience, most have been very receptive to this reminder and none of the students receiving the responses have been offended or hostile in their reactions.

Conclusion to Model

The final section to the GMRC model is the conclusion. Although this seems to be an obvious part and perhaps not a too important section of the model, it must be defended. The model is incomplete without an appropriate conclusion. This conclusion is the last opportunity to summarize the contents of the "message" section of the e-mail. It provides the last opportunity for the person to understand the importance of the e-mail and act on it or respond appropriately to the content of the material.

The professor who authored the response should not be expected to repeat the entire e-mail the way we often do in conversation as we say "well the reason I came by was to..." then later we indicate several times again "I just wanted to come by and..." The e-mail conclusion in the GMRC model must not be repetitive, rather it must be reinforcing. It is a brief reinforcement of the important message the professor has provided. It can be reinforcing in several ways that can be reviewed.

First, the conclusion can be utilized to reinforce the date some assignment is due or when the work as indicated in the message is supposed to be online. An example of the response could be, "Mr. Simon, don't forget that I will send the test online on Thursday, January 15th." Second, the conclusion can be an excellent reinforcer of a student who is achieving more in the class, such as, "Ms. Kellogg...I wanted to say again that your latest work is exactly what I am seeking in the course."

Ultimately the conclusion can provide two other types of information as the e-mail concludes. If the student has tremendously excelled in the course or the last assignments, it may be such a statement as "I will post some additional material on the Web this week. I wanted to assure you that you were doing excellent work and I am very glad you are in the class." On the other hand, the message could end with a little less positive tone, "I am very sorry you have had trouble meeting the deadlines on the system, I am confident you will get back on track."

All of these conclusions are clearly requiring the professor to analyze the student's individual needs. With large advising loads or classes that are more than 25 members, it is very difficult to address a personal matter in the conclusion. Some professors have been led to believe that teaching online or advising online will be easier and will not take the time that face-to-face advisement or teaching will require. This is certainly not true if the students are given the attention and concern as discussed in this model.

USING THE GMRC

Example I

The GMRC model is designed to begin using immediately and can be included in a professor's e-mail habits very quickly. However, it is understood that elements may be left out without practice and when the pressure of the semester becomes difficult. The model's benefits will not totally be seen until the practice becomes second nature to the professor using the model. This section will take a number of common scenarios and apply the model's steps in order to demonstrate the utility of the model.

The initial example is of a professor who has a student who has demonstrated excellent communication online and who has been attentive to the work, yet who has not been able to provide strong papers. The occasion is that the professor of history is trying to contact the student after the student questioned the grade of 70 on the last e-mail assignment.

The professor responds to the student as listed below:

"Mr. ______,
I am glad to hear from you in regard to your recent grades and the work you are doing. It is always helpful to know where you stand in the class. I looked at your score and the work that you submitted and found two reasons you were having trouble with the piece why the North won the Civil War. First, you did not examine the industrial might the North had in comparison to the South. Second, although you mentioned the British and potential recognition of the Southern government, you did not search out some of the reasons the British were reluctant to recognize the South. Leaving out both of these reasons made your paper less complete. As you know, we have two more reflection papers coming up so you can bring up your grades. Stay in contact and revisit the reasons the North won which will help you on the final."

Looking at the e-mail sent by the professor we see all four parts of the GMRC utilized. First, in the "greeting" the professor acknowledged the student's difficulty and the student's concern over the grade. In addition, the professor was supportive of communication. This made the greeting a little longer than it may take in some cases. However, it also gave the student some positive feedback in a negative situation.

In regard to the "message" in this example, the professor was succinct in what was said and the problems were clearly provided. Instead of mentioning that the student simply had problems with the work, specific shortcomings were listed. This type of response gives the student something to research and also to go back and review.

The "reminder" section of the message gave the student both information that there were two more assignments due and also hope for the student that the student could bring up the overall grades for the course. This is an important part of the "reminder" in this type of situation. The professor may perceive that the student is overwhelmed with school and the chance of late papers may be an issue. In addition, the idea of bringing the grade up was something the student may seriously need to hear.

Obviously, the "conclusion" in the e-mail revisits the problem the e-mail addressed. The student is reminded of the problems that were in existence on the paper and was told that these issues will again be addressed on the test. Therefore, the student now had reason to acknowledge the problems in the paper. In addition, the professor made it clear that the student could ask more questions when it was said to "stay in contact."

Example II

Many times professors may simply send out e-mail to those who are having problems with the course. This is not unusual since those who are doing well require less time and may learn "in spite of the professor." However, there is merit in trying to contact those who are doing well and encouraging them to strive even harder to learn not only the material for the course but other issues related to the subject. Perhaps this type of solid reinforcement should be used more often, especially in an electronic environment. This can help to bond the professor and the students and build rapport for the next course.

This professor-initiated e-mail could be written in the following manner. Notice that this e-mail continues to address the student by the student's last name. This is something each faculty member needs to fully implement or ignore. It is not equitable to call some students by their last names and then call others by their first names. We must have a personal policy that is consistent. This informational e-mail is unsolicited and would go to the student whenever the professor wished to commend the students. The following is an example of an unsolicited e-mail with all elements of the GMRC.

"Ms. ______,
I hope you are doing well this week! I wanted to take this opportunity to let you know that your work on our thought papers has been outstanding. You have hit the nail on the head in regard to the beliefs that Dr. Glasser has proposed in his new book. Your critiques of his work and the other two were very strong. Registration is coming up and I hope you consider more advanced counseling courses. Again, I am pleased with your work in the course...keep it going!"

The analysis of this communication demonstrates a friendly and shorter "greeting." Although any e-mail may have a friendly greeting as demonstrated by this model, it is evident in this e-mail that the professor wants to put the student at ease. Certainly some students who are high achievers may be "Type A" individuals who may immediately spring to action with contact from a professor. Therefore, this greeting is one that is designed to put the person at ease.

The "message" in the model is clearly stated. The student knows from the outset that the student has achieved and surpassed the expectations of the class. The student also knows that the professor was particularly interested in the work on William Glasser since his name was chosen above the other two. This also helps the student to realize that if the work on Glasser was named then the student should model the work on that assignment for the future. It was perhaps a hidden message, yet it can be a strong one.

The "reminder" part of the GMRC was going in a completely different direction than in the first example. This reminder is recruitment for solid students by the department. It is also an indicator that the student could consider counseling as a major course of study. The strength of this "reminder" following the other positive strokes can only endear the student to the department and peak the student's interest.

The "conclusion" section of the message reinforced the "message" and also made it clear that success was doing what the student was already doing. This was a short and to the point conclusion that still demonstrated warmth.

Example III

In a distance learning course it may be difficult to determine if a student has some personal problems that are impeding progress in the class. Because the professor does not see the student on a regular basis and sometimes never sees the student, it is difficult to read the body language that may cluė in the educator to some serious problems. On the other hand, it is not uncommon that some students may report more than expected about their situation. The next several examples will deal with self-reporting and also when the professor suspects there may be some issues that the student is confronting.

In the arena of self-report, a student may review an assignment much like the counseling assignment that was discussed in Example II. After reading and reviewing the assignment, the student may begin to self-diagnose and self-disclose. The author remembers one student who became so distraught in writing her life history that she indicated she could not continue. Therefore, intervention was required. The problem is compounded when the student is thousands of miles away in an electronic-based course and the professor knows the student only by electronic messaging and occasional phone calls. This makes it difficult to deal with the issues and offer sound professional advice.

The following is an example where a student e-mailed the professor about a serious personal problem of spousal abuse. This was the response:

"Ms. ____,
I am glad you shared your situation with me and I want you to continue to feel free to share this difficult situation with me. Let me first indicate that I am not a licensed counselor or health professional. However, with that said I still have some concrete suggestions for you. First, your local police department has a unit that deals with spousal abuse. They can provide you information and a place to stay to get away from the situation. Second, there are also counselors that your city has on staff to work with individuals like yourself. I checked the Internet for your city and noted this information and am providing you these phone numbers ____________, __________. Third, it is also good to contact close friends so you will have a support system that may include the religious community if you are members of a particular religious group. Finally, I believe you need to act now to check into these situations. In the event you cannot finish this course on time, I will certainly offer an "incomplete grade." Please keep me informed and I hope you will take my advice as quickly as possible."

This example was a difficult situation and one that may warrant a phone call. However, because of the difficult situation the professor may reject the phone call idea in the event the spouse was there and the student could not talk. Certainly there is an e-mail risk; however, the student must have felt safe since the topic was raised in the e-mail. The "greeting" was an inviting statement that let the student know that it was "ok" to share the information. In fact, it was more than "ok" in that the professor said the student could continue to correspond. Therefore, the "greeting" was both an indicator of help and of security to the student.

The "message" in this situation was a strongly worded step-by-step guide for the student. The professor went out of the way to even provide phone numbers for local help obtained from an Internet search. This is very useful since sometimes those hurting have a difficult time actually making the first call. The message also was clear to indicate that the professor had no special credentials that made the professor able to provide solid specific advice. Instead, the professor was referring some place specific for the student rather than saying "go get some help."

In regard to the reminder, this was not as clear in the message. However, it could be said that the sense of urgency was a reminder and the student could receive an "incomplete" grade in the event that it was too stressful or difficult to finish the course. This is something a student may already know due to policy. However, it is always positive to mention this during such a crisis.

The "conclusion" in the GRMC model reaffirmed that the student should stay in contact and also that the advice given should be followed. Certainly this conclusion enables the professor to continue the dialog in the future messages. If the student continues to complain, the professor can continue to go back to the initial premise in regard to what steps need to be taken. This is a decisive conclusion and provides clear direction on what the student should do.

Example IV

Another difficult situation, as mentioned earlier in the chapter, is where the professor in the course is suspicious that something is going on in the student's life that is serious or could bring harm to the student or others. Some students ramble in their answers to questions and often reveal some issues that are masked in essays or information that they provide. This can create a major difficulty for the faculty member since the course may not have anything to do with mental health issues. All the student's responses are incidental and directly or indirectly related to the course. However, the professor may believe there is something that is causing some serious problems for the student. Should the professor act on these thoughts that there is something wrong?

There are policies in existence in some institutions that address referral of students who are suspected of having serious issues. However, because time may be of the essence, the faculty has a unique opportunity to address the student's covert message or sometimes-blatant message. The author would recommend addressing it to the best of our ability as quickly as possible.

The following example is communication using the GMRC method to deal with a student's perceived problem.

"Mr.______,
Hope this e-mail finds you doing well and having a good week! I have been reading the reports you have been sending and wanted to share with you a couple of observations. First, you seem to relate to the problem "Bill" has in our reading. You provided a lot of information about your current life as relates to "Bill." Second, you have made some of these observations and comments on several occasions when you have responded. In looking at these two things, I wanted to offer my help or direction in finding an answer for these concerns you have expressed. I am willing to make some contacts or talk to you to be able to help with these. By the way, I did want you to remember that there are some online resources from our campus that we talked about in the first class that can also help you. At any rate, I wanted to let you know that even though I am your professor this semester that I am also concerned about your well being outside of the class. Let me know how I can help."

This communication would be difficult for the faculty member to write since it is engulfed in legal issues. The professor did not want to ignore the problem and continue reading the material the student sent. Yet, the professor did need to be careful not to indict or obligate the student. The "greeting" in the communication was friendly and noncommittal. It at least gave the opportunity to set a pleasant tone for the "message" which was to be difficult at the least.

The "message" part of the model was written in more of a clinical manner. The professor was making an attempt to demonstrate what the student had been doing in a nonjudgmental manner. Although we could continue to write and rewrite the "message" portion of the e-mail, this was at least an attempt to contact the student.

The "reminder" section of the e-mail simply made mention of previous resources that online students have. It was a fairly weak "reminder"; however, it was still in line with the model. In addition, it was certainly positive information for a negative situation.

It is clear that the "conclusion" reaffirms the issue and places the professor ready to assist the student. Although the conclusion is a risk, it is also a strong affirmation of what lengths a professor may travel to assist students electronically.

CONCLUSION

The GMRC is a model that is practical and an experienced-based model that attempts to provide the personal feelings and touch within a technical and impersonal environment. It can be utilized in virtually all e-mail situations. This may include the class and individual students. Each element of the GMRC may be larger or smaller depending on the information that is being conveyed. The examples provided are simply ways to adapt the model to situations that may arise as a faculty member negotiates communication with the distance learner. The author has found there has been a positive reception to the model; however, it needs to be empirically tested among a larger audience at several universities. Perhaps through such research future educators will be able to more completely address the gulf between technical delivery and personal relationships.

FUTURE RESEARCH DIRECTIONS

There remain several areas that future research may be directed. Initially, the first area revolves around the time that the professor must spend in answering e-mails using the model. Will the professors consistently utilize the model in their expanded correspondence during peak semester times? Perhaps even more difficult to consider is the overall positive or negative effect on student retention within a university setting. Will the GMRC be proven to positively impact on retention? We know that the weaker students will have difficulty in distance learning options as they do in the traditional learning environment. Can the GMRC provide the critical support difference needed to help these weaker students remain engaged? Finally, does the GMRC work as well for various ages, different ethnic groups, and genders? This study would assist in transforming the GMRC in the event that a significant difference is identified in regard to the various group responses to the GMRC. In retrospect, the area is replete with options for future research.

REFERENCES

Alden, J. (1998). *A trainer's guide to Web-based instruction*. Alexandra, VA: American Society for Training and Development.

Bowles, M. S. (2004). *Relearning to e-learn*. Carlton, Victoria, Australia: Melbourne University Press.

Bower, B. L., & Hardy, K. P. (2004). From correspondence to cyberspace: Changes and challenges in distance education. *New Directions for Community Colleges, 128*, 5-12.

Brewer, E. W., Dejonge, J. O., & Stout, V. J. (2001). *Moving to online*. Thousand Oaks, CA: Corwin Press, Inc.

Cole, R. (Ed.). (2000). *Issues in Web-based pedagogy*. Westport, CT: Greenwood Press.

Duran, R., Kelly, L., & Keaten, J. (2005). College faculty and use and perceptions of electric mail to communicate with students. *Communication Quarterly, 53*, 159-176.

Finch, C., Keaten, J., & Kelly, L. (2004). Reticent and non-reticent college students' preferred communication channels for interacting with faculty. *Communication Research Reports, 21*(2), 197 209.

Mason, R., & Rennie, F. (2006). *E-learning the key concepts*. New York: Routledge.

Payne, D. A. (2005). Succeeding in graduate school online: Tips from successful students. *College Student Journal, 39*(1), 117-28.

Polson, C. J. (2003). Adult graduate students challenge institutions to change. *New Directions for Student Sevices, 102*, 59-68.

Shih, T., & Hung, J. C. (Ed.). (2007). *Future directions in distance learning and communication technologies*. Hershey, PA: Information Science Publishing.

Singh, P. (2004). Online education: Lessons for administrators and instructors. *College Student Journal, 38*(2), 302-308.

Wagner, L. (2001). Virtual advising: Delivering student services. *Online Journal of Distance Learning Administration, 4*(3). Retrieved February 14, 2008, from www.westga.edu/%7Edistance/ojdla/fall43/wagner43.html

Williams, K. C. (2006). Active learning and quality in online courses. *NACTA Journal*. Retrieved February 14, 2008, from http://findarticles.com/p/articles/mi_qa4062/is_200612/ai_n17194101

ADDITIONAL READING

Dabbagh, N., & Bannan-Ritland, B. (2005). *Online learning concepts, strategies, and applications.* Upper Saddle River, NJ: Pearson Education, Inc.

De Figueiredo, A. D., & Afonso, A. P. (2006). *Managing learning in virtual settings: The role of context.* Hershey, PA: Information Science Publishing.

Discenza, R., Howard, C., & Schenk, K. (2002). *The design and management of effective distance learning programs.* Hershey, PA: IGI Global, Inc.

Howard, C., Schenk, K., & Discenza, R. (2004). *Distance learning and university effectiveness: Changing educational paradigms for online learning.* Hershey, PA: Information Science Publishing.

Khan, B. H. (2005). *Managing e-learning: Design, delivery, implementation and evaluation.* Hershey, PA: Information Science Publishing.

O'Neil, H. F. (2005). *What works in distance learning: Guidelines.* Greenwich, CT: Information Age Publishing.

Simpson, O. (2002). *Supporting students in online, open and distance learning.* London: Kogan Page Limited.

Section IV
Course Design and Classroom Teaching

Chapter XII
On the Design and Application of an Online Web Course for Distance Learning

Y.J. Zhang
Tsinghua University, Beijing, China

ABSTRACT

Web course design and implementation are very important in effective distance learning. Such a work is related to a number of issues and needs considerable attentions. In this paper, a feasible framework for developing Web courses and some of our experimental results along the design and application of a particular online course are discussed. It first addresses several major designing considerations to match the expected features with different factors in developing online Web courses for distance learning, takeing into consideration various distinctions between online courses and traditional courses; it then introduces the structure and components more suitable for self-learning. In addition, the tools for developing online Web courses play a significant role and have important influences over the developed courses. Based on some analytical discussions and real experimental results, different developing tools are compared in speed of loading, the file size generated, as well as security and flexibility. The comparison results are not only employed in the present development for improving the flexibility of the new Web course, but also usable for developing other courses. Finally, the principles proposed and the tools selected have been concretely integrated in the implementation of a particular Web course, which has been conducted with satisfactory results.

INTRODUCTION

With the fast progress of computers and network techniques, online distance learning has become a feasible learning tool, especially for our university study. On the other side, with the continuous evolvement of modern technology and society, continuous education turns out to be an indispensable part of present-day life. For example, in Tsinghua University, Beijing (a top one in China) alone, there are already more than ten thousand students (around the country) who pursue their Master's degrees through distance learning with registration to the Institute of Continuous Education. Though these students are all working people and already obtained their Bachelor or Engineering diploma a number of years ago, the new challenge takes them back to study.

Compared to the normal students inside the university campus, the students following the continuous education have some particularities. They are usually working at different regions, sometimes more than a thousand kilometers apart from each other. They often have sparse time to follow the course lecture or have few possibilities to contact teachers though they may not live too far from a university. Some of them even have obtained their diplomas in other disciplines than the new subjects they would like to follow as they are facing the challenge of new techniques. It is clear that due to these particular problems, the courses they need and their learning styles would be quite different from regular ones. In practice, a number of challenges for the design and implementation of online courses are faced (Pohjola, 1999).

In this paper, some general principles for designing considerations in developing online courses are first discussed; both expected properties and performance factors are included. Then, the components and structures of online courses are proposed and specified to distinguish the online courses from normal courses. One important ingredient of online courses is the ability to provide human-machine interaction through interactive demonstrations. How to select the development tools to increase course comprehension is discussed in detail with the help of both analytical and experimental results. As an example of the application of those general principles, a particular Web course has been practically developed and conducted within a network education platform. Both the course implementation and application environment are presented, and some effect evaluations are provided.

GENERAL DESIGNING CONSIDERATIONS

Properties Expected and Factors Considered

As a new means of education, a Web course has some properties/features of its own. Alternatively, the design of a Web course has significant influence on the course performance and education effects. Table 1 notes some expected properties of Web courses and several factors to be considered in the design of Web courses. An "X" indicates that the corresponding factor should take the corresponding property to design the Web course.

A brief discussion is given to Table 1 is given below (more details can be found in Zhu, 2001; Zhang, 2002). It can be seen from Table 1 that both in choosing information resources and in organizing course contents, the current trend in the development of discipline should be reflected. The structure and content of Web courses should be easily extended, not only for supporting association of related course units in a dynamic and layered way, but also for facilitating the selection of information resources and organizing the contents. In addition, the selection of information resources and the manner to navigate inside the course should be helpful for pushing the student's motivation in active and self learning. As the nature of a Web course is imposed, the course

Table 1. Properties of Web courses and factors should be considered in the designing of Web course

Property \ Factor	Information resources	Content organization	Course structure	Course navigation	Content representation
Reflecting the current trend in the development of discipline	X	X			
Structure and content of course could be easily extended	X	X	X		
Prompt student's motivation	X			X	
Different users' sharing of the resources		X			
Easy interaction			X	X	X
Endorses collaboration					X

structure should permit the users' sharing of the resources inside and/or outside the class, as well as the easy interactions between human-machine, teacher-student and teaching-learning. The latter is also very important for course navigation and content representation. Finally, the representation of course content should support the collaboration so learning task could be fulfilled by cooperation.

COMPONENTS AND STRUCTURE OF WEB COURSE

Multi Components

The contents of a Web course are related to the subject of the course, while the components of a Web course depend on the learning styles adopted by the course. Modern theories in educational psychology emphasize that use of different learning styles helps greatly in knowledge acquisition and retention. There are four important learning styles: (1) concrete experience, (2) abstract conceptualization, (3) reflective observation and (4) active experimentation. They can be simply stated as (1) feeling, (2) thinking, (3) watching and (4) doing, respectively. Depending on the personality and environment, a learner may prefer one learning style to another, but learning will be greatly enhanced if one is actively involved in all of them (Feldman, 1997). In addition, this would be some assistance for turning the drudgery of learning into fun and introducing students to impressive adventures of exploring a new domain of knowledge.

According to the above stated theory, computer-based educational programs should strive to offer their learners, if possible, all of the four learning styles. This can be done by providing a collection of different components in addition to normal text notes, such as illustrated examples (with formula, tables, drawing, pictures and video), interactive or animated demonstrations, quick quizzes or self-tests, as well as references and resource links, etc. These components will help the learning in different ways. For example, the animated demonstrations could attract students to watch, the illustrated examples would give students more concrete feeling of complicated concepts, the self-tests could push students for deeper thinking and the interactive demonstrations would fulfill the needs of students for processing tasks practically. In making different combinations, the possibilities for learning will grow. With these components, the course would permit students using different styles of learning to enhance the study.

Course Structure

The organization of course components heavily influences the performance and operation of the course. According to the above discussions, the course structure should support different learning styles to enhance the leaning efficiency. From the leaning point of view, the course is better divided into a number of modules (associated modules can be grouped in chapter), where each module refers to one separate study unit (SU). Each study unit is concentrated with several related concepts, and should be composed of various components (such as indicated in the above sub-section) that have embedded the teacher's teaching experience.

The proposed structure for the online Web course is sketched in Figure 1. When the course started, five lists are accessible. The SU title list provides a list of SUs, which serves as the content of the course. The other four lists are composed of a number of pop-up content modules, where each of them has a particular function. Look at Figure 1 horizontally; each SU is made of component modules from all four lists. The course structure is a combination of tree structure and graph structure. The tree structure provides a fast and logical way to access different branches of the course. On the other side, the graph structure with cross-reference makes the navigation inside the course quite easy and nature. This mixed data structure has similar performance to that of the hierarchical hyper-concept map (HHCM) proposed by Sung (2001).

The Web course is physically composed of a number of Web pages. Since the Web course is accessible from the Internet, people could use net browser tools (for example, IE) to get into these pages. The Web page is the natural unit of the Web course; each studying unit corresponds to an individual Web page. The Web page also provides the framework of the course and serves as a platform for embedding separate component modules. The navigation among different SUs, such as between the former page and the next page can be easily accomplished using functions provided by the browser.

In addition, the proposed structure can be efficiently implemented by using the hyper-link among different Web pages. Using the graph structures inherently embedded in the hyper-link connection, the physical location of different component modules is no long an obscurity to the users of the course. Users can easily follow the structure connection in Figure 1 to select the appropriate course contents.

Organization of Study Unit

The organization for each SU is illustrated in Figure 2. SU is accessible through selecting the appropriate SU title. It is composed of text, tables and formulae for explanation, as well as illustration examples, interactive demos and self-test functions. Examples and demos are made of pop-up windows, so the representation in the SU unit is clear and concise. Reference is provided for each

Figure 1. Structure of the course

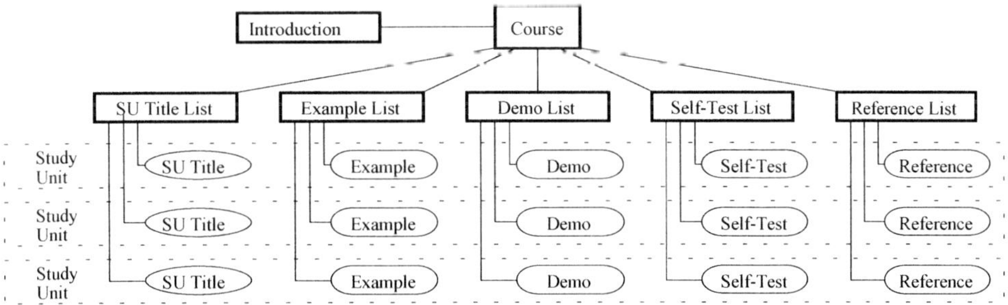

Figure 2. Organization of study

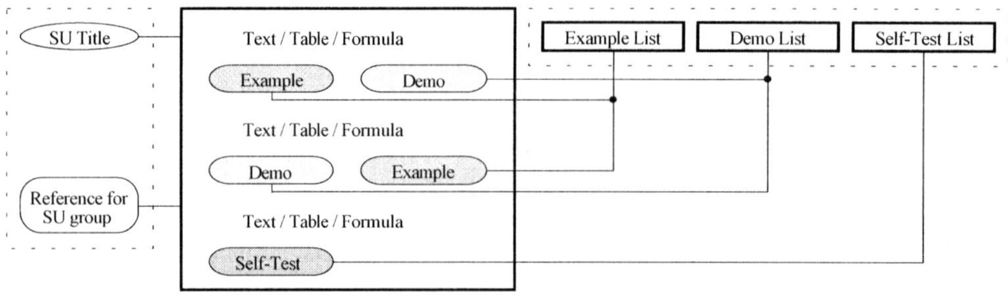

SU, however, to help the reader to concentrate on the course content first, references for related SU are grouped together, and are attainable from the table of contents. While examples, demos and self-tests are directly accessible from the SU unit, they can also be selected from their respective lists, as shown by the connections in Figure 2. This organization can help new users and old users to select different learning styles; as will be discussed.

SELECTION OF DEVELOPMENT TOOLS

In Figure 1, five lists are indicated. All the components of these lists are structured in modules. The title and reference modules are made only with text, so the implementation should be very straightforward. The example module is made of text plus graphics and pre-computed image, so it is still similar to a test module. On the other side, the modules for demonstration and self-test are more complicated; they should provide interactive functions and perform some processing and/or analysis tasks.

Some Existing Developmental Tools

In implementing demonstrations and self-tests, some developmental tools should be employed. In order to free the developers from purely technological concerns, we have confined our choice with tools using sophisticated technology and having widespread usage. Each tool, in fact, has its own advantages and drawbacks, so different tools are selected to serve a specific task in the development. We now give some comments of the five available developmental tools we have had an experience with, in the hope of providing future developers with some useful guidelines and references.

1. **Mathworks Matlab** provides a number of powerful toolboxes for different domains of scientific and/or engineering calculation. Such kinds of devoted toolboxes are most suitable to stay in the back yard and serve as a computing center to process raw data needed in course demonstrations.
2. **Macromedia Flash** can be used for creating and editing animated demonstrations. Flash provides seamless incorporation with Web pages, and compiled Flash Movies are generally small in terms of file size. This is especially desired since many learners need to access a Web course through a modem with limited bandwidth and slow downloading speed.
3. **Microsoft Visual J++** is for programming Java Applets, which are platform-independent, compact in size and capable of doing rather complex computations immediately. The only drawback is that a Web page containing Java Applets generally takes longer

time to load. However, once loaded, it can run very quickly on the local machine.

4. **Microsoft Visual C++** is for developing ActiveX controls, which can be downloaded and installed on the local machine. ActiveX controls are also capable of doing any complicated computation tasks, and can access image files on the local machine.
5. **Macromedia Dreamweaver** is an integrated developmental environment for Web site construction and maintenance. It provides powerful assistance to combine all relevant media components together seamlessly. It also serves as a handy editor of JavaScript, a simple script language for adding controllable functions to the Web pages (e.g., to pop up a new window, to expand or collapse directory trees).

In developing a particular Web course, these tools can be integrated as shown in Figure 3. Matlab toolboxes can serve as a backstage process center to produce resultant pictures or data further used by Macromedia Flash to create visually appealing multimedia, animated demos. Java Applets (programming using Microsoft Visual J++) and ActiveX Controls (compiled using Microsoft Visual C++) can provide dynamic computing power and show calculated results according to user-specified parameters. These two tools emphasize more the visual impact of the Web course. Macromedia Dreamweaver can combine all the results together into a user-friendly Web course interface on HTML pages.

COMPARISON OF DEMONSTRATIVE MODULES

Different implementation tools lend to different styles in the resulting modules. Some discussions and experimental comparison of three categories of demonstrative modules (Flash Movies, Java Applets and ActiveX controls) are given below.

Flash Movies

Flash Movies help learners to understand a new concept by demonstrating a sample process of certain ideas through animation. In this way, Flash Movies would provide learners with a vivid, intuitive interpretation of an abstract conception, and thereby enhance the learners' comprehension. As a common media on the Web, they shall also appeal psychologically to our online readers, who are already accustomed to, and expect always, the fun and excitement of Web surfing. A third virtue is that as images created in Flash are represented as graph vectors, complied Flash Movies are generally compact in size, which facilitates the process of page transmission over today's band limited Internet.

Figure 3. The organization of various tools for developing Web course

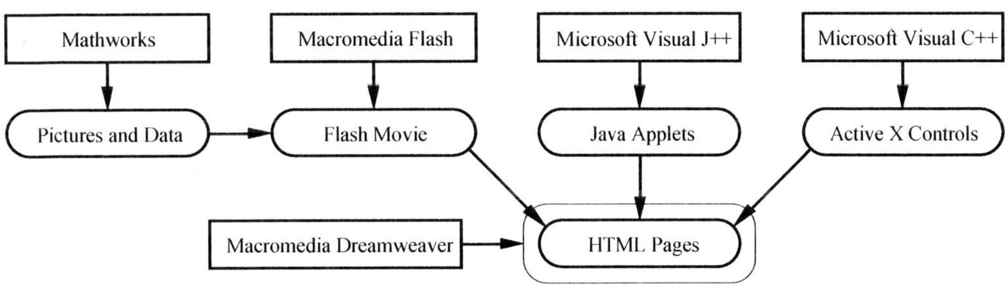

Java Applets

The application of Java technology could help greatly in making the course truly portable and less dependent on slow Internet connection (Ahluwalia, 2000). Each Java applet can serve certain functions with user-customized parameters, or in other words, it allows the user to choose operating parameters according to their own wishes.

ActiveX Controls

With a registered class id, ActiveX controls can process any image file stored in the user's own computer, if it is of the legible/permitted type. The resultant image can also be locally saved. Obviously, this offers the learner much greater flexibility. In addition, as ActiveX controls are programmed with C++ and pre-compiled before running, they also have of greater calculating power and higher runtime efficiency.

Comparison

Aside from the qualitative comments on each kind of produced components, an evaluation and comparison experiment of the features and performances of the three categories of demonstration modules in a numerical manner is also conducted. In this experiment, 16 Flash Movie files, 12 Java Applets and 12 ActiveX Control modules are tested for their average page load time of the local machine with a virtual stopwatch. Using the campus network, these files are downloaded from the server on course site. In Table 2, the average file sizes (including accompanying images) are listed with the experimental results to give a hint on the possible transmission time. These experimental results are obtained with a Pentium II 800 machine. However, the platform is not critical because what is interesting and important here is not the absolute time, but the relative ranking, as the selection of tools is based on their relative performance.

From the information above, it seems that Flash Movies are most suitable for network transmission. Java Applets are moderate in size but take quite a while to be loaded by the browser. ActiveX Controls are pre-registered on the local machine and will virtually take no time to start up once approved by the user. Although it is of remarkable size, it can be downloaded beforehand. In some sense, these three kinds of modules rank in an upward sequence from Flash to ActiveX in terms of interactivity, flexibility and process power.

However, components with greater computational power and runtime flexibility generally score less in terms of access security or stability, and vise versa. For instance, an ActiveX Control component has the greatest computational power, and allows the user to manipulate image files on his own machine, whereas a Java Applet will only allow the user to select from a number of pre-limited pictures and a Flash Movie gives no more than a pre-recorded demonstration. On the other hand, Flash Movies will do no harm to

Table 2. Comparison of development tools

Component Type	File Number	Avg. Load Time	Avg. File size (Including images)
Flash Movie	16	0.466 sec.	38.3 kB
Java Applet	12	1.319 sec	93.5 kB
ActiveX Control	12	< 0.4 sec	163 kB

Figure 4. Comparison of three kinds of modules used in IP&A-Web

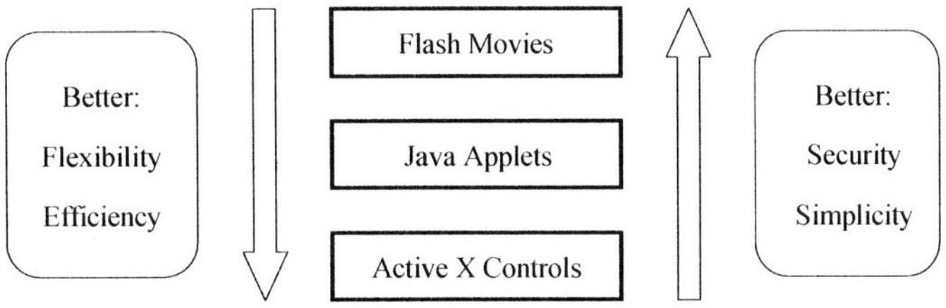

your local machine, and Java Applets will work on image files temporarily downloaded to your virtual machine, but there exists the danger of damaging your own files if they are of the illegal type and are improperly processed by the ActiveX controls. The comparison of these three kinds of component, and thus three tools, is summed up in Figure 4.

IMPLEMENTATION AND APPLICATION

Course Implementation

Based on the above discussions, a new Web course named "Fundamentals of Image Processing and Analysis", which is targeted specifically to the continuous education program for working people, has been implemented. Different from some other image processing CAI systems, which are only structured according to different categories of image operations (Ahluwalia, 2000), this course aims at providing multiple options to help learners with each new conception in image processing and analysis. It is a concrete course, in which theory and practice are closely related. Therefore, the interactive demonstrations and manipulations are thus very useful for capturing the meaning of different concepts. In addition, a number of practical image processing techniques and algorithms are discussed in the course. To give students a deep impression of related theories, showing the results obtained by applying these techniques is mandatory and can help the understanding of the principles.

In course implementation, the content has been divided into 60 study units. Each unit has a few numbers of examples and demonstrations linked to the related concepts in this unit. For each study unit, at least three quick quizzes or self-tests are provided near the end of the unit. All these examples, demonstrations and self-tests are convoked by keywords to appear in new windows, so the main stream of text can be kept concise. New users can simply follow the structure of the study unit for progressive learning.

On the other side, as indicated in Figure 2, different modules are also organized into several lists. From the example list, the user can get access to 91 examples with several hundreds of pictures and graphics. From the demonstration list, the user can select 36 interactive programs for image manipulation. From the self-test list, the user can pick up one of 219 quizzes with hints and explanatory answers to test himself. Note that the hints are important as the user can get more insight from it for better understanding the problem and can find the best way to give the solution. In addition, this can split a test into two levels, so it would be more efficient and flexible. Explanatory answer is also important as the learner can deeply capture the principle of problems. It helps learner to learn more things from making exercises. Reading the explanation is also an education phase.

Some interface examples can be found in Zhang (2002).

Application Environment

The above course has been conducted with the help of "Tsinghua WebSchool" (Web, 2002). "Tsinghua WebSchool" is a Web-based learning system made by Tsinghua University. It is a platform for conducting the tele-learning. The main functions provided include make announcement, help the query and answer between instructor and students, and some courseware management, as well as material download. It can be seen that this platform provides a suitable environment for continuous education (Collis, 1999).

Among the several thousands of registered students who pursue their Master's degrees through distance learning by using "Tsinghua WebSchool", more than 700 students followed the course "Fundamentals of Image Processing and Analysis" in 2002. As they are distributed in more than 50 cities around the country, with some of them as far apart several thousand kilometers from each other, such a Web course would be quite convenient.

Effect Investigation

An effect investigation/performance evaluation using the following 10 questions was carried out in order to get some overview of the course for credibility and effectiveness. The questions are: (1) Is it simple to access? (2) Do you think its interface is friendly? (3) Is it easy to manipulate? (4) Are the functions comprehensive? (5) Is it closely related to the course? (6) Does it clearly illustrate the concepts of the textbook? (7) Is it heuristic for your learning? (8) How do you think about its interactivity? (9) Does it provide assistance for your learning? (10) Are you satisfied with its running conditions? These 10 questions are designed on the basis of an early investigation work for a stand-alone courseware (Zhang, 1999) with some adaptations and improvements. Among these 10 questions, the first five are more for judging the functionality of this Web course and the last five are more for evaluating the competency of using this Web course in education.

Students are asked to select only one from four levels of ranks (Inferior, O.K., Good, and Excellent) for each of the above 10 questions. This would not be a cumbersome task, as most of these questions are quite concentrated only on a particular aspect of the Web course, so the answer can be simply made. Alternatively, with the coverage of different aspects, the statistic based on all the answers together could show a common opinion and provide some useful insights into suitability of the course.

The statistics of answers for these questions are depicted in Figure 5. It seems large numbers of students are quite satisfied as seen from the statistical results. Roughly speaking, for each question, around one third of students selected the "Excellent" answer, and more then two thirds selected the "Excellent" answer or the "Good" answer. This is quite encouraging. In fact, more than 90% of answers are confirmative (for questions 5, 6, 9 and 10, 100%), this indicate that this course in general has been developed successfully.

The statistical results in Figure 5 also indicated that a more user-friendly interface is required. One reason for this problem would be that as three modules (developed with Flash Movie, Java Applets and ActiveX Control, respectively) are used in one course, the interfaces of different modules have some variant appearances. According to the study and comparison made earlier, development tools–with each of them having its particular strength and weaknesses–should be used for modules of different purposes. Therefore, the solution for this problem would be to make the interfaces of different modules more similar, which would be taken into account for the next version. Another point should be discussed is related to question 7, for which the "Inferior" answer number is higher than that of other questions.

Figure 5. Results of evaluation

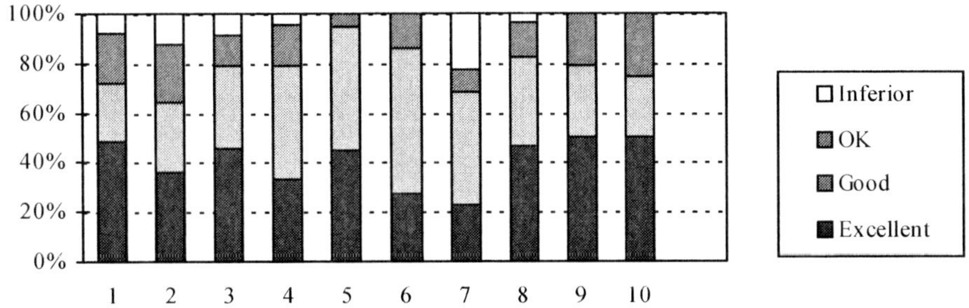

Though the Web course provides many modules for showing visual information and for prompting student's motivation, this is merely to make each study step more efficient; the whole strategy for learning is barely changed. To make the learning more heuristic, we believe, some further works on using some more suitable strategies for distance learning should be conducted.

CONCLUDING REMARKS

In this paper, a number of rules of thumb for designing online Web courses are discussed. These considerations serve as guidelines in the design of a particular Web course for distance learning. They have also been incorporated into the real implementation of this Web course that is highly scalable, and with suitable interactivity and portability.

In addition, different learning components to provide different learning styles and to increase learning performance have been created and integrated into a manageable course structure, which provides some flexible access ability and easy navigation capability. This module-based structure, with the organization of study units, also reduces the possible disorientation and cognitive overload in learning practice.

Finally, suitable development tools are compared and selected for different demonstration and interaction purposes. Three types of modules, each with appropriate properties for pre-defined functions, have been developed. Their combination makes the course with more attractive and comprehensible features, as indicated by the primary evaluation results.

ACKNOWLEDGMENT

This work has been supported by The Ministry of Education (NENC-2000-29).

Several students, especially W.J. Liu, X.Q. Zhu, S. Y. Dai, F. Jiang, H.J. Hu, and D. Xu, have contributed to the implementation of this Web course.

REFERENCES

Ahluwalia, A.K., Jonker, P.P & Young, I.T. (2000). An interactive image processing course for the Web, Proc. *First International Conference on Image and Graphics*, Y.J. Zhang ed., 589~593.

Collis, B. (1999). Design, development and implementation of a WWW-based course-support system, *Proc. International Conference on Computer in Education*, 11~18.

Feldman, L.J & Hofinger, R J. (1997). Active participation by sophomore students in the design of experiments, *ASEE/IEEE Frontiers in Education Conference*, 1526~1527.

Pohjola, V.J. (1999). Knowledge integration as a challenge for future education, *Proc. International Conference on Computer in Education*, 2: 299~306.

Sung, Y.T, Chiou, S.K & Chang, K.E. (2001) Use of hierarchical hyper-concept map in Web-based courses. *Proc. International Conference on Computer in Education, 1*133~1137.

Web (2002): *http://www.itsinghua.edu.cn* or *http://www.itsinghua.com*

Zhang, Y J & Xu, Y. (1999). Effect investigation of the CAI software for "Image Processing and Analysis". *Proc. of International Conference on Computer in Education'99,* 858-859.

Zhang, Y.J & Liu, W.J. (2002). A new Web course:'Fundamentals of Image Processing and Analysis'", *Proc. 6th Global Chinese Conference on Computer in Education,* 1: 597-602.

Zhu, X Q, Zhang, Y J & Liu, W J. (2001). IP&A-Web: an online course of image processing and analysis. *Proc. of International Conference on Computer in Education'01*, 729-734.

This work was previously published in International Journal of Distance Education Technologies, Vol. 2, No. 1, edited by T. Shih, pp. 31-41, copyright 2004 by IGI Publishing, formerly known as Idea Group Publishing (an imprint of IGI Global).

Chapter XIII
Teaching Information Security in a Hybrid Distance Learning Setting

Michael E. Whitman
Kennesaw State University, USA

Herbert J. Mattord
Kennesaw State University, USA

ABSTRACT

This chapter provides a case study of current practices and lessons learned in the provision of distance learning (DL)-based instruction in the field of information security. The primary objective of this case study was to identify implementations of distance learning techniques and technologies that were successful in supporting the unique requirements of an information security program that could be generalized to other programs and institutions. Thus the focus of this study was to provide an exemplar for institutions considering the implementation of distance learning technology to support information security education. The study found that the use of lecture recording technologies currently available can easily be used to record in-class lectures which can then be posted for student use. VPN technologies can also be used to support hands-on laboratory exercises. Limitations of this study focus on the lack of empirical evidence collected to substantiate the anecdotal findings.

INTRODUCTION

Information security (InfoSec) is an academic discipline that represents a teaching discipline that is distinct from the traditional fields of information systems, computer science, or information technology. As InfoSec programs are designed and implemented in institutions throughout the country, many instructors are struggling to develop programs to educate students in this new and exciting area. While teaching InfoSec does share some of the same challenges as those of other information technology topics, those not familiar with the specifics of the information security profession will find it difficult to develop curriculum without specialized outside support. With the shortage of established programs, many students interested in studying in this field are pressuring those institutions with established programs to provide distance learning (DL) options. This phenomenon is coupled with past experience that finds many InfoSec students are seasoned IT professionals, usually returning to academia for some specialized education when on-the-job experience is not available or is insufficient. These IT professionals usually maintain their current employment, further demanding alternative educational experiences that are flexible enough to deal with the irregularities of business travel, emergency demands on the employee time, and ongoing business change. The result is that many academic institutions, even those that have barely established coursework in the discipline at all, are beginning to evaluate distance learning support to provide service to a wider student base.

InfoSec education curriculum includes many topics, some technical and some managerial. Most of the specialty topics within the broader InfoSec domain will have improved learning outcomes when the theoretical elements of the subject delivered to the students as reading assignments and lectures are reinforced with additional interactive learning opportunities. The optimum learning environment will combine the best elements of theoretical instruction, using reading assignments, lectures, seminar discussion, and research assignments, reinforced with interactive modules made up of lab tutorials, lab exercises, electronically mediated content such as videos and Web-based seminars, and lab demonstrations. The combination of passive and active learning approaches will prepare the student for the integration of the theoretical material into the students experience with real-world opportunities when cooperative studies and/or internships are available. Many of these components lend themselves to distance learning, while others require substantial investigation into how to best meet student needs and modifications to current practices in order to sustain academic rigor yet provide the asynchronous distance support demanded by most students.

In 2004, Kennesaw State University began the Bachelor of Science in Information Security and Assurance, only the second such program in the U.S. at a public institution, and the first in the Southeast. Having pioneered the development and offering of undergraduate programs in information security in the Southeast since 2000, Kennesaw State is recognized nationwide for the quality of its programs and the expertise of its faculty. The faculty members teaching in this program have published a number of textbooks on the subject, have conducted an annual conference on information security curriculum development, and have made numerous presentations internationally on the subject.

This chapter provides a case study of current practices and lessons learned in the provision of distance learning-based instruction in the field of information security. The primary objective of this case study was to identify implementations of distance learning techniques and technologies that were successful in supporting the unique requirements of an information security program that could be generalized to other programs and institutions. A secondary objective was to identify the limitations of the program that other institutions must address before implementing.

PREVIOUS WORK

The previous work published in the area of distance learning is far too wide and varied to completely summarize here. By way of a focused synopsis with direct influence on this chapter, a few salient references in the literature are briefly reviewed. The reader is encouraged to explore these resources more fully.

In Spring 2006, Tallent-Runnels, Thomas, Lan, and Cooper (2006) did prepare a review of the research related to online distance learning. The authors found that the existing body of research was readily organized into four topics: course environment, learners' outcomes, learners' characteristics, and institutional and administrative factors. One of their most interesting findings was that most institutions currently performing distance learning have no written policies, guidelines, or technical support for faculty and students engaged in these activities. The absence of such support essentially puts those engaged in distance learning on their own, without guidance or support, setting them up for failure, saved for those talented and hard working faculty who find themselves providing their own support.

Figure 1. Location and time as instructional variables

		Location	
		Same	**Different**
Time	**Same**	**I.** Traditional Classroom Instruction	**II.** Synchronous–Satellite Campus Distance Learning Instruction
	Different	**III.** Asynchronous "Time-shifted" Classroom Instruction	**IV.** Asynchronous Distance Learning Instruction

ASYNCHRONOUS VS. SYNCHRONOUS LEARNING OPTIONS

When considering distance learning support for an academic program, one typically turns to a common 2x2 examination of academic educational delivery systems. As shown in Figure 1, this 2x2 matrix compares location and time as the two variables that can be considered.

Traditional Classroom Synchronous Instruction

Quadrant I from Figure 1 is the traditional classroom with a fixed position in both space and time. Students arrive at a predetermined location at a predetermined time, and typically receive instruction directly from the instructor. This is considered synchronous instruction, not distance learning. If the instructor were to be providing the lecture from a remote location, the scenario would fall under the next category.

Synchronous–Satellite Campus Distance Learning Instruction

Quadrant II in Figure 1 is the synchronous-satellite campus model. This model, popular in the early days of distance learning, involves student remotely, usually via video conferencing where a live instructor provides the course materials. An alternative version of the model is the provision of a distance or guest instructor, who was physically remote from the classroom. Originally, this model used broadcast television to provide distance learning options for students. Many students enrolled in classes at a main campus, and then watched distance learning lectures transmitted via local closed circuit or cable television broadcasts.

Advances in Internet-based distance learning applications are increasing the number of options faculty have in the provisioning of the synchronous model. As the experiences of the authors are not in the synchronous category, this option will not be explored in this case. However, it is possible to incorporate synchronous DL into any program, so long as the bandwidth issues discussed in later sections are considered. Solutions in this category typically require service contracts, and even external service providers to ensure quality of service of the live audio and video streams.

Asynchronous "Time Shifted" Classroom Instruction

Another instantiation of the distance learning classroom is the asynchronous "time shifted" classroom as indicated by Quadrant III of Figure 1. In this implementation, students arrive in the same classroom where instruction was previously delivered, but at a later time, to observe a recorded version of the instructional material. To date, the authors are not familiar with any implementation of this model in the academic arena. While theoretically possible, the fourth and final quadrant is far more common, and thus preferred.

Asynchronous Distance Learning Instruction

The most commonly implemented version of distance learning is the asynchronous distance learning instruction where students attend instruction at different locations and different times than the instruction was captured. The implementation of this type of instruction varies widely from packaged distance learning courses to the use of recorded lectures. Each of these differ mainly in the methodology used to store and then present the material, as well as the amount of live human interaction the student experiences. Several of the more common implementations are presented next.

Paper-Like Media

One of the sparest forms of information distribution is the organization of subject matter into a series of documents, computer screen images, Web pages, or slides that students read and absorb at their own pace. In its simplest form, paper-like media is essentially a collection of text with supporting images containing a straight-forward presentation of the information. This form of delivery is common in commercial distance learning self-paced courses, and provides very limited student engagement. The material may contain diagrams, illustrations, or figures, and may be in hard-copy or electronically delivered by CD or the World Wide Web. Few institutions of higher education support the exclusive use of this type of material, using it as a supplement to other forms of instructional delivery. The most common use of this type of content delivery is as course support, as in Blackboard or WebCT systems, designed to provide remediation for students having difficulty in the primary instructional delivery.

Multimedia Hypertext Media

The basic methods used for paper-like media can be enhanced to improve instruction and learning, though the integration of multimedia content. By adding computer animations, audio and video clips, and the like, the instructor can increase the content retention by the students and improve their learning. The content can still be delivered via the WWW, or by CD/DVD, providing portable, widely accessible content, which can be viewed at the student's leisure.

Recorded Lecture

The most widely used method to provide distance learning is the recorded lecture. In this model, the instructor records a delivered lecture from a studio or classroom, and then makes the recording available to the student. The use of recorded video lectures has been around since the availability of video recording and viewing equipment. Many executive MBA programs began with a lecture-by-mail model, where the instructor would record either a live in-class lecture or stage a studio-based presentation. The institution would then mass produce the video media and distribute them. Variations of this model continue today.

With the advent of digital audio and video recording, and publishing on the WWW, more and more institutions are following variants of this model. The case in this chapter elaborates on this approach using modern audio and screen-capture technology to create the lecture source files, which are then converted into a widely accepted Web streaming movie format. While the technical details may vary between proprietary and standardized formats, the methodology is essentially the same: capture the source lecture once, and render into a format viewable by the student.

From a commercial standpoint, awareness and training related education opportunities are often delivered using a Web-based seminar. Using both synchronous and asynchronous models, industrial marketing professionals are scheduling live seminars promoting various product lines, and inviting potential buyers to participate in a live teleconference that makes use of video and slide materials. During this event, the guest presenters can answer questions provided by phone, e-mail, or chat. Once the event has concluded, the recorded presentations are posted on the vendor's Web site for viewing by other potential clients.

One of the newest variations on the asynchronous record-and-post methodology is the capitalization on the iPod™ phenomenon. Apple Computer's portable audio and video players have become ubiquitous, and their competitive counterparts are also widely available. Once a source lecture has been captured, rendering it into a wide variety of formats is simply an extension of the technological capabilities of the software. If many students have the ability to listen, or even view recorded material specifically composed for their portable audio/video devices, then it is only natural that innovative faculty will realize this opportunity and convert their instructional material to take advantage. In addition, many working professionals use recorded lectures during commute trips to and from work and school, providing an increase in efficiency and productivity for the working student.

Enhanced Recorded Lecture

The final category of asynchronous lecture delivery is rapidly gaining momentum in the academic community. Taking the foundation of the recorded lecture, and applying additional capabilities through software, allows the student to use an enhanced recorded lecture to increase their learning experience. The traditional lectures is recorded, usually by computer-based capture software, then revisited by the instructor who threads value-added components to the material, such as incorporating active Web links, hypertext documents, imbedded supplemental content, animations, simulations, threading audio and video clips, and adding computer-based assessment tools creates a true enhanced learning opportunity for the student. While this does create additional work for the faculty, students experience an elevated sense of involvement in the instruction, as opposed to simply listening or watching a recorded presentation.

Media-richness theory is a well researched IS arena, and generally finds that the "richer" the communications, the higher the degree of information absorption (e.g., Daft & Lengel, 1986). In the arena as applied to distance learning, Shepherd and Martz (2006) found that when considering three key evaluation criteria, that is, "reported satisfaction with the DL course or program, more reported communication, and higher valuation of the course delivery platform, the studies found that the more media used effectively in a distance program, the greater the satisfaction with that program. Effective use of the technology further enhances the communication and ultimately the satisfaction with the program."

Student Interaction

One concern voiced by faculty and student alike in the development and conduct of distance learning is the level of student interaction with the instructor. The delivery of asynchronous content can result in a feeling of disconnection by the student which is offset by the degree of convenience that it offers. The level of interaction with the instructor becomes paramount to keeping the student engaged in the class. Even in synchronous content delivery, it is difficult to keep students engaged. This requires that synchronous and asynchronous delivery methods should plan for direct and/or indirect interaction between instructor and student.

Many faculty members accomplish this feedback by combining e-mail, posted discussion lists, chat sessions, and extended office hours. With synchronous content delivery, most software-based solutions provide mechanisms whereby students can type their questions in a chat box, or raise their electronic "hands" and be granted the opportunity to transmit questions to the entire class. This requires the faculty member to gain additional skill in the assessment of online content. Marra (2006), for example, provides a review of methodologies for assessing content of computer-mediated discussion forums, useful in this situation.

Even in the absence of this type of technology, instructors frequently keep open e-mail clients to look for and respond to student questions. For those instructors that use video conference-based systems, the technology should permit two-way communications in order to be most effective. This causes problems when the number of remote satellite classes exceeds two or three locations, as the video conferencing technology typically then requires the use of a multiplexer to handle the multiple streams. As indicated earlier, Web seminars often couple a one-way Web-based streamed presentation with a disconnected telephone conference call. Many institutions report that these calls can be prohibitively expensive for use in ongoing classroom instruction. While a marketing department may be able to afford a call once per month or per week for a product sales pitch, an academic institution could not afford such a conference once per class, with the typical class meeting 2-3 times per week for 16 weeks (in a typical semester).

In their summary of previous research on distance learning, Tallent-Runnels et al. (2006) found that "asynchronous communication seemed to facilitate in-depth communications (but not more than in traditional classes), students like to move at their own pace, learning outcomes appeared to be the same as in traditional courses, and students with prior training in computers were more satisfied with online courses" (p. 93).

Blended Learning: The Hybrid Approach

There is of course another option, that is, an approach that blends elements of each of the major strategies. In this option, referred to as the hybrid, blended, or partial DL approach, faculty members select the best of one style and combine it with the best of others. For example, a faulty member could lecture in class to students who prefer to attend live, face-to-face lectures, yet record the content for use by other students unable, or unwilling to attend these time-constrained meetings. The blended approach is generally considered the most palatable to faculty and students alike. The strengths of the best approaches are capitalized upon, while the weaknesses arebmitigated. In a review of perspectives on blended learning in higher education, Vaughn (2007) finds that:

Students indicate that a blended learning model provides them with greater time flexibility and improved learning outcomes but that initially they encounter issues around time management, taking greater responsibility for their own learning, and using sophisticated technologies. Faculty suggest that blended courses create enhanced opportunities for teacher-student interaction, increased student engagement in learning, added flexibility in the teaching and learning environment, and opportunities for continuous improvement. (p. 81)

Knowledge of these challenges goes a long way toward allowing faculty and students alike to avoid these issues. In many of the blended courses taught in the InfoSec program discussed in the case that follows, an advisory warning is given to students warning them of the challenges they will face, especially in time management.

ADDITIONAL CONSIDERATIONS FOR DISTANCE LEARNING

There are several issues to consider in the implementation of the distance learning classroom. Bates (1999) asserts that the selection of technology should be based on the academic needs of the students rather than the tendency to chase the latest technology trends. Britain and Liber (1999) further assert that distance learning platforms must lower teacher's administrative load, supporting efficient work in the institution and allowing teachers to respond to individual student educational needs.

Chickering and Ehrmann (1996) developed seven general principles concerning the use of technology in the support of education:

1. *Stimulate student-teacher contact.*
2. *Stimulate cooperation among students.*
3. *Stimulate active learning.*
4. *Offer a fast feedback with students.*
5. *Highlight the time invested in the assignment.*
6. *Transmit high expectations.*
7. *Respect the different learning abilities and styles* (p. 3).

Consideration of these principles, coupled with additional considerations of the structure of the DL model, the general purpose of the program, the expected outcomes of the students, and the capabilities of the institution should influence the model for distance learning selected (Mena, 2007).

The Information Security Classroom: DL Support

In this section we transition from the pedagogical approach to distance learning instruction covered in the previous section to the specifics of the technology tested and refined for use in a specific classroom, using the KSU experiences as a foundation. When designing the information security classroom, there is little difference from the challenges facing the creation of any distance learning classroom. The first foray into distance learning for the newly created information security courses was in 2000, when a distance learning "studio" was created. A server using Real™ technologies was set up running Real Presenter™, allowing the instructor to record lectures and post using the Real™ proprietary format. This allowed the faculty member to cover lost classes, but really did not reduce the work for the instructor. This studio was abandoned in 2003.

In 2004, when the newly approved BS-information security and assurance (ISA) was approved, the faculty decided to make distance learning an integral component of the program. The first task faced in the design of the information security distance learning component was the selection of the DL quadrant/model to be used. After extensive discussion, the faculty decided that the asynchronous model would be best, as synchronous interaction between faculty and student was not essential to the assimilation of the information. At the time, there were still few turnkey applications available for synchronous instruction at a reasonable expense. Once the asynchronous DL model was selected, the design of the classroom became the next priority.

In each of the following sections, a specific component of the distance learning program is presented, beginning with the requirements to convert a standard instructional classroom to a distance learning-capable instructional platform. Subsequent sections address the specifics of laboratory exercise support, and server-side file sharing.

Classroom Support

Building on a standard classroom equipped with a data projector and laptop connection capabilities, the design team began to evaluate available market for instructional components. After looking and experimenting with a number of solutions, the faculty selected Camtasia (http://www.techsmith.com/camtasia.asp), a program typically used for creating application computer-based training courses. The tool contains the ability to record screens ranging from a specific application window to full-screen, and integrate audio from the system or external sources. The next step was to select a platform to host the DL instruction. Based on the instructional methodology currently employed, the faculty desired to capture a live lecture, thus optimizing their work effort. The faculty selected Tablet PCs to incorporate both the Camtasia software, and the integral Windows Journal allows the importation of a slideshow, yet allows the instructor to add blank slides to use as whiteboards on the fly. With the combination of Camtasia and Windows Journal running on the Tablet PC, the instructor essentially has a mobile instructional workstation they can create and use slideshows, whiteboards, Web sites, and applications. Sound was integrated with the incorporate of an external wireless microphone solution, converting the ¼" output to 1/8" and connecting to the microphone port on the Tablet PC. By connecting this solution to the in-class data projector, the instructor eliminates the need to use the whiteboards and can both lecture and record at the same time.

Once a lecture is recorded with Camtasia, it can be rendered into a number of formats. The faculty selected Windows Movie (.wmv) format since it is widely compatible, and offers reasonably good compression keeping the size of media files to a manageable size. The Camtasia application allows for granular configuration of both audio and video rates to balance file size and quality of recordings. The lectures were then posted on a password-protected Web site for student download.

At one point the faculty considered media-based distributed recordings, and even pilot tested one application. This application allowed the protection of recorded media (CD or DVD) integrating an application allowing the use of password restrictions. Unfortunately this application required a per-copy license (of approximately $.25) making it unusable for mainstream courses. Since that time the emergence of freeware encryption utilities like TrueCrypt (http://www.truecrypt.org/) make this option more feasible. Instructors create a password protected TrueCrypt volume, insert the lecture, and then copy it to CD/DVD for distribution. The TrueCrypt volume creates an encrypted container securing the content from unauthorized access.

Course Support

The next phase in the design of the DL support for the information security program was to rely heavily on the university course support application, WebCT and eventually WebCT Vista, for student assignments, submissions, and notifications. With the ability to archive and upload subsequent sections, the solution, while far from ideal, was the best available. The optimal solution would be one that integrated the recorded lectures with the course support materials in the enhanced recorded lecture format discussed earlier.

Project and Team Function Support

Many postsecondary educational programs seek to engage the student in group work. This is a valuable way for students to gain experience in using the theory taught in the lecture components of classes. It is also an important extension to student capabilities as students learn to work well in group tasks, a setting that is used in many business tasks. This practical exposure to project work, almost universally valued by future employers, is less easily attained when student take courses that rely on DL techniques. Some mechanism is necessary in the DL approach to offer students these course experiences in teamwork.

Replicating the project team experience is possible using distance learning approaches, but requires the course instructor to create an environment that encourages student engagement and participation. Discussion boards and groupware support tools that enable file sharing can be included by the instructor into an approach that brings students together in a virtual way to collaborate on completing assignments and performing project work.

Legal Issues

An important consideration for the migration of information security classes to online variants, and equally a concern for other programs, is the management of protected intellectual property. While current copyright law allows for more liberal use of information for educational support, once that material is placed in a publicly accessible location, it is considered republished and different rules apply. The protection of IP requires faculty and students alike to ensure that any information distributed as part of the educational process is restricted to those involved in the curriculum. The use of password protected storage locations is essential. The use of course-support sites like Blackboard or WebCT provides a mechanism that assists the faculty in retaining control of the IP, and thus avoiding potential legal issues. Nemire (2007) provides additional considerations.

DISTANCE LEARNING AND THE INFORMATION SECURITY LABORATORY

A valuable extension to the learning in information security education occurs when students experience hands-on laboratory exercises (labs). The creation and operation of information security labs have sufficient challenges of their own. The needs and approaches used to deliver lab experiences and the physical makeup of information security lab facilities has been documented extensively elsewhere. It is possible to replicate all aspects of the lab experience in a remote access environment, when it is necessary. The availability of the requisite bandwidth and the technological infrastructure to make it a success will be dependent on the resources and circumstances of each situation.

As evidenced by Duan, Hosseini, Ling, and Gay (2006), integration of hands-on laboratories with distance learning infrastructures requires considerable examination. Once labs are removed from the local, physical requirement to a global, online availability, the ability to offer virtually any InfoSec course online becomes feasible. As we consider how to bring lab experiences to distance learning students, five options have been identified. These options are not mutually exclusive, rather they offer the instructor a range of possibilities that can be combined as needed based on the learning outcomes being sought and technical infrastructure that are available. These options are: physical presence, student's facilities, loaned facilities, VPN to physical lab, and VPN to virtual lab. These options are explained below and shown in Table 1. A fourth option is to use VPN access to lab LAN and then remote operation of the lab's physical devices. The final option is to use VPN access to lab LAN and then use of virtual lab systems.

Hardware Options

When implementing distance labs, the hardware environment will dictate what can be accomplished. When Option 1 is used and students must attend a physical lab, shared use labs are often in place. In this case, the instructor must use what is available. When possible, attempt to make the lab systems capable of network segregation (see next section).

Option 2 will benefit from specifying capabilities for students to acquire rather than specific devices. For instance, an instructor may instruct students to have a broadband Internet connection, a broadband router, a Microsoft Windows PC, and a Linux PC among other requirements. The more flexibility that is built into the specifications, the more likely it will be that students can comply.

When Option 3 is chosen, equipment should be specified to make transportation and configuration as straightforward as possible. The use of small networking devices and laptop computer systems can make is possible to transport an entire network configuration in a relatively small container.

Option 4 can be accomplished with a medium- or high-speed network connection and a moderately priced VPN concentrator. Many small office/home office (SOHO) grade routers also offer inbound VPN functionality that can support this capability.

In order to facilitate remote virtual operation, one or more server systems will be required. These systems can be configured with a product like Microsoft Virtual Server or VMWare GSX server. This will enable the instructor to preposition operating system virtual containers or require students to build their own operating system images.

Network Options

Once again, while legacy and shared lab designs are not incompatible with the needs of the InfoSec lab curriculum, some factors in network implementation can make operating an InfoSec lab easier. The strongest case for specialized design is in the area of network architecture since Infosec lab assignments have a unique nature. A certain

Table 1. Distance learning lab options

Option	Description	Advantages	Disadvantages
1. Physical Presence	Students must attend lab sessions in a physical lab.	Useful when sensitive topics are part of the curriculum or specialized devices requiring hands-on interaction are required.	Not a distance learning mode.
2. Student's Facilities	Students provide specified computer and networking devices.	Many students have broadband connections from a home LAN.	Limits the labs that can be performed as most students will not be able to provide specialized equipment.
3. Loaned Facilities	Institution provides computer and network equipment to student for use at the student's location.	Useful in hybrid settings where students have periodic physical contact.	Higher than average loss of equipment; high degree of instructor effort.
4. VPN To Physical Lab	Students access lab LAN using a secure tunnel. Remote connections are made to specific lab computers.	Maximizes use of physical lab investment.	Limited by physical lab capabilities and requires high degree of coordination by instructor.
5. VPN To Virtual Lab	Students access lab LAN using a secure tunnel. Remote access to computers and network devices is enabled including virtual server access.	Highly flexible.	High startup cost and complex management requirements.

minimum degree of isolation is required for some InfoSec exercises, and keeping the use of some software tools away from the campus network may be advisable. This is equally applicable for physical as well as virtual labs.

Default Route to Campus

The usual circumstance on many campuses is to use default routing to the Internet. This is where every client can get to the Internet for outbound requests. Whether or not outside traffic is allowed to connect to campus servers or is allowed only to specific addresses and/or ports using a firewall is a matter of campus policy. For lab purposes, the inbound situation is not usually an issue.

Using Network Address Translation

One common practice is to take the lab computers to local addressing. By using a locally hosted dynamic host control protocol (DHCP) server that uses network address translation (NAT), the lab computers can be kept somewhat isolated from the campus local network and still allow access to the Internet for browser access.

When making the lab subnet a nonroutable network using local addressing, you preclude lab students' activities using lab computers from inadvertently spilling over into the campus network. This will not necessarily preclude purpose misadventures by students, but will certainly make the miscreants identifiable to campus network administrators, since all traffic coming from the lab will have the lab router address assigned to it.

The equipment used to isolate the lab network can be one of three grades: residential, SOHO, or commercial. Along this continuum, the expense increases along with reliability and robustness of the devices. At KSU, initial attempts to use a server computer using Microsoft 2000 Server to provide DHCP and NAT found it was less reliable than required and took more than expected effort to configure. This solution was replaced using a residential grade cable/DSL router. This worked and was easy to configure, but required occasional rebooting and experienced unreliable performance. By increasing the asset from one with a cost of about $45 to one with a cost of about $800, a very reliable mainstream SOHO router can give near-commercial reliability within the reach of most lab budgets. An added benefit is that devices in this price range will provide the inbound VPN capabilities needed to support Options 4 and 5 for distance learning.

DMZ with Hardware Appliances

If the budget permits (or the vendor makes a donation) the lab designer may want to consider putting in all of the essential devices for a full blown screened subnet such as is used for commercial Web enterprises. Since this is an option that will provide all of the capabilities needed in any lab setting, the only comment is that if you can afford it, or have it donated, get the expert

Table 2. Reserved nonroutable address ranges

Class	From	To	CIDR Mask	Decimal Mask
Class "A" 24 Bit	10.0.0.0	10.255.255.255	/8	255.0.0.0
Class "B" 20 Bit	172.16.0.0	72.31.255.255	/12 or /16	255.240.0.0 or 255.255.0.0
Class "C" 16 Bit	192.168.0.0	192.168.255.255	/16 or /24	255.255.0.0 or 255.255.255.0

help you will require to get it fully configured and ready for use.

Internal Lab Subnets

Within the lab subnet, the distribution options are many. One option is to provide a home wiring run for each computer to a central hub or switch. Another option, which can be used to good effect for group assignments on firewalls and intrusion detection systems, is to have clusters of client computers use small (4-8 port) hubs or switches to enable day-to-day individual connectivity and ad hoc group segmentation when necessary for project assignments.

Best Practice Recommendations

The KSU experience has been that using a SOHO router that then has a tier of true 8-port hubs for clusters of three or four client computers provides a workable mix of reliability and flexibility without exceeding our available budget.

Operating System Options

When exploring some areas in InfoSec, especially in the area of vulnerability assessment, it is very useful to have multiples operating systems, perhaps multiple configurations of multiple operating systems available to the each student in the lab. This can be done in a number of ways including use of removable disk drives, using multiboot operating system features, and the use of virtual computer images.

Using Removable/Selectable Drives

When there is a need to use multiple operating systems (OS) in serial fashion (there is no need to have a student use more than one OS at a time) using removable or selectable hard drives is a viable solution. This applies to the boot drive, not the use of a removable expansion drive. There are a number of vendors that provide receiver and carrier combinations that permit the removal and replacement of a hard drive. Using this is as a boot drive does not present any serious issues. One problem that may arise is when the lab is administered using the Ghost image management software. This gives rise to a problem where each computer is identified to the Ghost server and not the drive that is mounted. It means that the automated image management features are not useable and manual image copying must be used, creating a bit more work.

Using Multiboot Systems

Another option is to create multiple boot partitions on the system drive of the client computer. This gives the student or lab administrator the ability to select the OS to run at startup time. Once again this enables the client computer to run one OS at a time. The advantage is that each OS is fully native and has full and complete native access to all of the hardware. There are additional approaches in this broader category, one such being HyperOS (http://www.hyperos2002.com/) that enables multiple Windows images to be swapped about in real time. The authors have not used this tool, but testimonials and vendor claims appear make this appear to be a viable and realistic option.

Using Virtual Images

When a lab student needs to access multiple computers with different (or the same) running operating systems, the choices are multiple computer systems, each booted to run the needed OS or the use of virtual OS images. Two widely used products are available for this purpose: VMWare and Microsoft Virtual PC. VMWare (http://www.vmware.com/) is a software application that runs in Windows or a Unix variant. Once the VMWare application is running the user can then activate as many other images as the client computer's resources can support. With a stout client configuration (emphasis on large RAM), it is possible to run three to five images in addition to the base image running VMWare itself. VMWare images can be built for any operating system that supports the Intel 32-bit architecture. VMWare provides bridged network access to the host computers network interface card and also allows managed control of all of the host OS resources and devices. VMWare tools are provided for select guest OS choices that enable improved information sharing between the host and guest and also easier operation of the interface as the user switches between running guest systems.

A competitor to VMWare is Microsoft Virtual PC or VPC (http://www.microsoft.com/windows/virtualpc/default.mspx). Virtual PC is a software virtualization solution that allows users to run multiple PC-based operating systems simultaneously on one computer. The package runs on Windows XP Professional and has many similarities to VMWare in how it functions. One advantage offered to the academic community is that VPC is available under the Microsoft Academic Alliance licensing program (http://msdn.microsoft.com/academic/program/factspage/). If the institution is already associated with this program, it is possible to outfit all labs for no additional cost. If not, this program alone can justify the annual cost per department. VPC and VMWare allow the client computer to have one stable OS that supports the virtual computer application. The application in turn can operate multiple simultaneous different OS images, making them all available to a bridged local network inside the host computer and also to the lab network.

When Option 5 is in use, the virtualization is best accomplished using a server virtualization platform such as VMWare GSX or Microsoft Virtual Server. These tools use the same approach as noted in the previous paragraphs, but have scaled them up for multiple simultaneous users.

BEST PRACTICE RECOMMENDATIONS

The KSU experience has been that using a stable and consistent lab computer platform such as Windows XP, configured in such a ways that student reconfiguration is not easily done, gives rise to more reliable lab experiences for all students. Using virtual images with a product such as VMWare or Virtual PC allows the student to select from the necessary operating system for each learning element, and in those situations where warranted allows simultaneous execution of different operating systems (such as a Linux server and a Windows client) under the direct control of the student and/or the instructor. This includes establishing multiple security targets on a single client workstation or classroom server for examination and assessment by students. KSU has used both VMWare and VPC; however, current licensing cost considerations mandate our use of VPC.

Software Options

When it is time for the students to use the lab computer systems, very few lab exercises are possible without additional software. Sure, a few drills with OS commands can be useful, but using ping and nbtstat in a Windows command line window has limited room for growth. There are many, many software packages that have value in the teaching of InfoSec labs. The major grouping, based on licensing approach, is freeware, demo- and shareware, and commercial software.

Freeware

Freeware is software that has been created by an author or a group of authors, perhaps using open source licensing (such as GNU General Public Licensing found at http://www.gnu.org/licenses/licenses.html#GPL) or by simply giving the right to use the software away. There are many freeware titles available and, in fact, lab manuals that make heavy use of freeware are available (e.g., the *Hands-on Information Security Lab Manual* by Whitman, Mattord & Shackleford © 2005 Course Technology). Use caution when using freeware, since not all of these applications will have the same degree of quality or reliability, and it is possible that some freeware may conceal malicious intent opening up your systems to backdoor exploitation. So long as freeware is limited to use in academic labs and kept from administrative and sensitive systems there should not be any problems.

Demoware and Shareware

Shareware (sometimes called demoware) is not free. Either the program stops working after a trial period, or, the capabilities are limited (giving this software another synonym of crippleware) or both. To get full functionality and/or to operate after the trial period, a license must be purchased. Lots of software is distributed as shareware. If managed properly, these programs can add a great deal to your student's lab experiences. Demonstration versions of commercial grade software are often available for 14 or 30 day trials and the capability limits will not usually preclude their use in the lab. The lab manual mentioned above also makes use of shareware as a valuable extension to the freeware available fro the Internet.

Commercial Software

When the budget permits (or the generous vendor donates licenses) it is a real treat for students and teachers alike when they can use full featured, unlimited use of a market-leading software product.

Best Practice Recommendations

Naturally, the use of industry-proven commercial software tools would be optimum, but the realities of the lab budget means that the perfectly capable and quite useful freeware and demoware titles are often found in most labs.

Lab Presentation Methods

Within the lab setting, students may be presented with learning opportunities in different ways depending on the learning outcomes sought by the teacher. Among those mechanisms that are widely used in the lab setting are tutorials, exercises, demonstrations, simulations, Webinars, and films and videos. Before getting into a description of the various elements, note that a central repository of InfoSec educational content (curriculum, tutorials, exercises, videos and many other items) is available from National Information Assurance Training and Education Center (NIATEC) (http://niatec.info/curriculum.htm). According to NIATEC, "this site brings together a series of education and training modules prepared to teach introductory material with other modules contributed by their authors to the NIATEC project" (Schou, 2006).

Tutorials

A tutorial is a set of step-by step instructions with explanations that are used to give a student some skill in a specific technical area. Tutorials are often provided for software applications to enable new users to get up to speed quickly in using the application. Many applications have prepared tutorials that can be used in the InfoSec lab to quickly get started. On occasion the teacher may create a tutorial to give students a critical skill for use in the lab.

Lab Exercise

A lab exercise, like a tutorial, is a set of step-by-step instructions that guides the student. Unlike the tutorial, the intent is to demonstrate the capability of a tool or to show the results of a sequence of activities. A certain amount of skill and familiarity will also be transferred by the exercise, but the intent is to show a result or to lead the student to the conclusion of a process.

Demonstration

When it is not practical for each student to perform the tutorial or exercise individually, it is useful to show the student how it is done. A demonstration of a technology or application enables the student to be exposed to the learning opportunity without the expense of preparing all lab systems, or because it is not practical for the student to perform the exercise due to licensing expense, technical limitations, security concerns, or other issues.

Simulation

A simulation uses a software package to permit the student to experience conditions and outcomes that are not practical to implement in the real world. A simulation can allow the student to experience the effects of network attack or worm outbreak without the expense of setting up an infrastructure to be attacked. Some events are not possible to recreate and others are too expensive to set up for the lab experience.

Web-based Seminars

Many vendors and professional organizations broadcast seminars and training events over the Internet. Many organizations will capture these events and provide them via their Web pages. The sessions available range from product marketing materials that include very nice explanations of key concepts in Infosec to current events news of interest to the InfoSec student. The video and animated appearance of these items can provide a welcome alternative to traditional content delivery in the classroom.

Films and Videos

Students appreciate the appropriate use of films and videos in the lab and classroom. Many sources of video content are available and can be acquired at little or no cost. Once source that offers useful licensing options for academia is the Public Broadcasting Service (http://www.pbs.org/) which has a few good titles within its documentary series *Nova* and *Frontline*.

Best Practice Recommendations

The KSU experience has been that it requires a mixture of each of these lab presentation models to create the most useful learning environment. Some activates are best learned by doing the actual steps on a real computing systems and can be offered in that manner. And, while every activity might be best done in a real-life fashion, some activities require the preparation and skill learning that only tutorials can deliver. Some activities are too complex to prepare and deliver in a multicomputer lab and must be shown to the student using a single real-life demonstration and some other activities are so restrictive (dangerous to the student or the academic institutions IT infrastructure) that they must be shown only through simulation. Often economic considerations also create a justification for demonstration and simulation. For instance, a high-end forensic analysis tool may be too expensive for every student to be provided with, but one copy for use by the instructor in a demonstration is possible.

Lab Content Sourcing Options

When a teacher prepares to take a class into the information security lab, they must have content for the learning experience. By selecting a balanced mix of tutorials and exercises using software applications, lightened up with demonstrations, simulations, Webinars and videos, the teacher can keep the student engaged. But, it is not always practical to create or even locate all of the content needed for a class. Fortunately, the availability of content continues to improve as does the overall quality.

Published Sources

Course technology and other publishers continue to bring new information and computer security titles to market. Some lab support guides are also available.

Web Sources

NIATEC (Schou, 2006) offers some lab content materials as do some of the National Security Agency Centers of Academic Excellence in Information Assurance Education (NSA, 2007). A review of the Web sites for these institutions may yield some useful content.

Best Practice Recommendations

The KSU experience has been that drawing lab content from any and all sources makes for a diverse learning experience but is accomplished at some cost to the lab instructor. Another issue when compiling from multiple sources is uneven quality of the resources (both appearance and content) and some challenges in making the labs function for students of varying abilities.

Student Teaming

When working in the lab setting, the instructor can have the student work in a variety of modes depending on a number of factors including:

- Limited equipment
- Varying skill levels of students
- Limits in the number or skill of lab assistants
- Distance learning constraints
- Requirements of other learning objectives (such as writing or teamwork objectives)

Individual Assignments

Many lab units are best done by individual students, giving each more control over the factors in the assignments, but limiting the degree of assumed collaboration.

Ad hoc Teams

Sometimes it is advantageous to group students for a single assignment or learning opportunity. For instance, if performing a firewall configuration exercise, it is useful to use small groups since there are many steps in the process and it usually requires operations across several computer systems.

Persistent Teams

Some lab instructors prefer to have teams that last for most or all of the duration of the class. This will improve productivity and learning for sound teams, but can pose special interpersonal management problems for some dysfunctional teams.

Best Practice Recommendations

The KSU experience has been to assign the bulk of the lab tutorials and exercises as individual assignments. Some assignments including activities with specialized equipment and project work are done using both ad hoc and persistent teams at the discretion of the instructor.

Lab Management

Client System Management & Configuration

The InfoSec lab is like any lab or small business network in that it needs to have certain routine system management activities completed on a regular basis. Using a centrally administered antivirus solution (such as Symantec Norton Antivirus Corporate Edition) along with an image management package (such as Ghost) will make these tasks less burdensome and require fewer passes through the lab configuring client computers.

Resetting Labs

When a lab is used for multiple sections of similar courses, there is an issue of system state to consider. For instance, if a tutorial requires that a student installs a software application as a step, that package is then installed on that computer. If a later class needs to do the same activity, that will become an issue. Depending on how the hardware is built and the OS is deployed this can be managed by the lab instructor:

- When using removable drives, each class section is assigned to a drive set and remounts the applicable drive set for use in the lab.
- When using partitioned multiboot drives, multiple versions of the same OS can be built and the image assigned to one of the class sections.
- If Ghost is being used to manage images, the image in use can be reset to a known condition prior to each lab meeting.
- If virtual OS images are used, each student or class section can be assigned a virtual image to use.

Instructed Labs vs. Self-Paced Labs

Some lab assignments are well suited to individual effort and can be performed by the student using a manual or handout. Other topics are better handled when the instructor leads the lab in a *click-and-check* model of instruction where the action is shown using a projector and an instructor's workstation and then the instructor and/or lab assistants circulate to check progress and help the students. Once that section is complete, the instructor covers the next segment and so on. The lab instructor will have to make the determination regarding which approach is right for each unit of lab instruction.

Theoretical Preparation

A question usually arises when heading to the lab about how much theoretical instruction in lecture should be completed on a specific topic before taking the students into the lab. Few will argue that some degree of theory is needed for the student to make sense out of the lab assignments. Some may argue that too much theory will reduce the feeling of discovery that the student will achieve in the lab if every aspect of the lab assignment is fully dissected before the lab itself. As in most discussions like this, each lab instructor will have to asses the degree of theoretical preparation needed to maximize student learning.

Building Vulnerability Assessment Targets

The InfoSec lab assignments that student seem to enjoy the most are vulnerability assessment and penetration testing against lab targets. This type of activity will give the student the feeling that they are *learning to hack* while minimizing the risk that the lab instructor and the educational institution would face were they actually preparing hackers for a life of crime. A challenge for the lab instructor is to prepare the targets for these assignments. After having done this a few times, here are some recommendations:

- Avoid using stacks of old computers as targets unless this is the only option. The hardware issues alone are a distraction and the care and feeding of four to seven aging PC systems is in fact a challenge.
- Use a reasonably capable server using a virtual OS server (such as VMware or VPC) to bring up your targets on the lab network.

CONCLUSION

Using the methods described here, the BS-ISA program has hosted a limited number of exclusive

DL students. These students were hand-selected as capable of handling the pilot testing of these students. Over the course of these individual pilot tests, students were interviewed as to the lessons learned. Overall all student comments on the use of the recorded lectures have been extremely positive. Students are enthusiastic about the ability to both make up missed lectures, and review key material for exams. The only disadvantage noted by the students is the drop in written feedback on the part of the instructors. Using electronic submissions and electronic evaluations, inevitably results in a decrease in the number and quality of written comments. Thus a learning process for both the faculty and students is inevitable, requiring periodic voice or chat-session interactions to provide additional feedback and interaction.

Theoretical Implications of the Study

The intent of this study was not to examine fundamental theoretical constructs in distance learning, rather to provide a successful demonstration of current practices and techniques in hybrid distance learning techniques that promise to provide solid support. As such the use of the tools and techniques described here were found to be consistent with the theoretical foundations as follows.

When considering Vaughn's assertion (2007), students supported the assertions of the availability of greater flexibility and better time management. While no formal surveys were performed specifically for this study, a review of end-of-term student comment forms reinforced this finding. The faculty teaching the course also indicated strong support for added flexibility in the teaching and learning environment.

Bates' statement (1999) on guiding the technology by the student needs and work were also reinforced by student and faculty member comments. In this learning environment, there are many sophisticated (and expensive) alternatives to the technologies indicated, however students indicated that their largest concern was for the ability to transfer the learning material to mobile technology (i.e., laptops, PDAs, portable MP3 players), and thus be able to listen and watch the lectures on their time-table and location.

Managerial Implications of the Results

From an instructional perspective this paper presents an easily implemented methodology with accompanying technologies that can be implemented in a typical academic institution to support distance learning education. Some of the overriding concerns with any DL program are that it does not substantially impact the pedagogical style of the instructor, it provides value-added learning for the students, and that the quality of instruction does not suffer as a result of the use of DL techniques. In this example we found that the use of this technology and methodology compliments most instructional styles, especially those that favor hybrid in-class/distance learning methods. The availability of recorded lectures for make-up and remedial review was highly regarded by students as educational support materials, and as methods for handling absences. The quality of the instruction as commented on by students and as evidenced in overall course grades did not suffer. General assessments of student performance in sections using the technologies and technique were as good as or better than those that did not. Finally, student satisfaction with the program increased significantly, with far more favorable comments on end-of-term student surveys.

Limitations of the Study and Future Research Directions

The dominant drawbacks of this study and the technologies indicated are presented here. The primary drawback to this study is the lack of empirical data collection to support the assertions made. In future studies, validated survey instruments will be created to collect and report student satisfaction with the study, using popular methods such as the technology acceptance model, and related technology diffusion assessments. The purpose of this study was to serve as an exemplar for institutions considering the implementation of this technology, more than an extension of the current research on the subject.

With regard to the tools and techniques implemented, one drawback discussed by faculty members involved in the project was the failure to realize the drop in teacher's administrative load, and increase in student learning outcomes as a result of the use of the technology. While there was much support for improved time management as indicated earlier, the instructors felt the need to manage the DL technology, plus managing the problems inherent in capturing, converting, and posting the recorded content, provided a small but noteworthy increase in administrative overhead for each course. In addition, the need to completely prepare an online support tool complete with online examinations, assignments, and instructional support materials requires a substantial time investment prior to the conduct of the class, which was hitherto spread out over the semester. While this case study did not incorporate or address self-paced programs, the extension of this technology to these types of courses would compound the need for prior preparation before the conduct of the course.

At the current time, the BS-ISA is seriously considering an online-only DL option. The ramifications of this type of program obviously extend beyond the courses offered within the program. As the university continues to roll out online-only options for general education courses, and within other support programs, this will be more feasible.

ACKNOWLEDGMENT

This material is based upon work supported by the National Science Foundation under Grant No. 0516192. Any opinions, findings, and conclusions or recommendations expressed in this material are those of the author(s) and do not necessarily reflect the views of the National Science Foundation.

REFERENCES

Bates, T. (1999). *Managing technological change: Strategies for academic leaders*. San Francisco: Jossey Bass.

Britain, S., & Liber, O. (1999). A framework for pedagogical evaluation of virtual learning environments. *Report to JISC Technology Applications Programme*. Retrieved May 10, 2007, from http://www.jisc.ac.uk/uploaded_documents/jtap-041.doc

Chickering, A., & Ehrmann, S. (1996). Implementing the seven principles: Technology as lever. *American Association for Higher Education Bulletin, 49*(2), 3-6.

Daft, R., & Lengel, R. (1986). Organizational information requirements, media richness and structural design. *Management Science, 32*(5), 554-571.

Duan, B., Hosseini, H., Ling, K., & Gay, R. (2006). An architecture for online laboratory e-learning system. *Journal of Distance Education Technologies, 4*(2), 87-101.

Marra, R. (2006). A review of research methods for assessing content of computer-mediated discussion forums. *Journal of Interactive Learning Research, 17*(3), 243-267.

Mena, M. (2007). E-learning quality: A look towards the demands of its good practices *Journal of Cases on Information Technology, 9*(2), 1-11.

Nemire, R. (2007). Intellectual property development and use for distance education courses: A review of law, organizations, and resources for faculty. *College Teaching, 55*(1), 26-30.

NSA. (2007). *Centers of academic excellence.* Retrieved June 21, 2007, from http://www.nsa.gov/ia/academia/caemap.cfm?MenuID=10.1.1.2

Schou, C. (2006). Curriculum. *National Information Assurance Training and Education Center.* Retrieved June 21, 2007, from http://niatec.info/(S(nfkpbb55vg5kpp45mlebbe55))/index.aspx?page=102

Shepherd, M., & Martz, W. (2006). Media richness theory and the distance education environment. *The Journal of Computer Information Systems, 47*(1), 114-122.

Tallent-Runnels, M., Thomas, J., Lan, W., & Cooper, S. (2006). Teaching courses online: A review of the research. *Review of Educational Research, 76*(1), 93-135.

Vaughn, N. (2007) Perspectives on blended learning in higher education. *International Journal on E-Learning, 6*(1), 81-94.

ADDITIONAL READING

International Journal on E-Learning

Journal of Distance Education Technologies

Managing Technological Change: Strategies for Academic Leaders

Chapter XIV
A Hybrid and Novel Approach to Teaching Computer Programming in MIS Curriculum

Albert D. Ritzhaupt
University of North Florida, USA

T. Grandon Gill
University of South Florida, USA

ABSTRACT

This chapter first discusses the opportunities and challenges of computer programming instruction for management information systems (MIS) curriculum, which includes the development of survey instruments and the meaningful integration of information and communication technology. Second, the chapter describes a unique and hybrid computer programming course for MIS curriculum that embraces an assignment-centric design, self-paced assignment delivery, low involvement multimedia tracing instructional objectives, and online synchronous and asynchronous communication. Third, the development and use of a survey is employed as a method to monitor and evaluate the course, while providing an informative discussion with descriptive statistics related to the course design and practice of computer programming instruction. Tests of significance show no differences on overall student performance or satisfaction using this instructional approach by gender, prior programming experiences, or work status. This chapter aims to provide generalizable knowledge to influence the practice in computer programming instruction in MIS curriculum.

INTRODUCTION

Graduates of management information systems (MIS) programs should possess a variety of organizational and technical skills, including a strong foundation in computer programming. While a majority (78%) of 1,250 information technology managers surveyed in a large national study suggested full-time study as the most effective way to gain the necessary skills and knowledge, only 20% of this same group reported that undergraduates were "equipped for work" (Brandon, Pruett, &

Wade, 2002). Research sponsored by the National Science Foundation (NSF) has reported that the U.S. has an inability to generate well-prepared new graduates in the information systems-centric disciplines (Lidtke, Stokes, Haines & Mulder, 1999). The study found that graduates lack computer programming knowledge and skills necessary to succeed in business and industry. These reports are of paramount concern to MIS educators and a signal to improve the quality of instruction in our programs.

To heighten the quality of instruction, educators place emphasis on the development and dissemination of best practices that intersect pedagogy, content domain, and information and communication technology (ICT) for instruction. Publication venues like the *Journal of Information Systems Education* or the *Journal of Information Technology Education* collect and share experiences about pedagogy in information-centric programs. These best practices are, perhaps, the necessary elements to equip our educators, and consequently, our graduates to compete in a global economy. This shared value highlights the importance of a *Handbook of Distance Learning for Real-Time and Asynchronous Information Technology Education.*

In the spirit of this important tradition, this chapter addresses the concerns by: 1) providing a rich description of the pedagogical context used in a novel, hybrid computer programming course in an MIS curriculum; 2) providing empirical evidence that demonstrates the instructional value of those elements found within this course; and 3) providing reliable and valid evidence of an instrument designed to monitor this course. The chapter first briefly examines the challenges and opportunities of teaching computer programming in MIS curriculum, and then examines the specific course under investigation. This chapter aims to provide generalizable knowledge to influence the practice in computer programming instruction in MIS curriculum.

The Challenges

Computer programming instruction in MIS curriculum poses many serious problems to educators, starting with the inherent difficulty of the content domain. Computer programming students have to learn to analyze problems critically, implement robust solutions in a programming language, debug code, and make enhancements to existing computer programs, and repeat this process several times in multiple programming assignments over the duration of a quarter or semester. All of this must be done while learning programming concepts, a programming language, and principles of software design. There is little surprise students are often challenged by one or more aspects of a computer programming course.

Empirical studies confirm that students struggle with computer programming. The most troubling numbers are from the introductory computer programming courses where failure and withdrawal rates exceed 50% (Woszczynski, Guthrie, & Shade, 2005). One study found that the probability of passing an introductory undergraduate programming course the first time was 40% across all majors, with an initial failure rate of 19.5% and a withdrawal rate of 40.5% (Beise, Myers, VanBrackle, & Chevli-Saroq, 2003). During a period of high enrollment growth, this may not have been such a problem. However, during a period of low enrollment, this problem can threaten the sustainability of an academic program.

Graduates of MIS programs are required to possess a strong foundation in computer programming. Yet, research suggests the degree of interest in learning computer programming is highly variable in MIS curriculum because many graduates pursue careers in the field where computer programming is not a required job activity (Gill, 2005a). In opposition to degree programs like computer science, with a more computer programming focused curriculum, MIS students may only be exposed to a single programming course in their entire program of study.

To add to the complexity of computer programming instruction, there is also increased emphasis on offering access to higher education at a distance. Taking classes at a distance poses a different set of challenges for students who are used to taking on-campus classes in terms of studying, time management, and autonomy (Moore & Thompson, 1998). More troubling, educational research on distance learning suggests that retention rates tend to be significantly lower in distance education classes (Carr, 2000; Garrison, 1987; Zajkowski, 1997).

Finally, developing MIS curriculum is an ongoing, dynamic process. Government and industry needs, along with fundamental changes in information technology are constantly driving the field of MIS to revise curricula (Al-Rawi, Lansari, & Bouslama, 2005). For instance, in 1987, Microsoft released Visual Basic 1.0 ©. This was followed by five more major releases of the Visual Basic language © (version 2.0 – 6.0). In 2001, Microsoft then released the Visual Basic .NET © language, which transformed the language into a fully object-oriented programming language. Subsequently, the company issued two more major releases of the product and associated .NET framework. Even before MIS educators are fully comfortable with one technology, they have to change it to keep up with practice.

The Opportunities

Due to the complex nature of computer programming instruction in MIS curriculum, a door of opportunity opens for educators and researchers to address the challenges and impact of both MIS educational research and practice. Two dimensions come to light in facing these challenges: 1) the integration of ICT to heighten the quality of instruction in face-to-face, distance and hybrid modality, and 2) the development, validation, and use of instruments designed to measure a number of factors influencing a student's perspective of a course. Both of these opportunities intersect in the goal of improving instruction.

The integration of ICT into MIS curriculum for instructional purposes is a clear fit for the discipline, since MIS faculty tend to be knowledgeable both about how to use ICT and the process surrounding its integration. ICT speaks of the infusion of tools to store, retrieve, and manipulate information with tools for communication. Integration initiatives range from using multimedia CD-ROM course delivery (Doube, 1998) to fully online instruction using specially developed course management systems (Molstad, 2001) to hybrid courses combining both face-to-face and distance technologies (Gill, 2006). The initiatives also integrate a number of different instructional strategies, like language independent approaches or cooperative learning (Lehman & Naumann, 1986; Nosek, 1998; Williams & Kessler, 2001). The opportunities are endless, yet the empirical characterization of these initiatives should demonstrate student achievement and satisfaction equal to or above traditional methods (Stansfield, McLellan, & Connolly, 2004).

In speaking of student satisfaction, the primary means that most instructors have for assessing their courses are word-of-mouth and university-wide course evaluations. While these two forms of feedback are helpful, they tend to be either anecdotal or very general in nature. To gain the sufficiently detailed information necessary to evaluate individual course elements, MIS educators may need to develop and tailor instruments to capture information that directly pertains to their courses.

Of course, the development of instruments to assess course design and student perspectives immediately raises concerns about reliability and validity. The development, validation, and use of instruments to assess these areas therefore present a fruitful research opportunity. As pointed out in the literature (Straub, 1989), confirmatory empirical findings will be strengthened in information systems research when "instrument validation precedes both internal and statistical conclusion validity" (Straub, 1989, p. 147).

MIS PROGRAMMING COURSE

The programming course for which this research is based is offered at a large university ("very high research activity") in a metropolitan city in the southeast United States (Carnegie Classification, 2007). The course is a required introductory programming course for undergraduate MIS majors, taught using the C++ programming language. The course is generally taken during a student's junior year, and is one of the first courses taken in the MIS major. The course historically enrolls anywhere from 80 to 100 students in the spring and fall semesters. The purpose of the course is to teach all students the basics of procedural and object-oriented thinking so students can pursue any career in MIS. The course used to be delivered in a face-to-face format, but later was transformed into a hybrid format, incorporating many different technologies and pedagogical strategies. Lectures were entirely removed from the course.

The course curriculum includes a variety of conceptual, algorithmic, and practical elements, such as data representation, flowcharting, functions, elementary algorithms, debugging techniques, memory organization, and input/output streams. Students are also introduced to classes, polymorphism, encapsulation, and inheritance. The course was supported by one instructor and up to five teaching assistants in any given semester. Table 1 shows the seven programming assignments students should complete in this course. The course description is:

Business Application Development*—Presentation of business application development using a modern programming language. Topics include data structures, indexing, file processing, and user interfaces. Good program design techniques are emphasized. Business applications are developed.*

The course diverges from traditional programming courses in that it is essentially self-paced, heavily supplemented with multimedia materials, assignment-centric, and makes extensive use of both synchronous and asynchronous communication technology to support the learning process. These characteristics should not be viewed as independent. Rather, they are part of an integrated approach with the distinct elements intersecting at many different levels. The follow-

Table 1. Course programming assignments

Assignment	Description
Assignment 1: Compiler Exercises	Compiler installation and simple compiles (Hello, World! and simple multifile project).
Assignment 2: Numbering Systems	Conversions between decimal, hex, and binary. Twos complement representation. Simple bitwise logical operations. Credit for assignment will be dependent on the results of an online exam conducted in the lab.
Assignment 3: Logic and Flowcharting	Creating flow charts for simple processes. Converting code to flow charts. Converting flow charts to code. Credit for assignment will be dependent on the results of an oral exam.
Assignment 4: Debugging and Pointer Arithmetic	Taking a program with a variety of compiler, linker and runtime errors and finding/removing the bugs. Using a memory grid to locate items in memory. Credit for assignment will be dependent on the results of an online exam conducted in the lab.
Assignment 5: Function Exercises	Creating a series of functions that perform simple string tasks. Credit for assignment will be dependent on the results of an oral exam.
Assignment 6: Structured CGI Application	Creating Web-based application that takes input from a Web form and returns it to a browser. Credit for assignment will be dependent on the results of an oral exam.
Assignment 7: OOP CGI Application	Rewriting Web-based CGI application using C++ classes. Credit for assignment will be dependent on the results of an oral exam.

Figure 1. Technology and pedagogy used in course

Asynchronous technology	Pedagogical strategy
• Discussion boards • Online assessment system • Animated screen captures w/ narration • Animated presentations w/ narration • Course management system • E-mail	• Self-paced instruction • Web-based instruction • Cooperative learning • Assignment-centric design • Self-regulated progress monitoring • Oral assessments • Proctored assessments
Synchronous technology	
• Virtual office hours • Online lab sessions	

ing sections provide more depth regarding these particular characteristics. Figure 1 summarizes the technologies and pedagogical strategies used in this course.

Assignment-Centric Design

Assignment-centric design can be characterized by the following principles: completion of the course assignments meets the course requirements; all assessment is directed toward validating students properly to complete the assignments; and the primary role of instruction is helping students complete the assignments (Gill, 2005a). This approach is similar to a project-driven approach, in which the programming assignments are designed to trace all learning objectives in the curriculum and drive the instruction in the course (Ritzhaupt & Zucker, 2006). However, traditional quizzes and examinations are removed from the curriculum, leaving only programming assignments in place with assessment directed at validating student completion of the assignments.

The integration of an assignment-centric design requires substantial modifications to a traditional course. These changes include providing access to learning materials in a flexible format, directing lab sessions towards assignment completion, and providing technical support 7-days a week to accommodate diverse schedules (Gill, 2005a). A Web-based content delivery system was developed combining both a course management system (Blackboard67©) for asynchronous communication and a Web site with all the instructional materials organized in assignment modules.

Validation of the exams involves two methods: 1) proctored online exams supervised by teaching assistants; and 2) oral exams with either a teaching assistant or instructor. The oral exams include specific questions to probe whether students completed and understood the material they submitted. Questions are typically open-ended, such as "What does this line of code do?" or "What would happen if this line of code were removed?" Any code included in an assignment submission is open to inclusion in the oral exams (Gill, 2005a).

Two other important characteristics surrounding the validation of an assignment include: 1) having no specific limit on the number of attempts a student can make on validating the assignment; and 2) the degree of the validation being directly proportional to the performance on the assignment. "Thus, a 95% score on an assignment meant a far tougher validation exam than a 60%" (Gill, 2005a, p. 342).

Self-Paced Delivery

This course embraces a self-paced approach. The development of a self-paced approach included the integration of three different, interdependent systems: a content delivery system, peer support system, and progress monitoring system. The content delivery system, as previously noted, included a Web site with all instructional materials, a course management system for group discussions, and some validation assessments, and later included an Elluminate Live!© component for conducting online lab sessions and a virtual office (Gill & Holton, 2006).

The peer support system was implemented by encouraging students to cooperate with each other, and teaching assistants (all of whom had previously completed the course as students) holding office hours and answering questions posted to the discussion board or via e-mail. Though traditional programming classes do not allow students to cooperate on their assignments, in the real-world, software development is rarely a solitary activity. Peopleware (DeMarco & Lister, 1999) reports that software developers generally spend 30% of their time working alone, 50% of their time working with one other person, and 20% of their time working with two or more people.

Cooperative strategies like pair-programming as an instructional method have seen tremendous growth in computer programming courses (Nosek, 1998; Williams, & Kessler, 2001). At the same time, the obvious drawback of cooperation is the potential loss of academic rigor that can result from the "free rider" effect (Gill & Holton, 2006). In practice, the assignment validations address this problem because students completing an assignment receive no credit for doing so until the assignment is validated.

The progress monitoring system was modeled after nuclear submarine training (Gill, 2005b). Each student was provided a validation card for progress monitoring (validation assessments and programming assignments) and was paired with one teaching assistant. Further, a system of credit for each assignment was instituted in which students had to update their assigned teaching assistant on their progress and post to an asynchronous, online progress report form. These various elements worked in concert to encourage students to complete the activities at an even pace (Gill & Holton, 2006). While the self-paced approach allowed more flexibility for students, the instructor did have suggested turn-in dates for each assignment and hard deadlines specifying when all assignments would be due, typically right before the exam period at the institution.

Low Involvement Multimedia

Developing high quality multimedia resources for distance learning, face-to-face, or hybrid courses is a resource intensive process, drawing from faculty time and institutional monies. This high-involvement process can quickly serve as a deterrent to faculty members with split research, teaching, and service responsibility. Consequently, there is a need for user-friendly, low budget multimedia authoring tools. This has led to the development of authoring tools like Camtasia© (2007) and Articulate© (2007), and explains their explosive growth in the higher education market.

For this course, multimedia tutorials have been developed and mapped to the instructional objectives of each assignment by the instructor. The multimedia materials included animated screen captures and presentations with the instructor's voice as guided narration (Gill, 2007). The primary advantage of these multimedia resources to both students and instructors is that a tremendous amount of information can be communicated in a relatively short amount of time. Resources can be developed as planned lectures or on-the-fly productions to answer student questions (Gill, 2007).

The multimedia resources developed using these authoring tools can be seamlessly integrated into a course management system or simply uploaded to Web space for efficient delivery. The resources could be burned to CDs and distributed early in the semester (an approach previously used in this course). More recently, the pervasiveness of broadband Internet connections has made

Web-based delivery the near-universal choice of students.

A critical lesson from the experiences of this course is that faculty members do not have to be multimedia development specialists to wield this powerful technology for their own activities. Given minimal support and tools, faculty members can generate materials for entire courses with a few months of preparation and implementation. Often, academic institutions will also provide support personnel to aid faculty in this development process.

Synchronous Online Interaction

One of the unfortunate drawbacks to most forms of multimedia is that students cannot ask questions and receive immediate feedback. Until recently, higher education did not have access to robust and scalable tools for synchronous, online communication aside from basic chat rooms and whiteboards without audio capability. Over the past 5 years, we have seen tremendous market growth in tools like Adobe Breeze© or Elluminate Live! to support online, synchronous interaction. However, synchronous communication tools like Elluminate Live! have not been thoroughly investigated in research literature (Johnson, 2006) and are often cost-prohibitive, which complicates the integration process.

Elluminate Live! is a voice-over Internet protocol (VoIP) package that has been particularly successful in penetrating the higher education market. The software package has many features making it well-suited for computer programming instruction. Of particular note in this regard is application sharing. This feature allows the instructor to give live demonstrations of writing source code, compiling, debugging, and even running programs. The students see exactly what the instructor sees and hear exactly what the instructor says. Students can ask the instructor questions during live demonstrations and receive immediate feedback as the tool is built to emulate a virtual classroom environment.

Elluminate Live! was not integrated into the course until the spring of the 2004-05 school year because the software was previously unavailable to the institution. The software package was used by the instructor and teaching assistants to provide virtual office hours and to hold online lab sessions to demonstrate skills and knowledge specific to assignments, such as writing functions. Students were not required to join the sessions, but rather the sessions served as supplementary instructional support. Additionally, those students that could not attend an online session could view the recorded session later in an asynchronous modality.

Course Grading Scale

The grading scale, presented in Table 2, was fixed throughout the study, although the weight given to individual assignments and participation did change over time. Because of the assignment-centric approach, the instructor instituted a system in which only the first four assignments had to be satisfactorily completed in order to receive a C grade, the first five assignments for a grade of a B, and six of the seven assignments to receive an A grade. Completing the requirements of an A grade resulted in depth of coverage far beyond what would normally be expected in an introductory computer programming course.

Table 2. Grading scale in summer of 2004

Numeric grade range	Letter grade
80-100	A
60-79	B
40-59	C
20-39	D
<20 F	F

METHOD

Instrumentation

The primary instrument used in this analysis was developed to monitor this complex programming course, and to collect useful information about a student's previous background and career expectations. The instrument was developed by the instructor of the course and contains over 300 items organized into five separate sections: student background information, assignments, opinions, student assessment of learning gains, and (added in spring 2005-06) Elluminate usage. The satisfaction and assessment of learning gains sections were, in turn, derived from three NSF-sponsored instruments, the "Student Opinion Survey," the "Computer Programming Survey," and the "Student Assessment of Learning Gains (SALG)."

The university-wide course evaluation instrument has eight items and uses a modified Likert scale in the range of 1 to 5 (1=poor, 2=fair, 3=good, 4=very good, 5=excellent), with an additional area for free-form comments. This instrument served as a secondary instrument in the analysis. The instrument was administered concurrently with the instructor-developed instrument, and is designed to measure a student's overall level of satisfaction with various areas of the course.

Procedure

The instrument was electronically released to students at the end of the 15th week of a 16-week semester, along with the university-wide course evaluations. Students were instructed to complete the survey in an electronic format (spreadsheet) and to e-mail the results to an administrative e-mail address that the course instructor could not access. To encourage students to complete the survey, they were provided extra credit points toward their final grades, which led to typical response rates of approximately 70%. The instrument was not anonymous for those students desiring extra credit, which allowed for the responses to be linked to responses on the university-wide course evaluation instrument, and student grades for the course. Students are instructed that they can request a small project as an alternative to this extra credit opportunity. However, this request has not been made since the inception of this procedure.

Prior to submission of course grades, a designated departmental administrative assistant provided the course instructor with a list of students who had responded, so each student's grade could be adjusted accordingly. Copies of the completed surveys, on a CD, were then made available to the instructor only after all course grades had been officially submitted to the registrar. This procedure was designed to prevent student concerns that their responses might be considered in assigning their final course grade. During 4 years of collecting data in this fashion, not a single complaint was raised—either to the course instructor or to the department chair—regarding the mechanics of the process.

Participants

This chapter reports on students (N=254) enrolled in the aforementioned course from the spring of the 2002-03 school year to the spring of 2004-05 school year on a semester system. The instrument's first section, labeled *student background information*, included a variety of background information about the student population. Approximately 69% of the participant students were male. Ages varied with a mean of 26.16 (SD=7.48) years old. Approximately 13% of the students were non-US citizens. Forty-seven percent of the students had never taken a programming course and 59% indicated this as their first C/C++ programming experience.

Nearly 40% of the students were working part-time and 35% full-time, while enrolled in the course. Interestingly, while the majority of the students indicated working either full- or part-time, nearly 78% also indicated full-time status as a student. Eighty-two percent of the students in the sample were MIS majors with the remaining being either other business majors or nondegree seeking students.

Of particular interest was the attractiveness of MIS-related and non-MIS-related careers to students. One item in the background section of the instrument asked how attractive a particular career was using a 5-point scale. Internal consistency for the scale is less than desirable with α=.49. The response frequencies for these items are shown in Table 3. As can be gleaned, 50% of the students indicate computer programming as an unattractive career and 17% had a neutral response. Seventy-one percent of the students are attracted to MIS project management, shortly followed by 65% attracted to both general management and network management as a career goal.

Data Analysis

Recognition of the need to provide valid and reliable measures served as a key component of the analysis. Quantitative analyses of the data included descriptive analysis of response frequencies and measures of variation and central tendency, internal consistency reliability analysis (Cronbach's alpha), an exploratory factor analyses (EFA) for the student level of satisfaction subscale, and statistical inferences using analysis of variance (ANOVA) and Pearson correlations for the data that are continuous. EFA and reliability analysis were conducted to explore the underlying structure of these data (satisfaction subscale only) and to demonstrate the reliability of the measures prior to statistical inference.

RESULTS

In this chapter, internal consistency reliability equal to or above a 0.7 threshold is assumed to be an acceptable measure (Nunnaly, 1978). Statistical significance is set at a 0.05 level for all statistical tests. When using 5-point Likert scales, responses and central tendency measures greater than 3 (central points) are considered favorable. Results including performance measures (assignment performance) include only those students that responded to the survey.

Table 3. Student career expectations

Career area	M	SD	VU	U	N	SA	VA
General mgmt.	3.65	1.10	6%	9%	20%	43%	22%
Programming	2.68	1.33	25%	25%	17%	24%	9%
Database mgmt.	3.48	1.14	6%	15%	22%	39%	19%
Network mgmt.	3.64	1.11	6%	10%	19%	44%	21%
Sales	2.44	1.32	33%	24%	18%	17%	8%
MIS project mgmt.	3.83	1.10	4%	10%	15%	41%	30%
CPA	2.09	1.24	47%	17%	21%	10%	5%
Lawyer	2.26	1.43	48%	12%	17%	13%	10%
Medical Doctor	2.00	1.27	53%	15%	15%	12%	5%
Game developer	3.23	1.40	19%	10%	21%	29%	21%

Note. M= mean response, SD = standard deviation, VU=very unattractive, U=unattractive, N=neutral/have no opinion, SA=somewhat attractive, VA=very attractive.

The instrument did not force students to respond to any items on the instrument, as doing so is considered poor instrument design. Because the number of surveys missing large numbers of responses was minimal (less than 10 responses to any survey item), it was decided that if a student did not respond to some items and responded to others, the data would be included in the analysis for those items where responses were provided.

Instruction Quality and Involvement

The second part of the instrument, labeled *assignments*, included a number of items examining the student's perspective of the assignments, the instruction supporting the assignments, and the validation assessments of the assignments. The area also attempts to measure the involvement in each of the assignments and supporting areas. Because the course uses an assignment-centric design and greatly relies on the use of multimedia resources, this type of information is especially important.

The items response scale ranged from 1 to 5 (1=strongly disagree, 2=mildly disagree, 3=neutral, 4=mildly agree, 5=strongly agree) with a not applicable option for each assignment. The internal consistency reliability of the categorical items demonstrates high internal consistency reliability at α=.93. Table 4 illustrates the response frequencies, mean, and standard deviations by assignment for these items. Assignment 7 was not included in the analysis because only a small fraction of the students completed the assignment each semester.

When examining whether students found each of the assignments to be "a helpful learning activity," whether "the multimedia resources related to the assignment were very helpful," and whether students could complete the assignments "with little or no difficulty," all the mean responses are above the central point for all six assignments. In terms of the programming assignment validations, students had favorable responses to all validation exams, both online and oral.

Understanding the amount of time a student invests into the course is also a pertinent measure. If the time a student invests in a particular assignment is too great, the instructor should modify the assignment or the supporting instruction to reduce the workload on a student. Table 5 provides student responses to the number of hours invested in the completion of each assignment. The results are clear in indicating the logic and flowcharting (Assignment 3, 18.76, *SD*=16.82) and structured common-gateway interface applications (Assignment 6, 17.3, *SD*=17.9) occupied the most amount of time in the course.

Another way to view the amount of time a student invests in the course is by activity or resource area. Students were asked to indicate the amount of time spent participating in a number of areas per week. The results, shown in Table 6, demonstrate the majority of the time students are completing assignments at approximately 11.16 (*SD*=10.1) hours per week, which is expected since the course embraces an assignment-centric design. The amount of time devoted to the other areas varied, but all appear to be, on average, below 4 hours per week.

One concern is that students are investing too much time into a specific area. One item stated, "The assignments required too much time," and used a 5-point agreement scale. Students indicated a mean of 3.93 (*SD*=1.15), suggesting that, on average, more students agreed with this statement than disagreed. This section of the instrument also provided quality measures associated with the teaching assistants supporting the course. While these measures are vital to the evaluation of the course (and instructional support staff), they were purposefully excluded from the chapter.

Student Level of Satisfaction

The level of satisfaction subscale included 16 items under the third section of the instrument labeled *opinions*. The response scale for these items uses a modified Likert scale in the range

Table 4. Assignment-centric measures

Assignment	M	S.D.	SD	MS	N	MA	SA	NA
Assignment 1								
Helpful	4.02	1.23	4%	5%	12%	36%	43%	0%
Multimedia	4.13	1.26	3%	6%	12%	24%	52%	3%
Independence	4.22	1.34	5%	6%	8%	18%	62%	1%
Assignment 2								
Helpful	3.68	1.40	5%	12%	16%	32%	34%	1%
Multimedia	3.64	1.40	5%	10%	23%	23%	32%	7%
Independence	4.02	1.33	3%	10%	10%	30%	47%	1%
Fairness	4.16	1.05	3%	7%	10%	31%	47%	2%
Assignment 3								
Helpful	4.04	1.22	4%	6%	12%	31%	47%	0%
Multimedia	3.69	1.35	7%	10%	18%	27%	35%	3%
Independence	3.46	1.38	10%	14%	16%	32%	27%	1%
Fairness	4.28	0.95	2%	3%	11%	29%	50%	4%
Assignment 4								
Helpful	4.09	1.14	3%	3%	14%	35%	42%	3%
Multimedia	3.78	1.29	6%	7%	19%	27%	35%	7%
Independence	4.00	1.21	3%	8%	14%	30%	43%	2%
Fairness	4.09	1.09	3%	5%	17%	22%	42%	12%
Assignment 5								
Helpful	3.86	1.42	5%	3%	15%	26%	34%	18%
Multimedia	3.43	1.47	7%	10%	19%	22%	22%	21%
Independence	3.01	1.45	11%	17%	19%	23%	13%	17%
Fairness	4.05	1.12	4%	2%	16%	17%	33%	29%
Assignment 6								
Helpful	3.58	1.71	4%	3%	15%	14%	24%	41%
Multimedia	3.21	1.71	5%	8%	16%	12%	16%	43%
Independence	3.01	1.73	8%	9%	15%	14%	13%	41%
Fairness	3.93	0.99	1%	1%	15%	12%	16%	55%

Note. M=Mean, S.D.=Standard Deviation, SD=strongly disagree, MD=mildly disagree, N=neutral, MA=mildly agree, SA=strongly agree, Helpful=The assignment was a helpful learning activity, Multimedia=The multimedia resources related to the assignment were very helpful, Independence=I could now complete the assignment by myself with little or no difficulty, Fairness=The oral or online validation exam I took for the assignment provided a fair assessment of my knowledge at the time.

Table 5. Student time (hours) investment by assignment

Assignment	Mean	S.D.
Assignment 1	3.97	4.17
Assignment 2	6.61	10.24
Assignment 3	18.76	16.82
Assignment 4	10.25	8.99
Assignment 5	7.42	9.44
Assignment 6	17.30	17.90
Assignment 7	9.78	15.35

Table 6. Student time (hours) investment by activity per week

Activity	Mean	S.D.
Doing course assignments	11.16	10.10
Reading the text	3.21	4.63
Viewing multimedia resources	3.08	4.32
Reading and interacting with Blackboard	3.34	4.32
Attending lab sessions	3.73	6.94
Interacting with TAs and instructors	3.60	6.40

of 1 to 5 (1=not at all satisfied, 2=somewhat dissatisfied, 3=neutral/don't know, 4=somewhat satisfied, 5=very satisfied), and also included a not applicable option. Students were instructed, "Indicate your level of satisfaction with the various elements/aspects of the course that are listed" (see Table 8).

The Cronbach Alpha for the 16-item instrument was α=0.89, a sign of high internal consistency for these items and data. This particular subscale lends itself to further investigation for these data because of its generic nature and consistent response scale. Consequently, this subscale is used to demonstrate the initial process of validating a portion of the instrument. As pointed out by Straub (1989), instrument validation should be done prior to drawing upon any statistical inferences.

Bartlett's test of sphericity for these data has a chi-square of 1225.99 ($p<.001$), which suggests the intercorrelation matrix contains adequate common variance. The Kaiser-Meyer-Olkin measure of sampling adequacy is 0.85, far above the 0.5 recommended threshold (Kaiser, 1974). Finally, the participants to items ratio is approximately 15:1, well above the 10:1 ratio for factor analysis suggested (Kerlinger, 1974). Thus, these data appeared to be well-suited for EFA.

An EFA was executed using principal axis factoring and an oblique (promax) rotation since the factors were anticipated to be correlated. The resulting correlation matrix showed that all items correlated in the range of r=.03 to r=.70, as they should since the items are all positively stated. An investigation of the item-to-total correlations showed that reliability would not be advantageously increased with the removal of an item.

Table 7 shows the Eigenvalues and variance explained by factor. A review of the Scree plot showed that little variability was added after the inclusion of the fourth factor, which explained approximately 63% of the cumulative variation. For subsequent factors, the Eigenvalues were below Kaiser's criterion at 1.0. Consequently, we chose parsimony over complexity to focus the results on a four factor model to explain a student's level of satisfaction.

To uncover the underlying factors of the instrument, the maximum factor loadings for each item were used to assign each item to a factor. The factor and item statistics are shown in Table 8. The results did not exhibit a truly simple structure, but the authors felt the loadings are reasonable for this type of instrument. The factors correlated in the range of r=.39 to r=.60, suggesting moderately strong relationships between the extracted factors.

The extracted factors all demonstrate acceptable internal consistency reliability with the exception of the fourth at α = .47. Since the items measure elements important to the particular course, no items were removed. The other factors' internal consistency reliability measures were .7, .85, and .82 for factors one to three, respectively. Future adjustment to the scale may be necessary to increase reliability.

Table 7. Eigenvalues, variance, cumulative variance by factor

Measurement	Factor 1	Factor 2	Factor 3	Factor 4
Eigenvalues	6.01	1.60	1.30	1.09
Variance (%)	37.58	10.02	8.17	6.80
Cumulative variance (%)	37.58	47.60	55.77	62.57

Table 8. Item response frequencies and statistics by factor

Item/factor	M	S.D.	NAS	SD	N	SA	VA
Instructional Materials	3.66	1.13					
The textbook that was used in the course	3.15	1.19	11%	22%	16%	41%	9%
Other written handouts that were distributed	3.90	1.01	4%	6%	14%	46%	28%
The use of multimedia to supplement lectures	3.95	1.18	6%	7%	11%	35%	39%
Course Structure and Composition	3.46	1.16					
The number of homework assignments	3.58	1.08	3%	17%	21%	38%	21%
The type of homework assigned	3.36	1.13	6%	18%	25%	34%	16%
The in-lab activities and exercises	3.47	1.17	7%	13%	23%	33%	20%
The use of problems from other areas	3.47	1.11	6%	11%	32%	31%	19%
The amount of material that was covered	3.32	1.24	6%	11%	32%	31%	19%
The content of the course materials	3.40	1.13	6%	17%	21%	40%	15%
The split lecture/lab format	3.62	1.22	5%	15%	21	27	30
Instructor and Student Interaction	3.23	1.34					
Instructor's use of different ways of teaching	3.20	1.29	13%	18%	22%	29%	17%
The interaction between the instructor/students	3.10	1.41	17%	18%	22%	17%	22%
The availability of the instructor	3.41	1.31	11%	12%	27%	21%	27%
Ancillary Support	4.25	0.96					
The group work between students	3.99	1.08	3%	4%	19%	24%	35%
The use of Blackboard and the Internet	4.42	0.82	1%	3%	5%	35%	56%
The availability of the teaching assistants	4.33	1.00	2%	6%	7%	25%	59%

Note. CA=Cronbach Alpha, M=Mean, S.D.=Standard Deviation, NAS = Not at all satisfied, SD= Somewhat dissatisfied, SA=Somewhat satisfied, VA=Very Satisfied.

Content Validity

To demonstrate content validity of the constructs, three MIS faculty members not involved in the course or the instrument's design reviewed the items in each factor and were instructed to determine the underlying factor by providing a title, description, and justification for each factor and its items. These responses were analyzed for similarity in semantics and language. The analysis resulted in four meaningful dimensions for the instrument: course structure and composition, instructional materials, instructor and student interaction, and ancillary support. The items and the titled factors are also shown in Table 8.

Construct Validity

Construct validity is provided by correlating the subscale with the university-wide course evaluation. The instrument positively correlated r=.70 (n=214, $p < .001$) with the university-wide course evaluation. The correlation between the level of satisfaction and the university-wide course evaluation was anticipated, since logically, a course evaluation is a theoretical measure of student satisfaction with a course.

Relationships Among Factors

A strong relationship is present between course structure and composition, and the instructional materials (r=.59) provided in the course. This may indicate that the instructional materials and their alignment with the number and type of assignments are critical to a student's level of satisfaction with the course. Further, the instructional materials, and instructor and student interaction also exhibit a strong relationships (r=.60), which may speak to a teacher effectively using the instructional material while interacting with students. Finally, a relatively strong relationship between instructor and student interaction, and ancillary support (r=.43) is present. This relationship may indicate the need for student and instructor interaction in relation to the additional support, such as tutors.

Student Assessment of Learning Gains

The fourth section of the survey is labeled *student assessment of learning gains*, and includes a number of items pertaining to a student's self-assessment of their learning. Particularly, the section attempts to capture a student's reflective assessment of particular skills and knowledge that are developed as a result of completing the course. The items in this section use two 5-point scales (1=No help, 2=little help, 3=moderate help, 4=much help, 5=very much help; and 1=Not at all, 2=A little, 3=Somewhat, 4=A lot, 5=A great deal) with not applicable options. The internal consistency reliability for the items in this section is more than acceptable at α=.95.

The course involves students in many activities. Some of these activities are shown in Table 9, along with the student responses. This item asked students "How did each of the following class activities help your learning?" The highest mean responses are teamwork in labs at 4.24 (SD=1.36) and hands-on lab activities at 3.95 (SD=1.3). These results indicate that students feel these areas are most helpful in learning the content. The discussions in section and in Blackboard appear to have the lowest positive responses at 29% and 33%, respectively. This indicates students may have not perceived asynchronous discussions as a helpful learning tool in this course, perhaps because of the lack of immediacy of the medium.

Another area of interest is a student's perception of the influence of course structure, such as the type of assignments, on their learning. The response statistics are shown in Table 10. The mean response for the self-paced approach is 2.90 (SD=1.37), suggesting students may not find a self-paced approach that helpful as an in-

Table 9. Student assessment of activity area

Activity	M	S.D.	NH	LH	MH	H	VH	NA
Discussions in sections	3.21	1.51	15%	15%	22%	18%	11%	8%
Hands-on lab activities	3.95	1.30	4%	9%	17%	21%	29%	7%
Teamwork in labs	4.24	1.36	4%	7%	13%	17%	33%	14%
Blackboard discussions	3.23	1.49	13%	18%	19%	13%	20%	4%

Note. M=mean, S.D.=standard deviation, NH=No help, LH=little help, MH=moderate help, H=much help, VH=very much help, NA=not applicable.

structional method for this type of course. Some students find themselves procrastinating when using a self-paced approach, which attests to the importance of the instructor frequently monitoring students to encourage timely completion of the work. Sixty-nine percent of the students reported the grading scale as helpful to their learning, which suggests students have a positive response to having control, based on how much effort they are willing to expend, over their final grade.

Being a computer programming course, it is important for students to be able to assess whether they have acquired the desired computer programming skills from the course. Specifically, this items asks students, "How much has this class added to your skills in each of the following areas?" The responses in Table 11 indicate a high degree of perceived skill development with all mean responses above the central point. More than 50% of the respondents indicated the course added a lot to their ability to create flowcharts (54%), reading and understanding the C++ programming language (53%), and the ability to work effectively with others (51%). Interestingly, only 39% positively indicated the course added to their ability write C++ computer programs. This finding may be attributable to the vast array of computer programming concepts, such as flowcharts and structured problem-solving skills, covered in this course, limiting students' syntax skills. Further, some aspects of the programming assignments involved the deciphering and modification of existing computer programs as oppose to students starting from scratch.

Initial learning in the context of a programming course is important; however, educators are more concerned with the transfer of learning, in which the skills and knowledge developed during a course can be applied in a different setting by a student (Brandon et al., 2002). Students were asked, "How much of the following do you think you will remember and carry with you into your other classes or aspects of your life?" The responses are shown in Table 12. The mean student responses to both "the core programming concepts" and "how to solve problems" are above the central point, suggesting students believe they have gained skills from the course that can be used in their future academic and professional careers. This is especially important since the goal of the course is to provide students problem-solving skills and the ability to think programmatically.

Synchronous Software Usage

During the semester Elluminate Live! was integrated, the software had been quickly accepted by students as it was offered as a supplementary resource for virtual office hours and online lab sessions to support students in completing their programming assignments. Though the use of the software was not a requirement in this semester, 72% of the students in semester attended at least one synchronous session.

Table 10. Student assessment of structure composition

Organization	M	S.D.	NH	LH	MH	H	VH	NA
Class activities each week	3.28	1.34	8%	21%	28%	25%	9%	8%
How parts of the class relate to each other	3.11	1.17	9%	21%	32%	24%	12%	1%
The grading system for the class	3.92	1.18	4%	11%	15%	29%	40%	0%
The self-paced structure	2.90	1.37	16%	30%	20%	15%	19%	0%

Note. M=mean, S.D.=standard deviation, NH=No help, LH=little help, MH=moderate help, H=much help, VH=very much help, NA=not applicable.

Table 11. Student assessment programming skills gained

Item	M	S.D.	NAT	ALI	SW	ALO	AGL	NA
Solving problems	3.15	1.11	10%	14%	38%	27%	10%	0%
Creating flowcharts	3.51	1.02	4%	12%	29%	38%	16%	0%
Reading and understanding C++	3.51	1.13	3%	18%	25%	30%	23%	0%
Writing C++ programs	3.08	1.20	12%	19%	30%	27%	12%	0%
Debugging software	3.42	0.99	2%	16%	34%	33%	14%	0%
Testing software	3.16	1.08	6%	21%	35%	28%	10%	1%
Working effectively with others	3.77	1.45	10%	10%	20%	24%	27%	10%
Explaining how programs work	3.44	1.19	4%	20%	31%	20%	24%	1%

Note. M=mean, S.D.=standard deviation, NAT=Not at all, ALI=A little, SW=Somewhat, ALO=A lot, ALG=A great deal, NA=not applicable.

Table 12. Student assessment learning transfer

Item	M	S.D.	NAT	ALI	SW	ALO	AGL	NA
Core concepts of programming	3.42	1.15	4%	14%	28%	30%	23%	1%
How to solve problems	3.57	1.14	4%	19%	31%	25%	20%	1%

Note. M=mean, S.D.=standard deviation, NAT=Not at all, ALI=A little, SW=Somewhat, ALO=A lot, ALG=A great deal, NA=not applicable.

Table 13. Elluminate usage response statistics

Item	M	S.D.	SD	MD	N	MA	SA	NA
Using Elluminate was helpful to my learning	4.17	1.40	6%	3%	25%	25%	19%	22%
I wish the course had used Elluminate more	4.22	1.49	6%	8%	22%	6%	39%	19%
I had difficulty setting up Elluminate	2.61	1.84	42%	17%	14%	11%	0%	17%
Elluminate was similar to face-to-face	3.97	1.40	3%	14%	19%	31%	14%	19%
Most courses would benefit from Elluminate	4.36	1.31	6%	0%	19%	22%	33%	19%

Note. M=mean, S.D.=standard deviation, SD=strongly disagree, MD=mildly disagree, N=neutral, SA=mildly agree, and SA=strongly agree.

The course instructor added six additional items to the instrument in a section labeled *Elluminate usage* prior to its administration to gather relevant information from students as to whether the technology added value to their learning experience. One item simply asked whether students had participated in a synchronous session, while the remaining five items used a 5-point scale (1=strongly disagree, 2=mildly disagree, 3=neutral, 4=mildly agree, and 5=strongly agree) with a not applicable option. The scale demonstrates an acceptable degree of internal consistency reliability at $\alpha = .84$.

Though the descriptive statistics shown in Table 13 only highlight the usage of the tool during one academic semester, the results are promising. Fifty-nine percent of the respondents indicated the tool was easy to set up, and 44% indicated it was helpful in learning. Forty-five percent of the respondents indicated the tool was similar to a face-to-face experience, and 55% indicated most courses would benefit from the tool's use.

Outcomes on Satisfaction and Performance

Inferential statistical tests were only performed on measures perceived to be valid and reliable. This includes the assignment performance scores, level of satisfaction subscale, and final course grade.

Overall Performance and Retention

Table 14 reports the mean performance on each assignment and the overall grade point average for the students that responded to the survey (not including Assignment 7). It is important to note that students that did not complete the assignments received a zero grade, but were included in the mean averages for each assignment. The data did not exhibit normality for all performance measures; however, there were not any severe departures from normality and ANOVA is robust to violations of this assumption.

Table 14. Outcomes by assignment and final grade

Performance measure	N	M	S.D.
Final course GPA	245	3.00	0.95
Assignment 1	250	95.81	14.48
Assignment 2	250	88.44	18.78
Assignment 3	250	87.61	23.00
Assignment 4	250	81.33	26.32
Assignment 5	250	65.37	43.79
Assignment 6	230	48.78	47.68

From the fall of 2003-04 to the spring of 2004-05, the course averaged a 58% retention rate with 23% of the students withdrawing from the course and approximately 19% receiving a failing grade (D or F). The course retention rates improved from approximately 50% the first semester to approximately 65% in the last semester included in the analysis. Of the students passing the course, approximately 25% of the students earned an A, 22% earned a B, and the remainder earned a C grade.

Gender

To detect statistically significant differences on performance and satisfaction between males and females, one-way ANOVAs were constructed for numeric grades on each assignment, the level of satisfaction subscale, and the final grade in the course with gender serving as a between-subject condition. Results were statistically insignificant on every account as shown in Table 15.

Gender was also statistically insignificant when examining the satisfaction subscale $F(1,231)=1.09$, $p=.297$. Males reported a satisfaction scale of .681 ($SD=.131$) and females .705 ($SD=.124$). These are both positive findings in that they attest to the course not being characterized as gender bias on performance or student satisfaction. Both males and females appear to perform nearly the same and have a similar outlook on this course.

Table 15. Performance outcomes by gender

Performance measure	Male	Female	F	p
Final course GPA	2.98	3.04	0.19	.66
Assignment 1	95.36	95.95	0.07	.79
Assignment 2	87.26	90.29	1.30	.26
Assignment 3	87.37	90.09	0.72	.40
Assignment 4	79.38	83.96	1.55	.21
Assignment 5	63.14	64.17	0.03	.87
Assignment 6	43.49	41.79	0.06	.81

Programming Experience

To detect statistically significant differences on performance based on prior programming experience, one-way ANOVAs were constructed for numeric grades on each assignment, the level of satisfaction subscale, and the final grade in the course with prior programming experience as a between-subjects condition. Again, no statistically significant differences were detected as shown in Table 16.

Table 16. Performance outcomes by prior programming experience

Performance measure	Prior	No prior	F	p
Final course GPA	3.04	3.08	0.30	.59
Assignment 1	93.89	96.95	2.15	.14
Assignment 2	89.71	86.76	1.31	.25
Assignment 3	88.57	88.43	0.00	.96
Assignment 4	81.83	80.18	0.23	.64
Assignment 5	65.98	61.53	0.59	.44
Assignment 6	42.11	44.44	0.13	.72

Prior programming was also not statistically significant when examining the satisfaction subscale $F(1,231)=1.01, p=.317$. Those individuals with prior programming experience reported a satisfaction scale of .676 ($SD=.128$), while those without reported .703 ($SD=.127$). Again, both findings are positive in that they suggest the course content is appropriate for both novices and intermediate students. Both students performed nearly the same and were equally satisfied.

Employment Status

To detect statistically significant differences on performance and satisfaction among students employed full-time, part-time, or unemployed, one-way ANOVAs were constructed for numeric grades on each assignment, the level of satisfaction subscale, and the final grade in the course with employment-status serving as a between-subject condition. This demographic resulted in a statistically significant difference on Assignment 2—$F(1,243)= 4.36, p=.01$—in favor of full-time employed students. All other measures were not significant, as shown in Table 17.

One might observe more variation in the final course grade based on employment status as it approaches significance $p=.07$. Unemployed students appear to achieve a higher final course grade than those students that are currently employed. This is a logical finding in that it suggests unemployed

Table 17. Performance outcomes by employment status

Performance measure	Unemployed	Part-time	Full-time	F	p
Final course GPA	3.19	3.03	2.82	2.72	.07
Assignment 1	95.49	95.76	95.79	0.01	.99
Assignment 2	88.15	84.87	92.95	4.36	.01
Assignment 3	86.22	87.84	87.79	0.11	.90
Assignment 4	81.86	80.60	81.51	0.05	.95
Assignment 5	61.17	70.66	63.35	1.07	.34
Assignment 6	47.85	52.51	45.99	0.42	.66

students, perhaps, have more time to devote to their academics. Employment status was not statistically significant when examining the satisfaction subscale $F(1,231)=0.07$, $p=.93$, indicating both employed and unemployed students had a similar degree of satisfaction with the course.

Computer Programming Interest

The final course grade and satisfaction subscales were correlated with the degree of a student's interest in computer programming. There is a statistically significant, positive correlation ($r=.18$, $p=.004$) between the degree of interest in computer programming and overall performance and satisfaction ($r=.16$, $p=.014$) in the course. This is an expected outcome in that one would anticipate those individuals more interested in the content domain would achieve better performance and be more satisfied with the course.

Satisfaction and Performance

The final course grade and the satisfaction subscale mildly correlated ($r=.15$, $p=0.05$). Although statistically insignificant, the low correlation between the instrument and final course grade was somewhat surprising. One explanation could be a result of expectations. At the beginning of the Spring and Fall 2005-06 semesters, students were surveyed regarding their targeted course grades. The results were that 43% were targeting As, 26% were targeting Bs, and 31% were targeting Cs. Since the instrument was administered at the end of the academic semester, at a time when most students had a good estimate of their final grade, it is plausible that satisfaction was related to performance relative to target grade, rather than relative to actual grade. The results also provide strong evidence that students were willing to provide critical feedback about the course that was independent of their grade.

DISCUSSION

What can be concluded from this course and analysis, and what will be contributed to the best practices for computer programming instruction? The chapter has illustrated a number of challenges facing MIS computer programming instruction, and two key opportunities to impact MIS research and practice: the integration of ICT for instructional purposes, and the development, use, and validation of instruments designed to monitor our courses.

Of course, the results of this analysis must be interpreted in light of the limitations. This analysis has been conducted using data that were collected during different academic semesters. Technology, students, and curriculum has changed over the last 3 years, thus the design of the instrument has changed to collect relevant information. Movement of survey items within the survey, and modification or addition of items may have changed the constructs used in the analysis. In addition, combining responses from survey items to make composite variables may not adequately measure the constructs.

The degree of accuracy of these measures may also be questionable since the items are self-reported measures. Students, though informed their responses would have no bearing on their final grades, may have feared retaliation, and consequently, responded more favorably to the instructional methods. Further, since the instrument is completed by students at the end of an academic semester, experiences regarding assignments and methods earlier in the semester may be skewed. Though reliability and validity evidence could not be provided for all aspects of the instrument and some scales demonstrated less than acceptable internal consistency reliability, the authors wish to emphasize that instrument design is an on-going process.

In light of these limitations, this chapter has demonstrated the value of instruments to monitor our courses. Because university-wide course evaluations only summarize a course at large, it is often difficult for educators to specifically monitor assignments and activities at a level of granularity that will inform instruction. The instrument demonstrated a level of granularity (assignment) that was useful for informing instruction. The chapter has attempted to demonstrate the validity of scores to measure students' level of satisfaction with an MIS computer programming course, and in doing so, provided statistical conclusions demonstrating gender, prior programming experience, and employment status were not statistically different on student satisfaction in this course.

The instrument also provided some interesting descriptive results. Overall descriptive statistics indicate a positive outlook on learning computer programming skills and the transfer of these skills to different settings. Teamwork appears to be a favorable characteristic, from a student perspective, in computer programming instruction. And the use of synchronous communication tools shows great promise for computer programming instruction. This chapter has also described a novel and hybrid approach for the delivery of computer programming instruction in MIS curriculum using a variety of ICTs and pedagogical strategies. Some unique characteristics of the course include an assignment-centric design, self-paced completion, and the use of multimedia resources to replace lectures and provide flexible delivery of instruction. Analysis of student performance showed no differences in gender or prior programming experience, though a student's employment status did demonstrate some variation.

In closing, the authors believe MIS faculty and administration should be mindful of a student's perspective of their courses, and should take careful steps in the integration of novel ICTs and pedagogical strategies. We also feel that educators should not solely rely on university-wide course evaluations to efficiently and effectively capture this critical information from students. More relevant information can be collected by developing instruments tailored closely to our courses to aid in the decision-making process.

FUTURE RESEARCH DIRECTIONS

This research has documented the value of developing, validating, and using survey instruments to monitor and evaluate courses, demonstrated the use of novel pedagogical strategies (e.g., assignment-centric design), and provided many different forms of ICTs that can be gracefully integrated into the curriculum. While this chapter has provided evidence to demonstrate the instructional value of the technology and method, more work needs to be executed in this area. For instance, replicating the instructional methods and use of technology in hybrid classroom environments, both inside and outside of computer programming, is necessary to generalize findings and develop best practices to inform practice.

Educational research in computing- and information-centric disciplines also needs a stronger connection to other areas of educational research, an interdisciplinary approach. Of particular interest is the discipline of instructional technology. In recent years, researchers in instructional technology-related areas have been criticized for holding a one-directional view of the connection between theory and practice in which basic research questions or theory precede and gives rise to investigations having an applied focus or practice. The concept of developmental research or "design experiments" offers an exciting and useful alternative, whereby practical instructional interventions are rigorously studied for their usefulness in solving authentic problems (Reeves, 2000). Educational research without a strong grounding in a context and real-world connection, such as MIS curricula, lacks the necessity to inform practice.

Research in the realm of MIS curricula is open-ended with many opportunities for improvement and innovation, especially in computer programming instruction. Future research efforts should embrace an interdisciplinary approach, and include contexts to solve authentic instructional problems. The additional readings section includes several readings that can serve as a starting place for this suggested and needed future research.

REFERENCES

Al-Rawi, A., Lansari, A., & Bouslama, F. (2005). A holistic approach to develop IS curricula: Focusing on accreditation and IT certification. *Journal of Information Techonology Education*, 4, 307-327

Articulate. (2007). *Articulate e-learning software*. Retrieved on February 10, 2007, from http://www.articulate.com/

Beise, C., Myers, M., VanBrackle, L, & Chevli-Saroq, N. (2003). An examination of age, race, and sex as predictors of success in the first programming course. *Journal of Informatics Education and Research, 5*(1), 51-64.

Brandon, D., Pruett, J., & Wade, J. (2002). Experience in developing and implementing a capstone course in information technology management. *Journal of Information Technology Education, 1*(2), 92-102.

Camtasia. (2007). *TechSmith*. Retrieved on February 10, 2007, from http://www.techsmith.com/

Carnegie Classification. (2007). *The Carnegie classification of institutions of higher education*. Retrieved on February 10, 2007, from http://www.carnegiefoundation.org/classifications/

Carr, S. (2000). As distance education comes of age, the challenge is keeping the students. *Chronicle of Higher Education, 46*(22), 39-A41.

DeMarco, T., & Lister, T. (1999). *Peopleware: Productive projects and teams* (2nd ed.). New York: Dorset House Publishers.

Doube, W. (1998). Multimedia delivery of computer programming subjects: Basing structure on instructional design. In *Proceedings of the 3rd Australasian conference on Computer Science Education,* Sydney, Australia, (pp. 85-91).

Garrison, D. R. (1987). Researching dropout in distance education. *Distance Education, 8*(1), 95-101.

Gill, T. G. (2005a). Assignment-centric design: Testing the assignments not the lectures. *Decision Sciences Journal of Innovative Education, 3*(2), 339-346.

Gill, T. G. (2005b). Teaching C++ submarine style. *IEEE Transactions on Education, 48*(1), 150-156.

Gill, T. G. (2006) The mystery of a self-paced course. *Informing Faculty, 1*(3), 95-105.

Gill, T. G. (2007). Quick and dirty multimedia. *Decision Sciences Journal of Innovative Education, 5*(1), 197-206.

Gill, T. G., & Holton, C. (2006). A self-paced introductory programming course. *Journal of Information Technology Education, 5*, 95-105.

Johnson, G. M. (2006). Synchronous and asynchronous text-based communication in educational contexts: A review of recent research. *TechTrends, 50*(4), 46-53.

Lehman, J. A., & Naumann, J. D. (1986). A language independent course in program design and programming for MIS students. *ACM SIGCSE Bulletin, 18*(4), 32-37.

Lidtke, D. K., Stokes, G. E., Haines, J., & Mulder, M. C. (1999). *ISCC '99: An information systems-centric curriculum '99 program guides for educating the next generation of information systems specialists, in collaboration with industry. Supported by the National Science Foundations grants.*

Molstad, L. (2001). Teaching computer programming using distance education technology. *Journal of Computing Sciences in Colleges, 17*(1), 265-277.

Moore, M. G., & Thompson M. M. (1997). *The effects of distance learning* (Rev. ed. ACSDE Research Monograph No. 15). University Park, PA:

Pennsylvania State University, American Center for the Study of Distance Education.

Nosek, J. T. (1998, March). The case for collaborative programming. *Communications of the ACM, 41*(3), 105-108.

Nunnaly, J. (1978). *Psychometric theory.* New York: McGraw Hill.

Reeves, T. C. (2000). Socially responsible educational technology research. *Educational Technology, 40*(6), 19-28.

Ritzhaupt, A. D., & Zucker, R. (2006). Teaching object-oriented concepts using visual basic .NET. *Journal of Information Systems Education, 17*(2), 163-170.

Stansfield, M., McLellan, E., & Connolly, T. (2004). Enhancing student performance in online learning and traditional face-to-face class delivery. *Journal of Information Technology Education, 3.*

Straub, D. W. (1989). Validating instruments in MIS research. *MIS Quarterly, 13*(2), 147-169.

Williams, L. A., & Kessler, R. R. (2001). Experiments with industry's "pair-programming" model in the computer science classroom. *Computer Science Education, 11*(1), 7-20.

Woszczynski, A., Guthrie, T., & Shade, S. (2005). Personality and programming. *Journal of Information Systems Education, 16(4)*, 293-299.

Zajkowski, M. E. (1997). Price and persistence in distance education. *Open Learning, 12*(1), 12-23.

ADDITIONAL READING

Aiken, J. (2004, September). Technical and human perspectives on pair programming. *ACM SIGSOFT Software Engineering Notes, 29(5)*, 1-14.

Baddely, A. D. (1986). *Working memory.* Oxford, England: Oxford University Press.

Baddeley, A. D., & Hitch, G. J. (1974). Working memory. In G.A. Bower (Ed.), *The psychology of learning and motivation: Advances in research and theory* (Vol. 8, pp. 47-89). New York: Academic Press.

Barron, A. E. (2004) Auditory instruction. In D. H. Jonassen (Ed.), *Handbook of research on educational communications and technology* (pp. 949-978). Mahwah, NJ: Lawrence Erlbaum.

Barron, A. E., & Kysilka, M. (1993). The effectiveness of digital audio in computer-based training. *Journal of Research on Computing in Education, 15*, 277-289.

Beck, K. (1999, October). Embracing change with extreme programming. *Institute of Electrical and Electronics Engineers Computer, 32(10)*, 70-77.

Chandler, P., & Sweller, J. (1991). Cognitive load theory and format of instruction. *Cognition and Instruction, 8*(4), 293-332.

Clark, J. M. (1983). Reconsidering research on learning from media. *Review of Educational Research, 53*(4), 445-459.

Clark, J. M., & Paivio, A. (1991). Dual coding theory and education. *Educational Psychology Review, 3*(3), 149-170.

Johnson, D. W., & Johnson, R. T. (2004). Cooperation and the use of Technology. In D. Jonassen (Ed.), *Handbook of research for educational communications and technology* (2nd ed.) (pp. 785-812). Mahwah, NJ: Lawrence Erlbaum Associates.

Kullhavey, R. W., Lee, B. J., & Caterino, L. C. (1985). Conjoint retention of maps and related discourse. *Contemporary Educational Psychology, 10*, 28-37.

Mayer, R. E. (2001). *Multimedia learning.* New York: Cambridge University Press.

Mayer, R. E., & Anderson, R. B. (1991). Animations need narrations: An experimental test of a dual-coding hypothesis. *Journal of Educational Psychology, 83*, 484-490.

Mayer, R. E., & Gallini, J. K. (1990). When in an illustration worth ten thousand words? *Journal of Educational Psychology, 82*(4), 715-726.

Moreno, R., & Mayer, R. E. (2002). Verbal redundancy in multimedia learning: When reading helps listening. *Journal of Educational Psychology, 94*(1), 156-163.

Paivio, A. (1986). *Mental representations.* New York: Oxford University Press.

Slavin, R. E. (1996). Research on cooperative learning and achievement: What we know, what we need to know. *Contemporary Educational Psychology, 21*, 43-69.

Schnotz, W. (2005). An integrated model of text and picture comprehension. In Mayer (Ed.), *The Cambridge handbook of multimedia learning* (pp. 49-69). New York: Cambridge University Press.

Schnotz, W., & Bannert, M. (1999, October 27-30). Supports and interference effects in learning from multiple representations. In S. Bangera (Ed.), *European Conference on Cognitive Science*, Instituto di Psicologia, Nazionale delle Ricerche, Rome, (pp. 447-452).

Sweller, J. (1988). Cognitive load during problem solving: Effects on learning. *Cognitive Science, 12*, 257-285.

Sweller, J., & Chandler, P. (1994). Why some material is difficult to learn. *Cognition and Instruction, 12*, 185-233.

Williams, L. A., & Kessler, R. R. (2000, May). All I really need to know about pair programming I learned in kindergarten. *Communications of the Association of Computing Machinery, 43(5)*, 108-114.

Williams, L. A., & Kessler, R. R. (2001, March). Experiments with industry's "pair-programming" model in the computer science classroom. *Computer Science Education, 11*(1), 7-20.

Williams, L. A., Kessler, R. R., Cunningham, W., & Jeffries, R. (2000). Strengthening the case for pair programming. *IEEE Software, 17*(4), 19-25.

Yi, J. (2005). Effective ways to foster learning. *Performance Improvement Journal, 44*(1), 34-38.

Chapter XV
Delivering Online Asynchronous IT Courses to High School Students:
Challenges and Lessons Learned

Amy B. Woszczynski
Kennesaw State University, USA

ABSTRACT

As high schools begin to offer more distance learning courses, universities have an opportunity to establish partnerships to deliver online IT courses. Delivering online courses at the high school level, however, means overcoming obstacles that may not be faced at the university level. In particular, establishing partnerships with high schools requires politically savvy navigations of bureaucratic roadblocks while ensuring the integrity of course content and delivery. This chapter provides a primer on establishing relationships with high schools to deliver college-level IT curriculum to high school students in an asynchronous learning environment. We describe the curriculum introduced and discuss some of the challenges faced and the lessons learned.

INTRODUCTION

In this chapter, we describe the CyberTech I program, a National Science Foundation (NSF) funded initiative which delivers university level introduction to IT curriculum online to nine schools in a large metropolitan area in the southeastern United States. Delivering online curriculum to U.S. high school students (grades 9-12, with approximate ages between 14 and 18 years old) provides university educators with unique challenges. Unlike the college environment, in which professors have local autonomy over curriculum delivery and instruction, public high school curriculum has rigid standards that must be achieved, along with guidelines on methods of delivery. Forming a politically savvy team aware of how to navigate the high school environment is a must for ensuring successful establishment of partnership endeavors.

The applicability of this type of program to similarly structured university-high school partnerships is obvious. However, lessons learned in the CyberTech I program may also assist university faculty members who coordinate introductory online courses with large numbers of students and teaching assistants.

BACKGROUND

In U.S. public school systems, IT-related courses often fall under "Career and Technology Education" or "Business Education" departments. This positioning gives IT courses a distinctly different and typically lower regarded position than other mathematics and science courses. Often, students encouraged to take career and technology education courses select a noncollege-prep track. Moreover, since many of the teachers in these departments have a primary background in business, they may find it difficult to teach an advanced IT course, leaving students unable to take more advanced courses, such as advanced placement (AP) computer science, at their own schools.

In January 2005, we partnered with nine schools in a large metropolitan area to deliver an online introduction to IT course to high school students. University initiatives to deliver IT-related courses in the high schools have proven successful in Finland (Grandell, 2005), where five different college-level courses were offered at the high school level. Moreover, a community college in Pennsylvania successfully partnered with high schools to offer college-level computer information systems courses with significant success (Harvey, 2004). In addition to the goodwill established between high schools, students, and universities, programs such as these offer the ability to expand curriculum offerings at high schools significantly (Donlevy, 2003). Since 25% of public high schools have distance learning alternatives, and 19 states have virtual high schools (Mupinga, 2005), partnerships between colleges and high schools may become more commonplace. Distance learning has even expanded into the elementary level in some cases (Anastasiades, 2003), with a great deal of success.

After establishing a partnership with local school systems, we then had to decide which online learning management system to use. We considered WebCT, Blackboard, and Moodle. We needed to use a flexible, Web-based learning management system for several reasons. First, some of the schools were unable to offer the course during the regular school day. Therefore, we needed an asynchronous method to communicate with students on a 24/7 schedule. Second, those schools who did offer the online class during the regular school day met at different times and in different locations. Obviously, Web-based solutions offer the ability to reach a geographically dispersed population at different times.

CyberTech I PROGRAM

The CyberTech I program seeks to attract women, minorities, first-generation college students, and other underrepresented groups into technology related fields. CyberTech I is the first course that selected freshman high school students take, culminating with AP computer science in their senior year. In the middle years, students participate in SummerTech, which teaches basic programming skills and logic using VB.net, followed by Weekend Academy, an adventure game programming seminar that meets once per month throughout the student's junior year of high school. Finally, students are encouraged to take AP computer science in their senior year, and the grant pays all testing fees. This chapter will focus on the issues surrounding successful implementation of the 75-hour, one semester (18 weeks) CyberTech I initiative, including the lessons learned and challenges.

Funding for the CyberTech I program was provided through an NSF grant (#0423576). NSF funded the grant under its Information Technology for Students and Teachers (ITEST) program. NSF awards ITEST grants to programs that show promise in increasing the number of students and teachers who learn about, experience, and use IT. The grant award included stipends for high school teachers who assisted the students (site facilitators), course releases for participating faculty members, funding for graduate teaching assistants, and miscellaneous fees for students. Students participating in the program received all materials and services free of charge, as did the schools involved. Each school agreed to provide a site facilitator and, where possible, a classroom devoted to CyberTech I students. Before turning to a discussion of the important participants in the program, we begin with an overview of the online tools used in the CyberTech I program.

Online Learning Tools

Online learning tools may be broadly classified as either synchronous or asynchronous. Synchronous tools allow much greater interaction and faster feedback as compared to their asynchronous counterparts. However, these advantages come with the caveat that everyone has to be online at the same time. In general, we did not use synchronous tools for the entire student group due to the varying schedules of the participating schools. CyberTech I classes met as early at 8:00 am and as late as 3:30 pm. Some classes met everyday, while others met two or three times weekly. One school even offered CyberTech I as a year-long course, meeting twice a week for 36 weeks.

In addition, the participating public schools in our program had slightly different break schedules, including different starting and ending dates. For these reasons, we only made limited use of synchronous tools. In particular, each teaching assistant offered online office hours and was available for live chat sessions at specified times. We also gave the students the opportunity to use instant messaging, but few used this option. In fact, students only made very limited use of the chat and instant messaging options, preferring instead to use e-mail or discussion board postings.

We did, however, make extensive use of asynchronous tools to deliver the course content. Students overwhelmingly reported that the discussion board activities were helpful in keeping them connected to their peers. We included multiple discussion board assignments to engage the students. We also created a student lounge and a question and answer (Q&A) discussion board. The student lounge gave participants an opportunity to talk with other students about interests outside the classroom. In fact, we required that all students post an introduction in the student lounge at the beginning of the course. Many students also posted a picture of themselves, which helped everyone put a name and face together, much as students do in a regular classroom setting. Since some CyberTech I sections included students from multiple schools, the student lounge gave an opportunity for participants to meet other students with whom they had no physical contact.

The Q&A discussion board gave students the ability to ask classroom-relevant questions. We encouraged all students to check the Q&A discussion board frequently and to post responses if they had the answer. Of course, teaching assistants monitored this discussion board to ensure accuracy of postings.

Requiring that students post to discussion boards provided several additional advantages. First, students taking the class at different times and/or in different schools could respond at the most appropriate time for them. Second, since all students were required to make postings, the loudest or brightest students did not dominate the discussion as they might in a traditional classroom environment. Rather, everyone had an opportunity to express an opinion. Therefore, the discussion board postings reflected the diversity of our group, with constructive discussions on multiple topics, particularly those topics related to ethics. We found that many students posted more times than required, signaling student interest and involvement, characteristics that are difficult to find in most high school and university classrooms.

However, the discussion boards also presented some challenges. Since we were dealing with young, often immature students, the posters would occasionally flame another student or use inappropriate language. The teaching assistants were encouraged to quickly stop the inappropriate postings, remind the student of appropriate discussion board etiquette, and remove any offensive postings promptly.

Students extensively used e-mail communication tools. We quickly learned that we needed to specify e-mail response times and post availability for teaching assistants. Otherwise, students would send an e-mail in the middle of the night and expect a response immediately. If students had made more extensive use of the chat opportunities, they may have overcome some of the disadvantages of asynchronous communications.

We did face some additional challenges with e-mail communications. In particular, students using this type of communication tool often used less formal communication strategies. Professors and teaching assistants might receive e-mail messages addressed to them using first names only, as if the teaching assistants and students were close friends. E-mail messages using all lower-case type or multiple abbreviations were frequently received. In addition, students would forget to sign their names, thus leaving the instructor to determine which student matched to a particular (often unrelated) e-mail address. As we developed and improved the program, we made more formal written requirements for e-mail etiquette, but we still encountered similar difficulties throughout the CyberTech I program. Next we turn to a discussion of the important components of the CyberTech I program.

Project Manager/Director

We learned very quickly that we needed a project manager/director to oversee all facets of the program and who could interact well with university faculty, high school teachers, principals, counselors, parents, and students. We recommend the appointment of a project manager/director with an education background, both at the kindergarten through 12th grade (K-12) level and at the university level. This person should have excellent organizational skills, the ability to balance competing needs simultaneously, an understanding of how the K-12 system operates, and the ability to work with all levels involved in the project, from county superintendents to teaching assistants (TAs), students, and teachers. The next section describes the CyberTech I curriculum and the delivery method selected.

Curriculum and Delivery

As the students' first IT course, the curriculum focused on the basics of computing and how the technology is used. The curriculum for CyberTech I spanned the following general subject areas: 1) The computer (hardware, software, and how computers execute software); 2) problem solving, 3) algorithm design; 4) introduction to programming languages; 5) abstract data types; 6) operating systems; 7) artificial intelligence; 8) networks; 9) simulation; 10) software (including spreadsheets and databases); and 11) the Internet. Also, we introduced students to legal and ethical issues in computing as well as information security and forensics. The course materials included a very readable textbook (Dale & Lewis, 2003) and a computing laboratory manual (Meyer, 2003) as supplements to the online curriculum. Jones & Bartlett Publishers graciously donated the textbooks and the accompanying laboratory manual and CDs to support the program.

The CyberTech I program mapped to the quality core curriculum (QCC) standards established by the state's Department of Education for IT Foundations 11.412, an existing high school course. Therefore, students were eligible to receive high school credit for the course. Further, students who successfully completed the CyberTech I curriculum and later enroll at our university will be able to receive college credit for a comparable introduction to IT course that many computer science and information systems majors must take. For students who do not enroll at our university, they have the option of taking an advanced standing test to receive credit for the introductory course.

Students successfully completing the 75-hour CyberTech I course are eligible to participate in the 75-hour SummerTech VB.net programming course, which builds on concepts learned in CyberTech I. Upon completion of the summer programming course, students have the opportunity to attend Weekend Academy, an adventure game programming seminar that meets once per month throughout the student's junior year of high school. Finally, students are encouraged to take AP computer science in their senior year, and the grant pays all testing fees.

During the first offering of CyberTech I, we made all course materials available to students using the Blackboard delivery option. We selected Blackboard as the delivery vehicle for CyberTech I since all K-12 schools in the state use Blackboard for online courses, and the students and teachers are likely very familiar with the look and feel of the system. The county managing the Blackboard licenses required us to pay a per-student fee for usage of the service, and the NSF grant covered all costs. In the second and subsequent semester offerings of CyberTech I, we moved to Moodle, an open source alternative, saving approximately $20,000 per semester over the costs associated with using Blackboard. We developed online training for students using Moodle and offered an in-person training session for all high school teachers assisting in the CyberTech I classroom.

Assignments were divided into units, with each unit encompassing a chapter or similar grouping of material. Within every unit, we integrated ethics components. We included a substantial number of video clips, hands-on activities, and Web exercises to stimulate students. Each unit further included interactive discussion boards for critical evaluation of issues relevant to the students. Discussion board topics included subjects such as software piracy, reliability of information on the Web, software licensing and patents, careers in computing, and Grace Hopper's impact on computing. We particularly recommend the inclusion of topics highlighting the contributions to IT of underrepresented groups, such as women and minorities.

Based on feedback from teachers and students, we designed a student agreement after the initial offering of CyberTech I. This agreement specified student expectations and outlined rules of conduct for the class. Specifically, the students agreed to the following:

- To complete unit assignments in a timely manner, by 11:45 pm on the date specified, and to inform the instructors promptly if I am unable to complete the work in a timely manner.
- To stay in close contact with the teacher at my school, the CyberTech I instructor, and the TA.
- To submit only work that is my own, and not to share my work with anyone else, although I understand that I may work together on homework and laboratory assignments.
- To regularly visit the course site for updates and assignments, checking at least every 3 business days.
- To interact with classmates and the course instructors in a professional and mature manner.
- To follow the instructions for assignments carefully, making sure to answer the questions posed.
- To spell-check and proofread all discussion board postings, laboratory exercises, and e-mail messages before submitting them.
- To submit assignments in the appropriate manner and format as specified online.
- To contact my teacher, the CyberTech I instructor, or my TA with problems or questions.
- To keep my e-mail address updated and to check and respond to e-mail regularly, at least every 3 business days.
- To spend approximately 10-14 hours per week on in-class and out-of-class activities.

We provided the students with a copy of the syllabus and the student agreement on the first day of class. They had to sign the agreement and have a parent/guardian sign as well and return the form the next day. Students, site facilitators, and parents all reported significantly less confusion once we developed a student agreement, and we would recommend that others embarking on a similar project develop comparable guidelines.

Schools

Before submitting a grant proposal to NSF, we secured multiple letters of support for the CyberTech I program from schools in the surrounding areas. When we received funding and began to implement the program, we believed we would encounter little difficulty getting schools to participate. After all, we were offering teachers the option of facilitating a class with no grading requirements and no lesson plan development. Further, we were offering the schools funding to cover all books and miscellaneous fees and the potential ability to expand their curriculum to offer a course not typically offered. Although we thought schools would be excited to participate in the program, we ran into a few problems. Before even considering whether to implement the CyberTech I program, the schools wanted to ensure that the curriculum met state guidelines for IT courses. Therefore, we mapped all course learning outcomes to corresponding objectives in IT 11.412, IT Foundations. After one of the largest school districts in the state approved the online course as meeting the requirements for IT 11.412, all other districts approved the course as well.

Further, we found that school participation was contingent upon securing additional support from administrative levels as high in the hierarchy as possible. In school districts where the superintendent was supportive and excited about the program, we found principals and teachers correspondingly supportive. We also learned that it was essential to get the full support of principals, counselors, teachers, parents, and students at each school. The more support we received, the better the student participation in the program. To secure support at all levels, a project manager/director visited school district offices, local schools, and individual teachers, counselors, and principals on a regular basis. We found that it was important to keep the level of excitement high, and personal contact with the project manager/director was necessary to sustain this passion for the program.

Course Scheduling

Once we received support from the school systems, we then had to find a way to get CyberTech I on the school schedule. Unfortunately, we ran into several problems in this regard. High school schedules are often created a year or more in advance. NSF funding for our proposal, however, was not awarded until October. That made it challenging to schedule the course the following January, a mere three months to get the course on the schedule and the online materials in place. For the initial offering of CyberTech I, we developed several workarounds to compensate for the schedule timing issues.

First, we offered students the option of taking the CyberTech I course before or after school, as an additional class to the load they were already taking. Second, we offered students the opportunity to take the course at their homes, on their own computers. Both of these options presented problems.

The withdrawal process for high school students is different than the process normally encountered in universities. High school students are not allowed to withdraw from classes that meet during the day. However, for additional classes students take beyond the school day, they can withdraw almost until the end of the semester. Since CyberTech I was an additional course, high school students had the option to drop the course.

Several of the high schools with which we dealt had very lenient drop dates almost at the end of the semester, which caused us to lose a number of students. It is very important to establish clear guidelines for withdrawal that do not conflict with individual high school policies.

In addition, it was difficult to keep the students motivated when they only met an afternoon or two a week in a lab, or when they worked from home. We also found it quite hard to maintain contact with the students—whose e-mail addresses were often invalid—and difficult to keep these students involved, engaged, and interested in the CyberTech I class.

As a second option, we put CyberTech I students in other computer classrooms that had extra capacity. So while one class was being taught, the CyberTech I students could access their assignments on a computer in a regular classroom. The obvious drawbacks to this situation are that teachers do not have time to assist the CyberTech I students, and the CyberTech I students found it difficult to concentrate when another class was being taught.

After offering alternative scheduling arrangements out of necessity the first semester, we learned that students in the regularly scheduled classes that met every day in the classroom with a teacher did much better than students who worked before or after school, in other classrooms, or at home. We would highly recommend that others undertaking similar endeavors schedule CyberTech I as part of the school day to achieve greater levels of success.

The rules for changing grades also differ in the high schools and universities. High school students can appeal grades more than a year after the class is completed. In some cases, they can request withdrawal from a course well after the class has been completed. High school teachers—and sometimes counselors—have much more flexibility in changing grades than do professors in most university settings. Establishing clear guidelines for grade appeals is also necessary before starting any curriculum endeavor in the high schools.

CyberTech I Instructor

The CyberTech I instructor selected for this program had years of experience teaching the introductory course material with excellent student evaluations of performance. Further, the instructor had several years of experience coordinating many sections of the course at two different universities. We found that it is critical to select an instructor who works well with students at all levels and promptly responds to student inquiries. In particular, we suggest that the chosen instructor has a passion for working with young students, since the high school students have a different maturity level than their college counterparts. Instructors with experience teaching college-level freshmen and lower-level courses might be most capable of teaching these adolescents. Moreover, we recommend that others undertaking similar programs avoid the temptation to assign adjunct or new faculty to this endeavor. The chosen instructor needs experience teaching the course, excellent organizational skills, the ability to juggle conflicting needs seamlessly, and superior supervision skills to monitor the performance of the teaching assistants.

Site Facilitators

After choosing the faculty member to oversee the course, we then worked with the participating school systems to select the high school teachers. Selecting the site facilitators presented challenges significantly different from the selection of the CyberTech I instructor. While we had total control of instructor selection, we had only limited input into the selection of a site facilitator. The school principals assigned the site facilitators, often selecting volunteer teachers, who may or may not have had the necessary background. In fact, we learned that if a volunteer did not come forward to facilitate CyberTech I, the principal would simply force a teacher to take the responsibility, leading to potential resentment. We found that giving the principal advance notice of the need for instructors helped in the assignment of a site facilitator with the necessary skills. Ideal site facilitators would have extensive experience teaching advanced IT courses, a passion to teach AP courses, and excellent teaching skills. Moreover, the site facilitator needs to be organized and able to work well in online environments. Since we learned that some site facilitators were unfamiliar with online teaching environments, we recommend requiring that all site facilitators complete a special training course in the online learning management system to ensure that they can provide basic assistance to students when needed. Previous studies have noted the importance of allocating sufficient training time to teachers (Donlevy, 2003).

We found that principals were more likely to assign excellent site facilitators when: 1) they received ample advance notice, 2) they understood the goals of the program, and 3) they understood the importance of teacher involvement on student success in the CyberTech I classroom. This program targets underrepresented groups, including minorities, women, disabled, and first-generation college students. Since these students typically may not perform well on state-mandated tests, standards upon which the principal is graded, we used anticipated improved student performance as another selling point. If we can convince the principal that the program is going to help these students perform better in the classroom, we are more likely to receive assistance in the selection of an excellent site facilitator.

We also found it very difficult to motivate the site facilitators to perform well in the classroom, even though we stressed to them the benefits of participating in the program. Specifically, we provided all lesson plans, completed all grading, and answered student questions. The site facilitator only had to answer questions for which students might need immediate help; for example, assistance with logging into the learning management systems (i.e., Blackboard or Moodle). The site facilitators received their regular pay for teaching the CyberTech I course, although they did not have to put together lesson plans, grade assignments, or calculate final grades. If teachers were not teaching the CyberTech I course during this time period, they would receive their regular pay—but no stipend—to teach another course, including developing lesson plans, grading all assignments, and assigning final grades. We also gave all site facilitators a modest stipend of $400 but still found it difficult to motivate the teachers. Since we had no input into the teachers' evaluations, we could not motivate them to perform better or worse. We were unable to find a suitable workaround for the problem of teacher motivation. Indeed, K-12 administrators face a constant struggle of motivating teachers, and we simply experienced one small part of the problem.

Students

To recruit students for the program, we offered several information sessions in the evenings, where parents could learn about the program and ask questions. We learned that it was important to secure the support and excitement of at least one teacher or school counselor to ensure that a large number of participants attended the information sessions. To that end, we involved counselors and principals early and often. In particular, school counselors have enormous influence on which courses students will take, so we made sure that the project manager/director for the CyberTech I project met with all counselors on a regular basis, stressing the benefits of the program.

In addition to open recruiting presentations, we solicited input from teachers and counselors to select appropriate students. Rather than simply selecting students who were on the noncollege-preparation track, we asked teachers and counselors to identify potentially high-performing students, particularly students from our targeted groups.

Once students were identified—either from the information sessions or through teacher identification—we then interviewed the students. We set grade-point average (GPA) guidelines, setting a cutoff level of 3.0 on a 4.0 scale. In our case, a grade of A received 4 points, a grade of B received 3 points, a grade of C received 2 points, a grade of D received 1 point, and a grade of F received 0 points. To earn an A, a student had to have an average of 90% or higher for the course; for a B, an average of 80% or higher; for a C, an average of 70% or higher; for a D, an average of 60% or higher; and for an F, any average grade below 60%. We further considered additional desired traits such as good attendance records. In some cases, based on teacher and/or counselor recommendations, we selected students whose GPA might have been a little lower than desired, but who had excellent attendance and great potential in the classroom. We also learned that it is important for students to be excited about the program so that they will talk positively to their peers who might be eligible for the program the following year. Word of mouth advertising was an inexpensive method to recruit students as the project continued.

Teaching Assistants

Teaching assistants were critical to the program's success, since they were the first line of contact for the students. We have a Master of Science in Information Systems (MSIS) program at our institution, so we had a ready group of qualified graduate students who could serve the high school students well. The NSF grant allowed us the opportunity to reward well-qualified students with tuition remission and a small stipend. Like the instructor and the site facilitators, it is critical to select teaching assistants who are organized, fast to respond to student questions, and able to work well in an online environment. Since the TAs are often the first line of contact for the students, their ability to handle problems in a tactful manner is of paramount importance. Moreover, their communication skills, both written and verbal, must be superlative. They need to work well with the CyberTech I instructor and the site facilitators. Finally, previous teaching or training experience is highly valued, and if certified teachers are available, they should receive priority consideration.

We completed a competitive application process for the TA positions, which were highly desired by the MSIS students. The CyberTech I faculty member interviewed each potential TA after reviewing the application and resume. In addition, we relied on faculty input to make the best choices. We believe it is critical to form a supportive group of faculty members who can provide input into the selection process. In addition, if a school undertakes a similar project, and does not have a graduate program from which to select TAs, we recommend establishing a partnership with a nearby university to recruit qualified and capable TAs. We cannot overstate the importance of good TAs for a successful classroom experience.

Before hiring the TAs, we estimated the optimal student load that each TA could handle. In initial semesters, one TA could handle about 25-30 students. That time includes grading approximately one discussion board posting and one laboratory assignment per week for each student. Weekly quizzes, the midterm, and the final exam were all graded automatically. In subsequent semesters working in CyberTech I, the same previously trained TAs could handle up to 50 students each. We generally kept the TA-student ratio at about 1:25 and found that the TAs, students, teachers, and faculty members were all satisfied with performance at those levels.

Once a TA was hired, we began an extensive preparatory program. TAs went through the same exercises that students would complete, as well as an online training course. In addition, TAs were required to submit weekly status reports detailing the accomplishments of the week. In the status report, the TA had to include names of students who were not making satisfactory progress and steps taken to improve student performance. In addition, the TAs reported which specific assignments had been graded and which assignments were pending. We believe it is critically important for the CyberTech I faculty member to stay current with TA performance so that problems can be resolved quickly.

We did learn early in the program that TAs should not be the first line of contact for parents. Unlike the college environment, high school teachers often have to deal with irate parents who demand immediate answers. When the parents contacted the TAs directly, the TAs were always told to forward that information to the CyberTech I Instructor. In addition, although the TAs graded all student work submitted, the CyberTech I Instructor, in consultation with the site facilitators, assigned final grades.

Further, we learned that the K-12 calendar does not correspond to many college calendars. Specifically, the college fall semester begins after many of the K-12 schools have been in session for some time, and the college spring semester ends before K-12 schools complete their year. In addition, spring breaks and other holidays may not match. Therefore, we recommend hiring at least one teaching assistant who is available to work during breaks. We hired this TA as a temporary employee, and the position was renewable for a 1-year time period.

We also learned that the TA could serve an increasingly important role if one of the high school teachers was not certified to teach IT-related courses. In that case, we recommend finding a TA who is able to travel to the school and provide help sessions about once or twice per month. We further recommend that the TA who has to travel off-site to help the students should be compensated with a reduced teaching load (10-15 students as opposed to 25-30 typical for other TAs).

Advisory Board

The advisory board provides the final piece for our team of players needed for a successful outcome. Advisory board members should include representatives from university personnel, K-12 employees experienced in administration, publishers and authors of textbooks appropriate for use in the program, graders for the AP computer science exam, and people familiar with assessment strategies for large-scale, longitudinal projects. We recommend meeting with the advisory board once or twice a year and updating them on strategies used in the project. With the input of the CyberTech I advisory board, we were able to develop meaningful assessment instruments and appropriate curriculum for the high school level.

ASSESSMENT

Formative evaluation of each student's progress was conducted through daily quizzes and weekly assessment of work products by the CyberTech I instructor and teaching assistants. In addition to feedback from instructors and TAs, students received guidance from a site facilitator at their local high school. Moreover, students were required to complete an initial evaluation before enrolling to determine their suitability for taking an online course. We used a test from one of the county systems that was given to all high school students who were considering enrollment in online learning. If the answers indicated that the student might not perform well in an online course, the student was given online advice and encouraged to visit the school counselor and/or talk with the instructor for further information. We strongly recommend that universities undertaking similar projects use a test to identify whether a student has the focus and initiative to succeed in an online course.

We completed several assessment measures to determine the success of CyberTech I. First, we measured the number of students who enrolled in CyberTech I and the number of students who successfully completed the course. Over three semesters and with approximately 250 enrollees, students posted a successful completion rate (defined as earning an A, B, C, or D) of 78%. Considering that we were dealing with at-risk, transient, and underrepresented groups, we feel that this success rate was excellent. We would encourage others completing similar projects to measure and report success rates for thoughtful comparison purposes and to establish a benchmark.

Second, we asked all students to take a preassessment test. This test was a 45-item multiple-choice exam matched to the stated goals and objectives of the course. We compared the student's preassessment score to the score on a similarly-matched 45-item multiple-choice final exam to determine if learning occurred throughout the semester. Our results showed statistically significant differences ($p<0.05$) between the pretest and the final exam scores, which we used as an indication that learning had occurred.

We also conducted attitudinal surveys to gauge student satisfaction with the course. Our initial results indicated that students were generally satisfied with the CyberTech I course. However, we learned that high school students need a significant amount of interaction and activities to accomplish their online course assignments, more so than their university student counterparts. Therefore, we plan to include more online activities in future endeavors and recommend that others do the same.

Our study did face a number of challenges in regards to gathering data. Since high school students are not old enough to sign a contract, they cannot agree to complete surveys. Therefore, we had to secure parent permission for student participation in assessment activities. We learned that we should give the permission forms to the students on the first day of class and have them take the forms home and return them promptly.

FUTURE RESEARCH

Although the CyberTech I program discussed in this chapter received funding from an external agency, clearly similar endeavors may need to find alternate funding sources. Corporate sponsors may provide another option for universities that are not fortunate enough to receive some of the available limited funding. Before beginning the CyberTech I program, we offered a similar summer session that taught high school students how to program in a nonthreatening, enriching environment. We acquired corporate funding for these initiatives, along with support from the participating high schools, and we would encourage other universities to seek funding from multiple sources.

University-high school partnerships have only recently emerged as a viable option for students, teachers, and administrators. Much more research is needed to determine what types of online learning are most effective in different situations, with different age levels, and in countries outside of the U.S. Will the proliferation of virtual high schools across the U.S. provide better options for joint online delivery of courses at the high school and university levels? As infrastructure improves around the world, will international partnerships between universities and high schools in different countries be offered? If so, what impact does culture have on the success of online course delivery? Researchers could analyze whether international partnerships offer students a more enriching learning experience than partnerships offered in one country only. In addition, researchers may study whether factors such as different time zones and languages affect the learning experience.

Although our study found that students did not use instant messaging or chat options very frequently, future research should determine the applicability of such tools in university-high school partnerships. Since instant messaging and chat have recently become mainstream communication alternatives, perhaps future research will find more student use of these options.

Future research could also examine how to educate students on the appropriate methods of using e-mail, instant messaging, and chat. Since

Table 1. Challenges and key lessons learned by component

Component	Challenges and Key Lessons Learned
Project Manager/ Director	• Hire one person to oversee all aspects of the project. • Visit principals and schools regularly.
Curriculum and Delivery	• Work with local school districts to match course guidelines to appropriate course outcomes. • Ensure that students receive high school credit. • Use an online delivery option that schools accept. • Require that students and parents read and sign a student agreement form.
Schools	• Secure support from schools before pursuing university-high school partnerships. • Maintain personal contact with principals, counselors, and teachers.
Course Scheduling	• Plan course offerings well in advance. • If no in-class options are available, consider offering the class before/after school, in another classroom, or as a study at home course. • Set policies on withdrawals and grade changes.
CyberTech I Instructor	• Select a faculty member with supervisory and/or coordination experience. • Choose a faculty member who enjoys working with younger students. • Do not assign part-time faculty members.
Site Facilitators	• Work with principals to select site facilitators. • Let the principal know in advance the site facilitator qualities needed. • Encourage the principal to select the best teacher for the job, rather than just the one who volunteers. • Find innovative ways to motivate site facilitators.
Students	• Offer multiple information sessions at various schools, and market to parents and students. • Solicit input to get the best students. • Interview students, and select those most likely to succeed based on GPA, attendance, and so forth.
Teaching Assistants	• Select graduate students to fill TA positions. • Look for previous teaching experience when hiring TAs; use certified teaching applicants if available. Estimate the amount of work a TA can handle within the time allotted. • Train TAs to complete the student work, answer questions, and use the online delivery tool. • Hire at least one TA to help during breaks.
Advisory Board	• Ask representatives from important constituents to participate in an advisory capacity. • Meet 1-2 times per year.

e-mail, instant messaging, and chat by their very nature lead to informal communication methods, teachers may need to use different strategies to help students realize how to use communication methods in professional and appropriate ways.

CONCLUSION

Proper planning and effective project management are essential to the success of university-high school partnerships to deliver online courses. In this chapter, we presented details of an asynchronous online learning class delivered by university faculty and teaching assistants to high school students. Some of the lessons learned may be applied to similar university-high school partnerships or to large, introductory online college classrooms that may have many teaching assistants. In addition, we presented essential considerations to improve the chances of successfully completing a major endeavor. In particular, we discussed challenges and key lessons learned for the project manager/director, curriculum and delivery, schools, course scheduling, CyberTech I instructor, site facilitators, students, teaching assistants, and advisory board, as shown in Table 1.

We recommend that anyone who is considering a similar endeavor should ensure that an appropriate team is formed, including all of the components discussed in this chapter and summarized in Table 1. In addition to the CyberTech I instructor, we enlisted the assistance of a full-time project manager/director who served as a liaison to all of the high schools. The high school principals and counselors appreciated the presence of a single point of contact they could call with questions. Ultimately, we learned that a synergistic team including university faculty, teaching assistants, site facilitators, principals, counselors, project managers, and, especially, the high school students themselves, worked together to make the university-high school partnership a success.

REFERENCES

Anastasiades, P. S. (2003). Distance learning in elementary schools in Cyprus: The evaluation methodology and results. *Computers & Education, 40*, 17-40.

Dale, N., & Lewis, J. (2003). *Computer science illuminated* (2nd ed.). Boston: Jones and Bartlett Publishers.

Donlevy, J. (2003). Teachers, technology and training. *International Journal of Instructional Media, 30*(2), 117-121.

Grandell, L. (2005). High school students learning university level computer science on the Web: A case study of the DASK-model. *Journal of Information Technology Education, 4*, 207-218.

Harvey, S. (2004). Bridging the digital divide: How technology can change higher education delivery for high school students. *Community College Journal of Research and Practice, 28*, 73-74.

Meyer, R. M. (2003). *Explorations in computer science: A guide to discovery.* Boston: Jones and Bartlett Publishers.

Mupinga, D. M. (2005, January-February). Distance education in high schools. *The Clearing House, 78*(3), 105-108.

ADDITIONAL READINGS

Carswell, L., Thomas, P., Petre, M., Price, B., & Richards, M. (2000). Distance education via the Internet: The student experience. *British Journal of Educational Technology, 31*(1), 29-46.

Gal-Ezer, J., & Lupo, D. (2002). Integrating Internet tools into traditional CS distance education: Students' attitudes. *Computers & Education, 38,* 319-329.

Inman, E., Kerwin, M., & Mayes, L. (1999). Instructor and student attitudes toward distance learning. *Community College Journal of Research and Practice, 23,* 581-591.

Katz, Y. J. (2002). Attitudes affecting college students' preferences for distance learning. *Journal of Computer Assisted Learning, 18,* 2-9.

Ponzurick, T. G., France, K. R., & Logar, C. M. (2000). Delivering graduate marketing education: An analysis of face-to-face versus distance education. *Journal of Marketing Education, 22*(3), 180-187.

Section V
Economic Analysis and Adoption

Chapter XVI
Motivators and Inhibitors of Distance Learning Courses Adoption:
The Case of Spanish Students

Carla Ruiz Mafé
University of Valencia, Spain

Silvia Sanz Blas
University of Valencia, Spain

José Tronch García de los Ríos
University of Valencia, Spain

ABSTRACT

The main aim of this chapter is to present an in-depth study of the factors influencing asynchronous distance learning courses purchase decision. We analyse the impact of relations with the Internet, distance course considerations, and perceived shopping risk on the decision to do an online training course. A convenience sample of 111 students attending classroom-taught postgraduate and management training programmes were used in March 2005 to obtain the information necessary to test in the Spanish market the conceptual model proposed by applying logistical regression. The results show that perceived course utility, lack of mistrust in the organising institution (service considerations), and satisfaction with the use of Internet when doing this type of training (relations with the medium) determine the asynchronous distance learning course purchase intention. Finally, the authors consider a set of recommendations for company managers.

INTRODUCTION

Information and communication technologies (ICT) are bringing radical changes to the world of work, culture, interpersonal relationships, the way knowledge is shared, and teaching-learning processes (Pagliarello, 2007). This, in turn, presents many challenges: challenges for employment because new jobs and new qualifications are required; challenges for culture as Internet-centred developments of new services which affect cultural practices are sometimes seen as a threat and sometimes as an opportunity for Europe's cultural and linguistic diversity; and challenges for education, especially to ensure that technological innovations are really serving education and are relevant from the pedagogical point of view and educational in very different learning contexts, respecting linguistic, cultural, and social diversity (Escarré & Barros, 2007; Hennecke & Cerny, 2007).

In the sphere of teaching-learning, ICT provide new methodological and didactic opportunities, encouraging universities, companies, and institutions in the training sector to develop alternative pedagogical methods to the traditional classroom. Training must aim to develop skills which favour flexible adaptation to the changes which are occurring in the information society (Bruder, Beyerlein, & Blessing, 2007; Butrimiene & Danilevicius, 2007; Hasan & Dunn, 2007).

Distance learning has been an alternative form of education for years for people whose personal circumstances (because they are working, have little time to attend class, etc.), geographical location (training centres are far away and, therefore, involve travel and in some cases a change of residence, which is a significant cost in terms of time and money) lead them to choose this option because they feels it suits them better than traditional training.

Originally, the focus was on autonomous, independent learning, with very little participant-teacher relations and none between participants (Williams, Paprock, & Covington, 1999). Distance learning is currently considered equivalent to "no classroom attendance," "online," or "e-learning," and is based on technological networks which contribute to participants' intellectual development (Macknight, 2000), providing the stimulus to create work teams, encouraging critical thought (Muilenberg & Berge, 2002), and mutual support for collaborative work (Salter, 2000). This teaching methodology is based therefore on a didactic dialogue between the teacher and the student who learns independently but also cooperatively and collaboratively. The dialogue takes place through a technological bidirectional communication system which can be addressed to a mass audience and replaces personal interaction in the classroom by the joint, systematic action of various didactic resources and the support of an organisation and tutoring which provide the students with independent and flexible learning.

This new method makes learning more convenient, eliminating space and distance barriers and providing cost savings (E-Leusis, 2004; Favretto, Caramia, & Guardini, 2005; Hannum, 2001). It also offers other methods of communicating with the teachers which are more flexible, graphic, and fast and compensate for the shortfalls in the classical method of distance learning. E-learning has new interactive tools which make it convenient and simple for the student to assimilate the content, building a more appealing form of training.

In this way, e-learning, conceived as the use of new multimedia and Internet technologies to improve the quality of learning, emerges not only as a channel for facilitating learning processes but also as a privileged mechanism for articulating information, management, and training systems. It makes it possible to integrate the different components which influence continuous training issues and the factors arising from the use of the new technologies, contributing to promote the teaching-learning process and also favouring the acquisition of skills for working in virtual environments (Cenich, 2006).

The advantages in terms of flexibility, interactivity, and accessibility of distance courses are turning e-learning into one of the main methods of training in higher education (Graff, 2003). In this context, the perspectives point to exponential growth opportunities in the distance course market. Since the University of Phoenix offered the first distance course in 1989, the distance course market has grown spectacularly. If we look at Spain, the results of the "5th Panel of results on E-learning application in large companies" (Grupo Doxa, 2005) show that e-learning represented 6.2% of business training, rising to 8.4% in large companies and 11% including blended learning. The study forecasts that business e-learning courses in Spain will grow faster than classroom taught courses, with the latter showing a mere 8% increase in 2004 while e-learning grew by 20%. It is forecasted that over the next few years there will be a purging of the e-learning market followed by natural growth which will reach maturity around 2010 (E-Leusis, 2004).

There is no unanimously agreed definition of the e-learning concept. The common denominator among specialists reveals it as a mode of training that is independent of time and place thanks to the use of new technologies such as CD-ROMs, Internet, video conferencing, DVDs, and Intranet. Bellier (2001) distinguishes between four types of e-learning methods: (a) the wholly distance method without tutorial intervention, (b) the wholly distance method with tutorial intervention, (c) the mixed distance/face-to-face with distance self-training, and (d) the mixed distance/face-to-face without distance self-training. In this chapter we consider e-learning as "a wholly distance method of training with tutorial intervention that is independent of time and place thanks to Internet."

Previous research into distance learning has mainly focused on the introduction of virtual learning environments in different markets (Alaoutinen &Voracek, 2004; Wilcox, Petch, & Dexter, 2005), the differences between traditional classroom instruction and online courses (Favretto, Caramia, & Guardini, 2005), students self-discipline (Eom & Reiser, 2000; Mc Manus, 2000), the advantages and limitations of distance learning (Graff, 2003; Lindh & Soames, 2004), and evaluation methods in virtual learning environments (Dyson & Barretto, 2003; Woods & Keeler, 2001), but there are no in-depth studies of the key drivers in the adoption decision of an e-learning system.

Despite the growing importance of distance learning courses, there are still not enough studies to provide a holistic view of motivators and inhibitors of distance learning courses purchase decision. It is also crucial for managers of educational institutions to understand which aspects students value the most, the synergies between the different channels (traditional, Internet, mobile), and the barriers to adoption in order to assign resources effectively to obtain competitive advantages.

Distance learning is not immediately adopted, instead consumers have to overcome a set of barriers, some of them cognitive, before doing distance courses instead of live training. Among other factors, distance learning previous experience, student attitudes and perceptions of distance learning, and perceived risk are key factors in the speed of distance learning adoption (Lee, 2006). The future commercial success of distance learning depends to some extent on whether companies also use virtual environments for management training.

Therefore, reflection on e-learning should pay attention to the most important capital of the training experience: the learners. They are believed to be proactive actors and customers, and their perceptions determine their predisposition to accept or refuse such courses (Triki & Ouejden, 2007). This chapter offers an empirical insight into distance learning in Spain from the learner perspective, which has not previously been investigated.

It is clearly important to understand the success factors contributing to learners' acceptance of the Web-based e-learning system. The chapter aims to present an in-depth study of the factors influencing the distance learning course purchase decision. The chapter's specific goals are to:

1. Identify motivators and inhibitors of distance course adoption among consumers, focusing

on the impact of relations with the medium, service considerations and perceived purchase risk.
2. Provide empirical research on the Spanish market that analyses the influence of perceived purchase risk, consumer mistrust of distance learning institutions, perceived utility, perceived ease-of-use, perceived customer service, perceived satisfaction, and previous experience as a distance learning student in the future asynchronous distance learning course purchase decision.
3. Analyse the motivators and barriers that encourage and discourage Spanish students to adopt asynchronous distance learning courses.

This chapter will give managers and students insight into the distance learning industry and the different factors that influence e-learning courses adoption. In addition, these factors can be applied to the specific context of the Spanish market. Table 1 shows the theoretical and empirical contribution of this chapter.

The chapter is divided in two parts. The first part includes the literature review. The second part includes the methodology used in the empirical study of a sample of 111 Spanish students attending classroom-taught postgraduate and management training programmes and the data analysis. We analyse the impact of relations with the Internet, service considerations, and perceived shopping risk on the decision to do an online training course.

Table 1. Contribution of this chapter

Theoretical contribution	Empirical contribution (focused on the Spanish market)
To identify reasons why universities and business schools are moving to an online model and why they can integrate their marketing channels (physical, Internet, and mobile).	Provide a reference framework for comparative studies with other countries with different rates of: (i) Internet adoption and (ii) distance learning adoption.
Provide a holistic view of factors influencing distance learning courses adoption, from the consumer point of view.	Identify segments of consumers more likely to purchase asynchronous distance learning courses. Identify the perceived benefits and inhibitors that influence asynchronous distance learning courses purchase decision by Spanish customers and use the study's findings to develop strategies for managers on how to maximise the adoption rate.

THEORETICAL BACKGROUND: MOTIVATIONS AND INHIBITORS ON DISTANCE COURSE ADOPTION

Past research has identified a number of motivations and barriers influencing distance learning courses adoption by consumers.

Most studies have examined e-learning on the Internet by investigating the relationship between instructional materials and their structure, teaching strategies, learner personalities, and student self-discipline. Eom and Reiser (2000) found that younger students needed a more organised structure of course materials and ongoing help. McManus (2000) concluded that learner personalities, the structure of the materials, and teaching strategies had some influences on the ways in which students self-regulated their learning behaviour. Mason and Weller (2000) found that Web-creation skills, previous computing experience, group collaboration, and time input are important factors affecting students' acceptance of the long distance education system. Lee (2006) found that an e-learning system should be developed to target changes in perceived usefulness, perceived ease-of-use, and perceived network externality.

Previous research also has indicated that perceived content quality is important in determining users' satisfaction with the system (Delone & McLean, 1992; Katerattanakul & Siau, 1999; McKinney, Yoon, & Zahedi, 2002), which in turn leads to system utilisation. There are two dimensions of content quality: "content richness" and "update regularity" (Lee, 2006). Internet offers content richness far beyond other technologies since the special characteristics of Internet hyperlinks and interactivity allow students and teachers to access complementary course content. Content richness positively affects learners' level of satisfaction with the course (Arbaugh, 2000; Burns, Clift, & Duncan, 1990). Moreover, learner satisfaction would be enhanced significantly if they could obtain updated e-learning content on a regular basis. Updated content may lead students to feel that the system is a useful means of gaining new knowledge and learning.

Boticario and Gaudioso (2000) posit that an appropiate framework for developing e-learning (1) should include an interactive and online resource model that consider the stakeholders at various levels including lecturers, students, and tutors in the distance process; (2) should stimulate student participation in the use of the different resources (3); should promote new ways of communication to facilitate working groups of students and lecturers with common interests; and (4) should stimulate the use of technological resources available among users.

This section shows a description of the impact of the influence of relations with the Internet (i.e., perceived ease of use, satisfaction with Internet use in distance learning courses and previous distance learning experience), service considerations of the distance course to be chosen (i.e., customer service during the course, perceived utility_of the course and mistrust in the organising institution), and perceived risks (i.e., financial purchase risk, perceived psychosocial risk, and perceived waste-of-time risk) on the future asynchronous E-learning course intention.

Relations with the Medium

Perceived Satisfaction

When a consumer is satisfied with a product/service, the customer's short term behaviour is expected to be coherent with that satisfaction and consequently the customer will show greater willingness to buy that product/service in the future (Bolton, 1998).

Internet offers consumers many benefits which help to increase their perception of the satisfaction this type of training can bring; it offers the possibility of asynchronic collaboration and encourages interaction between group members (Dillenbourg, 1999; Kearsley, 2000). Internet permits not only teacher-student interaction and student-student interaction but also interaction with information sources which allow students to accumulate knowledge and develop skills. So, it is expected that consumers who think that the advantages of using Internet in distance courses provide a high level of satisfaction will adopt online training in the future.

Perceived Ease-of-Use

New technological advances in some direct shopping methods like online shopping need to take into account the individual's capacity to understand the changes and complexities in the new technologies. To succeed in distance courses students must have adequate computer skills (Dupin & Du-Charme, 2005). These computer skills include basic computer operation, file management, Web browsing, and e-mail operation. So, it is important that teachers try to compensate for participants' limited technical skills to avoid perceived difficulties in managing virtual environments (Hannafin, Hill, Oliver, Glazer, & Sharma, 2003).

Many students encounter barriers associated with the need to use the computer, so it is worth noting that while the initial sensation of difficulty disappears after the pupil uses the computer and discovers that it is possible to obtain very satisfactory academic results, learners initially believe that the technology will make learning more difficult. (Tesouro & Puiggalí, 2004)

Previous research focused on online shopping shows perceived difficulty in using computers and technophobia as inhibitors of online shopping adoption (Mattila, Karjaluoto, & Pento, 2003; Suganthi, Balachandher, & Balachandran, 2001). The need to use the new technologies (a computer, Internet, etc.) to access distance learning can be considered a dissuasive factor for certain consumers, especially low-educated older ones (Bruder et al., 2007). Therefore, in the e-learning context, perceived Internet ease-of-use is expected to have a positive influence on the decision to opt for distance learning.

Previous Experience

The attitude of Internet users to e-shopping changes as their use and experience of Internet grows. Web-users who spend more time surfing the net and who have more online use experience will be more familiar with the opportunities Internet offers and will have more knowledge about using it (Mason & Weller, 2000). Research by Swaminathan, Lepkowska-White, and Rao (1999) shows there are different levels of experience and consequently of the perception of the usefulness of Internet in different scenarios of Internet use. In other words, there is a positive relation between online experience and perceived e-shopping benefits.

Furthermore, preliminary studies have shown that the use of Internet increases students' confidence and capacity to learn as satisfactory results are obtained (Tesouro & Puiggalí, 2004). Prior experience with Internet has a positive influence on learning, cognition, attitudes, interactivity, personalisation, means of research, and individualised learning.

The literature review also shows that having some prior experience of distance learning has a positive influence on future e-learning adoption (Dupin & Du-Charme, 2005; Hannafin et al., 2003; Mason & Weller, 2000). Prior experience and success in the use of computer-based applications is important to succeed in technologically-mediated learning environments (Hannafin et al., 2003). So, it is to be expected that experienced users of this system will value its potential more highly and therefore develop a greater future purchase intention than nonusers.

Service Considerations

Quality of Service

Nowadays, there is increasing demand for personalised service. Distance training methodology makes it possible to adapt more closely to the customer's needs and consequently offer greater service quality (Graff, 2003). First, these methods offer flexible learning which means students can progress at their own pace. Second, online methods make participants responsible for their own training, as they focus on the student's activity and not on the teacher's explanations. Finally, the teacher can facilitate access to study material at any time and place or it can be accessible online for students to use it at their own convenience.

On the basis of a precise diagnosis of the students' competencies, the learner can choose the modules that develop the aspects the student wishes to reinforce (Triki & Ouejden, 2007). E-learning allows the learner to become the principal actor of training; the student builds personalised courses, sets objectives, and controls the training process. E-learning enables the student to manage the student's own personal career and guarantee the student's own employability over the long run (Booker, 2000).

E-learning also offers great flexibility. It allows the learner to plan a training path more easily, to better reconcile the time devoted to training, and the requirements of the learner's activity (Booker, 2000). Students learn better in an online program because they can learn according to their own rate, in their own environments, and at the moment of the day which is most appropriate to them.

Perceived Utility

A positive attitude to the benefits provided by the Internet influences the decision to purchase online. The TAM model (Davis, 1989, 1993) establishes the fact that intention to use a technology is determined by the individual's attitude towards using that technology and its perceived utility. In turn, that attitude is determined by the perceived utility and perceived ease-of-use. Therefore, it is to be expected that the perceived utility of virtual learning environments has a positive influence on the decision to opt for distance learning (Lee, 2006).

Trust/Mistrust of Distance Learning Institutions

One of the characteristics of e-learning is the separation of teachers and learners, distinguishing it from face-to-face interaction. The influence of an educational organisation differentiates e-learning from self-study and private tutoring, as the use of a computer network to distribute educational contents and two-way communication via a computer network enables students to benefit from communication with other students, teachers, and staff.

Previous studies (e.g., Black & Lin, 2003; Picardo, 2002) show that the reputation of an institution offering distance courses has a significant influence on reducing the risk perceived by the student. Lee and Tan (2003) show that consumers are more willing to acquire products and services through Internet from companies with a good reputation. Stigler (1961) states that a company's reputation indicates persistence of quality and makes it possible to establish a higher price because it reduces the information search effort. Reputation is achieved through recommendations from other consumers, advertising, and brand image (Bolton, 1998).

It seems logical to expect that a prestigious institution for classroom taught courses would use this positioning to launch distance course programmes. However, other authors (e.g., Cookson, 2001; Jones & Pritchard, 2000) state that one of the attributes that a distance learning institution should have is precisely the non coexistence of distance courses and classroom taught courses. For example, the successful Open University in the United Kingdom and the Universitat Oberta de Catalunya in Spain, which are dedicated exclusively to distance courses and the increased demand for online courses from both institutions, can only be due to their prestige, academic rigour, and professionalism (Jones & Pritchard, 2000) as neither of them offer classroom taught courses.

Perceived Shopping Risk

We can define the perceived e-shopping risk as "the Internet user's expectation of losing in a particular electronic transaction" (Forsythe &bShi, 2003; Ko, Jung, Kim, & Shim, 2004). The choice of a purchasing channel is heavily influenced by the perceived shopping risk, which becomes a barrier that may prevent consumers from doing online courses.

In distance shopping, perceived risk is greater than when shopping in traditional environments (Ko et al., 2004; Lee & Tan, 2003). If we concentrate on the sphere of distance learning, the perceived risk is a consequence of several factors: belief that distance learning is of lower quality than attended courses and therefore should cost less, rejection due to mistrust of the institution offering this form of learning, perception of insufficient acceptance in society of online training qualifications, and the perceived need to make greater effort to achieve the same teaching objectives (Black & Lin, 2003; Cookson, 2001).

Financial risk is the perception that the value of the course is less than its selling price (Schiffman & Kanuk, 2003). In the sphere of distance learning, financial risk is associated with the perception that doing this type of course should cost less than doing a classroom taught course. Previous research (e.g., Forsythe & Shi, 2003;

Tan, 1999) has shown that the perceived financial risk is an important predictor of the e-shopping decision. A money-back guarantee may reduce this type of risk although the higher the cost of the product, the greater this type of risk (Tan, 1999), especially if the consumer is not familiar with the product.

The perceived psychosocial risk is the perception of society's lack of acceptance of the qualification to be obtained with distance learning courses (Black & Lin, 2003). With the aim of minimising this type of risk in the countries where distance learning is more widespread (i.e., USA, Australia, Scandinavian Countries, etc.) universities and institutions offering e-learning do not make distinctions between their classroom taught diplomas and distance diplomas (Jones & Pritchard, 2000), since there are no significant differences between the academic results obtained through either system.

Online training provides greater access to learning resources such as texts, images, audio and video sequences, online databases, virtual libraries, e-books, and so forth than traditional classroom instruction (Harasim, 2000). However, this apparent advantage may become another obstacle in the decision to opt for distance learning, as for certain people it may translate into a greater perceived difficulty of accessing specific information, making learning tasks more difficult and increasing the perceived risk of time wasting (Harasim, 2000). For example, it takes more in-depth reflection to compose an answer when participating in a computer conference than it does in face-to-face discussion, as the reply is written, and thus requires more time than simple oral expression (Harasim, 1995). If we focus on the teacher's point of view, more planning is required in a virtual environment to provide authentic answers to student questions, repetition, and feedback at the same time in comparison with the common verbal exchanges in a traditional class.

THE CASE OF SPANISH STUDENTS

After identifying the key drivers of distance learning adoption, the second part of the chapter presents an empirical study of the Spanish market.

The use of Internet in Spain began around 1997 with 1.6% penetration of the population. According to data from the General Mass Media Study (AIMC, 2006), at present Internet penetration of the population in Spain is at 37.2%, far exceeding the penetration of other media such as newspapers, magazines, and the cinema. If we focus on the evolution of online purchases, in recent years, Spain has seen an increase in the e-commerce adoption rate from 0.47 million euros in 1997 to 2,143 million euros in 2005 (AECE, 2006).

According to the Bioeduca 2006 (Spanish Internet Observatory [OEI], 2006) around 350,000 Spanish students are currently following various online courses and 50% of them are university educated. Distance pupils represent 5% of total student numbers at present, and this share is forecast to rise to 10%.

The profile of an e-learning course user is a consumer between the ages of 30 to 45, with one or more children and in full-time work and with insufficient time to attend a training centre (i.e., university, academy or business school) and who studies at times when family commitments are not pressing. It is worth noting that this type of training is becoming increasingly popular with executives and middle management.

The results of the "5th Panel of results on E-learning application in large companies" (Grupo Doxa, 2005) show that in 2004, employers invested around 410 million euros in outsourcing continuous training, 40 million of which were for e-learning (including blended learning). Spanish companies are beginning to include more e-learning in their training plans in some sectors of the economy such as banking, energy, telecommunications, and other advanced services, while transport, industry, and construction are the minority sectors for this type of training.

Figure 1. Determining variables in the e-learning decision

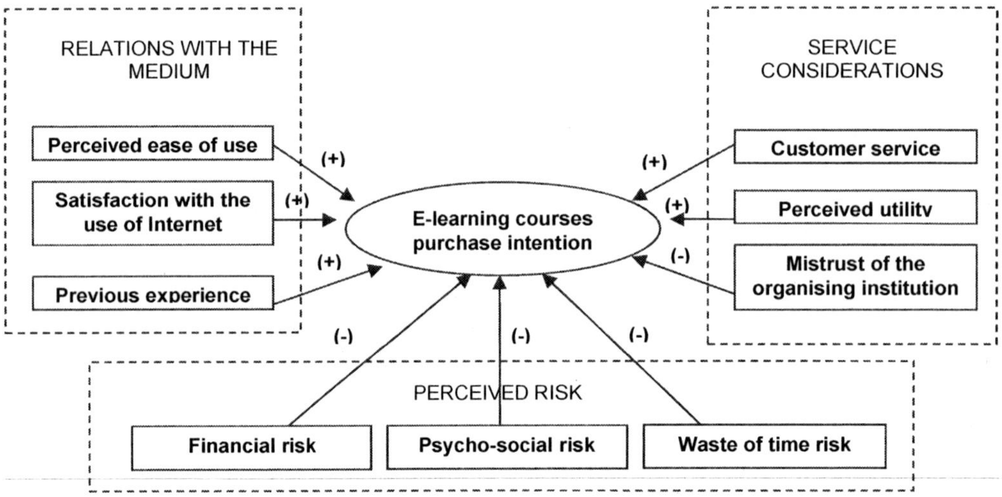

The conceptual model of distance learning courses adoption which will be contrasted in the Spanish market (see Figure 1) is an outcome of the literature review presented above.

The quantitative analysis provides answers to the following research questions:

1. What are the main perceived benefits and barriers in distance learning courses for Spanish consumers?
2. How does mistrust of distance learning institutions and perceived shopping risk influence e-learning course adoption?
3. Do relations with the Internet influence the future distance learning course purchase decision?
4. What are the impact of perceived customer service and perceived course utility on the distance learning course purchase decision?

Methodology

A convenience sample of 111 students attending classroom-taught postgraduate and management training programmes and self-administered questionnaires were used in March 2005 to obtain the information.

This study uses both quantitative and qualitative analysis techniques. The questionnaire was developed and tested with 9 focus groups to examine the dynamics of distance learning course users and institutions (managers of business schools, university students and professional students) with different levels of previous experience with distance learning courses. The qualitative research helped us to identify Spanish consumer standards of behaviour and attitudes in the adoption of this type of learning and to focus the study on asynchronous distance learning because the use of synchronous learning technologies is still scarce in Spain. Based on the information provided by focus group meetings, the questionnaire was modified and finalised.

A research instrument with close-ended questions was used for this study. Questionnaires were delivered to and collected from volunteer participants over the age of 18. A total of 202 Spanish students were contacted during the survey; 157 agreed to participate in this study. Among the questionnaires received, 111 were completed and analysed. Only 27% of the sample has previous experience of asynchronous distance courses.

Demographics of the sample are shown on Figures 2 to 5 (see below). The average age of the sample is 34 years old, with 55.3% men and 44.7% women. Of the sample, 73.9% have university studies, and 26.1% have completed secondary school. Also, 74.7% of the participants were employed.

The different types of perceived purchase risks were measured on a 5-point Likert scale (1 "Strongly disagree" to 5 "Completely agree"). Thus, the financial risk was measured by the degree to which consumers agreed with the statement, "The price of asynchronous distance courses should be lower than the price for classroom taught courses," the psychosocial risk with the statement, "Society places more value on qualifications obtained from classroom taught courses than those from asynchronous distance learning courses," and perceived waste-of-time risk was measured with the statement, "With traditional classroom instruction you have to work less than in asynchronous distance learning to achieve the same teaching objectives."

Prior experience as user was measured with a dichotomous variable (Yes/No) where consumers indicated whether they had previously done any type of asynchronous distance course.

The perceived ease-of-use (the need to use Internet and computers does not mean any additional effort in asynchronous distance learning over traditional classroom instruction), perceived satisfaction (I think that the advantages of using Internet in asynchronous distance courses provide a high level of satisfaction), customer service (asynchronous distance learning provides greater service quality than classroom taught courses), mistrust in the organising institutions (the prestige of the offering institution is more important in the

Figure 2. Sample description (gender)

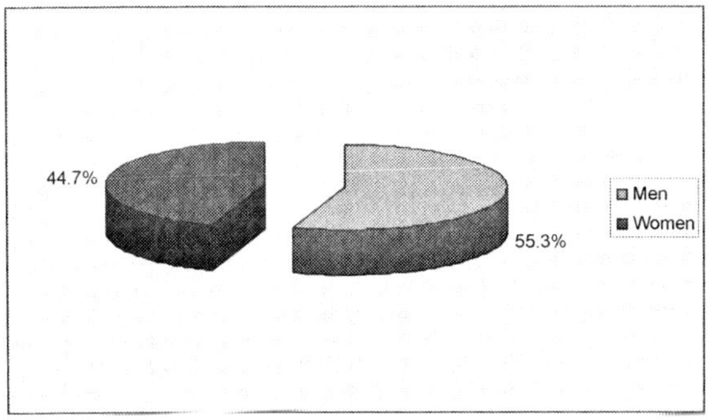

Figure 3. Sample description (age)

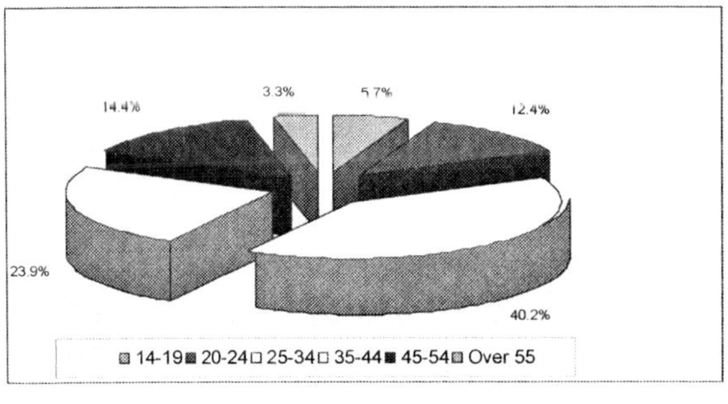

Figure 4. Sample description (education)

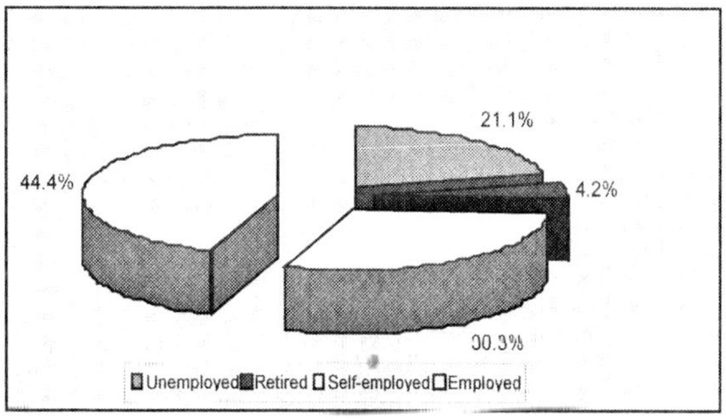

Figure 5. Sample description (occupation)

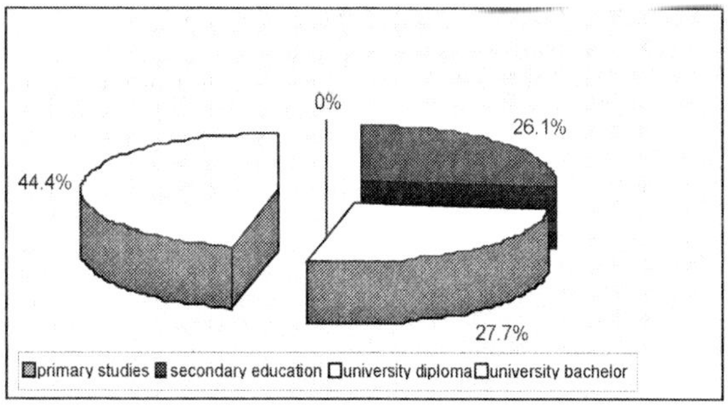

decision to do asynchronous distance courses than classroom taught ones), and perceived utility (an asynchronous distance course is more useful than a classroom taught course) were also measured on a 5-point Likert scale.

The intention of doing distance courses in the future (purchase intention) was measured on a dichotomous scale with two possible response categories (Yes/No). The future purchase intention as a variable to be explained has already been used in recent studies on consumer behaviour in virtual environments (Bhattarcherjee, 2001; Goldsmith, 2002).

In the quantitative analysis, we first describe the behavioural and attitudinal profiles of "future e-learning adopters" (i.e., consumers who intend to do asynchronous distance courses in the future) and "nonfuture e-learning adopters" (i.e., consumers who will never do asynchronous distance courses) (see Table 1). Second, we used logistic regression to empirically contrast the model proposed in Figure 1.

Table 2. Behavioural and attitudinal profile of respondents

Characteristic	Future E-learning intention		Mean (N=111)
	Yes (N=51)	No (N=60)	
Satisfaction with the use of Internet	2.53	1.84	2.14
Perceived ease-of-use	2.79	2.25	2.49
Perceived utility	2.19	1.61	1.86
Mistrust of the organis-ing institution	3.39	4.06	3.68
Perceived customer service	4.50	4.40	4.45
Psychosocial risk	3.08	3.15	3.19
Financial risk	4.13	4.16	4.16
Perceived waste-of-time risk	2.96	3.11	3.04

Data Analysis

Table 2 shows the description of the sample. It shows that 46% of Spanish consumers have future e-learning intention, while the rest would prefer to do face-to-face courses.

The sample shows high levels of perceived financial risk (4.16) and mistrust of institutions organising asynchronous distance courses (3.68). Evaluation of the psychosocial risk (3.19) and waste–of-time risk (3.04) is also high although slightly less so. Those interviewed showed medium agreement (2.49) on the difficulty of using the new technologies in asynchronous distance courses. However, perceived utility in relation to classroom attendance is low (1.86), although the perception that satisfaction with the use of Internet in asynchronous distance courses is more likely than with classroom attendance is medium (2.14), with a higher perception of the quality of the service offered by distance course institutions (4.45).

If we focus on the segment of consumers with future purchase intention, the descriptive analysis shows medium levels of utility (2.19), satisfaction (2.53), and perceived ease-of-use (2.79) in doing asynchronous distance courses. The perception of service quality is very high (4.50) for this collective. Consumers with a future purchase intention for asynchronous distance courses award importance to the prestige of the organising institution (3.39), with high levels of perceived financial risk (4.13) and psychosocial risk (3.08), with a slightly lower perceived waste-of-time risk (2.96).

A logistic regression (N=111 Spanish students) was used to test the proposed model. We will try to explain the potential effects on e-learning asynchronous course purchase intention of relations with the Internet (i.e., perceived ease of use,

Table 3. Regression results

Variable	B	SE	Wald	Df	Sig.	Exp (B)
Satisfaction with the use of Internet	0.535	0.255	4.401	1	0.041	1.707
Perceived ease of use	0.229	0.182	1.583	1	0.213	1.257
Mistrust of institution	0.528	0.198	7.111	1	0.008	0.589
Perceived utility	0.569	0.309	3.390	1	0.046	1.766
Perceived customer service	0.109	0.215	0.257	1	0.718	1.115
Psychosocial risk	0.145	0.233	0.387	1	0.534	0.865
Waste-of-time risk	0.082	0.228	0.129	1	0.788	0.921
Previous experience	0.235	0.501	0.220	1	0.641	0.790
Financial risk	0.209	0.208	1.009	1	0.314	0.811
Intercept	2.375	1.332	3.179	1	0.076	0.093

satisfaction with Internet use in asynchronous distance learning courses, and previous experience as a asynchronous distance learning user), considerations of the service (i.e., the customer service during the course, the perceived utility of the course, and mistrust of the organising institution), and perceived risks of the distance course (i.e., financial purchase risk, psychosocial risk, and waste-of-time risk) using logistic regression analysis, with future e-learning purchase intention as the dependent variable.

Hypothesis testing of the significance of the regression coefficients (β) gave the following results (see Table 3):

There are six variables with nonsignificant coefficients ($p>0.05$) according to the Wald statistic: EXPERIENCE, QSERVICE, FINANCR, PSYCHOSOCR, WOTR AND PERCEIVED EASE OF USE. Therefore, experience as a asynchronous distance course user, the possible difficulties in managing virtual learning environments, and perceived service quality do not influence the intention to continue using this training methodology. The belief that the price of distance courses should be lower than the same course taught in the classroom, the feeling that with traditional classroom instruction the students have to work less than in asynchronous distance learning to achieve the same teaching objectives, and the perception of better acceptance by society of classroom taught diplomas have no significant influence on future purchase intention either.

The variables satisfacation and perceived utility have a positive estimated coefficient. This means that if the other variables remain constant, a one point increase in consumer perception of satisfaction with the use of Internet in asynchronous distance courses will produce a more than proportional increase (1.707) in the future purchase intention. Similarly, a unit increase in the perceived utility of doing asynchronous distance courses will cause a more than proportional increase (1.766) in the intention to do an asynchronous distance course in the future.

The variable mistrust has a negative estimated coefficient. This means that if the other variables remain constant, a one point increase in the degree of consumer agreement that it is more important for institutions providing asynchronous distance courses to have prestige in the market than it is for institutions which provide classroom taught courses, will cause a less than proportional decrease (0.589) in the future purchase intention. Consequently, consumers who mistrust the institutions which provide distance training and need them to be well positioned in the market develop less future purchase intention than those who are not so wary.

Having checked the statistical significance of the estimated logistic regression coefficients, we proceed to verify the overall of the model using the Hosmer and Lemeshow test. The Chi-Square value is equal to 6.774, with 8 degrees of freedom (significance=0.561>0.05). It may be stated, therefore, that model fit is good. Furthermore, the model presents very good predictive capacity: 77.25% of the cases are correctly classified given a cut-off value of 0.5.

CONCLUSION

The revolution in computers and telecommunications networks, along with the global explosion in knowledge and ready access to powerful communications tools, are creating unprecedented changes in business, commerce, science, and education.

Computers are constantly redefining the way we live and work. So, in an environment like the present where developments in information and communication technologies influence practically all facets of our lives, universities, companies, and institutions in the training sector are addressing the task of developing alternative teaching methods to the traditional classroom. The computer age means classrooms will never be the same again. Computers are already introducing and even forcing new developments in the way students are taught and learn.

E-learning is a training method based on ICT. This new method of teaching via ICT is regarded as an innovation at technological, pedagogical, organisational, social, and cultural levels. Distance learning is considered to be a flexible, interactive, and accessible training option, making it one of the methods with the greatest potential for higher education (Favretto et al., 2005). In short, the interest aroused by the application of technological networks in the field of training, together with technological progress, has promoted many teaching-learning experiments based on these networks.

This chapter's main objective is to identify the predictive variables for the decision to do a distance course. In terms of the academic contribution of this study, it should be noted that while in the literature there are descriptive studies on the evolution of the distance training sector and its advantages and drawbacks, there are hardly any studies which propose purchase behaviour models for distance training.

This chapter extends the current knowledge base by providing insight into the different factors that influence the decision to do an asynchronous online training course. Specifically, this study will help to improve managers' understanding of the influence of relations with the Internet medium, service considerations, and perceived shopping risk and their relation to the asynchronous e-learning purchase intention. In addition, the application of these factors to the Spanish market allows comparisons to be made with studies done in other countries with different Internet adoption rates.

The results show that perceived course utility, trust in the organising institution, and satisfaction with the use of Internet when doing this type of training positively influence the asynchronous e-learning course purchase intention.

Financial, waste-of–time, and psychosocial risks show no significant influence, indicating that these obstacles do not determine the consumer's purchase decision. Consequently the price of the training activities, the time and effort dedicated to doing asynchronous distance courses, and acceptance of the qualifications by society are not determining factors in the intention to do this type of distance course in the future.

The study has also allowed us to verify that user experience and the perceived ease-of-use do not significantly influence the purchase decision. This confirms that prior experience is not a sufficient condition to develop a favourable future purchase intention, but rather the consumer needs to perceive that the user will be more satisfied than with classroom-taught courses, and the utility of the asynchronous course, as these variables do have a significant influence. The perceived ease-of-use does not influence the future purchase intention, possibly due to the homogeneity of the sample in terms of the high level of education, familiarity with the new technologies, and the fact that distance learning tools are becoming increasingly easier to use and interactive environments can resolve any doubt or difficulty the user may have.

Despite the fact that customer service quality has not proved to have a significant influence on the future purchase intention, the descriptive analysis shows that students perceive high levels of service quality.

The results obtained in our study make it possible to present the following set of recommendations for company management:

- The importance of the relation between satisfaction and the decision to do an asynchronous distance course suggests that companies in the sector should analyse the factors which generate consumer satisfaction continuously and systematically in order to include them in their Web pages (wide range of courses, interactivity, etc.).
- It is also recommended that companies integrate their marketing channels in order to reduce the feeling of mistrust in consumers. In short it is a question of good positioning in Internet to create positive synergies, increasing in turn the sales of other types of courses in the physical world. For example, an attractive, reliable Web site makes it possible to obtain quality information on the company's training activities both classroom and distance courses, while good positioning in the physical world would reduce consumer wariness of doing distance courses.
- Developing customer service centres to provide systematic communication with consumers both through conventional channels and online communication tools and attend their doubts, problems, and suggestions may prove to be a good option for reducing mistrust and increasing perceived utility.
- Asynchronous distance courses are tools with great potential for companies in the training sector due to the benefits they provide users, making it possible to reach new market niches. For this reason, companies in the training sector should not focus only on knowing the system and developing valid pedagogical models; an important part of their efforts should be directed at studying the barriers which slow down adoption by the market and defining the profile of potential users who are the ones, in the last instance, who will determine the success or failure of this training option.

One of the limitations of this study is the homogeneity of the sample formed by consumers with a high level of education because all of them were doing postgraduate or management training programs. Despite this limitation, the conclusions can be generalised to the business sector because most of the participants (74.7%) were employed (30.3% self-employed).

Another limitation would be that, due to the speed at which the new technologies change, it would be advisable to repeat the study periodically to verify whether the results obtained remain valid.

FUTURE TRENDS AND FUTURE RESEARCH DIRECTIONS

Five future trends can be observed in the education sector which is undoubtedly favoured by the introduction of Internet in the learning sphere.

- **Increased collaborative learning:** Virtual meeting systems (e.g., videoconferencing, chats, discussion forums, etc.) will reduce costs and avoid displacements.
- **Increased "live training" distance courses:** Live training sessions on the Internet complemented by the telephone or voice over IP. This system makes it possible to access training immediately for urgent matters and address audiences throughout the world from a single geographical point.
- **Personalised training:** Advances in artificial intelligence will make it possible to detect training lacunae in individuals and adapt training to fill in these gaps; it will be possible to adapt training to each employee's particular profile.
- **Return on the investment by the students:** Companies will look for indicators to evaluate and quantify the return in the form of greater worker productivity and future savings of time and money.
- **Training portals:** Companies with large numbers of employees can consider developing their own training platforms through which they can offer all types of courses. This strategy makes it possible to increase employee loyalty and reduce company training costs. Developing portals by training areas will be an important field of growth in future years, with finance institutions as the pioneering sector.

The consumer's cultural background is one of the aspects that can influence the creation of a favourable climate for developing and consolidating electronic transactions worldwide. A significant aspect is adapting the content and design of the virtual learning environment and the style of communications (formal or informal) to each country's culture. So, we propose as a future line of research to compare the results of our Spanish study with the perceptions of English speaking students.

This study has focused on asynchronous distance learning. It would also be interesting to analyse the key drivers of the synchronous distance learning purchase decision.

Another future line of research would be to empirically contrast an integrating model for the influence of sociodemographic, behavioural, and attitudinal characteristics in the distance course decision and apply them to a probabilistic sample of distance course users.

REFERENCES

AECE (Asociación Española de Comercio Electrónico). (2006). *Estudio sobre comercio.* Paper presented at the Electrónico B2C 2006, Madrid. Retrieved November 14, 2005, from http://www.aece.es

AIMC (Asociación de Investigación de los Medios de Comunicación). (2006). *Estudio general de medios.* Pape presented at the Tercera ola 2005, Madrid. Retrieved September 12, 2006, from http://www.aimc.es

Alaoutinen, S., & Voracek, J. (2004). Combining traditional and virtual teaching techniques in cross-border higher education. *Electronic Journal of E-learning, 2*(1), 1-10.

Arbaugh, J. (2000). Virtual classroom characteristics and student satisfaction with Internet based MBA. *Journal of Management Education, 24*(1), 32-54.

Bellier, S. (2001). *The EL.* Paris: Editions Connections.

Bhattarcherjee, A. (2001). An empirical analysis of the antecedents of electronic commerce service continuance. *Decision Support Systems, 32*, 201-214.

Black, I., & Lin, Y. (2003). *Perceptions of risks in e-commerce, a comparison of perceived risk in on-line degree courses versus campus based learning*. Paper presented at the 33rd EMAC Conference, Glasgow.

Bolton, R. (1998). A dynamic model of the duration of the customer's relationship with a continuous service provider: The role of satisfaction. *Marketing Science*, *17*(1), 45-65.

Booker, D. (2000). *Getting to grips with online delivery*. Adelaide: NCVER.

Boticario, J., & Gaudioso, E. (2000). Adaptive Web site for distance learning. *Campus-Wide Information Systems*, *17*(4) , 120-128.

Bruder, C., Beyerlein, T., & Blessing, L. (2007). Transfer training success into everyday use. In L.Gómez, D. Martí, & I. Candel (Eds.), *International Association of Technology, Education and Development Conference Proceedings*. Valencia: IATED.

Burns, J., Clift, J., & Duncan, J. (1990). Understanding of understanding: Implications for learning and teaching. *British Journal of Educational Psychology*, *61*, 276-289.

Butrimiene, E., & Danilevicius, E. (2007). Prerequisites for a sistemic employment of e-learning: Adequacy between the opportunities and student's needs. In L. Gómez, D. Martí, & I. Candel (Eds.), *International Association of Technology, Education and Development Conference Proceedings*. Valencia: IATED.

Cenich, G. (2006). Hipertexto y nuevas tecnologías: Su aporte al elearning. *Edutec. Revista Electrónica de Tecnología Educativa*, *20*, 1-14.

Cookson, P. (2001). La práctica de la educación superior a distancia: El ejemplo de la Universidad de Athabasca–Universidad Abierta de Canadá. *Revista Electrónica de Tecnología Educativa*, *14*. Retrieved October 15, 2001, from http://www.uib.es/depart/gte/edutec-e/Revelec14/cookson.html

Davis, F. (1989). Perceived usefulness, perceived ease of use and user acceptance of information technology. *MIS Quaterly*, *13*(3), 319-340.

Davis, F. (1993). User acceptance of information technology: System characteristics, user perceptions and behavioral impacts. *International Journal of Man-Machine Studies*, *38*, 475-487.

Delone, W., & McLean, E. (1992). Information systems success: The quest for the dependent variable. *Information Systems Research*, *3*(1), 60-95.

Dillenbourg, P. (1999). *Collaborative learning: Cognitive and computational approaches*. Amsterdam: Pergamon–Elsevier.

Dupin, B., & Du-Charme, H. (2005). Assesing student needs in Web-based distance education. *Journal of instructional technology and distance learning*, *2*(1), 39-48.

Dyson, M., & Barretto, S. (2003). Evaluating virtual learning environments: What are we measuring. *Electronic Journal of E-learning*, *1*(1), 11-20. Retrieved November 10, 2003, from http://www.ejel.org/volume-1-issue-1/issue1-art2-dyson-campello.pdf

E-Leusis. (2004). *El futuro del eLearning: Análisis del mercado y del contexto actual del eLearning*. Informe realizado en el marco del Proyecto Espacios de Excelencia Transfronterizos (E.E.T.) cofinanciado por la Unión Europea. Iniciativa Comunitaria INTERREG III A España, Portugal.

Eom, W., & Reiser, R. (2000). The effects of self-regulation and instructional control on performance and motivation in computer-based instruction. *International Journal of Instructional Media*, *27*(3), 247-261.

Escarré, R., & Barros, F. (2007). Pihe network. Best practice guide on university cooperatin between Europe and Lationamerica. In L. Gómez, D. Martí, & I. Candel (Eds.), *International Association of Technology, Education and Development Conference Proceedings*. Valencia: IATED.

Favretto, G., Caramia, G., & Guardini, M. (2005). E-learning measurement of the learning differences between traditional lessons and online lessons. *European Journal of Open Distance and E-learning, 2*. Retrieved December 23, 2005, from http://www.eurodl.org/materials/contrib/2005/Giuseppe_Favretto.htm.

Forsythe, S., & Shi, B. (2003). Consumer patronage and risk perceptions in Internet shopping. *Journal of Business Research, 56*, 867-875.

Goldsmith, R. (2002). Explaining and predicting consumer intention to purchase over the Internet: An exploratory study. *Journal of Marketing, 66*, 22-28.

Graff, M. (2003). Cognitive style and attitudes towards using online learning and assessment methods. *Electronic Journal of E-learning, 1*(1), 21-28

Grupo Doxa. (2005). *Panel de E-learning en las grandes empresas*. Retrieved October 27, 2005, from http://www.grupodoxa.com

Hannafin, M., Hill, J. R., Oliver, K., Glazer, E., & Sharma, P. (2003). Cognitive and learning factors in Web-based distance learning environments. In M. G. Moore & W. G. Anderson (Eds.), *Handbook of distance education* (pp. 245-260). Mahwah, NJ: Lawrence Erlbaum Associates.

Hannum, W. (2001). Web-based training: Advantages and limitations. In B. Khan (Ed.), *Web-based training* (pp. 13-20). NJ: Educational Technology Publications.

Harasim, L. (1995). *Learning networks: A field guide to teaching and learning online*. MA: MIT Press.

Harasim, L. (2000). *Redes de aprendizaje: Guía para la enseñanza y el aprendizaje en red*. Barcelona: Gedisa/EDIUOC.

Hassan, H., & Dunn, I. (2007). Exploring engineers employability competencies through interpersonal skills and enterprise skills. In L. Gómez, D. Martí, & I. Candel (Eds.), *International Association of Technology, Education and Development Conference Proceedings*. Valencia: IATED

Hennecke, A., & Cerny, L. (2007). Bridging the gulf between technology and intercultural communication. In L. Gómez, D. Martí, & I. Candel (Eds.), *International Association of Technology, Education and Development Conference Proceedings*. Valencia: IATED.

Jones, D., & Pritchard, L. (2000, November-December). The distance education debate: The Australian view. *Change, 32*(6), 32-33.

Katerattanakul, P., & Siau, K. (1999). Measuring information quality of Web-sites: development of an instrumen. In *Proceedings of International Conference on Information Systems*, Charlotte, (pp. 279-285).

Kearsley, G. (2000). *Learning and teaching in cyberspace*. Retrieved September 5, 2000, from http://home.sprynet.com~gkearsley/chapts.htm

Ko, H., Jung, J., Kim, J., & Shim, S. (2004). Cross-cultural differences in perceived risk of online shopping. *Journal of Interactive Advertising, 4*(2). Retrieved October 2, 2004, from http://jiad.org

Lee, Y. (2006). An empirical investigation into factors influencing the adoption of an e-learning system. *Online Information Review, 30*(5), 517-541.

Lee, K., & Tan, S. (2003). E-retailing versus physical retailing. A theoretical model and empirical test of consumer choice. *Journal of Business Research, 56*, 877-885.

Lindh, J., & Soames, C. (2004). A dual perspective on an online university course. *Electronic Journal of E-learning, 2*(1), 129-134. Retrieved December 9, 2004, from http://www.ejel.org/volume-2/vol2-issue1/issue1-art16-lindh-soames.pdf

Macknight, C. (2000). Teaching critical thinking through online discussions. *Educative Quarterly, 4*, 38-41.

Mason, R., & Weller, M. (2000). Factors affecting students' satisfaction on a Web course. *Australian Journal of Educational Technology, 16*(2), 173-200.

Mattila, M., Karjaluoto, H., & Pento, T. (2003). Internet banking adoption among mature customers: Early majority or laggards? *Journal of Service Marketing, 17*(5), 514-528.

McKinney, V., Yoon, K., & Zahedi, F. (2002). The measurement of Web-customer satisfaction: an expectation and disconfirmation approach. *Information Systems Research, 13*(3), 296-315.

McManus, T. (2000). Individualizing instruction in a Web-based hypermedia learning environment: nonlinearity, advance organizers and self-regulated learners. *Journal of Interactive Learning Research, 11*(3), 219-251.

Muilenberg, L., & Berge, Z. (2002). *A framework for designing question for online learning.* Retrieved August 22, 2002, from http://www.ssn.flinders.edu.au/innovations/ teacherresources.html

Pagliarello, M. C. (2007). Reluctant teachers and traditional learning methods: The multimedia teaching as cross-roads between the old and the new. In L. Gómez, D. Martí, & I. Candel (Eds.), *International Association of Technology, Education and Development Conference Proceedings.* Valencia: IATED

Picardo, O. (2002). Enseñar a aprender en la sociedad del conocimiento. *Revista Electrónica de Tecnología Educativa, 15.* Retrieved November 1, 2002, from http://www.uib.es/depart/gte/edutec-e/revelec15/oscarpicardo.htm

Salter, G. (2000). Making use of online discussion groups. *Journal of the Australian Council for Computers in Education, 15*(2), 5-10.

Schiffman, L., & Kanuk, L. (2003). *Consumer behaviour* (8th ed.). NJ: Prentice Hall.

Stigler, G. (1961). The economics of information. *The Journal of Political Economy, 49*(3), 213-225.

Suganthi, B., Balachandher, K., & Balachandran, K. (2001). Internet banking patronage: An empirical investigation of Malaysia. *Journal of Banking and Commerce, 6*(1). Retrieved May 22, 2001, from http://www.arraydev.com/commerce/jibc/0103_01.htm

Swaminathan, V., Lepkowska-White, E., & Rao, B. (1999). *Browsers or buyers in cyberspace? An investigation of factors influencing electronic exchange.* Retrieved October 3, 2001, from http://www.ascusc.org/jcmc/vol5/issue2/swaminathan.html

Tan, S. (1999). Strategies for reducing consumer's risk aversion in Internet shopping. *Journal of Consumer Marketing, 16*(2), 163-180.

Tesouro, M., & Puiggalí, J. (2004). Beneficios de la utilización del ordenador en el aprendizaje: un diseño experimental. *Revista Electrónica de Tecnología Educativa, 17.* Retrieved November 15, 2005, from http://www.uib.es/depart/gte/edutec-e/revelec17/tesouro_16a.htm

Triki, A., & Ouerghi, O. (2007). The tunisian e-learning experience: A focus on the main actors perceptions. In L. Gómez, D. Martí, & I. Candel (Eds.), *International Association of Technology, Education and Development Conference Proceedings.* Valencia: IATED.

Wilcox, P., Petch, S., & Dexter, H. (2005). Towards an understanding of UKeU business processes within an e-learning lifecycle model. *Electronic Journal of E-learning, 3*(1), 77-86. Retrieved December 11, 2005, from http://www.ejel.org/volume-3/v3-i1/v3-i1-art8-wilcox.pdf

Williams, M., Paprock, K., & Covington, B. (1999). *Distance learning: The essential guide.* London: SAGE Publications.

Woods, R., & Keeler, S. (2001). The effect of instructor's use of audio e-mail messages on student participation in and perceptions of online learning: A preliminary case study. *Open Learning, 16*(3), 263-278.

ADDITIONAL READING

Arends, R. I. (1997). *Classroom instruction and management.* New York: The McGraw-Hill

Battezzati, L., Coulon, A., Gray, D., Mansouri, I., & Ryan, M. (2004). *E-learning for teachers and trainers: Innovative practices, skills and-competences.* Luxembourg: Office for Official Publications of the European Communities.

Beasley, J. E. (1995). Determining teaching and research efficiencies. *Journal of the Operational Research Society, 46*, 441-452.

Blass, E., Ettinger, A., & Holton, V. (2006). Recognising differences in achievement. The case for a separate classification for qualifications undertaken by e-learning. In F. Li (Ed.), *Social implications and challenges for e-business.* Pending: Idea Publishing Group.

Burdea, G. C. (2004). Teaching virtual reality: Why and how? *Presence: Teleoperators and Virtual Environments, 13*(4), 463-483.

Carlson, R., Chandler, P., & Sweller, J. (2003). Learning and understanding science instructional material. *Journal of Educational Psychology, 95,* 629-640.

Correas, J. M., Correas, I., & López, P. (2006). An open source approach in designing third-generation systems for distance learning. *WSEAS Trans. Inform. Science & Applic, 3*(12), 2398-2402.

Ettinger, E., Holton, V., & Blass, E. (2005) E-learner experiences: Learning from the pioneers. *Industrial and Commercial Training, 37*(6/7).

Ettinger, E., Holton, V., & Blass, E. (2006). E-learner experiences: A lesson on in-house branding. *Industrial and Commercial Training, 38*(1), 22-37.

Felder, R. M., & Brent, R. (2001). Effective strategies for cooperative learning. *Cooperation & Collaboration in College Teaching, 1*(2), 69-75.

García-Aracil, A. (2006). Overview of regional differences in the Spanish higher education system. In A. Bonaccorsi & D. Cinzia (Eds.), *Universities and strategic knowledge creation* (pp. 376-404). UK: Edward Elgar Publishing Ltd.

Grandío, A., Agut, S., & Peris, R. (2006, July 16-21). *Perceived efficacy and achievement in virtual learning environments.* Paper presented at the 26th International Congress of Applied Psychology, Athens, Greece.

Grandío, A., Peris, R., Pinazo, D., & Jiménez, A. (2004). Features of the interaction lecturer-student in e-learning. *Current developments in technology-assisted education.* Badajoz: Formatex.

Grandío, A., Pinazo, D., & Gimeno, M. A. (2006). *Emerging patterns within 'blended learning strategies' in education.* Paper presented at the 26th International Congress of Applied Psychology, Athens.

Kearsley, G. (1997). *A guide to online education.* Retrieved November, 1, 2006, from http:// home.sprynet.com/~gkearsley/online.htm

Keegan, D. (2002). *The future of learning: From e-learning to m-learning.* Retrieved September 7, 2002, from http://learning.ericsson.net/leonardo/thebook/ chapter4.html#milearn

Khan, B. (1997). A framework for Web-based learning. In B. Khan (Ed.), *Web-based training* (pp. 5-98). NJ: Educational Technology Publications.

Ko, L. (2004). *Teaching interpreting by distance mode: A revealing experience.* Paper presented at 4th International Conference on Interpreting in the Community, Stockholm, Sweden.

Ko, L. (2006). Teaching interpreting by distance mode: Possibilities and constraints. *Interpreting, 8*(1), 67-96.

Paas, F., & Van Gog, T. (2006). Optimising worked example instruction: Different ways to increase germane cognitive load. *Learning and Instruction, 16*, 87-91.

Pastor, D., Ortega, B., Martinez, A., Capmany, J., Sales S., & Munoz, P. (2004). *Next generation of teaching optical communications: Interactive multimedia CD, tele-measuring and advanced simulation tools.* Paper presented at the Workshop on Education in Networks and Optical Communications, NOC 2004 (pp. 290-297).

Rubio, E., Gallego, D. J., & Alonso, C. (2003). E-learning in distance education and in the new cooperative environments. *UPGRADE, 4*(5), 39-46.

Scheiter, K., Gerjets, P., & Schuh, J. (2004). The impact of example comparisons on schema acquisition: Do learners really need multiple examples?. In Y. B. Kafai, B. Sandoval, N. Enyedy, A. S. Nixon, & F. Herrera (Eds.), *Proceedings of the 6th International Conference of the Learning Sciences* (pp. 457-464). Mahwah, NJ: Erlbaum.

Sein-Echaluce, M., Fidalgo, A., & Gil, J. J. (2004). DSED: A new technological platform for e-learning, collaborative work and knowledge management. *World Conference on Educational Multimedia, Hypermedia and Telecommunications, 1*, 1611-1616. Retrieved March 13, 2005, from http://dl.aace.org/15625

Slavin, R. (1990). *Cooperative learning.* NJ: Prentice-Hall.

Teo, C. B. & Gay, R. K. L. (2005). *Content authoring system to personalize e-learning.* Paper presented at 5th WSEAS Int. Conf. On Dist. Learning and Web Engin. Corfu., Greece, (pp. 105-110).

APPENDIX 1: QUESTIONNAIRE

Indicate how strongly you agree or disagree with the following statements (1= Strongly disagree; 5= Completely agree). Please respond to all the statements EVEN IF YOU HAVE NEVER DONE A DISTANCE COURSE.

0.1	Have you ever done any asynchronous distance course?	YES	NO

How strongly to you agree or disagree with the following statements?	(1 = Strongly disagree; 5 = Completely agree)				
1. I think that the advantages of using Internet in asynchronous distance courses provide a high level of satisfaction	1	2	3	4	5
2. Asynchronous distance learning provides greater service quality than classroom taught courses	1	2	3	4	5
3. The need to use Internet and computers does not mean any additional effort in asynchronous distance learning over traditional classroom instruction	1	2	3	4	5
4. An asynchronous distance course is more useful than a classroom taught course	1	2	3	4	5
5. The price of asynchronous distance courses should be lower than the price for classroom taught courses	1	2	3	4	5
6. Society places more value on qualifications obtained from classroom taught courses than those from asynchronous distance learning courses	1	2	3	4	5
7. With traditional classroom instruction you have to work less than in asynchronous distance learning to achieve the same teaching objectives.	1	2	3	4	5
8. The prestige of the offering institution is more important in the decision to do asynchronous distance courses than classroom taught ones	1	2	3	4	5

0.2	Would you be willing to do an asynchronous distance course in the future?	YES	NO

Classification data

Gender: _______________

Age: _________________

Education (mark the correct response with an x):

Primary studies	
Secondary education	
University diploma	
University bachelor	

Occupation (mark the correct response with an x):

Unemployed	
Retired	
Self-employed	
Employed	

Chapter XVII
ICT Impact on Knowledge Industries: The Case of E-Learning at Universities

Morten Falch
Technical University of Denmark, Denmark

Hanne Westh Nicolajsen
Technical University of Denmark, Denmark

ABSTRACT

This chapter analyzes e-learning from an industry perspective. The chapter studies how the use of ICT-technologies will affect the market for university teaching. This is done using a scenario framework developed for study of ICT impact on knowledge industries. This framework is applied on the case of e-learning by drawing on practical experiences.

INTRODUCTION

This chapter analyses e-learning from an industry perspective. The chapter studies how the use of information and communication technologies (ICT) will affect the market for university teaching. This is done by the use of a scenario framework developed for the study of ICT impact on knowledge industries. This framework is applied on the case of e-learning by drawing on practical experiences made at the Center for Information and Communication Technologies (CICT) at Technical University of Denmark.

The impact of ICT on knowledge services such as e-learning relates to production processes, content, and delivery. Production of knowledge services can, as the production of goods and of other services, make use of ICT in order to increase efficiency. This can for instance be done through processing, sharing, and reuse of data. A special feature for information services is that electronic delivery can be used both in the production process and in delivery to end users. Electronic delivery is, however, not just a new way to deliver an existing service. Electronic delivery changes the content of the service delivered. Provision of online access to information is a service, which is different from provision of the same information through a weekly newsletter.

Knowledge services using ICT in either production or delivery can be termed e-knowledge services (Sundbo, 2006). Professional e-knowledge services have been studied by, for example, Haukness (1999) and Miles (1994). But services directed towards private citizens have been less common. However, with improved network access for private citizens, use of the Internet or mobile networks as delivery channels are becoming more widespread.

First a scenario framework for analysing ICT impact on knowledge services is outlined. Second, different types of e-learning are discussed and the experiences made at CICT are presented. Thereafter follows an analysis of the market for e-learning and the possibilities for universities to address this market.

SCENARIOS FOR ICT IMPACT ON KNOWLEDGE SERVICES

We will in this section present two sets of service scenarios, which later will be used to discuss the implications of ICT-based learning methods for university teaching. These scenarios have been developed as part of the research project 'E-services—Knowledge Services, Entrepreneurship and the Consequences for Business Customers and Citizens.'[1]

The first set of scenarios describes customer relationships. Here two dimensions are defined:

- Codification and openness
- Social relationships

A high level of codification and openness implies that information can be made available on the Internet and it is possible for consumers to get easy access to a low price. Services, where information and knowledge are more difficult to codify or where access for other reasons is limited, are more expensive and available for a limited audience only.

The other dimension relates to interaction between producers and users. Some services involve intensive communication (e.g., coaching), while services like cash dispensing demand very little direct interaction.

Use of ICT may imply that the characteristics of a particular service are moving from one point to another in this two-dimensional continuum.

Scenario I deals with highly codified information, where ICT is used to expand coverage and user interaction. Wikepedia is a typical example of this of service. Information is easily available, and both users and producers are able to interact. The success depends on the reliability of the information, and usability for users (Christensen, 2006).

Scenario II deals with highly specialised information used in close interaction with users, which is difficult to codify. This implies that although ICT may be used in part of the process, it is difficult to deliver all parts of the service without one-to-one communication. Most often service delivery will necessitate at least some face-to-face communication. Many consultancy firms are found in this scenario. Management consultants for instance will, in spite of intensive use of ICT, need to interact with customers through personal communication. ICT is mainly used for standardisation and modulation of production (Baark, Falch, et al., 2002).

In Scenarios III and IV interaction with users are less necessary. ICT may here be used to develop self-service concepts. This is particularly relevant in scenarios where information is codified more easily. Electronic payment systems are an example of a service where the self-service concept has been developed in full (Scenario IV). But also, more complicated services, where codification is more difficult (Scenario III), can be provided through development of expert systems.

Figure 1. Customer relationship scenarios

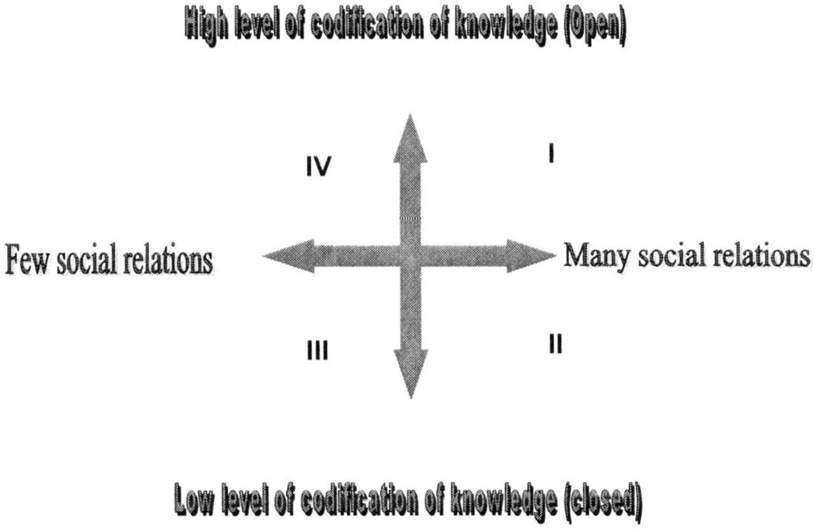

Figure 2. Industry organisation scenarios

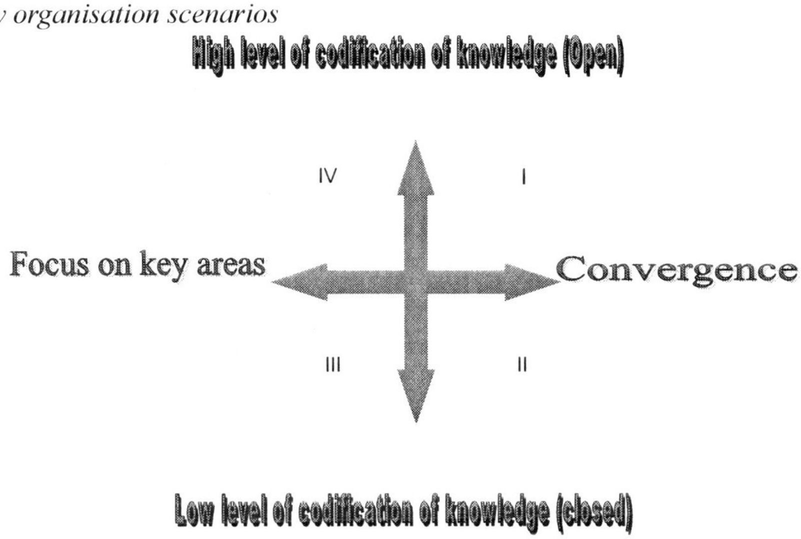

A similar set of scenarios can be defined for the organisation of the industry (Figure 2). Here the second axis is replaced with a dimension describing whether the industry tends to converge or to split into different specialised service areas. Convergence implies that different service areas are combined by the use of ICT in order to develop new service concepts and business areas. Focussing implies that ICT is used to expand expertise and upgrade service products in a particular area.

In the first scenario ICT is used to combine services in new ways. One example is new media houses providing their services on several platforms such as TV broadcast, Internet, and mobile phones. Other examples are various types of e-shops. E-shops may not need to keep stocks of the products they offer. Therefore an e-shop may be able to offer a very wide range of products.

Banks may be in the second scenario as they use ICT to expand their product range into other financial services such as insurance and sales of real estate. Local affiliates are used for communication with customers, but the information and expertise may be provided from other locations.

ICT can expand the geographical reach of companies, and in this way create sufficient volume for more specialised services. Consultancy firms are able to focus on their core competences and still obtain a sufficient volume to cover their costs (Scenario III). Wikepidia is also benefiting from addressing the global market in a limited niche (Scenario IV).

WHAT IS E-LEARNING?

The idea of using computers as a learning tool is almost as old as the computer, and e-learning is one out of several concepts which are used for describing a host of new learning methodologies using e-learning in parts of or in the entire learning process. Concepts like flexible learning, distance learning, telelearning, and computer supported learning cover to a wide extent use of similar learning methodologies.

The EU e-learning action defines e-learning as 'the use of new multimedia technologies and the Internet to improve the quality of learning by facilitating access to resources and services as well as remote exchanges and collaboration' (CEC, 2001). This definition is rather broad as it neither specifies the kind of learning methodologies nor the kind of technologies supporting it (multimedia technologies may cover almost any kind of computer based applications). It does however distinguish itself from distance learning, which can be done without the use of ICT. Moreover distance is not a necessary condition for application of e-learning, although one of the most important advantages by e-learning is the flexibility it offers with regard to distance.

E-learning is facilitated by different types of communication technologies where especially the use of online access of the Internet provides unique possibilities to deliver e-learning across space and support interaction-based learning types, which, for example, CD-ROMS do not.

E-learning is not confined to any particular part of the educational system, rather the contrary. One of the advantages of e-learning is that it makes it possible to extend the reach of educational and training systems into new areas. Thus e-learning can be applied both in the formal educational system (i.e., public schools, colleges, universities, etc.), as well as for vocational training. It can be used both for private use as well as in the public and the private sector.

A report from the Danish Ministry of Science, Technology and Innovation operates with four different types of e-learning methodologies, which illustrate the wide spectrum learning methodologies covered by the concept of e-learning (Box 1):

- **Model A:** E-learning where the teacher and the students never meet physically, and where no dialogue between students or students and the teacher takes place.
- **Model B:** E-learning where the teacher and the students never meet physically, but where the dialogue between the participants is supported by use of IT-based communication services.
- **Model C:** E-learning where parts of the learning takes place in a classroom and parts of the learning is done elsewhere, where the students work on a computer on their own (e.g., at home or at work).
- **Model D:** E-learning where all teaching is done in a classroom, and where computers

Box 1. Types of e-learning methodologies

Model A: E-Learning without Presence and without Communication

This type of e-learning can be done entirely off-line as all information can be stored on a CD-ROM or on a hard disk. Continuous or occasional online connection will however enable updates of the teaching material.

The user is provided with information on a certain topic, and may thereafter be given training through a number of exercises. The user may also be tested through a number of multiple choice tests. The user may seek guidance through a help function or similar.

The main advantage by this type of e-learning is its flexibility. The learning can take place everywhere and at all times if the requested equipment is available. This enables use of this type of e-learning exactly where and when there is a need to acquire a certain type of competence. On the other hand it is difficult to design the learning process according to the needs of the individual user, and the user cannot seek guidance beyond what is included in the e-learning system beforehand. The users must be able to work independently and solve unexpected problems by themselves.

This type of e-learning is mainly used for teaching in very specific competences such as use of a particular IT system, training in a new sales concept, and so forth. But the method is less suitable for teaching in general competences and is therefore difficult to apply in teaching at universities.

Model B: E-Learning without Presence and with Communication

This type of model demands some type of connectivity. Communication can either be asynchronous (e.g., e-mail communication) or synchronous (e.g., chat rooms). Communication can either be with a tutor or with fellow students. The model is almost as flexible as Model A. As a tutor is involved in the learning process, use of an e-learning system will often demand the user to register as a participant if the user wants to receive advice from the tutor.

The use of a tutor enables use of less-automated training exercises and tests. The model can therefore be used for teaching where reflection and dialogue are important for the learning process. The model is often used in situations where flexibility in time and space is important. For instance, the model is used for cross-border teaching by American universities.

Model C: E-Learning Combined with Occasionally Presence (Blended Learning)*

In this model e-learning is combined with traditional classroom teaching. A wide spectrum of models is here possible. The 'electronic' part can be with or without communication and it can either be a minor supplement to the traditional teaching or the traditional teaching can be a minor supplement to the 'electronic' part of the course.

Use of classroom teaching adds to the economic costs, but it also helps to make e-learning more efficient as it facilitates a dialogue between students and between students and the tutor also outside the classroom.

As explained in further detail below, CICT uses this model in most of its courses. In most courses, e-learning is used to supplement traditional classroom teaching. However, in our teaching in supplementary training e-learning is used more intensively.

Another example is the international shipping company Maersk who uses e-learning as part of their Maersk International Shipping Education. This is a 2-year education with 600 students from 80 different countries.

Model D: E-Learning used as a Tool in Class-Room Teaching

E-learning can also be used as a tool in the traditional classroom teaching. The major advantage here is that this enables use of modern pedagogic teaching methods (e.g., use of games and scenarios in realistic settings).

*This concept is discussed further by Derntl and Motschnig-Pitrik (2005)

are used as a learning tool.("Perspectives," 2003)

E-LEARNING AT CICT

The Center for Information and Communication Technologies (CICT), located at Technical University of Denmark (DTU), has for many years used ICT as an integral part of its teaching. This section outlines how ICT and e-learning are used for providing teaching to students outside the campus of the Technical University of Denmark. The section focuses on e-learning used for postgraduate training, as e-learning is deemed to have the largest potential in this area. Postgraduate students will most often be full-time employed elsewhere, and find it inconvenient to make frequent travels to CICT.

Current Use in Regular Courses

As many other universities, the Technical University of Denmark has developed an Intranet called 'CampusNet,' which can be accessed by both students and teachers. This facility is used for all course information to students. Each course maintains its own Web site only available to students registered to this particular course. These Web sites are used for posting information on lectures, group work, and so forth. Course material such as literature, exercises, PowerPoint presentations, and so forth can also be made available. Tools for provision of interactive training is provided, but only a few teachers use this opportunity, as it is very resource demanding to develop good quality teaching material for this purpose. Another possibility could be to podcast lectures. This has been considered, but has not yet been implemented.

CICT is also working on the development of an interface enabling the use of mobile devices and transmission through a campus-wide DVB-H system.

Course in Hypermedia

Back in 1997-98, CICT established videoconferencing facilities in order cooperate with Intermedia at University of Aarhus (350 km from CICT). The facilities included two–way video transmission and the possibility of transmission of a PowerPoint presentations. The system was based on a 2 Mbit video link provided via the Nordic university network, Nordunet. A video link established by a telecom operator would at that time be far too costly. Therefore the system could only be used for communication with other universities.

This system was used for providing a course in hypermedia to DTU students. All lectures were conducted in Aarhus for students from Aarhus University and transmitted to DTU, so students from DTU were able to follow the teaching. Even oral examinations were done by the use of the video link, so the same examiner could examine students from both Aarhus and DTU.

The technical performance of the system was not entirely stable, and it was necessary to have a technician in both ends to monitor the system during the lectures. The only reason that this way of teaching was acceptable to the students was that they had an interest in the technology itself.

Although this course was provided by use of e-learning technologies, content and teaching methods remained unaffected. The course was developed as a traditional course with lectures and report writing, followed by an oral examination. E-learning was used only to make the course available to more students.

mMIC: Master in Mobile Internet Education

When CICT created a new master education in 2001 in collaboration with Aalborg University, e-learning was a part of the concept from the very beginning. The master was designed especially for students working in the flourishing mobile communication sector. As most students had a full-time job and were located in different parts of the country, it was not possible to demand students to come to DTU or to Aalborg in order to attend course lectures every week.

Therefore teaching was concentrated in 15-weekend seminars during a period of 2 years. The students were requested to deliver a short report 1-2 month after each seminar, and thereafter to present this report in an oral examination.

In order to reduce accomodation costs for the participants, the first day of the seminars was transmitted via a video link. In this way students from Aalborg only had to spend one night in Copenhagen.

Before each seminar, students were demanded to prepare themselves by reading selected textbooks and carrying out a number of partly interactive exercises provided by the use of an e-learning tool.

The system applied was not particularly advanced compared to what is available today. At that time DTU did not yet have a fully developed e-learning tool. Therefore a trial version of a system called Uniflex developed by researchers at Aalborg University was used.

Remote teaching during the seminars was done by use of a video conferencing system from Polycom. The conference system worked in real time, that is, the students in the remote classroom could on one display see the students in other location. The lecturer was made visible at another display. The teacher was provided with a display, where the teacher was able to see the students in the remote classroom.

Apart from the live video connection, we used a data connection for the PowerPoint presentations. We tried by using these technologies to create a situation where the geographical distance did not matter, and the lectures were given as in a traditional classroom environment.

In this way the students at both locations could see each other and participate in the discussions. There was a huge participation in the teaching from the remote site.

The same concept was used during the final examinations, so students could avoid travelling to DTU in order to participate in examinations. Most students used the video-facilities available at Aalborg University for this purpose, but one student made his presentation from Taiwan.

The teachers had to go through a learning and adaptation process before they felt comfortable with the new teaching environment. Although it was possible to see all students, it was much more difficult to secure a lively interaction with the students located at the remote site than with the students located just in front of the teacher.

All students were very enthusiastic about the program. They felt like the video barrier was almost absent, but suggested at the same time to shorten lectures conducted on video. This indicates that lectures on video are more difficult to follow than ordinary lectures. On the other hand the students appreciated the flexibility enabled by the use of e-learning.

It was clear from the responses that the quality of both video and sound was essential for the learning. Also the lightening of the classroom was important. A few lectures were transmitted by use of an ordinary ADSL line. Both video and sound went through without technical difficulties, but a lower picture resolution and a poorer voice quality affected the learning.

The students felt that it was important that only a part of the training was done by the use of video. They argued that it was mportant to talk directly to the teacher at least a few times as this increases the efficiency of the training, as well as subsequent training made by video.

It is suggested that CICT in cooperation with Ghana Telecom University College (GTUC) offer a remote teaching program in Ghana, using the same concept as the above mentioned master program. However, this probably creates other challenges, technical, cultural, and language wise. The longer distances make it difficult to meet 15 times over 2 years and their might be differences in understanding and expectations to teaching.

SCENARIOS FOR E-LEARNING

The four different models for e-learning (A, B, C, and D) can be inserted in the customer relationship scenario framework described initially (see Figure 3). Model A learning without presence and without communication is most suitable for learning processes where little interaction is needed. Use of Model A is most suitable for learning of knowledge which can be codified in a systematic way. Model A may provide individual exercises, but in many areas students will need at least some interaction with a tutor in order to discuss solutions (Model B); in particular, if a problem has more than one correct solution. Models B (e-learning without presence and with communication), C (e-learning combined with occasionally presence), and D

Figure 3. Customer relationship scenarios for e-learning

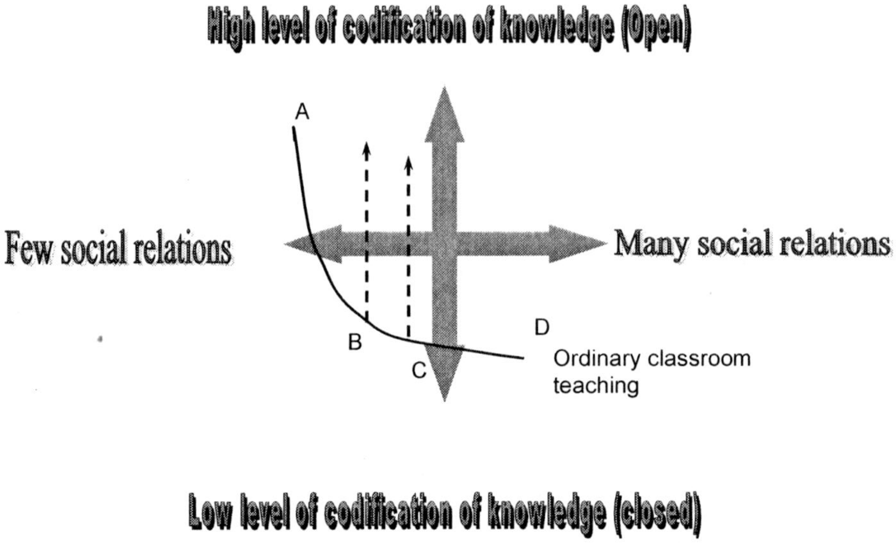

Box 2. Bloom's taxonomy of cognitive skills

- Knowledge (simple knowledge of facts terms and theories)
- Comprehension (understanding of the meaning of this knowledge)
- Application (the ability to apply this knowledge and comprehension in practice)
- Analysis (ability to break material down into its constituent parts and to se the relationship between them)
- Synthesis (ability to reassemble these parts into new and meaningful relationship)
- Evaluation (ability to judge and value material using coherent and explicit criteria)

Bloom. B.S. (1956). *Taxonomy of educational objectives, handbook 1: The cognitive domain.* London: Longmans Green. (adapted from (Beardwell & Holden, 1994)

(e-learning used as a tool in class-room teaching) do not necessarily imply more codification, but experiences from CICT show that teaching using videoconferencing and e-mail communication works better if more codified teaching methodologies are used. For instance, is there a need for more detailed PowerPoint presentations, as it is difficult to supplement with other learning tools such as whiteboard, demonstrations requiring physical interaction, and so forth? E-learning will thus motivate teachers to codify the learning process more than if traditional classroom teaching is used.

The usefulness of the different models depends on the environment, the kinds of users, and the type of competence that the learning process is aimed to develop (see Box 2). Blooms taxanomy of skills refers to the development starting with the knowledge of certain models and facts without understanding. Later on, comprehension of the models is developed and afterwards the level of being able to use the knowledge is aquired based on the former two levels of skills. The later levels imply more reflection and critical assesment of the knowledge refining existing knowledge.

Learning of different types of skills involves different types of learning processes. Some of these are more suitable for adaption of e-learning methodologies than others. The lowest levels of skills such as simple knowledge are the easiest to codify and hence to implement in an ICT-based learning tool. The learning processes can be categorised according to the skills to be developed through the learning process (Attwell & others, 2003):

1. Learning as a process for acquiring information
2. Learning as a process for acquiring information and processing experience.
3. Learning as a process for acquiring information and processing experience that effects a long-term change in the consciousness of the learner.
4. Learning as a process for acquiring information and processing experience in which the learner integrates new information and experience into the user's current knowledge base.
5. Learning as a process for acquiring information and processing experience in which the learner perceives, selects, and integrates new information and experience into the user's current knowledge base, thereby changing it.
6. Learning as a process for acquiring information and processing experience, in which the learner selects and constructs knowledge that is useful and appropriate for the user and in turn uses this to drive and determine the user's own continuous learning process.
7. Learning that becomes an individual process of interaction between the individual and the individual's environment, in which the subjective reality of the learner is actively constructed.

The first of these learning processes are the least demanding with regard to interaction between student and teacher. These processes are therefore more suitable for the use of e-learning Model A than Processes 3-5. Model A may also provide individual exercises, but in many areas students will need at least some interaction with a tutor in order to discuss the student's solutions (Model B). A long-term change in consciousness (3) may be difficult to obtain without any social interaction with fellow students. This points towards use of Model C or Model D. All models will however be able to include examples and exercises, which will be difficult to provide in a non-IT environment.

Learning Processes 6 and 7 will benefit from taking place in a praxis-related environment, for example, a working place. E-learning can therefore play an important role here, although codification of all elements of the learning process will be difficult.

IS THERE A MARKET FOR E-LEARNING?

In this section we discuss how e-learning expands universities' potential for offering educations and supplementary training.

Official statistics measuring the level of e-learning activities are rather scanty, but a number of indicators for use of e-learning for supplementary training exist. They all indicate that the market for e-learning is growing rapidly and that e-learning will become widespread in most types of private and public institutions engaged in training.

One of the most substantive surveys has been made by the British consulting firm Alphametrics on behalf of the EU Commission. Alphametrics has in cooperation with Bizmedia made surveys

Figure 4. Users of training by time spent on e-learning, blended learning, and classroom tuition by size of organisation in EU15 (Source: Massy, Harrison, & Ward, 2002)

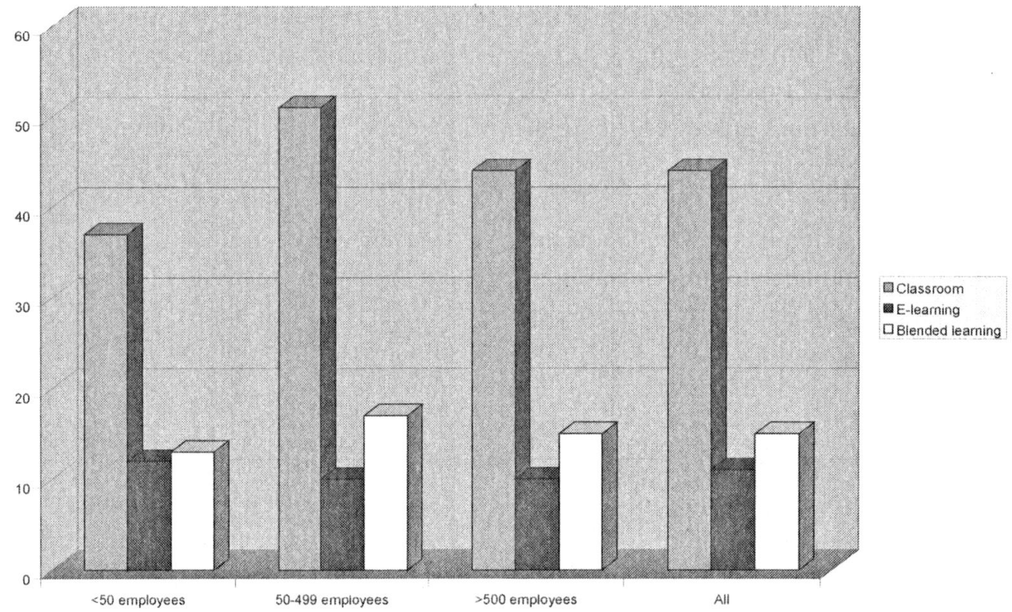

on the use of e-learning in Europe in 2001 and 2002 (Massy, Harrison, et al., 2002). In 2002, 638 organisations involved in training (538 from Europe) responded to the survey. As many as 83% of the European respondents reported that they have used e-learning in some way as part of their training. Although organisations with an interest in e-learning will be more inclined to respond, this indicates that experiences with e-learning are widespread in Europe among institutions involved in training.

It must however be emphasized that the study by Alphametrics focuses on institutions involved in training. The penetration among private companies in general is much lower. For instance, only 6 out of 27 SMEs included in an Austrian survey use e-learning as part of their training (Attwell, 2003).

According to the survey prepared by Alphametrics in 2002, 45% of the time employees in the EU spent on training was spent in the classroom, 12% was spent on e-learning following Model A or Model B outlined above, and 15% was spent on blended solutions following Model C.

One interesting result is that the share of nonusers is about the same in small and large organisations. This contrasts experiences made on diffusion patterns from a number of other new ICT-based applications, where large organisations dominates the population of early adopters, while SMEs dominate the population of late adopters.

This indicates that e-learning is not only a technology, which provides economies of scale for large organisations, but also a technology which can benefit small organisations. The survey indicates that classroom training is used less in small enterprises (37% of total time spent on training) than in middle sized (51%) and large enterprises (44%).

Although experiences with e-learning are widespread this does not indicate that the market for e-learning has matured. First of all, the above-

Table 1. Ranking of uses of e-learning by time spent on e-learning as % of total time spent on training (Source: http://www.eurolearn.net/docs/CEDEFOP_ELEARNING.PPT)

IT/computing	1
Technical (non IT)	2
Languages	3
Management	4
Process/production	5
Sales/marketing	6
Teamwork/communication	7
Quality	8
New products	9
Other	10

mentioned survey only includes organisations involved in training. But the potential of e-learning goes far beyond such organisations. E-learning is a flexible tool that can be used by any company or organisation in their daily business.

Second there is a potential to a more intense use of e-learning among those organisations already using this technology. And third, there is room for qualitative changes in the technology and the opportunities it offers.

An indicator for the growth potential of e-learning is the growth in expenditures related to e-learning. According to the Alphametrics/Biz-media survey users of e-learning report that their growth in e-learning expenditures was more than 70% in 2001 and just under 50% in 2002. Suppliers of e-learning are reporting even higher rates of growth. This may be due to more optimistic market expectations than on the user side, but it may also be an indicator of increasing use of e-learning outside the traditional population of training organisations.

Connectivity is an important precondition for use of most types of e-learning, and provision of infrastructure and equipment was the first action line of the EU eLearning Action Plan (CEC 2001). Since then, much work has been done to ensure connectivity to educational institutions.

Use of e-learning is not equally widespread among industries. According to a survey conducted by the Danish Technology Institute, e-learning is used most intensively in business services including consultancy firms, auditors, and traders in real estate, while usage is least intensive in the building industry ("E-learning in practice," 2003). A similar conclusion can be drawn from a later study made by E-learning Circuits (Ellis, 2005). According to their survey finance and investment management is the largest user. However, there is a clear trend towards increasing use in other industries, for example, healthcare.

Looking at the subjects in which e-learning is used, all surveys indicate that e-learning is used most intensively for training in IT and computing, while other important areas are teaching technical (non-IT) and teaching in languages. E-learning is mainly used by professionals—particularly IT professionals—and technicians, while blue collar workers' use of e-learning still is very limited. However, e-learning tools are also developed in these areas, for example, a training programme for shop stewards offered by the trade union, but these kind of e-learning progammes are of less interest to the universites as they build on other types of qualifications.

In general the European market for e-learning is a very segmented market split into a large number of regions. One reason for this is language and cultural differences. There are, however, signs of change particularly in working-place related e-learning, where the market is becoming more international oriented.

A preliminary study prepared by the Danish Technology Institute in cooperation with Alphametrics indicate that the majority of suppliers of e-learning are small businesses or even micro businesses without cash reserves and with limited growth potentials. There is however also a small number of large suppliers, mainly with a U.S. parentage (e.g., publishers, universities, and

Figure 5. E-learning scenarios for universities

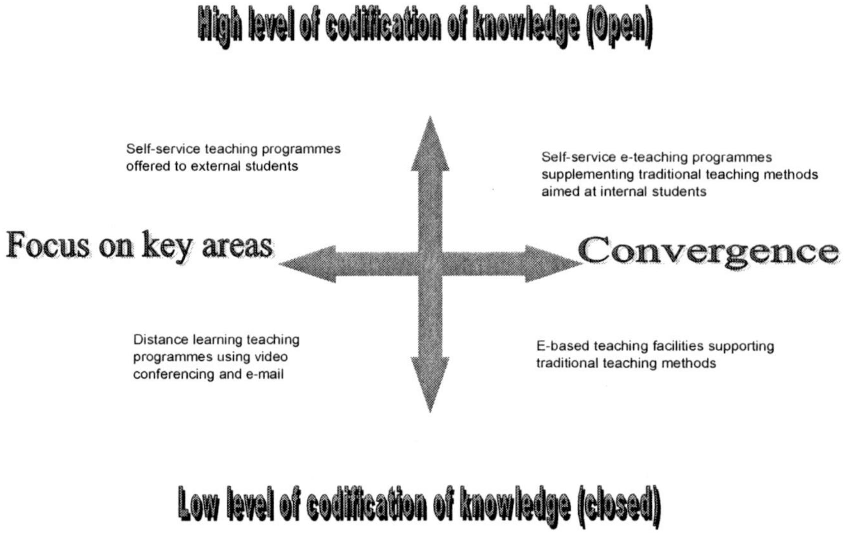

Box 3. E-learning provided by publishers

Publishers of university textbooks offer e-learning material, which supplements the textbooks. For instance, Prentice Hall has produced a number of tests and exercises connected to each chapter in a textbook, which students can use for testing their understanding of the texts. These texts are made available on their Web-site and can be used as part of the teaching at universities using books from Prentice Hall in their courses.*

* see http://www.prenhall.com/blanchard/

broadcasters) providing their services in several countries.

A number of university institutions like CICT are engaged in providing e-learning to external students. Many of these institutions are teaching subjects related to ICT, as they are the ones having the technical capability to set up e-learning facilities. However, others are very reluctant to go into this business. They lack both funding and short term economic incentives. This is one of the reasons for setting up public funded e-learning programmes, as it is done in a number of countries. The organisation of university teaching using e-learning can take different forms (see Figure 5).

As described in the previous section, e-learning does not necessitate increasing codification of learning processes. Videoconferencing, podcasting, and e-mail can be used without changes in course content. On the other hand, use of e-learning will stimulate codification as the benefits become more obvious.

E-learning can be used both for supporting existing teaching activities and as an enabler for new types of training, where the flexibility offered by e-learning is important. When e-learning is used to support the existing teaching activities, it may improve efficiency and performance, but it will not directly affect the overall organisation of the market for university teaching.

E-learning may however help universities to broaden their range of courses into new areas outside their core competences, as they can rely on input from other sources by the use of complete e-learning course packages through the combining of reusable e-learning objects (Muzio, Heins, et al., 2002). Publishing houses offer already e-learning material supplementing their textbooks (see Box 3). This trend may develop to provision of complete course packages either by publishing houses or other international institutions. By use of such facilities, it becomes easier to develop teaching programmes covering a wide range of disciplines.

The most obvious opportunity for expansion by use of e-learning is in the market for postgraduate training. But in the long term the international market for provision of university degrees may be more important than the market for supplementary training. In particular, English training institutions have offered distance learning in the form of correspondence courses. Use of e-learning enables the use of the same concept for more intensive education, in particular if combined with intensive seminars provided at regular intervals.

It will be easier for students to follow courses and take degrees in other countries. This may lead to a higher degree of specialisation among universities. Utilisation of the market potentials offered by e-learning will in particular benefit universities with special capabilities in certain areas.

On the other hand it may become less important for each university to cover all subjects, as students with particular interests are able to follow courses or a full programme elsewhere. CICT provides an example of a centre aiming at broadening the geographic coverage and in this way—in spite of a limited home market—attract sufficient resources to maintain world class expertise in a specialised area.

CONCLUSION: E-LEARNING AT UNIVERSITY INSTITUTIONS

E-learning offers a wide range of possibilities for universities. These relate to course content, teaching methodologies, as well as extension of the population of students. Particularly in supplementary training flexibility is important.

Moreover e-learning has helped in the shift from a teacher-centred model (i.e., lecture, notes, examination) towards a learner-centred model (i.e., problems, literature, information, investigation, discussions). This paradigm shift started before the introduction of e-learning, however e-learning supports this change of direction.

Flexibility offered by use of e-learning tools can be used to offer supplementary training to students located either abroad or in other parts of the country. The most important lesson from the experiences learned so far is that although substantial part of the learning can be done by use of ICT, it is essential for the students to meet occasionally. Once personal contacts to students and fellow teachers are established, interactive learning by use of online communication can be performed much more efficient.

Another experience is that preparation of an e-learning course demands substantial resources, even if the technology is in place. This limits the use of e-learning for specialised university courses with a limited target audience.

A third experience is that technology plays an important role. The quality of video and sound is more important than expected in order to ensure efficient learning.

There is a fast growing market for supplementary training also in areas where universities possess the relevant competences. However, substantial resources are needed if this potential is to be utilised. If successful, universities can use this opportunity, not only to expand their business, but also to upgrade teaching of their regular students.

In the long term e-learning used for teaching of university students may become an important market for universities. Self-service teaching programmes may be developed and help universities to improve efficiency and productivity. This may lead to more specialisation between universities, as it will become easier for students to follow an education at a university in a different region or country. It may however also enable small universities to broaden the teaching programme, as they can rely on the use of e-learning material to supplement their own teaching.

Such trends will be most visible in areas where it is possible to make use of codified teaching methods, but electronic communication such as videoconferencing and e-mail can be used to support other teaching methods as well.

FUTURE DIRECTIONS FOR RESEARCH

E-learning offers new opportunities for universities, but more research is needed to develop the right business models for how universities can apply e-learning. One problem is whether university teaching should be made publicly available at the Internet. Concerns have been raised on a possible negative impact of this on the quality of the teaching, but no studies on this impact have been made so far. Another aspect is the use of e-learning as a tool for distance learning. To what extend can distance learning be used, and to what extend is local presence necessary? So far distance learning has been applied in different contexts, but experiences in the use of distance learning in regular bachelor or master programs are limited. Further research on the potentials in this area is needed.

REFERENCES

Attwell, G. (2003). The challenge of e-learning in small enterprises: Issues for policy and practise in Europe. *Cedefop Panorama Series, 82*. Luxembourg: Cedefop Office for Official Publications of the European Communities.

Attwell, G., Dirckinck-Holmfeld, L., Fabian, P., Kárpáti, A., & Littig, P. (2003). *E-Learning in Europe results and recommendations: Thematic monitoring under the LEONARDO DA VINCI-Programme*. Bonn, Germany: Nationale Agentur Bildung für Europa beim Bundesinstitut für Berufsbildung.

Baark, E., Falch, M., Henten, A., & Skouby, K.E. (2002). *The tradability of consulting services*. New York/Geneva: UNCTAD.

Beardwell, I., & Holden, L. (1994). *Human resource management: A contemporary perspective*. London: Pitman Publishing.

CEC. (2001). The eLearning action plan: Designing tomorrow's education. COM(2001)172 final

Christensen, L. G. (2006). *E-service: Knowledge services, entrepreneurship, and the consequences for business customers and citizens*. Lyngby, Denmark: CICT DTU.

Derntl, M., & Motsching-Pitrik, R. (2005). The role of structure, patterns, and people in blended learning. *The Internet and Higher Education, 8*, 111-130.

E-learning in practice. (2003). Aarhus, Denmark: Danish Technology Institute.

Ellis, R. K. (2005). E-learning trends 2005. *Learning Circuits*. Retrieved April 18, 2008, from http://www.learningcircuits.org/

Haukness, J. (1999). *Services in innovation: Innovation in services*. Oslo: STEP Group.

Massy, J., Harrison, T., et al. (2002.). *The European e-learning market*. London: BizMedia.

Miles, I. (1994). *Knowledge intensive business services*. Manchester: PREST University of Manchester.

Muzio, J. A., Heins, T., et al. (2002). Experiences with reusable objects: From theory to practice. *The Internet and Higher Education, 5*, 21-34.

Perspectives for competence development: Report on e-learning. (2003). Copenhagen: Danish Ministry of Science and Innovation.

Sundbo, J. (2006). Customer-based innovation of e-knowledge services. *E-services - knowledge services, entrepeurship, and the consequences for business customers and citizens* (pp. 1-19). Lyngby, Denmark: CICT Technical University of Denmark.

ADDITIONAL READING

Bang, J, (2006). *eLearning reconsidered. Have e-learning and virtual universities met the expectations?* Retrieved September 17, 2007, from http://www.elearningeuropa.info/index.php?page=doc&doc_id=7778&doclng=6

Bang, J., Dalsgaard, C., & Kjaer, A. (2007). *Beyond blended learning! Undiscovered potentials for e-learning in organizational learning.* Retrieved September 17, 2007, from http://www.elearningeuropa.info/index.php?page=doc&doc_id=7778&doclng=6

Cordon, O., Anaya, K., Gonzalez, A., & Pinzon, S. (2007). Promoting the use of ICT for education in a traditional university: The case of the virtual learning center of the University of Grandada. *Journal of Cases on Information Technology, 9*(1), 90-107

Dalsgaard, C. (2006). Social software: E-learning beyond learning management systems. *European Journal of Open, Distance and E-Learning.* Retrieved September 17, 2007, from http://www.eurodl.org/materials/contrib/2006/Christian_Dalsgaard.htm

Falch, M. (2004). *A study on practical experiences with using e-learning methodologies and cooperative transnational development methodology* (working papers). Lyngby: Technical University of Denmark, Center for Information and Communication Technologies.

Falch, M., & Tadayoni, R. (2006, August 23-25). *Practical experiences with using e-learning methodologies at CICT* (e-service working papers No. 10). Paper presented at the Digital Learning India 2006 Conference, New Delhi.

Fidas, C., Kapsalis, V., Tranoris, C., & Avouris, N. (2006). Developing a blended-learning community in a university of setting. *Campus-Wide Information Systems, 23*(3), 138-148. Emerald Group Publishing Limited.

Gunga, S. O., & Ricketts, I. W. (2007). Facing the challenges of e-learning initiatives in African universities. *British Journal of Educational Technology, 38*(5), 896-906.

OECD. (2005). *E-learning in tertiary education. Where do we stand?* Author.

Perez, M. J. V. (2007). E-learning university networks: an approach to a quality open education. *Journal of Cases on Information Technology, 9*(2), 2-25. Idea Group Publishing.

Sherman, W. H., & Beaty, D. M. (2007). The use of distance technology in educational leadership preparation programs. *Journal of Educational Administration, 45*(5), 605-620. Emerald Group Publishing Limited.

Volpentesta, A. P., & Frega, N. (2007). Developing a blended-learning community in a university of setting. *International Journal of Web Based Communities, 3*(2), 134-150.

ENDNOTE

[1] http://www.eservice-research.dk/

Chapter XVIII
Economies of Scale in Distance Learning (DL)

Sudhanva V. Char
Life University, USA

ABSTRACT

As per conventional wisdom, the larger the size of the distance learning (DL) educational facility in terms of student enrollments, lower also would be the unit capital and unit operating costs. Looking at empirical evidence, the correlation between the two variables of enrollments and average total costs is unmistakable, even if not significant. In this chapter the nature and strength of such relationship is of more interest. This work discusses ramifications of scale-related economies for public policy, such as a mega or open university and so forth, for cost effectiveness of tax dollars, if any, spent on a DL unit. However, the scope of the chapter is limited to scale-related economies and it does not encompass the nitty-gritty of social cost benefit analysis. Subsequently, DL costs of a mega university are looked into to identify and quantify scale-related economies. The last part suggests what would make it possible to achieve minimum efficient scale (MES) size so that scale-related economies are achieved or diseconomies of scale are surmounted.

INTRODUCTION

One of the reasons that distance learning (DL) is alluringly attractive to some educationists at least is that it lends itself capably and economically to reach higher education to a very large body of students on a scale inconceivable for a campus-based institution. The number of students in such a modern DL facility runs into hundreds of thousands. In 1971, Walter Perry, the first administrator of the UK Open University ignored the advice of several experts to start DL on a pilot scale and let it prove its merits before expanding it into a large educational unit. In the stimulating words of John Daniel (2003), "Walter Perry ignored this advice. I believe he had two reasons. Even in those early days he understood that one of the great virtues of distance learning was the potential to operate at scale. He could already see that starting an open university required a big investment, but he could also see that if it were able to operate at scale the marginal cost of serving each additional student could be lower than in conventional institutions. He knew therefore, that if he started with a small pilot project of a few hundred students the cost per student would be enormous and people would ridicule the whole idea." That is a fit and functional elucidation of the economies of scale in DL.

Fixed costs as well as variable costs to a lower extent tend to decline as the volume of output expands. Such costs keep falling up to an optimum point beyond which the costs start climbing due to diseconomies of scale. The classic long run average cost curve (LRACC) is parabolic, although on account of the dynamics of current production systems, there are LRACCs that are not U-shaped but are L-shaped. Distance learning is a service industry and it would be interesting to explore if such economies are sizeable and the nature of such economies. The Appendix at the end shows the LRACC curve for a DL facility and it is L-shaped. We will discuss the implications later.

When up-front investment costs are spread over a larger output, average total costs, average fixed costs in particular, as well as marginal costs tend to decline. In terms of a DL facility this assertion would mean that as the number of students (or student credit hours) enrolled in distance education increases, the per student (credit hour) costs would keep declining at first, remain steady at the optimum level with further increases in enrollment, and eventually start rising slightly as still more students enroll. This fact is brought out in the estimates of the economics of a hypothetical DL unit at a southern location presented in Table 1. Note the significant change that the economies of scale make between 6,000 credit hours and 30,000 credit hours to both the costs and the net revenues. The data therein are of *ex ante* planning value, but not all encompassing to include social cost benefit analysis (SCBA), environmental benefits, and other topics beyond the scope of this study. Also such an analysis is location-specific. It is undertaken after the all conjectures and numbers are firmed up. Because of the economies (spread of fixed costs over a larger number of credit hours) in Table 1 the break-even for such credit hours is slashed substantially from 15,800 to 5,300 credit hours. Also observe that semivariable costs (also called semifixed costs) such as marketing, student support, faculty salary, and such others constitute as much as 67% of total costs at 30,000 student credit hours, going up from 55% at 6,000 credit hours. Whereas fixed costs decline from 41% of total costs at 6,000 credit hours to 23% at 30,000 credit hours. The dramatic fall in cost per credit hour is obvious. This evidence validates the presence of economies of scale in DL. In Table 4 as well, the same proof is available.

Economies of scale in DL are a hot issue in contemporary higher education, notably where there is capital deepening by making use of heavy-duty IT infrastructure consisting of cutting edge technology to deliver DL and otherwise manage it. Economies of scale and heavy investments work in step and style with each other. And in order to meet the "exploding demand" for DL courses, many universities have already launched DL programs or are contemplating them. Enrollment increases are in the range of 20 to 30% per annum (Carnevale, 2006).

The Secretary of Education, Margaret Spellings, thinks that "nearly two-thirds of all high-growth, high-wage jobs created in the next decade will require a college degree; a degree only one-third of Americans have" (Spellings, 2006). This vast hiatus between demand and supply cannot be bridged by mortar and brick institutions alone and DL has a role cut out for it in this milieu.

The government repealed early last year the federal rule mandating that colleges provide at least one half of the instruction on campus. The repeal would further augment enrollment in the online environment. According to an estimate by Eduventures in Boston, by early 2008, 1 out of 10 college students is expected to be enrolled in an online degree program. The demand for online programs will outstrip supply during the next 5 years. (Golden, 2006)

Table 1. Economics of a hypothetical DL firm (Estimates by S.V. Char2007)

Revenues:			
Student Enrollment	200	500	1000
Credit Hours per student per year	30	30	30
Total Credit Hours	6,000	15,000	30,000
Fees per Credit Hour $	150	150	150
Total Revenues $	900,000	2,250,000	4,500,000
Expenditures:	$	$	$
A. Variable Expenditures:			
Instructional Materials: E-Books, E-Library and other on-line Materials per student	200	200	200
Total Cost of Instructional Materials	40,000	100,000	200,000
	(3.3)	(6.4)	(9.3)
B. Semivariable Costs:			
a. Marketing Costs	150,000	200,000	250,000
b. Student Support and Assistance	400,000	600,000	1,000,000
c. DL Faculty Salary	120,000	150,000	200,000
Total Semivariable Costs	670,000	950,000	1,450,000
	(55.4)	**(61.3)**	**(67.4)**
C. Fixed Costs:			
a. Course Production Costs:			
i. Course Development Faculty	**50,000**	**50,000**	**50,000**
ii. E-System Management Staff	**50,000**	**50,000**	**50,000**
Total Course Production Costs	100,000	**100,000**	**100,000**
b. Administrative Overheads:			
i. Director of DL, Instruction Designer and others	**250,000**	**250,000**	**250,000**
ii. Registrar, Admissions Personnel, PR, and so forth	**150,000**	**150,000**	**150,000**
Total Administrative Costs	**400,000**	**400,000**	**400,000**
Total Fixed Costs	500,000	**500,000**	**500,000**
	(41.3)	**(32.3)**	**(23.3)**
Total Variable + S.variable + Fixed Costs (A+B+C)	**1,210,000**	**1,550,000**	**2,150,000**
Net Revenues	**(310,000)**	**700,000**	**2,350,000**
Cost per Credit Hour $	**202**	**103**	**72**
BE Level Cr. Hours = FC/Fees - (V+SV)	15,787.81	**6250.0**	**5263.2**
Rounding up Cr. Hrs #	**15788**	**6250**	**5263**

Figures in parenthesis are % of costs to total costs

DL VIRTUES AND CONCERNS

The UNESCO portal listed in the reference section (UNESCO, 2007) of this chapter provides a description of a mega university (MU) as institutions that have "enrolments in excess of 100,000 learners per year" and "provide a combination of media to accommodate learners." The term "mega university" was coined in 1993 by the Vice-Chancellor of the UK Open University, John

Daniel (Daniel, 1996). Initially it meant any institution that had a DL component in an otherwise regular brick and mortar school. Later the term came to mean a predominantly DL unit, with a teaching campus, if any, merely of secondary importance, like it happened to UKOU. In this chapter the term is used only in this expansive sense in tune with the UNESCO (2007) portrayal as any institution that provides distance learning primarily, and only incidentally, has a brick and mortar campus learning facility. Fine tuning of this concept is difficult in view of a wide range of both on and off campus education made available by institutions under DL. Also the MU could use any communications technology, including the humble snail mail to qualify as a MU. Allama Iqbal OU in Pakistan is considered the world's largest DL facility with 1.8 million students, including mainly the off-campus students in both graduate and undergraduate courses (Vice-Chancellor, UK Open University, 2007)

Student enrollment in "mega iniversities" is massive. UK's Open University, dedicated to delivering online learning, has over 150,000 undergraduate students and 30,000 postgraduate students. Some 10,000 students of the university suffer disabilities. China and India have millions of students, in aggregate terms, in "open universities" (OU). During 2005, India alone had 11 open universities and some 70 distance education units in that many universities with hundreds of thousands of students in each. In America, the for-profit University of Phoenix has 145,000 online students seeking online degrees. University of Maryland has 51,450, Troy University has 19,000, University of Massachusetts (UMass) has 9200, Pennsylvania State University has 5,700, and so forth. (Data for number of students in different universities are published in issues of *Chronicle of Higher Education* and *Inside Higher Education-Daily Updates*).

OUs are now commonplace in about 50 countries around the world. UMass experienced a quadrupling of its 2001 enrollment. DL appears to be so accessible, affordable, and versatile that increasing faith is being put into it to meet the demand for higher education and for "cyber scholars." It is also flexible in the sense that students can learn at their own comfortable pace and just-in-time. Online colleges turned out to be resourceful and inventive in the delivery of educational courses to Katrina victims and provided continuity in education.

What is not common in the enrollments at the various American and non-American distance education providers is that with the exception of a few eminent DL schools such as the University of Phoenix, it is not unreasonable to suggest that by and large the enrollments are originating in the same state where the DL facility is located. More often the students are in the same town, if not the same zip code area. This fact has a bearing on scalability. A single city or county or even a state by themselves are not likely to fill the e-seats of a mega university that can service 100,000 or more students. And DL facilities, as a result, will not lend themselves well for being scaled up. In the case of the UKOU there was a phenomenal growth from 25,000 to 40,000 in just a year and by 1980 the total student number was 70,000 and there were 6,000 graduations a year (UK Open University, 2007.) One of the features that helped this growth was the fact the UKOU was only one of its kind, a virtual monopoly in the UK and with a strong presence all over Europe and elsewhere. Some of these propitious circumstances may not be available in the USA. There may not be any takers for the e-seats, at least in the initial phases of the facility. This does not suggest that scale-related economies cannot be realized by nonmega units. The nifty advances of modern management do enable the setting up of viable, but reasonably efficient size units that could attain 60 to 70% the economies of scale (Louisiana Department of Education, 2003), if not 80% or more as a mega unit would.

Profitability is a function of revenues and costs. Online colleges that are for-profit are now numerous. However, in terms of enrollments, they are not anywhere close to those of some Asian giants such as China TV University (CTVU) and the Indira Gandhi National Open University (IGNOU). At IGNOU enrollment is now around 1.5 million and is present in 35 countries (IGNOU, 2007) and that could be true of CTVU too. In a sense, they could be dominant units in the domain of DL. The demand *in general* for DL is burgeoning and will continue to support the growth of new relatively smaller units. But considering that though the expectation is that enrollment would be global, in fact it has been largely local will the incremental market be large enough to support the new mega or huge (enrollment-wise) units? This is the main concern of this study.

On account of the exploratory nature of this study, all DL units, regardless of whether in the private or the public (government) sector, have been treated alike. The scope of this study does not allow the undertaking of a full-fledged social cost benefit analysis (SCBA) of DL projects which would be virtually mandatory wherever tax dollars are drawn on. Also such an SCBA would take into account environmental factors such as minimizing carbon emissions and factor in other good externalities of DL as well.

It will be increasingly tricky to find, *ab initio,* mega enrollments such as 100,000+ because of market segmentation in the USA, unless substantial population increases occur. Later we suggest after a study of the scale-related costs that modular increments in DL gear and equipment, and also staff, to match enrollments would perhaps be a solution.

The phenomenal growth in DL raises several controversial questions that relate to the quality of education (Gibson, 1998) that is delivered predominantly by part-time instructors. Such instructors are not compensated at the same rate as full-time instructors and as such there could be differences in the quality of instruction. It is equally possible that part-timers do a much better job too, making it a great learning experience and an effective outcome for one's efforts. The South African National Council on Higher Education found in 1996 that DL was more effective in learning outcomes than "face-to-face" education and the costs were lower too. (For conclusions quite the contrary see Ketterer & George, 2006.)

In response to an overall rapidly expanding market DL facilities at times have been overdimensioned at substantial costs in the hope of cheapening future expansion of the DL facility. Not knowing the ramifications of such investments and economics of such decisions would be tantamount to rolling the die in regard to the economic viability of DL units. In-depth analysis of the options in this study could help moderate the risks. A review of the economics of scalability of DL facilities is thus an urgent and real need.

FACTORS INFLUENCING SCALE-RELATED COSTS

Empirical data clearly establish that the technology employed to deliver DL weighs in heavily on average costs and scale-related cost savings. This is true wherever there is capital deepening with heavy upfront investment in technology. In the area of DL, for instance, while computer conferencing has the lowest costs, broadcast television has the highest. Audio cassette, videoconferencing, computer-based learning, and others fall in between (Harrison, 1999). All media costs keep declining even if insignificantly or flatten out, without curving up for diseconomies. The same fact has been observed at other mega universities.

Technology calls the shots in the scale of costs. Besides technology, economies of scale in DL depend on a variety of factors and no two locations or DL institutions seem to be the same. The variety begins with whether the DL school is for-profit or a charitable or state unit. The latter is likely to be an established unit already endowed with its campus and buildings and other facilities for course development. Second, for cost purposes, it does matter if it is a designated DL unit or an on-going university that is diversifying into DL. Third is the nature of the course: whether it is a highly technical one such as a C++ or an Oracle programming course, or a non-technical English literature course. The more technical or novel the course, the more expensive would be the cost of course development and delivery. Fourth, course development costs are also dependent on materials that would be made available to the DL student. Table 2 clearly brings out the very wide spread between a bare bones on line course and an all-input providing sumptuous course. The difference in costs with each additional material is striking.

Table 2 brings out that the more technological the medium of delivery, the more expensive also is the course development cost. Instructional technology impinges on the up-front investment costs which in turn influence scale-related economies or diseconomies. Distance education is multifaceted offering a multiplicity of courses from accounting to zoology through an assortment of Internet technology that is an intermix of snail-mail, voice-mail, immersive 3-D chatroom, blackboard, e-mail, customizable bulletin board, print materials, CD-ROMs, e-libraries and e-databases, instant message, videoconference, teleconference (Sims, 2006) and so forth. In some instances it includes on-campus learning activities or testing. Additionally, there is also electronic networking through threaded discussions (Walls, 2006). And yet another option is synchronous or asynchronous environment in which to deliver education. The permutations possible with these several IT and non-IT media/techniques, say ten, are 10! (ten factorial) or 3,628,800. This diversity of choices in to reach DL to the student makes it obvious that costs too would be varied depending on the choice of instructional technology. No two DL projects are likely to have identical costs. Other critical factors impinging on DL costs are:

1. **Instructor/tutor salaries:** This expense varies widely from a mere 10% in developing countries such as India and China to a heavy 37% in the U.S. (PRG Inc., p. 115).
2. **Faculty-student ratios:** The larger the number of students per faculty person, the lower would be the costs and vice versa.
3. **Course development costs:** This has been discussed earlier. There are overlaps with instructor/tutor salaries above.
4. **Teaching loads of faculty and whether faculty is full-time, or adjunct, or even graduate assistants.**
5. **Sacrifice of faculty time for research and service:** Designated DL institutions may not have faculty undertaking research. However, at other campuses difficult choices would have to be made in this regard.

Table 2. Cost of developing a three-unit Internet course (US$) (Source: Rumble, 2001)

(Arizona Learning Systems, 1998)	
Course Outlines and Assignments	6,000
Text	12,000
Text with Reference Material	18,000
Text with Ref. Material and Images	37,500
Audio and Video	120,000
Simulations	250,000
Virtual Reality	1,000, 000

VIABILITY OF MEGA ONLINE UNIVERSITY

Very large DL schools, one would speculate, have been hugely successful, going by their proliferation, both from the educational and business points of view. However, currently what is germane is whether in the coming years new online units, let alone mega units, can repeat such success. Will these trends continue? There is need for circumspection. The market eventually, sooner than later at least in America, may not generate a large enough demand to justify new mega open universities of the IGNOU kind, or even of the University of Phoenix type. They may be significantly smaller, impacting on the viability of DL units.

The new qualms about the economic viability of DL are triggered, *inter alia*, by the emerging constraints on the scale of operations. Mega universities like the ones listed above cannot be replicated anymore and there would have to be garden variety, relatively smaller units that cater to local or at the most state-level populace. It is true that during the last academic year UMass enrollments jumped 23% to 23,682 despite there being no dearth of higher learning institutions in the state. What is noteworthy nevertheless, is that just 28% of students at UMass are out of state. University of Illinois launched its online programs on the same lines with a goal of 10,000 students to be enrolled in 5 years and 50,000 in 10 years (Stokes, 2006; *Inside Higher Education – Daily Update*). The strategic point at this phase of distance education (DE) is that "most enrollments are local" (Primary Research Group, 1997). If that is so, even if a DL college or institution has world-wide capabilities, it would be a "college without boundaries" more in name and much less in student enrollment, unless it can be claimed that any reputed school can have a world-wide enrollment eventually raising it to the status of mega university. Essentially, they would be local to a geographical area and not at all global. In this sense, there would be a paradigm shift with a negative impact on their economics.

The sobering findings of the Primary Research Group, Inc. call for a new analysis of the economics of DL. The group stated that "40% of the distance education programs operate at a loss. 25% of such programs operate with a profit margin of less than 10% above costs while another 25% operate with a profit margin between 11% and 30% above costs. 10% of distance education programs earn between 31% and 50% above costs" (PRG Inc., 1997a). Thanks to what appears to be an overcrowding of the DL business area, 40% of the institutions are in the red, and possibly more may join their ranks. While the University of Phoenix is a success, the New York University Online is considered not a success. The public policy implication would then be that when it becomes harder for DL units to resort to mass education and scale, and thereby minimize costs, the economics of brick and mortar campuses become more appealing albeit the fact that they have much leeway to cover to catch up with DL units in costs and overall viability.

CAPITAL DEEPENING AND OVER-DIMENSIONING

If the growth of mega universities is thus predicated by waning numbers of out-of-state online students, let alone international students, and with state and local populace constituting the bulk of the online enrollments, new questions would arise about the size of such units, and their costs and revenues. Economies of scale account for a significant part of the relatively lower costs of online education, in particular for new institutions entering DL. If the numbers of students enrolling into them do not turn out to be large enough (e.g., the numbers required under a minimum efficient scale [MES] size unit such as 25,000 students), economies of scale may be lost and over-dimensioning of facilities is bound to occur. The choice of 25,000 as the MES, if sounds arbitrary or random, the explanation why it is not so, is simple. In 1971, when the UKOU was set up, the Open University started with 25,000 students and grew to 40,000 the second year (Daniel, 2003). Also in his 2004 study Anthony Dean mentions a threshold size of 20,000. Since then cost inflation would have been substantial. Accordingly, for planning purposes, this author has increased the MES by 25% to 25,000, a ballpark quantity. Capital intensity, not later justified by the number of students registering for courses at the DL institutions would result in heavier per capita capital costs and therefore higher per student average total costs, negatively impacting the bottom line of the DL providers. The savings that emerge from better utilization of mortar-brick campuses for DL projects would be lost.

As shown in Table 1, one could use the break even point (BEP) method to estimate the student enrollment necessary to break even. If, say, up-front investment is $50 million and the fee for a one-credit hour is $125, and the variable costs are $25 per student per hour, the contribution margin would be $100 (125-25) and the DL unit would have to vend 500,000 ($50,000,000/100) credit hours. If on an average, a typical student accepts a 25 credit hour instruction load during a year, there would have to be 20,000 students merely to break even. If the credit hours are less, say 20 hours per year, the students on the roll would have to be 25,000, and so forth. If a 20% return is expected on the upfront investment of $50 million, other things being the same, that is, the fee is $125, the contribution margin is $100, and 25 credit hour instruction per student is typical, there would have to be 24,000 {($50 million + 20%) / (100 * 25)} students on the rolls.

A minimum enrollment of 20,000 students is valid under the above assumptions relating to the upfront investment, minimum credit hours, the fees, and the returns. Even if any one of the assumptions is altered, the estimates need to be redone. The costs and course fees for different locations would indeed vary widely. The BEP could be as low as 5,000 and as high as 30,000 fee-paying students. It is unrealistic to postulate that larger the capacity of a DE facility, lower also would be the per student costs under all conditions. A look at the L-shaped average cost curve (Appendix A) should bring this point home. Keeping this fact in mind, key items of equipment or technology for DL are identified below in order to pinpoint capital costs that would contribute to savings arising out of scale of operations.

THE PRG SURVEY OF TECHNOLOGY PREFERENCE

The big-ticket items of expenditure are listed in the survey of 44 DL programs by the Primary Research Group (1998). The group states that four technological approaches dominate the DL market according to the technology selected: a) The Internet b) live or tape broadcast/wire technology based on satellite, television or cable, and c) snail mail and phone.

Pareto analysis (arranging items in terms of percentage share) of the information on Table 3 would highlight that Internet (27.1% of the 44 units surveyed use as the Primary Medium for their DE programs) followed by interactive video (23.95%) would be the main tech-based media that may offer opportunities for savings. Mail correspondence, ranking as the third largest medium, would offer little, if any, scale-related economy in view of its variable cost nature, the total of such costs being a function of volume. Opportunities for economy may exist in a few other technology-based media such as tape video, video cassette, tape and cable broadcast, and similar media, but again the scope for cutback on costs would not be large, in view of their semifixed nature.

IGNOU Case Study

Mega universities mentioned elsewhere such as The China TV University System with about a million student admissions per year, The Centre National d'Enseignement a Distance in France, with 350,000 students, Universitas Terbuka in Indonesia with 350,000 students, Payame Noor University in Iran with 120,000 students, University of South Africa with 150,000 students, and several others obviously enjoy economies of scale and economic viability. The relevant data for one such university, Indira Gandhi National Open University (IGNOU), is presented in Table 4 and in the charts in Appendix A.

Table 3. Technology preferences of 44 units (Compiled PRG Inc., 1997)

Technology	% using as Primary Medium	% Intending Greater Use	% Intending Lesser Use
Mail Correspondence	19.95	23	18
Fax	5.73	38.5	7.7
Fax Broadcasting	2.78	7.9	7.9
E-mail Correspondence	17.61	75	2.5
Internet	27.10	92.5	0
Audio Cassette	3.90	7.7	12.8
Live Audio	4.41	12.8	2.6
Live Radio	0.05	2.6	5.1
Telephone	10.27	20.5	7.7
Voice Mail	7.17	43.6	2.5
Interactive Voice Response	5.00	22.5	2.5
Videocassette	14.31	36.8	7.9
Tape Video	17.80	23	7.7
Tape Broadcast	9.07	30.8	5.1
Tape Cablecast	9.76	35.9	2.6
Tape Satellite	2.49	23.1	2.5
Live Video	9.12	28.9	2.6
Live Broadcast	2.93	20.5	5.1
Live Cablecast	3.46	20.5	2.5
Live Satellite	2.51	23.7	2.6
Interactive Video	23.95	64.1	5.1
Computer to Computer Video-conferencing	0.05	59	2.6
CD-ROM	0.61	56.4	0

In the initial stages the decline in costs is shown to be substantial, although at later stages with larger numbers such as 40,000 students and more particularly after 70,000 students, the decline in costs is small though not insignificant. The economies associated with scale are well demonstrated in the IGNOU study. The cost curve based on this study slopes downward. Savings in costs are based mostly on the modular build up of the technical architecture for DE. Size related economies of the firm are important too. The LRACC is L-shaped with little incremental scale-related economies after 100,000 students.

The least squares regression equation is $Y = 4312.4 - 0.018X$ and the plot is shown in Appendix A ($p < 0.0910$). The good fit of the negative sloping regression line indicates that it is safe to work on the basis of overtones of significant association between the two variables under study. The L shape of the average cost curve could mean that after 100,000 student enrollments, there is not much of scale-related economies, the costs becoming somewhat constant. There needs to be no fears of increasing costs or diseconomies due to excess students either, considering DL units with more than a million students, too, are economically viable today.

Table 4. Scale economies: IGNOU (Source: Rumble, 1997)

# of Students	Total Costs	Average		% Decline in	Marginal Cost
	Rupees (Millions)	Costs (Rupees)	Decline in Ave. Costs	Ave. Costs	Per Student
5000	59.522	11, 904.40			
10000	62.508	6,250.80	5,653.60	47.49	597.2
20000	68.478	3,423.90	2,826.90	45.22	597.0
30000	74.449	2,481.63	942.27	27.52	597.1
40000	80.421	2,010.53	471.11	18.98	597.2
50000	87.521	1,750.42	260.11	12.94	710.0
60000	96.218	1,603.63	146.79	8.39	869.7
70000	104.916	1,498.80	104.83	6.54	869.8
80000	113.615	1,420.19	78.61	5.25	869.9
90000	122.313	1,359.03	61.15	4.31	869.8
100000	131.011	1,310.11	48.92	3.60	869.8
200000	217.993	1,089.97	220.15	16.80	869.8
300000	304.975	1,016.58	73.38	6.73	869.8

The relationship between costs and students is not necessarily linear and therefore calls for looking at other related variables, a conclusion this chapter has been promoting. The Pearson plot based on data in columns 1 and 3, linking average costs to number of students, does indicate cost economies. However, the computed r value of -0.4899 being lower than the critical r value of 0.5529 indicates that the correlation coefficient is not significant for $n = 13$. And yet the coefficient of determination (r^2) being 21%, a sizeable part of the variation in costs, is explained by scaling.
The economies of scale witnessed in IGNOU may not be emblematic of DL elsewhere, given the vast differences in capital costs as well as in capital deepening. Most DL facilities are purposely over dimensioned with the objective of reducing future expansion costs. By the same token, the economies gained by IGNOU may be relatively larger at other locations depending upon how much more capital intensive the DL operations are and also how much more pricey the cutting edge DL technology is.

CONCLUSION

Technology would perhaps be the major capital cost for a unit that already has the brick and mortar configuration such as a building, organization, and staff. If DL is an extension of such a university edifice, a second major cost would be the cost of developing courses for online delivery. If such an online institution has a mother institution like the one described above, the costs of developing a course per student can be vastly reduced by means of online delivery of the course to thousands of students over several years.

Besides size-capital cost, or volume-capital cost relationship, that brings about much of the scale economies, there are also cost savings associated with the size of the operations. However, recent studies of Fortune 500 companies using simple ordinary least squares techniques to look into nonlinearities have established that there are no perceptible advantages to large firm size in terms of return on equity, return on assets, or in terms of profit margin. Quality of management, adopting best practices, and so forth could undercut diseconomies. For an interesting and yet edifying narration of scale-related issues in Fortune 500 companies see Jerry Useem's work (2007) as well as On-Line Game Services (2006).

There are several other factors that lend a modicum of validity to size-related savings in costs. For instance, environmental costs of on-campus education tilt the overall costs against such education. The Open University, UK study was concerned about greening of higher education by means of reducing environmental impacts such as savings in energy and lower emissions of CO_2 arising out of mainly a reduction in commuting (Roy, Potter, Yarrow, & Smith, 2005) Thus insufficient student registration for DL can prove to be doubly negative, deleteriously affecting education, and also making it a less sustainable means of higher education for both institutions and students.

While economies of scale are indeed valid and do materialize in DL, the diversity in this regard makes it hard to oversimplify in cost and revenue matters. One hat does not fit all units and qualitative variables such as basic design and environmental considerations, video-based systems, audio-based systems, computer-based systems, integrated-tech systems, educational standards complied with, compensation to instructors, course development costs, infrastructure and technology and equipment fixed costs, labor-intensive or capital-intensive modes, and others have to be factored into the modus operandi. There is much uncertainty and only a nuts and bolts evaluation of each DL project would make available information that would enable decision making regarding profitability. Here Table 1 would serve as a template to derive *ex ante* planning data.

FUTURE RESEARCH

In view of the conclusion of this chapter that economies of scale may not be achieved in the future thanks to factors inhibiting enrollments across boundaries, students and researchers should be encouraged to come up with Plan A, such as pragmatic case studies including cost estimates as per Table 1 for different locations, technological infrastructure, courses, and of course nonmega size hypothetical units. A bunch of such studies could then be summarized into a metastudy that could provide answers to questions about realizing economies of scale, minimum efficient size, and viability of DL units. There could be quite a few wild card revelations.

Alternatively, Plan B would be the more arduous task: scout for cost data for existing units all over the USA, or globally, and examine the different technologies used and make them comparable and then sort them out for matching likes with likes, process them, and draw conclusions. Plan A would still trump with better answers.

REFERENCES

Carnevale, D. (2006, September 27). Distance education: Keeping up with exploding demand. *The Chronicle of Higher Education*. Retrieved February 13, 2008, from www.chronicle.com

Daniel, J. S. (1996). *Mega-universities and knowledge media: Technology strategies for higher education*. London: Kogan Page.

Daniel, J. S. (2003). *Mega-Universities: Mega-Impact on access cost and quality*. Retrieved February 13, 2008, from http://portal.unesco.org/education/en/ev.php-URL_ID=26277&URL_DO=DO_TOPIC&URL_SECTION=201.html

Dean, F. A. (2004). *Australian universities in the information economy: Electronic commerce and the business of distance education*. Unpublished doctoral dissertation, University of Wollongong. Retrieved February 13, 2008, from www-library.uow.edu.au/adt-NWU/public/index.html

Gibson, C. C. (Ed.). (1998). *Distance learners*

in higher education: Institutional responses for quality outcomes. Madison, WI: Atwood Publishing.

Golden, D. (2006). *Degrees @StateU.edu.* Retrieved May 9, 2006, from Wall Street Journal Online. Retrieved from http://online.wsj.com/article/SB114713782171047386.html?modegooglenews_wsj

Harrison, N. (1999). *How to design self-directed and distance learning programs.* McGraw-Hill.

IGNOU. (2007). Retrieved February 13, 2008, from http://www.ignou.ac.in/

Inside Higher Ed -- Daily Update (on-line edition) frequently carries interesting information about DE and data regarding UMass and U of Illinois are compiled from different issues. *http://newsroom@insidehighered.com*

Ketterer, J. J., & George, E. M., II. (2006, Spring). Re-conceptualizing intimacy and distance in instructional models. *Online Journal of Distance Learning Administration, 9*(1).

Louisiana Department of Education. (2003) *Small school districts and economies of scale.* Retrieved February 13, 2008, from www.doe.state.la.us/LDE/uploads/3475.pdf

On-Line game services. (2006). Retrieved February 13, 2008, from http://www-306.ibm.com/software/success/cssdb.nsf/CS/MCAG-6WNQRG?OpenDocument&Site

Primary Research Group (PRG), Inc. (1997a). *Ibid* (p. 18). Author.

Primary Research Group (PRG), Inc. (1997b). *The survey of distance learning programs in higher education* (p. 97). Author.

Primary Research Group (PRG), Inc. (1998). *Profiles of college and university distance learning programs.* Author.

Roy, R., Potter, S., Yarrow, K., & Smith, M. (2005, March). *Final report on towards sustainable higher education: Environmental impacts of campus-based and distance higher education systems.* The Open University, Design Innovation Group.

Rumble, G. (1997). *The costs and economics of open and distance learning.* London: Kogan Page.

Rumble, G. (2001). The costs and costing of networked learning. *Journal of Asynchronous Networks.* Sloan Publications. Retrieved February 13, 2008, from www.sloan-c.org/publications/jaln

Sims, D. (2006). *TMC net, teleconferencing and distance learning: An inside look.* Retrieved February 13, 2008, from http://www. tmcnet.com/channels

Spellings, M. (2006). *Secretary of education while addressing the National Press Club on the commission on higher education in October 2006.* Retrieved February 13, 2008, from http://www.ed.gov/news/speeches/2006

Stokes, P. (2006). *Consultant for the Illinois State on-line education project quoted in Inside Higher Education – Daily Update.* Retrieved February 13, 2008, from *http://newsroom@insidehighered.com*

UK Open University. (2007). Retrieved February 13, 2008, from http://www.open.ac.uk/about/ou/p7.shtml

UNESCO. (2007). *Description of mega university.* Retrieved February 13, 2008, from

http://portal.unesco.org/education/en/ev.php URL_ID=42857&URL_DO=DO_TOPIC&URL_SECTION=201.html

Useem, J. (2007). *Is scale an asset or liability? The case of Fortune 500 companies.* Retrieved February 13, 2008, from http://money.cnn.com/magazines/fortune/fortune_archive/2007/04/30/8405390/index.htm

Vice-Chancellor, UKOU. (2007). Retrieved

February 13, 2008, from http://www.open.ac.uk/vice-chancellor/News_3a00_Insights/Insights-2007-April/Our_founding_fathers_would_be_proud_of_the_achievements_of_the_OU_around_the_world.html

Walls, C. M. (2006). *Some strategies for balancing economies of scale and interaction in online/distance education courses.* Oregon State University. Retrieved February 13, 2008, from drwalls@orst.edu

ADDITIONAL READING

Bates, A. W. (2005). *Technology, e-learning and distance education, studies in distance education.* Routledge Falmer.

Bates, A. W., & Poole, G. (2003). Effective teaching with technology in higher education: Foundations for success. *The Jossey-Bass higher and adult education series.*

Bodain, Y., & Robert, J.-M. (2000). *Investigating distance learning on the Internet.* Retrieved February 13, 2008, from http://www.isoc.org/inet2000/cdproceedings/6a/6a_4.htm

Claude, G. (Ed). (2003). *Usability evaluation of online learning programs.* Hershey, PA: Information Science Pub.

Compora, D. P. (2003, Summer). Current trends in distance education: An administrative model. *Online Journal of Distance Learning Administration, 6*(2).

Duffy, T. M., & Kirkley, J. R. (Eds.). (2004). *Learner-centered theory and practice in distance education: Cases from higher education.* L. Erlbaum.

Jones, R. (2001). *A recommendation for managing the predicted growth in college enrollment at a time of adverse economic conditions.* Retrieved February 13, 2008, from http://www.westga.edu/~distance/ojdla/spring61/jones61.htm

Ko, S., & Rossen, S. (2001). *Teaching online: A practical guide.* Houghton Mifflin.

Mohammed, R., Fisher, R. J., Jaworski, B. J., & Cahill, A. (2002). *Internet marketing* (1st ed.). New York: McGraw-Hill.

Moore, M. G., & Anderson, B. (2003). *The handbook of distance education.* Routledge.

Morgan, B. M. (2000). *Determining the costs of online courses.* Retrieved February 13, 2008, from http://www.marshall.edu/distance/

Palloff, R. M., & Pratt, K. (2001). *Lessons from the cyberspace classroom: The realities of online teaching.* Jossey-Bass.

Valentine, D. (2002, Fall). Distance learning: Promises, problems, and possibilities. *Online Journal of Distance Learning Administration, 5*(3).

APPENDIX A: REGRESSION ANALYSIS OF I.G. NATIONAL OPEN UNIVERSITY COST DATA IN TABLE 4

	N = 13				
R	0.4899				
R^2	0.24				
Adjusted R^2	0.17				
SE	2790.8823				
Term	**Coefficient**	**SE**	**p**	**95% CI of Coefficient**	
Intercept	4312.4	1103.6	0.0024	1883.42	to 6741.4
Slope	-0.0180	0.0097	0.0910	-0.0393	to 0.0034
Source of variation	**SSq**	**DF**	**MSq**	**F**	**P**
Due to regression	26,723,727.2	1	26,723,727.2	3.43	0.0910
About regression	85,679,263.4	11	7,789,023.9		
Total	112,402,990.6	12			

*Economies of scale based on IGNOU data *Note. The curve with rings is the Average Cost Curve and has an L shape. The thick middle line is the forecast regression line.*

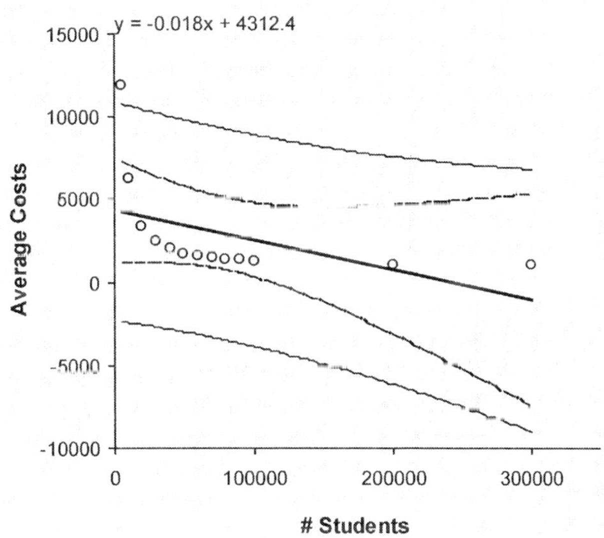

Compilation of References

Abbott, L. et al. (1993). *Videoconferencing and distance learning. Faculty of education and department of adult and continuing education.* Ireland: University of Ulster.

Abler, R. T., & Wells, I. G. (2005). Distributed engineering education: Evolution of the telecollaboration stations for individualized distance learning. *IEEE Transactions on Education, 48*(3), 490-496.

Abraham, T. (2002). Evaluating the virtual management information systems classroom. *Journal of Information Systems Education, 13*(2), 125-133.

Academic impressions building learning communities: Strategies for collaborative learning on-line conference overview. (2006). Retrieved October 12, 2006, from https://www.academicimpressions.com/conferences/1006-collaborative-communities.php

Adams, M., Tenney, Y., & Pew, R. (1995). Situation awareness and the cognitive management of complex systems. *Human Factors, 37*, 85-104.

ADEC Guiding Principles for Distance Learning and Teaching. (1999). *American Distance Education Consortium.* Retrieved February 21, 2008, from http://www.adec.edu/admin/papers/distanceteaching_ principles.html

AECE (Asociación Española de Comercio Electrónico). (2006). *Estudio sobre comercio.* Paper presented at the Electrónico B2C 2006, Madrid. Retrieved November 14, 2005, from http://www.aece.es

Ahluwalia, A.K., Jonker, P.P & Young, I.T. (2000). An interactive image processing course for the web, Proc. *First International Conference on Image and Graphics*, Y.J. Zhang ed., 589~593.

Ahrens, J. D., & Sankar, C. S. (1993). Tailoring database training for end users. *MIS Quarterly, 17*(4), 419-439.

Ahuna, C. (2006). *Online game communities are social in nature.* Retrieved February 12, 2006, from http://switch.sjsu.edu/v7n1/articles/cindy02.html

AIMC (Asociación de Investigación de los Medios de Comunicación). (2006). *Estudio general de medios.* Pape presented at the Tercera ola 2005, Madrid. Retrieved September 12, 2006, from http://www.aimc.es

Alaoutinen, S., & Voracek, J. (2004). Combining traditional and virtual teaching techniques in cross-border higher education. *Electronic Journal of E-learning, 2*(1), 1-10.

Alavi, M. (1994). Computer-mediated collaborative learning: An empirical evaluation. *MIS Quarterly, 18*, 159-174.

Alavi, M., & Leidner, D. E. (2001). Research commentary: Technology mediated learning-a call for greater depth and breadth of research. *Information Systems Research, 12*(1), 1-10.

Alavi, M., Marakas, G. M., & Yoo, Y. (2002). A comparative study of distributed learning environments on learning outcomes. *Information Systems Research, 13*(4), 404-417.

Alavi, M., Wheeler, B. C., & Valacich, J. S. (1995). Using IT to reengineer business education: An exploratory investigation of collaborative telelearning. *MIS Quarterly, 19*, 293-312.

Alavi, M., Yoo, Y., & Vogel, D. R. (1997). Using information technology to add value to management education. *Academy of Management Journal, 40*(6), 1310-1333.

Alden, J. (1998). *A trainer's guide to Web-based instruction*. Alexandra, VA: American Society for Training and Development.

Allen, G. K., Wedman, J. F., & Folk, L. C. (2001). Looking beyond the valley: A five year case study of course innovation. *Innovative Higher Education*, *26*(2), 103-119.

Allen, M., Mabry, E., Mattrey, M., Bourhis, J., Titsworth, S., & Burrell, N. (2004). Evaluating the effectiveness of distance learning: A comparison using meta-analysis. *Journal of Communication*, *54*(3), 403-420.

Al-Rawi, A., Lansari, A., & Bouslama, F. (2005). A holistic approach to develop IS curricula: Focusing on accreditation and IT certification. *Journal of Information Techonology Education*, 4, 307-327

Anastasiades, P. (2002). The theory of information reversal. *Computers and Society, ACM Special Interest Group on Computers and Society (ACM SIGCAS)*, *32*(2), 10-16.

Anastasiades, P. (2002). *Towards the global information society: The enactment of a regulatory framework as a factor of transparency and social cohesion* (LNCS 2510, pp. 527-535). Berlin/Heidelberg: Springer-Verlag.

Anastasiades, P. (2003). Distance learning in elementary schools in Cyprus: The evaluation methodology and results. *Computers & Education, 40*(1), 17-40. Elsevier Science.

Anastasiades, P. (2003, March 24-29). Synchronous distance learning courses: Evaluation methodology. In *Proceedings of the International Conference SITE 200: Society for Information Technology and teacher Education*, Nashville/ New Mexico.

Anastasiades, P. (2003, April 1-5). Building up a pedagogical model for synchronous distance learning courses (selected paper). In *Proceedings of the 13th International Conference on College Teaching and Learning*, Florida.

Anastasiades, P. (2004, June 27-30). *From the conventional classroom to the new hybrid learning environment: Steps and methodology*. Paper presented at the Eleventh Literacy and Education Research Network Conference on Learning, Cojímar Pedagogical Convention Centre. Havana, Cuba: Common Ground.

Anastasiades, P. (2005, November 11-13). Lifelong e-learning and digital divide. In *Proceedings of the Third International Conference on Open and Distance Learning*. Patra, Greece: Greek Open University.

Anastasiades, P. (2005, July 14-17). Web based education and digital divide towards the European lifelong learning society. In *Proceedings of the 3rd International Conference on Education and Information Systems: Technologies and Applications*, Orlando, FL.

Anastasiades, P. (2005, June 27-July 2). Synchronous vs asynchronous learning? Principles, methodology and implementation policy of a blended learning environment for lifelong learning, at the University of Crete. In *Proceedings of the ED-MEDIA 2005 World Conference on Educational Multimedia, Hypermedia and Telecommunications, (AACE)*, Montreal. Association for the Advancement of Computing in Education.

Anastasiades, P. (2005, October 24-28). Lifelong learning and Greek diaspora towards the global information society: Suggested e-learning monitoring methodology (e-LMM). In *Proceedings of the E-Learn 2005--World Conference on E-Learning in Corporate, Government, Healthcare, and Higher Education*, Vancouver, Canada.

Anastasiades, P. (2006, July 5-7). Interactive videoconferencing in lifelong learning: Methodology and implementation policy at the University of Crete (E.DIA.M.ME). In *Proceedings of the Diverse 2006, 6th International Conference on Video and Videoconferencing in Education*. Scotland, Glasgow: Caledonian University.

Anastasiades, P. (2006, July 5-7). Interactive videocon ferencing in K-9 education: "ODUSSEAS 2000-2004" a case study in elementary schools in Greece and Cyprus. In *Proceedings of the Diverse 2006 1 6th International Conference on Video and Videoconferencing in Education*. Scotland, Glasgow: Caledonian University.

Anastasiades, P. S. (2003). Distance learning in elementary schools in Cyprus: The evaluation methodology and results. *Computers & Education*, *40*, 17-40.

Anastasiades, P., & Spantidakis, J. (2006, June 26-30). Advanced learning technologies and the new hybrid learning environment towards the knowledge society: Steps and implementation policy. In *Proceedings of the*

ED-MEDIA World Conference on Educational Multimedia, Hypermedia and Telecommunications (ED MEDIA 2006). Orlando, FL: Association for the Advancement of Computing in Education (AACE).

Anderson, T., Rourke, L., Garrison, D., & Archer, W. (2001). Assessing teaching presence in a computer conferencing context. *Journal of Asynchronous Learning Networks, 5*(2), 1-17.

Angel, R. B., Sanders, R. L., & Tashner, J. H. (2005, March). *Constructing learning communities through Web-based environments: Problem based learning in cross-disciplinary social constructivist frameworks.* Phoenix, AZ: Society for Information Technology and Teacher Education.

Aniebonam, M. C. (2000, October). Effective distance learning methods as a curriculum delivery tool in diverse university environments: The case of traditional vs. historically black colleges and universities. *Communications of the Association for Information Systems, 4*(8), 1-35.

Arbaugh, J. (2000). Virtual classroom characteristics and student satisfaction with Internet based MBA. *Journal of Management Education, 24*(1), 32-54.

Armitt, G., Slack, F., Green, S., & Beer, M. (2002, January). *The development of deep learning during a synchronous collaborative on-line course.* Paper presented at the CSCL 2002, Boulder, Colorado.

Armstrong, T. (2000). *Multiple intelligences in the classroom.* Alexandria, VA: Association for Supervision and Curriculum Development.

Articulate. (2007). *Articulate e-learning software.* Retrieved on February 10, 2007, from http://www.articulate.com/

Attwell, G. (2003). The challenge of e-learning in small enterprises: Issues for policy and practise in Europe. *Cedefop Panorama Series, 82.* Luxembourg: Cedefop Office for Official Publications of the European Communities.

Attwell, G., & others (2003). *E-Learning in Europe results and recommendations: Thematic monitoring under the LEONARDO DA VINCI-Programme.* Bonn, Germany: Nationale Agentur Bildung für Europa beim Bundesinstitut für Berufsbildung.

Baark, E., Falch, M., et al. (2002). *The tradability of consulting services.* New York/Geneva: UNCTAD.

Baldwin, T.T., & Ford, J.K. (1988). Transfer of training: A review and directions for future research. *Personnel Psychology, 41,* 63-105.

Bandura, A. (1977). *Social learning theory.* Morristown, NJ: General Learning Press.

Barrie, T. (1996). *Spiritual path, sacred place: Myth, ritual, and meaning in architecture.* Boston: Shambhala.

Barrows, H. S. (1986). A taxonomy of problem-based learning methods. *Medical Education, 20,* 481-486.

Bastiaens, T., & Martens, R. (2000). Conditions for Web-based learning with real events. In B. Abbey (Ed.), *Instructional and cognitive impacts of Web-based education* (pp. 1-32). London: Idea Group.

Bates, A. W. (1995). *Technology: Open learning and distance education.* New York: Routledge.

Bates, T. (1999). *Managing technological change: Strategies for academic leaders.* San Francisco: Jossey Bass.

Bayman, P., & Mayer, R. E. (1988). Using conceptual models to teach BASIC computer programming. *Journal of Educational Psychology, 80*(3), 291-298.

Beardwell, I., & Holden, L. (1994). *Human resource management: A contemporary perspective.* London: Pitman Publishing.

Beise, C., Myers, M., VanBrackle, L, & Chevli-Saroq, N. (2003). An examination of age, race, and sex as predictors of success in the first programming course. *Journal of Informatics Education and Research, 5*(1), 51-64.

Bellier, S. (2001). *The EL.* Paris: Editions Connections.

Bender, T. (2003). *Discussion-based online teaching to enhance student learning; Theory, practice, and assessment.* Sterling, VA: Stylus Publishing.

Benjamin, A. (2005). *Differentiated instruction using technology.* Larchmont, NY: Eye on Education.

Berger, K. A., & Topol, M. T. (2001). Technology to enhance learning: Use of a Web site platform in traditional classes and distance learning. *Marketing Education Review, 11*(3), 15-26.

Bhattarcherjee, A. (2001). An empirical analysis of the antecedents of electronic commerce service continuance. *Decision Support Systems, 32*, 201-214.

Bielefield, A., & Cheeseman, L. (1997). *Technology and copyright law*. New York: Neal-Schuman Publishers, Inc.

Bishop, M. J., & Cates, W. M. (1996). A door is a big wooden thing with a knob: Getting a handle on metaphorical interface design. In *Proceedings of Selected Research and Development Presentations at the 1996 National Convention of the Association for Educational Communications and Technology* (pp. 80-88). Indianapolis, IN (ERIC Document Reproduction Service No. ED397779).

Black, I., & Lin, Y. (2003). *Perceptions of risks in e-commerce, a comparison of perceived risk in on-line degree courses versus campus based learning*. Paper presented at the 33rd EMAC Conference, Glasgow.

Bolton, R. (1998). A dynamic model of the duration of the customer's relationship with a continuous service provider: The role of satisfaction. *Marketing Science, 17*(1), 45-65.

Booker, D. (2000). *Getting to grips with online delivery*. Adelaide: NCVER.

Booth, S., & Hulten, M. (2004). Opening dimensions of variation: An empirical study of learning in a Web-based discussion. In P. Goodyear, S. Banks, V. Hodgson, & D. McConnell (Eds.), *Advances in research on networked learning* (Vol. 4, pp. 153-174). MA: Kluwer Academic Publishers.

Borning, A., & Travers, M. (1991). Two approaches to causal interaction over computer and video networks. In *Proceedings of the Conference on Human Factors in Computing Systems* (pp. 13-19). New York: ACM.

Bostrom, R. P., Olfman, L., & Sein, M. K. (1990). The importance of learning style in end-user training. *MIS Quarterly, 14*(1), 101-109.

Boticario, J., & Gaudioso, E. (2000). Adaptive Web site for distance learning. *Campus-Wide Information Systems, 17*(4), 120-128.

Bower, B. L., & Hardy, K. P. (2004). From correspondence to cyberspace: Changes and challenges in distance education. *New Directions for Community Colleges, 128*, 5-12.

Bowles, M. S. (2004). *Relearning to e-learn*. Carlton, Victoria, Australia: Melbourne University Press.

Bradburn, E.M., & Zimbler, L. (2002). *Distance education instruction by postsecondary faculty and staff: Fall 1998*. National Center for Education Statistics. Retrieved from http://nces.ed.gov/pubs2002/2002155.pdf

Brandon, D., Pruett, J., & Wade, J. (2002). Experience in developing and implementing a capstone course in information technology management. *Journal of Information Technology Education, 1*(2), 92-102.

Brewer, E. W., Dejonge, J. O., & Stout, V. J. (2001). *Moving to online*. Thousand Oaks, CA: Corwin Press, Inc.

Brewer, P. D. (2004). An examination of alternative instructional methods. *Delta Pi Epsilon Journal, 46*(2), 92-104.

Britain, S., & Liber, O. (1999). A framework for pedagogical evaluation of virtual learning environments. *Report to JISC Technology Applications Programme*. Retrieved May 10, 2007, from http://www.jisc.ac.uk/uploaded_documents/jtap-041.doc

Broady, E. (1996). Learner attitudes towards self-direction. E. Broady & M. M. Kenning (Eds.), *Promoting learner autonomy in university language teaching* (pp. 215-236). London: CILT.

Bronack, S., Riedl, R., & Tashner, J. (in press). Learning in the zone: A social constructivist framework for distance education in a 3D virtual world. *Journal Interactive Learning Environments*.

Brookfield, S. D. (1986). *Understanding and facilitating adult learning*. San Francisco: Jossey-Bass.

Brookfield, S. D. (1987). *Developing critical thinkers*. San Francisco: Jossey-Bass

Brookfield, S. D. (1990). *The skillful teacher*. San Francisco: Jossey-Bass.

Brookfield, S. D. (1991). Discussion. In M. W. Galbraith (Ed.), *Adult learning methods* (pp. 187-204). Malabar, FL: Krieger Publishing Company.

Brookfield, S. D. (1991). The development of critical reflection in adulthood. *New Education, 13*(1), 39-48.

Brookfield, S. D. (1992). Developing criteria for formal theory building in adult education. *Adult Ed. Q. 42*(2), 79-93.

Brookfield, S. D. (1994). Adult learning: An overview. In A. Tuinjman (Ed.), *International encyclopedia of education.* Oxford: Pergamon Press.

Brookfield, S. D. (1995). *Becoming a critically reflective teacher.* San Francisco: Jossey-Bass.

Brookfield, S. D. (1996). Adult learning an overview. In A. Tuijnman (Ed.), *International encyclopedia of adult education and training* (pp.175, 380). Oxford: Pergamon.

Brookfield, S. D. (2001). Repositioning ideology critique in a critical theory of adult learning. *Adult Education Quarterly, 52*(1), 7-22.

Brookfield, S. D. (Ed). (1985). The continuing educator and self-directed learning in the community. In self-directed learning: From theory to practice. *New Directions for Continuing Education* (25). San Francisco: Jossey-Bass.

Brooks, J. G., & Brooks, M. G. (1993). *The case for constructivist classrooms.* Alexandria, VA: ASCD Publications.

Brown, S. (2001). *Views on videoconferencing higher education and research opportunities in the UK (HERO).* Retrieved February 21, 2008, from http://www.hero.ac.uk/inside_he/archive/views_on_videoconferencin883.cfm?

Bruder, C., Beyerlein, T., & Blessing, L. (2007). Transfer training success into everyday use. In L.Gómez, D. Martí, & I. Candel (Eds.), *International Association of Technology, Education and Development Conference Proceedings.* Valencia: IATED.

Bruner, J. (1996). *Toward a theory of instruction.* New York: Norton.

Bruner, J. S. (1990). *Acts of meaning.* Cambridge: Harvard University Press.

Brunner, D. L. (2006). The potential of the hybrid course vis-a-vis online and traditional courses. *Teaching Theology and Religion, 9*(4), 229-235.

Brusilovsky, P. (1996). Methods and techniques of adaptive hypermedia. *User Modeling and User-Adapted Interaction, 6*(2/3), 87-129.

Brusilovsky, P. (1999). Adaptive and intelligent technologies for Web-based education. *Kunstliche Intelligenz: Special Issue on Intelligent Systems and Teleteaching, 4*, 19-25.

Brusilovsky, P. (2001). Adaptive hypermedia. *User Modeling and User-Adapted Interaction, 11*(1/2), 111-127.

Burge, E. J. & Carter, N. M. (1997). *It's building, but is it designing? Constructing Internet-based learning environments.* Paper presented at the World Conference of the International Council for Distance Education, University Park, PA (ERIC Document Reproduction Service No. ED412333).

Burns, J., Clift, J., & Duncan, J. (1990). Understanding of understanding: Implications for learning and teaching. *British Journal of Educational Psychology, 61*, 276-289.

Butrimiene, E., & Danilevicius, E. (2007). Prerequisites for a sistemic employment of e-learning: Adequacy between the opportunities and student's needs. In L. Gómez, D. Martí, & I. Candel (Eds.), *International Association of Technology, Education and Development Conference Proceedings.* Valencia: IATED.

Camtasia. (2007). *TechSmith.* Retrieved on February 10, 2007, from http://www.techsmith.com/

Carmel, E. (1999). *Global software development.* Prentice Hall.

Carnegie Classification. (2007). *The Carnegie classification of institutions of higher education.* Retrieved on February 10, 2007, from http://www.carnegiefoundation.org/classifications/

Carnevale, D. (2006, September 27). Distance education: Keeping up with exploding demand. *The Chronicle of Higher Education.* Retrieved February 13, 2008, from www.chronicle.com

Carr, S. (2000). As distance education comes of age, the challenge is keeping the students. *Chronicle of Higher Education, 46*(22), 39-A41.

Carr, W., & Kemmis, S. (2002). *For a critical educational theory. Education, knowledge and action-research.* Athens: Kodikas.

Casini, M., & Vincino, A. (2003). The automatic control telelab: A user-friendly interface for distance learning. *IEEE Transactions on Education, 46*, 252-257.

Cates, W. M. (1994). Designing hypermedia is hell: Metaphor's role in instructional design. In *Proceedings of Selected Research and Development Presentations at the 1994 National Convention of the Association for Educational Communications and Technology* (pp. 95-108). Indianapolis, IN (ERIC Document Reproduction Service No. ED373706).

Cates, W. M. (1996). Towards a taxonomy of metaphorical graphical user interfaces: Demands and implementations. In *Proceedings of Selected Research and Development Presentations at the 1996 National Convention of the Association for Educational Communications and Technology* (pp. 101-110). Indianapolis, IN (ERIC Document Reproduction Service No. ED397781).

CEC. (2001). The eLearning action plan: Designing tomorrow's education. COM(2001)172 final

Cenich, G. (2006). Hipertexto y nuevas tecnologías: Su aporte al elearning. *Edutec. Revista Electrónica de Tecnología Educativa, 20*, 1-14.

Chan, T., & Rapman, J. (1999). Flow in Web based instructional activity: An exploratory research project. *International Journal of Educational Communications, 5*(3), 225-237.

Chang, S. (2004). High tech vs. high touch in distance education. *International Journal of Distance Education Technologies, 2*(2), i-iii.

Chen, Y., & Willits, F. (1998). A path analysis of the concepts in Moore's theory of transactional distance in a videoconferencing environment. *Journal of Distance Education, 13*(2), 51-65.

Chickering, A., & Ehrmann, S. (1996). Implementing the seven principles: Technology as lever. *American Association for Higher Education Bulletin, 49*(2), 3-6.

Chou, C. (2002). *A comparative content analysis of student interaction in synchronous and asynchronous learning networks.* Paper presented at the 35th Annual Hawaii International Conference on System Sciences, Hawaii.

Christensen, L. G. (2006). *E-service: Knowledge services, entrepreneurship, and the consequences for business customers and citizens.* Lyngby, Denmark: CICT DTU.

Chu, G.-L. (1999). *The relationships between cultural differences among American and Chinese university students and the design of personal pages on the World Wide Web.* Unpublished doctoral dissertation, University of Georgia.

Clark, H. (1996). *Using language.* Cambridge, England: Cambridge University Press.

Clark, J. (2002). A product review of WebCT. *Internet and Higher Education, 5*, 79-82.

Cobb, P. (1994). Where is the mind? Constructivist and sociocultural perspectives on mathematical development. *Educational Researcher, 23*(7), 13-20.

Cogburn, D. L., & Hurup, D. (2006, April 13). The world is our campus: Synchronous collaboration software lets universities unite colleagues, students, and researchers from all over the globe. *Network computing for IT by IT* (pp. 57-68). Retrieved August 6, 2006, from http://www.networkcomputing.com/showArticle.jhtml?articleID=184428959&pgno=1

Cole, R. (Ed.). (2000). *Issues in Web-based pedagogy.* Westport, CT: Greenwood Press.

Colis & Moonen (2001). *Flexible learning in a digital world: Experiences and expectations.* London: Kogan Page.

Collis, B. (1999). Design, development and implementation of a WWW-based course-support system, *Proc. International Conference on Computer in Education*, 11~18.

Compeau, D. R., & Higgins, C. A. (1995). Application of social cognitive theory to training for computer skills. *Information Systems Research, 6*(2), 118-143.

Connell, R. W. (1994). Poverty and education. *Harvard Educational Review, 64*, 125-149.

Connick, P. (1999). *The distance learner guide.* Upper Saddle River, NJ: Prentice

Cookson, P. (2001). La práctica de la educación superior a distancia: El ejemplo de la Universidad de Athabasca–Universidad Abierta de Canadá. *Revista Electrónica de Tecnología Educativa, 14*. Retrieved October 15, 2001, from http://www.uib.es/depart/gte/edutec-e/Revelec14/cookson.html

Cornford, H. J., & Pollock, N. (2003). *Putting the university online: Information, technology and organisational change*. Buckingham, UK: The Society for Research into Higher Education and Open University Press.

Coventry, L. (2000). *Videoconferencing in higher education*. Heriot Watt University, Edinburg: Institute for Computer Based Learning. Retrieved February 21, 2008, from http://www.man.ac.uk/MVC/SIMA/video3/contents.html

Cox, G., Carr, T., & Hall, M. (2004). Evaluating the use of synchronous communication in two blended courses. *Journal of Computer Assisted Learning, 20*, 183-193.

Crook, C., & Light, P. (2002). Virtual society and the cultural practice of study. In S. Woolgar (Ed.), *Virtual society? Technology, cyberbole, reality*. Oxford: Oxford University Press.

Cropley, A. J., & Kahl, T. N. (1983). Distance education and distance learning: Some psychological considerations. *Distance Education, 4*(1), 27-39.

Cutler, N. E., & Danowski, J. A. (1980). Process gratification in aging cohorts. *Journalism Quarterly, 57*, 269-77.

CyberAtlas. (2003). E-Learning market expanding beyond IT training. Jupiter Research. Retrieve July 28, 2006, from *http://cyberatlas.internet.com/markets/education/article/0,,5951_914901,00.html*

Daft, R. L., & Lengel, R. H. (1984). Information richness: A new approach to managerial behavior and organizational design. In L. L. Cummings & B. M. Staw (Eds.), *Research in organizational behavior 6* (pp. 191-233). Homewood, IL: JAI Press.

Daft, R., & Lengel, R. (1986). Organizational information requirements, media richness and structural design. *Management Science, 32*(5), 554-571.

Dagada, R., & Jakovljevic, M. (2004). Where have all the trainers gone? E-learning strategies and tools in the corporate training environment. In *Proceedings of the 2004 Annual Research Conference of the South African Institute of Computer Scientists and Information Technologists on IT Research in Developing Countries* (pp. 194-203). Stellenbosch, Western Cape, South Africa.

Dale, N., & Lewis, J. (2003). *Computer science illuminated* (2nd ed.). Boston: Jones and Bartlett Publishers.

Damanakis & Anastasiades (2005). Life long & distance learning and the diaspora: Implementing a virtual learning environment at the University of Crete. *Themes in Education: Special Issue Information & Communication Technologies in Diaspora, 6*(1), 83-96.

Daniel, J. S. (1996). *Mega-universities and knowledge media: Technology strategies for higher education*. London: Kogan Page.

Davis, F. (1989). Perceived usefulness, perceived ease of use and user acceptance of information technology. *MIS Quaterly, 13*(3), 319-340.

Davis, F. (1993). User acceptance of information technology: System characteristics, user perceptions and behavioral impacts. *International Journal of Man-Machine Studies, 38*, 475-487.

De Bra, P., Eklund, J., Kobsa, A., Brusilovsky, P., & Hall, W. (1999). *Adaptive hypermedia: Purpose, methods and techniques*. Paper presented at the 10th ACM Conference on Hypertext and Hypermedia (pp. 199-200).

De Laat, M., & Lally, V. (2004). Complexity, theory and praxis: Researching collaborative learning and tutoring processes in networked learning community. In P. Goodyear, S. Banks, V. Hodgson, & D. McConnell (Eds.), *Advances in research on networked learning* (Vol. 4, pp. 11-42). MA: Kluwer Academic Publishers.

Dean, F. A. (2004). *Australian universities in the information economy: Electronic commerce and the business of distance education*. Unpublished doctoral dissertation, University of Wollongong. Retrieved February 13, 2008, from www-library.uow.edu.au/adt-NWU/public/index.html

Dearing, R. (1997). Higher education in the learning society. *Report of the Committee of Inquiry into Higher Education*. Retrieved September 15, 2005, form http://www.ncl.ac.uk/ncihe/index.htm

Decker, P. J., & Nathan, B. R. (1985). *Behavior modeling training.* New York: Praeger.

Delone, W., & McLean, E. (1992). Information systems success: The quest for the dependent variable. *Information Systems Research, 3*(1), 60-95.

DeMarco, T., & Lister, T. (1999). *Peopleware: Productive projects and teams* (2nd ed.). New York: Dorset House Publishers.

DeNeui, D. L., & Dodge, T. L. (2006). Asynchronous learning networks and student outcomes: The utility of online learning components in hybrid courses. *Journal of Instructional Psychology, 33*(4), 256-259.

Denzin, N., & Lincoln, Y. (2000). Introduction: The discipline and practice of qualitative research. In N. Denzin & Y. Lincoln (Eds.), *Handbook of qualitative research* (2nd ed., pp. 1-28). London: Sage Publications.

Derntl, M., & Motsching-Pitrik, R. (2005). The role of structure, patterns, and people in blended learning. *The Internet and Higher Education, 8*, 111-130.

Dick, W., & Carey, L. (1996). *The systematic design of instruction.* New York: Harper Collins College Publishers.

Dillenbourg, P. (1999). *Collaborative learning: Cognitive and computational approaches.* Amsterdam: Pergamon–Elsevier.

Dillenbourg, P., & Traum, D. (1999). Does a shared screen make a shared solution? In C. Hoadly & J. Roschelle (Eds.), *Proceedings of the Third Conference on Computer Supported Collaborative Learning* (pp. 127-135). California: Stanford University.

Dillenbourg, P., Baker, M., Blaye, A., & O'Malley, C. (1996). The evolution of research on collaborative learning. In E. Spada & P. Reiman (Eds.), *Learning human and machine* (pp. 189-211). Oxford: Elsevier.

Donati, J. (2003). Least squares deconvolution of Stellar Spectra. Retrieved July 28, 2006, from http://webast.ast.obs-mip.fr/people/donati/multi.html

Donlevy, J. (2003). Teachers, technology and training. *International Journal of Instructional Media, 30*(2), 117-121.

Dori, Y. J., Barak, M., & Adir, N. (2003). A Web-based chemistry course as a means to foster freshmen learning. *Journal of Chemical Education, 80*(9), 1084-1092.

Doube, W. (1998). Multimedia delivery of computer programming subjects: Basing structure on instructional design. In *Proceedings of the 3rd Australasian conference on Computer Science Education,* Sydney, Australia, (pp. 85-91).

Dourish, P., & Belotti, V. (1992). Awareness and coordination in shared workspaces. In *Proceedings of the Conference on Computer-Supported Cooperative Work* (pp. 107-114). New York: ACM.

Duan, B., Hosseini, H., Ling, K., & Gay, R. (2006). An architecture for online laboratory e-learning system. *Journal of Distance Education Technologies, 4*(2), 87-101.

Duin, H., & Hansen, C. (1994). Reading and writing on computer networks as social construction and social interaction. In C. Selfe & S. Hilligoss (Eds.), *Literacy and computers: The complications of teaching and learning with technology* (pp. 89-112). New York: The Modern Language Association.

Dunderstadt, J. (1997). The future of the university in an age of knowledge. *Journal of Asynchronous Learning Networks, 1*(1), 78-88.

Dunlap, T. (2007). *IBM, Cisco team up on chat, video platform.* Retrieved March 2007, from http://www.intranetjournal.com/articles/200703/ij_03_07_07a.html

Dunn, K., & Dunn, R. (1987). Dispelling outmoded beliefs about student learning. *Educational Leadership, 44*, 6.

Dupin, B., & Du-Charme, H. (2005). Assesing student needs in Web-based distance education. *Journal of instructional technology and distance learning, 2*(1), 39-48.

Duran, R., Kelly, L., & Keaten, J. (2005). College faculty and use and perceptions of electric mail to communicate with students. *Communication Quarterly, 53*, 159-176.

Dykes, M., & Schwier, R. (2003, Spring). Content and community redux: Instructor and student Interpretations of online communication in a graduate seminar. *Canadian Journal of Learning and Technology, 29*(2).

Dyson, M., & Barretto, S. (2003). Evaluating virtual learning environments: What are we measuring. *Electronic Journal of E-learning, 1*(1), 11-20. Retrieved November 10, 2003, from http://www.ejel.org/volume-1-issue-1/issue1-art2-dyson-campello.pdf

Eastman, J. K., & Swift, C. O. (2001). New horizons in distance education: The online learner-centered marketing class. *Journal of Marketing Education, 23*(1), 25-34.

Eastman, J., & Iyer, R. (2004). The elderly's uses and attitudes towards the Internet. *Journal of Consumer Marketing, 21*(3), 208-220.

Eighmey, J., & McCord, L. (1998). Adding value in the information age: Uses and gratifications of sites on the World Wide Web. *Journal of Business Research, 41*, 187-194.

E-Leusis. (2004). *El futuro del eLearning: Análisis del mercado y del contexto actual del eLearning.* Informe realizado en el marco del Proyecto Espacios de Excelencia Transfronterizos (E.E.T.) cofinanciado por la Unión Europea. Iniciativa Comunitaria INTERREG III A España, Portugal.

Ellis, C., Gibbs, S., & Rein, G. (1991). Groupware: Some issues and experiences. *Communications of the ACM, 34*(1), 38-58.

Ellis, R. K. (2005). E-learning trends 2005. *Learning Circuits.*

Endsley, M. (1995). Toward a theory of situation awareness in dynamic systems. *Human Factors, 37*(1), 32-64.

Engestrom, Y. (1987). *Learning by expanding: An activity theoretical approach to developmental research.* Helsinki, Finland: Orienta Konsultit Oy.

Eom, W., & Reiser, R. (2000). The effects of self-regulation and instructional control on performance and motivation in computer-based instruction. *International Journal of Instructional Media, 27*(3), 247-261.

Escarré, R., & Barros, F. (2007). Pihe network. Best practice guide on university cooperatin between Europe and Lationamerica. In L. Gómez, D. Martí, & I. Candel (Eds.), *International Association of Technology, Education and Development Conference Proceedings.* Valencia: IATED.

Espinosa, A. J., Lerch, J. F., & Kraut, E. R. (2004). Explicit versus implicit coordination mechanisms and task dependencies: One size does not fit all. In E. Salas & S. M. Fiore (Eds.), *Team cognition: Understanding the factors that drive performance and process* (pp. 107-129). Washington, D.C.: American Psychological Association.

Evans, J. R. (1986). Creative thinking and innovative education in the decision sciences. *Decision Sciences, 17*(2), 250-163.

Evans, J. R. (2001). The emerging role of the Internet in marketing education: From traditional teaching to technology-based education. *Marketing Education Review, 11*, 1-14.

Favretto, G., Caramia, G., & Guardini, M. (2005). E-learning measurement of the learning differences between traditional lessons and online lessons. *European Journal of Open Distance and E-learning, 2.* Retrieved December 23, 2005, from http://www.eurodl.org/materials/contrib/2005/Giuseppe_Favretto.htm.

Feldman, L.J & Hofinger, R J. (1997). Active participation by sophomore students in the design of experiments, *ASEE/IEEE Frontiers in Education Conference*, 1526~1527.

Ferber, M. (1994). *Multi-agent systems: An introduction to distributed artificial intelligence.* Addison-Wesley.

Fillion, G., Limayem, M., & Bouchard (1999). Videoconferencing in distance education: a study of student perceptions in the lecture context. *Innovations in Education and Training International, 36*, 302.

Finch, C., Keaten, J., & Kelly, L. (2004). Reticent and non-reticent college students' preferred communication channels for interacting with faculty. *Communication Research Reports, 21*(2), 197-209.

Finn, K. E., Sellen, A. J., & Wilbur, S. B. (Eds.). (1997). *Video-mediated communication.* Mahwah, NJ: Lawrence Erlbaum Associates.

Fischer, F., Kollar, I., Mandl, H., & Haake, J. M. (Eds.). (in press). Scripting computer-supported collaborative learning. *Cognitive, computational and educational perspectives.* New York: Springer.

Fisher, C., Berliner, D., Filby, N., Marliave, R., Cahen, L., & Dishaw, M. (1980). Teaching behaviors, academic learning time, and student achievement: An overview. In C. Denham & A. Lieberman (Eds.), *Time to learn* (pp. 7-32). Washington, DC: National Institutes of Education.

Flowers, L., Pascarella, E. T., & Pierson, C. T. (2000). Information technology use and cognitive outcomes in the first year of college. *The Journal of Higher Education, 71*(6), 637-667.

Forsyth, I. (2001). *Teaching & learning materials & the Internet.* London: Kogan Page.

Forsythe, S., & Shi, B. (2003). Consumer patronage and risk perceptions in Internet shopping. *Journal of Business Research, 56*, 867-875.

Freire, P. (1984). *The importance of reading and the liberation process.* Mexico: Siglo XXI.

Fulk, J. (1993). Social construction of communication technology. *Academy of Management Journal, 36*, 921-950.

Gagne, R. M. (1992). *Principles of instructional design.* New York: Holt, Rinehart and Winston, Inc.

Gagné, R., Briggs, L., & Wager, L. (1994). *Principles of instructional design.* Fort Worth, TX: HBJ College Publishers.

Gao, T., & Lehman, J. D. (2003). The effects of different levels of interaction on the achievement and motivational perceptions of college students in a Web-based learning environment. *Journal of Interactive Learning Research, 14*(4), 367-387.

Gardner, H. (1991). *The unschooled mind: How children think and how schools should teach.* New York: Basic Books.

Garnham, C., & Kaleta, R. (2002, March 20). Introduction to hybrid courses. *Teaching with Technology Today, 8*(6), 1-3. Retrieved May 5, 2007, from http://www.uwsa.edu/ttt/articles/garnham.htm

Garrison, D. (2003). Cognitive presence for effective asynchronous online learning: The role of reflective inquiry, self-direction and metacognition. In J. Bourne & J. Moore (Eds.), *Elements of quality online education: Practice and direction* (Vol. 4). Needham, MA: The Sloan Consortium.

Garrison, D. R. (1987). Researching dropout in distance education. *Distance Education, 8*(1), 95-101.

Garrison, D. R. (2006). Online collaboration principles. *Journal of Asynchronous Learning Networks, 10*(1), 25-34.

Garrison, D., Anderson, T., & Archer, W. (2000). Critical inquiry in a text-based environment: Computer conferencing in higher education. *Internet and Higher Education, 11*(2), 1-14.

Garrison, D., Anderson, T., & Archer, W. (2001). Critical thinking, cognitive presence, and computer conferencing in distance education. *American Journal of Distance Education, 15*(1), 7-23.

Gaver, W. (1991). Sound support for collaboration. In L. Bannon, M. Robinson, & K. Schmidt (Eds.), *Proceedings of the Second European Conference on Computer-Supported Cooperative Work* (pp. 293-308). Amsterdam: Klewer.

Gibson, C. C. (Ed.). (1998). *Distance learners in higher education: Institutional responses for quality outcomes.* Madison, WI: Atwood Publishing.

Gibson, C., & Cohen, S. (Eds.). (2003). *Virtual teams that work: Creating conditions for effective virtual teams.* San Francisco: Jossey-Bass/Wiley.

Gill, T. G. (2005). Assignment-centric design: Testing the assignments not the lectures. *Decision Sciences Journal of Innovative Education, 3*(2), 339-346.

Gill, T. G. (2005). Teaching C++ submarine style. *IEEE Transactions on Education, 48*(1), 150-156.

Gill, T. G. (2006) The mystery of a self-paced course. *Informing Faculty, 1*(3), 95-105.

Gill, T. G. (2007). Quick and dirty multimedia. *Decision Sciences Journal of Innovative Education, 5*(1), 197-206.

Gill, T. G., & Holton, C. (2006). A self-paced introductory programming course. *Journal of Information Technology Education, 5*, 95-105.

Ginsberg, R. B., & Foster, K. R. (1998). The wired classroom. *IEEE Spectrum, 35*(8), 44-51.

Gist, M. E., Schwoerer, C., & Rosen, B. (1989). Effects of alternative training methods on self-efficacy and performance in computer software training. *Journal of Applied Psychology, 74*, 884-891.

Gloster, C., Jr., & Doss, C. (2000). A distance education course in computer engineering at NC State University. *Computers in Education Journal, 10*, 22-26.

Golden, D. (2006). *Degrees@StateU.edu*. Retrieved May 9, 2006, from Wall Street Journal Online.

Goldsmith, R. (2002). Explaining and predicting consumer intention to purchase over the Internet: An exploratory study. *Journal of Marketing, 66*, 22-28.

Graff, M. (2003). Cognitive style and attitudes towards using online learning and assessment methods. *Electronic Journal of E-learning, 1*(1), 21-28

Graham, D., McNeil, J., & Pettiford, L. (2000). *Untangled Web: Developing teaching on the Internet*. Harlow, UK: Pearson Education.

Grandell, L. (2005). High school students learning university level computer science on the Web: A case study of the DASK-model. *Journal of Information Technology Education, 4*, 207-218.

Gregory, G., & Chapman, C. (2002). *Differentiated instructional strategies – one size doesn't fit all*. Thousand Oaks, CA: Corwin Press.

Grigoriadou, M., Papanikolaou, K., Cotronis, Y., Velentzas, C., & Filokyprou, G. (1999). Designing and implementing a Web-based course. In *Proceedings of International Conference of Computer Based Learning in Science*, Enschede, Netherlands.

Grupo Doxa. (2005). *Panel de E-learning en las grandes empresas*. Retrieved October 27, 2005, from http://www.grupodoxa.com

Gürer, D., Kozma, R., & Millán, E. (1999). Impact of shared applications and implications for the design of collaborative learning environments. In S. Lajoie (Ed.), *Proceedings for the ninth world conference on artificial intelligence in education* (AI-ED99) (pp. 439-445). Amsterdam: IOS Press.

Gutwin, C., & Greenberg, S. (2004). The importance of awareness for team cognition in distributed collaboration. In E. Salas & S. M. Fiore (Eds.), *Team cognition: Understanding the factors that drive performance and process* (pp. 177-201). Washington, D.C.: American Psychological Association.

Guzdial, M., & Turns, J. (2000). Effective discussion through a computer-mediated anchored forum. *Journal of the Learning Sciences, 9*, 437-469.

Häkkinen, P., Järvelä, S., & Byman, A. (2001). Sharing and making perspectives in Web-based conferencing. In P. Dillenbourg, A. Eurelings, & K., Hakkarainen (Ed.), *European perspectives on computer-supported collaborative learning* (pp. 285-292). In Proceedings of the 1st European Conference on Computer-Supported Collaborative Learning, Maastricht University, Maastricht.

Hamer, L. O. (2001). Distance learning technologies as facilitators of learning and learning-related student activities. *Marketing Education Review, 11*, 55-67.

Hannafin, M., & Land, S. (1997). The foundations and assumptions of technology enhanced student-centered learning environments. *Instructional Science, 25*, 167-202.

Hannafin, M., Hill, J. R., Oliver, K., Glazer, E., & Sharma, P. (2003). Cognitive and learning factors in Web-based distance learning environments. In M. G. Moore & W. G. Anderson (Eds.), *Handbook of distance education* (pp. 245-260). Mahwah, NJ: Lawrence Erlbaum Associates.

Hannum, W. (2001). Web-based training: Advantages and limitations. In B. Khan (Ed.), *Web-based training* (pp. 13-20). NJ: Educational Technology Publications.

Hanson, D., Maushak, N., Schlosser, C., Anderson, M., Sorensen, C., & Simonson, M. (1997). *Distance education: Review of the literature* (2nd ed.). Washington, D.C./Ames, IA: Association for Educational Communications and Technology and Research Institute or Studies in Education.

Hara, N. (1998, October 14-17). Students' perspectives in a Web-based distance education course. In *Proceedings of the Mid-Western Educational Research Association*. Retrieved February, 2007, from http://php.ucs.indiana.edu/~nhara/paper/mwera98.htm

Hara, N., & Kling, R. (1999). Students' frustrations with a Web-based distance education course. *First Monday, 4*(12).

Hara, N., Bonk, C. J., & Angeli, C. (2000). Content analysis of online discussions in an applied educational psychology course. *Instructional Science, 28*, 115-152.

Harasim, L. (1995). *Learning networks: A field guide to teaching and learning online*. MA: MIT Press.

Harasim, L. (2000). *Redes de aprendizaje: Guía para la enseñanza y el aprendizaje en red*. Barcelona: Gedisa/EDIUOC.

Harasim, L., Hiltz, S. R., Teles, L., & Turoff, M. (1995). Network learning: A paradigm for the twenty-first century. *Learning networks: A field guide to teaching and learning online* (pp. 271-278). Cambridge. MA: MIT Press.

Harley, D. (2001). Higher education in the digital age: Planning for an uncertain future. *Syllabus: New Dimensions in Education Technology, 15*(2), 10-12.

Harrison, N. (1999). *How to design self-directed and distance learning programs*. McGraw-Hill.

Hartman, J. (2004). *The horizontal university: E-learning as a catalyst for organizational transformation*.

Harvey, S. (2004). Bridging the digital divide: How technology can change higher education delivery for high school students. *Community College Journal of Research and Practice, 28*, 73-74.

Hassan, H., & Dunn, I. (2007). Exploring engineers employability competencies through interpersonal skills and enterprise skills. In L. Gómez, D. Martí, & I. Candel (Eds.), *International Association of Technology, Education and Development Conference Proceedings*. Valencia: IATED

Haukness, J. (1999). *Services in innovation: Innovation in services*. Oslo: STEP Group.

Hayden, K. L. (1999). *Videoconferencing in K-12 education: A Delphi study of characteristics and critical strategies to support constructivist learning experiences*. Retrieved February 21, 2008, from http://hale.pepperdine.edu/~kahayden/dissertation.html

Haythornthwaite, C., Kazmer, M., Robins, J., & Shoemaker, S. (2000). Community development among distance learners: Temporal and technological dimensions. *Journal of Computer Mediated Communication, 6*(1).

He, A., Zhang, G., & Cheng, Z. (2004). A design of real-time and interactive distance education. *International Journal of Distance Education Technologies, 2*(2), 1-12.

Hearnshaw, D. (1999). Capitalising on the strengths and availability of desktop videoconference. *Active Learning, 9*, 52-59.

Hennecke, A., & Cerny, L. (2007). Bridging the gulf between technology and intercultural communication. In L. Gómez, D. Martí, & I. Candel (Eds.), *International Association of Technology, Education and Development Conference Proceedings*. Valencia: IATED.

Henry, A., & Crawford, C. M. (2001). Creating a collaborative Web-based environment through the inclusion of metaphorically enhanced graphics. In *Proceedings of WebNet 2001: World Conference on the World Wide Web and Internet* (pp. 1-8). Orlando, FL (ERIC Document Reproduction Service No. ED462914).

Herring, S. (1999). Interactional coherence in CMC. *Journal of Computer Mediated Communication, 4*(4).

Herzog, H. (1944). What do we really know about daytime serial listeners? In P. Lazarsfeld & F. Stanton (Eds.), *Radio research 1942-1943*. New York: Duel, Sloan and Pearce.

Hill, J. R., & Hannafin, M. J. (2001). Teaching and learning in digital environments: The resurgence of resource-based learning. *Educational Technology Research and Development, 49*(3), 37-52.

Hiltz, S. R. (1994). *The virtual classroom: Learning without limits via computer networks*. Norwood, NJ: Ablex.

Hodges, C. B. (2005). Self-regulation in Web-based courses: A review and the need for research. *The Quarterly Review of Distance Education, 6*(4), 375-383.

Hoffman, D. L., & Novak, T. P. (1996). Marketing in hypermedia computer-mediated environments: Conceptual foundations. *Journal of Marketing, 60*, 50-68.

Hofstede, G. (1997) *Cultures and organizations: Software of the mind.* New York: McGraw-Hill.

Holmberg, B. (1986). *Growth and structure of distance education.* London: Croom Helm.

Horton, S. (2000). *Web teaching guide: A practical approach to creating course Web sites.* London: Yale University Press.

Horton, W. (2000). *Designing Web-based training: How to teach anyone anything anywhere anytime.* New York: John Wiley & Sons.

Huang, H., & Lu, C. (2003). Java-based distance learning environment for electronic instruments. *IEEE Transactions on Education, 46*(1), 88-94.

IDC. (2002, September 30). While corporate training markets will not live up to earlier forecasts, IDC suggests reasons for optimism, particularly e-learning. Retrieved July 28, 2006, from http://www.idc.com/getdoc.jhtml?containerId=pr2002_ 09_17_150550

IDE. (1996). Three models of distance education. *University of Maryland University College.* Retrieved February 21, 2008, from http://www.umuc.edu/IDE/modeldata.html

IGOU. (2007). Retrieved February 13, 2008, from http://www.ignou.ac.in/

Illeris, K. (2002). *The three dimensions of learning: Contemporary learning theory in the tension field between the cognitive, the emotional, and the social.* UK: Leicester.

Illeris, K. (2003) Towards a contemporary and comprehensive theory of learning. *International Journal of Lifelong Education, 22*(4), 396-406.

Ives, B., Olson, M., & Baroudi, S. (1983). The measurement of user information satisfaction. *Communications of the ACM, 26,* 785-793.

Jakobsson, M., & Taylor, T. L. (2003). The Sopranos meets EverQuest: Social networking in massively multiplayer online games. *Melbourne DAC.* Retrieved February 9, 2007, from http://hypertext.rmit.edu.au/dac/papers/Jakobsson.pdf

Jarvis, P. (2004). *Adult and continuing education. Theory and practice.* Athens: Metaixmio.

Jensen, E. (1998). *Teaching with the brain in mind.* Alexandria, VA: ASCD.

Jensen, E. (2000). *Different brains, different learners.* San Diego: The Brain Store. Retrieved February 9, 2008, from www.thebrainstore.com

Jiang, M., & Ting, E. (2000) A study of factors influencing students' perceived learning in a Web-based course environment. *International Journal of Educational Telecommunications, 6*(4), 317-338.

Johnson, G. M. (2006). Synchronous and asynchronous text-based communication in educational contexts: A review of recent research. *TechTrends, 50*(4), 46-53.

Johnson, J. D. (2000). Levels of success in implementing information technologies. *Innovative Higher Education, 25*(1), 59-75.

Johnson, T. & Johnson, R. (1999). *Learning together and alone: Cooperative, competitive and individualistic learning.* Boston: Allyn and Bacon.

Jolliffe, A., Ritter, J., & Stevens, D. (2001). *The online handbook: Developing and using Web based learning.* London: Kogan Page.

Jonassen, D. (1997). Instructional design models for well-structured and ill-structured problem-solving learning outcomes. *Educational Technology Research & Development, 45,* 65-94.

Jonassen, D. (1997, Spring). *INSYS 527 designing constructivist learning environments.* Retrieved October 12, 2006, from http://www.coe.missouri.edu/~jonassen/INSYS527.html

Jonassen, D. H. (1992). Objectivism versus constructivism: Do we need a new philosophical paradigm? *Educational Technology Research and Development, 39*(3), 5-14.

Jonassen, D., Howland, J., Moore, J., & Marra, R. (2003). *Learning to solve problems with technology, a constructivist perspective.* Upper Saddle River, NJ: Merrill Prentice Hall.

Jones, D., & Pritchard, L. (2000, November-December). The distance education debate: The Australian view. *Change*, *32*(6), 32-33.

Jung, Carl (1923). *Psychological types* (H.G. Baynes, Trans.). New York: Harcourt, Brace & Co.

Kagan, S. (1994). *Cooperative learning*. San Juan: Capistrano.

Karakaya, F., Ainscough, T. L., & Chopoorian, J. (2001). The effects of class size and learning style on student performance in a multimedia-based marketing course. *Journal of Marketing Education*, *23*(2), 84-90.

Karpov, Y. V., &. Haywood, H. C. (1998, January) Two ways to elaborate Vygotsky's concept of mediation. Implications for instruction. *American Psychologis,* 27-36.

Katerattanakul, P., & Siau, K. (1999). Measuring information quality of Web-sites: development of an instrumen. In *Proceedings of International Conference on Information Systems,* Charlotte, (pp. 279-285).

Katz, E. (1950). Mass communication research and the study of popular culture: An editorial note on a possible future for this journal. *Studies in Public Communication*, *2*, 1-6.

Kear, K. (2004). Peer learning using asynchronous discussion systems in distance education. *Open Learning*, *19*(2), 151-164.

Kear, K., & Heap, N. (1999). Technology-supported group work in distance learning. *Active Learning, 10,* 21-26.

Kearsley, G. (2000). *Learning and teaching in cyberspace*. Retrieved September 5, 2000, from http://home.sprynet com~gkearsley/chapts.htm

Keegan, D. (1986). On defining distance education. In D. Sewart, D. Keegan, & B. Holm-berg (Eds.), *Distance education: International perspectives* (pp. 6-33). London: Routledge.

Keegan, D. (1995). *Distance education technology for the new millennium: Compressed video teaching* (ERIC Document Reproduction Service No. ED 389 931).

Keegan, D. (1996). *The foundations of distance education*. London: Croom Helm.

Kemmis, S. (1985). Action research and the politics of reflection. In D. Boude et al. (Eds.), *Reflection turning experience into learning*. London: Kogan Page.

Kerlinger, F. N., & Lee, H. B. (2000). *Foundations of behavioral research*. New York: Harcourt Brace College Publishers.

Kern, R. (1995). Restructuring classroom interaction with networked computers: Effects on quantity and characteristics of language production. *The Modern Language Journal, 79*, 457- 476.

Ketterer, J. J., & George, E. M., II. (2006, Spring). Reconceptualizing intimacy and distance in instructional models. *Online Journal of Distance Learning Administration, 9*(1).

Kiesler, S., Siegel, J., & McGuire, T. (1984). Social psychological aspects of computer-mediated communication. *American Psychologist, 39*(10), 1123-1134.

Kirpatrick, D. L. (Ed.). (1967). *Evaluation of training: Training and development handbook*. New York: McGraw-Hill.

Kirschner, A. (2005). Alma mater in the time of TiVo. *Chronicle of Higher Education*, *52*. Retrieved March, 2007, from http://chronicle.com/weekly/v52/i16/16b00601.htm

Klapper, J. T. (1963). Mass communication research: An old road resurveyed. *Public Opinion Quarterly*, *27*, 515-527.

Kneser, C., Pilkington, R., & Treasure-Jones, T. (2001). The tutor's role: An investigation of the power of exchange structure analysis to identify different roles in CMC seminars. *International Journal of Artificial Intelligence in Education, 12*, 63-84.

Knowles, M. S. (1970). *The modern practice of adult education: Andragogy versus pedagogy.* New York: Association Press.

Ko, H., Jung, J., Kim, J., & Shim, S. (2004). Cross-cultural differences in perceived risk of online shopping. *Journal of Interactive Advertising*, *4*(2). Retrieved October 2, 2004, from http://jiad.org

Kokkos, A. (2007). Special characteristics and aims of adult education. *Adult Education*, **9**, 4-19.

Kokkos, A., & Lionarakis, A. (1998). *Open and distance learning, relations teachers and students* (Vol. B). Hellenic Open University Press.

Kortti, H. (1999). *On some similarities between discourse in the IRC and the conventions of spoken English.* Retrieved 9 November, 2004, from http://www.student.oulu.fi/~hkortti/proseminar-final.html

Kostoula, N., & Makrakis (2006). *Intercultural and education.* Athens.

Kuhl, D. (2002). *Investigating online learning communities.* U.S. Department of Education Office of Educational Research and Improvement (OERI).

Langa, M. & Yost, J. (2007). *Curriculum mapping for differentiated instruction.* Corwin Press: Sage Publications. Thousand Oaks, CA.

Lapadat, J. (2002). Written interaction: A key component in online learning. *Journal of Computer-Mediated Communication, 7*(4).

Lavooy, M. J., & Newlin, M. H. (2003). Computer mediated communication: Online instruction and interactivity. *Journal of Interactive Learning Research, 14*(2), 157-165.

Lawyer-Brook, D. (1991). *The use of distance education to deliver, chapter I services.* Denver: Region 5 R-TAC.

Lederer, A. L., Maupin, D. J., Maupin, M. P., Sena, M.P. & Zhuang, Y. (2000). The technology acceptance model and the World Wide Web. *Decision Support Systems, 29*, 269-282.

Lee, K., & Tan, S. (2003). E-retailing versus physical retailing. A theoretical model and empirical test of consumer choice. *Journal of Business Research, 56*, 877-885.

Lee, Y. (2006). An empirical investigation into factors influencing the adoption of an e-learning system. *Online Information Review, 30*(5), 517-541.

Lehman, J. A., & Naumann, J. D. (1986). A language independent course in program design and programming for MIS students. *ACM SIGCSE Bulletin, 18*(4), 32-37.

Leidner, D. E., & Jarvenpaa, S. L. (1995). The use of information technology to enhance management school education: A theoretical view. *MIS Quarterly, 19*, 265-291.

Leinonen, P., Järvelä, S., & Lipponen, L. (2003). Individual students' interpretations of their contribution to the computer-mediated discussions. *Journal of Interactive Learning Research, 14*(1), 99-122.

Lemak, D. L., Montgomery, J. C., & Reed, R. (2003). Instructor effectiveness in distance education: The case of technology and transactional distance. In *Proceedings of the 2003 Academy of Management Conference.*

Lemak, D. L., Shin, S. J., Reed, R., & Montgomery, J. C. (2005). Technology, transactional distance and instructor effectiveness: An empirical investigation. *Academy of Management Learning & Education, 4*(2), 150.

Levene, H. (1960). In I. Olkin et al. (Eds.) *Contributions to probability and statistics: Essays in honor of Harold Hotelling.* (pp. 278-292). Stanford University Press.

Levin, D. S., & Ben-Jacob, M. G. (1998, November). *Using collaboration in support of distance learning.* Paper presented at the WebNet 98 World Conference of the WWW, Internet and Intranet Proceedings, Orlando, FL (ERIC Document Reproduction Service No. ED 427 716).

Levy, M. R., & Windahl, S. (1984). Audience activity and gratifications: A conceptual clarification and exploration. *Communication Research, 1*, 51-78.

Lewin, K. (1951). *Field theory in social science: Selected theoretical papers.* New York: Harper and Row.

Lewis, L., Snow, K., Farris, E., Levin, D., & Greene, B. (1999). *Distance education at postsecondary education institutions: 1997-98.* National Center for Education Statistics. Retrieved from http://nces.ed.gov/pubs2000/2000013.pdf

Lidtke, D. K., Stokes, G. E., Haines, J., & Mulder, M. C. (1999). *ISCC '99: An information systems-centric curriculum '99 program guides for educating the next generation of information systems specialists, in collaboration with industry. Supported by the National Science Foundations grants.*

Light, P., Nesbitt, E., Light, V., & White, S. (2000). Variety is the spice of life: Student use of CMC in the context of campus based study. *Computers & Education, 34*, 257-267.

Lim, H. L. (2006). *Constructing learning conversations: A study of the discourse and learner experiences of online synchronous discussions.* Unpublished doctoral thesis, Murdoch University, Perth, Australia.

Lin, N. (1977). Communication effects: Review and commentary. In B. Ruben (Ed.), *Communication yearbook 1*. New Brunswick, NJ: Transaction Books.

Lincoln, D. J. (2001). Marketing educator Internet adoption in 1998 versus 2000: Significant progress and remaining obstacles. *Journal of Marketing Education, 23*(2), 103-116.

Lindh, J., & Soames, C. (2004). A dual perspective on an online university course. *Electronic Journal of E-learning, 2*(1), 129-134. Retrieved December 9, 2004, from http://www.ejel.org/volume-2/vol2-issue1/issue1-art16-lindh-soames.pdf

Lionarakis, A. (1998). Polymorfic education. A pedagogical framework for open and distance learning in universities in a digital era. *Transformation, Innovation and Tradition – Roles and Perespectives of Open and Distance Learning, 2*, 499-504. Italy: University of Bologna.

Lionarakis, A. (2001). Qualitative approaches of the design and production ofdistance, polymorphic educational material. In Makrakis (ed), *Proceedings of the Pan-Hellenic Conference with International Participation for New Technologies in Education and in Distance Education* (Vol. B). Atrapos Editions.

Lionarakis, A. (2003, June 15-18). A *preliminary framework for a theory of open and distance learning: The evolution of its complexity.* Paper presented at the 12th Conference of the European Distance Education Network, Rhodes.

Lionarakis, A. (2006). *Open and distance learning, theory and practice*. Athens: Probobos.

Loh, L. L., & Heng-Yu, K. (2003). Development of a Web-based template for active learning. *Quarterly Review of Distance Education, 4*(3), 213-227.

Louisiana Department of Education. (2003) *Small school districts and economies of scale.* Retrieved February 13, 2008, from www.doe.state.la.us/LDE/uploads/3475.pdf

Macdonald, J. (2003). Assessing online collaborative learning: Process and product. *Computers and Education, 40*, 377-391.

Macknight, C. (2000). Teaching critical thinking through online discussions. *Educative Quarterly, 4*, 38-41.

Maier, P., & Warren, A. (2000). *Integrating technology in learning and teaching*. London: Kogan Page.

Maki, R. H., Maki, W. S., Patterson, M., & Whittaker, P. D. (2000). Evaluation of a Web-based introductory psychology course: I. Learning and satisfaction in online versus lecture courses. *Behavior Research Methods, Instruments, and Computers, 32*(2), 230-239.

Manninen, T. (2001). Rich interactions in the context of networked virtual environments: Experiences gained from the multi-player games domain. In A. Blandford, J. Vandersoickt, & P. Gray (Eds.), *Joint Proceedings of HCI 2001 and IHM 2002 Conference* (pp. 383-398). Springer-Verlag.

Marra, R. (2006). A review of research methods for assessing content of computer-mediated discussion forums. *Journal of Interactive Learning Research, 17*(3), 243-267.

Marshall, G. (2000). Models, metaphors and measures: Issues in distance learning. *Educational Media International, 37*(1), 2-8.

Martin, M. (2000). *Videoconferencing in teaching and learning: Case studies and guidelines.* Western Education and Library Board.

Marzano, R. (1992). A different kind of classroom: Teaching with dimensions of learning. Alexandria, VA: Association for Supervision and Curriculum Development.

Marzano, R., Pickering, D., & Pollock, J. (2001). *Classroom instruction that works: Research-based strategies for increasing student achievement.* Alexandria, VA: Association for Supervision and Curriculum Development.

Mason, R., & Bacsich, P. (1998). Embedding computer conferencing into university teaching. *Computers in Education, 30*(3/4), 249-258.

Mason, R., & Rennie, F. (2006). *E-learning the key concepts.* New York: Routledge.

Mason, R., & Weller, M. (2000). Factors affecting students' satisfaction on a Web course. *Australian Journal of Educational Technology, 16*(2), 173-200.

Massialas, B. (1989). *Claims and counterclaims of the educational uses of microcomputers.* Paper presented at the Third International Pedagogical Conference, Athens, Greece.

Massy, J., Harrison, T., et al. (2002.). *The European e-learning market.* London: BizMedia.

Mattila, M., Karjaluoto, H., & Pento, T. (2003). Internet banking adoption among mature customers: Early majority or laggards? *Journal of Service Marketing, 17*(5), 514-528.

McGehee, W., & Tullar, W. (1978). A note on evaluating behavior modification and behavior modeling as industrial training techniques. *Personal Psychology, 31*, 477-484.

McKinney, V., Yoon, K., & Zahedi, F. (2002). The measurement of Web-customer satisfaction: an expectation and disconfirmation approach. *Information Systems Research, 13*(3), 296-315.

McLoughlin, C., & Luca, J. (1999). *Lonely outpourings or reasoned dialogue? An analysis of text-based conferencing as a tool to support learning.* Paper presented at the ASCILITE 99, Brisbane, Australia.

McManus, T. (2000). Individualizing instruction in a Web-based hypermedia learning environment: nonlinearity, advance organizers and self-regulated learners. *Journal of Interactive Learning Research, 11*(3), 219-251.

McNeese, M., Salas, E., & Endsley, M. (2001). *New trends in cooperative activities: Understanding system dynamic in complex environments.* San Diego: Human factors and Ergonomics Society Press.

McTighe, J. & Wiggins, G. (2004). *Understanding by design professional development workbook.* Alexandria, VA: Association for Supervision and Curriculum Development.

Mena, M. (2007). E-learning quality: A look towards the demands of its good practices *Journal of Cases on Information Technology, 9*(2), 1-11.

Mennecke, B. E., Crossland, M. D., & Killingsworth, B. L. (2000). Is a map more than a picture? The role of SDSS technology, subject characteristics, and problem complexity on map reading and problem solving. *MIS Quarterly, 24*(4), 601-627.

Mercer, D. (2003). *Using synchronous communication for online social constructivist learning.* Paper presented at the 2003 CADE-ACED Conference, St Johns, Newfoundland.

Meyer, K. (2003). Face-to-face versus threaded discussions: The role of time and higher-order thinking. *Journal of Asynchronous Learning Networks, 7*(3), 55-65.

Meyer, K. (2004). Evaluating online discussions: Four different frames of analysis. *Journal of Asynchronous Learning Networks, 8*(2), 101-114.

Meyer, R. M. (2003). *Explorations in computer science: A guide to discovery.* Boston: Jones and Bartlett Publishers.

Mezirow, J. (1980). How critical reflection triggers transformative learning. In J. Mezirow (Ed.), *Fostering critical reflection in adulthood: A guide to transformative and emancipatory learning.* San Francisco: Jossey-Bass.

Mezirow, J. (1981). A critical theory of adult learning and education. *Adult Education, 32*(1), 3-24.

Mezirow, J. (1985). A critical theory of self-directed learning. In self-directed learning: From theory to practice. *New Directions for Continuing Education*, (25). San Francisco: Jossey-Bass.

Mezirow, J. (1991). *Transformative dimensions of adult learning.* San Francisco: Jossey-Bass.

Mikropoulos, T. A. (2000). Design, development and evaluation of advanced learning environments. An overall approach, HERMES. *Advanced systems for teaching and learning over the world wide Web* (pp. B42-B52), Samos.

Miles, I. (1994). *Knowledge intensive business services.* Manchester: PREST University of Manchester.

Miller, J. W., Martineau, L. P., & Clark, R. C. (2000). Technology infusion and higher education: Changing teaching and learning. *Innovative Higher Education, 24*, 227-241.

Millis, B.J., & Cottell Jr., P.G. (1997). *Cooperative learning for higher education faculty*. Oryx Press.

Mitchell, J. et al. (1993). *Video-conferencing in higher education in Australia*. Canberra: AGPS.

Mohammed, R., Fisher, R. J., Jaworski, B. J., & Cahill, A. (2002). *Internet marketing* (1st ed.). New York: McGraw-Hill.

Molstad, L. (2001). Teaching computer programming using distance education technology. *Journal of Computing Sciences in Colleges, 17*(1), 265-277.

Moon, J., & Kim, Y. (2001). Extending the TAM for a World Wide Web context. *Information & Management, 38*, 217-230.

Moore, M. G. (1989). Editorial: Three types of interaction. *American Journal of Distance Education, 3*(2), 1-16.

Moore, M. G., & Kearsley, G. (1996). *Distance education: A systems view*. Wadsworth Pub Co.

Moore, M. G., & Thompson M. M. (1997). *The effects of distance learning* (Rev. ed. ACSDE Research Monograph No. 15). University Park, PA: Pennsylvania State University, American Center for the Study of Distance Education.

Moreland, R. L., & Levine, M. J. (1992). Problem identification in groups. *Group processes and productivity* (pp. 17-47).

Morris, D., Shaw, B., & Perney, J. (1990). Helping low readers in grades 2 and 3: An after-school volunteer tutoring program. *The Elementary School Journal, 91*, 133-150.

Mouza, C., Kaplan, D., & Espinet, I. (2000). *A Web-based model for online collaboration between distance learning and campus students* (IR020521). Office of Educational Research and Improvement, U.S. Department of Education.

Muilenberg, L., & Berge, Z. (2002). *A framework for designing question for online learning*. Retrieved August 22, 2002, from http://www.ssn.flinders.edu.au/innovations/ teacherresources.html

Muniz, A. M., Jr., Billingsley, W., & Brill, T. (2002). The GoReader launch: Developing marketing strategy for an innovative education technology. *Journal of Interactive Marketing, 16*(1), 67-88.

Mupinga, D. M. (2005, January-February). Distance education in high schools. *The Clearing House, 78*(3), 105-108.

Murphy, K., & Collins, M. (1997). Communication conventions in instructional electronic chats. *First Monday, 2*(11).

Muzio, J. A., Heins, T., et al. (2002). Experiences with reusable objects: From theory to practice. *The Internet and Higher Education, 5*, 21-34.

Myers, I. (1985). *Manual: The Myer-Briggs Type Indicator*. Palo Alto, Ca: Consulting Psychologist Press.

Nachmias, R. & Mioduser, D. (2000). Web-supported emergent-collaboration in higher education courses. *Educational Technology & Society, 3*(3), 94-104.

National Research Council (1999). *How people learn: Brain, mind, experience, and school*. Washington, DC: National Academy Press.

Negash, S., & Wilcox, M. V. (2007). Synchronous hybrid e-learning: Teaching complex information systems classes online. In *Proceedings of the 18th Annual International Information Resources Management Association Conference*, Vancouver, British Columbia, Canada.

Nelson, T. (1990). The right way to think about software design. In B. Laurel (Ed.), *The art of human computer interface design* (pp. 235-243). Reading, MA: Addison-Wesley.

Nemire, R. (2007). Intellectual property development and use for distance education courses: A review of law, organizations, and resources for faculty. *College Teaching, 55*(1), 26-30.

Newcomer, J., & Stafford, T. F. (2001). Teaching simultaneously in dual classrooms: Tips for when you're asked to use videoconferencing as an instructional tool. In *Proceedings of the Southwest Academy of Management*, SWFAD, Houston.

Newman, D., Johnson, C., Webb, B., & Cochrane, C. (1997). Evaluating the quality of learning in computer supported co-operative learning. *Journal of the American Society for Information Science, 48*(6), 484-495.

Ngwenya, J., Annand, D., & Wang, E. (2004). Supporting asynchronous discussions among online learners. In T. Anderson & F. Elloumi (Eds.), *Theory and practice of online learning* (pp. 319-347). Canada: Athabasca University.

Nicholson, J., & Sarker, S. (2002). Unearthing hidden assumptions regarding on-line education: The use of myths and metaphors. In *Proceedings of the International Academy for Information Management (IAIM) Annual Conference: International Conference on Informatics Education Research (ICIER)* (pp. 298-306). Barcelona, Spain (ERIC Document Reproduction Service No. ED481748).

Norman, D. (1993). *Things that make us smart.* Reading, MA: Addison-Wesley.

Norris & Pippa. (2001). *Digital divide: Civic engagement, information poverty and the Internet worldwide.* Cambridge: University Press.

Northey, S. (2005). *Handbook on differentiated instruction for middle and high schools.* Larchmont, NY: Eye on Education.

Norton, B. (2001). Non-participation, imagined communities and the language classroom. In P. Breen (Ed.), *Learner contributions to language learning: New directions in research* (pp. 159-171). Essex: Pearson Education Limited.

Nosek, J. T. (1998, March). The case for collaborative programming. *Communications of the ACM, 41*(3), 105-108.

NSA. (2007). *Centers of academic excellence.* Retrieved June 21, 2007, from http://www.nsa.gov/ia/academia/caemap.cfm?MenuID=10.1.1.2

Nunnaly, J. (1978). *Psychometric theory.* New York: McGraw Hill.

O'Malley, C. (1995). *Computer supported collaborative learning.*, Berlin: Springer Verlag.

O'Malley, C. E. (Ed.). (1995). *Computer supported collaborative learning.* Springer-Verlag.

On-Line game services. (2006). Retrieved February 13, 2008, from http://www-306.ibm.com/software/success/cssdb.nsf/CS/MCAG-6WNQRG?OpenDocument&Site

Pagliarello, M. C. (2007). Reluctant teachers and traditional learning methods: The multimedia teaching as cross-roads between the old and the new. In L. Gómez, D. Martí, & I. Candel (Eds.), *International Association of Technology, Education and Development Conference Proceedings.* Valencia: IATED

Palloff, R., & Pratt, K. (2003). *The virtual student: A profile and guide to working with online learners.* San Francisco: Jossey-Bass.

Paris, S. G., Cross, D. R., & Lipson, M. Y. (1984). Informed strategies for learning: A program to improve children's reading awareness and comprehension. *Journal of Educational Psychology, 7*, 1239-1252.

Parnell, J. A., & Carraher, S. (2003). The management education by Internet readiness (MEBIR) scale: Developing a scale to assess personal readiness for Internet-mediated management education. *Journal of Management Education, 27*(4), 431-446.

Patton, M. Q. (2002). *Qualitative research and evaluation methods* (3rd ed.). Thousand Oaks, CA: Sage.

Paulsen, M. F. (1997). *Teaching methods and techniques for computer mediated communication.* Paper presented at the 18th ICDE World Conference. Pensylvania: State University.

Paulsen, M. F. (2003). *Online education and learning management systems. Global e-learning in a Scandinavian perspective.* Oslo: NKI Forlaget.

Payne, D. A. (2005). Succeeding in graduate school online: Tips from successful students. *College Student Journal, 39*(1), 117-28.

Pea, R. (1995). *Practices of distributed intelligence and designs for education, distributed cognitions: Psychological and educational considerations* (pp. 47-87). Cambridge, England: Cambridge University Press,.

Perspectives for competence development: Report on e-learning. (2003). Copenhagen: Danish Ministry of Science and Innovation.

Phillips, D. C. (1995). The good, the bad, and the ugly: The many faces of constructivism. *Educational Researcher, 24*(7), 5-12.

Picardo, O. (2002). Enseñar a aprender en la sociedad del conocimiento. *Revista Electrónica de Tecnología Educa-*

tiva, 15. Retrieved November 1, 2002, from http://www.uib.es/depart/gte/edutec-e/revelec15/oscarpicardo.htm

Piccoli, G., Ahmadand, R., & Ives, B. (2001). Web-based virtual learning environments: A research framework and a preliminary assessment of effectiveness in basic IT skills training. *MIS Quarterly, 25*(4), 401-426.

Pilkington, R., & Walker, S. (2004). Facilitating debate in networked learning: Reflecting on online synchronous discussion in higher education. In P. Goodyear, S. Banks, V. Hodgson, & D. McConnell (Eds.), *Advances in research on networked learning* (Vol. 4, pp. 67-90). MA: Kluwer Academic Publishers.

Pilkington, R., Bennett, C., & Vaughan, S. (2000). An evaluation of computer mediated communication to support group discussion in continuing education. *Educational Technology and Society, 3*(3), 349-359.

Pitcher, N., Davidson, K., & Goldfinch, J. (2000). Videoconferencing in higher education. *Innovations in Education and Training International, 37*, 199.

Pohjola, V.J. (1999). Knowledge integration as a challenge for future education, *Proc. International Conference on Computer in Education*, 2: 299~306.

Polson, C. J. (2003). Adult graduate students challenge institutions to change. *New Directions for Student Sevices, 102*, 59-68.

Rafii, F. (1995). How important is physical co-location to product development success. *Business Horizons, 38*(1) 78-84.

Ramsden, P. (Ed.). (1988). Context and strategy: Situational influences on learning. In *Learning strategies and learning styles*. New York: Plenum Press.

Ranganathan, S., Negash, S., & Wilcox, M. V. (2007). Hybrid learning: Balancing face-to-face and online class sessions. In *Proceedings of the Tenth Annual Conference of the. Southern Association for Information Systems*, Jacksonville, FL.

Raptis & Rapti (2004). *Teaching and learning towards the information age*. Athens.

Reed, J., & Woodruff, M. (1995). An introduction to using videoconferencing technology for teaching. The Distance Educator Newsletter, Fall, Scott, T D and Pitcher, N (1997) SUMSMAN: Scottish, from: http://www.kn.pacbell.com/wired/vidconf/Using.html

Reeves, T. C. (2000). Socially responsible educational technology research. *Educational Technology, 40*(6), 19-28.

Rehg, J. A. (1999). Developing Web-based courses using an online development guide and templates. *Computers in Education Journal, 9*, 51-55.

Reich College of Education – Appalachian State University, Boone, NC. (2005). *Conceptual framework*. Retrieved November 6, 2006, from http://ced.appstate.edu/about/conceptualframework.aspx

Reiser, R. (2001). A history of instructional design and technology. Part 2: A history of instructional design. *Educational Technology Research & Development, 49*, 57-67.

Rheingold, H. (2000). *The virtual community: Homesteading on the electronic frontier* (2nd ed.). Cambridge, MA: MIT Press.

Richter, M. (1999). Overseas Internet users protest high cost of access. *The New York Times*, p. G3.

Riedl, R., Bronack, S., & Tashner, J. (2005, January). *3D Web-based worlds for instruction*. Phoenix: The Society for Information and Teacher Education.

Ritzhaupt, A. D., & Zucker, R. (2006). Teaching object-oriented concepts using visual basic .NET. *Journal of Information Systems Education, 17*(2), 163-170.

Roberts, L. H. (1995). *A template for converting classroom courses to distributed, asynchronous courses*. ITS, Center for Instructional Technology, The University of North Carolina at Chapel Hill. Retrieved February 21, 2008, from http://www.unc.edu/cit/iat-archive/publications/roberts/template.html

Robinson, A. (1997). Adult learning by videoconferencing. In J. Field (Ed.), *Electronic pathways. Adult learning and the new communication technologies*. Leicester: National Organisation for Adult Learning (NIACE).

Rogers, A. (2003). *What is the difference? A new critique of adult learning and teaching*. Leicester: NIACE.

Rogers, C. R. (1969). *Freedom to learn*. Columbus, OH: Merrill.

Rogoff, B. (1990). *Apprenticeship in thinking.* New York: Oxford University Press.

Rohfeld, R. W., & Hiemstra, R. (1995). Moderating discussions in the electronic classroom. In Z. Berge & M. Collins (Eds.), *Computer mediated communication and the online classroom (Vol. 3): Distance learning* (pp. 91-104). Cresskill, NJ: Hampton Press.

Rohrer, T. (1995). *Metaphors we compute by: Bringing magic into interface design.* Retrieved August 4, 2004, from University of Oregon, Department of Philosophy Web site: http://philosophy.uoregon.edu/metaphor/gui-4web.htm

Rosbottom, J. (2001). Hybrid learning: A safe route into Web-based open and distance learning for the Computer Science teacher. *ACM Association for Computing Machinery Sigcse Bulletin, 33*(3), 89-92.

Rosendahl-Kreitman, K. (1990). The challenge of interface design: Creating quality experience for the user. *Multimedia interface design.* Santa Clara, CA: Multimedia Computing Corporation.

Rovai, A. (2001). Building classroom community at a distance: A case study. *Educational Technology Research and Development, 49,* 1042-1629.

Roy, R., Potter, S., Yarrow, K., & Smith, M. (2005, March). *Final report on towards sustainable higher education: Environmental impacts of campus-based and distance higher education systems.* The Open University, Design Innovation Group.

Rumble, G. (1997). *The costs and economics of open and distance learning.* London: Kogan Page.

Rumble, G. (2001). The costs and costing of networked learning. *Journal of Asynchronous Networks.* Sloan Publications. Retrieved February 13, 2008, from www.sloan-c.org/publications/jaln

Russel, T. (2003). *The no significant difference phenomenon.* TeleEducation New Brunswick. Retrieved from http://teleeducation.nb.ca/nosignificant difference/

Russel, T. (2003). *The significant difference phenomenon.* TeleEducation New Brunswick. Retrieved from http://teleeducation.nb.ca/significant difference/

Ryan, S., Scott, B., Freeman, H., & Patel, D. (2000). *The virtual university: The Internet and resource-based learning.* London: Kogan Page.

Sacks, H., Schegloff, E., & Jefferson, G. (1974). A simplest systematics for the organization of turn-taking for conversation. *Language, 50*(4), 696-735.

Salas, E., & Fiore, M. S. (2004). Why team cognition: An overview. *Team cognition: Understanding the factors that drive process and performance* (pp. 3-8). Washington, D.C.: American Psychological Association.

Salomon, G. (1992). New challenges for educational research: Studying the individual within learning environments. *Scandinavian Journal of Educational Research, 36*(3), 167-182.

Salomon, G. (1992). What does the design of effective CSCL require and how do we study its effects? *SIGCUE Outlook, Special Issue on CSCL, 21*(3), 62-68.

Salomon, G. (1995). *Distributed cognitions: Psychological and educational considerations.* In B. B. Seels & R. C. Richey (Eds.), *Instructional technology: The definitions and domains of the filed.* Washington D.C.: Association for Educational Communications and Technology.

Salomon, G., & Perkins, D. N. (1988). Teaching for transfer. *Educational Leadership, 46*(1), 22-35.

Salter, G. (2000). Making use of online discussion groups. *Journal of the Australian Council for Computers in Education, 15*(2), 5-10.

Sanders, R. L., & Angel, R. B. (2005, August). *Shared decision-making: Case study analysis to promote cross-program dialogue between administrators and media coordinators.* Paper presented at the International Conference on Computers and Advanced Technology in Education, Oranjestad, Aruba.

Sanders, R. L., & McKeown, L. (2007, January). *Promoting reflection through action learning in a 3D virtual world.* Paper presented at the Association of Library and Information Science Educators Annual Conference, Seattle, WA.

Sanders, R. L., Bronack, S., Cheney, A., Tashner, J., Reidl, R., & Gilman, R. (2007, February). *Education in the zone: Dynamic learning communities in a 3D virtual world.*

Paper presented at the IADIS International Conference of Web Based Communities 2007, Salamanca, Spain.

Santhanam, R., & Sein, M. K. (1994). Improving end-user proficiency: Effects of conceptual training and nature of interaction. *Information Systems Research, 5*(4), 378-399.

Sauers, D., & Walker, R. C. (2004). A comparison of traditional and technology-assisted instructional methods in the business communication classroom. *Business Communication Quarterly, 67*(4), 430-442.

Schiffman, L., & Kanuk, L. (2003). *Consumer behaviour* (8th ed.). NJ: Prentice Hall.

Schiller, J., & Mitchell, J. (1993). Interacting at a distance: Staff and student perceptions of teaching and learning via video conferencing. *Australian Journal of Educational Technology, 9*(1), 41-58.

Schniederjans, M.J., & Kim, E.B. (2005). Relationship of student undergraduate achievement and personality characteristics in a total Web-based environment: An empirical study. *Decision Sciences Journal of Innovative Education, 3*(2), 205-221.

Schou, C. (2006). Curriculum. *National Information Assurance Training and Education Center*. Retrieved June 21, 2007, from http://niatec.info/(S(nfkpbb55vg5kpp45mlebbe55))/index.aspx?page=102

Schroeder, R. (2002). Social interaction in virtual environments: Key issues, common themes, and a framework for research. In R. Schroeder (Ed.), *The social life of avatars: Presence and interaction in shared virtual environments*. London: Springer.

Schroeder, R. (2006). Being there together and the future of connected presence. *Presence: Teleoperatores and Virtual Environments, 15*(4), 438-454.

Schroeder, R. (Ed). (2002). *The social life of avatars: Presence and interaction in shared virtual environments* (pp. 1-18). Great Britain: Springer-Verlag/London Limited.

Schroeder, R., Steed, A., Axelsson, A., Heldal, I., Abelin, A., Widestrom, J., et al. (2001). Collaborating in networked immersive spaces: As good as being there together? *Computers and Graphics, 25*, 781-788.

Schwier, R., & Balbar, S. (2002, Spring). The interplay of content and community in synchronous and asynchronous communication: Virtual communication in a graduate seminar. *Canadian Journal of Learning and Technology, 28*(2).

Scnurr, C., & Smith, C. (1995). *Video conferencing in education: Meeting teachers and learners support and training needs a report to the advisory group on computer graphics* (SIMA Report Series ISSN 1356-5370). Retrieved February 21, 2008, from http://www.agocg.ac.uk/mmedia.htm

Seepanski, J. M., & von Wahlde, B. (1998). Megasystem collaboration: Cross-continent consortial cooperation. *Information Technology and Libraries, 17*(1), 30-35.

Semper, R. (1990). Hypercard and education: Reflections on the hyperboom. In S. Ambron & K. Hooper (Eds.), *Learning with interactive multimedia: Developing and using multimedia tools in education* (pp. 52-67). Redmond, WA: Microsoft Corporation.

Seng, L. C., & Al-Hawamdeh, S. (2001). New mode of course delivery for virtual classroom. *Aslib Proceedings, 53*(6), 238-242.

Shale, D. (1988). Toward a reconceptualization of distance education. *The American Journal of Distance Education, 2*(3), 25-35.

Shepherd, M., & Martz, W. (2006). Media richness theory and the distance education environment. *The Journal of Computer Information Systems, 47*(1), 114-122.

Sheremetov, L., & Arenas, A. G. (2002). *EVA: An interactive Web based collaborative learning environment Computers & Education, 39*, 161-182.

Shih, T., & Hung, J. C. (Ed.). (2007). *Future directions in distance learning and communication technologies*. Hershey, PA: Information Science Publishing.

Sikora, A., & Carrol, D. (2002). *A profile of participation in distance education: 1999-2000*. National Center for Education Statistics. Retrieved from http://nces.ed.gov/pubs2003/2003154.pdf

Silver, H., Hanson, J., Strong, R., & Schwartz, P. (2003) *Teaching styles and strategies*. Ho-Ho-Kus, NJ: Thoughtful Education Press.

Silver, H.F., Strong, R.W., & Perini, M.J. (2000). *So each may learn: Integrating learning styles and multiple intelligences*. Alexandria, VA: Association for Supervision and Curriculum Development.

Simon, S. J., & Werner, J. M. (1996). Computer training through behavior modeling, self-paced, and instructional approaches: A field experiment. *Journal of Applied Psychology, 81*(6), 648-659.

Simon, S. J., Grover, V., Teng, J. T. C., & Whitcomb, K. (1996). The relationship of information system training methods and cognitive ability to end-user satisfaction, comprehension, and skill transfer: A longitudinal field study. *Information Systems Research, 7*(4), 466-490.

Simon, S. J., Grover, V., Teng, J. T., & Whitcomb, K. (1996). The relationship of information systems training methods and cognitive ability to end-user satisfaction, comprehension, and skill transfer: A longitudinal field study. *Information Systems Research, 7*(4), 466-490.

Simonson, M., Smaldino, S., Albright, M., & Zvacek, S. (2000). *Teaching and learning at a distance: Foundations of distance learning*. Upper Saddle River, New Jersey: Merrill.

Simonson, M. (2002). *Teaching and learning at a distance: Foundations of distance education*. Prentice Hall.

Sims, D. (2006). *TMC net, teleconferencing and distance learning: An inside look*. Retrieved February 13, 2008, from http://www. tmcnet.com/channels

Singh, P. (2004). Online education: Lessons for administrators and instructors. *College Student Journal, 38*(2), 302-308.

Skinner, B. F. (1938). *The behavior of organisms: An experimental analysis*. New York: Appleton-Century Company, Incorporated.

Smith, P. L., & Ragan, T. J. (1993). *Instructional design*. Upper Saddle River, NJ: Prentice-Hall.

Smyth, R. (2005). Exploring the usefulness of broadband videoconferencing for student-centred distance learning in tertiary science. In C. McLoughlin & A. Taji (Eds.), *Student centred teaching in science*. New York: Howarth.

Solomon (1987). Social influences on construction of pupil's understanding of science. *Studies in Science Education, 14*, 63-82.

Sonwalkar, N. (2003). Logging in with Nishikant Sonwalkar: Online education must capitalize on students' unique approaches to learning. *Chronicle of Higher Education – Distance Education*.

Sousa, D. (2003). How the gifted brain learns. Thousand Oaks, CA: Corwin Press Incorporated.

Spellings, M. (2006). *Secretary of education while addressing the National Press Club on the commission on higher education in October 2006*. Retrieved February 13, 2008, from http://www.ed.gov/news/speeches/2006

Spencer, D., & Hiltz, S. (2003). *A field study of use of synchronous chat in online courses*. Paper presented at the 36th Annual Hawaii International Conference in System Sciences (HICSS 03), Big Island, Hawaii.

Stafford, T. F. (2000). Internet as metamedium: Emerging uses of the World Wide Web – a tutorial. In H.M. Chung (Ed.), *Proceedings of the Association for Information Systems Conference, AIS*, Long Beach, CA.

Stafford, T. F. (2005). Understanding motivations for Internet use in distance education. *IEEE Transactions on Education, 48*(2), 301-306.

Stafford, T. F., & Simon, J. C. (2002). High-tech adjuncts: Using technology-mediated virtual partnerships to facilitate the delivery of information systems course content. In *Proceedings of the 2002 Decision Sciences Institute Conference, DSI*, Atlanta.

Stafford, T. F., Stafford, M. R., & Schkade, L. L. (2004). Determining uses and gratifications for the Internet. *Decision Sciences, 35*(2), 259-288.

Stake, R. (1995). *The art of case study research*. Thousand Oaks, CA: Sage Publications.

Stansfield, M., McLellan, E., & Connolly, T. (2004). Enhancing student performance in online learning and traditional face-to-face class delivery. *Journal of Information Technology Education, 3*.

Stellar. (2003). *Course management system*. Retrieved October 25, 2005, from http://stellar.mit.edu/

Stiefelhagen, R., Chen, X., & Yang, J. (2005). Capturing interactions in meetings with omnidirectional cameras. *International Journal of Distance Education Technologies, 3*(3), 34-47.

Stigler, G. (1961). The economics of information. *The Journal of Political Economy, 49*(3), 213-225.

Stilborne, L., & Lindy, W. (1996). *Meeting the needs of adult learners in developing courses for the Internet.* Retrieved November 2007, from http://www.isoc.org/isoc/whatis/conferences/inet/96/proceedings/c4/c4_2.htm

Stokes, P. (2006). *Consultant for the Illinois State on-line education project quoted in Inside Higher Education – Daily Update.* Retrieved February 13, 2008, from *http://newsroom@insidehighered.com*

Stöttinger, B., & Schlegelmilch, B. B. (2002). Information and communication technologies in tertiary education: A 'customer' perspective. *Marketing Education Review, 12,* 63-72.

Straub, D. W. (1989). Validating instruments in MIS research. *MIS Quarterly, 13*(2), 147-169.

Straub, D. W. (1994). The effect of culture on IT diffusion: E-mail and FAX in Japan and the U.S. *Information Systems Research, 5*(1), 23-47.

Straub, D., Keil, M., & Brenner, W. (1997). Testing the technology acceptance model across cultures: A three country study. *Information and Management, 33,* 1-11.

Strijbos, J. W., Martens, R. L., & Jochems, W. M. G. (2004). Designing for interaction: Six steps to designing computer-supported group-based learning. *Computers & Education, 42*(4), 403-424. Elsevier.

Sudweeks, F. (2003). *Promoting cooperation and collaboration in a Web-based learning environment.* Paper presented at the 2003 Informing Science and Information Technology Education Conference, Informing Science Institute, Santa Rosa, CA.

Sudweeks, F. (2003). Connecting students with group work. In C. Constantinou & Z. Zacharia (Eds.), *Computer-based learning in science* (Vol. 1, pp. 173-183). Nicosia, Cyprus: University of Cyprus.

Sudweeks, F. (2003). *The reflective learner: A framework for reflective e-learning.* Paper presented at the ICIER03, Seattle, WA.

Sudweeks, F. (2004). *Development and leadership in computer-mediated collaborative groups.* Unpublished doctoral thesis, Murdoch University, Perth, Australia.

Sudweeks, F. (2005). *Unit outline.* Murdoch University, School of Information Technology.

Sudweeks, F., & Simoff, S. (2000). *Participation and reflection in virtual workshops.* Paper presented at the 3rd Western Australian Workshop on Information Systems Research, Perth, Australia.

Sudweeks, F., & Simoff, S. (2005). *Leading conversations: Communication behaviour of emergent leaders in virtual teams.* Paper presented at the 38th Hawaii International Conference on System Sciences (HICSS05), Hawaii.

Suganthi, B., Balachandher, K., & Balachandran, K. (2001). Internet banking patronage: An empirical investigation of Malaysia. *Journal of Banking and Commerce, 6*(1). Retrieved May 22, 2001, from http://www.arraydev.com/commerce/jibc/0103_01.htm

Sullivan, M., Jolly, D., Foster, D., & Tompkins, R. (1994). *Local heroes: Bringing te le communications to rural, small schools.* Austin, TX: Southwest Educational Development Laboratory.

Sumner, M., & Hostetler, D. (2002). A comparative study of computer conferencing and face-to-face communications in systems design. *Journal of Interactive Learning Research, 13*(3), 277-291.

Sundbo, J. (2006). Customer-based innovation of e-knowledge services. *E-services - knowledge services, entrepeurship, and the consequences for business customers and citizens* (pp. 1-19). Lyngby, Denmark: CICT Technical University of Denmark.

Sung, Y.T, Chiou, S.K & Chang, K.E. (2001) Use of hierarchical hyper-concept map in Web-based courses. *Proc. International Conference on Computer in Education,* 1133-1137.

Suthers, D. (2001, January 3-6). Collaborative representations: Supporting face to face and online knowledge-building discourse. In *Proceedings of the 34th Hawai`i International Conference on the System Sciences (HICSS-34)*, Maui, Hawaii [CD-ROM]. Institute of Electrical and Electronics Engineers.

Svastisinha, R. W. (1999). *Wahhn: Web-based design. Wind and human comfort for Thailand.* Unpublished doctoral dissertation, University of Southern California.

Swaminathan, V., Lepkowska-White, E., & Rao, B. (1999). *Browsers or buyers in cyberspace? An investigation of factors influencing electronic exchange.* Retrieved October 3, 2001, from http://www.ascusc.org/jcmc/vol5/issue2/swaminathan.html

Tallent-Runnels, M., Thomas, J., Lan, W., & Cooper, S. (2006). Teaching courses online: A review of the research. *Review of Educational Research, 76*(1), 93-135.

Tan, S. (1999). Strategies for reducing consumer's risk aversion in Internet shopping. *Journal of Consumer Marketing, 16*(2), 163-180.

Tapscott, D. (1995). *Digital economy. Promise and peril in the age of networked intelligence.* McGraw-Hill.

Tashner, J., Bronack, S., & Riedl, R., (2005, March). *Virtual worlds: Further development of Web-based teaching.* Paper presented at the Hawaii International Conference on Education, Honolulu.

Tesouro, M., & Puiggalí, J. (2004). Beneficios de la utilización del ordenador en el aprendizaje: un diseño experimental. *Revista Electrónica de Tecnología Educativa, 17.* Retrieved November 15, 2005, from http://www.uib.es/depart/gte/edutec-e/revelec17/tesouro_16a.htm

Thomas, B., Jones, P., Packham, G., & Miller, C. (2004, April 5-7). *Student perceptions of effective e-moderation: A qualitative investigation of E-college Wales.* Paper presented at the Networked Learning Conference 2004, Lancaster University, England.

Thorndike, R. L. (1949). *Personnel selection: Test and measurement techniques.* New York: John Wiley & Sons.

Tobin, K. (1990). Social constructivist perspectives on the reform of science education. *Australian Science Teachers Journal, 36*(4), 29-35.

Tomlinson, C. A. (1995). *How to differentiate instruction in mixed ability classrooms.* Alexandria, VA: ASCD.

Tomlinson, C. A. (1999). *The differentiated classroom: Responding to the needs of all learners.* Alexandria, VA: ASCD.

Tomlinson, C. A. (2000). *Leadership for differentiating schools and classrooms.* Alexandria, VA: ASCD.

Tomlinson, C. A., & McTighe, J. (2006). *Integrating differentiated instruction and understanding by design.* Alexandria, VA: ASCD.

Trevino, L. K., Daft, H. R., & Lengel, L. R. (1987). Media symbolism, media richness, and media choice in organizations. *Communication Research, 14*(5), 87-96.

Triki, A., & Ouerghi, O. (2007). The tunisian e-learning experience: A focus on the main actors perceptions. In L. Gómez, D. Martí, & I. Candel (Eds.), *International Association of Technology, Education and Development Conference Proceedings.* Valencia: IATED.

Twigg, C. (2002). Improving quality and reducing costs: Designs for effective learning using information technology. *The Observatory on Borderless Higher Education.* Retrieved March, 3, 2003, from http://www.obhe.ac.uk/products/reports/

U.S. Department of Education. (2005). Toward a new golden age in American education: How the Internet, the law and today's students are revolutionizing expectations. *National Education Technology Plan 2004.* Retrieved February 9, 2008, from www.ed.gov/about/offices/list/os/technology/plan/2004/index.html

Uiterwijk, J., Seoane, D., Mitchell, L., & Welch, J. (1998). The virtual classroom. *InfoWorld, 20*(47), 64-70.

UK Open University. (2007). Retrieved February 13, 2008, from http://www.open.ac.uk/about/ou/p7.shtml

UNESCO. (2007). *Description of mega university.* Retrieved February 13, 2008, from

United Nations. (2003). *Address by UN Secretary-General to the World Summit on the Information Society, Geneva, Switzerland.* Retrieved April 5, 2005, from www.itu.int/wsis/geneva/coverage/statements/opening/annan.html

Unruh, D. L. (2000). Desktop videoconferencing the promise and problems of delivery of Web-based training. *Internet and Higher Education, 3,* 183-99.

Useem, J. (2007). *Is scale an asset or liability? The case of Fortune 500 companies.* Retrieved February 13, 2008, from http://money.cnn.com/magazines/fortune/fortune_archive/2007/04/30/8405390/index.htm

Van Berlo, M. P. W. (2000). Empirical validation of team training ID-guidelines. In *Proceedings of the 44th Annual Meeting of the Human Factors and Ergonomics Society*, San Diego, (Vol. 2, pp. 394-397).

Vaughn, N. (2007) Perspectives on blended learning in higher education. *International Journal on E-Learning, 6*(1), 81-94.

Vertelney, L., Arent, M., & Lieberman, H. (1990). Two disciplines in search of an interface: Reflections on the design process. In B. Laurel (Ed.), *The art of human computer interface design* (pp. 45-55). Reading, MA: Addison-Wesley.

Vosniadou, S. & Kollias, V. (2001). Information and communications technology and the problem of teacher training: Myths, dreams and the harsh reality. *Themes in Education, 2*(4), 341-365.

Vygotsky, L. (1962). *Thought and language*. Cambridge, MA: MIT Press.

Vygotsky, L. (1978). Interaction between learning and development. In E. Souberman (Ed.), *Mind in society.* Cambridge, MA: Harvard University Press.

Vygotsky, L. S. (1978). *Mind in society: The development of higher psychological processes.* Cambridge, MA: Harvard University Press.

Wagner, L. (2001). Virtual advising: Delivering student services. *Online Journal of Distance Learning Administration, 4*(3). Retrieved February 14, 2008, from www.westga.edu/%7Edistance/ojdla/fall43/wagner43.html

Walls, C. M. (2006). *Some strategies for balancing economies of scale and interaction in online/distance education courses*. Oregon State University. Retrieved February 13, 2008, from drwalls@orst.edu

Wang, X., & Teles, L. (1998) Online collaboration and the role of the instructor in two university credit courses. In. T. W. Chan, A. Collins, & J. Lin (Eds.), *Global Education on the Net, Proceedings of the Sixth International Conference on Computers in Education* (Vol. 1, pp. 154-161). Beijing: China High Education Press

Warschauer, M. (1996). Comparing face-to-face and electronic discussion in the second language classroom. *CALICO Journal, 13*(2-3), 7-26.

Wedemeyer, C. (1981). *Learning at the backdoor.* Madison, WI: University of Wisconsin Press.

Wegner, E. (2001). *Supporting communities of practice. A research and consulting report.*

Wenger, E. (1998, June). Communities of practice: Learning as a social system. *Systems thinker.* Retrieved October 16, 2006, from http://www.co-i-l.com/coil/knowledge-garden/cop/lss.shtml

Werry, C. (1996). Linguistic and interactional features of Internet relay chat. In S. Herring (Ed.), *Computer-mediated communication* (pp. 47-64). Philadelphia: John Benjamins Publishing Company.

Wilcox, P., Petch, S., & Dexter, H. (2005). Towards an understanding of UKeU business processes within an e-learning lifecycle model. *Electronic Journal of E-learning, 3*(1), 77-86. Retrieved December 11, 2005, from http://www.ejel.org/volume-3/v3-i1/v3-i1-art8-wilcox.pdf

Wilkes, R. B., Simon, J. C., & Brooks, L. D. (2006). A comparison of faculty and undergraduate students' perceptions of online degree programs. *Journal of Information Systems Education, 17*(2), 131-140.

Williams, K. C. (2006). Active learning and quality in online courses. *NACTA Journal.* Retrieved February 14, 2008, from http://findarticles.com/p/articles/mi_qa4062/is_200612/ai_n17194101

Williams, L. A., & Kessler, R. R. (2001). Experiments with industry's "pair-programming" model in the computer science classroom. *Computer Science Education, 11*(1), 7-20.

Williams, M., Paprock, K., & Covington, B. (1999). *Distance learning: The essential guide*. London: SAGE Publications.

Williams, S., & Pury, C. (2002). Student attitudes toward participation in electronic discussions. *International Journal of Educational Technology, 3*(1), 1-15.

Wilson, B., & Ryder, M. (2006). *Dynamic learning communities: An alternative to designed instructional systems.* Retrieved October 6, 2006, from http://carbon.cudenver.edu/~mryder/dlc.html

Winograd, D. *A medium for collaborative learning.* Retrieved February 9, 2008, from www.Emoderators.com

Woods, R., & Keeler, S. (2001). The effect of instructor's use of audio e-mail messages on student participation in and perceptions of online learning: A preliminary case study. *Open Learning, 16*(3), 263-278.

Woszczynski, A., Guthrie, T., & Shade, S. (2005). Personality and programming. *Journal of Information Systems Education, 16(4)*, 293-299.

Wu, A. (2003). Supporting electronic discourse: Principles of design from a social constructivist perspective. *Journal of Interactive Learning Research, 14*(2), 167-184.

Yi, M. Y., & Davis, F. D. (2001). Improving computer training effectiveness for decision technologies: Behavior modeling and retention enhancement. *Decision Sciences, 32*(3), 521-544.

Yi, M. Y., & Hwang (2003). Predicting the use of Web-based information systems: Self-efficiency, enjoyment, learning goal orientation, and the technology acceptance model. *International Journal of Human–Computer Studies, 59*(4), 431-449.

Yilmaz, L., & Phillips, J. (2006). Organization-theoretic perspective for simulation modeling of agile software processes. In *Proceedings of the SPW/ProSim2006 Workshop,* Shanghai, China, (LNCS 3966, pp. 234-241). Springer-Verlag.

Yin, R. (1994). *Case study research: Design and methods* (Vol. 5). Thousand Oaks, CA: Sage Publications.

Young, M. R. (2001). Windowed, wired and webbed: Now what? *Journal of Marketing Education, 23*(1), 45-54.

Zajkowski, M. E. (1997). Price and persistence in distance education. *Open Learning, 12*(1), 12-23.

Zhang, Y J & Xu, Y. (1999). Effect investigation of the CAI software for "Image Processing and Analysis". *Proc. of International Conference on Computer in Education'99,* 858-859.

Zhang, Y.J & Liu, W.J. (2002). A new web course:'Fundamentals of Image Processing and Analysis'", *Proc. 6th Global Chinese Conference on Computer in Education,* 1: 597-602.

Zhao, C. M., & Kuh, G. D. (2004). Adding value: Learning communities and student engagement. *Research in Higher Education, 45*(2), 115-138.

Zhu, E., Payette, P., & DeZure, D. (2003). *An introduction to teaching online* (CRLT Occasional Papers No. 18). University of Michigan.

Zhu, X Q, Zhang, Y J & Liu, W J. (2001). IP&A-Web: an online course of image processing and analysis. *Proc. of International Conference on Computer in Education'01*, 729-734.

About the Contributors

Solomon Negash specializes in e-learning, business intelligence, and information and communications technology (ICT) for developing economies. He is the 2007 Distinguished Graduate Teaching Award recipient from his university and the 2005 recipient of the distinguished e-learning award from his department. His work is published in *Information & Management*, *Communication of the ACM*, *Psychology and Marketing*, *Communication of AIS*, *International Journal in ICT Education,* and at conference proceedings in the U.S., Canada, Spain, Ethiopia, Kenya, and Malaysia. Professor Negash is the program coordinator for the Bachelor of Science in information systems (BSIS) program at Kennesaw State University. With an engineering, management and information systems background, his over 20 years of industry experience include consulting, entrepreneurship, management, and systems analysis. His teaching area includes system analysis and design, project management, information systems policy, and information technology management.

Michael E. Whitman, PhD CISM, CISSP, is a professor of information systems at Kennesaw State University, Kennesaw, Georgia, where he is also the director of the KSU Center for Information Security Education and the coordinator of the Bachelor of science in information security and assurance program, which is the first program of its kind in the Southeast. Dr. Whitman is an active researcher and author in information security policy and curriculum development, ethical computing, and information systems research methods. He currently teaches graduate and undergraduate courses in information security, local area networks, and data communications. Dr. Whitman has five information security textbooks currently in print and has published articles in *Information Systems Research*, *Communications of the ACM*, the *Journal of International Business Studies*, *Information and Management*, and the *Journal of Computer Information Systems*. Dr. Whitman earned his PhD and MBA at Auburn University, Al, and has taught previously for the University of Nevada, Las Vegas and Auburn University.

Amy B. Woszczynski is MSIS director and an associate professor of information systems at Kennesaw State University. She completed a bachelor's in industrial engineering from Georgia Tech, an MBA from Kennesaw State University, and a PhD in industrial management from Clemson University. Dr. Woszczynski's current research interests include diversity in IT and distance learning pedagogy and curriculum initiatives. She has published papers on these and other topics in *Journal of Information Systems Education*, *Journal of Computer Information Systems*, *Computers in Human Behavior*, and *Industrial Management & Data Systems*. She also coedited *The Handbook of Information Systems Research.*

Herbert J. Mattord, MBA, CISM, CISSP recently completed 24 years of IT industry experience as an application developer, database administrator, project manager, and information security practitioner before joining faculty as a full time tenure-track instructor. During his career as an IT practitioner, he has been an adjunct professor at a number of universities throughout the South for over 20 years. He currently teaches courses in information security, data communications, local area networks, database technology, project management, and systems analysis and design. He is the coauthor of *Principles of Information Security, Management of Information Security, Principles of Incident Response and Disaster Recovery, Readings and Cases in the Management of Information Security*, and *The Hands-On Information Security Lab Manual*. He was formerly the manager of corporate information technology security at Georgia-Pacific Corporation.

* * *

Panagiotes Anastasiades is currently assistant professor on lifelong and distance learning in the Department of Education at the University of Crete. He is also tutor counsellor at the Hellenic Open University (postgraduate level, master's in education, Module: EKP65 Open and Distance Learning). He has been visiting assistant professor in the Department of Computer Engineering and Informatics at the Polytechnic School of the University of Patras (1999-2002), and also visiting faculty in the Department of Computer Science at the University of Cyprus (2000-2002). His current research emphasis focuses on lifelong and distance learning via advanced learning Internet technologies and interactive videoconferencing, social and educational informatics, and information society theories. He has additional papers published in *Journal of Computers and Education, Computers in the Social Studies Journal, ACM Special Interest Group on Computers and Society (ACM SIGCAS), Lecture Notes in Computer Science, International Journal of Communication, International Journal of Learning, International Journal of the Humanities,* and so forth.Home Page: http://www.edc.uoc.gr/~panas/index.html

Roma Angel is an assistant professor in the Department of Leadership and Educational Studies and currently serves as assistant dean in the Reich College of Education, Appalachian State University, Boone, NC. She received her EdD from the University of North Carolina at Greensboro and a master's degree from Wake Forest University. Dr. Angel has held various positions in higher education including serving as developer/director of the teacher education program at Lees-McRae College. Her research interests include examination of leadership roles in developing effective environments for communities of learners.

Silvia Braidic serves as an assistant professor in the Administrative Program for Principals at California University of Pennsylvania. Currently she is teaching online for the Principal's Program. Previously she served as an assistant professor and director of graduate and undergraduate secondary education at Duquesne University. Her research interests include teaching online, online professional development, instructional strategies/differentiation, and principal/teacher leadership. In addition to her work at the university level, she has experience as a principal and assistant principal in the Mt. Lebanon School District in Pittsburgh, Pennsylvania. She also served as the district's coordinator for strategic planning. Prior to her work in administration, she taught middle school mathematics. She holds a Doctorate of Education in administrative and policy studies-educational administration from the University of Pittsburgh, K-12 Principal Certification from Carnegie Mellon University, a MSEd in elementary education, and a B.S.Ed in secondary education-mathematics from Duquesne University.

Steven C. Bronack is an associate professor of instructional technology in the Department of Leadership and Higher Education, in the Reich College of Education at Appalachian State University Boone, NC. His research interests include teaching and learning in virtual worlds, social constructivist learning environments, and case-based approaches to education. Dr. Bronack holds BS and MA degrees from Appalachian State and a PhD in education from the University of Virginia.

Sudhanva Char has a Master of Commerce degree from Bombay University and a Doctor of Philosophy degree in industrial economics from the Gokhale Institute of Economics, Poona University. He has diplomas in C+, Oracle, and data base management. He has taught business courses on distance learning programs at Chattahoochee Technical College. Earlier Dr. Char served as executive-in-charge in the Corporate Information Services Department of a steel conglomerate. He is currently an associate professor in the Business Department at Life University, Georgia. Dr. Char has published research papers in refereed, trade, and e-journals, as well a book on farm income taxation.

Amelia Cheney is an assistant professor in the instructional technology program at Appalachian State University in the AETZone. Prior to joining the faculty, she worked in K-12 for more than 13 years in teaching and CTO capacities for two school districts in North Carolina. She holds BA and MEd degrees from Wake Forest University, and an EdD from Appalachian State.

Morten Falch is an associate professor in the Center for Information and Communication Technologies at the Technical University of Denmark. He holds a PhD degree from DTU, a master's degree in economics, and a bachelor's in math. His research interests include a wide range of issues related to information and communication technologies such as cost analysis of telecom networks, e-government, regulation of the telecom sector, ICT industry policy, the role of competition in innovation of new services, use of ICT in knowledge services, and tele-based community centres. He has participated in many EU funded research projects and conducted a large number of consultancies for national and international organisations such as ITU, UNCTAD, the World Bank, and the National Telecom Agencies in Denmark, Norway and Sweden.

Grandon Gill is an associate professor in the Information Systems and Decision Sciences Department at the University of South Florida. He holds a doctorate in management information systems from Harvard Business School, where he also received his MBA. His principal research focus is in the area of information systems (IS) education, and he has published many articles describing how technologies and innovative pedagogies can be combined to increase the effectiveness of teaching across a broad range of IS topics. Currently, he teaches programming, database, and managerial courses to both undergraduate and graduate students.

Regis M. Gilman recently became associate dean of educational outreach at Western Carolina University, Cullowhee, NC. Prior to joining WCU, she was assistant professor of instructional technology in the Department of Leadership & Educational Studies at Appalachian State University. Before joining the faculty at ASU, she directed the Appalachian Transition to Teaching Program, a $1.6 million US Department of Education initiative. She holds a BSEd from Northwest Missouri State University, a MS in adult education from Drake University, an EdD in educational policy–higher education from the University of Kansas, and a postdoctorate MA in ed media–instructional technology.

Keith Lindsey is an assistant professor of business administration at Trinity University in San Antonio, Texas. He holds a PhD in management information systems from the University of Memphis. His research interests include the use of technology in higher education, business intelligence and knowledge management, applying information strategies in small and medium sized enterprises, and accounting information systems.

Lim Hwee Ling is an assistant professor in the Communication Department at The Petroleum Institute (Abu Dhabi, UAE). She has bachelor's (English language, literature), master's (English language) degrees, and a professional diploma in education from The National University of Singapore. Dr. Lim has a PhD (information technology) from Murdoch University (Perth, Western Australia). Her areas of research interest are educational technology, computer-mediated communication, and distance learning.

Richard E. Riedl is a professor in the instructional technology program at Appalachian State University and assistant department chair of the Leadership and Educational Studies Program. He received his PhD in curriculum and instruction from Arizona State University.

Albert D. Ritzhaupt is an instructor in the School of Computing at the University of North Florida. He has a BS in computer and information sciences and an MBA from the University of North Florida. His research focuses on the meaningful integration of information and communication technology in higher education and computing education, and has been published in several venues. Albert is a PhD candidate at the University of South Florida, and has taught in the areas of operating systems, database systems, computer programming, multimedia applications, and computer networking to undergraduate students.

Carla Ruiz Mafé is an assistant professor in the Department of Marketing at the University of Valencia (Spain) and coordinator of postgraduate training programs in the Chamber of Commerce of Valencia. She received her PhD from the University of Valencia. She is the author of international publications on e-commerce and e-learning. Her primary research interests include e-commerce, communication, interactive marketing, and consumer behaviour.

Silvia Sanz Blas is an assistant professor in the Department of Marketing at the University of Valencia (Spain). She received her PhD from the University of Valencia. She is the author of international publications on distance shopping and e-learning. Her primary research interests include communication, sales, e-commerce, interactive marketing and consumer behaviour.

Robert L. Sanders coordinates the Appalachian State University Library Science program. Prior to this, Dr. Sanders served as the president of *para instructional designs*, an e-learning design company based in Cincinnati, OH. These roles have provided numerous opportunities to explore applications of information and instructional technologies and the impact of these technologies on student and patron behaviors, perspectives, and interactions. Dr. Sanders' current research is focused on the utilization of action learning in 3D immersive learning environments, and students sense of presence and the role of serendipitous interactions in virtual worlds.

Stephen B. Springer, LPC, CPM is the program chair and associate professor in occupational education at Texas State University-San Marcos. He has taught elementary through graduate college and has held administrative positions in both public education and at the university level. He has developed courses for online as well as DVDs for classes at the university. The current university program he chairs is a nontraditional adult program that has been in existence since 1973. Dr. Springer also had a private counseling practice and has served in public office as well as an officer in the State Guard.

Tom Stafford is Suzanne Downs associate professor of management information systems for the Fogelman College of Business and Economics at University of Memphis, and editor of *ACM Data Base for Advances in Information Systems*. He holds doctorates in MIS from University of Texas–Arlington and in marketing from University of Georgia. Stafford's research spans issues of human computer interaction and technology adoption, and has appeared in journals such as *Decision Sciences*, *Communications of the ACM*, and *IEEE Transactions on Engineering Management*.

Fay Sudweeks is a senior lecturer at the School of Information Technology (Murdoch University). She has bachelor's (psychology, sociology) and master's (cognitive science) degrees from the University of New South Wales, and a PhD from Murdoch University. Her current research interests are social, cultural, and economic aspects of CMC and CSCW, group dynamics, and e-learning. She has published six edited books and more than 50 papers in journals, books, and conference proceedings. She is on the editorial review board of *International Journal of e-Learning*, *Journal of CMC*, and *Journal of Electronic Commerce in Organizations*.

John H. Tashner is a professor and coordinator of the instructional technology program at Appalachian State University with 30+ years experience in university teaching. Prior to this, John served in various roles as a central office school administrator, an assistant principal, and a public school science teacher. He received his BS in biology and an MS in science education, both from Old Dominion University. He earned his EdD in curriculum and instruction-science education from the University of Virginia in 1973. Current research interests involve the creation and study of viable pedagogies for use in 3D immersive worlds for education.

José Tronch García de los Ríos is a part-time professor in the Department of Marketing at the University of Valencia (Spain) and expert for International Cooperation Programs in the Foreign Trade Department of the Chamber of Commerce of Valencia. He is the author of Spanish publications on e-learning. His primary research interests include e-commerce, e-learning, service quality, and consumer behaviour.

Hanne Westh Nicolajsen is an assistant professor in the Center for Information and Communication Technologies at the Technical University of Denmark. Her research interests include organizational implementation and use of IT, knowledge management, and computer-mediated communication. Her current research focuses on the use of information and communication technology for innovation in the service sector. Nicolajsen holds a PhD from the Technical University of Denmark.

Marlene Wilcox is an assistant professor in the Department of Business Management and Administration at Bradley University. Marlene is a graduate of Claremont Graduate University's (CGU) School of Information Systems and Technology. She holds a PhD and MS in management of information systems from CGU and an MBA and BS in management from Pepperdine University. Marlene's research interests include knowledge management, organizational and interorganizational learning, and online education.

Levent Yilmaz is assistant professor of computer science and software engineering at the College of Engineering at Auburn University. Dr. Yilmaz earned his PhD and MS degrees from Virginia Polytechnic Institute and State University (Virginia Tech). His research focuses on (1) improving cognition in modeling and design education, (2) advancing the theory and methodology of simulation modeling via novel modeling and simulation formalisms, and (3) agent-directed simulation. Dr. Yilmaz is a member of ACM, IEEE Computer Society, Society for Computer Simulation International, and Upsilon Pi Epsilon.

Index

E

F

G

H

I

J

K

L

$210.00

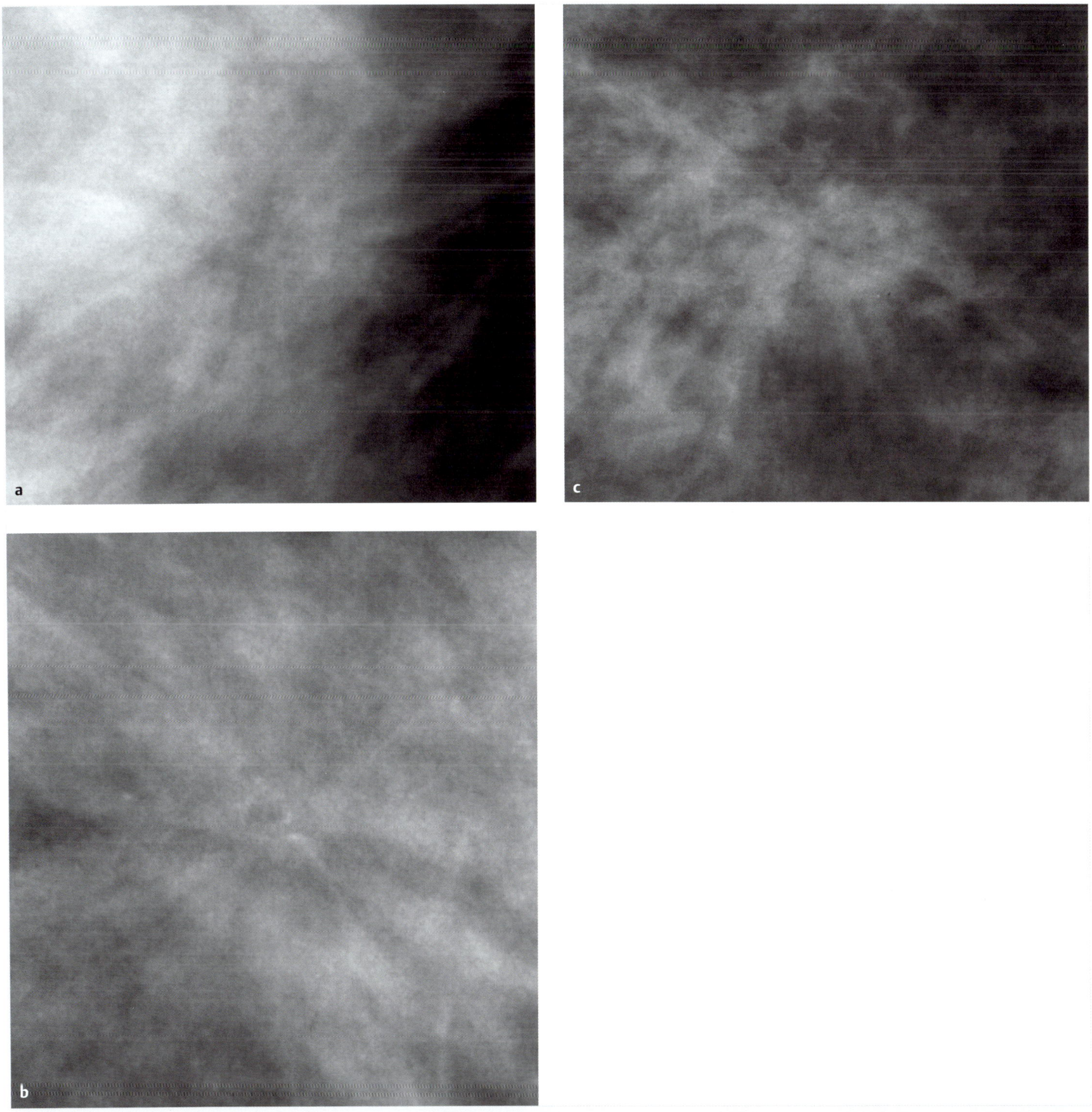

Fig. 5.50 Examples of architectural distortion in the mammogram. (a) Radial scar. (b) Radial scar with associated small carcinoma. (c) Tubular breast carcinoma.

are required. This may be the case, for example, when an additional magnification mammographic view is required for the further work-up of indistinct microcalcifications, or when a spot compression image is needed for the work-up of an architectural distortion.

The MX BI-RADS category 0 is almost invariably reserved for an examination within the framework of a screening program. On completion of the conventional mammographic screening, the patient leaves the screening center; because the mammograms are read and assessed at a later time, the patient is no longer present when it is judged that supplementary views or examinations are necessary for the final assessment. In the screening situation, a patient whose mammogram is classified as BI-RADS 0 will have to be recalled for further diagnostic procedures. But in the circumstances of individualized examination, the BI-RADS 0 category should not have to be used at all. Further diagnostic work-up of an ambiguous lesion, including any supplementary imaging, can be performed promptly when the need is determined because the patient will still be present at the time of assessment of the mammograms. Thus it possible to assign each mammogram a BI-RADS category 1 to 5.

MX BI-RADS 1

Assignment to the BI-RADS 1 category indicates that the mammogram in question can be conclusively evaluated and that no

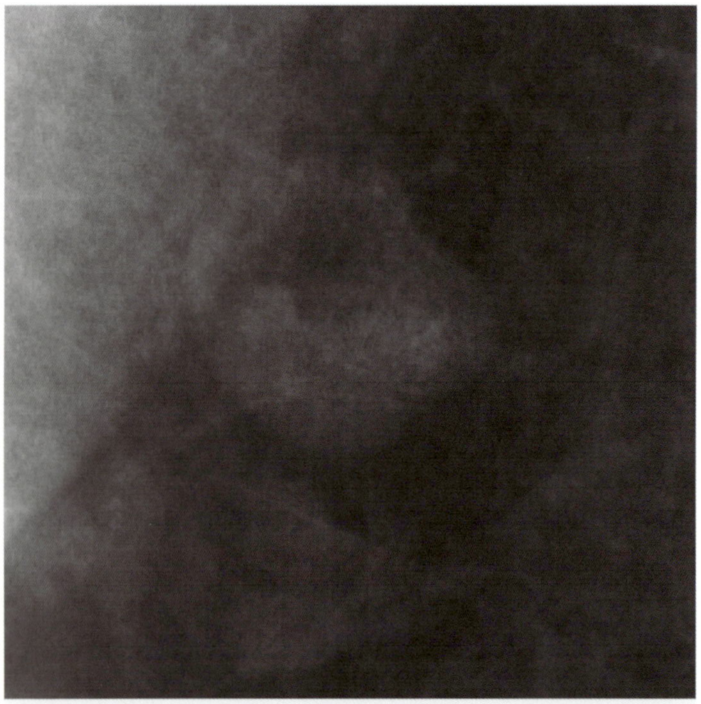

Fig. 5.51 **Intramammary lymph node.** A special case in mammography.

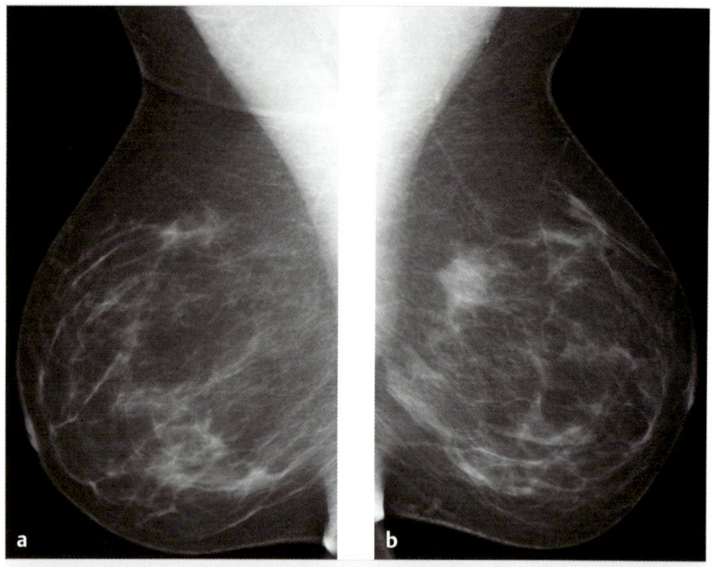

Fig. 5.52 **Focal asymmetry in bilateral comparison.** A special case in mammography. **(a)** Right breast. **(b)** Left breast.

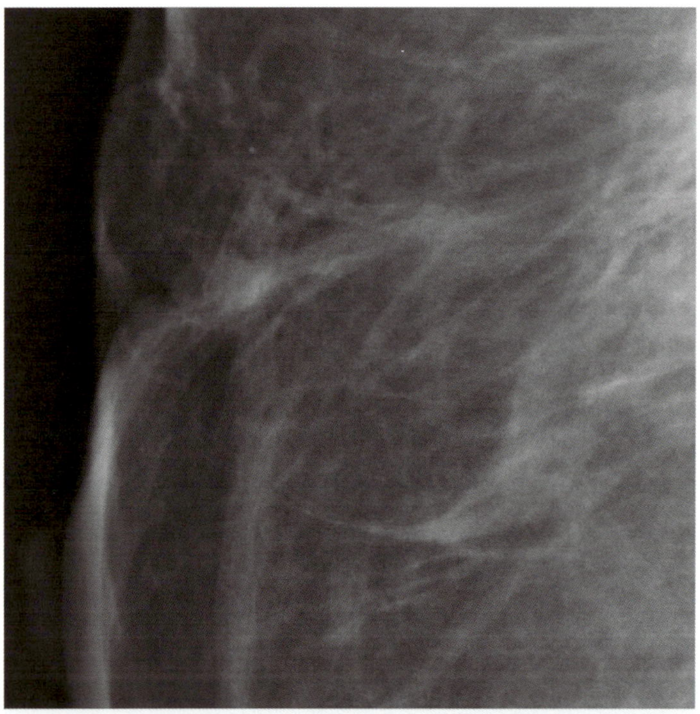

Fig. 5.53 **Skin retraction.** An associated finding in the mammogram.

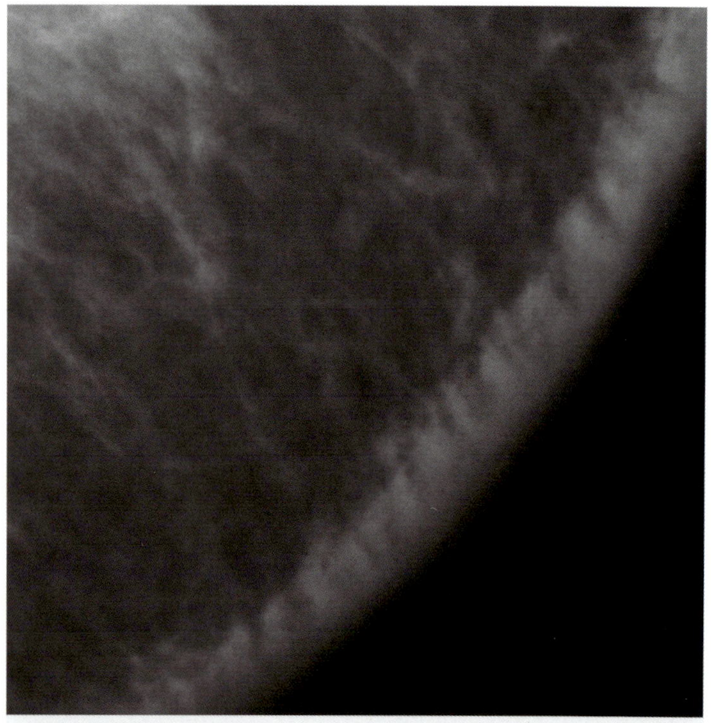

Fig. 5.54 **Skin thickening.** An associated finding in the mammogram.

lesions worth describing are present—i.e., a normal finding. It also implies that the breasts are symmetrical and that no densities, mass lesions, architectural distortions, or suspicious microcalcifications are present. It is implicit in the BI-RADS 1 category that—depending on the parenchymal density—the sensitivity for detection of breast cancer can be reduced down to approximately 40% in spite of normal findings (▶ Fig. 5.56). The next examination should normally be recommended at the usual screening interval, but this will depend on the individual breast cancer risk.

MX BI-RADS 2

Assignment as BI-RADS 2 also conveys that the mammogram in question can be conclusively evaluated and that ultimately the findings are normal. In this case, however, there is a finding that is considered noteworthy but can be reliably assessed as benign. Examples of such lesions in the mammogram that can be interpreted as unequivocally benign are:

- Calcified fibroadenomas (▶ Fig. 5.57a).
- Multiple secretory calcifications.
- Findings with lipomatous inclusions, such as a lipoma, oil cysts, and hamartomas (▶ Fig. 5.57b).
- Galactoceles.

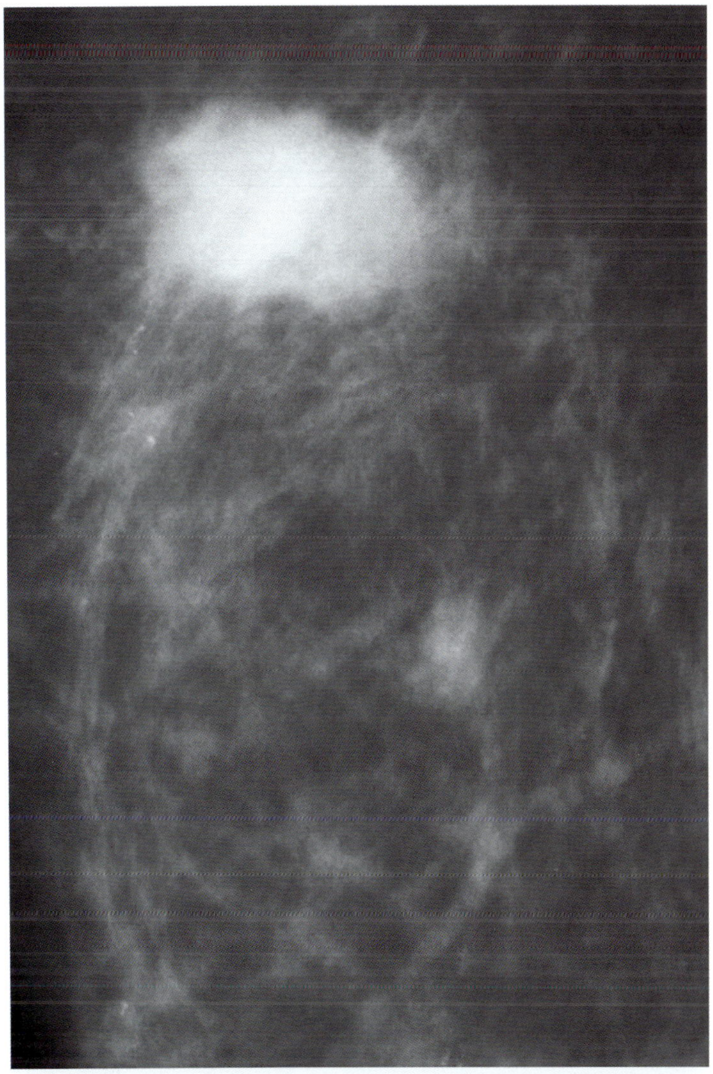

Fig. 5.55 **Calcifications in direct proximity to a breast carcinoma as indication of possible extensive intraductal components (EIC).** An associated finding in the mammogram.

Table 5.4 BI-RADS classification for mammography (MX BI-RADS). Description of findings and risk of the presence of a malignant tumor according to the BI-RADS classification

MX BI-RADS category	Description	Risk of carcinoma (%)
1	Normal	0
2	Benign	0
3	Probably benign	≤2
4	Possibly malignant	>2–>95[a]
5	Highly suspicious	≥95

[a]Because of to the large range of the MX BI-RADS category 4 (2–4%), subdivisions may be used: MX BI-RADS 4A (>2–≤10%), 4B (>10–≤50%), and 4C (>50–<95%).

All of these entities exhibit a characteristic appearance that obviates their being mistaken for a malignant tumor. The probability that such a finding represents breast cancer is effectively zero. Accordingly the recommendation for the next examination is at the usual screening interval.

Table 5.5 BI-RADS classification and management. Recommended course of action according to MX BI-RADS category

MX BI-RADS category	Recommended management
1	Follow-up within the usual screening interval
2	Follow-up within the usual screening interval
3	Follow-up in 6 months
4	Diagnostic work-up by percutaneous biopsy
5	Pretherapeutic diagnostic work-up by percutaneous biopsy

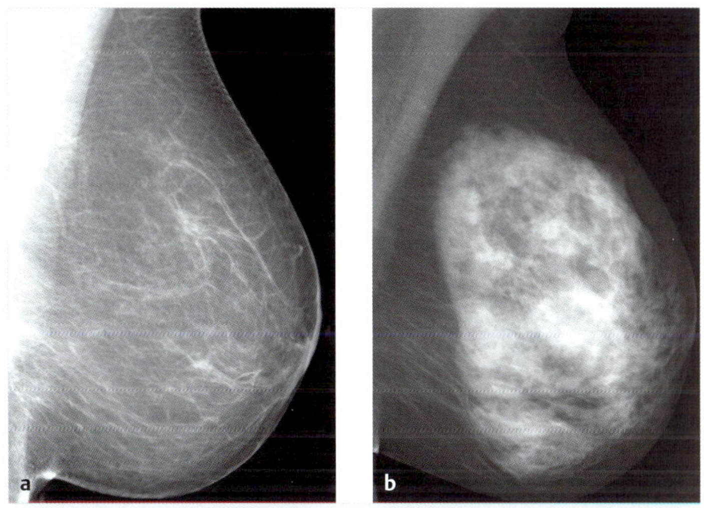

Fig. 5.56 **BI-RADS 1 in a mammogram.** Typical examples of mammograms in the MX BI-RADS 1 category. **(a)** Normal findings with very high sensitivity (ACR density type I). **(b)** Normal findings with considerable limitation of the sensitivity for the detection of breast cancer (ACR density type IV).

MX BI-RADS 3

BI-RADS 3 represents probably the most difficult and challenging evaluation category for a lesion seen in a mammogram that can be conclusively evaluated. It describes the presence of a lesion that has a very high probability of being benign, but where there is a minimal remnant of uncertainty (≤2%). The BI-RADS lexicon describes the following three concrete BI-RADS 3 findings:

- Noncalcified, circumscribed, solid mass lesions (▶ Fig. 5.58a).
- Focal asymmetry.
- Cluster of round (punctiform) microcalcifications (▶ Fig. 5.58b).

All diagnostic imaging possibilities should be exhausted before a lesion is classified as BI-RADS 3.

Usually, a short-term follow-up interval of 6 months is recommended for MX BI-RADS 3 findings. Further follow-up may be continued until the lesion is proven to be stable over a period of 2 years. Alternatively, if the patient wishes, histological verification can be obtained by percutaneous biopsy. If the lesion shows progression during the short-term follow-up, then it is reclassified as BI-RADS 4 and a percutaneous biopsy is performed. In retrospective monitoring, the rate of carcinoma in the BI-RADS 3 group within the recommended follow-up period should be ≤2%. It should explicitly recognized that for those rare cases in which a

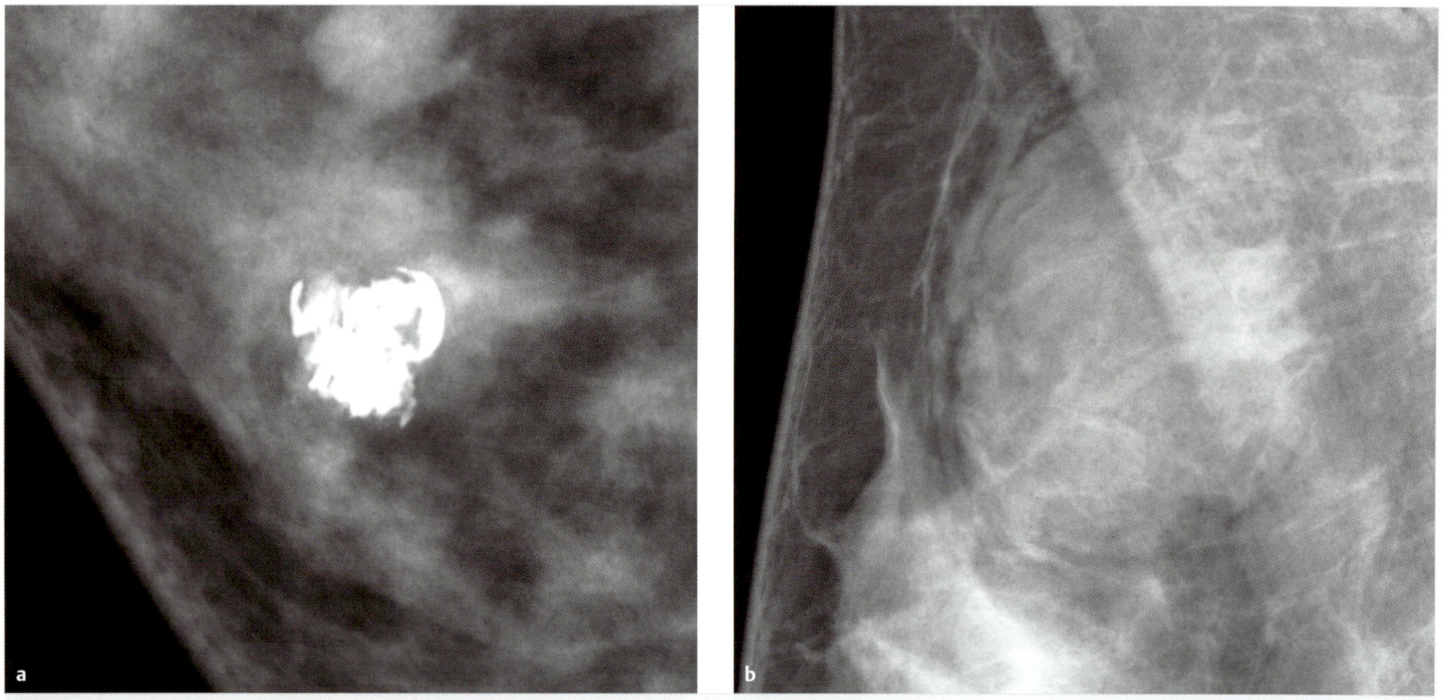

Fig. 5.57 BI-RADS 2 in a mammogram. Typical examples of mammographic findings in the MX BI-RADS 2 category. **(a)** Popcorn-shaped calcifications in a fibroadenoma. **(b)** A hamartoma.

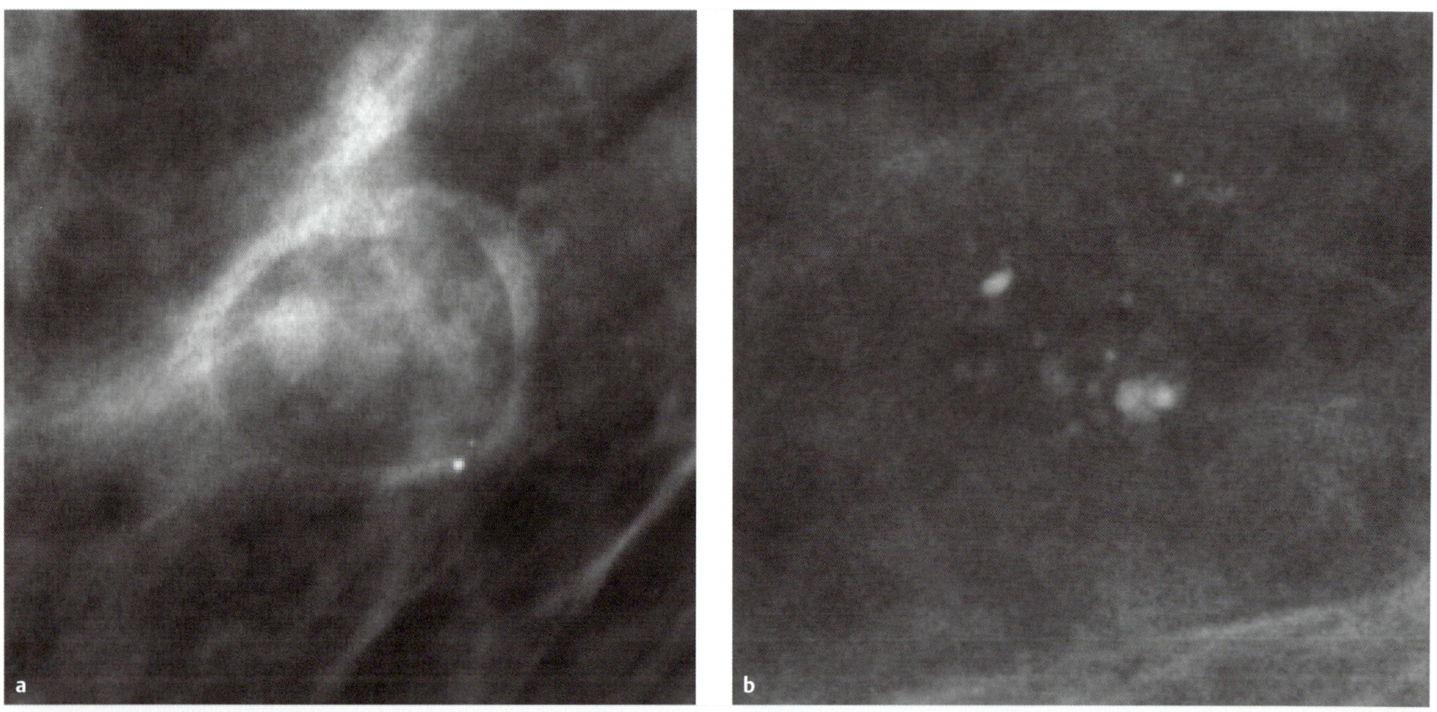

Fig. 5.58 BI-RADS 3 in a mammogram. Typical examples of mammographic findings in the MX BI-RADS 3 category. **(a)** A smooth mass lesion with internal fat-equivalent density and associated microcalcifications. **(b)** Clustered, round microcalcifications.

carcinoma is found in the course of the follow-up period, neither the staging allocation of the tumor nor the prognosis for the patient is worsened.

MX BI-RADS 4

The BI-RADS 4 category indicates that the mammogram in question can be conclusively evaluated and that changes are present that do not exhibit the typical characteristics of a carcinoma but do have a definite probability of being malignant (> 2– < 95%). An interventional diagnostic work-up in the form of a percutaneous biopsy is recommended for findings in this category.

Because of the broad range of malignancy risk in this category (> 2– < 95%), BI-RADS 4 is divided into subgroups BI-RADS 4A to 4C, differing in the acuteness of the indication for a biopsy.

- *BI-RADS 4A* describes an abnormality in the mammogram with a minor risk of malignancy (> 2– ≤ 10%). Examples of these are palpable lumps that are only partially well-defined and have the sonographic criteria of a fibroadenoma, a

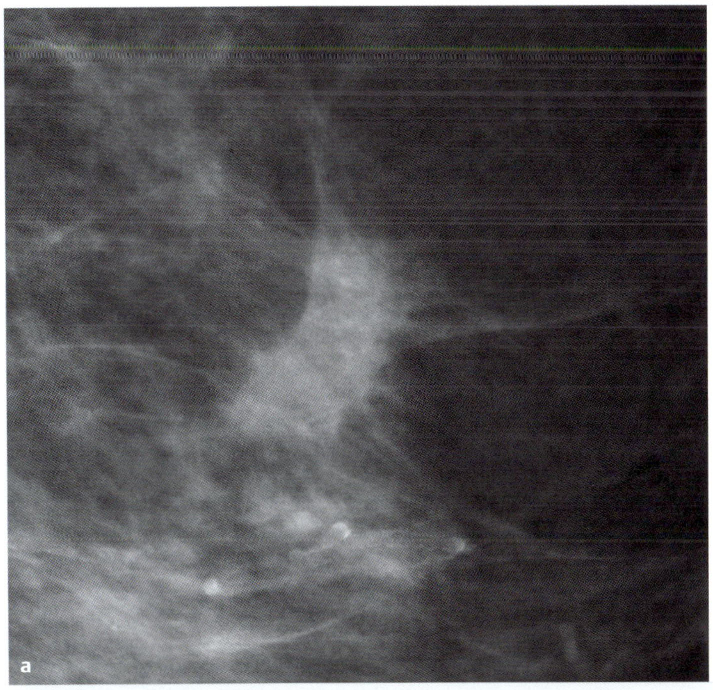

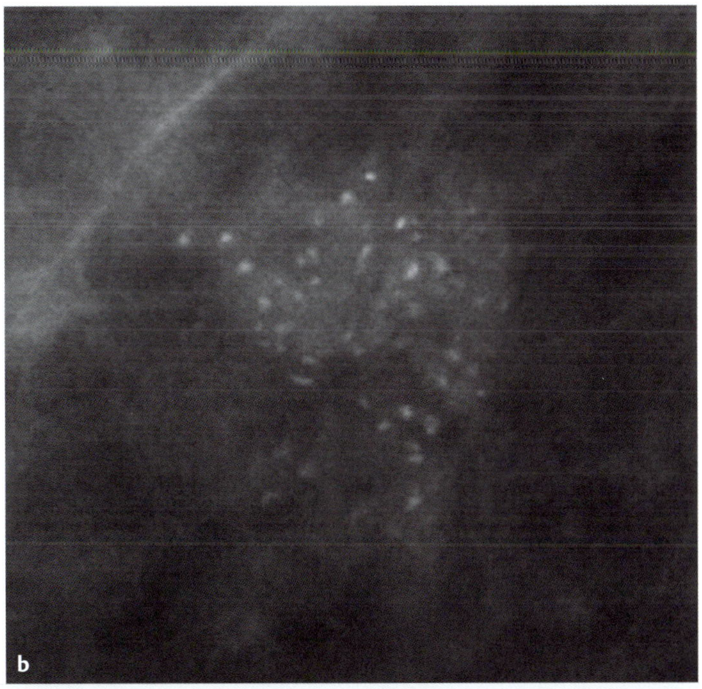

Fig. 5.59 BI-RADS 4 in a mammogram. Typical examples of mammographic findings in the MX BI-RADS 4 category. **(a)** Mass lesion with partially ill-defined margins. **(b)** A new occurrence of pleomorphic microcalcifications.

palpable complex cyst, and a probable abscess. Results of a percutaneous biopsy are expected to be benign. In such a case, follow-up examination after 6 months or at the routine interval is sufficient.

- *BI-RADS 4B* describes an abnormality in the mammogram that carries a moderate risk of malignancy (> 10– ≤ 50%). One example of this is a focal lesion that is partially well-defined and partially indistinct (▸ Fig. 5.59a). When a percutaneous biopsy yields histologically benign results, it must be decided case by case whether repeat biopsy, open biopsy, or wait-and-see management with a follow-up examination after 6 months is indicated.
- *BI-RADS 4C* describes an abnormality in the mammogram that carries a very high risk of malignancy (> 50– ≤ 95%) but does not exhibit the classic malignancy criteria of a BI-RADS 5 finding. Examples include a poorly defined, irregular mass lesion, and a newly occurring cluster of fine, pleomorphic microcalcifications (▸ Fig. 5.59b). The results of a percutaneous biopsy are expected to be malignant. If histology yields benign findings, the pathologist should be motivated to investigate further. It is important to discuss repeat biopsy or open biopsy as the appropriate course of further action.

MX BI-RADS 5

The BI-RADS 5 category is intended for those lesions that exhibit the typical radiological appearance of breast cancer. The probability that it is indeed a malignancy lies statistically between 95 and 100%. The ACR specifies the following classic examples for this category:

- Spiculated, irregular, very dense masses.
- Segmental or linearly arranged microcalcifications (▸ Fig. 5.60a).
- Irregular, spiculated masses with pleomorphic calcifications (▸ Fig. 5.60b).

In accordance with established oncologic treatment guidelines, a finding in the BI-RADS 5 category should be histopathologically verified by percutaneous biopsy before the onset of therapy. If histology yields benign findings, one must consider the possibility of having missed the lesion, and a repeat biopsy or an open biopsy should be seriously considered.

MX BI-RADS 6

The BI-RADS 6 category describes those findings that appear mammographically suspicious and have been histologically verified as malignant tumors by percutaneous biopsy. They now correspond to B5a (DCIS) or B5b (invasive carcinoma), or, where applicable, to B5c or B5d lesions (see Chapter 3.2) in the B-categorization of the National Coordinating Group for Breast Screening Pathology. Allocation a lesion to the BI-RADS 6 category is thus decoupled from morphological criteria in the imaging. The designation as BI-RADS 6 is removed when the lesion has been surgically excised and appropriately classified (pT staging) according to the TNM system of the AJCC (American Joint Committee on Cancer) and the UICC (Union for International Cancer Control).

Note

- In the overall assessment, the final diagnosis stands at the end of an examination chain (clinical examination, mammogram, ultrasonography, breast MRI). Taking the specific reliability of each method into consideration, a comprehensive evaluation is expressed in the form of a final overall BI-RADS assessment category 1, 2, 3, 4, or 5.
- Assuming that the histological result and the mammographic image findings are compatible, a short-term follow-up examination should be done 6 to 12 months after each core needle biopsy of a BI-RADS category 4 or 5 lesion.
- The designation as BI-RADS 6 ends on the day the carcinoma is surgically excised.

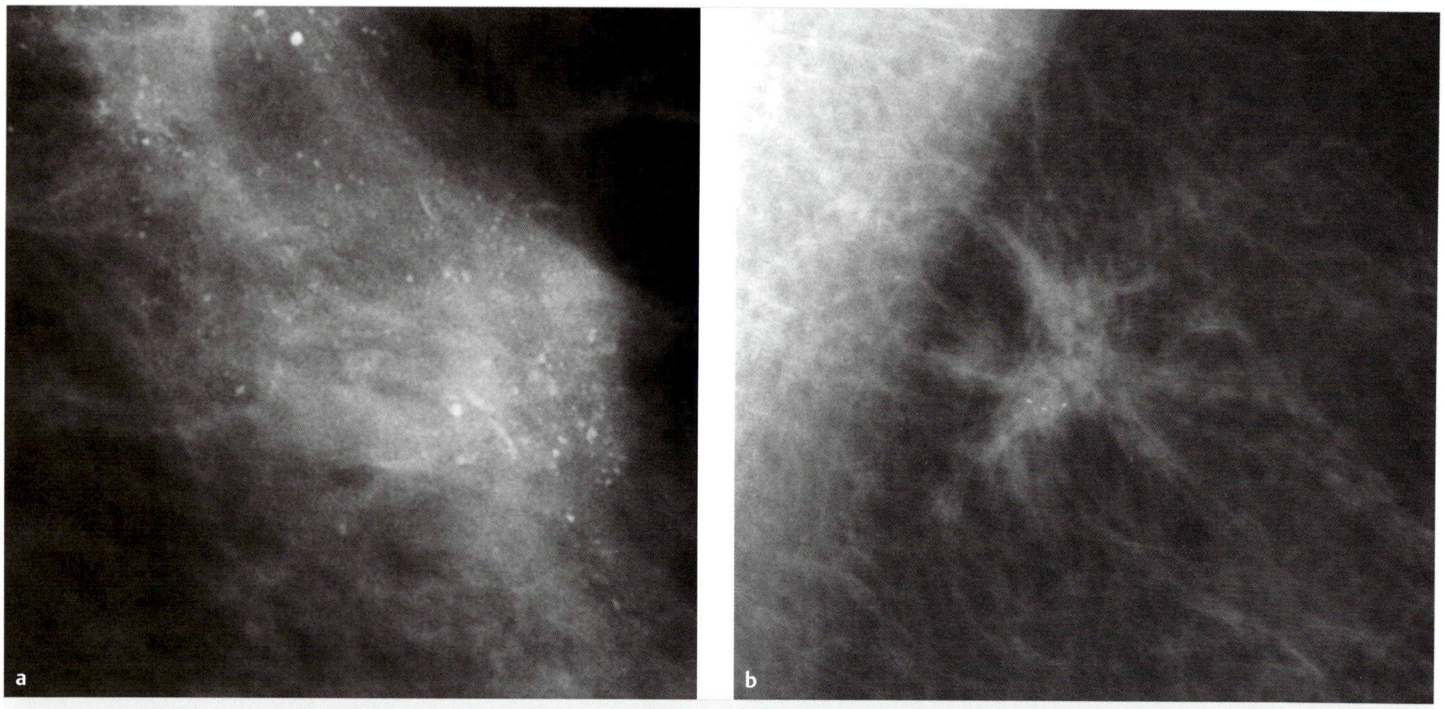

Fig. 5.60 BI-RADS 5 in a mammogram. Typical examples of mammographic findings in the MX BI-RADS 5 category. **(a)** A new occurrence of pleomorphic microcalcifications with segmental distribution. **(b)** Mass with spiculated margins and associated microcalcifications.

5.3.5 Normal Findings in the Mammogram

The extent of glandular tissue and adipose tissue essentially determines the mammographic appearance of the female breast. The composition of both of these tissue components depends primarily on age, hormonal stimulation, and nutritional status. Typically, a younger woman exhibits a prominent parenchymal body, which atrophies after about age 40 years. The adipose tissue component commonly increases during the perimenopausal and postmenopausal periods. Skin, subcutaneous adipose tissue, Cooper's ligament, veins, arteriosclerotic artery changes, and intramammary and prepectoral lymph nodes may also be distinguished in a normal mammogram.

Bibliography

▶ Technique and Methods

[1] Bloomquist AK, Yaffe MJ, Pisano ED, et al. Quality control for digital mammography in the ACRIN DMIST trial: part I. Med Phys. 2006; 33(3):719–736

[2] Cole EB, Pisano ED, Kistner EO, et al. Diagnostic accuracy of digital mammography in patients with dense breasts who underwent problem-solving mammography: effects of image processing and lesion type. Radiology. 2003; 226 (1):153–160

[3] Diekmann F, Diekmann S, Taupitz M, et al. Use of iodine-based contrast media in digital full-field mammography—initial experience. Rofo. 2003; 175 (3):342–345

[4] Diekmann F, Diekmann S, Jeunehomme F, Muller S, Hamm B, Bick U. Digital mammography using iodine-based contrast media: initial clinical experience with dynamic contrast medium enhancement. Invest Radiol. 2005; 40 (7):397–404

[5] Di Nubila B, Cassano E, Origgi D, et al. Analogic versus digital mammographic examination a radiological study of mammary microcalcifications on 52 surgical samples. Radiol Med (Turin). 2003; 106(4):297–304

[6] Pisano ED, Cole EB, Major S, et al. The International Digital Mammography Development Group. Radiologists' preferences for digital mammographic display. Radiology. 2000; 216(3):820–830

[7] Fischer U, Baum F, Obenauer S, et al. Comparative study in patients with microcalcifications: full-field digital mammography vs screen-film mammography. Eur Radiol. 2002; 12(11):2679–2683

[8] Fischmann A, Siegmann KC, Wersebe A, Claussen CD, Müller-Schimpfle M. Comparison of full-field digital mammography and film-screen mammography: image quality and lesion detection. Br J Radiol. 2005; 78 (928):312–315

[9] Funke M, Obenauer S, Hermann KP, Fischer U, Grabbe E. Soft copy versus hard copy findings in digital mammography [in German]. Radiologe. 2002; 42 (4):265–269

[10] Gong X, Vedula AA, Glick SJ. Microcalcification detection using cone-beam CT mammography with a flat-panel imager. Phys Med Biol. 2004; 49 (11):2183–2195

[11] Jong RA, Yaffe MJ, Skarpathiotakis M, et al. Contrast-enhanced digital mammography: initial clinical experience. Radiology. 2003; 228(3):842–850

[12] Karssemeijer N, Hendriks JH. Computer-assisted reading of mammograms. Eur Radiol. 1997; 7(5):743–748

[13] Lai CJ, Shaw CC, Whitman GJ, et al. Visibility of simulated microcalcifications —a hardcopy-based comparison of three mammographic systems. Med Phys. 2005; 32(1):182–194

[14] Lewin JM, Hendrick RE, D'Orsi CJ, et al. Comparison of full-field digital mammography with screen-film mammography for cancer detection: results of 4,945 paired examinations. Radiology. 2001; 218(3):873–880

[15] Lewin JM, D'Orsi CJ, Hendrick RE, et al. Clinical comparison of full-field digital mammography and screen-film mammography for detection of breast cancer. AJR Am J Roentgenol. 2002; 179(3):671–677

[16] Lewin JM, Isaacs PK, Vance V, Larke FJ. Dual-energy contrast-enhanced digital subtraction mammography: feasibility. Radiology. 2003; 229(1):261–268

[17] NHS Connecting for Health. Picture Archiving and Communications System (PACS). Available at: http://webarchive.nationalarchives.gov.uk/20130502102046/http://connectingforhealth.nhs.uk/systemsandservices/pacs. Archived 05.03.2013. Accessed January 4, 2017

[18] Niklason LT, Christian BT, Niklason LE, et al. Digital tomosynthesis in breast imaging. Radiology. 1997; 205(2):399–406

[19] Obenauer S, Hermann KP, Schorn C, Fischer U, Grabbe E. Digitale Vollfeldmammographie: Dosisabhängige Detektion von simulierten Herdbefunden

und Mikrokalzifikationen [Full-field digital mammography: dose-dependent detectability of breast lesions and microcalcinosis]. Rofo. 2000; 172 (12):1052–1056

[20] Obenauer S, Luftner-Nagel S, von Heyden D, Munzel U, Baum F, Grabbe E. Screen film vs full-field digital mammography: image quality, detectability and characterization of lesions. Eur Radiol. 2002; 12(7):1697–1702

[21] Pisano ED, Cole EB, Kistner EO, et al. Interpretation of digital mammograms: comparison of speed and accuracy of soft-copy versus printed-film display. Radiology. 2002; 223(2):483–488

[22] Pisano ED, Gatsonis C, Hendrick E, et al. Digital Mammographic Imaging Screening Trial (DMIST) Investigators Group. Diagnostic performance of digital versus film mammography for breast-cancer screening. N Engl J Med. 2005; 353(17):1773–1783

[23] Pisano ED, Gatsonis CA, Yaffe MJ, et al. American College of Radiology Imaging Network digital mammographic imaging screening trial: objectives and methodology. Radiology. 2005; 236(2):404–412

[24] Rong XJ, Shaw CC, Johnston DA, et al. Microcalcification detectability for four mammographic detectors: flat-panel, CCD, CR, and screen/film). Med Phys. 2002; 29(9):2052–2061

[25] Schulz-Wendtland R, Aichinger U, Säbel M, Böhner C, Dobritz M, Bautz W. Experimentelle Untersuchungen zur Bildgüte konventioneller Film-Folien-Mammographie, digitaler Mammographie mit Speicherfolien in Vergrösserungstechnik und voll digitaler Mammographie in CCD-Technik [Experimental studies on image quality in conventional film screen system, digital phosphor storage plate mammography in magnification technique and digital mammography in CCD-technique]. Rofo. 2000; 172(12):965–968

[26] Skaane P, Young K, Skjennald A. Population-based mammography screening: comparison of screen-film and full-field digital mammography with soft-copy reading—Oslo I study. Radiology. 2003; 229(3):877–884

[27] Skaane P, Skjennald A. Screen-film mammography versus full-field digital mammography with soft-copy reading: randomized trial in a population-based screening program—the Oslo II Study. Radiology. 2004; 232(1):197–204

[28] Skaane P, Diekmann F, Balleyguier C, et al. Observer variability in screen-film mammography versus full-field digital mammography with soft-copy reading. Eur Radiol. 2008; 18(6):1134–1143

[29] Skarpathiotakis M, Yaffe MJ, Bloomquist AK, et al. Development of contrast digital mammography. Med Phys. 2002; 29(10):2419–2426

[30] Suryanarayanan S, Karellas A, Vedantham S, et al. Comparison of tomosynthesis methods used with digital mammography. Acad Radiol. 2000; 7(12):1085–1097

[31] Venta LA, Hendrick RE, Adler YT, et al. Rates and causes of disagreement in interpretation of full-field digital mammography and film-screen mammography in a diagnostic setting. AJR Am J Roentgenol. 2001; 176(5):1241–1248

[32] Webber RL, Underhill HR, Freimanis RI. A controlled evaluation of tuned-aperture computed tomography applied to digital spot mammography. J Digit Imaging. 2000; 13(2):90–97

[33] Wu T, Stewart A, Stanton M, et al. Tomographic mammography using a limited number of low-dose cone-beam projection images. Med Phys. 2003; 30 (3):365–380

[34] Yamada T, Ishibashi T, Sato A, et al. Comparison of screen-film and full field digital mammography: image contrast and lesion characterization. Radiat Med. 2003; 21(4):166–171

▶ Parameters and Positioning

[1] Barth V, Herrmann J, Koubenec HJ. Brustkrebs Info. Available at: http://www.brustkrebs.de, Version: 27.02.2013

[2] Boyce M, Gullen R, Parashar D, et al. Comparing the use of PGMI scoring systems used in the UK and Norway to assess the technical quality of screening mammograms: a pilot study. Breast Cancer Res. 2012; 14 Suppl. 1:P41

[3] Farria DM, Bassett LW, Kimme-Smith C, DeBruhl N. Mammography quality assurance from A to Z. Radiographics. 1994; 14(2):371–385

[4] Faulconer LS, Parham CA, Connor DM, et al. Effect of breast compression on lesion characteristic visibility with diffraction-enhanced imaging. Radiol Clin North Am. 1992; 30(1):21–53

[5] Fischer U. Röntgenmammografie. Verstehen. Anwenden. Optimieren. Stuttgart: Thieme; 2003

[6] Fischer U, Baum F. Direct Diagnosis in Radiology. Breast Imaging. Stuttgart: Thieme; 2007

[7] Geiser WR, Haygood TM, Santiago L, Stephens T, Thames D, Whitman GJ. Challenges in mammography: part 1, artifacts in digital mammography. AJR Am J Roentgenol. 2011; 197(6):W1023–:W10:30

[8] Heywang-Köbrunner SH. Planungen der Deutschen Röntgengesellschaft und des Berufsverbandes zur Qualitätssicherung bei Mammographie [Plans of the German Roentgen Society and the Professional Association for Quality Assurance in Mammography]. Radiologe. 2001; 41(4):352–358

[9] Heywang-Köbrunner SH. Bildgebende Mammadiagnostik. Stuttgart: Thieme; 2003

[10] Heywang-Köbrunner SH, Schreer I, Barter, S. Diagnostic Breast Imaging. Stuttgart: Thieme; 2014

[11] McLelland R, Hendrick RE, Zinninger MD, Wilcox PA. The American College of Radiology Mammography Accreditation Program. AJR Am J Roentgenol. 1991; 157(3):473–479

[12] Moreira C, Svoboda K, Poulos A, Taylor R, Page A, Rickard M. Comparison of the validity and reliability of two image classification systems for the assessment of mammogram quality. J Med Screen. 2005; 12(1):38–42

[13] Otto H. Mammadiagnostik für MTRA und Ärzte. Berlin: Springer; 2001

[14] Peters J, Jacobi V, Krapfl E, Kirchner J. Mammografie-Einstelltechnik: Wie betrifft sie den Gynäkologen? [Mammografic positioning: how does it affect the gynecologist?]. Zentralbl Gynakol. 2002; 124(6):326–330

[15] Pfandzelter R, Wülfing U, Boedeker B, Heywang-Köbrunner S. Die diagnostische Bildqualität von Mammografien in der vertragsärztlichen Versorgung Deutschlands [Diagnostic image quality of mammograms in German outpatient medical care]. Rofo. 2010; 182(11):993–1000

[16] Roth-Ganter G, Fischer U. Mammografie-Handbuch für die tägliche Praxis. Stuttgart: Thieme; 2002

[17] Spuur K, Hung WT, Poulos A, Rickard M. Mammography image quality: model for predicting compliance with posterior nipple line criterion. Eur J Radiol. 2011; 80(3):713–718

[18] Warren LM, Mackenzie A, Cooke J, et al. Effect of image quality on calcification detection in digital mammography. Med Phys. 2012; 39(6):3202–3213

▶ Evaluation

[1] American College of Radiology (ACR). Breast Imaging Reporting and Data System Atlas (BI-RADS Atlas). Reston, VA: American College of Radiology; 2003

[2] Balleyguier C, Ayadi S, Van Nguyen K, Vanel D, Dromain C, Sigal R. BIRADS classification in mammography. Eur J Radiol. 2007; 61(2):192–194

[3] Boyer B, Canale S, Arfi-Rouche J, Monzani Q, Khaled W, Balleyguier C. Variability and errors when applying the BIRADS mammography classification. Eur J Radiol. 2013; 82(3):388–397

[4] Deutsche Gesellschaft für Senologieee.V., Deutsche Krebshilfe e.V. Stufe-3-Leitlinie "Brustkrebs-Früherkennung in Deutschland". Munich: W. Zuckschwerdt; 2003

[5] European Commission. European Guidelines for Quality Assurances in Mammography Screening, 3rd ed. Cologne: Bundesanzeiger Verlag GmbH; 2001

[6] Fischer U. Röntgenmammografie. Verstehen. Anwenden. Optimieren. Stuttgart: Thieme; 2003

[7] Fischer U, Baum F. Mammography Casebook. Stuttgart: Thieme; 2006

[8] Fischer U, Baum F. Direct Diagnosis in Radiology. Breast Imaging. Stuttgart: Thieme; 2007

[9] Fischer U, Helbich T. Mammografiebefundung nach BI-RADS. Illustrierte Anleitung zur einheitlichen Befunderstellung von Mammografie, Mammasonografie, MR Mammografie. 2nd ed. Stuttgart: Thieme; 2006

[10] Fischer U, Hermann KP, Baum F. Digital mammography: current state and future aspects. Eur Radiol. 2006; 16(1):38–44

[11] Fischer U. Röntgenmammographie: Herdbefunde, Verkalkungen und Architekturstörungen [X-ray mammography: calcifications, masses and architectural distortions]. Radiologe. 2008; 48(8):785–797, quiz 798

[12] Heywang-Köbrunner SH, Schreer I, Barter S. Diagnostic Breast Imaging. Stuttgart: Thieme; 2014

[13] Hirunpat S, Tanomkiat W, Khojarern R, Arpakupakul N. Accuracy of the mammographic report category according to BIRADS. J Med Assoc Thai. 2005; 88 (1):62–65

[14] Lanyi M. Brustkrankheiten im Mammogramm. Berlin: Springer; 2003

[15] Leitlinienprogramm Onkologie der AWMF, Deutschen Krebsgesellschaft eV und Deutschen Krebshilfe eV. Leitlinienreport der S 3 Leitlinie fur die Diagnostik, Therapie und Nachsorge des Mammakarzinoms. AWMF Register-Nr. 032–045OL. Available at: http://www.awmf.org/leitlinien/detail/ll/032–045OL.html. Accessed February 9, 2017

[16] Müller-Schimpfle M, Wersebe A, Xydeas T, et al. Microcalcifications of the breast: how does radiologic classification correlate with histology? Acta Radiol. 2005; 46(8):774–781

[17] Otto H. Mammadiagnostikfür MTRA und Ärzte. Berlin: Springer; 2001

[18] Pfarl G, Helbich TH. Breast Imaging Reporting and Data System (BI-RADS)—German version [in German]. Rofo. 2002; 174(7):921–926

[19] Roth-Ganter G, Fischer U. Mammografie-Handbuch für die tägliche Praxis. Stuttgart: Thieme; 2002

5

6 Breast Ultrasonography

S. Luftner-Nagel

6.1 Technique and Methods

6.1.1 Basic Principles

Breast ultrasonography is a dynamic real-time examination in which image generation is based on the conduction and reflection of high-frequency mechanical sound waves (ultrasound) in the breast tissue. The technique is also known as sonography and is loosely but commonly referred to simply as *ultrasound* or ultrasound examination). Numerous physical and technical factors play an important role in the formation of an ultrasound image. A wide range of high-performance examination devices are available for everyday clinical practice (▶ Fig. 6.1a).

Physical Principles

Modern ultrasound probes (transducers) contain numerous piezoelectric elements or crystals that can produce ultrasonic impulses and receive the returning echo signals. An ultrasound impulse is a mechanical compression wave generated by applying AC voltage to the surface of a piezoelectric crystal, causing it to oscillate. Thus these crystals transform an electrical current into mechanical ultrasound waves. The number of oscillations per unit time is termed frequency, given in hertz. One hertz (1 Hz) equals one oscillation per second, and one megahertz (1 MHz) is thus one million oscillations per second. Medical ultrasound devices work at a sound frequency (f) of 3 to 15 MHz, well above the audible hearing range (which is up to 20 kilohertz).

Sound waves require an elastic medium in which to propagate. They are a continuous, longitudinal transmission of kinetic energy. Particles that are induced to oscillate transmit these oscillations to the neighboring particles, which in turn transmit them on to the next particle. The kinetic energy spreads out continuously in a sine wave manner, so that regions of particle compression and decompression develop at periodic intervals. The distance between two identical phases of a wave (locations of the same particle density) corresponds to the wavelength, λ (lambda).

The velocity with which the kinetic energy is transmitted depends on the density and compressibility of the medium through which the wave passes. Because the particle density in dense tissue is higher than that in soft tissue, the distance between one particle and the next is less and the sound propagation velocity is therefore higher. Sound waves travel faster in solids than in liquids or gases. The velocity (v) with which sound waves travel through the different soft human body tissues, however, is very similar and is calculated to be approximately 1540 meters per second (m/s) (international standard value).

The velocity of a wave is a function of the wavelength and frequency:

$$v = \lambda \times f$$

where v = velocity of propagation (m/s); λ = wavelength (m); f = frequency (s^{-1})

Thus, the higher the frequency, the shorter the wavelength. For the sound frequency applied in breast ultrasonography of 7.5 to 15 MHz, the wavelength is 0.1–0.2 mm. These short wavelengths enable good spatial resolution and the visualization of small structures.

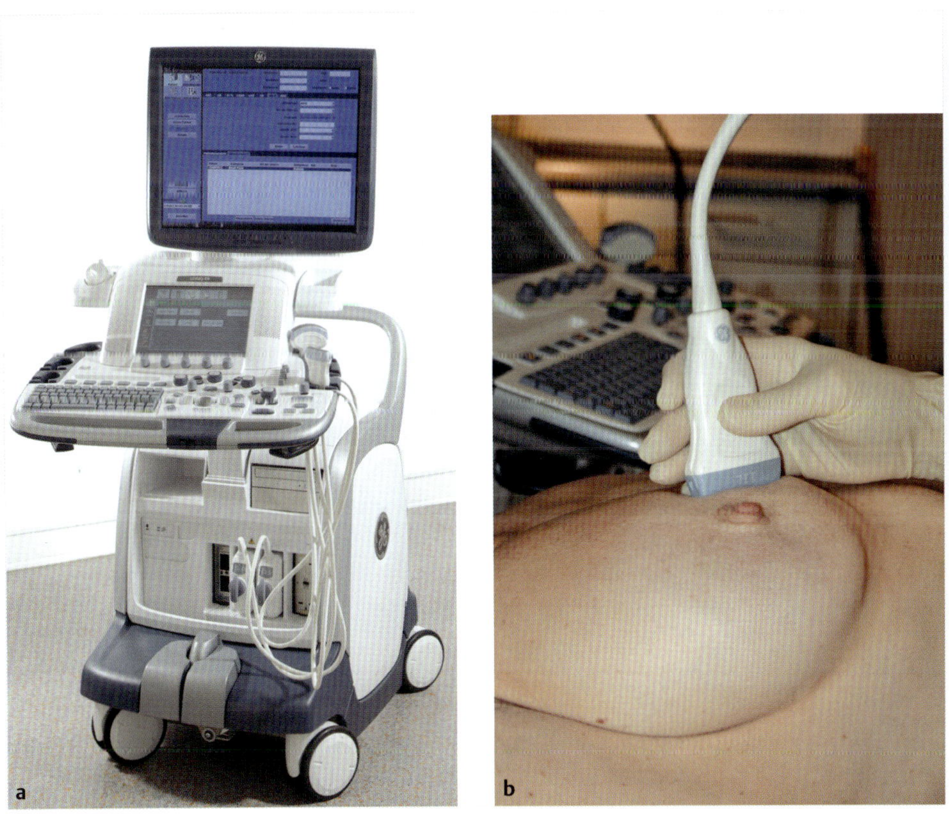

Fig. 6.1 Sonography unit and transducer position for an ultrasound examination of the breast. (a) Commercially available, high-definition ultrasound device (manufacturer: GE Healthcare, Milwaukee, WI, USA). **(b)** Typical handling of the transducer with fixation of the base. Movement is carried out from the wrist.

Generation and Transmission of Sound Waves

In breast diagnostics, virtually all ultrasound techniques are based on pulsed sound waves, the so-called pulse–echo principle. A short electrical impulse causes the piezoelectric crystals of the transducer to oscillate, producing the sound impulse. The acoustic pulse is terminated by damping and is of only short duration (ca. 1 microsecond [µs]). In the time between impulses, the piezo element is available to receive the returning ultrasound echoes.

When sound waves encounter the boundary between two different materials with different densities there is a change in the sound propagation velocity and acoustic impedance. The greater the difference in the acoustic impedance of the materials or tissues, the greater the proportion of sound wave energy that will be reflected at the interface rather than transmitted onward. In breast ultrasonography, the oscillations of the piezo element in the transducer are transmitted to the body over a coupling layer that compensates for the impedance (sound resistance) between the oscillator and the skin surface. This minimizes the energy loss involved in transmission of the sound to the breast. Because the impedance differences between tissues within the breast are slight, the sound waves propagate well and only a small proportion of the acoustic energy is reflected at breast structure interfaces at different depths. The mechanical energy of the reflected sound waves returning is transduced into electric current at the piezo element. Because the time between sending and receiving a signal is proportional to the distance between the transducer and the reflecting structure, this information can be converted into an image.

6

Note

Ultrasound systems send 3,000 to 5,000 ultrasonic impulses per second and receive their reflection as echoes. This information is converted and processed into images fast enough to enable an examination in real time.

Because of the very similar impedances of breast tissue structures, the larger part of the applied acoustic energy is transmitted through interfaces that are encountered, thus remaining available for undergoing further reflection and generating images of deeper structures. However, if the ultrasound meets an interface or a structure with a significantly smaller or larger impedance, such as air, macrocalcifications, or some tumors, then almost all of the sound wave energy is reflected. Posterior to these structures, the sound is extinguished and no image information can be gained there.

Note

Differences in impedance are slight in normal breast tissue. Therefore only a small proportion of the energy of the ultrasound is reflected at any interface and serves to generate the image.

Physical Phenomena

Sound waves propagate optimally if they strike a smooth interface at right angles. Depending on the impedance jump, a greater or lesser portion of the energy is reflected back to the transducer and the other portion travels further on in a straight line, i.e., at the same angle. This situation rarely occurs within the human body. Instead, ultrasonic waves undergo various interactions as they travel through tissue: absorption, reflection, refraction, and scattering.

Acoustic Attenuation

With increasing penetration, there is increased acoustic attenuation. One source of this energy loss is absorption, which is the conversion of acoustic energy into heat. In biological tissues, absorption is approximately linearly dependent on the sound frequency: the higher the frequency, the higher the absorption. Also, deflection of the sound wave occurs at the interfaces of tissue structures with varying density, depending on the difference in impedance, the angle of incidence, and the roughness of the interface surface (reflection, refraction, and scattering). The portions of sound energy that are not reflected back to the transducer and received as echoes are lost to image generation.

Note

The optimal transducer frequency is a compromise between the required penetration depth and the desired spatial resolution.

If the ultrasound strikes an interface at a slanting angle, a small portion of the acoustic energy (a function of amplitude) is reflected back toward the source (transducer) along the direction of incidence. The remaining ultrasound is transmitted into the second medium at an angle different from the angle of incidence (refraction). The angle of refraction depends on the incident angle (the more inclined, i.e., the greater the angle of incidence, the greater the angle of refraction) and on the sound wave velocity in the second medium. If tissue structure interfaces are highly oblique to the sound wave propagation direction, the refraction of the sound wave can be such that it does not penetrate the tissue surface, causing posterior acoustic attenuation—i.e., shadowing, within which there is no image information. This is of practical importance for breast ultrasonography. Using an appropriate examination technique, increasing the transducer pressure (i.e., compressing the breast) can often flatten the tissue interfaces, making them more horizontal to the transducer surface and thereby improving ultrasound transmission and the ultrasound image (see ▶ Fig. 6.4).

Practical Tip

When there is poor sound penetration (due to inhomogeneous parenchyma or scarring), the ultrasound transmission can often be improved by increasing the breast compression using the transducer probe.

Interference

Superimposition of two or more sound waves with different phases can strengthen (constructive interference) or weaken (destructive interference) the combined wave according to the superposition principle. The spatial arrangement of regions of weakening and strengthening is known as the "interference pattern." This has a significant influence on the appearance of the ultrasonic image impression (its homogeneity). In the ultrasonographic examination of the breast, an inhomogeneous interference pattern can significantly compromise the sensitivity of the examination. Because the interference pattern is determined primarily by the number and arrangement of oblique interfaces in the breast and not by the mammographic density of the parenchyma, the sensitivity of breast ultrasonography can vary, even in women with similarly structured parenchyma on mammography.

Posterior Acoustic Enhancement

Using optimal time gain compensation (TGC), there is a uniformly brighter tissue strip posterior to structures that have less acoustic attenuation than the surrounding parenchyma—as in the case of cysts, for example. This impression of posterior acoustic enhancement arises because the TGC compensates for the normal acoustic attenuation of the neighboring tissue, which increases with increasing depth. Thus the reflected sound signals from regions posterior to a nonattenuating structure are less weakened than the surrounding tissue and appear strengthened.

Reverberation Artifacts

Reverberation occurs when sound waves encounter two highly reflective parallel surfaces, such as the walls of a puncture needle. A portion of the sound wave energy is reflected back and forth between the two surfaces before returning to the transducer and being received as echoes. Because the probe detects prolonged traveling times after repeated reflections and treats these as longer distance of travel, the structure erroneously appears repeatedly at equally spaced intervals with decreasing brightness at deeper levels in the image. The echo that returns to the transducer after a single reflection corresponds to the correct position of the reflecting structure.

6.1.2 Device Adjustments

Display Unit and Documentation

Contrast and brightness can be adjusted according to a gray scale, which is usually displayed at the image border. Optimal adjustment of this parameter is important for being able to utilize the entire grayscale information and to enable a good comparison for later follow-up or for a second opinion.

Signal Enhancement

The brightness of the ultrasound image depends on the power output of the transmitter, amplitude of the received signal, and depth compensation. The higher the emitted sound amplitude and overall gain, the brighter the image. The base setting should be made in the center of the image, where most parenchyma is displayed. With an average power output, the dial or slide controls (TGC) are used to adjust the depth compensation so that superficial connective tissue fibers have the same level of brightness as those at a greater depth, i.e., throughout the image.

Focusing

Most transducers used for breast examinations provide multiple focal zones with variable focal positioning. When a single focus is used, the focus should be set at the level of the anterior to middle third of the breast. In the supine position, most breasts have a depth of 2 to 4 cm. When using multiple foci with two or three zones, covering the area from 0.5 to 3 or 4 cm depth is usually optimal.

Image Scaling

The penetration depth should be set just large enough so that the chest wall is visible on the bottom of the monitor. If the scale is smaller, the screen width will not be optimally utilized. If the scale is larger, the deep tissue structures will be incompletely displayed. Zooming (detail enlargement) is advantageous for targeted imaging of small lesions.

6.1.3 Examination Technique

Positioning

For the ultrasound examination of the breast, the patient should lie supine with the hands folded behind the head (▶ Fig. 6.2). In this position, the breast is distributed over the chest wall and slightly immobilized, facilitating a systematic examination. Acoustic penetration is improved by the flattening of the tissue structures and reduction of the tissue layer thickness. With large breasts that fall to the side, elevating the ipsilateral chest wall with a pillow properly placed under the patient can enable better positioning of the breast. In individual cases, a supplementary examination can be carried out in other positions (for example, in the sitting position) for targeted examination of a tumor that is not palpable in the supine position. Ideally, the ultrasound machine should be positioned across from the examiner, in the direct line of sight (▶ Fig. 6.2).

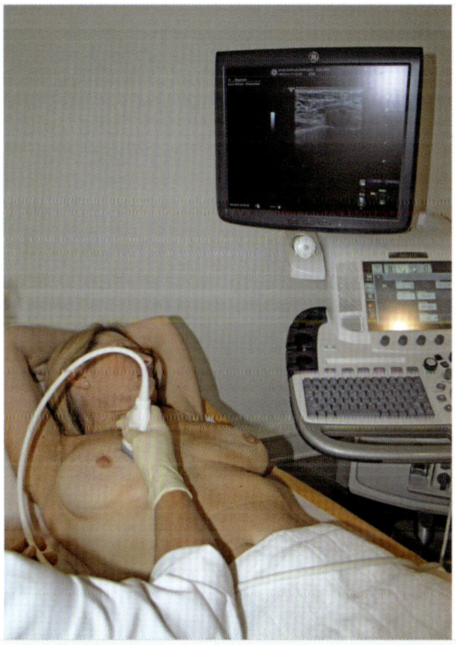

Fig. 6.2 Orientation of the ultrasound device. Ideally, the ultrasound unit should be in a position that is in the direct line of sight of the examining physician and such that it is possible for the patient to follow the examination on the monitor.

Coupling Gel

To facilitate a systematic and uninterrupted ultrasound examination, the entire breast should be amply coated with coupling gel. This serves to effectively transfer the sound waves from the transducer to the breast by preventing the formation of a layer of air, which would cause an immediate reflection of the sound energy due to the large impedance discontinuity.

Holding the Transducer

The examiner should hold the transducer at the base and guide it using the wrist. For the systematic examination of the breast, the transducer is placed perpendicular to the skin surface (▶ Fig. 6.1b). Normally the weight of the transducer and the hand is sufficient coupling pressure. A greater pressure with compression and flattening of the breast structures is applied selectively as necessary, for areas in which the sound penetration is insufficient. On the other hand, if the coupling pressure is too high, the transducer can displace structures and push them ahead of the transducer. It is even possible to overlook tumors as a result of displacing them out of the image plane. It is best for the comfort of the examiner for them to sit next to the patient at a height that allows the examiner to lay the forearm lightly on her body. In this way, the sense of space and controlled transducer guidance is optimized (▶ Fig. 6.2).

Examination Procedure

Proper technique with systematic guidance of the transducer is important for the reliable detection of small breast tumors. The entire breast is examined with overlapping transducer paths, moving in a back-and-forth pattern with care not to miss any areas. The examiner is free to choose which sectional plane to use, but usually the sagittal plane is preferred. An examination in the transverse plane is usually performed as a supplementary procedure for additional assurance that the examination has been thorough, without any areas having been overlooked (for example, by novice sonographers or in a "second-look" ultrasound examination). The radial plane is usually used specifically as a diagnostic work-up of milk duct changes. To evaluate the retroareolar region, commonly restricted by the development of acoustic shadowing, it is often helpful to tilt the transducer in the appropriate direction and to increase the transducer contact pressure in a controlled manner.

In the dynamic examination, changing the image plane allows assessment of masslike breast structures in three dimensions. A true mass lesion, which can be demarcated from the normal breast structures on all sides, can easily be distinguished from a pseudotumor (▶ Fig. 6.3). In addition, the elasticity of structures and tissue can be assessed by increasing the transducer contact pressure. Soft tissues such as parenchyma and adipose tissue deform when compressed. Malignant tumors, on the other hand, have a firm consistency and can be compressed only little, if at all.

Practical Tip

Because a lesion in the breast is displaced when the transducer contact pressure is increased, one must take care to examine and document a lesion with and without compression in the same plane.

Decreased ultrasound penetration (attenuation) in normal, inhomogeneous parenchyma or posterior to scar tissue can often be improved by increasing the transducer contact pressure. This leads to flattening of the oblique interfaces and a decrease in sound refraction, facilitating the assessment of the ultrasound examination and the differential diagnosis from malignant changes (▶ Fig. 6.4). Carcinomas often have inhomogeneous internal structure and show a posterior acoustic shadow that is not diminished by increasing the transducer contact pressure because of the limited compressibility of the tumor.

Image Documentation

The stored image should exhibit an image quality that allows evaluation of the findings by a third party. It must include data concerning patient identification, examination date, and the institution where the examination was performed, as well as technical data about the device settings (▶ Table 6.1). As a matter

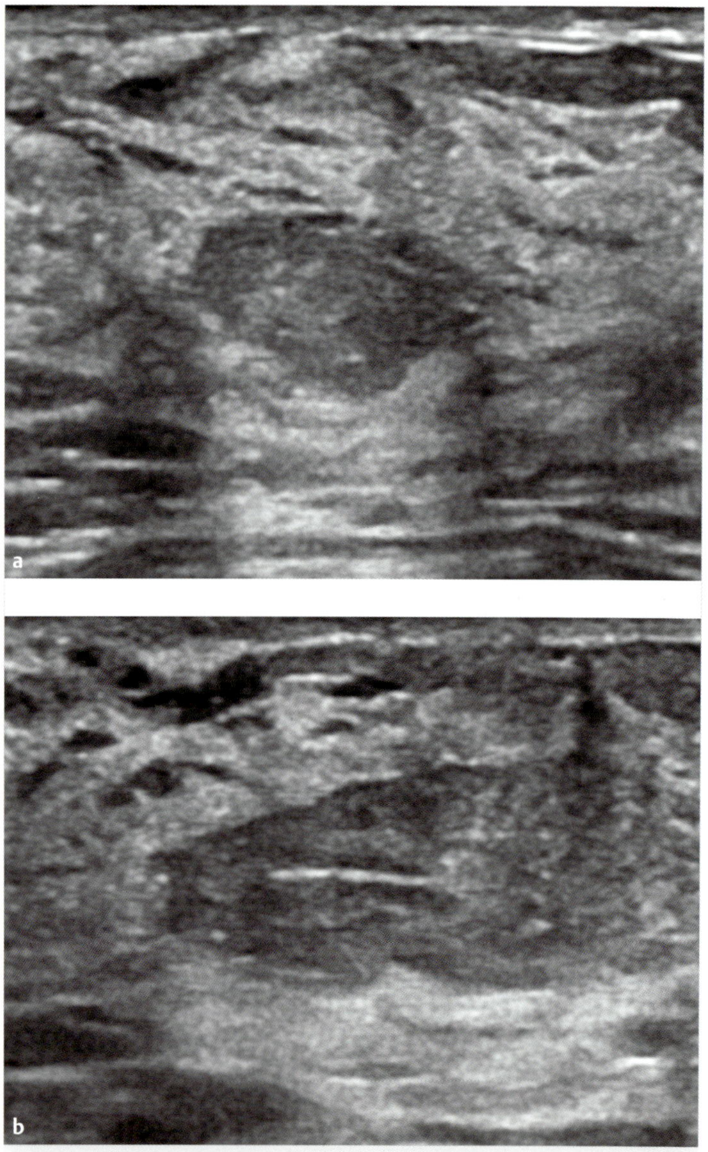

Fig. 6.3 Changing the plane to differentiate a true mass lesion from a pseudotumor. (a) Hypoechic mass lesion in the sagittal transducer position. **(b)** Turning the transducer shows that the lesion is not circumscribed in three dimensions (pseudotumor).

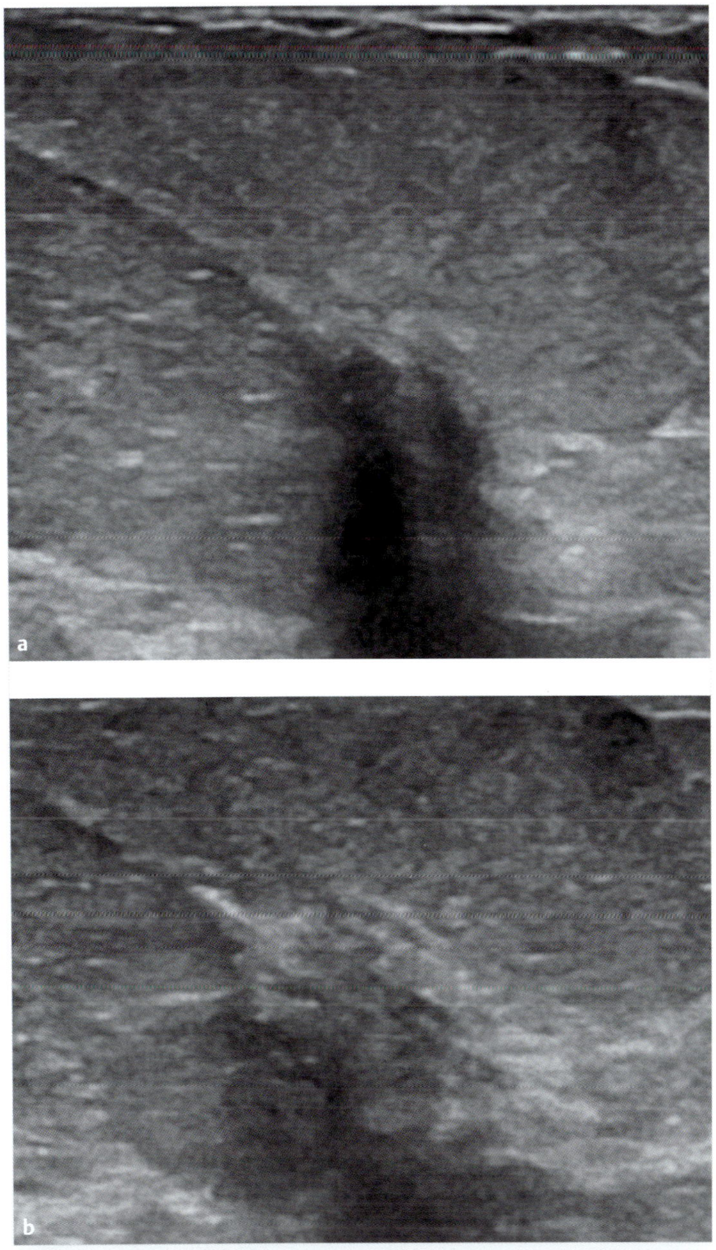

Fig. 6.4 **Effect of increased transducer pressure on the sound penetration and the image quality in the case of inhomogeneous parenchyma. (a)** Normal examination without additional compression using the transducer. **(b)** Same image section with increased breast compression, showing a visible improvement in the acoustic penetration.

of principle, every focal lesion should be documented in two perpendicular planes. In the case of multiple, similar, benign-appearing lesions, such as simple cysts, a representative sample should be documented. If there are no focal lesions, bilateral representative image sections of the parenchyma and axilla should be documented. These images should convey an impression of the assessability and reliability of the examination.

Practical Tip

It is a good idea to include sonographic image documentation of areas that are suspicious on clinical examination, mammography, or breast MRI, even if no sonographic lesion is verified there. This both documents the absence of a sonographic correlate and demonstrates that adequate care and attention were taken.

Table 6.1 Image documentation in breast ultrasonography

Documentation	Data
Written Information on the ultrasound image report	Name and location of the institution
	Examination date
	Patient's name and birth date and/or identification number
	Initials of the examiner
	Device adjustment (frequency, performance, focal position)
Image acquisition	Transducer orientation or plane on a diagram
	In the case of unremarkable findings: bilateral representative cross sections of the parenchyma + axilla
	In the case of a suspicious lesion: documentation images with and without measurements in two perpendicular planes
	Lesion localization on a diagram (with indication of the distance to the nipple)

Written Documentation

The written documentation of an ultrasound examination should contain both a statement as to the assessability of the examination, as well as a description and assessment of conspicuous findings using standardized evaluation criteria and nomenclature.

6.1.4 Ultrasound Techniques

Brightness Mode

For the production of an ultrasound image, short ultrasonic impulses are transmitted that spread out into the tissue and are reflected back from different depths. Because the time lapse between sending and receiving a signal is proportional to the distance between the reflecting interface and the transducer, the depth of these structures can be calculated. If the amplitude of a reflection is depicted as a point of corresponding brightness, the reflections from multiple ultrasound beam scans can be visualized next to each other to produce a two-dimensional image of the examined organ (▶ Fig. 6.5a). In modern devices, 256 gray-scale levels are usually represented. The stronger the signal (greater amplitude), the brighter the corresponding image point. The high image refresh rate enables real-time visualization.

Color-Coded Duplex Sonography (Doppler-Sonography)

Measurement of tissue perfusion can be helpful in differentiating between benign and malignant tumors because rapidly growing malignant tumors by their nature usually have increased vascularization. But because proliferative benign changes can also exhibit increased vascularization such measurement cannot allow definitive differentiation.

The Doppler effect is the velocity-dependent change in the received frequency (frequency shift) of the wave from a transmitter or a reflecting object (such as blood cells) that is moving relative to the receiver. In duplex sonography, B-mode sonography (to detect vessels) and pulsed Doppler imaging (to

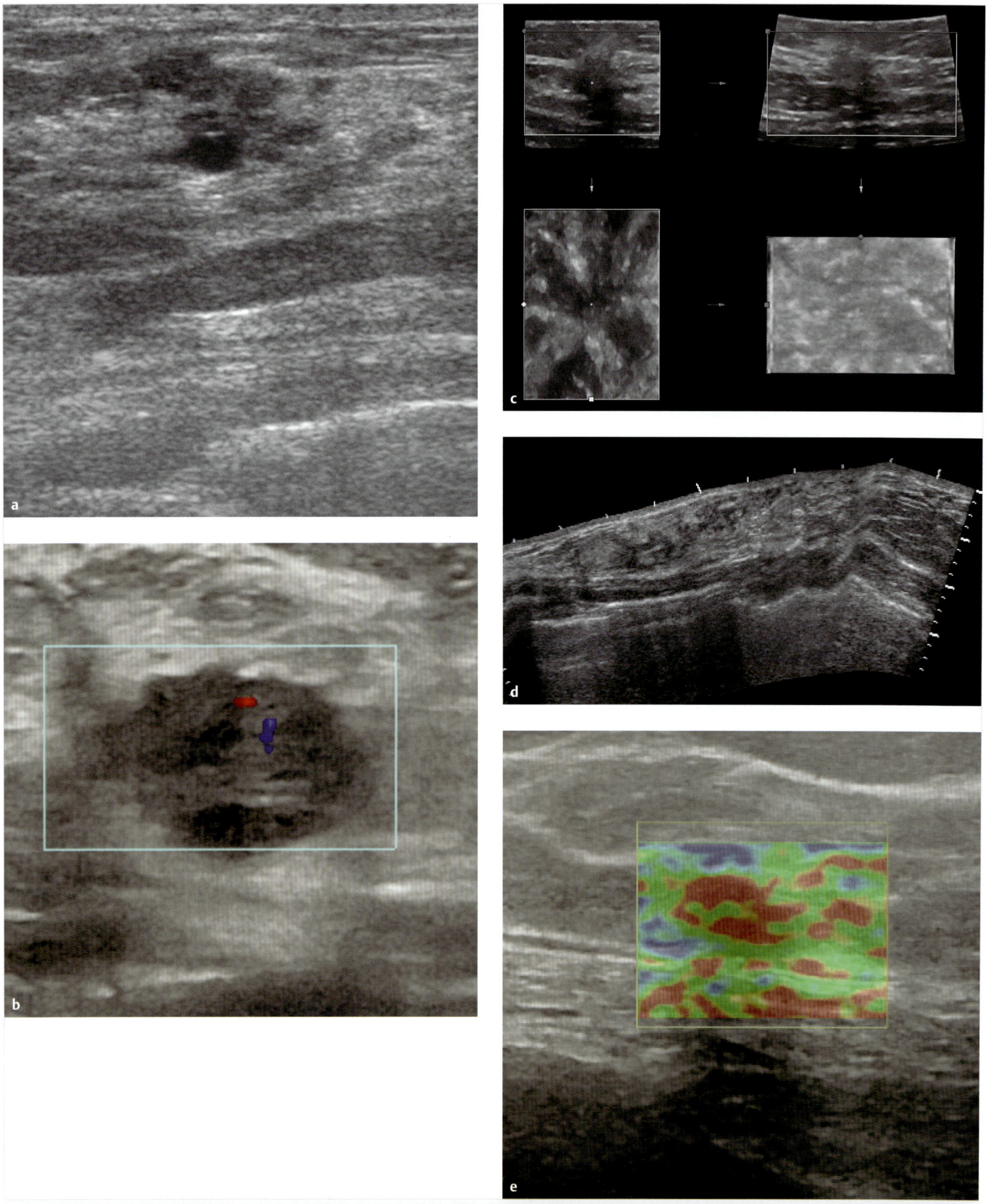

Fig. 6.5 Ultrasound techniques. (a) B-mode. **(b)** Color-coded duplex sonography. **(c)** 3D sonography. **(d)** Panorama image. **(e)** Elastograph.

analyze flow in the detected vessels) are combined. Because tumor blood vessels are very small, they are not themselves discernible by sonographic imaging, so this method by itself has no significant utility in tumor diagnosis. However, in color-coded Doppler sonography (or color Doppler sonography), the frequency shifts (due to flow movements) are visualized as colored areas in a defined image section of the B mode image. The color shade encodes the flow direction and the color brightness encodes the velocity of flow. This method enables two-dimensional visualization of blood flow in real-time imaging and the localization of small vessels, which are of importance in tumor diagnosis (▶ Fig. 6.5). Addition of the duplex mode makes it possible to calculate the flow velocity within the visualized vessels and analyze the frequency profile. Because of the limited sensitivity of color Doppler in locating microscopically small vessels and the angle-dependency of the Doppler principle, another method of signal processing has been developed. The power Doppler method encodes only the intensity of the Doppler signal, without providing information on the *direction* of flow. The advantage of this technique is its high sensitivity: it can make the smallest vessels visible and detect very slow flows. As an example, it can differentiate a thickened cyst with internal echoes from a round, very hypoechoic solid process. Its disadvantages include susceptibility to artifacts caused by the high sensitivity to movement and the inability to quantify measurements.

Contrast-Enhanced Sonography

Echo contrast agents are substances that, after intravenous injection, raise the sonographic reflectivity of blood. After approximately 20 to 40 seconds the contrast material reaches the arterial blood vessels and produces improved visualization of these blood vessels in the color Doppler examination. Like noncontrast Doppler diagnostics, contrast-enhanced sonography is not well suited as a method of detection. Also, because of the short window of opportunity before the contrast medium has passed through the arterial vessels, this method—which is still experimental—only comes into use if needed as a diagnostic work-up in selected problem cases.

Doppler sonography requires the examiner to have extensive experience and sufficient practice. Because it does not provide reliable differentiation between benign and malignant changes, it is advisable to carry out an ultrasound-guided core needle biopsy, if there is any doubt.

3D Sonography

Three-dimensional (3D) ultrasonography is based on the acquisition and processing of numerous, closely spaced two-dimensional (2D) image sections. For this, a specialized probe affords automatic guidance of the transducer and ensures a geometrically correct 3D reconstruction.

3D visualization of breast tumors facilitates the spatial presentation and image interpretation. Its particular advantage, however, is the ability to document the 3D findings and communicate them to other attending physicians, such as the surgeon. In addition, this is the only method that allows reconstruction of the horizontal sectional plane, in which the retraction pattern of malignant tumors is often well visualized. The 3D visualization made possible by volume rendering provides good orientation in cases of multifocal tumor foci and for the exact needle guidance in all planes during interventional procedures (▶ Fig. 6.5c).

Panorama Scan Mode

A panorama scan is a 2D wide image reconstruction in one image plane. The transducer is guided over the region of interest in an appropriate sectional plane, blending multiple images together to form one long image of an entire breast section. This gives a clear graphic survey of the breast anatomy and indicates the full dimensions of any lesions present and their spatial relationship(s) to each other in the imaged plane (▶ Fig. 6.5d).

Real-Time Compound Scan

This image reconstruction technique, also known as Sono CT, combines images that have been generated with various ultrasound incidence angles. In this way, creation of artifacts due to increased or diminished sound transmission is suppressed. Because these artifacts can also be important diagnostic criteria (acoustic enhancement, acoustic attenuation), the compound scan should only be used in conjunction with a simple scan to evaluate suspicious findings.

Automated Volume Scan

An automated system generates continuous ultrasound section images through the entire breast in this examination technique. The examination is performed using a dedicated ultrasound device (ABVS, automated breast volume scanner); the patient must be positioned comfortably, and the transducer, which is attached to a swivel arm, is placed on the breast. Sonographic section images are automatically acquired and can then be used to reconstruct a full-field volume image. For most breast sizes, two to four such volume scans suffice. The automated examination has the advantages that it is a physician-independent examination procedure compatible with a screening situation; that it requires a relatively short imaging time; that it provides coronal section reconstruction and volume visualization; and that it provides standardized documentation. The ultrasound images generated are sent to a workstation, where they can be processed and evaluated at any time by one or more examiners; they are thus easily accessible for a second opinion. Disadvantages include the uncertain inclusion of lateral parts of the breast and the axillary tail due to inadequate placement of the transducer on the breast, and the inability to evaluate the individual physical characteristics of tissues, such as lesion compressibility and plasticity.

Ultrasound Elastography

Malignant tumors usually have a harder consistency and are less compressible than benign tumors and normal breast structures: elastography uses real-time acquisition of the differences in tissue elasticity to distinguish these changes. Two different methods are under test:

- *Compression by manual transducer movement:* Some systems compress the breast tissue by manual oscillation of the transducer toward and away from the chest wall with quick, uniformly intense movements over the lesion being examined. The deformations this causes produce detectable changes in the ultrasound signal that are represented by color coding. To date, only limited reproducibility of the findings has been shown, because of the difficulty of carrying out the procedure.
- *Shear-wave method:* The shear-wave method uses a different technique to measure elasticity. Here, in addition to the ultrasound waves emitted to produce the image, a separate ultrasound impulse is sent out into the tissue, which generates a transverse mechanical (shear) wave. The velocity at which this transverse wave is propagated is dependent on the tissue elasticity, which can be measured and portrayed using color coding. Because this examination is not dependent on the examiner and renders absolute tissue density values in different measurement windows, one would expect

the study results to be reproducible. However, it is not uncommon to find benign tumors that are soft, and malignant tumors that are hard; for this reason, preference should be given to the liberal use of ultrasound-guided core needle biopsy, which is simple, cost-effective and reliable (▶ Fig. 6.5e).

6.1.5 Quality Assurance

The German National Association of Statutory Health Insurance Physicians has stipulated the following requirements for the performance of the ultrasound examination of the regional lymph nodes (2012):

- *Mode:* B mode.
- *Transducer:* linear array.
- *Transducer frequency:* ≥ 7.0 MHz.
- *Transmit focal setting:* electronically variable focal zone for optimal lateral resolution.
- *Transmission aperture:* variable, depending on the chosen focal zone.
- *Reception amplification:* variable time gain compensation (TGC).
- *Dynamic range:* ≥ 60 dB.
- *Field of view:* penetration depth ≥ 6 cm, width ≥ 8 cm.
- *Image refresh rate:* ≥ 15 images/second.
- *Bit depth:* ≥ 8 bits.

The German Society for Ultrasound in Medicine (DEGUM: Deutsche Gesellschaft für Ultraschall in der Medizin) has formulated concrete requirements for devices and technical specifications:

- Device requirements: image quality in clinical use:
  - 128 gray levels.
  - Good differentiation of breast tissue components (soft tissue differentiation).
  - Depiction of structural irregularities in breast tissue.
  - Depiction of tumor border irregularities.
  - Distinct depiction of cysts larger than 4 mm (higher standard: 2 mm).
  - Distinct depiction of solid tumors larger than 10 mm (higher standard: 5 mm).
  - Penetration depth of at least 4 cm with good spatial resolution.
  - Depiction of a 20-gauge needle in breast tissue along the image plane.
- Technical requirements:
  - Documentation: digital or hardcopy system (multiformat film, video printer, Polaroid).
  - Transducer frequency > 5 MHz or multifrequency transducer (must fulfill quality assurance requirements on phantom tests).
  - Image refresh rate > 12 images/second.
  - Field of view > 3.8 cm.
  - Incorporated acoustic standoff, if near-field resolution is not sufficient.
  - Image monitor display: patient identification number, date, transducer, measuring caliber, body marker (or other localization specification, e.g. text), capacity, time gain compensation (TGC), preset, depth scale.
  - Device safety in accordance with the regulations for the device.
  - Freehand biopsy or puncture guidance with image screen display (5 mm lesions must be accessible at all depths).

Device performance is tested on a phantom and on the basis of breast examinations. Quality control by technical experts should be carried out every 6 months.

6.2 Evaluation

6.2.1 Terminology

The most recent edition of the ACR BI-RADS lexicon incorporated breast sonography and breast MRI alongside mammography and it is recommended that this terminology be used to describe sonographic breast findings.

6.2.2 Tissue Type in Sonography

The sensitivity of any breast examination method will depend on the individual anatomical and/or physiological conditions. Just as for the specifications of density types according to the ACR in X-ray mammography, the sensitivity of the breast ultrasound examination should be assessed and documented for each examination. The ACR established systematic and standardized criteria for X-ray mammography findings in its BI-RADS lexicon, first published in 1999. In the second edition, published in 2003, the ACR established corresponding criteria for the categorization of ultrasound findings, and thereby created a uniform international terminology as a basis for a standardized evaluation. The German Society for Ultrasound in Medicine (DEGUM) has modified this US BI-RADS lexicon for the German-speaking European regions, incorporating its years of experience with breast sonography.

The echotexture—that is, the echogenicity and homogeneity—of the parenchyma in the sonographic image is the fundamental factor in the reliability of the sonographic examination for the detection of carcinoma of the breast (▶ Table 6.2, ▶ Fig. 6.6). With high echogenicity, a clear, sharp image with good penetration of the parenchyma can usually be acquired. When there is additional fibrosis, however, the acoustic penetration can be impaired to a greater or lesser degree, and the result is an inhomogeneous-appearing image with regions of acoustic attenuation, which can considerably impair the assessability and sensitivity of the examination. In an involuted breast, the ultrasound echogenicity of the breast tissue is lower. It is easier to overlook small, hypoechoic mass lesions without posterior attenuation within such an echotexture than within homogeneous, hyperechoic parenchyma. Fortunately, it is particularly for the involuted breast that

Table 6.2 Correlation between the composition of the mammary gland and the sensitivity of breast ultrasonography (Goettingen 4-level model)

Level	Echotexture	Sensitivity
1	Homogeneous, parenchyma equivalent	Very high
2	Homogeneous, mixed lipomatous/ parenchyma equivalent	High
3	Homogeneous lipomatous	Limited
4	Inhomogeneous, fibrocystic changes	Significantly limited

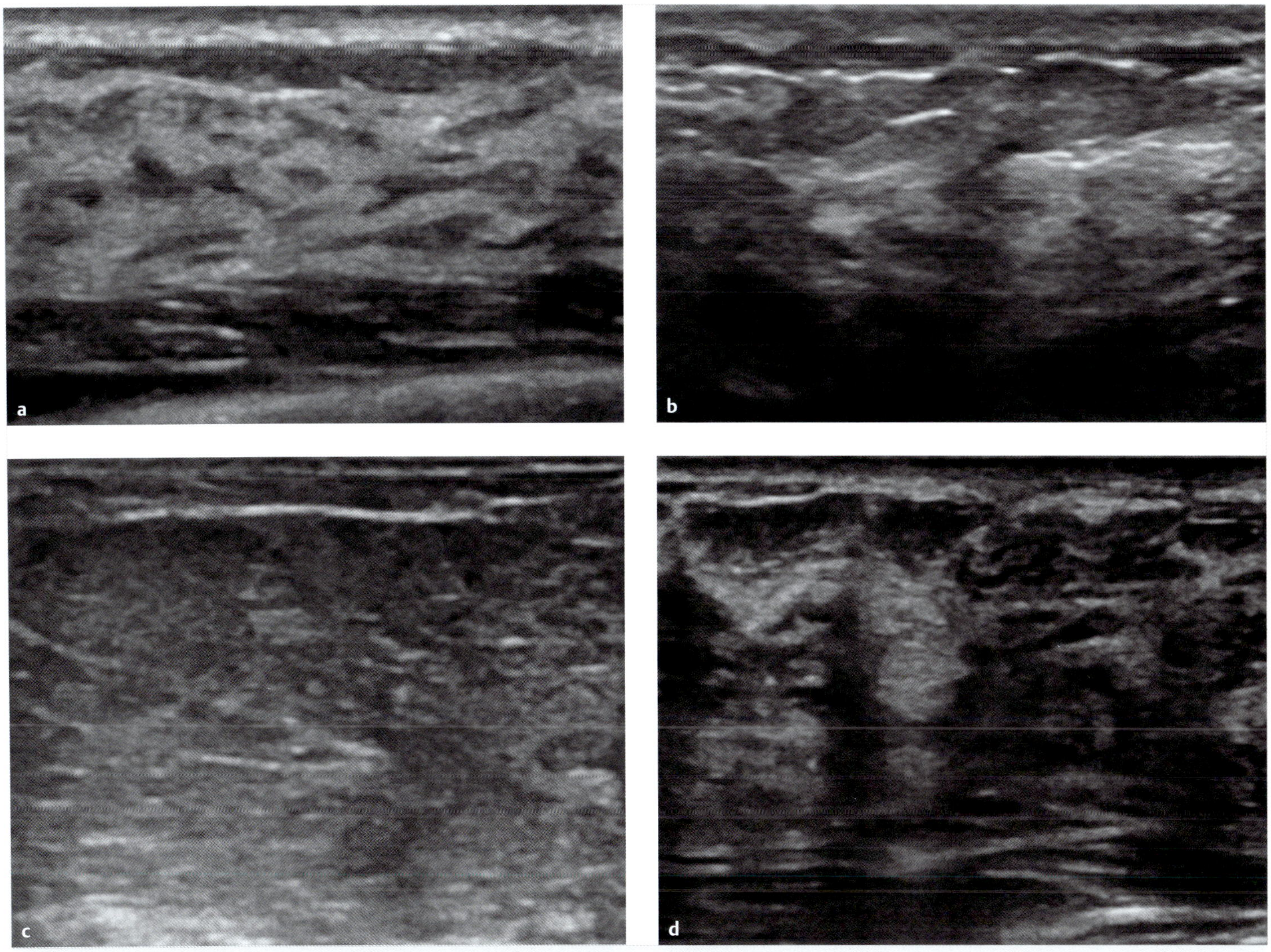

Fig. 6.6 Typical examples of different echo textures in breast sonography. (a) Glandular homogeneous. **(b)** Mixed lipomatous/parenchyma equivalent. **(c)** Homogeneous lipomatous. **(d)** Fibrocystic.

X-ray mammography has the highest sensitivity for the detection of breast cancer. The sonography examination report should always include a succinct description of the breast ultrasound composition pattern; this indicates the extent to which the sensitivity of the examination may have been compromised and a lesion missed. When breast cancer cannot be reliably ruled out in cases with reduced sensitivity, further diagnostic examinations may be indicated.

Note

- The written report of an ultrasound examination of the breast should contain annotation regarding the homogeneity of the echotexture, since an inhomogeneous echotexture may significantly reduce the sensitivity of the examination.
- Sonographic detection of small tumors in an involuted, fatty breast is often difficult.

6.2.3 Evaluation Criteria

Breast ultrasound allows reliable detection and characterization of mass lesions about 5 mm in size and larger. In addition, macrocalcifications about 5 mm in size and larger can easily be recognized by their strong anterior reflex with intense posterior shadowing/extinction, especially if the mammography results are known. However, reliable verification of microcalcifications is not possible sonographically; that remains the domain of X-ray mammography. When scanning the breast, one should pay special attention to architectural distortions (e.g., lobular breast carcinoma) and changes or discontinuities in the normal anatomical structures (e.g., milk ducts, Cooper's ligaments, the chest wall).

Note

Three main findings are distinguished in breast sonography: "mass," "architectural distortion" and "disruption of the normal anatomical structures."

Mass Lesion

A mass lesion is a three-dimensional, space-occupying, circumscribed structure. By rotating the transducer, one can demonstrate and document the spatial dimensions of a lesion and thus differentiate it from a pseudolesion, such as an adipose tissue island, which is not circumscribed in all planes.

The following criteria are used to assess the benign or malignant nature of a mass lesion:

- Shape.
- Orientation.
- Margin.
- Encircling border.
- Echo pattern.
- Internal structure.
- Ultrasound transmission.
- Calcifications.
- Elasticity assessment/compressibility.
- Mobility assessment.
- Architectural distortion or disruption of the surrounding structures.

All relevant findings should be documented in images and in a written report, including information on the position and size of a lesion (▶ Table 6.3). Whenever possible, comparison with previous images is obligatory. If the breast ultrasound is performed as a diagnostic work-up for a clinical, mammographic, or breast MRI finding (second-look sonography), an image of the area of interest is acquired for documentation, even in the absence of ultrasound findings.

Information about the position of a lesion should include the side (R, L), the clock face position, and the distance from the nipple. This information can also be mapped onto a diagram of the breast. The size of a mass lesion is usually expressed as the maximal dimensions in two perpendicular planes (see ▶ Table 6.3).

Practical Tip

One can use the length of the transducer head (e.g., 3.5 cm) as a "measuring stick" to estimate the distance between a sonographic finding and the nipple.

The terms used to describe the findings in terms of the criteria "shape" and "margin" are the same as those used for description of lesions in X-ray mammography:

- *Shape:* A benign mass lesion is most often oval in shape (it may include two or three undulations, i.e., be macrolobulated) or round. An irregular shape is neither round nor oval and is more likely to indicate malignancy.
- *Margin:* The margin signifies the delineation of the mass from the surrounding tissue. This can be circumscribed, i.e., well- or sharply-defined (more likely benign), or not circumscribed, i.e., indistinct, microlobulated, angular, or spiculated (more likely malignant).

Some descriptors are unique to the description of a lesion in breast ultrasonography (▶ Table 6.4, ▶ Fig. 6.7):

- *Orientation:* The orientation of a mass lesion is defined with reference to the skin surface. A mass oriented with the long axis parallel to the skin ("wider than tall") usually has a rather low risk of carcinoma, while a mass with the long axis running more perpendicularly to the skin line (not parallel) is suspicious for malignancy.
- *Border:* The presence of an echogenic, relatively wide border is a criterion that is strongly suspicious for malignancy. A distinct interruption of the normal connective tissue running parallel to the skin is also usually apparent here.

Table 6.3 Description of the position and size of a mass lesion in breast ultrasonography

Criterion	Information
Position	Clock face position and distance from the nipple
Size	Maximal dimensions in two perpendicular planes

Table 6.4 Sonographic lesion criteria

Criterion	Low risk of malignancy	High risk of malignancy
Shape	Oval, round	Irregular
Orientation	Parallel to the skin	Not parallel to the skin
Margin	Circumscribed	Not circumscribed: indistinct, angular, microlobulated, spiculated
Border	N/A	Hyperechoic rim
Echo pattern	Anechoic, hyperechoic	Hypoechoic
Internal structure	Homogeneous	Inhomogeneous
Posterior features	Posterior enhancement	Posterior attenuation/ shadowing
Edge shadowing	Both sides of lesion	N/A
Calcifications	Macrocalcifications	N/A
Surrounding tissues	Compression and displacement	Disruption or architectural distortion
Elasticity assessment	Soft	Hard
Lesion mobility	Good	Absent or slight
3D criteria	Compression pattern	Retraction pattern
Vascularity	Absent	Internal vascularity Vessels in rim
Lymph nodes	Echogenic fatty hilum	Hypoechoic with inconspicuous hilum Rounded shape, possibly with irregular margins
Milk ducts	Normal (< 2 mm diameter)	Irregularities in caliber, abrupt cutoff, solid internal structures

- *Echo pattern:* The echogenicity of a mass lesion is defined with reference to the echogenicity of the mammary fat. A homogeneous, anechoic or hyperechoic mass has a high probability of being benign.
- *Internal structure:* A hypoechoic internal structure is seen in both benign and malignant masses, thus the homogeneity of the internal lesion structure is a weak diagnostic criterion. Because both homogeneous malignant tumors and inhomogeneous benign tumors are not rare, echogenicity alone has little specificity.
- *Ultrasound transmission:* Posterior enhancement is seen when sound transmission is unimpeded in its passage through a mass, which is typically the case with cysts. Homogeneous solid

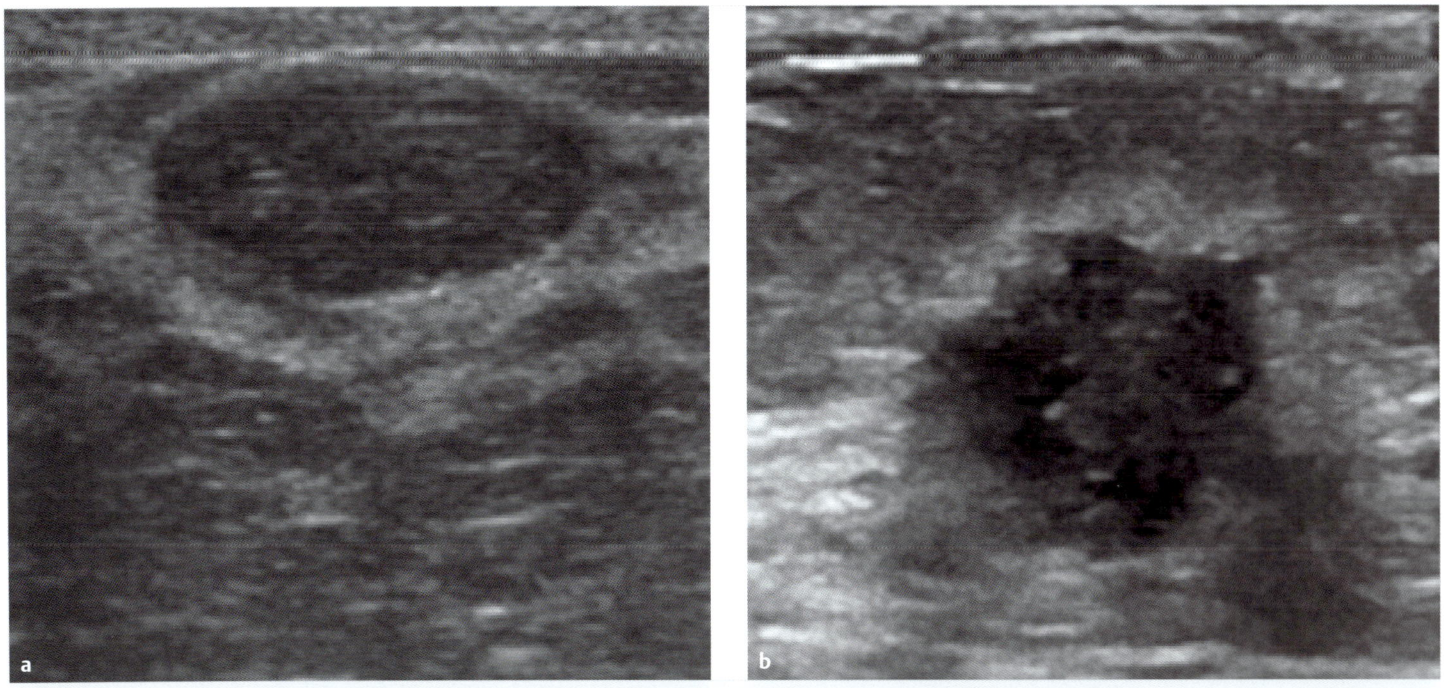

Fig. 6.7 **Mass lesions in breast ultrasonography. (a)** Benign mass lesion with the criteria "oval" and "circumscribed," and oriented parallel to the skin. **(b)** Malignant mass lesion with the criteria "irregular," "hypoechogenic," "hyperechogenic margin," "posterior acoustic attenuation," and "disruption of surrounding structures."

tumors with a high fluid content (e.g., myxoid fibroadenomas) may also show enhancement, which generally suggests for benignity. However, high-grade, homogeneous carcinomas may also exhibit enhancement or no change in posterior features. Posterior acoustic attenuation, i.e., shadowing or extinction, is a characteristic of many carcinomas. The differential diagnosis here includes primarily scars, but also fibrocystic changes of the mammary gland. Macrocalcifications, as seen for example in areas of fat necroses, also often cause posterior extinction.
- *Compressibility and mobility:* These are dynamic lesional criteria that can only be ascertained during the performance of the ultrasound examination.
- *3D imaging* of a lesion requires the use of a special transducer and is optional, as are color Doppler and perfusion analysis.
- *Architectural distortion or disruption of the surrounding structures.*

No one of these criteria by itself should be used as the sole feature to assess the benign or malignant nature of a tumor. There is a large overlap in the appearance of benign und malignant tumors. A round, circumscribed, homogeneous, hypoechoic mass with no posterior acoustic features can still be malignant, and an irregular-appearing, inhomogeneous, hypoechoic mass with indistinct margins can be benign.

Note

Because the morphological appearances in the ultrasound image of benign and malignant tumors overlap, no diagnostic criterion alone substantiates the diagnosis. In case of doubt, an ultrasound-guided core needle biopsy is recommended for histological confirmation.

Architectural Distortion

Architectural distortions in breast ultrasonography are usually recognized as regionally circumscribed, relatively hypoechoic distortions in otherwise homogenous breast structures, often with a relatively abrupt transition to the surrounding tissues. By their nature, they are difficult to detect in a breast with a sonographically inhomogeneous echotexture. Especially when coinciding with an area of increased palpable consistency, architectural distortions are particularly likely to indicate the presence of a lobular breast carcinoma. The differential diagnosis includes the following findings, which may also be accompanied by architectural distortion (▸ Fig. 6.8):
- Postoperative scars.
- Radial scars.
- Fibrocystic changes.
- Invasive tubular or ductal breast carcinomas.

Special Cases

Findings that cannot reasonably be described with the standard assessment criteria and have a unique diagnosis are called special cases. The following findings fall into this category:
- Clustered microcysts (▸ Fig. 6.9).
- Simple and complicated cysts.
- Skin lesions.
- Foreign bodies, including implants.
- Intramammary lymph nodes (▸ Fig. 6.10).

Ordinarily these are harmless findings without pathologic significance.

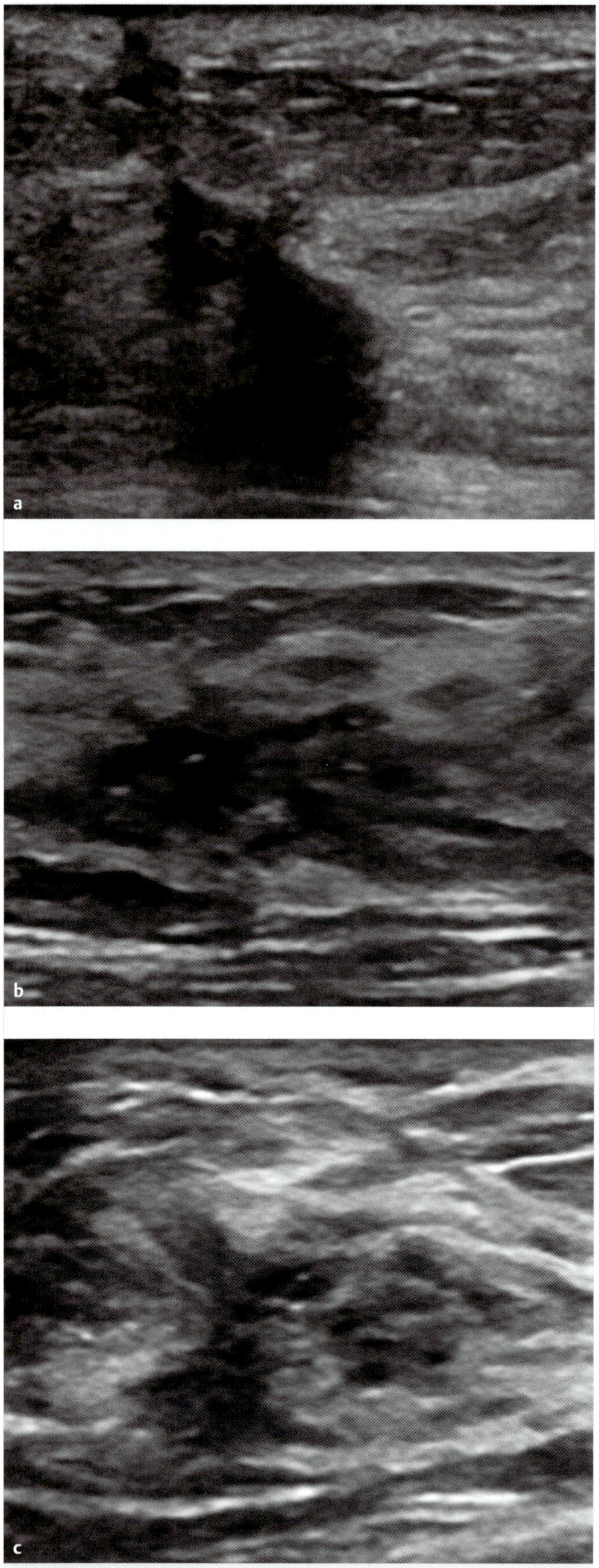

Fig. 6.8 **Architectural distortion in the sonogram. (a)** Postoperative scarring. **(b)** Radial scars. **(c)**Tubular breast carcinoma.

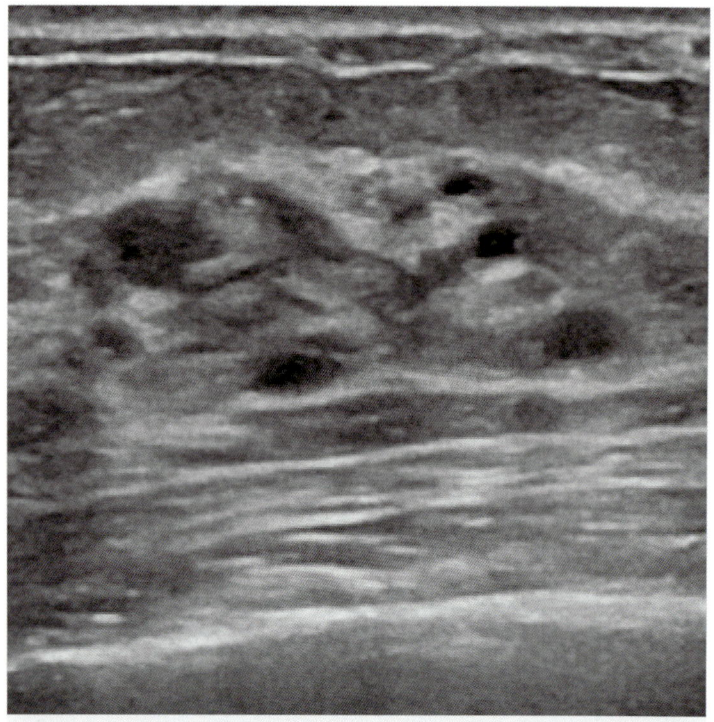

Fig. 6.9 **Grapelike grouped microcysts.** A special case in ultrasonography.

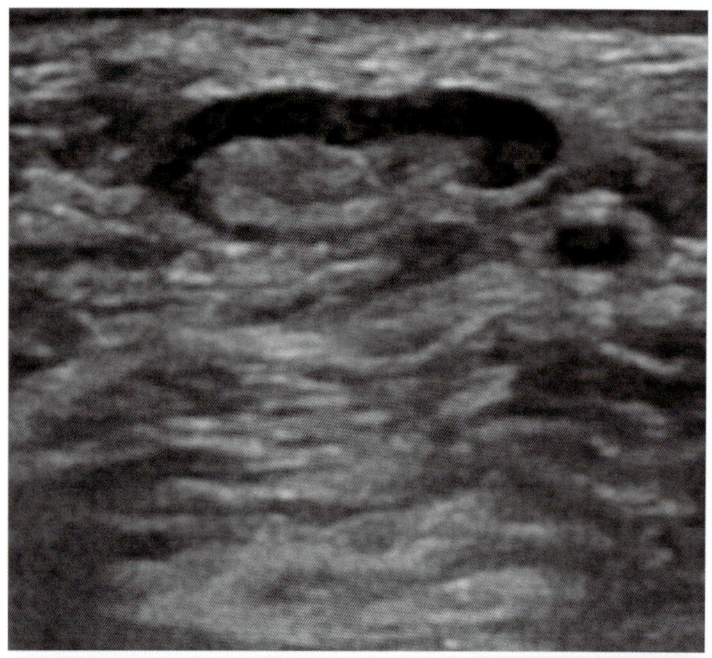

Fig. 6.10 **Intramammary lymph nodes.** A special case in ultrasonography.

Associated Features

Findings that are seen in connection with malignant tumors are termed associated features. Examples include

- Localized skin thickening (▶ Fig. 6.11) or retraction.
- Architectural distortion.
- Edema.

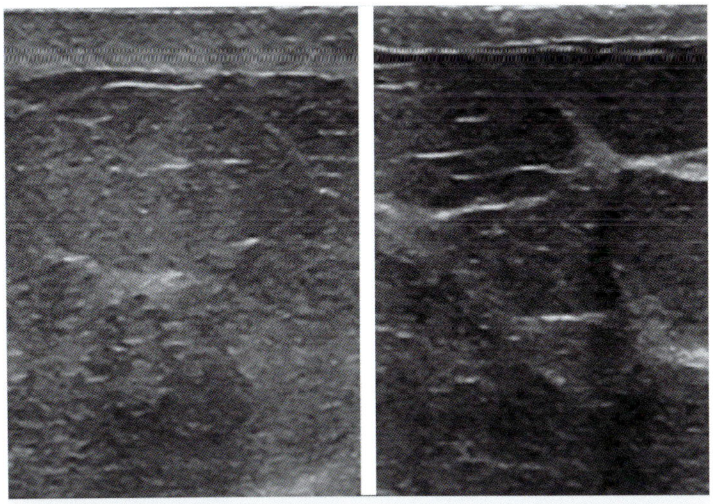

Fig. 6.11 Skin thickening. Associated findings in ultrasonography. *Left:* skin thickening and radiation-induced edema of the intramammary tissue. *Right:* normal structures with unremarkable skin thickness.

- Chest wall infiltration.
- Axillary lymph node enlargement or with internal structural changes (▶ Fig. 6.12).

6.2.4 BI-RADS Classification of Breast Ultrasonography

The final assessment of pertinent sonographic findings entails the assignment of a US BI-RADS category from the ACR BI-RADS lexicon. Seven categories are distinguished, wherein the categories BI-RADS 0 and BI-RADS 6 are of special significance. The BI-RADS assessment categories 1 to 5 (▶ Table 6.5) prescribe concrete management recommendations (▶ Table 6.6).

US BI-RADS 0

BI-RADS category 0 is assigned when no conclusive assessment of the ultrasound examination is possible. This can be the case, for example, if pronounced fibrosis is present and the acoustic penetration is not sufficient for visualization of all parts of the breast. Occasionally, fresh scar changes with wide acoustic shadowing can also be assigned to the US BI-RADS 0 category. Because a very large breast (macromastia) may be difficult to examine reliably and completely with sonography, the US BI-RADS 0 category may also be assigned in such cases.

Depending on the situation, either further diagnostic work-up using an alternative imaging technique (such as mammography or breast MRI), a core needle biopsy, or expectant management is indicated. Especially in cases of expectant management, the patient should be informed in detail of the clearly limited sensitivity of her ultrasound examination.

US BI-RADS 1

BI-RADS 1 is assigned when an examination is determined to be complete and adequate for assessment. Findings are normal and without abnormalities on which to comment. Mass lesions and architectural distortions are not present. The risk of cancer is assessed to be 0%.

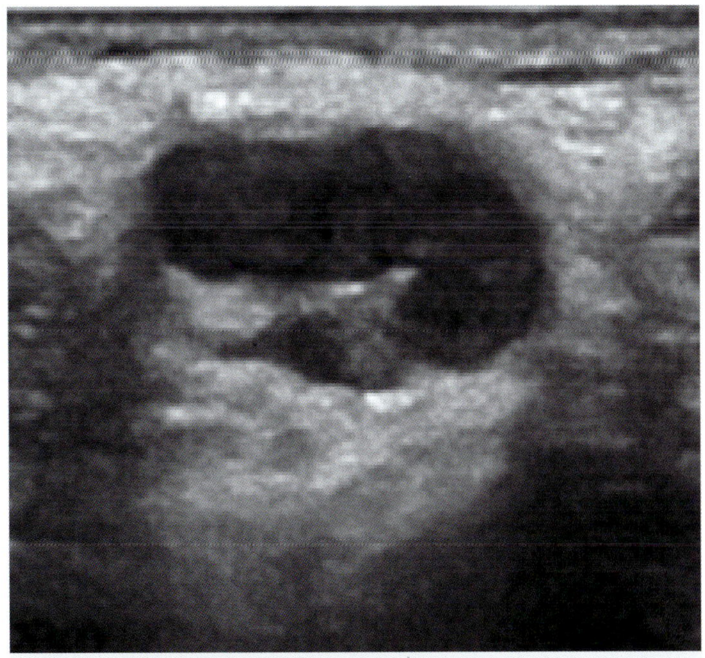

Fig. 6.12 Lymphadenopathy. Associated findings in ultrasonography.

Table 6.5 BI-RADS classification for breast ultrasonography findings (US BI-RADS); correlation of description of findings and risk of malignancy

US BI-RADS category	Description	Risk of malignancy
1	Normal	0
2	Definitely benign	0
3	Probably benign	> 0% – ≤ 2%
4	Suspicious	> 2% – < 95%[a]
5	Highly suggestive of malignancy	≥ 95%

[a]Because of the large range of risk within the category US BI-RADS 4 (> 2– < 95%), subdivisions may be used: subcategory US BI-RADS 4A (> 2– ≤ 10%), 4B (> 10 ≤ 50%) and 4C (> 50– < 95%).

Table 6.6 Recommended management according to the US BI-RADS category

US BI-RADS category	Recommended management
1	Next examination within the routine screening interval
2	Next examination within the routine screening interval
3	Follow up after 6 months; ultrasound core needle biopsy when appropriate
4	Histological work-up by percutaneous biopsy, typically with ultrasound guidance
5	Pretherapeutic histological work-up by percutaneous biopsy, typically with ultrasound guidance

US BI-RADS 2

The BI-RADS 2 category is assigned when an examination is determined to be complete and adequate for assessment, and there is a finding that is worth describing but clearly benign. In most cases, this refers to cysts (▶ Fig. 6.13a), typical lymph nodes, or masses that are typical of fibroadenomas and have not changed during surveillance over 2 to 3 years (▶ Fig. 6.13b). Breast implants can also be classified in this category. Scars that can be relatively well penetrated by ultrasound and that have remained unchanged while under surveillance should also be classified as BI-RADS 2. Findings in this category have no special management recommendation. The risk of cancer is assessed to be 0%.

US BI-RADS 3

The BI-RADS 3 category is assigned if an examination is determined to be complete and adequate for assessment, and a lesion is present that has a high probability of being benign. By definition, the residual risk of malignancy is ≤ 2%.

This category includes mass lesions typical of a fibroadenoma (solid, oval, long axis parallel to the skin, well circumscribed; ▶ Fig. 6.14a) that are seen for the first time because there has been no previous examination, complex cysts (▶ Fig. 6.14b), or clustered microcysts. Ultrasound-guided core needle biopsy as a diagnostic work-up of such findings is favored as a rule because it is simple, fast, relatively free of complications, and cost effective. Additionally, the histological work-up with the elimination of all "question marks" is usually a considerable psychological relief for the affected woman. If expectant management is desired, however, a short-term follow-up examination after 6 months is recommended. If the findings are unchanged, further follow-up

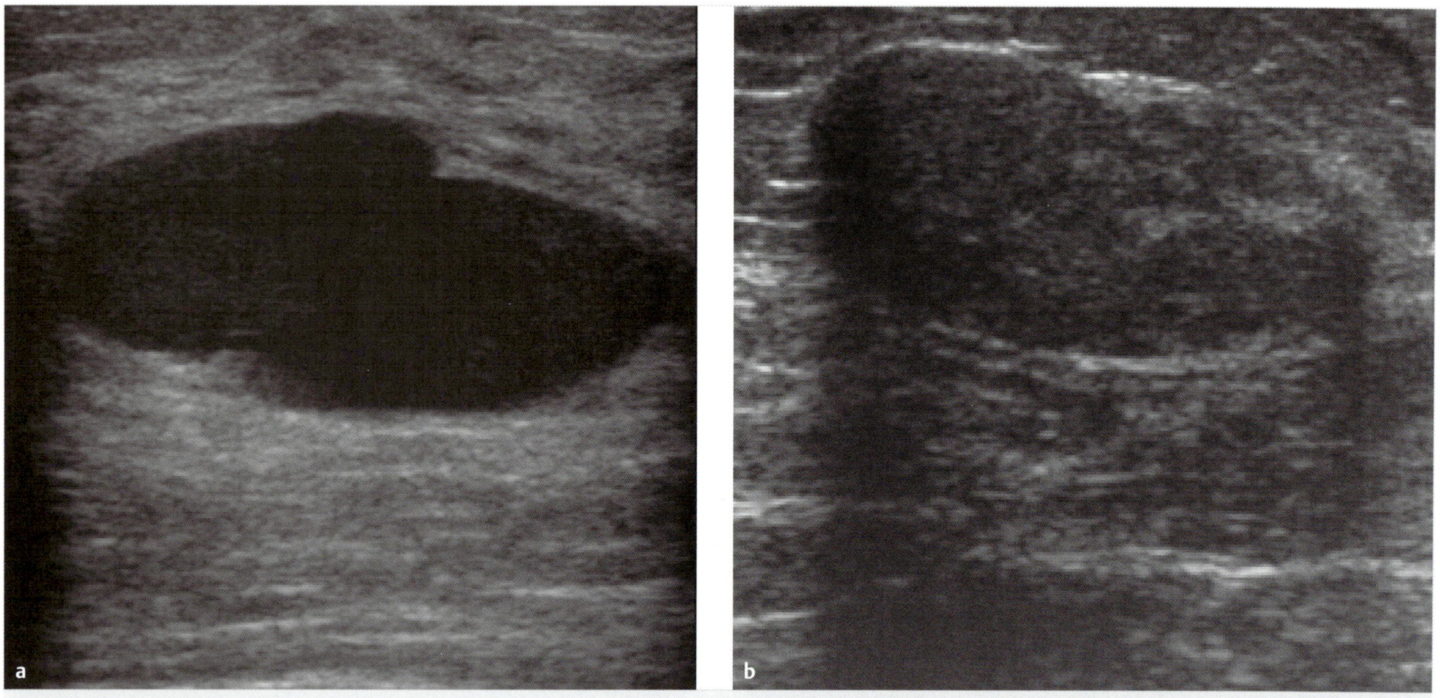

Fig. 6.13 US BI-RADS 2. Typical examples. **(a)** Macrocysts. **(b)** Fibroadenoma, constant finding for years.

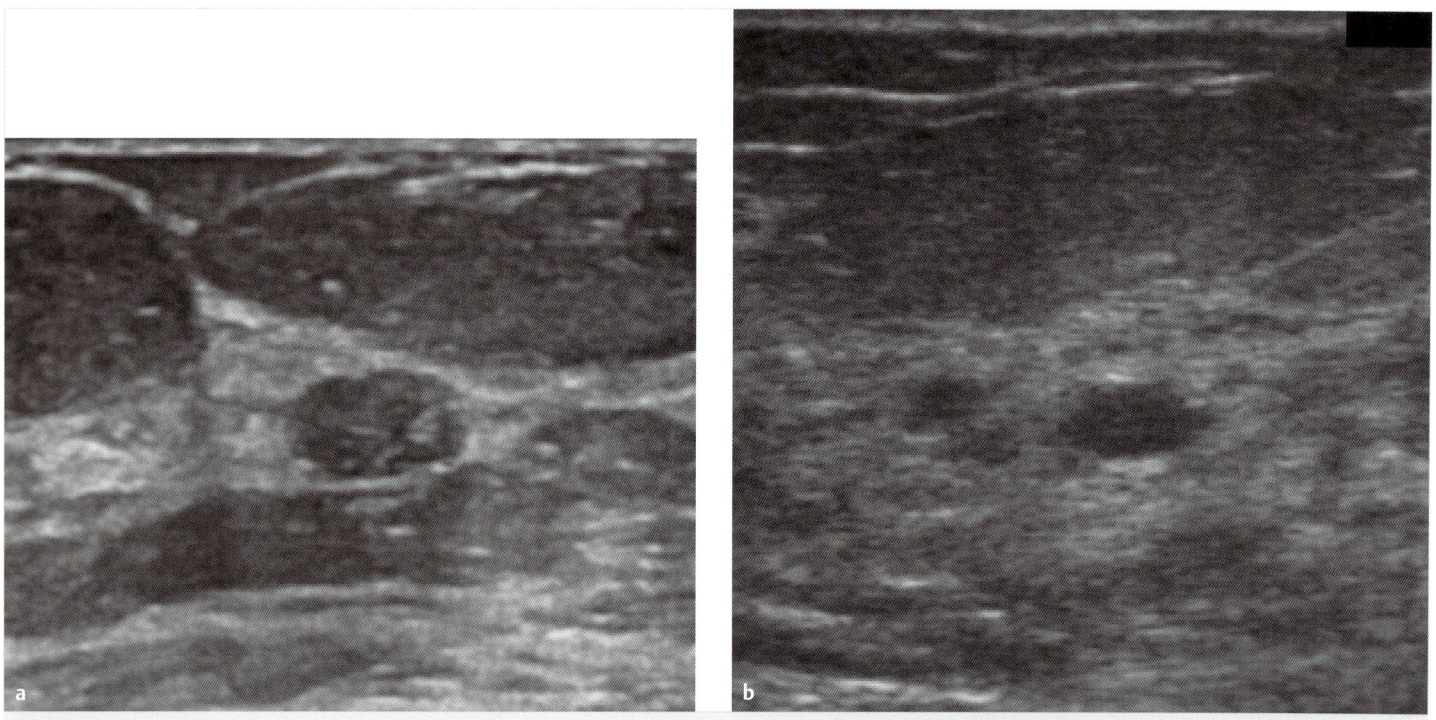

Fig. 6.14 US BI-RADS 3. Typical examples. **(a)** Findings typical of a fibroadenoma, detectable for the first time. **(b)** Complex cysts.

examinations at 6- to 12-month intervals are continued until stability of the findings has been documented for over 2 years. After 2 to 3 years of stability, reclassification as US BI-RADS 2 is warranted. If any follow-up results in reclassification of the lesion to a higher level (BI-RADS 4), a percutaneous biopsy is performed for histological evaluation.

US BI-RADS 4

The BI-RADS 4 category is assigned if an examination is determined to be complete and adequate for assessment, and a lesion is present that has sufficient probability of being malignant to warrant biopsy (▶ Fig. 6.15). These findings have neither the classic appearance of a benign lesion nor that of a malignant lesion. By definition, these findings carry a risk of malignancy of > 2 to < 95%. As a rule an ultrasound-guided core needle biopsy for histological diagnosis is recommended. Due to the broad range of likelihood of malignancy in this category, it can be subdivided into three subgroups (US BI-RADS 4A to 4C) to facilitate the making of informed decisions on the ultimate course of action.

US BI-RADS 4A

Findings falling into this category have a low risk of malignancy, between > 2 and ≤ 10%. Examples include:

- Hypoechoic mass lesions with only partially circumscribed margins and an oval shape.
- Palpable complex cysts.

Expectant management may be desired in exceptional cases. It is then prudent to recommend a short-term follow-up examination after 3 to 6 months. If the findings are unchanged, further follow-up examinations are continued until stability of the findings is documented for over 2 years. If there is progression, the probability of malignancy is greater and a percutaneous biopsy is indicated.

US BI-RADS 4B

Findings in this category have a moderate risk of malignancy of > 10 to ≤ 50%. Examples include.

- Inhomogeneous hypoechoic mass lesions with indistinct margins and a round shape.
- Palpable, hypoechoic areas with architectural distortion.

For findings in this category, histological work-up with ultrasound-guided core needle biopsy should always be recommended. If histological results are benign, then one should decide on a case-to-case basis whether these findings are compatible with the image findings, in which case a short-term follow-up examination after 6 months may be recommended. Otherwise, a repeat percutaneous core needle biopsy, a vacuum-assisted biopsy, or an open biopsy should be performed.

Intracystic and intraductal proliferation usually belong in this category (US BI-RADS 4B). In this special case, however, a primary open biopsy should be performed rather than a percutaneous core needle biopsy because, after percutaneous needle biopsy, fluid often no longer surrounds the biopsied lesion and it may then be difficult to distinguish it from the surrounding tissue. If biopsy reveals that the lesion should be excised, preoperative localization of the lesion may be problematic unless an ultrasound-guided clip has been placed marking the puncture site. This does not apply if the lesion is identifiable on X-ray mammography or breast MRI, which then offers an alternative approach for preoperative localization.

US BI-RADS 4C

Lesions belong in this category that have a high risk of malignancy (> 50– < 95%) but that do not exhibit the classic malignancy criteria of a BI-RADS 5 lesion. Examples include:

- Hypoechoic mass lesions with indistinct margins, a round shape, and posterior acoustic attenuation.
- Palpable, hypoechoic areas with distinct architectural distortion.

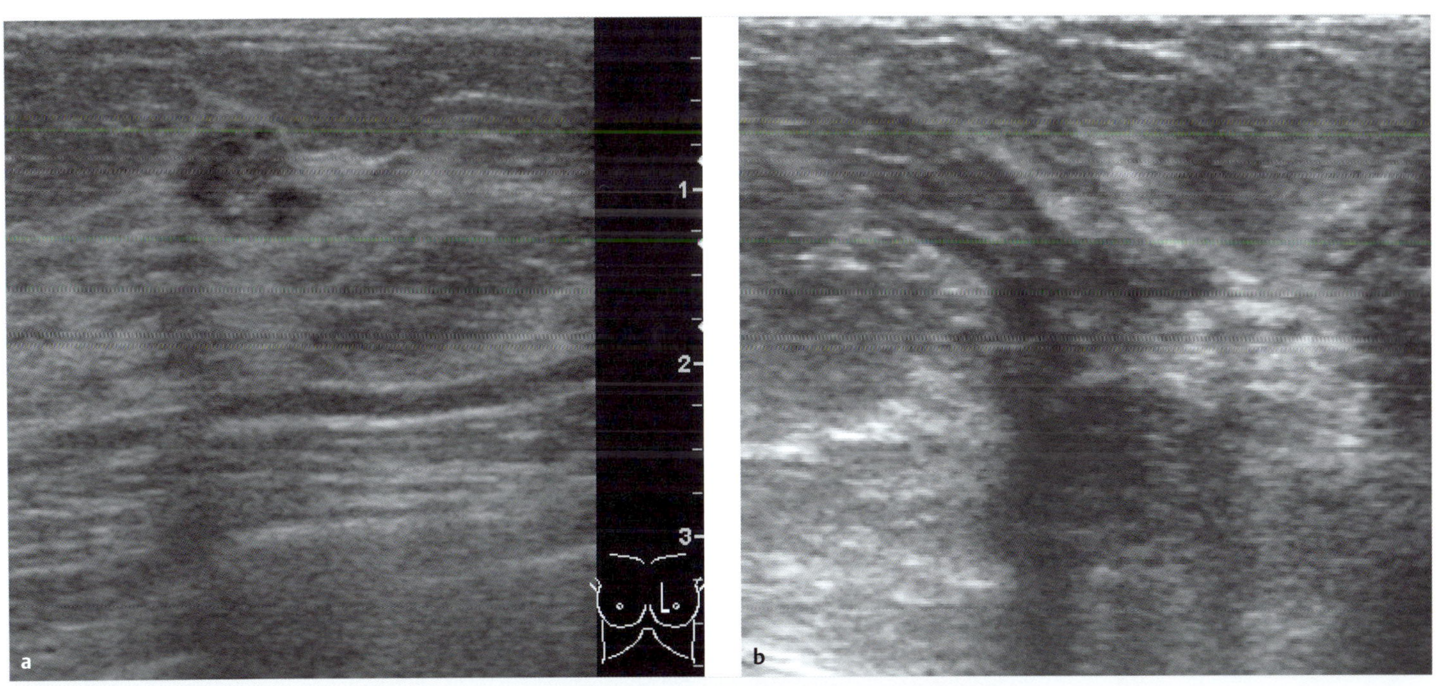

Fig. 6.15 US BI-RADS 4. Typical examples. **(a)** Irregular mass. **(b)** Palpable architectural distortion.

6

- Hypoechoic mass lesions in correlation with a newly occurring mass lesions in the mammogram or breast MRI.

For findings in this category, histological work-up with ultrasound-guided core needle biopsy should always be performed in the absence of clinical contraindications. If histological results are benign, it must be considered whether the lesion is compatible with the image findings and, if at all possible, further management should be discussed in an interdisciplinary conference. At the least, the examining pathologist should be consulted. In case of doubt, a repeat percutaneous core needle biopsy, a vacuum-assisted biopsy, or an open biopsy should be performed.

US BI-RADS 5

Lesions assigned to this category have a very high risk of malignancy of ≥ 95%. These findings have multiple characteristics typical of a malignant lesion (▶ Fig. 6.16). Classic examples include:

- Hypoechoic mass lesion with irregular shape, hyperechoic rim, interruption of the normal architecture, and posterior acoustic attenuation.
- Palpable, hypoechoic area with irregular shape, distinct architectural distortion, and posterior acoustic attenuation, with or without increased vascularization.
- Hypoechoic mass lesion with irregular shape, correlated with a newly occurring or suspicious mass lesion in the mammogram or breast MRI.

For findings in this category, histological work-up with ultrasound-guided core needle biopsy should always be performed in the absence of clinical contraindications. If histological results are benign, it is probable that the lesion was missed on biopsy. Histology is not compatible with the image findings, so that a repeat percutaneous core needle biopsy, a vacuum-assisted biopsy, or a diagnostic excision should be performed for clarification.

US BI-RADS 6

Lesions are placed in this category once histology has confirmed a malignancy, corresponding to a lesion in the histological B-classifications B5a, B5b, B5c, or B5d (see Chapter 3). The classification of a lesion to this category is thus independent of the morphological criteria in the image findings. The BI-RADS 6 category is only to be used until the malignant lesion has been surgically excised. Postoperatively, classification is according to the pT staging of the TNM system of the UICC (Union for International Cancer Control).

Note

- In the overall assessment, the final diagnosis stands at the end of an examination chain (clinical examination, mammogram, ultrasonography, breast MRI). Here, taking the specific reliability of each method into consideration, a summarizing evaluation of all image findings is assigned in the form of a final combined BI-RADS score to category 1, 2, 3, 4, or 5.
- Assuming that the histological result and the ultrasound image findings are compatible, re-biopsy or a follow-up examination after 6 to 12 months should be performed after every core needle biopsy with benign histology performed on US findings in a BI-RADS category 4 or 5.
- The designation as “BI-RADS 6” is reserved for the percutaneous biopsy–proven malignancy prior to surgery.

6.2.5 Normal Findings in Sonography

The breast ultrasound appearance generally depends on the breast composition, i.e., the extent of glandular tissue, adipose tissue, and fibrocystic changes. The major influencing factors are age, hormonal stimulation, and nutritional status. The following intramammary structures can be distinguished in a breast ultrasound: skin, subcutaneous adipose tissue, Cooper’s (suspensory) ligaments, veins and arteries (Doppler technique), and intramammary and prepectoral lymph nodes.

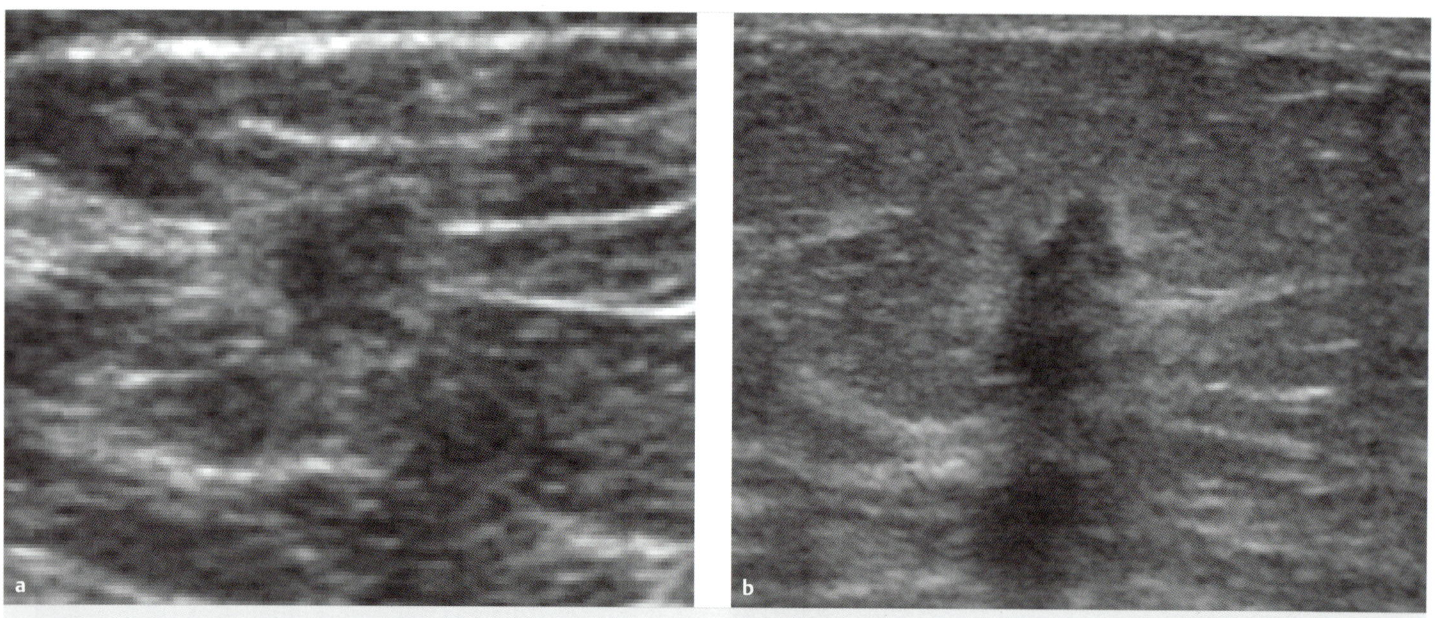

Fig. 6.16 US BI-RADS 5. Mass lesions typical of carcinoma with numerous criteria for malignancy. **(a)** Interruption of anatomical structures, hyperechoic border. **(b)** Irregular shape, vertical tumor axis, posterior acoustic extinction.

Typically, a younger woman exhibits a prominent parenchymal body that, if homogeneous, is usually reliably assessed with sonography. In this respect, sonography supplements X-ray mammography in an ideal way in women with high tissue density. The reliability of sonography declines to the extent that the adipose tissue component increases in the perimenopausal and postmenopausal years.

Take Home Point

Because the sensitivity of mammography is limited when breast parenchyma has a high mammographic density (ACR III and IV), supplemental breast ultrasonography should be done to increase overall sensitivity.

Breast ultrasonography is not suitable as the sole method for early breast cancer detection.

Second-look breast ultrasonography is especially suitable for the diagnostic work-up of nonpalpable mammographic or MRI findings (BI-RADS 0, III, IV, and V).

For women under 40 years with palpable findings, breast ultrasonography should be done as the imaging method of first choice.

Bibliography

▶ Technique and Methods

[1] Ärztliches. Zentrum für Qualität in der Medizin (ÄZQ). Evidenzbericht 2007 zur S3-Leitlinie "Brustkrebsfrüherkennung in Deutschland", Version 1.00. Berlin, Ärztliches Zentrum für Qualität in der Medizin (ÄZQ), Vol 32. Available at: http://www.aezq.de/mdb/edocs/pdf/schriftenreihe/schriftenreihe32.pdf; Accessed February 13, 2017

[2] Baker JA, Soo MS. Breast US: assessment of technical quality and image interpretation. Radiology. 2002; 223(1):229–238

[3] Barr RG, Destounis S, Lackey LB , II, Svensson WE, Balleyguier C, Smith C. Evaluation of breast lesions using sonographic elasticity imaging: a multicenter trial. J Ultrasound Med. 2012; 31(2):281–287

[4] Beissert M, Jenett M, Kellner M, Wetzler T, Haerten R, Hahn D. SieScape panorama imaging in radiologic diagnosis [in German]. Radiologe. 1998; 38(5):410–416

[5] Berg WA, Cosgrove DO, Doré CJ, et al. BE1 Investigators. Shear-wave elastography improves the specificity of breast US: the BE1 multinational study of 939 masses. Radiology. 2012; 262(2):435–449

[6] Cho N, Jang M, Lyou CY, Park JS, Choi HY, Moon WK. Distinguishing benign from malignant masses at breast US: combined US elastography and color Doppler US—influence on radiologist accuracy. Radiology. 2012; 262(1):80–90

[7] Deutsche Gesellschaft für Ultraschall in der Medizin e.V. (DEGUM). Available at: http://www.degum.de (Stand: 22.03.2013). Accessed February 10, 2017

[8] Fischer T, Peisker U, Fiedor S, et al. Significant differentiation of focal breast lesions: raw data-based calculation of strain ratio [in German]. Ultraschall Med. 2012; 33:372–379

[9] Gokalp G, Topal U, Kizilkaya E. Power Doppler sonography: anything to add to BI-RADS US in solid breast masses? Eur J Radiol. 2009; 70(1):77–85

[10] Huang YH, Chen JH, Chang YC, et al. Diagnosis of solid breast tumors using vessel analysis in three-dimensional power Doppler ultrasound images. J Digit Imaging. 2013; 26(4):731–739

[11] Kassenärztliche Bundesvereinigung. Vereinbarung von Qualitätssicherungsmaßnahmen nach § 135 Abs. 2 SGB V zur Ultraschalldiagnostik. Stand 2012

[12] Kelly KM, Dean J, Comulada WS, Lee SJ. Breast cancer detection using automated whole breast ultrasound and mammography in radiographically dense breasts. Eur Radiol. 2010; 20(3):734–742

[13] Madjar H, Mundinger A, Degenhardt F, Duda V, Hackelöer BJ, Osmers R. Quality control in breast sonography [in German]. Ultraschall Med. 2003; 24 (3):190–194

[14] Madjar H. Dopplersonografie in der Differentialdiagnose von Mammatumoren. Stuttgart: Thieme; 2005

[15] Madjar HK, Mendelson EB. The Practice of Breast Ultrasound. Techniques, Findings, Differential Diagnosis. 2nd ed. Stuttgart: Thieme; 2008

[16] Parris T, Wakefield D, Frimmer H. Real world performance of screening breast ultrasound following enactment of Connecticut Bill 458. Breast J. 2013; 19 (1):64–70

[17] Shen SJ, Sun Q, Xu YL, et al. Comparative analysis of early diagnostic tools for breast cancer [in Chinese]. Zhonghua Zhong Liu Za Zhi. 2012; 34(11):877–880

[18] Staren ED. Physics and principles of breast ultrasound. Am Surg. 1996; 62 (2):103–107

[19] Tozaki M, Fukuma E. Does power Doppler ultrasonography improve the BI-RADS category assessment and diagnostic accuracy of solid breast lesions? Acta Radiol. 2011; 52(7):706–710

[20] Weismann C, Hergan K. Current status of 3D/4D volume ultrasound of the breast [in German]. Ultraschall Med. 2007; 28(3):273–282

[21] Wenkel E, Heckmann M, Heinrich M, et al. Automated breast ultrasound: lesion detection and BI-RADS classification—a pilot study. Rofo. 2008; 180 (9):804–808

▶ Analysis

[1] Abdullah N, Mesurolle B, El-Khoury M, Kao E. Breast imaging reporting and data system lexicon for US: interobserver agreement for assessment of breast masses. Radiology. 2009; 252(3):665–672

[2] Ärztliches. Zentrum für Qualität in der Medizin (ÄZQ). Evidenzbericht 2007 zur S3-Leitlinie "Brustkrebsfrüherkennung in Deutschland", Version 1.00. Berlin, Ärztliches Zentrum für Qualität in der Medizin (ÄZQ), Vol 32. Available at: http://www.aezq.de/mdb/edocs/pdf/schriftenreihe/schriftenreihe32.pdf; Accessed February 13, 2017

[3] Alvarenga AV, Infantosi AF, Pereira WC, Azevedo CM. Assessing the combined performance of texture and morphological parameters in distinguishing breast tumors in ultrasound images. Med Phys. 2012; 39(12):7350–7358

[4] American College of Radiology. BIRADS - Breast Imaging Reporting and Data System. 3rd ed. American College of Radiology; 2003

[5] Baker JA, Soo MS. Breast US: assessment of technical quality and image interpretation. Radiology. 2002; 223(1):229–238

[6] Costantini M, Belli P, Ierardi C, Franceschini G, La Torre G, Bonomo L. Solid breast mass characterisation: use of the sonographic BI-RADS classification. Radiol Med (Turin). 2007; 112(6):877–894

[7] Friedrich M. Lehratlas der Mammasonographie. Stuttgart: Wissenschaftliche Verlagsgesellschaft mbH; 1999

[8] Heinig J, Witteler R, Schmitz R, Kiesel L, Steinhard J. Accuracy of classification of breast ultrasound findings based on criteria used for BI-RADS. Ultrasound Obstet Gynecol. 2008; 32(4):573–578

[9] Hong AS, Rosen EL, Soo MS, Baker JA. BI-RADS for sonography: positive and negative predictive values of sonographic features. AJR Am J Roentgenol. 2005; 184(4):1260–1265

[10] Lee HJ, Kim EK, Kim MJ, et al. Observer variability of Breast Imaging Reporting and Data System (BI-RADS) for breast ultrasound. Eur J Radiol. 2008; 65(2):293–298

[11] Leitlinienprogramm Onkologie der AWMF. Deutschen Krebsgesellschaft e.V und Deutschen Krebshilfe e. V. Leitlinienreport der S 3 Leitlinie fur die Diagnostik, Therapie und Nachsorge des Mammakarzinoms. AWMF Register-Nr. 032–045OL. Available at: http://www.awmf.org/leitlinien/detail/ll/032–045OL.html; Accessed February 9, 2017

[12] Levy L, Suissa M, Chiche JF, Teman G, Martin B. BIRADS ultrasonography. Eur J Radiol. 2007; 61(2):202–211

[13] Madjar HK, Mendelson EB. The Practice of Breast Ultrasound. Techniques, Findings, Differential Diagnosis. 2nd ed. Stuttgart: Thieme; 2008

7 Magnetic Resonance Imaging of the Breast

U. Fischer

7.1 Technique and Methods

7.1.1 Basic Principles

Magnetic resonance imaging (MRI) is based on the physical principle of nuclear magnetic resonance, first demonstrated in solids by Purcell and Bloch in 1946. Classically, electrically charged particles produce a local magnetic field when they have rotational motion. Atomic nuclei with an odd number of protons and neutrons have an intrinsic "spin" (analogous to classical angular momentum) and have a resultant magnetic moment. Because of their high concentration in biological tissues (in water and fat), protons are the primary particle used for imaging purposes. Furthermore, the hydrogen nucleus has the largest magnetic moment of all naturally occurring isotopes. When such nuclei are subjected to a strong external magnetic field, their spins align themselves with this static magnetic field either in a parallel orientation (low-energy state) or an antiparallel orientation (high-energy state). Only the spins oriented in an antiparallel direction ($1:10^5$) contribute to the signal generation. In addition, the external magnetic field exerts a torque on the individual magnetic moments of the protons, causing them to gyrate (precess) around the magnetic field direction—i.e., the longitudinal axis—at a certain frequency. This "Larmor frequency" (ω, omega) is proportional to the strength of the external magnetic field (B_0) according to the formula

$$\omega = y \times B_0$$

where the proportionality constant *y* is called the gyromagnetic ratio and represents a material-specific constant.

When an external electromagnetic pulse of high-frequency (a radiofrequency [RF] pulse) that corresponds to the Larmor frequency of the proton is applied, the individual magnetic moments are deflected out of alignment with the longitudinal axis of the external magnetic field. The amount of this deflection is characterized by the excitation pulse, measured in angular degrees (°). After the external electromagnetic pulse is turned off, the magnetic moments tend to return to their original state of longitudinal alignment and precession: this is the process of *relaxation*. During the relaxation process, the absorbed input energy is released in the form of radio waves. Further energy is lost through electromagnetic dipole–dipole interaction and thermal dissipation. The restoration of the magnetization vector to its original state of alignment is an exponential process characterized by the increase of magnetization in the longitudinal plane (T1 relaxation time) and the loss of transverse magnetization (T2 relaxation time or T2 decay). The decrease in the transverse magnetization is caused by local magnetic field inhomogeneities and temporally fluctuating magnetic fields, so-called spin–spin interactions. Due to these interactions, some the magnetic moments of protons will be caused to precess faster and some to precess slower, and thus to fall out-of-phase (dephasing). The T1 and T2 relaxation times are unique and intrinsic properties of the tissue examined. The emitted electromagnetic signal (magnetic resonance [MR] signal) can be detected as an induced voltage by an appropriate receiver coil and processed to produce an image. The relaxation times are the times that it takes for the RF signal to exponentially rise (T1) or decay (T2) to half its maximum value.

For image generation, an emitted signal must be identified with a specific location on a three-dimensional matrix using gradient fields in the three spatial axes (*x*, *y*, *z*). Spatial encoding can be done using 2D or 3D techniques.

- *2D technique:* Here spatial encoding is carried out by applying a linear magnetic field gradient along a selected plane (*x*, *y*, or *z* gradient), then applying an RF pulse with a selected center frequency (resonant frequency) that excites only spins located within a specific slice along that gradient (slice selection). Phase and frequency encoding within this slice is achieved by application of magnetic field gradients along the other two axes. This allows the spatial localization of MR signals from each voxel within the slice, which can then be used to create an image.
- *3D technique:* In contrast, for 3D acquisition, an entire tissue volume is excited. An additional phase-encoding gradient must therefore be applied along the third orthogonal dimension (e.g., the *z*-axis) to allow spatial encoding along the slice-select direction, thereby generating a set of contiguous slices within the volume.

Factors that influence contrast in an MR image include factors inherent in the tissue and also external variables. The most important of the inherent factors are the proton density and the T1 and T2 relaxation times. The external factors include hardware and software parameters such as slice thickness, angulation, and acquisitions; the type of pulse sequence chosen, such as SE (spin echo), GRE (gradient echo), and fat saturation; the field strength; and whether a contrast agent is administered.

A variety of different pulse sequences are available for MR imaging. The basic principles of the SE sequences and GRE sequences (of special significance for dynamic breast MRI) are explained below.

- *SE sequences:* For classical SE (spin-echo) sequences, excitation always occurs by application of a 90° RF pulse that completely abolishes the longitudinal magnetization and converts it into transverse magnetization. After half of the echo time (TE), a second, 180° RF pulse is applied. This flips the vectors so that precession now brings the vectors back to their starting positions; that is, it leads to a rephasing or refocusing of the spins, resulting in a maximum signal (spin echo) that is read out after the full TE. Successive repetitions of the entire sequence are performed after the lapse of a given repetition time (TR). Depending on the chosen TE and TR, resulting images that are either T1 weighted, proton-weighted, or T2 weighted.
- *GRE sequences:* Gradient-echo (GRE) pulse sequences use a reversal of gradient polarity to form an echo rather than a 180° RF pulse. Additionally, the initial 90° RF pulse can be replaced by a smaller pulse (flip angle, $\alpha < 90°$) that does not completely eliminate the longitudinal magnetization. This results in more

rapid image acquisition and a weaker signal yield. In comparison with SE sequences, GRE sequences show higher sensitivity for artifacts caused by susceptibility differences in tissues (bone, air) or through metal deposits. Correspondingly, the sensitivity of GRE sequences to paramagnetic contrast media is greater.

Signal intensity in MRI imaging can be modified by the administration of contrast materials. Paramagnetic substances are primarily used for this purpose, but superparamagnetic substances can also be used. Substances of both these groups exhibit the property of changing the relaxation times of the anatomical structures being imaged. Paramagnetic contrast media shorten the T1 and T2 relaxation times. Contrast-enhanced breast MRI takes advantage of this shortening effect on the T1 relaxation time to increase the signal intensity in the T1W sequences. Superparamagnetic substances have no utility for breast MR imaging.

7.1.2 Tumor Detection

MRI of the breast has high sensitivity for the detection of hypervascularized areas in the breast through the depiction of the intramammary structures in very thin layers before and after intravenous administration of a contrast agent suitable for MRI. Contrast-enhancing areas are characterized by increased vascularization, increased vessel permeability, and an enlarged interstitial space (tissue matrix). These tissue characteristics are found in almost all invasive cancers and in a majority of intraductal breast carcinomas (keyword: tumor neoangiogenesis). Some benign breast lesions, however, are also associated with an increased perfusion pattern, so that there is an overlap in the imaging characteristics of benign and malignant processes. Technical and methodological aspects of a high-quality breast MRI are discussed in the following sections.

7.1.3 Equipment

Breast MRI is generally performed in a whole-body magnet with field strengths of 1.5 and 3.0 T (tesla) (▸ Fig. 7.1a). It is no longer acceptable to use systems with 0.5 T. Special bilateral surface coils are required for breast MRI to achieve a sufficiently high spatial resolution. These are adapted to fit the shape of the female breast and should preferably be open to enable optimal support and positioning of the breast, and to allow for carrying out interventional procedures when required (▸ Fig. 7.1b). Dedicated compression plates can be integrated into such open breast coils to reduce the occurrence of motion artifacts (▸ Fig. 7.1c).

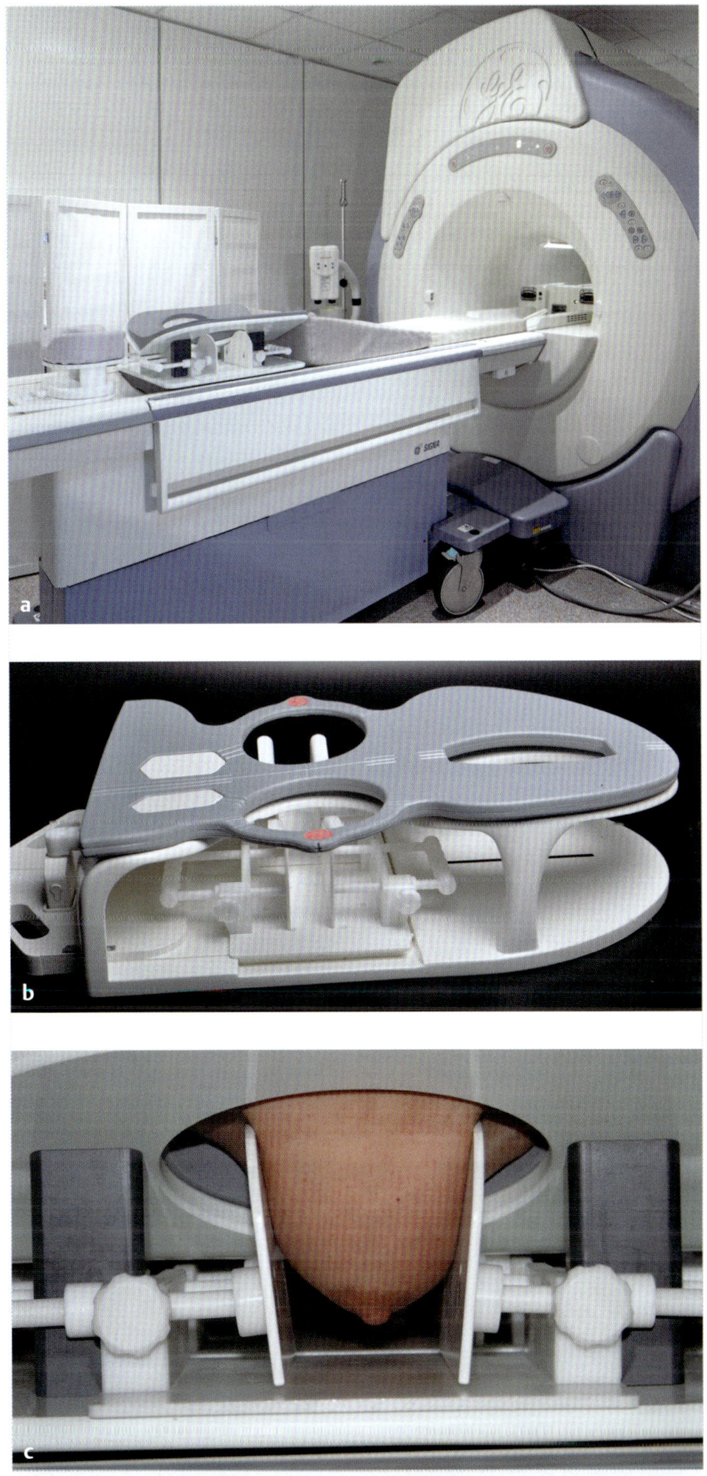

Fig. 7.1 Breast MRI. (a) 1.5 T Examination unit (CE Healthcare, Milwaukee, WI, USA) with the surface coil in place. **(b)** Open surface coil with integrated compression plates (Noras, Höchberg, Germany). **(c)** Compression device with superior-inferior fixation of the breast (Noras, Höchberg, Germany).

Note

Open breast surface coils with integrated compression plates are an essential requirement for excellent breast MRI quality.

7.1.4 Timing of the Examination

The parenchymal perfusion pattern, and thus the contrast enhancement in the mammary gland, fluctuates with the changing hormonal influences on the female breast during the menstrual cycle. The degree of disturbing signal enhancement on breast MRI imaging is least in the second week of the menstrual cycle. Thus, when circumstances allow, an examination should be scheduled between the 7th and 14th days of the cycle. Examinations performed in the third week of the menstrual cycle are generally still acceptable, but the first and fourth weeks should be avoided. If the imaging has to be done during this unfavorable period, the potential diagnostic disadvantage from increased enhancement must be taken into account.

Practical Tip

The best time to perform breast MRI is in the second week of the menstrual cycle.

After menopause, there are no concerns related to the menstrual cycle when planning a breast MRI examination appointment. Occasionally, however, there can be undesirable early parenchymal contrast enhancement due to hormone replacement therapy (HRT), necessitating a repeat examination after stopping HRT for 4–6 weeks. As a general principle, however, it is not justifiable to stop hormone substitution before performing a breast MRI examination.

An MRI examination can be performed without restriction after a previous fine needle puncture, and after a core needle biopsy without significant hematoma. After a vacuum-assisted biopsy, there is typically a large resection cavity with reactive, hyperemic changes around the edge and hematoma in the center. In general, it is better to undertake MRI before rather than after such an interventional procedure.

Subsequent to an open biopsy, no breast MRI should be performed within an interval of 6 months so as to avoid interference from increased enhancement associated with wound healing. After tumor resection (as breast-conserving therapy [BCT]) and radiation therapy, this interval is increased to about 12 months after termination of the radiotherapy. There is, however, pronounced variability between individuals with respect to the extent and duration of the reactive changes after a course of radiation therapy.

7.1.5 Patient Positioning

The examination is performed with the examinee in the prone position with the arms along the sides, preferably tucked into the belt of a bath robe (▶ Fig. 7.2). As an alternative, the arms can be crossed above the head, although in this position the breasts are often not completely visualized in the axial slice orientation. The use of a cushioned head support on which the forehead rests has proven value, allowing the woman to view a picture motif by way of an angled mirror during the procedure.

Note

The key to a high-quality breast MRI examination is patient comfort.

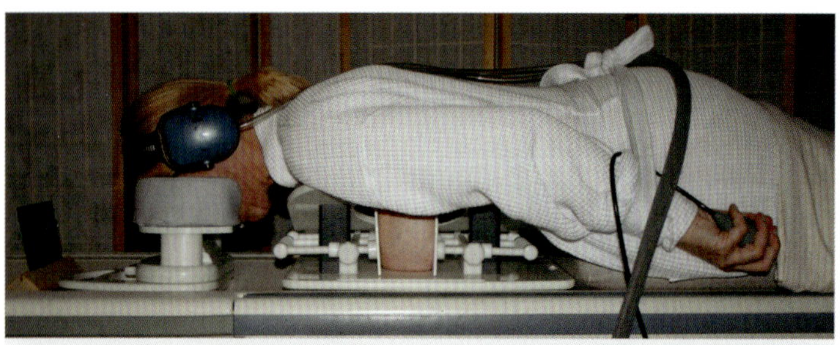

Fig. 7.2 Prone patient positioning for breast MRI. Frontal headrest with angled mirror and picture motif for viewing during the examination. Headphones for listening to recorded music. Compression plates with positioned and fixed breast. Arms along the body with the arms tucked into the belt of the bath robe and relaxed pectoral muscles. Alarm bulb to terminate the examination if necessary.

7.1.6 Measurement Parameters

Technique

Contrast-enhanced breast MRI can be performed with the 2D and 3D techniques. Each technique has advantages and disadvantages that do not significantly affect the assessability or sensitivity of the MRI examination. The decision as to which technique should be used should be based on the experience of the examiner; it is not advisable, however, to alternate techniques frequently. Recently, *parallel imaging* has optimized image acquisition speed and spatial resolution. This is essentially simultaneous acquisition of image data over two or more receiving coils with different spatial sensitivities.

Sequences

T1-weighted (T1W) GRE sequences still represent the heart of the breast MRI examination (▶ Fig. 7.3a). Essentially, they allow sensitive capture of an enhancement after administration of an MR-appropriate paramagnetic contrast agent. These T1W sequences are typically performed once before and several times after the peripheral venous administration of contrast medium. To allow clear visualization of contrast-enhancing areas, it is necessary to eliminate or suppress interfering intramammary fat signals. This can be accomplished in two ways:

- By the subtraction of identical slice images before and after administration of contrast material ("European school").
- By the acquisition of sequences with primary fat suppression, i.e., saturation (FAT-SAT) ("American school").

The additional acquisition of a T2 weighted (T2W) sequence is recommended in order to obtain a signal-intense visualization of water-containing or grossly edematous structures (▶ Fig. 7.3b). For this, breast MRI primarily employs T2W SE or TSE (turbo spin echo) sequences, or IR (inversion recovery) sequences, combined with simultaneous fat saturation. The T2W measurement is not for tumor detection, but rather to characterize suspicious findings found in the contrast-enhanced T1W measurement.

Notes

- In the T1W image, contrast media, fat, protein, and blood appear white (signal intense).
- In the T2W image, water-containing substances appear white (signal intense).

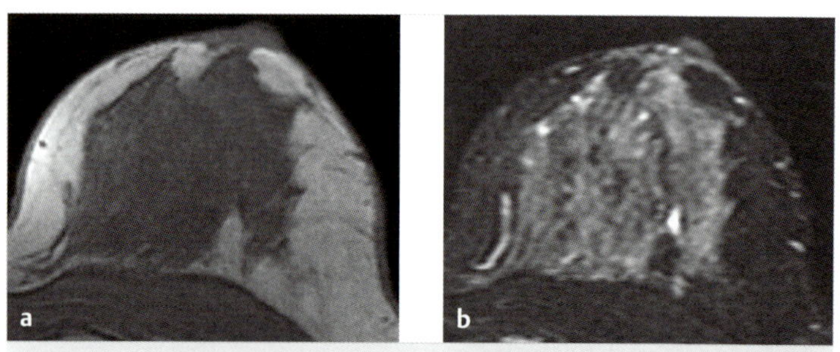

Fig. 7.3 T1W and T2W sequences in breast MRI. (a) Precontrast T1W image. **(b)** T2W image.

Orientation

MRI allows examinations in any slice orientation. In Europe, the transverse (i.e., axial) plane is usually preferred when contrast-enhanced breast MRI is used for tumor detection; in North America, the sagittal plane is preferred. For implant evaluation, at least one complementary sagittal angulation is expressly recommended.

Temporal Resolution

In the vast majority of cases, contrast enhancement occurs more rapidly in hypervascularized breast tumors than in the surrounding glandular parenchyma. This is particularly true for invasive breast carcinomas, in which the maximum signal intensity peak is usually reached approximately 3 minutes after the intravenous administration of the contrast agent. However, some carcinomas can exhibit their maximum signal intensity peak as early as 1 minute after the administration of the contrast agent, followed by a "washout" effect. In contrast, the signal intensity of parenchyma displays a continuous increase during the entire examination time of 8 to 10 minutes, and there is a considerable degree of variation between individuals. Because early and strong contrast enhancement of parenchyma may occur (e.g., in areas of adenosis or due to hormonal stimulation), pathological processes can be masked (masking effect). Generally, a temporal resolution of maximally 2 to 2.5 minutes/sequence over a period of at least 6 minutes after contrast agent administration is required for dynamic measurements.

Spatial Resolution

Various factors and setting parameters affect the spatial resolution of breast MRI. Specifically, these are the matrix (depending on the chosen field of view), the slice thickness, and the magnitude of the resulting volume. For examinations in axial orientation, the field of view is predetermined by the patient's proportions (usually 300–350 mm). Recommendations advise a matrix size of at least 256 × 256 pixels. Ideally, however, a matrix of 512 × 512 pixels should be used, which should not be the result of interpolation. Although general recommendations are for a slice thickness between 2 and 4 mm, breast MRI should be performed with single slice thicknesses of 1.5 to 2.5 mm. Indisputably, the best results are obtained by high resolution breast MRI with a matrix of 512 × 512 pixels and a single slice thickness of 1.5 to 2 mm.

Contrast Medium

Intravenous administration of an extracellular paramagnetic contrast agent allows visualization of the vascularization within the breast. Repetitive measurement after the administration of contrast media gives information about the uptake and clearance of the contrast material over time. For dynamic breast MRI (1.5 T), the recommended dose of contrast agent is 0.1 mmol Gd-DTPA (gadopentetate dimeglumine) per kg body weight. The contrast agent is mechanically injected at a rate of 2 to 3 mL/s via a venous access previously placed in the cubital fossa. Gd-DTPA is cleared quickly and completely by way of renal excretion. Tubular reabsorption has not been demonstrated. The half-life of Gd-DTPA in the blood is approximately 90 minutes. After 24 hours, more than 90% of the substance has been excreted. Gd-DTPA is very well tolerated. From available data, the rate of side effects is considered to be less than 2%, and approximately 80% of side effects are classified as mild. Adequate and clear information must be provides to the patient to allow written consent to be obtained before starting the examination.

The use of Gd-DTPA during pregnancy has not been established as innocuous, so contrast-enhanced breast MRI in pregnant women should be avoided on principle. The use of gadolinium-based contrast material can lead to nephrogenic systemic fibrosis in patients with renal insufficiency. This risk is especially high in patients with acute or chronic renal insufficiency and a glomerular filtration rate less than 30 mL/min per 1.73 m^3, and in patients with acute renal insufficiency in conjunction with hepatorenal syndrome or a history of liver transplantation.

Phase-Encoding Gradient

Spatial orientation in three-dimensions is achieved in MRI by way of phase encoding and frequency encoding of selected slice planes. For the axial view of breast MRI, as is commonly used in Europe, the phase-encoding gradient is typically positioned in a mediolateral direction. However, this occasionally impairs the ability to evaluate the axillary aspects of the breast, with the overlapping artifacts usually being worse in the vicinity of the heart (left axillary area) than on the right side. This problem has been resolved by modifying the examination protocol to feature alternation of the phase-encoding gradient. As signal curve analysis in high-resolution breast MRI becomes less significant, revision of the established examination protocols should take new considerations into account.

Zebra Protocol

The zebra protocol (▶ Fig. 7.4) involves sequentially alternating the phase-encoding gradient. Before administration of contrast medium, one measurement is made with the phase-encoding gradient along the mediolateral or horizontal direction, and one with it along the frontoposterior or vertical direction. These serve as the base measurements for the later image subtraction. After administration of contrast medium, two measurements are performed with the phase-encoding gradient along the mediolateral direction followed by one measurement with it along the frontoposterior direction. Subsequent image postprocessing yields three subtracted images:

- Earliest subtraction with mediolateral phase-encoding gradient.
- Early subtraction with mediolateral phase-encoding gradient.
- Late subtraction with frontoposterior phase-encoding gradient

The zebra protocol enables complete visualization of all the parenchymal segments, including the potential "dead spaces" in the left axillary tail region while preserving a very high temporal resolution.

7.1.7 Image Postprocessing

The multitude of images acquired in dynamic MR mammography necessitates additional postprocessing to help in the detection of contrast-enhancing lesions, the further characterization of findings, and the optimal presentation of lesions.

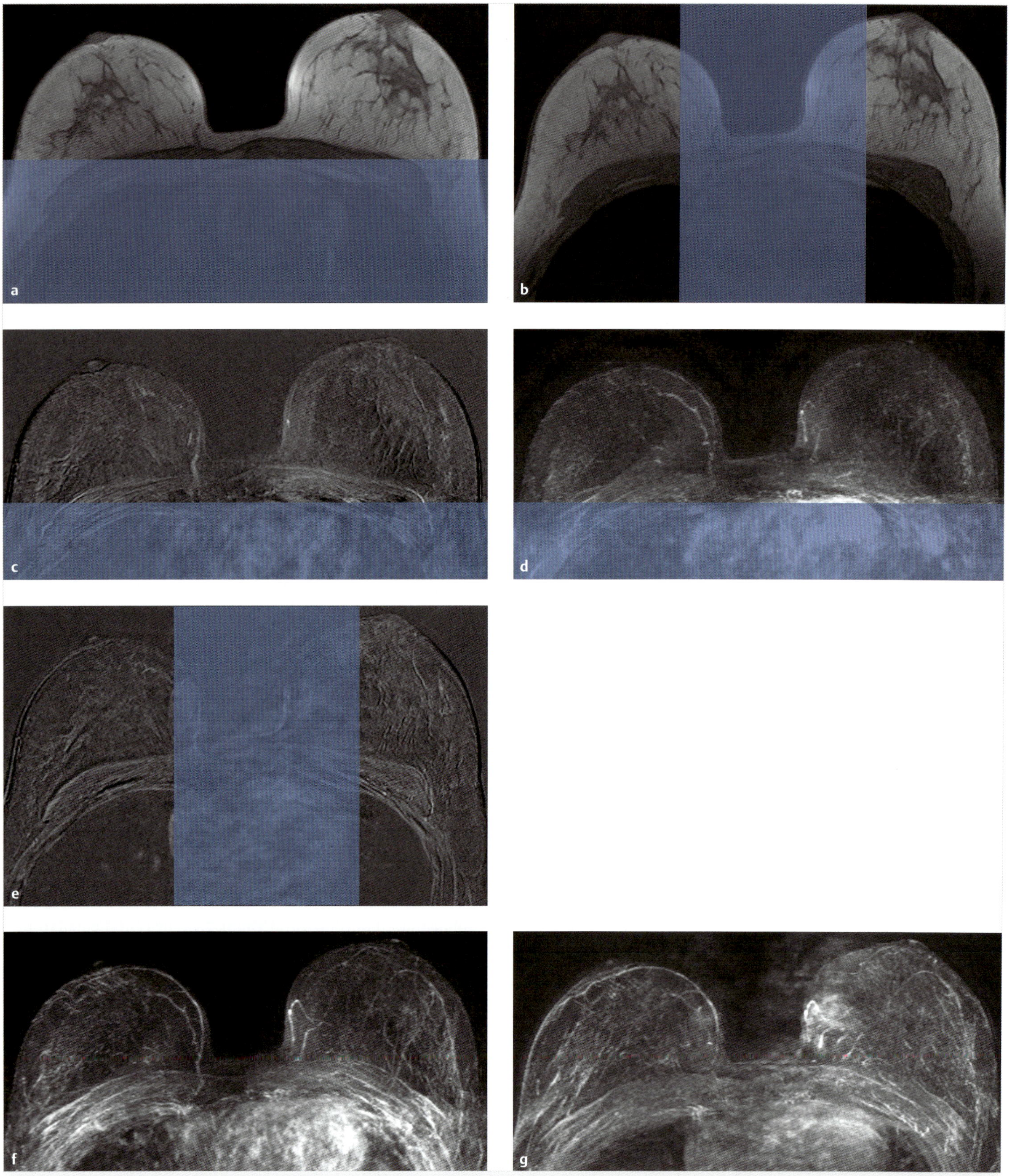

Fig. 7.4 Zebra protocol. Acquisition of dynamic T1 sequence slices with alternating phase-encoding gradients. **(a)** Precontrast examination with anteroposterior phase-encoding gradient. **(b)** Precontrast examination with mediolateral phase-encoding gradient. **(c, d)** After administration of contrast medium, two T1 image sequences with mediolateral phase-encoding gradients are acquired: first for performing the earliest subtraction **(c)**, and second for performing the early subtraction **(d)**. **(e)** After these postcontrast T1W images for the earliest and the early subtraction postprocessing are acquired, the third postcontrast T1W sequence is acquired with anteromedial phase-encoding gradient for the late image subtractions. **(f)** Maximum intensity projection with mediolateral phase-encoding gradient. **(g)** Maximum intensity projection with anteroposterior phase-encoding gradient.

Image Subtraction

The subtraction of precontrast images from postcontrast images with identical slice positions serves to eliminate the signal-intense adipose tissue and facilitates detection of hypervascularized areas of tissue (▶ Fig. 7.5a). It is mandatory to subtract the precontrast examination series from the first (earliest subtraction) and the second (early subtraction) postcontrast series on a pixel-to-pixel basis. Within the zebra protocol, an additional subtraction of the precontrast series from the third series after administration of contrast medium (late subtraction) is performed, allowing an evaluation of the tissue in the axillary tails (see ▶ Fig. 7.4).

Curve Analysis

The analysis of time–signal intensity curves (TICs) within representative areas of contrast enhancement has lost importance with the increasing optimization of the spatial resolution of breast MRI. The analysis remains useful for mass lesions with defined volumes, but it does not provide any additional information in the case of multiple foci or non-space-occupying lesions. TIC measurements should be made in defined measurement areas (ROIs, regions of interest), which are selected such that they include the greatest possible proportion of the maximally enhancing area while disregarding less vascularized areas (e.g., central necrotic tissue areas). The size of the field of view here should be between 2 and 5 pixels and the selection of three ROIs per lesion is recommended. The numerical results of the time–signal intensity measurements are expressed as percentage increase over the initial signal intensity values (▶ Fig. 7.5b).

Maximum Intensity Projection

The maximum intensity projection technique yields a comprehensive three-dimensional view of both breasts. Based on the subtraction images generated by postprocessing, which only take into account image pixels having a certain signal intensity (threshold algorithm), the representation of image information gives the impression of a "transparent breast" which can be viewed from different angles (▶ Fig. 7.5c).

Computer-Assisted Image Analysis

Various models of CAD (computer-assisted diagnosis) for dynamic MRI examinations of the breast have been presented in the literature. These range from color-coded parameter images (▶ Fig. 7.5d) over automatically defined regions of interest, and from pulsating playback sequences with a repetition rate of 1 or 2 images per second to the use of pharmacokinetic models.

▶ Table 7.1 shows the breast examination protocol used at the Göttingen Women's Health Care Center (BZG [Diagnostisches Brustzentrum Göttingen]).

7.1.8 Implant Evaluation

Examinations that target the detection of complications related to prosthetic implants differ fundamentally from examinations designed to rule out carcinoma. For one thing, implant diagnostics do not require a dynamic examination, so that IV administration of contrast medium is unnecessary. For another, special sequences come into use that allow the selective evaluation of the different prosthesis components (e.g., silicone or saline).

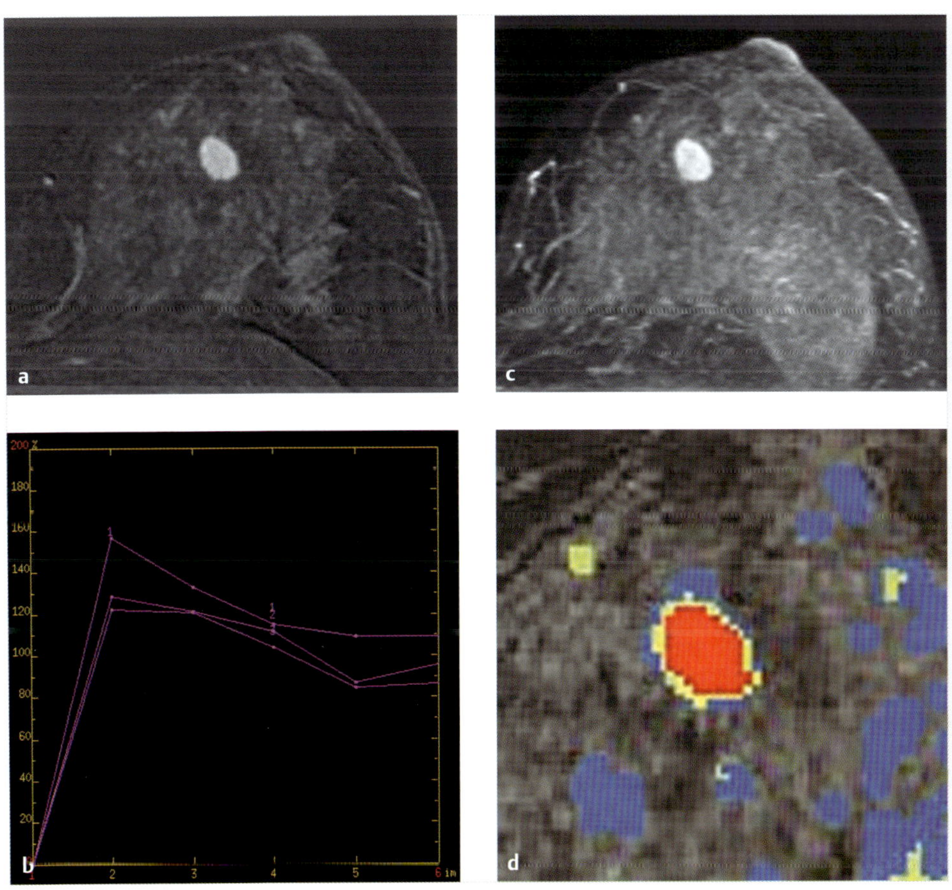

Fig. 7.5 Image postprocessing in breast MRI. **(a)** Subtraction. **(b)** Curve analysis. **(c)** Maximum intensity projection. **(d)** Computer-assisted diagnosis (CAD).

Table 7.1 Examination protocol at the BZG (Göttingen Diagnostic Breast Center, 2013)

Parameter	Setting
Field strength	1.5 T
Surface coil	Bilateral, open
Breast stabilization	Dedicated breast compression plates, superoinferior
Angulation	Axial
Technique	3D
Phase-encoding gradient	Zebra protocol (see ▶ Fig. 7.4)
Sequences	T1W GRE repetitive, T2W inversion recovery with fat saturation
Dynamics	Two measurements before, three measurements following administration of contrast media (zebra protocol)
Contrast agent	Gd-DTPA 0.1 mmol/kg body weight
Contrast agent injection	Cubital fossa; automated injection (3 mL/s); followed by 20 mL saline solution
Spatial resolution	1.5–2.5 mm
Field of view	300–350 mm
Matrix	512 × 512, noninterpolated
Temporal resolution	87 s/sequence
TR	8.4 ms
TE	4.1 ms (in-phase)

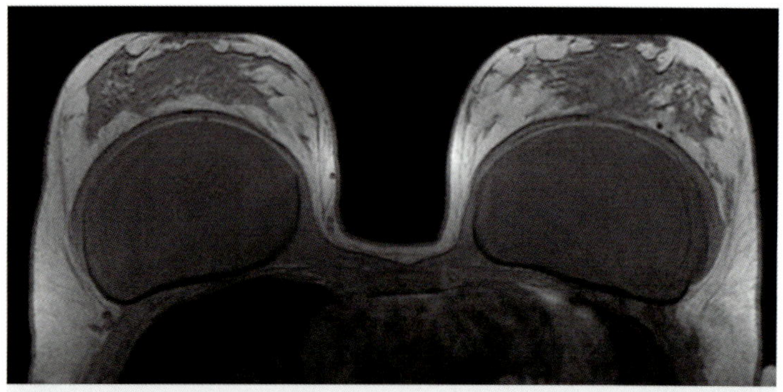

Fig. 7.6 Breast implant MRI. Visualization of an implant in the usual T1W sequence with normal findings.

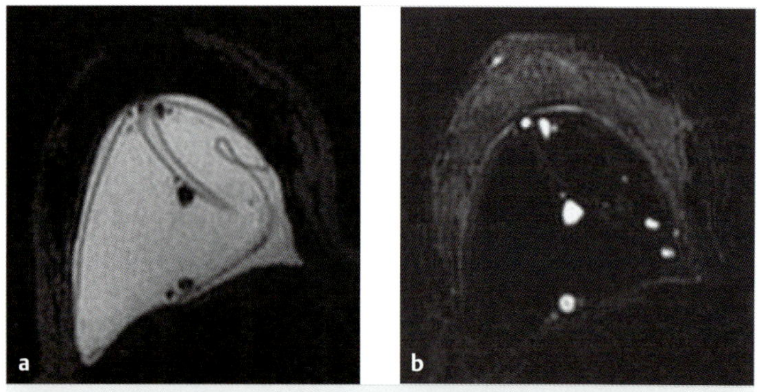

Fig. 7.7 Breast implant with intracapsular rupture on MRI. **(a)** Examination with silicone-sensitive sequences and suppression of the water signal. Evidence of rupture of the implant shell and collapse with incipient linguine sign. **(b)** Examination with water-sensitive sequence and silicone signal suppression. Visualization of aqueous inclusions within the silicone ("salad oil sign").

As opposed to contrast-enhanced breast MRI for early tumor detection, MRI examination of breast prostheses requires the acquisition of images in multiple planes. In addition to the usual axial slice orientation, sagittal images are acquired to detect any silicone located outside the superior and inferior circumference of the prosthesis, as well as to evaluate any deformation of the prosthesis in a superoinferior direction (▶ Fig. 7.6). A good prosthesis examination requires extremely high spatial resolution (1–2 mm), but this is unproblematic as there is no need for administration of contrast medium, and thus the measurement protocol has plenty of leeway regarding the temporal resolution.

Fat-suppressing inversion recovery sequences are usually used, because they are most effective in differentiating between the various prosthetic fluids. By additionally suppressing the water signal, such inversion recovery sequences allow a signal-intense visualization of silicone (silicone-sensitive measurement), while the signal of the saline components and the surrounding adipose tissue is suppressed (▶ Fig. 7.7). Conversely, inversion recovery sequences combined with silicone suppression allow the selective visualization of the saline components (water-sensitive measurement), so that, especially in the case of double-lumen prostheses (silicone/saline), the two prosthetic lumina can be visualized separately.

7.1.9 Nonestablished Examination Techniques

Additional examination techniques have been investigated and will continue to be investigated to further characterize conspicuous findings on MRI. These include MR spectroscopy, MR elastography, and diffusion-weighted imaging techniques. The usefulness of these methods has yet to be clearly established.

MR Spectroscopy

Magnetic resonance spectroscopy (MRS) investigates the interactions between molecules and electromagnetic fields. In addition to allowing the characterization of molecular properties such as bond lengths and bond strengths, it also enables the identification of atomic constituents. The observed molecular spectra differ from conventional atomic spectra by having very many more, usually overlapping spectral lines ("bands"). The choline peak has particular significance in tumor diagnosis and characterization, because carcinomas commonly exhibit a high concentration of choline. There continue to be problems with the placement of measuring volumes and the associated partial volume fraction contaminated by adjacent or endotumoral necroses, parenchyma, or adipose tissue.

MR Elastography

The goal of MR elastography (MRE) is to present a visual image of the viscoelastic tissue properties (hardness, elasticity, and cushioning effect) and derived parameters, such as anisotropy and elasticity tensor of the intramammary tissue structures. This makes it possible to distinguish between benign and malignant tumors on the basis of their differing elastic properties. Performance of this examination requires a breast surface coil with an integrated oscillator capable of sending low-frequency sound waves (e.g., 60 Hz) into the tissues. The wave pattern is then measured, and, through calculation, translated into a color scale that enables a slice-by-slice analysis of the elasticity and hardness of suspicious lesions.

Diffusion-Weighted MR Imaging

Diffusion-weighted MRI (DWI) produces an image representing the diffusional movement of water molecules in body tissue. The technique is used primarily to examine the brain, because for some diseases of the central nervous system the diffusion pattern is changed in characteristic ways. In the breast too, the diffusion rate of water molecules depends upon the nature and extent of the pathological changes at the cellular level. Thus diffusion measurements are made to obtain an additional criterion for differentiating between benign and malignant changes on the basis of such disturbances in the cellular structure of the tumor.

7.2 Evaluation

7.2.1 Terminology

Breast MR images are described and categorized according to the BI-RADS lexicon, which presents a standardized terminology for the description and characterization of MRI findings. Currently, the 5th edition of the BI-RADS lexicon (2013) is available from the American College of Radiology (ACR).

7.2.2 Perfusion Pattern

The sensitivity of the breast MRI examination is determined to a significant extent by the physiologic vascularization of the parenchyma. Early parenchymal enhancement hinders the detection of hypervascularized lesions. In analogy to the density types on X-ray mammography, a categorization of the density on breast MRI has also been devised (MRM density = magnetic resonance mammography density) to express the transparency of the breast (▶ Table 7.2, ▶ Fig. 7.8). The greater the transparency of the glandular structures, the more sensitive the method is for the detection breast lesions at a smaller size and at an earlier stage. The range of sensitivity here spans from about 50% (MRM density type IV) to almost 100% (MRM density type I), depending on the physiologic perfusion (see ▶ Table 5.1).

There are no defined criteria for assigning an MRM density level to a corresponding MRI examination. Ultimately, it is a subjective decision made by the evaluator. In any case, the reference parameter for this is the early-subtraction image, i.e., subtraction from the second measurement after administration of contrast medium.

Table 7.2 Density types on breast MRI. Description and sensitivity of breast MRI according to the level of enhancement in the early phase of the examination

ACR MRM density type	Description	Sensitivity of breast MRI
I	No early parenchymal enhancement	98%
II	Mild early parenchymal enhancement	~90%
III	Moderate early parenchymal enhancement	~70%
IV	Intense early parenchymal enhancement	~50%

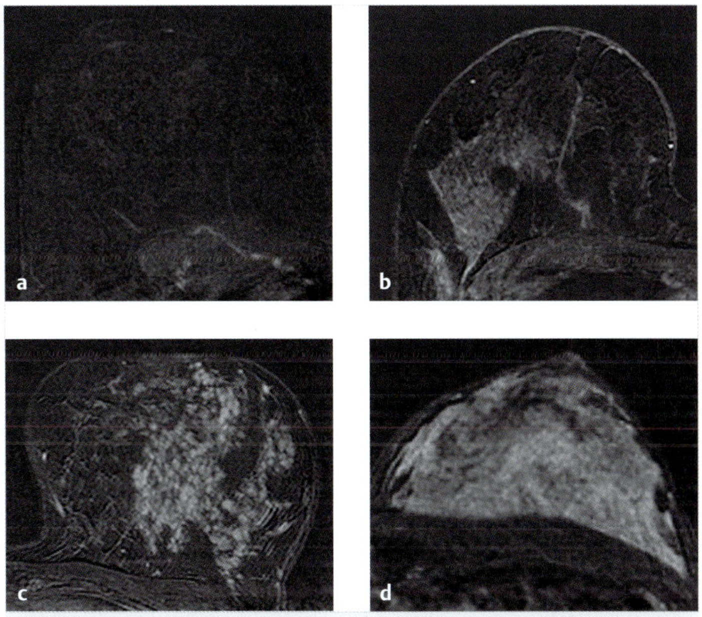

Fig. 7.8 Density types on breast MRI. Examples of MRI examinations (early subtraction) with different MRM density types: **(a)** type I; **(b)** type II; **(c)** type III; **(d)** type IV.

Note

The assignment of the breast MRI density type makes a statement about the reliability with which the particular breast MRI can detect breast cancer.

7.2.3 Findings in the T1-Weighted Precontrast Image

Important information that can be important in characterizing suspicious lesions and in evaluating an individual examinee's breast medical history or composition can be obtained from the T1W image findings before administration of contrast medium (T1W precontrast image). Thus the detection of adipose tissue within a mass lesion usually argues in favor of a benign condition, e.g., an intramammary lymph node (▶ Fig. 7.9a). The contents of cystic lesions (e.g., blood or proteinaceous fluid) can also be more

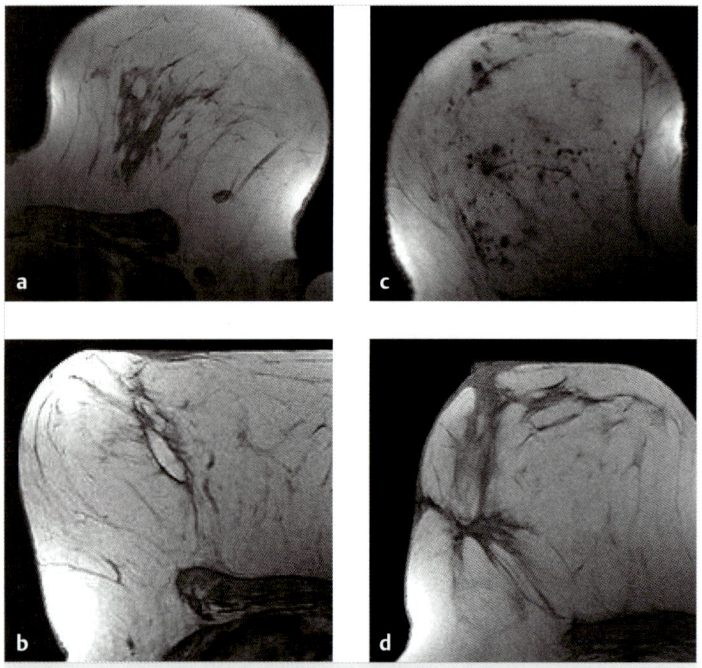

Fig. 7.9 Findings in the precontrast T1W MR image of the breast. Examples. **(a)** Lymph nodes (prepectoral and intramammary). **(b)** Postoperative hematoma. **(c)** Postoperative susceptibility artifacts. **(d)** Postoperative scarring.

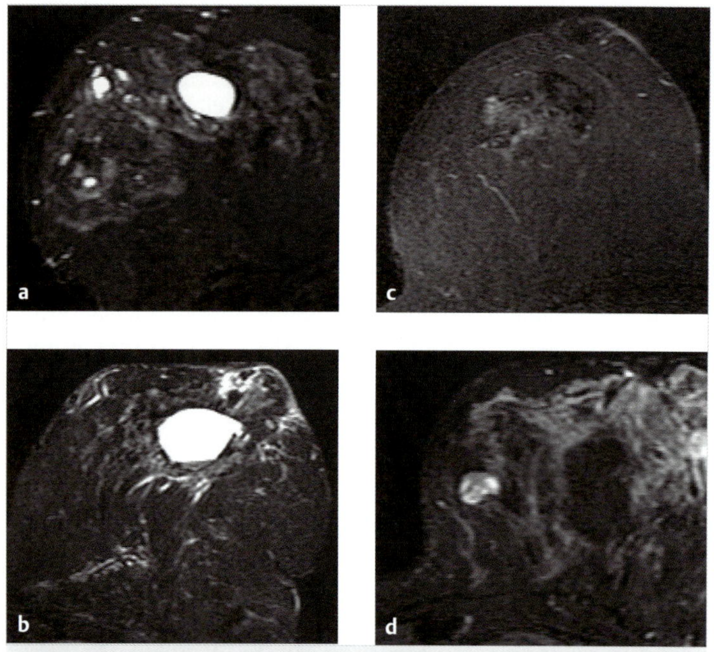

Fig. 7.10 Findings in the T2W MR image of the breast. Examples. **(a)** Cyst. **(b)** Seroma. **(c)** Area of adenosis. **(d)** Fibroadenoma.

precisely defined using the T1W information (▶ Fig. 7.9b). After previous surgical interventions, one often finds “susceptibility artifacts” as a consequence of electrocautery (▶ Fig. 7.9c). In addition, postoperative scars (▶ Fig. 7.9d), fat necrosis, and oil cysts can also be localized and evaluated on the precontrast images. However, the use of precontrast T1W image findings for the initial detection of breast cancer is appropriate only in exceptional cases, such as when a spiculated lesion lies within surrounding adipose tissue and can thus be easily identified based on its morphology.

Note

In the breast MRI examination, T1W images serve primarily to characterize suspicious findings; only occasionally do they aid in their initial detection.

7.2.4 Findings in the T2-Weighted Image

T2W images give a signal-intense visualization of fluid-containing structures. As a rule, they are not suitable for detection of carcinoma. However, information about the water content in a lesion can contribute important differential diagnostic information in the characterization of ambiguous hypervascularized masses or areas (▶ Fig. 7.10). Thus, invasive carcinomas frequently exhibit a lower endotumoral fluid content than, for example, myxomatous fibroadenomas or papillomas. (Invasive medullary or mucinous carcinomas are exceptions to this rule.) Nevertheless, assessment of the fluid content per se does not permit reliable differentiation between benign and malignant changes.

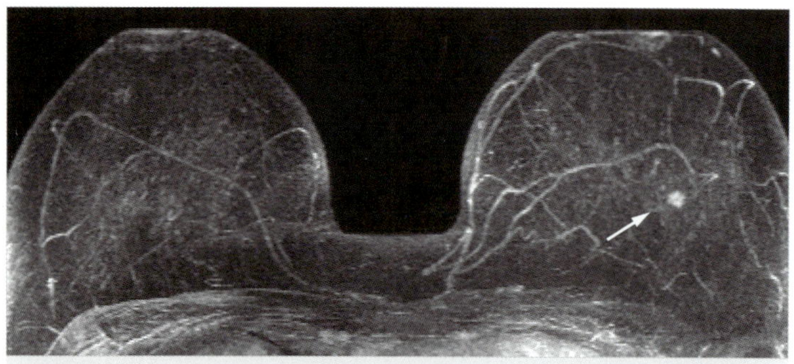

Fig. 7.11 Typical MRI subtraction image. Physiologic early contrast enhancement of the vessels and the nipple as well as suspicious enhancement within breast cancer in the left breast (arrow).

Note

In the breast MRI examination, T2W images serve primarily to characterize but not to detect suspicious findings.

Practical Tip

In the breast MRI, T2W- and T1W images should be performed with identical slice thickness and position.

7.2.5 Findings in the T1-Weighted Contrast-Enhanced Image

Suspicious findings on breast MRI are detected by T1W images after administration of contrast medium; the visualization of contrast-enhancing findings is best in the subtraction images of identical precontrast and postcontrast slices (▶ Fig. 7.11). As well as the conspicuous enhancement of a lesion, morphological criteria are of particular diagnostic importance for its characterization.

The significance of the kinetic curve assessment has declined due to the continuing improvement of the spatial resolutionof breast MRI and dynamic measurements.

Note

Contrast enhanced T1W breast MRI images, especially after subtraction postprocessing, serve the primary technique for the detection of suspicious mammary findings, but also in their characterization, based on morphological criteria.

7.2.6 Evaluation Criteria

The BI-RADS lexicon differentiates between three primary findings on breast MRI:

- Focus or foci.
- Mass lesion.
- Lesion without space-occupying characteristics (nonmasslike enhancement).

Foci

A focus is a tiny spot of enhancement within the parenchyma, exhibiting a maximum size of < 5 mm. A solitary focus can be an expression of hormonal stimulation (▶ Fig. 7.12a), but one must also include the very early stage of a malignant process in the differential diagnosis, so a follow-up MRI after 6 months is the strategic recommendation. The occurrence of multiple foci typically suggests a physiologic increase in perfusion (▶ Fig. 7.12b). In this case, no targeted follow-up is indicated.

Mass Lesions

Three-dimensional, space-occupying areas of increased enhancement of 5 mm and larger are designated "mass lesions." One should evaluate mass lesions based on the criteria "shape," "margins", and "endotumoral contrast media distribution," as well as certain dynamic criteria in representative regions of interest (ROIs): "initial rise" and "delayed phase" or "postinitial" uptake of contrast medium. To help translate findings based on these criteria into a BI-RADS assessment category, a system has been developed that assigns a points value to each criteria (the Goettingen score, ▶ Table 7.3). In this system, phenomena that strongly suggest malignancy—such as ring enhancement and washout effect—are given a higher score than findings that are more likely to be associated with benign findings, such as smooth margins and slight initial signal increase. The sum of all the points makes up the overall score for a mass lesion, which can range from 1 to 8 points. This overall score enables a fairly reliable evaluation of a mass lesion and its assignment to a BI-RADS assessment category (▶ Table 7.4, ▶ Fig. 7.13). Individual work groups have expanded the Goettingen Score to include the criteria "endotumoral water content" in the T2W measurement, and the presence of "endotumoral septations."

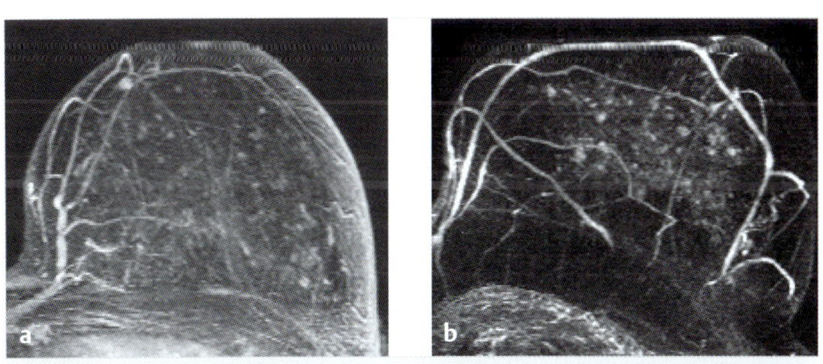

Fig. 7.12 Foci on breast MRI. (a) Solitary focus in the early subtraction image (maximum intensity projection). **(b)** Multiple foci in the early subtraction image (maximum intensity projection).

Table 7.3 Goettingen score for the evaluation of mass lesions on breast MRI. Assignment of points (0–2) for each criterion, depending on the specified findings. Maximal achievable point score is 8

Evaluation criterion	Score		
	0	1	2
Shape	Round, oval	Irregular, spiculated	-
Margin	Circumscribed	Indistinct	-
Contrast media distribution	Homogeneous	Inhomogeneous	Rim sign
Initial signal increase (1st–3rd minute)	Mild (< 50%)	Moderate (50–100%)	Strong (> 100%)
Post-initial signal sequence	Continuous increase	Plateau phase	Washout phenomenon

Table 7.4 Assignment of an MR BI-RADS category according to the Goettingen score. Mass lesions with an overall score of < 3 points are typically considered to be benign, while lesions with a sore of ≥ 4 points are considered to be potentially suspicious for malignancy. Mass lesions with a score of 3 points are assigned to MR BI-RADS 3

Total point count in the Goettingen score	MR BI-RADS category
1	1
2	2
3	3
4 and 5	4
6, 7, and 8	5

Note

Neither any individual criterion ("shape," "margins", "contrast media distribution") nor either of the dynamic criteria, considered alone, enables reliable assessment of a mass lesion in breast MRI.

Non-Space-Occupying Changes

Areas of increased enhancement that do not exhibit space-occupying characteristics are termed "nonmasslike enhancement." Such findings are typically associated with enhancement of the parenchymal or tumor matrix, whereas the intramammary adipose tissue does not enhance (▶ Fig. 7.14). The application of the Goettingen score here makes little sense, especially given that the ROIs used to determine the dynamic characteristics will often

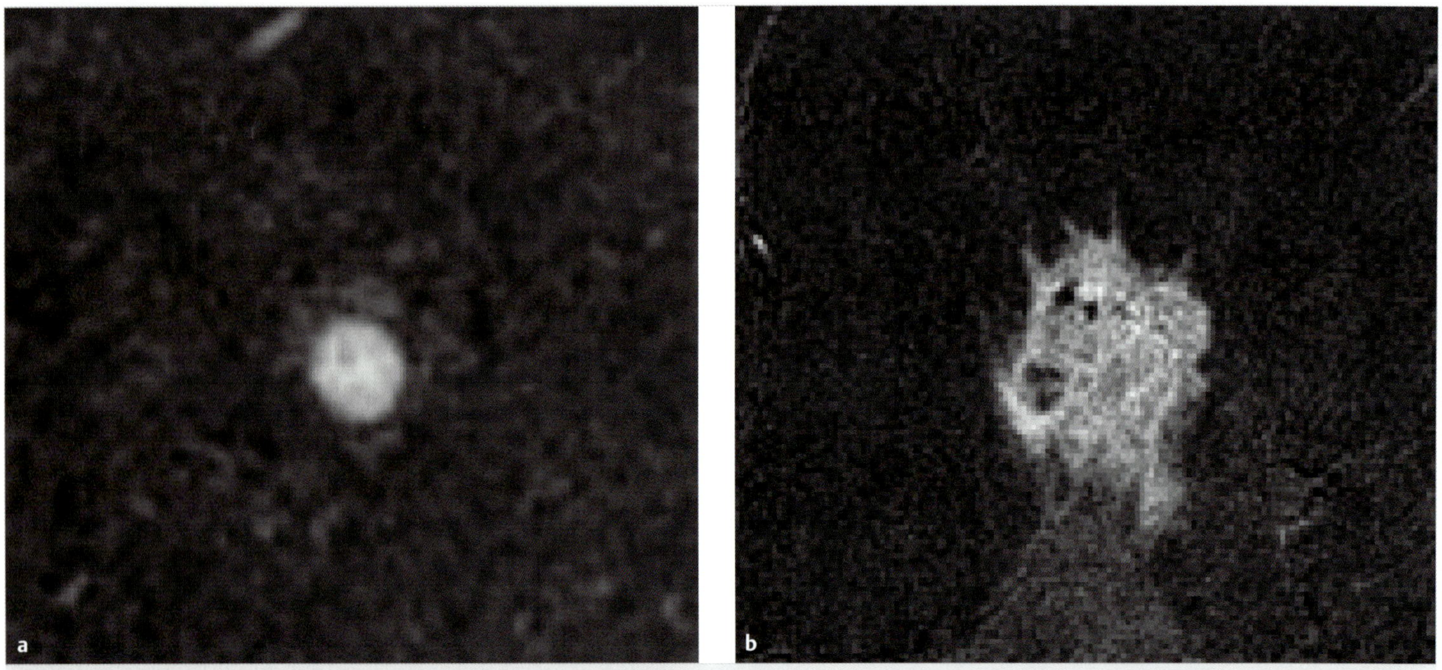

Fig. 7.13 Mass lesions on breast MRI. (a) Benign mass in the early subtraction image (fibroadenoma). **(b)** Malignant mass lesion in the early subtraction image (carcinoma).

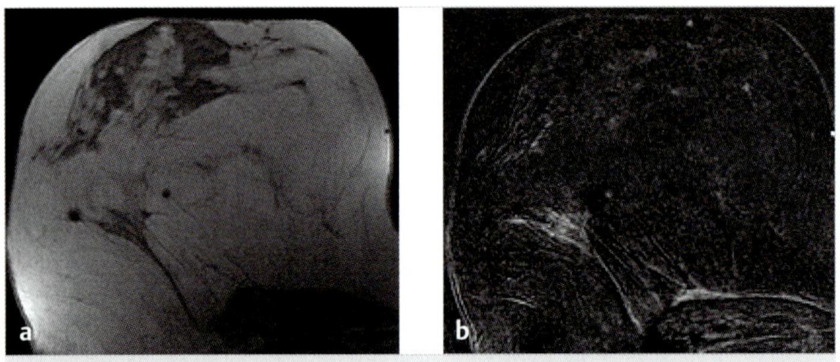

Fig. 7.14 Non-masslike lesions on breast MRI. A non-space-occupying process (invasive, lobular breast carcinoma): adipose tissue is not infiltrated by the tumor process (single-file infiltration pattern). Secondary characteristics: postoperative susceptibility artifacts from a previous surgery. **(a)** T1W precontrast image. **(b)** Subtraction image following administration of contrast medium.

include lipomatous regions, thus complicating the interpretation. Aside from a regional physiologic increase in perfusion (a normal finding), the differential diagnosis should consider intraductal tumors, invasive lobular carcinomas, radial scars, and inflammatory changes.

Note

Microcalcifications, whether carcinoma-associated or not, are not depicted on breast MRI. MRI is very sensitive, however, in the detection of the increased vascularization associated with proliferating tumor processes.

7.2.7 BI-RADS Classification of MRI of the Breast

The final assessment of a breast MRI examination must always be in accordance with the ACR MRI BI-RADS classification system. Seven categories are distinguished, whereby the categories BI-RADS 0 and BI-RADS 6 are of special significance. The BI-RADS assessment categories 1 to 5 are used to encode the probability that an abnormality is a malignant process (▶ Table 7.5); these prescribe concrete recommendations for the further course of action (▶ Table 7.6).

MR BI-RADS 0

As for the other modalities, the BI-RADS category 0 is assigned when no conclusive assessment of the MRI examination is possible and further diagnostic procedures are required. For breast MRI, this primarily involves examinations that are deemed to be technically or methodologically inadequate and must be repeated, e.g., as a consequence of artifacts caused by patient movement. The BI-RADS 0 category may also be assigned when it is necessary to acquire additional information from other imaging modalities (mammography, sonography) or when past medical history information must be obtained.

Table 7.5 BI-RADS classification for breast MRI findings (MR BI-RADS). Description of findings and corresponding risk of malignancy

MR BI-RADS category	Description	Risk of carcinoma (%)
1	Normal	0
2	Definitely benign	0
3	Probably benign	> 0% – ≤ 2%
4	Suspicious	> 2% – < 95%[a]
5	Highly suggestive of malignancy	≥ 95%

[a]Due to the large risk range within the category MR BI-RADS 4 (> 2– < 95%), subdivisions may be used: subcategory MR BI-RADS 4A (> 2– ≤ 10%), 4B (> 10– ≤ 50%) and 4C (> 50– < 95%).

Table 7.6 Recommended management according to the MR BI-RADS category

MR BI-RADS category	Recommended management
1	Next examination within the routine screening interval
2	Next examination within the routine screening interval
3	Follow-up after 6 months
4	Histological work-up by percutaneous biopsy
5	Pretherapeutic histological work-up by percutaneous biopsy

Within the concept of individualized examinations, the BI-RADS 0 category should not have to be used at all. Ambiguous findings can be conclusively evaluated by promptly carrying out appropriate supplementary examinations and assigning a BI-RADS category 1 to 5 while the patient is still on site.

MR BI-RADS 1

The BI-RADS 1 category is assigned when an examination is determined to be complete and adequate for assessment. Findings are normal and without abnormalities to comment on (▶ Fig. 7.15). This signifies that both breasts are visualized symmetrically and that there are no mass lesions with increased vascularization, architectural distortions, or suspicious areas of enhancement.

Depending on the individual breast cancer risk, further examinations are recommended at appropriate screening intervals.

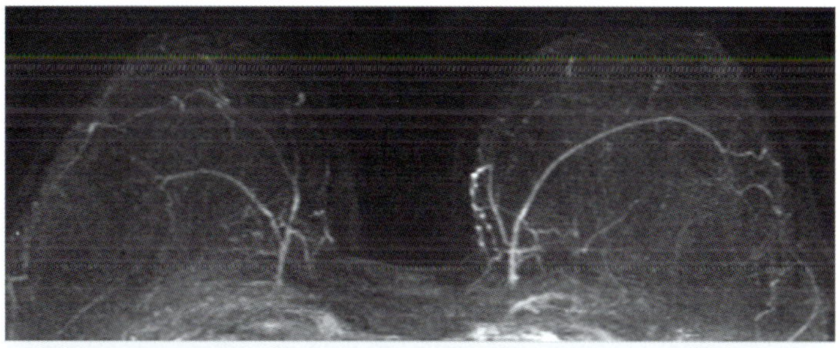

Fig. 7.15 MR BI-RADS 1. Unremarkable visualization of both breasts on breast MRI (maximum intensity projection).

MR BI-RADS 2

BI-RADS 2 categorization also conveys that the breast MRI examination is determined to be complete and adequate for assessment. In contrast to the MR BI-RADS 1 category, in BI-RADS 2 a finding is present that is worth describing but is clearly benign. Examples of such findings on MRI include:

- Fibrotic fibroadenoma without enhancement (▶ Fig. 7.16a).
- Cysts.
- Old scars without enhancement.
- Findings with lipomatous inclusions, such as lipomas, oil cysts (▶ Fig. 7.16b), galactoceles, or hamartomas.
- Implants with no indication of malignant changes.

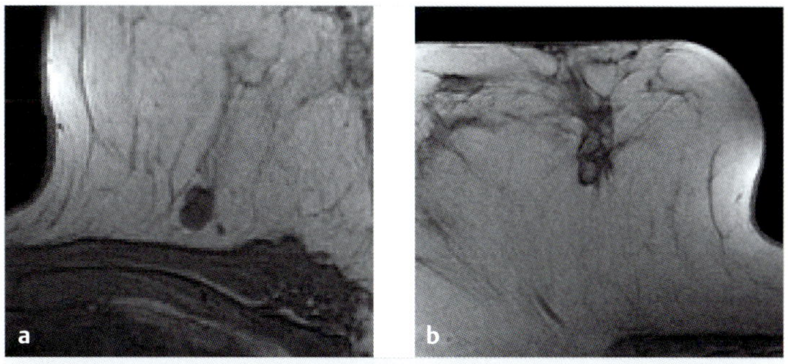

Fig. 7.16 MR BI-RADS 2. Typical examples. Both images show T1W precontrast examination. No enhancement is seen after administration of contrast material (not shown). (a) Fibrotic fibroadenoma. (b) Oil cyst.

All the itemized entities exhibit a characteristic appearance on MRI that excludes them from the differential diagnoses of malignant tumors. The risk of cancer is assessed to be 0%, and further examinations should be performed at appropriate screening intervals.

MR BI-RADS 3

The category BI-RADS 3 is assigned if an examination is determined to be complete and adequate for assessment, and a lesion is present that has a high probability of being benign. By definition, the residual risk of malignancy is ≤ 2%. The BI-RADS lexicon does not describe any concrete examples of this, so the classification of a lesion to the MR BI-RADS 3 category ultimately remains intuitive. On principle, it appears prudent to avoid using the BI-RADS 3 category, because this leads to a high false-positive rate for benign findings (overdiagnosis), while, on the other hand, there is a risk that the definitive diagnosis of a malignant lesion might be delayed. From the author's personal experience, however, there are two concrete constellations in which it makes sense to use the MR BI-RADS 3 category without risk of any appreciable worsening of prognosis if the lesion does correspond to a carcinoma:

- Solitary focus (▶ Fig. 7.17).
- Small mass lesion with 3 points on the Goettingen score (presupposing that a second-look ultrasound is unremarkable).

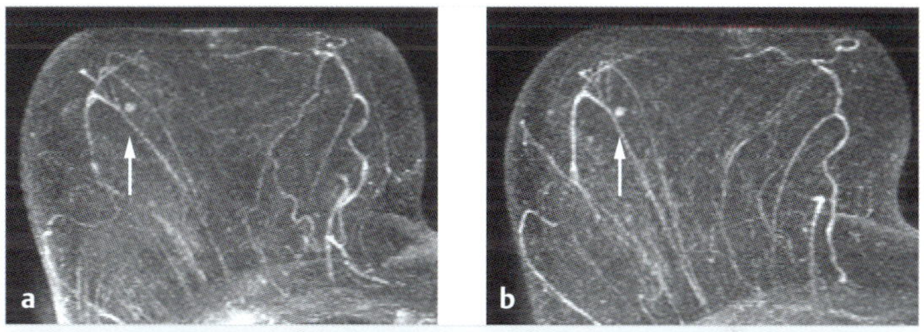

Fig. 7.17 MR BI-RADS 3. (a) Solitary focus in the right breast (arrow). (b) Findings initially categorized as MR BI-RADS 3. Follow-up MRI 6 months later showed identical findings (arrow).

In any case, as for the other modalities, all imaging possibilities should be exploited to clarify the finding before a lesion is classified as BI-RADS 3. This is especially true for the targeted second-look ultrasound, which is performed with exact information on the localization, size, and configuration of the MRI finding.

A short-term follow-up examination after an interval of 6 months is usually recommended for findings in the MR BI-RADS 3 category. As an alternative, if the patient desires, a prompt diagnostic work-up with a percutaneous biopsy can be performed. If short-term follow-up shows enlargement of the lesion being followed, then the lesion is upgraded and reclassified as an MR BI-RADS 4 lesion, and a percutaneous biopsy for diagnostic work-up is should be done. Here, too, the supplementary performance of a second-look ultrasound examination is indicated, in particular to clarify whether the finding can be biopsied by means of an ultrasound-guided core needle biopsy.

In retrospective monitoring, the rate of carcinoma found in the MR BI-RADS 3 group within the recommended follow-up examination period should be less than 2%.

MR BI-RADS 4

The BI-RADS 4 category (▶ Fig. 7.18) is assigned if an examination is determined to be complete and adequate for assessment, and a lesion is present that has sufficient probability of being malignant to warrant biopsy. These findings do not have the classic appearance of a malignant lesion but, by definition, carry a > 2 to < 95% risk of malignancy. For MRI, as for the other modalities, it is possible to subclassify MR BI-RADS 4 into the subcategories MR BI-RADS 4A (carcinoma risk > 2– ≤ 10%), MR BI-RADS 4B (> 10– ≤ 50%) and MR BI-RADS 4C (> 50– < 95%), but the BI-RADS lexicon does not present concrete examples for the corresponding subclassifications.

As a rule, an interventional diagnostic work-up, i.e., a percutaneous biopsy, is recommended for findings in the MR BI-RADS 4 category. With knowledge of the MRI findings, however, the mammographic images should be reviewed again and a second-look ultrasound performed. When no corresponding findings are seen on the mammogram and/or ultrasound, then diagnostic work-up is indicated in the form of an MRI-guided, vacuum-assisted biopsy. If, after the biopsy, the lesion is found to be histopathologically benign, then the compatibility between the histology and imaging must be reviewed. If it appears probable that the lesion was missed, re-biopsy or an open biopsy should be considered, in accordance with appropriate guidelines. If no re-biopsy is performed, a repeat MRI examination after 6 to 12 months is recommended.

MR BI-RADS 5

The BI-RADS 5 category designates those lesions that exhibit the typical MRI appearance of breast cancer (▶ Fig. 7.19). In the Goettingen score, these mass lesions typically receive 7 or 8 points. The probability of malignancy is statistically between 95 and 100%. Here also, the ACR does not specify any concrete examples for this category.

Findings in the MR BI-RADS 5 category are clarified by means of an MRI-guided, vacuum-assisted biopsy, as long as there are no corresponding findings on mammography or sonography. The course of action is essentially that for findings in the MR BI-RADS 4 category.

MR BI-RADS 6

Lesions are placed in this category once histology has confirmed a malignancy corresponding to a lesion in the histological B-classifications B5a (DCIS: ductal carcinoma in situ), B5b (invasive carcinoma), B5c, or B5d. The classification of a lesion to this category is thus independent of the morphological criteria in the image findings. The BI-RADS 6 category is only to be used until the malignant lesion has been surgically excised. Postoperatively, classification is made according to the pT staging of the TNM system of the UICC (Union for International Cancer Control).

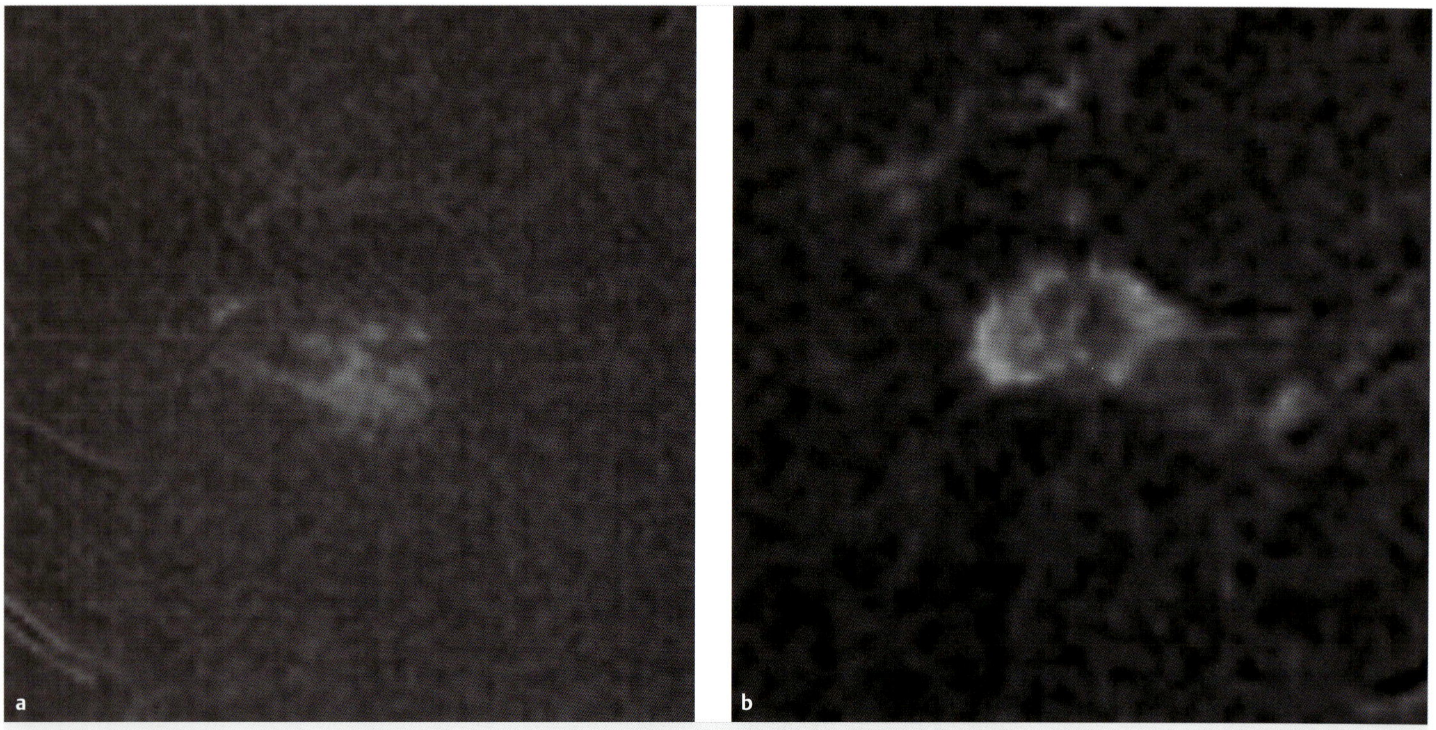

Fig. 7.18 MR BI-RADS 4. Typical examples. **(a)** Nonmasslike lesion (differential diagnosis: adenosis, mastitis, ductal carcinoma in situ [DCIS], invasive lobular carcinoma [ILC]). **(b)** Hypervascularized mass lesion (differential diagnosis: nodular adenosis, fibroadenoma, papilloma, invasive carcinoma).

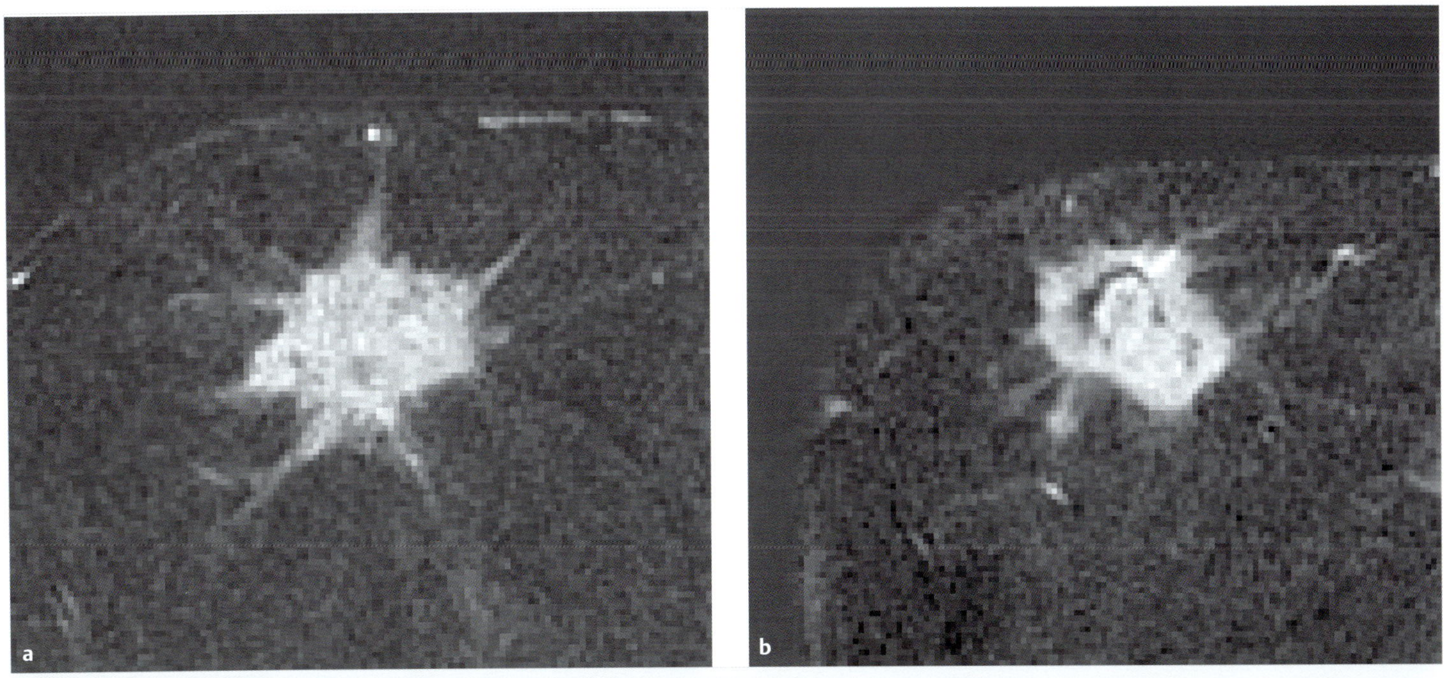

Fig. 7.19 MR BI-RADS 5. Classic carcinoma of the breast. **(a)** Irregular shape and spiculations. **(b)** Indistinct margins with ring enhancement and increased central enhancement.

Note

- In the overall assessment, the final diagnosis stands at the end of an examination chain (clinical examination, mammography, ultrasonography, breast MRI). Here, taking the specific reliability of each method into consideration, a summarizing evaluation of all image findings is assigned in the form of a final combined BI-RADS score to category 1, 2, 3, 4, or 5.
- Assuming that the histological result and the MRI image findings are compatible, re-biopsy or a follow-up examination after 6 to 12 months should be performed after every core-needle biopsy with benign histology performed on MRI findings in a BI-RADS category 4 or 5.
- The designation as "BI-RADS 6" is reserved for the percutaneous-biopsy–proven malignancy prior to surgery.

7.2.8 Normal Findings in MRI of the Breast

Changes seen in the T1W and T2W MR images are predominantly age-related physiological changes with respect to the composition of the breast, particularly regarding the distribution pattern and proportion of glandular and adipose tissue. The enhancement after the administration of contrast material correlates primarily with the extent of the hormonal stimulation of the glandular tissue, which, in interindividual comparisons, is more pronounced in premenopausal women than in postmenopausal women. Intraindividual fluctuations in premenopausal women depend primarily on the menstrual phase. Before (fourth week of the cycle) and during the menses (first week of the cycle), it is more pronounced than in the second and third weeks of the cycle. As expected, during pregnancy and in the postpartum lactation phase, there is generally extremely strong enhancement, so that the detection of contrast-enhanced findings is hardly, if at all, possible.

Bibliography

▶ Technique and Methods

[1] Ballesio L, Savelli S, Angeletti M, et al. Breast MRI: Are T2 IR sequences useful in the evaluation of breast lesions? Eur J Radiol. 2009; 71(1):96–101

[2] Baltzer PA, Freiberg C, Beger S, et al. Clinical MR-mammography: are computer-assisted methods superior to visual or manual measurements for curve type analysis? A systematic approach. Acad Radiol. 2009; 16(9):1070–1076

[3] Baltzer PA, Renz DM, Kullnig PE, Gajda M, Camara O, Kaiser WA. Application of computer-aided diagnosis (CAD) in MR-mammography (MRM): do we really need whole lesion time curve distribution analysis? Acad Radiol. 2009; 16(4):435–442

[4] Bhooshan N, Giger M, Lan L, et al. Combined use of T2-weighted MRI and T1-weighted dynamic contrast-enhanced MRI in the automated analysis of breast lesions. Magn Reson Med. 2011; 66(2):555–564

[5] Boetes C, Barentsz JO, Mus RD, et al. MR characterization of suspicious breast lesions with a gadolinium-enhanced TurboFLASH subtraction technique. Radiology. 1994; 193(3):777–781

[6] Buckley DL, Mussurakis S, Horsman A. Effect of temporal resolution on the diagnostic efficacy of contrast-enhanced MRI in the conservatively treated breast. J Comput Assist Tomogr. 1998; 22(1):47–51

[7] Damadian R. Tumor detection by nuclear magnetic resonance. Science. 1971; 171(3976):1151–1153

[8] El Yousef SJ, Alfidi RJ, Duchesneau RH, et al. Initial experience with nuclear magnetic resonance (NMR) imaging of the human breast. J Comput Assist Tomogr. 1983; 7(2):215–218

[9] Fischer U, von Heyden D, Vosshenrich R, Vieweg I, Grabbe E. Signal characteristics of malignant and benign lesions in dynamic 2D-MRT of the breast [in German]. Rofo. 1993; 158(4):287–292

[10] Fischer U, Vosshenrich R, Kopka L, Kahlen O, Grabbe E. Contrast medium assisted dynamic MR-mammography after diagnostic and therapeutic interventions on the breast [in German]. Bildgebung. 1996; 63(2):94–100

[11] Fischer U. Practical MR Mammography. 2nd ed. Stuttgart: Thieme; 2012

[12] Frahm J, Haase A, Matthaei D. Rapid three-dimensional MR imaging using the FLASH technique. J Comput Assist Tomogr. 1986; 10:363

[13] Haase A, Frahm J, Matthaei D, et al. FLASH-imaging: rapid NMR imaging using low flip angle pulses. J Magn Reson. 1986; 67:258–266

[14] Harms SE, Flamig DP, Hesley KL, et al. Fat-suppressed three-dimensional MR imaging of the breast. Radiographics. 1993; 13(2):247–267

[15] Harms SE, Flamig DP, Hesley KL, et al. MR imaging of the breast with rotating delivery of excitation off resonance: clinical experience with pathologic correlation. Radiology. 1993; 187(2):493–501

[16] Heywang SH, Fenzl G, Edmaier M, Eiermann W, Bassermann R, Krischke I. Nuclear spin tomography in breast diagnosis [in German]. Rofo. 1985; 143 (2):207–212

[17] Heywang SH, Fenzl G, Beck R, et al. Use of Gd-DTPA in the nuclear magnetic resonance study of the breast [in German]. Rofo. 1986; 145(5):565–571

[18] Heywang-Köbrunner SH, Beck R. Contrast-Enhanced MRI of the Breast. Berlin: Springer; 1995

[19] Huang W, Fisher PR, Dulaimy K, Tudorica LA, O'Hea B, Button TM. Detection of breast malignancy: diagnostic MR protocol for improved specificity. Radiology. 2004; 232(2):585–591

[20] Kaiser WA, Zeitler E. Nuclear spin tomography of the breast—1st clinical results [in German.]. Rontgenpraxis. 1985; 38:256

[21] Kaiser WA. MR-Mammography (MRM). Berlin: Springer; 1993

[22] Kelcz F, Furman-Haran E, Grobgeld D, Degani H. Clinical testing of high-spatial-resolution parametric contrast-enhanced MR imaging of the breast. AJR Am J Roentgenol. 2002; 179(6):1485–1492

[23] Kuhl CH, Bieling HB, Lutterberg G, et al. Standardization and acceleration of quantitative analysis of dynamic MR mammographies via parametric images and automatized ROI definition [in German]. RoFo. 1996; 164:475–482

[24] Kuhl CK, Klaschik S, Mielcarek P, Gieseke J, Wardelmann E, Schild HH. Do T2-weighted pulse sequences help with the differential diagnosis of enhancing lesions in dynamic breast MRI? J Magn Reson Imaging. 1999; 9 (2):187–196

[25] Kuhl CK, Gieseke J, von Falkenhausen M, et al. Sensitivity encoding for diffusion-weighted MR imaging at 3.0 T: intraindividual comparative study. Radiology. 2005; 234(2):517–526

[26] Kuhl CK, Jost P, Morakkabati N, Zivanovic O, Schild HH, Gieseke J. Contrast-enhanced MR imaging of the breast at 3.0 and 1.5 T in the same patients: initial experience. Radiology. 2006; 239(3):666–676

[27] Lauterbur PC. Image formation by induced local interactions: examples employing nuclear magnetic resonance. Nature. 1973; 242:190–191

[28] Mansfield P, Morris PG, Ordidge R, Coupland RE, Bishop HM, Blamey RW. Carcinoma of the breast imaged by nuclear magnetic resonance (NMR). Br J Radiol. 1979; 52(615):242–243

[29] Müller-Schimpfle M, Rieber A, Kurz S, Stern W, Claussen CD. Dynamic 3D MR mammography using a rapid gradient-echo sequence [in German]. Rofo. 1995; 162(1):13–19

[30] Müller-Schimpfle M, Ohmenhäuser K, Claussen CD. Effect of age and menstrual cycle on mammography and MR mammography [in German]. Radiologe. 1997; 37(9):718–725

[31] Nunes LW, Schnall MD, Siegelman ES, et al. Diagnostic performance characteristics of architectural features revealed by high spatial-resolution MR imaging of the breast. AJR Am J Roentgenol. 1997; 169(2):409–415

[32] Perman WH, Heiberg EV, Herrmann VM. Half-Fourier, three-dimensional technique for dynamic contrast-enhanced MR imaging of both breasts and axillae: initial characterization of breast lesions. Radiology. 1996; 200 (1):263–269

[33] Renz DM, Böttcher J, Diekmann F, et al. Detection and classification of contrast-enhancing masses by a fully automatic computer-assisted diagnosis system for breast MRI. J Magn Reson Imaging. 2012; 35(5):1077–1088

[34] Schorn C, Fischer U, Döler W, Funke M, Grabbe E. Compression device to reduce motion artifacts at contrast-enhanced MR imaging in the breast. Radiology. 1998; 206(1):279–282

[35] Schorn C, Fischer U, Luftner-Nagel S, Grabbe E. Diagnostic potential of ultrafast contrast-enhanced MRI of the breast in hypervascularized lesions: are there advantages in comparison with standard dynamic MRI? J Comput Assist Tomogr. 1999; 23(1):118–122

[36] Shi K, Zhou K, Niu X, et al. Investigation of motion artifacts associated with fat saturation technique in 3D flash imaging. Med Phys. 2011; 38 (8):4556–4562

▶ Evaluation

[1] Al-Khawari H, Athyal R, Kovacs A, Al-Saleh M, Madda JP. Accuracy of the Fischer scoring system and the Breast Imaging Reporting and Data System in identification of malignant breast lesions. Ann Saudi Med. 2009; 29(4):280–287

[2] Baker JA, Kornguth PJ, Floyd CE , Jr. Breast imaging reporting and data system standardized mammography lexicon: observer variability in lesion description. AJR Am J Roentgenol. 1996; 166(4):773–778

[3] Ballesio L, Savelli S, Angeletti M, et al. Breast MRI: Are T2 IR sequences useful in the evaluation of breast lesions? Eur J Radiol. 2009; 71(1):96–101

[4] Baltzer PA, Benndorf M, Dietzel M, Gajda M, Runnebaum IB, Kaiser WA. False-positive findings at contrast-enhanced breast MRI: a BI-RADS descriptor study. AJR Am J Roentgenol. 2010; 194(6):1658–1663

[5] Baum F, Fischer U, Vosshenrich R, Grabbe E. Classification of hypervascularized lesions in CE MR imaging of the breast. Eur Radiol. 2002; 12(5):1087–1092

[6] Buchbinder SS, Leichter IS, Lederman RB, et al. Computer-aided classification of BI-RADS category 3 breast lesions. Radiology. 2004; 230(3):820–823

[7] Dietzel M, Baltzer PA, Vag T, et al. Potential of MR mammography to predict tumor grading of invasive breast cancer. Rofo. 2011; 183(9):826–833

[8] Fischer DR, Wurdinger S, Boettcher J, Malich A, Kaiser WA. Further signs in the evaluation of magnetic resonance mammography: a retrospective study. Invest Radiol. 2005; 40(7):430–435

[9] Fischer U, von Heyden D, Vosshenrich R, Vieweg I, Grabbe E. Signal characteristics of malignant and benign lesions in dynamic 2D-MRT of the breas. Rofo. 1993; 158(4):287–292

[10] Girardi V, Carbognin G, Camera L, et al. Fischer's score criteria correlating with histopathological prognostic factors in invasive breast cancer. Radiol Med (Turin). 2010; 115(3):421–433

[11] Goto M, Ito H, Akazawa K, et al. Diagnosis of breast tumors by contrast-enhanced MR imaging: comparison between the diagnostic performance of dynamic enhancement patterns and morphologic features. J Magn Reson Imaging. 2007; 25(1):104–112

[12] Huang W, Tudorica LA, Li X, et al. Discrimination of benign and malignant breast lesions by using shutter-speed dynamic contrast-enhanced MR imaging. Radiology. 2011; 261(2):394–403

[13] Kim SJ, Morris EA, Liberman L, et al. Observer variability and applicability of BI-RADS terminology for breast MR imaging: invasive carcinomas as focal masses. AJR Am J Roentgenol. 2001; 177(3):551–557

[14] Kinkel K, Helbich TH, Esserman LJ, et al. Dynamic high-spatial-resolution MR imaging of suspicious breast lesions: diagnostic criteria and interobserver variability. AJR Am J Roentgenol. 2000; 175(1):35–43

[15] Kühl CK, Bieling H, Gieseke J, et al. Breast neoplasms: T2* susceptibility-contrast, first-pass perfusion MR imaging. Radiology. 1997; 202(1):87–95

[16] Kuhl CK, Klaschik S, Mielcarek P, Gieseke J, Wardelmann E, Schild HH. Do T2-weighted pulse sequences help with the differential diagnosis of enhancing lesions in dynamic breast MRI? J Magn Reson Imaging. 1999; 9(2):187–196

[17] Kuhl CK, Schild HH, Morakkabati N. Dynamic bilateral contrast-enhanced MR imaging of the breast: trade-off between spatial and temporal resolution. Radiology. 2005; 236(3):789–800

[18] Kuhl CK, Jost P, Morakkabati N, Zivanovic O, Schild HH, Gieseke J. Contrast-enhanced MR imaging of the breast at 3.0 and 1.5 T in the same patients: initial experience. Radiology. 2006; 239(3):666–676

[19] Kuhl C. The current status of breast MR imaging. Part I. Choice of technique, image interpretation, diagnostic accuracy, and transfer to clinical practice. Radiology. 2007; 244(2):356–378

[20] Kvistad KA, Rydland J, Vainio J, et al. Breast lesions: evaluation with dynamic contrast-enhanced T1-weighted MR imaging and with T2*-weighted first-pass perfusion MR imaging. Radiology. 2000; 216(2):545–553

[21] Lazarus E, Mainiero MB, Schepps B, Koelliker SL, Livingston LS. BI-RADS lexicon for US and mammography: interobserver variability and positive predictive value. Radiology. 2006; 239(2):385–391

[22] Levman JE, Causer P, Warner E, Martel AL. Effect of the enhancement threshold on the computer-aided detection of breast cancer using MRI. Acad Radiol. 2009; 16(9):1064–1069

[23] Liberman L, Morris EA, Lee MJ, et al. Breast lesions detected on MR imaging: features and positive predictive value. AJR Am J Roentgenol. 2002; 179 (1):171–178

[24] Liberman L, Morris EA, Dershaw DD, Abramson AF, Tan LK. Ductal enhancement on MR imaging of the breast. AJR Am J Roentgenol. 2003; 181(2):519–525

[25] Liberman L, Mason G, Morris EA, Dershaw DD. Does size matter? Positive predictive value of MRI-detected breast lesions as a function of lesion size. AJR Am J Roentgenol. 2006; 186(2):426–430

[26] Marklund M, Torp-Pedersen S, Bentzon N, Thomsen C, Roslind A, Nolsøe CP. Contrast kinetics of the malignant breast tumour—border versus centre enhancement on dynamic midfield MRI. Eur J Radiol. 2008; 65(2):279–285

[27] Matsubayashi R, Matsuo Y, Edakuni G, Satoh T, Tokunaga O, Kudo S. Breast masses with peripheral rim enhancement on dynamic contrast-enhanced MR images: correlation of MR findings with histologic features and expression of growth factors. Radiology. 2000; 217(3):841–848

[28] Meeuwis C, van de Ven SM, Stapper G, et al. Computer-aided detection (CAD) for breast MRI: evaluation of efficacy at 3.0 T. Eur Radiol. 2010; 20(3):522–528

[29] Moraes PdeC, Chala LF, Chang YS, et al. Observer variability in the application of morphologic and dynamic criteria according to the BI-RADS for MRI. Breast J. 2010; 16(5):558–560

[30] Morakkabati-Spitz N, Leutner C, Schild H, Traeber F, Kuhl C. Diagnostic usefulness of segmental and linear enhancement in dynamic breast MRI. Eur Radiol. 2005; 15(9):2010–2017

[31] Penn A, Thompson S, Brem R, et al. Morphologic blooming in breast MRI as a characterization of margin for discriminating benign from malignant lesions. Acad Radiol. 2006; 13(11):1344–1354

[32] Piccoli CW. Contrast-enhanced breast MRI: factors affecting sensitivity and specificity. Eur Radiol. 1997; 7 Suppl 5:281–288

[33] Sakamoto N, Tozaki M, Higa K, et al. Categorization of non-mass-like breast lesions detected by MRI. Breast Cancer. 2008; 15(3):241–246

[34] Sardanelli F, Boetes C, Borisch B, et al. Magnetic resonance imaging of the breast: recommendations from the EUSOMA working group. Eur J Cancer. 2010; 46(8):1296–1316

[35] Schnall MD, Blume J, Bluemke DA, et al. Diagnostic architectural and dynamic features at breast MR imaging: multicenter study. Radiology. 2006; 238 (1):42–53

[36] Sherif H, Mahfouz AE, Oellinger H, et al. Peripheral washout sign on contrast enhanced MR images of the breast. Radiology. 1997; 205(1):209–213

[37] Siegmann KC, Müller-Schimpfle M, Schick F, et al. MR imaging-detected breast lesions: histopathologic correlation of lesion characteristics and signal intensity data. AJR Am J Roentgenol. 2002; 178(6):1403–1409

[38] Siegmann KC, Moron HU, Baur A, et al. Diagnostic value of a breast MRI score for the prediction of malignancy of breast lesions detected solely with MRI [in German]. Rofo. 2009; 181(6):556–563

[39] Thomassin-Naggara I, Salem C, Darai E, et al. Non-masslike enhancement on breast MRI: interpretation pearls [in French]. J Radiol. 2009; 90(3 Pt 1):269–275

[40] Tokuda Y, Kuriyama K, Nakamoto A, et al. Evaluation of suspicious nipple discharge by magnetic resonance mammography based on breast imaging reporting and data system magnetic resonance imaging descriptors. J Comput Assist Tomogr. 2009; 33(1):58–62

[41] Tozaki M, Fukuda K. High-spatial-resolution MRI of non-masslike breast lesions: interpretation model based on BI-RADS MRI descriptors. AJR Am J Roentgenol. 2006; 187(2):330–337

[42] Tozaki M, Igarashi T, Fukuda K. Breast MRI using the VIBE sequence: clustered ring enhancement in the differential diagnosis of lesions showing non-mass-like enhancement. AJR Am J Roentgenol. 2006; 187(2):313–321

[43] Uematsu T. MRI findings of inflammatory breast cancer, locally advanced breast cancer, and acute mastitis: T2-weighted images can increase the specificity of inflammatory breast cancer. Breast Cancer. 2012; 19(4):289–294

[44] Van Goethem M, Schelfout K, Kersschot E, et al. Enhancing area surrounding breast carcinoma on MR mammography: comparison with pathological examination. Eur Radiol. 2004; 14(8):1363–1370

[45] Vassiou K, Kanavou T, Vlychou M, et al. Characterization of breast lesions with CE-MR multimodal morphological and kinetic analysis: comparison with conventional mammography and high-resolution ultrasound. Eur J Radiol. 2009; 70(1):69–76

[46] Wiener JI, Schilling KJ, Adami C, Obuchowski NA. Assessment of suspected breast cancer by MRI: a prospective clinical trial using a combined kinetic and morphologic analysis. AJR Am J Roentgenol. 2005; 184(3):878–886

[47] Williams TC, DeMartini WB, Partridge SC, Peacock S, Lehman CD. Breast MR imaging: computer-aided evaluation program for discriminating benign from malignant lesions. Radiology. 2007; 244(1):94–103

8 Imaging of Breast Lesions

U. Fischer and S. Luftner-Nagel

8.1 Benign Findings

8.1.1 Cysts

The simple cyst is a fluid-filled cavity with a lining consisting of an inner epithelial layer and an outer myoepithelial layer.

► **Characteristics**

- *Incidence:* very common.
- *Peak age:* all age groups; incidence decreases after menopause.
- *Multifocality:* frequent.
- *Bilateral occurrence:* frequent.
- *Complications:* none.
- *Risk of malignancy:* less than 1%.
- *Prognosis:* excellent.
- *Histological classification after biopsy:* B2.

► **Implications for work-up/treatment.** None if symptom-free. If there is pain, ultrasound-guided cyst aspiration, preferably by fine needle, may be performed.

► **Shape.** Round, oval.

► **Margins.** Circumscribed.

► **Internal composition.** Homogeneous fluid.

► **Clinical findings.** Small cysts normally have no correlative clinical findings. In the case of larger cysts, a circumscribed, firm-elastic mass may be palpable. Occasionally there is tenderness.

► **Imaging**

- *Sonography* (► Fig. 8.1a): Anechoic mass; posterior acoustic enhancement; unambiguous sonographic finding.
- *Mammography* (► Fig. 8.1b): mass with parenchyma-equivalent, homogeneous density; difficult or impossible to detect within dense breast tissue structures; good visibility within the lipomatous breast; occasional cyst wall calcifications.
- *Breast MRI* (► Fig. 8.1c): mass:
  - T1W precontrast image: hypointense.
  - T2W image: hyperintense.
  - After administration of contrast medium: no enhancement.

Note

Mammography cannot reliably distinguish between a cyst (fluid-filled) and a proliferative process (solid tissue).

► **Differential diagnosis.** Inflamed cyst, complex cyst, cyst with intracystic proliferation, triple-negative carcinoma.

8.1.2 Inflamed Cysts

The inflamed cyst is a simple cyst with inflamed—and thus hyperemic—cyst walls.

► **Characteristics**

- *Incidence:* rare.
- *Peak age:* all age groups; incidence decreases after menopause.
- *Multifocality:* rare.
- *Bilateral occurrence:* rare.
- *Complications:* occasional pain.
- *Risk of malignancy:* less than 1%.
- *Prognosis:* excellent.
- *Histological classification after biopsy:* B2.

► **Implications for work-up/treatment.** None if symptom-free. If there is pain, ultrasound-guided cyst aspiration, preferably by fine needle, may be performed.

► **Shape.** Round, oval.

► **Margins.** Predominantly circumscribed, occasionally slightly indistinct.

► **Internal composition.** Homogeneous fluid.

► **Clinical findings.** Small cysts normally have no correlative clinical findings. In the case of larger cysts, a circumscribed, firm-elastic, occasionally tender mass may be palpable.

► **Imaging**

- *Sonography* (► Fig. 8.2a): anechoic mass, cyst wall mildly thickened, possibly indistinct; posterior acoustic enhancement.
- *Mammography* (► Fig. 8.2b): mass with parenchyma-equivalent, homogeneous density; difficult or impossible to detect within dense breast tissue structures; good visibility within the lipomatous breast; occasional cyst wall calcifications.
- *Breast MRI* (► Fig. 8.2c): mass:
  - T1W precontrast image: hypointense.
  - T2W image: hyperintense.
  - After administration of contrast medium: cyst wall enhancement.

Note

When interpreting breast MRI, one should use the term "cyst wall enhancement" rather than "rim enhancement," because the term "rim enhancement" is closely associated with malignancy.

► **Differential diagnosis.** Complex cyst, cyst with intracystic proliferation, triple-negative carcinoma.

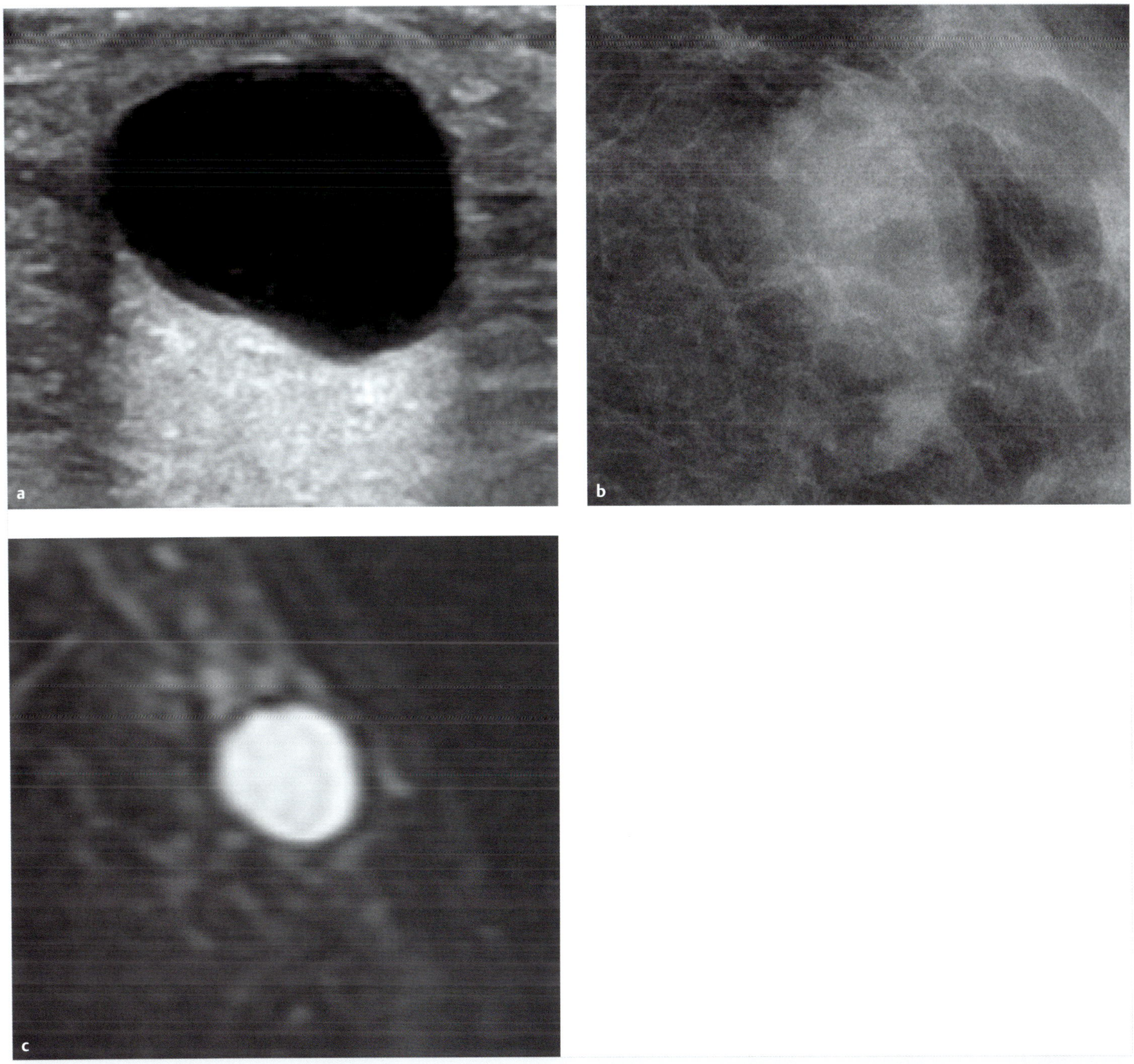

Fig. 8.1 Simple cyst. (a) Ultrasound image. (b) Mammogram. (c) T2W breast MR image.

8.1.3 Complex Cysts

When bleeding occurs into a simple cyst, it is referred to as a complex cyst (synonyms: hemorrhagic cyst, chocolate cyst). A cyst filled with highly proteinaceous fluid also justifies use of the term "complex cyst."

► **Characteristics**

- *Incidence:* rare.
- *Peak age:* all age groups; incidence decreases after menopause.
- *Multifocality:* rare.
- *Bilateral occurrence:* rare.
- *Complications:* none.
- *Risk of malignancy:* less than 1%.
- *Prognosis:* excellent.
- *Histological classification after biopsy:* B2.

► **Implications for work-up/treatment.** None if symptom-free. If there is pain, ultrasound-guided cyst aspiration using a larger gauge needle (10G) may be indicated.

► **Shape.** Round, oval.

► **Margins.** Circumscribed.

► **Internal composition.** Homogeneous to inhomogeneous due to fresh blood (with sedimentation effect), old blood, protein.

► **Clinical findings.** Small lesions normally have no correlative clinical findings. In the case of larger cysts, a circumscribed, firm-elastic mass may be palpable. The lesions are rarely tender to palpation.

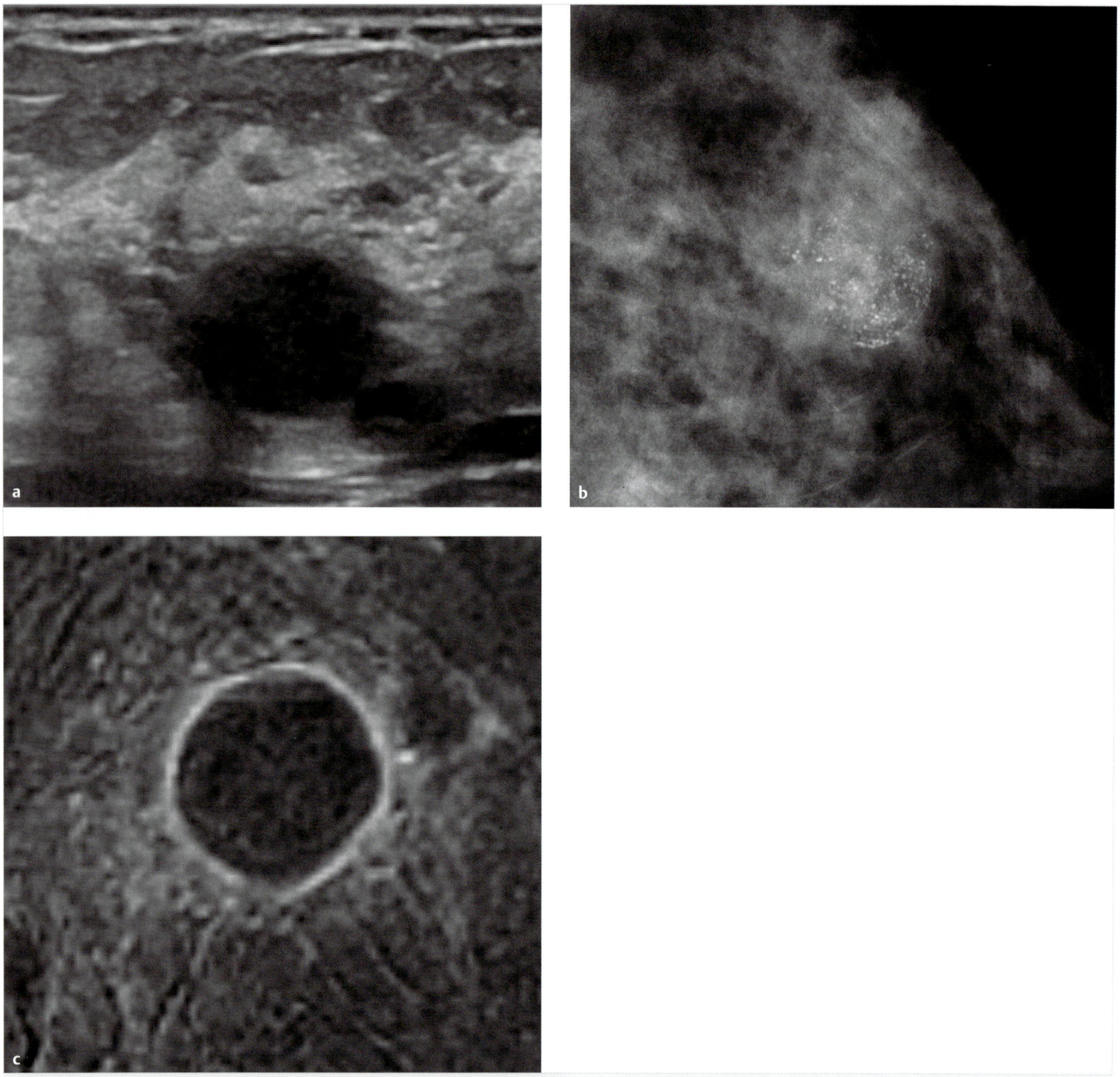

Fig. 8.2 Inflamed cyst. (a) Ultrasound image. **(b)** Mammogram. **(c)** Contrast-enhanced breast MRI subtraction image.

► **Imaging**

- *Sonography* (► Fig. 8.3a): hypoechoic mass; minor posterior acoustic enhancement (protein), moderate acoustic enhancement to acoustic extinction (old blood).
- *Mammography* (► Fig. 8.3b): mass with parenchyma-equivalent, homogeneous density; difficult or impossible to detect within dense breast tissue structures; good visibility within the lipomatous breast.
- *Breast MRI* (► Fig. 8.3c): mass:
  - T1W precontrast image: isointense (protein) to hypointense (old blood); sedimentation may be visible.
  - T2W image: isointense (protein) to hypointense (old blood); sedimentation may be visible.
  - After administration of contrast medium: no to slight cyst wall enhancement.

Note

When there is uncertainty in the differential diagnosis of a complex breast cyst, percutaneous biopsy should be undertaken.

► **Differential diagnosis.** Inflamed cyst, cyst with intracystic proliferation, triple-negative carcinoma, carcinoma with central degeneration.

8.1.4 Myxoid Fibroadenoma

Fibroadenomas are benign, mixed fibroepithelial tumors. In younger women, they usually present with a high epithelial content (myxoid fibroadenoma).

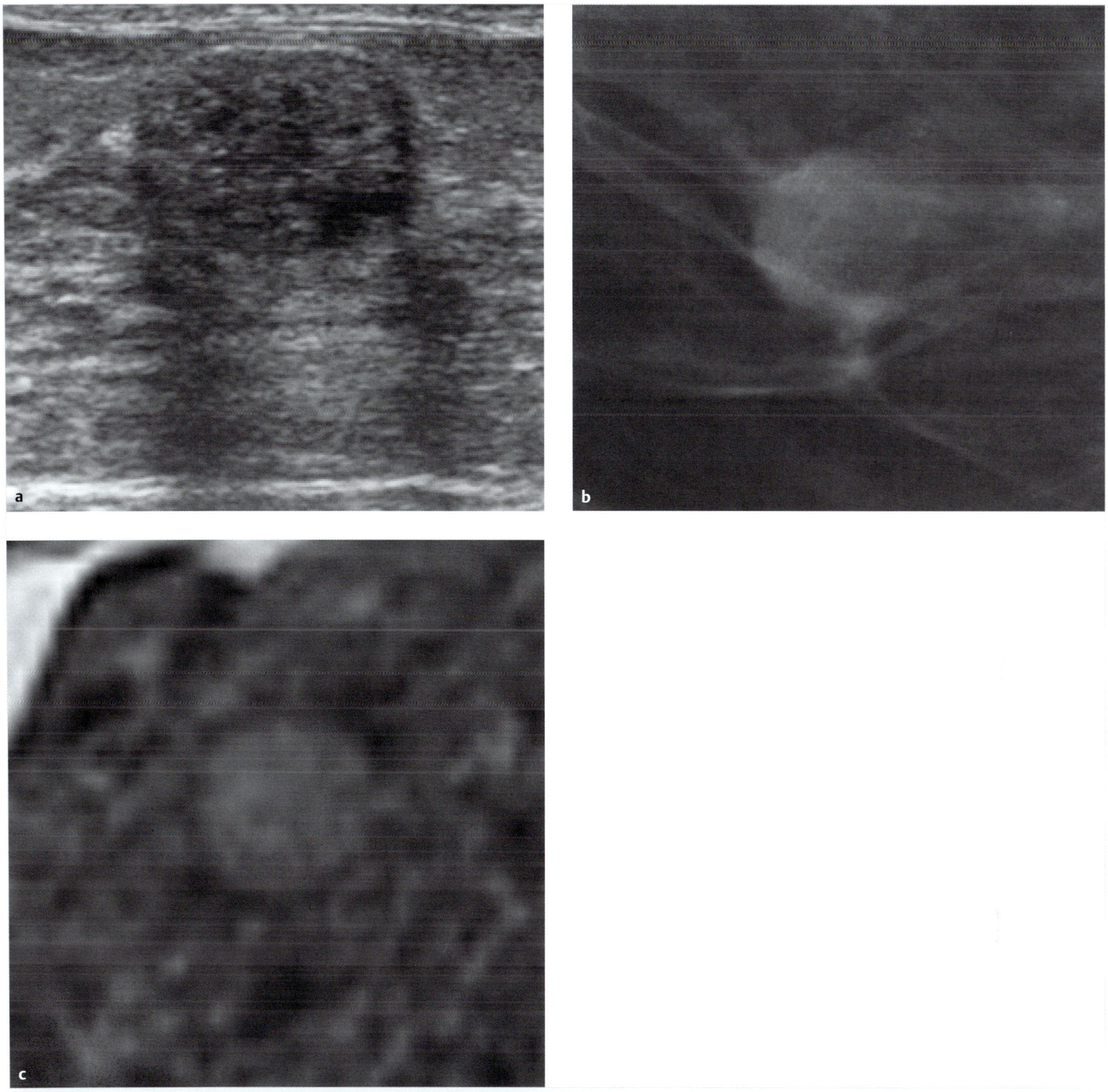

Fig. 8.3 **Hemorrhagic cyst. (a)** Ultrasound image. **(b)** Mammogram. **(c)** T2W breast MR image.

▸ **Characteristics**

- *Incidence:* frequent.
- *Peak age:* 18 to 50 years.
- *Multifocality:* frequent.
- *Bilateral occurrence:* frequent.
- *Complications:* usually none. The juvenile giant fibroadenoma is an exception due to its massive size (diameter up to 12–15 cm).
- *Risk of malignancy:* 0%.
- *Prognosis:* excellent.
- *Histological classification after biopsy:* B2.

▸ **Implications for work-up/treatment.** For smaller lesions in younger women (up to 30 years of age), no treatment is required. For smaller lesions in women aged 30 years and above, diagnosis should be histologically confirmed via percutaneous, preferably ultrasound-guided, core needle biopsy. Giant fibroadenomas should be surgically excised.

▸ **Shape.** Lobular, oval.

▸ **Margins.** Circumscribed.

▸ **Internal composition.** Homogeneous, proliferative tissue.

▸ **Clinical findings.** Small cysts normally have no correlative clinical findings. In the case of larger myxoid fibroadenomas, circumscribed lumps may be palpable.

► **Imaging**

- *Sonography* (► Fig. 8.4a): homogeneous, isoechoic mass; longitudinal axis parallel to the skin; mild elasticity; mild posterior acoustic enhancement.
- *Mammography* (► Fig. 8.4b): mass with parenchyma-equivalent, homogeneous density; difficult or impossible to detect within dense breast tissue structures; good visibility within the lipomatous breast; rarely displays endotumoral microcalcifications.
- *Breast MRI* (► Fig. 8.4c): focus, mass:
  - T1W precontrast image: hypointense.
  - T2W image: hyperintense; may display endotumoral hypointense septations with increasing fibrosis.
  - After administration of contrast medium: strong homogeneous enhancement; may display nonenhancing endotumoral septations with increasing fibrosis (dark septations).

Note

Myxoid fibroadenomas are the most common tumors in young women.

► **Differential diagnosis.** Adenoma, papilloma, phyllodes tumor, invasive carcinoma, intramammary metastasis.

8.1.5 Fibrotic Fibroadenoma

Fibroadenomas are benign, mixed fibroepithelial tumors. With increasing age, the fibrous component increasingly predominates (fibrotic fibroadenoma; synonyms: hyaline fibroadenoma, regressive fibroadenoma).

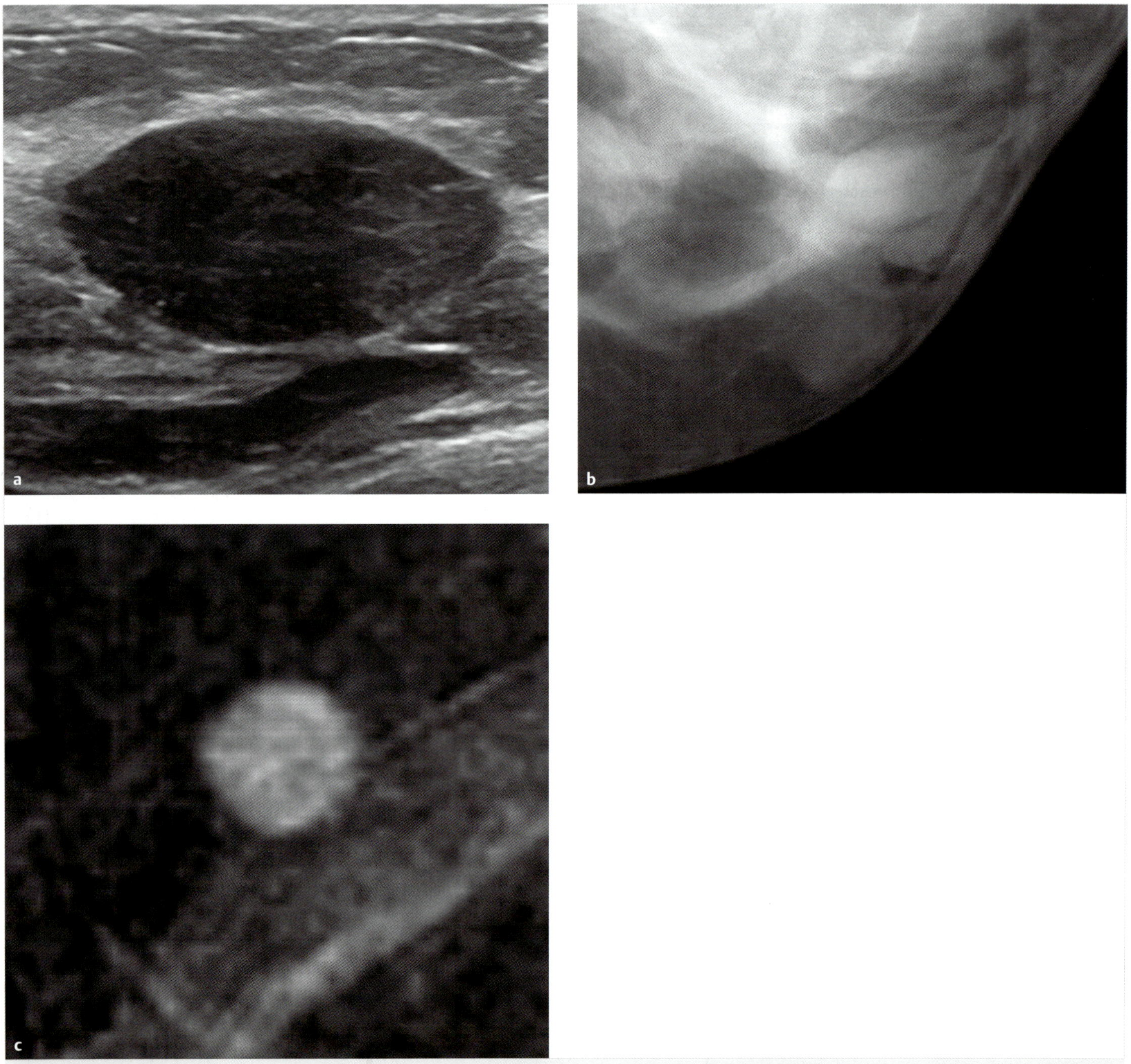

Fig. 8.4 Myxoid fibroadenoma. Fibroadenoma with high water content. **(a)** Ultrasound image. **(b)** Mammogram. **(c)** Contrast-enhanced breast MRI subtraction image.

► **Characteristics**

- *Incidence:* frequent.
- *Peak age:* 40 to 80 years.
- *Multifocality:* frequent.
- *Bilateral occurrence:* frequent.
- *Complications:* none.
- *Risk of malignancy:* 0%.
- *Prognosis:* excellent.
- *Histological classification after biopsy:* B2.

► **Implications for work-up/treatment.** When there is uncertainty in the differential diagnosis, percutaneous, preferably ultrasound-guided, core needle biopsy should be undertaken.

► **Shape.** Lobular, oval.

► **Margins.** Circumscribed.

► **Internal composition.** Inhomogeneous to homogeneous proliferative tissue.

► **Clinical findings.** Small lesions normally have no correlative clinical findings. In the case of a larger fibrotic fibroadenoma, a circumscribed, rather firm mass may be palpable.

► **Imaging**

- *Sonography* (► Fig. 8.5a): inhomogeneous or homogeneous, hypoechoic to isoechoic mass; longitudinal axis parallel to the skin; no elasticity; intermediate posterior echo pattern to acoustic extinction (acoustic extinction is particularly seen in the case of endotumoral macrocalcifications).

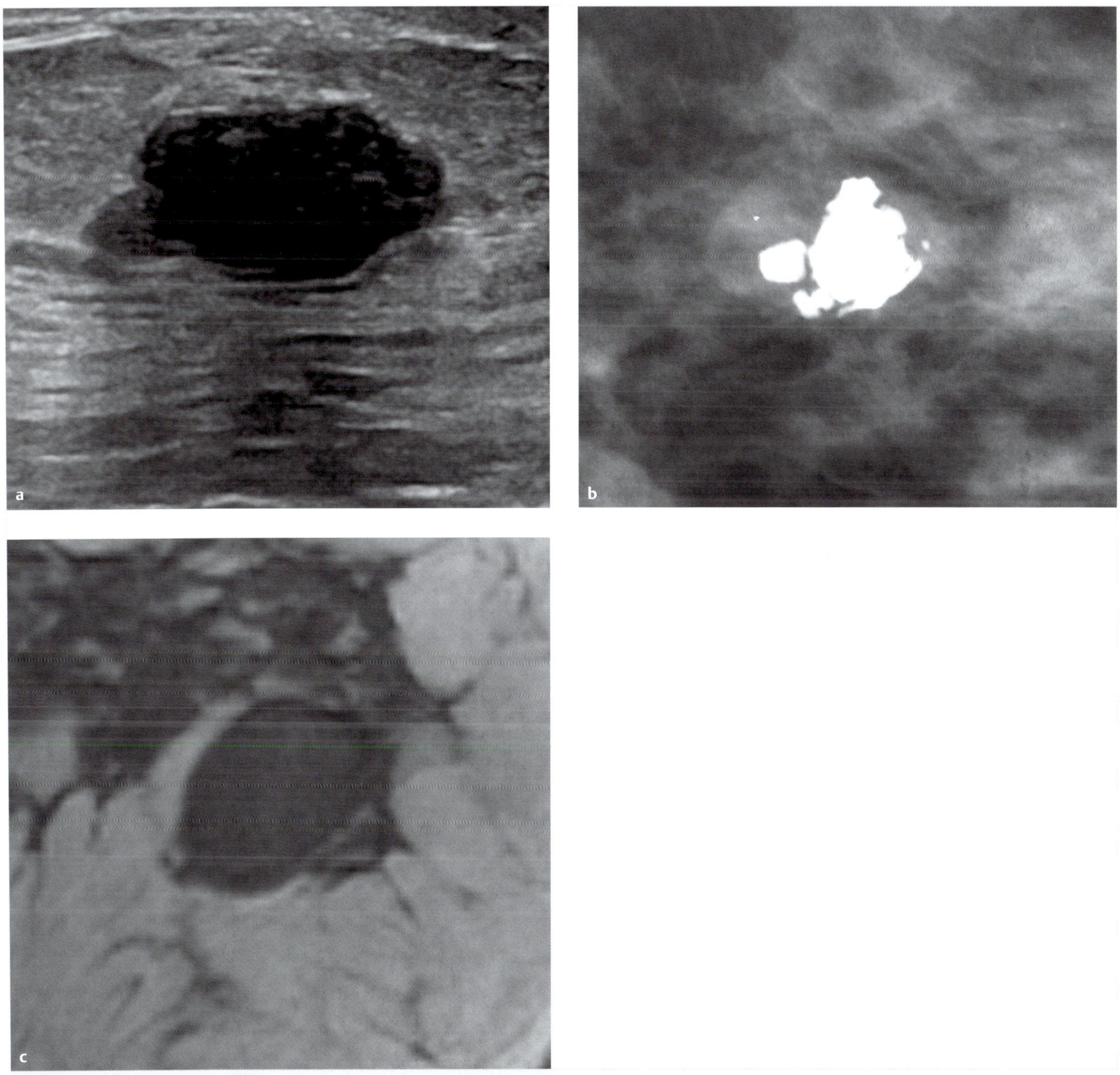

Fig. 8.5 Fibrous fibroadenoma. Fibroadenoma with regressive changes. **(a)** Ultrasound image. **(b)** Mammogram. **(c)** T1W precontrast breast MR image.

- *Mammography* (▸ Fig. 8.5b): mass with parenchyma-equivalent, homogeneous density; difficult or impossible to detect within dense breast tissue structures; good visibility within the lipomatous breast; frequently displays endotumoral macrocalcifications (popcornlike).
- *Breast MRI* (▸ Fig. 8.5c): focus, mass:
  - T1W precontrast image: isointense; endotumoral signal loss in the case of macrocalcifications.
  - T2W image: isointense; endotumoral signal loss in the case of macrocalcifications.
  - After administration of contrast medium: minor or no enhancement.

Note

Popcornlike endotumoral macrocalcifications on mammography are pathognomonic for a fibrotic fibroadenoma.

▸ **Differential diagnosis.** Adenoma, papilloma, phyllodes tumor, invasive carcinoma, intramammary metastasis.

8.1.6 Adenoma

The adenoma, as opposed to the fibroadenoma, exhibits only a very minor stromal component. It is a circumscribed tumor with tubular structures and a surrounding pseudocapsule.

▸ **Characteristics**

- *Incidence:* rare.
- *Peak age:* usually in younger women up to 50 years of age.
- *Multifocality:* rare.
- *Bilateral occurrence:* rare.
- *Complications:* none.
- *Risk of malignancy:* 0%.
- *Prognosis:* excellent.
- *Histological classification after biopsy:* B2.

▸ **Implications for work-up/treatment.** For smaller lesions in younger women (up to 30 years of age), no treatment is required. For smaller lesions in women aged 30 years and above, confirmation of the diagnosis via percutaneous, preferably ultrasound-guided, core needle biopsy should be undertaken.

▸ **Shape.** Round, oval, lobular.

▸ **Margins.** Circumscribed.

▸ **Internal composition.** Usually homogeneous proliferative tissue; may rarely display lipomatous inclusions.

▸ **Clinical findings.** Small lesions normally have no correlative clinical findings. In the case of a larger adenoma, a circumscribed, rather soft mass may be palpable.

▸ **Imaging**

- *Sonography* (▸ Fig. 8.6a): homogeneous, isoechoic mass; longitudinal axis parallel to the skin; good elasticity; mild posterior acoustic enhancement.
- *Mammography* (▸ Fig. 8.6b): mass with parenchyma-equivalent, homogeneous density; difficult or impossible to detect within dense breast tissue structures; good visibility within the lipomatous breast; may rarely display endotumoral microcalcifications.
- *Breast MRI* (▸ Fig. 8.6c): focus, mass:
  - T1W precontrast image: hypointense.
  - T2W image: hyperintense.
  - After administration of contrast medium: moderate to strong homogeneous enhancement.

Note

Adenoma of the nipple represents a variant form of adenoma. It grows in the terminal portion of the nipple and is preferably treated with surgical excision (differential diagnosis: eczema, Paget's disease).

▸ **Differential diagnosis.** Fibroadenoma, papilloma, phyllodes tumor, invasive carcinoma.

8.1.7 Hamartoma

The hamartoma (synonym: fibroadenolipoma, lipofibroadenoma, "breast within the breast") is a circumscribed mixed tumor with organoid composition and pseudocapsular margins.

▸ **Characteristics**

- *Incidence:* rare.
- *Peak age:* all age groups.
- *Multifocality:* rare.
- *Bilateral occurrence:* rare.
- *Complications:* none.
- *Risk of malignancy:* 0%.
- *Prognosis:* excellent.
- *Histological classification after biopsy:* B2.

▸ **Implications for work-up/treatment:** None. When there is uncertainty in the differential diagnosis, percutaneous, preferably ultrasound-guided, core needle biopsy, should be undertaken and should include portions of the pseudocapsule.

▸ **Shape.** Round, oval, lobular.

▸ **Margins.** Circumscribed; the lesion displays a pseudocapsule.

▸ **Internal composition.** Inhomogeneous parenchyma-equivalent tissue (fat, parenchyma, vessels).

▸ **Clinical findings.** Small lesions normally have no correlative clinical findings. In the case of a larger hamartoma, a circumscribed mass may be palpable. The consistency is dependent upon the composition.

▸ **Imaging**

- *Sonography* (▸ Fig. 8.7a): inhomogeneous mass with echogenicity depending upon the composition (fat and parenchymal components).

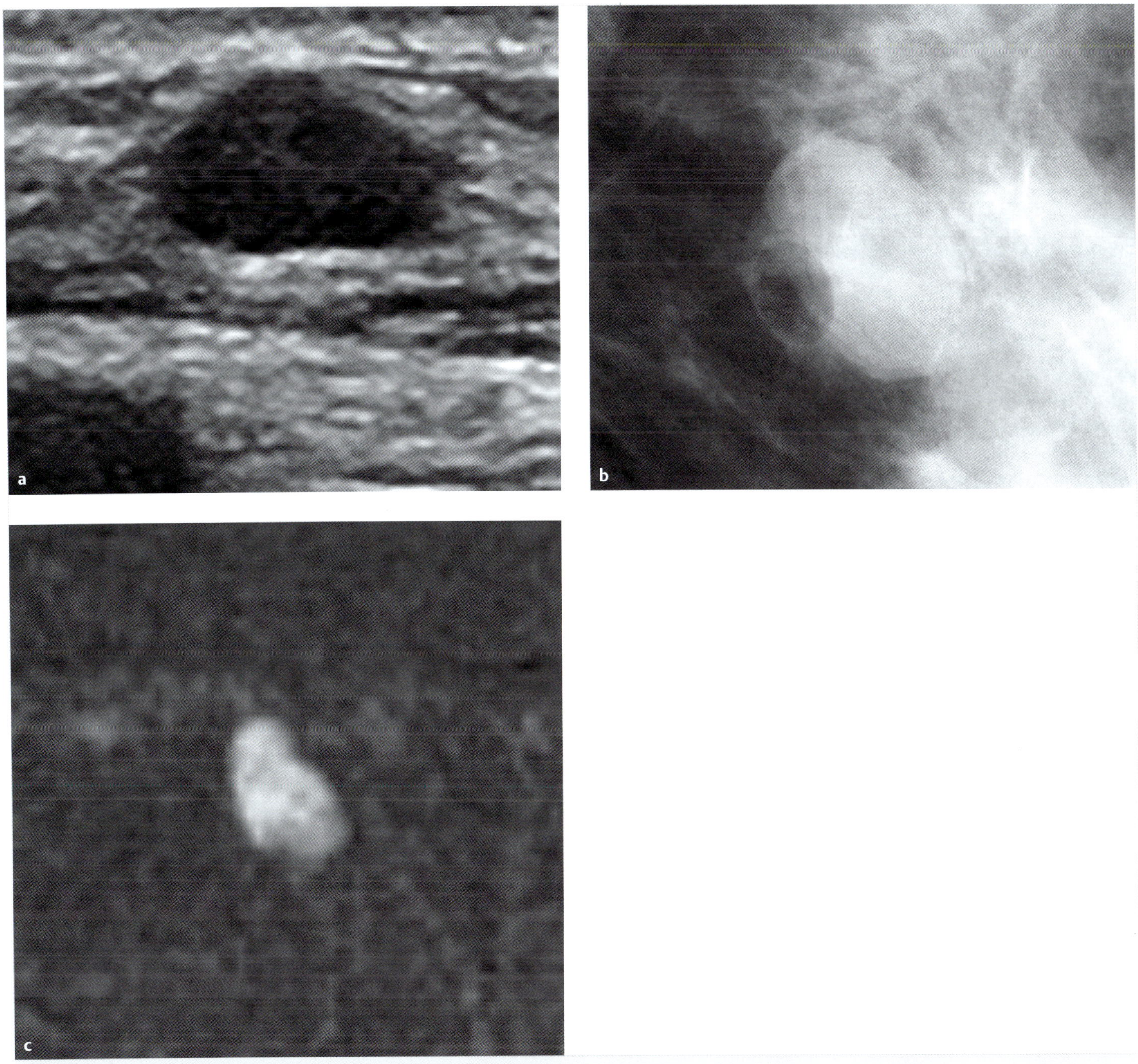

Fig. 8.6 Adenoma. (a) Ultrasound image. (b) Mammogram. (c) Contrast-enhanced breast MRI subtraction image.

- *Mammography* (▶ Fig. 8.7b): mass with hyperdense to hypodense tumor areas depending upon the endotumoral composition; impression of a "breast within the breast" (termed so by the pathologist Thomas Bässler); very rarely displays endotumoral microcalcifications.
- *Breast MRI* (▶ Fig. 8.7c): mass:
  - T1W precontrast image: mixed presentation (hypointense to hyperintense), depending upon composition.
  - T2W image: mixed presentation (hypointense to hyperintense), depending upon composition.
  - After administration of contrast medium: usually displays inhomogeneous enhancement of the parenchymal tumor areas.

Note

The composition and perfusion pattern of a hamartoma can be differentiated distinctly from the characteristics of the normal glandular tissue.

▶ **Differential diagnosis.** Phyllodes tumor.

8.1.8 Lipoma

The lipoma (synonym: adipose tumor) is an encapsulated tumor that contains exclusively mature adipose cells.

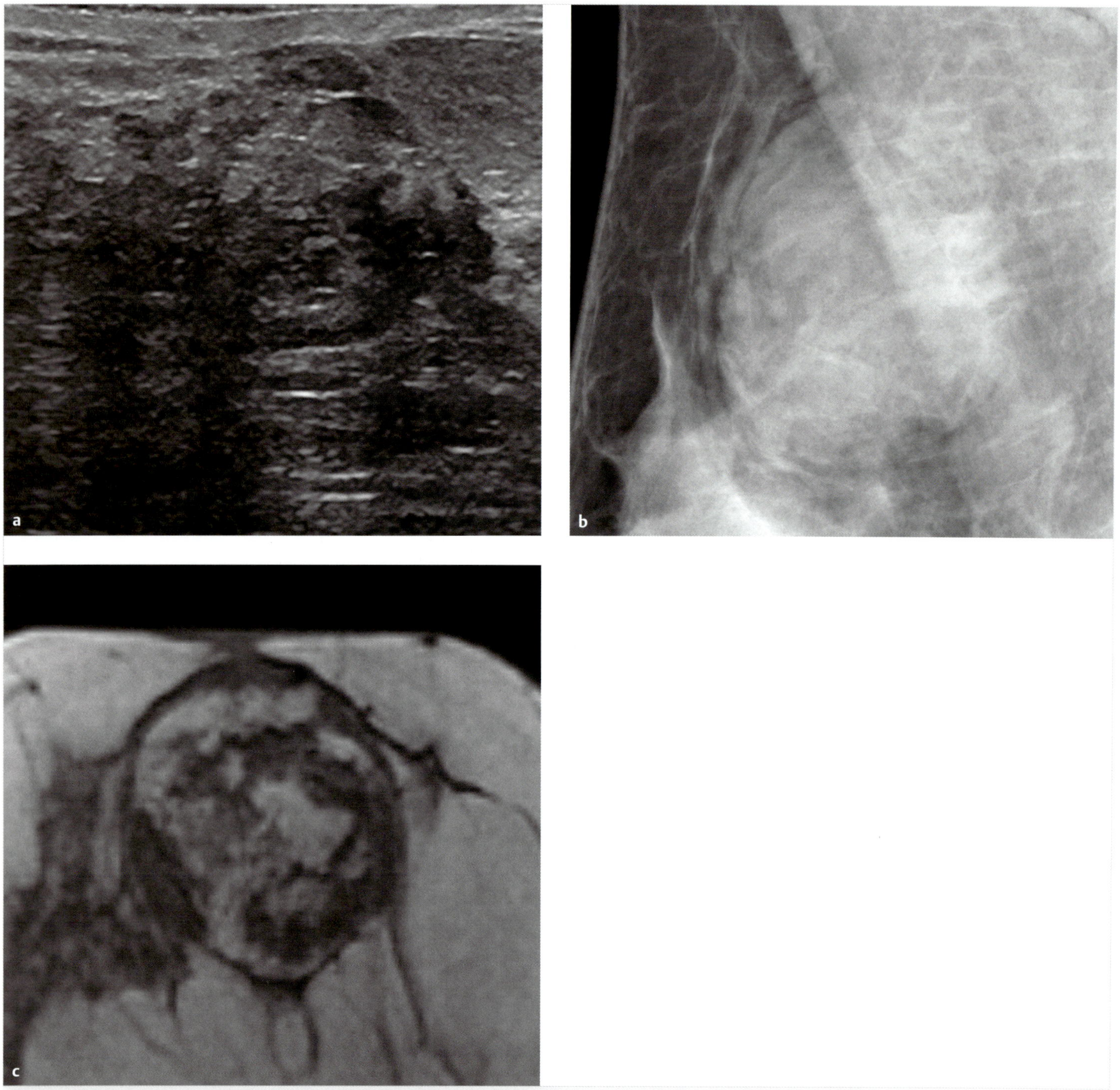

Fig. 8.7 Hamartoma. Mixed tumor. **(a)** Ultrasound image. **(b)** Mammogram. **(c)** T1W precontrast breast MR image.

► **Characteristics**

- *Incidence:* rare.
- *Peak age:* all age groups.
- *Multifocality:* rare.
- *Bilateral occurrence:* rare.
- *Complications:* none.
- *Risk of malignancy:* 0%.
- *Prognosis:* excellent.
- *Histological classification after biopsy:* B2.

► **Implications for work-up/treatment.** None.

► **Shape.** Oval.

► **Margins.** Circumscribed, displays a delicate capsule.

► **Internal composition.** Homogeneous fat.

► **Clinical findings.** Small lesions normally have no correlative clinical findings. In the case of a larger lipoma, a circumscribed, soft, cushionlike mass may be palpable.

► **Imaging**

- *Sonography* (► Fig. 8.8a): hyperechoic mass (due to the arrangement of the adipose cells, echogenicity is frequently higher than that of subcutaneous or intramammary adipose tissue).
- *Mammography* (► Fig. 8.8b): hypodense (fat-equivalent) mass with delicate surrounding capsule; no associated microcalcifications.
- *Breast MRI* (► Fig. 8.8c): mass:
  - T1W precontrast image: hyperintense.

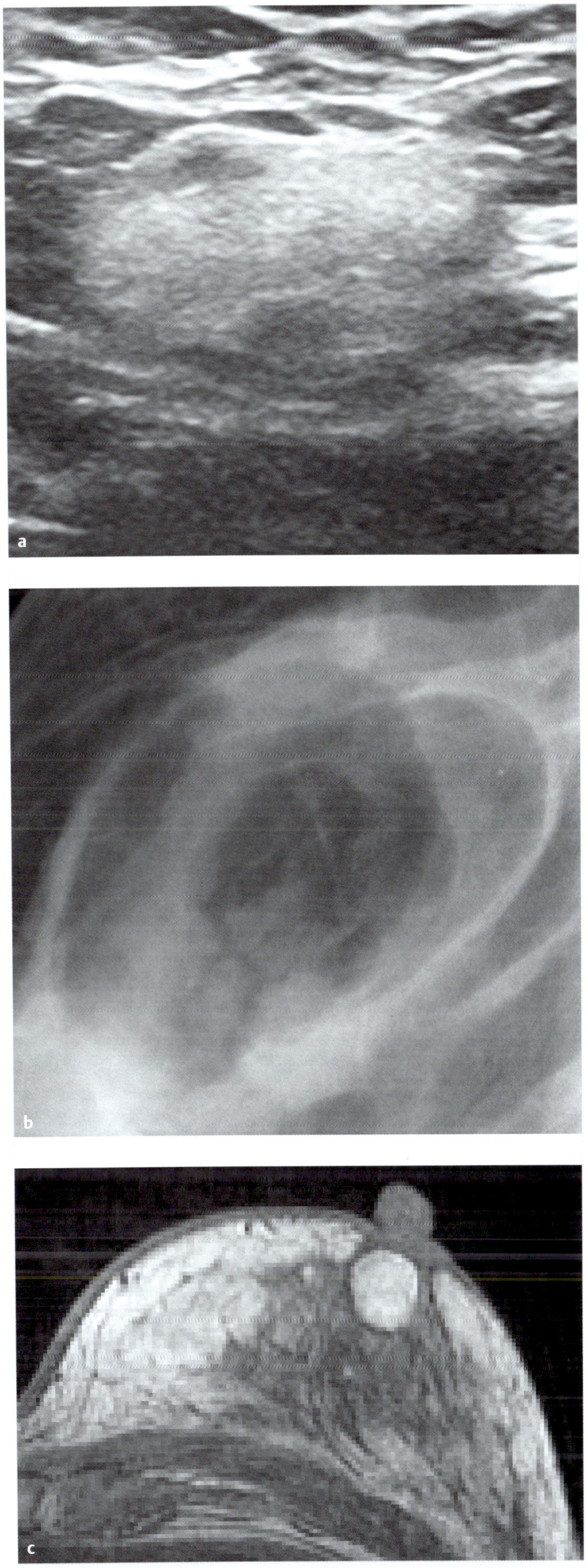

Fig. 8.8 **Lipoma.** Adipose tumor. **(a)** Ultrasound image. **(b)** Mammogram. **(c)** T1W precontrast breast MR image.

- ○ T2W image: hyperintense; hypointense in T2W fat-suppressed.
- ○ After administration of contrast medium: no enhancement.

Note

Liposarcoma does not develop from a breast lipoma.

▶ **Differential diagnosis.** Free intramammary adipose tissue.

8.1.9 Mammary Fibrosis

Mammary fibrosis is a regional or diffuse proliferation of the stroma with obliteration of the lactiferous ducts.

▶ **Characteristics**
- *Incidence:* frequent.
- *Peak age:* all age groups.
- *Multifocality:* frequent.
- *Bilateral occurrence:* frequent.
- *Complications:* none.
- *Risk of malignancy:* not increased.
- *Prognosis:* excellent.
- *Histological classification after biopsy:* B2.

▶ **Implications for work-up/treatment.** None. When there is uncertainty in the differential diagnosis, percutaneous biopsy should be undertaken.

▶ **Shape, distribution.** Irregular, regional.

▶ **Margins.** Indistinct.

▶ **Internal composition.** Inhomogeneous proliferative tissue.

▶ **Clinical findings.** Small lesions normally have no correlative clinical findings. An areas of extensive fibrosis may be palpable as a firm mass.

▶ **Imaging**
- *Sonography* (▶ Fig. 8.9a): nonspecific inhomogeneous, hypoechogenic area, frequently occult.
- *Mammography* (▶ Fig. 8.9b): inhomogeneous, hyperdense area, occasionally presenting as a focal asymmetry or density; nonspecific; rarely associated with microcalcifications.
- *Breast MRI* (▶ Fig. 8.9c): non-space-occupying lesion (nonmasslike lesion):
  - ○ T1W precontrast image: intermediary signal, nonspecific.
  - ○ T2W image: intermediary signal, nonspecific.
  - ○ After administration of contrast medium: mild to strong enhancement, nonspecific.

Note

Mammary fibrosis is usually visualized as ambiguous microcalcifications on the mammogram or as a nonmasslike area of increased enhancement in the MRI.

▶ **Differential diagnosis.** Adenosis, DCIS, ILC.

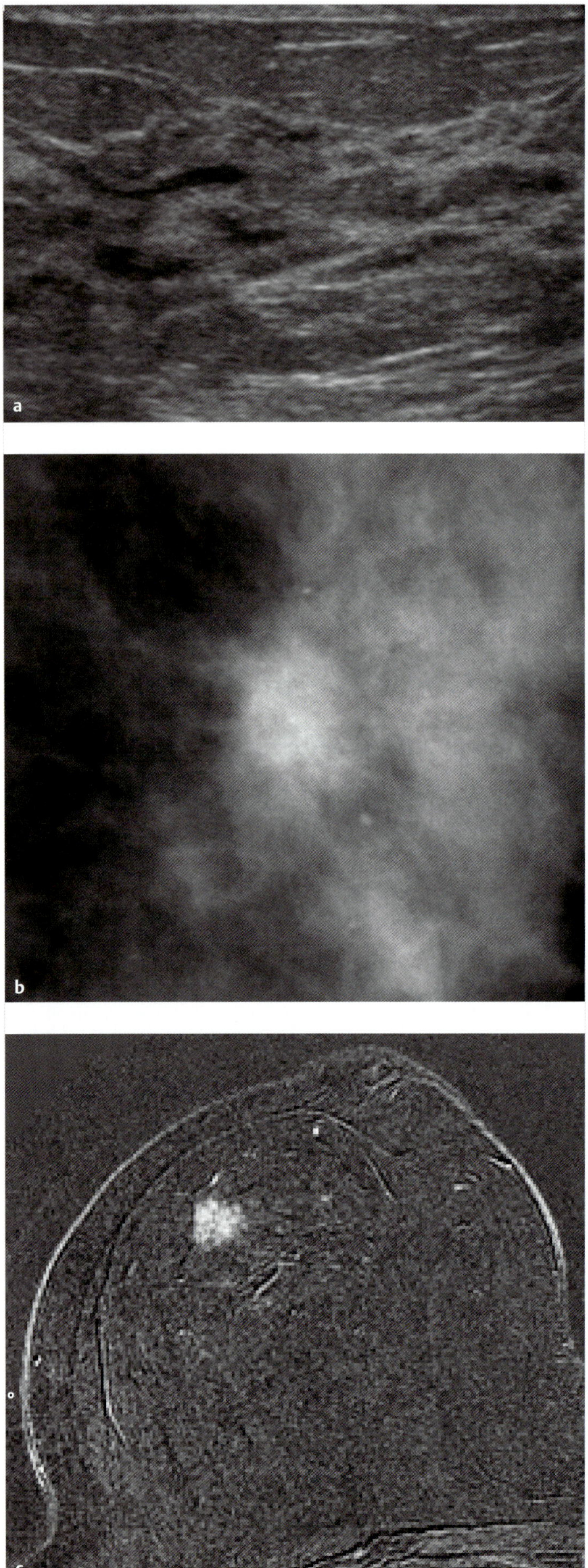

Fig. 8.9 **Mammary fibrosis. (a)** Ultrasound image. **(b)** Mammogram. **(c)** Contrast-enhanced breast MRI subtraction image.

8.1.10 Adenosis of the Breast

Adenosis (synonym: adenosis mammae) is a clustered proliferation of the small ductal segments and terminal ducts. Histopathology differentiates the following types: sclerosing adenosis, microcystic adenosis (blunt-duct adenosis), and the less common microglandular adenosis and the radial scar (see Chapter 8.2.2).

► **Characteristics**
- *Incidence:* frequent.
- *Peak age:* all age groups.
- *Multifocality:* frequent.
- *Bilateral occurrence:* frequent.
- *Complications:* none.
- *Risk of malignancy:* not increased.
- *Prognosis:* excellent.
- *Histological classification after biopsy:* B2.

► **Implications for work-up/treatment.** None. When there is uncertainty in the differential diagnosis, percutaneous biopsy should be undertaken.

► **Shape, distribution.** Irregular; regional or multiregional.

► **Margins.** Indistinct.

► **Internal composition.** Inhomogeneous proliferative tissue.

► **Clinical findings.** Small lesions normally have no correlative clinical findings. A larger area of adenosis may be palpable as an area of increased firmness.

► **Imaging**
- *Sonography* (► Fig. 8.10a): nonspecific, inhomogeneous, hypoechogenic area, frequently occult.
- *Mammography* (► Fig. 8.10b): inhomogeneous, hyperdense area, occasionally presenting as a focal asymmetry or density; nonspecific; frequently associated with microcalcifications (monomorphic, round); in the case of blunt-duct adenosis, often associated with amorphous calcifications in lobular distribution.
- *Breast MRI* (► Fig. 8.10c): non-space-occupying lesion (nonmasslike lesion):
  - T1W precontrast image: intermediary signal, nonspecific.
  - T2W image: intermediary signal, nonspecific.
  - After administration of contrast medium: often displays strong enhancement, nonspecific.

Note

Circumscribed adenosis is the most common cause of false-positive findings on breast MRI because it exhibits an irregular shape and is often associated with increased perfusion.

► **Differential diagnosis.** Mammary fibrosis, DCIS, ILC.

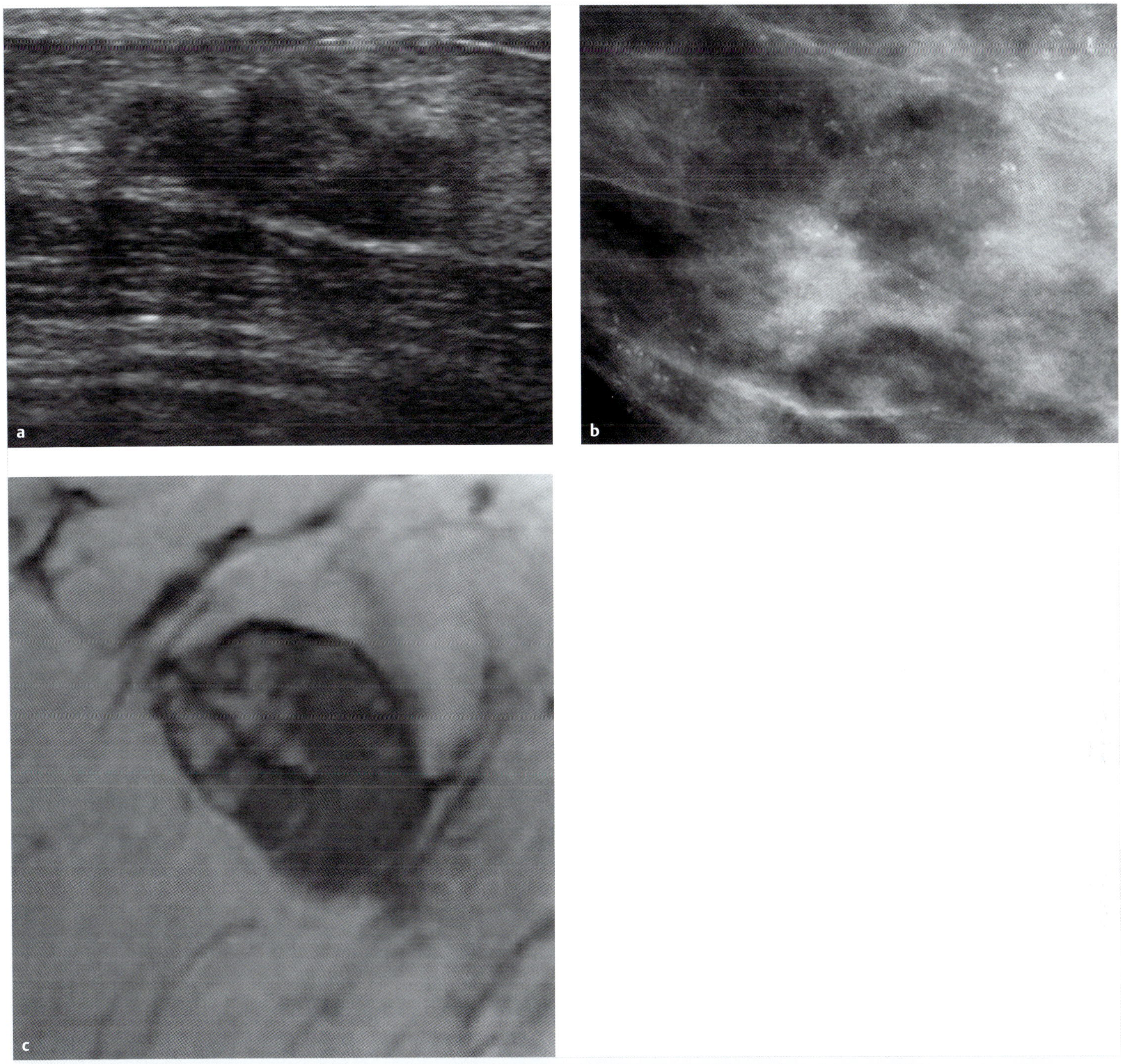

Fig. 8.10 **Adenosis.** Sclerosing adenosis. **(a)** Ultrasound image. **(b)** Mammogram. **(c)** Nodular adenosis in the T1W precontrast breast MR image.

8.1.11 Fibrocystic Condition of the Breast

Morphological components of the fibrocystic condition of the breast (also referred to as fibrocystic mastopathy, fibrocystic breasts, fibrocystic disease of the breast, fibrocystic changes) are: micro- and macrocysts in varying degrees; hyperplasia of the lobules; epithelial hyperplasia of the lactiferous ducts and acini; and stromal fibrosis.

► **Characteristics**
- *Incidence:* frequent.
- *Peak age:* all age groups.
- *Multifocality:* frequent.
- *Bilateral occurrence:* obligatory.
- *Complications:* none.
- *Risk of malignancy:* not increased.
- *Prognosis:* excellent.
- *Histological classification after biopsy:* B1.

► **Implications for work-up/treatment.** None. When there is uncertainty in the differential diagnosis, percutaneous biopsy should be undertaken.

► **Shape, distribution.** Irregular, diffuse, rarely regional.

► **Margins.** Indistinct.

► **Internal composition.** Inhomogeneous proliferative tissue + cysts.

► **Clinical findings.** Unremarkable to mild or moderately increased firmness.

▸ **Imaging**

- *Sonography* (▸ Fig. 8.11a): Generally hyperechogenic breast tissue; hypoechoic tubular changes (ectatic milk ducts); anechoic lesions with posterior acoustic enhancement (cysts); often bilaterally symmetric.
- *Mammography* (▸ Fig. 8.11b): circumscribed or diffuse areas of hyperdensity; frequently associated with micro- and macrocalcifications; often bilaterally symmetric.
- *Breast MRI* (▸ Fig. 8.11c): non-space-occupying lesion (non-masslike lesion):
  - T1W precontrast image: hypointense lesions (cysts); otherwise nonspecific.
  - T2W image: hyperintense lesions (cysts); otherwise nonspecific.
  - After administration of contrast medium: no/slight/strong enhancement, sparing the cystic lesions; nonspecific.

Note

In general, the term, "mastopathy," should be avoided and regarded as archaic. Currently it is only used in association with fibrocystic changes.

▸ **Differential diagnosis.** Fibrosis, adenosis, any diffuse tumor manifestation.

8.1.12 Adenomyoepithelioma

An adenomyoepithelioma is a circumscribed nodule consisting of a proliferation of myoepithelial cells.

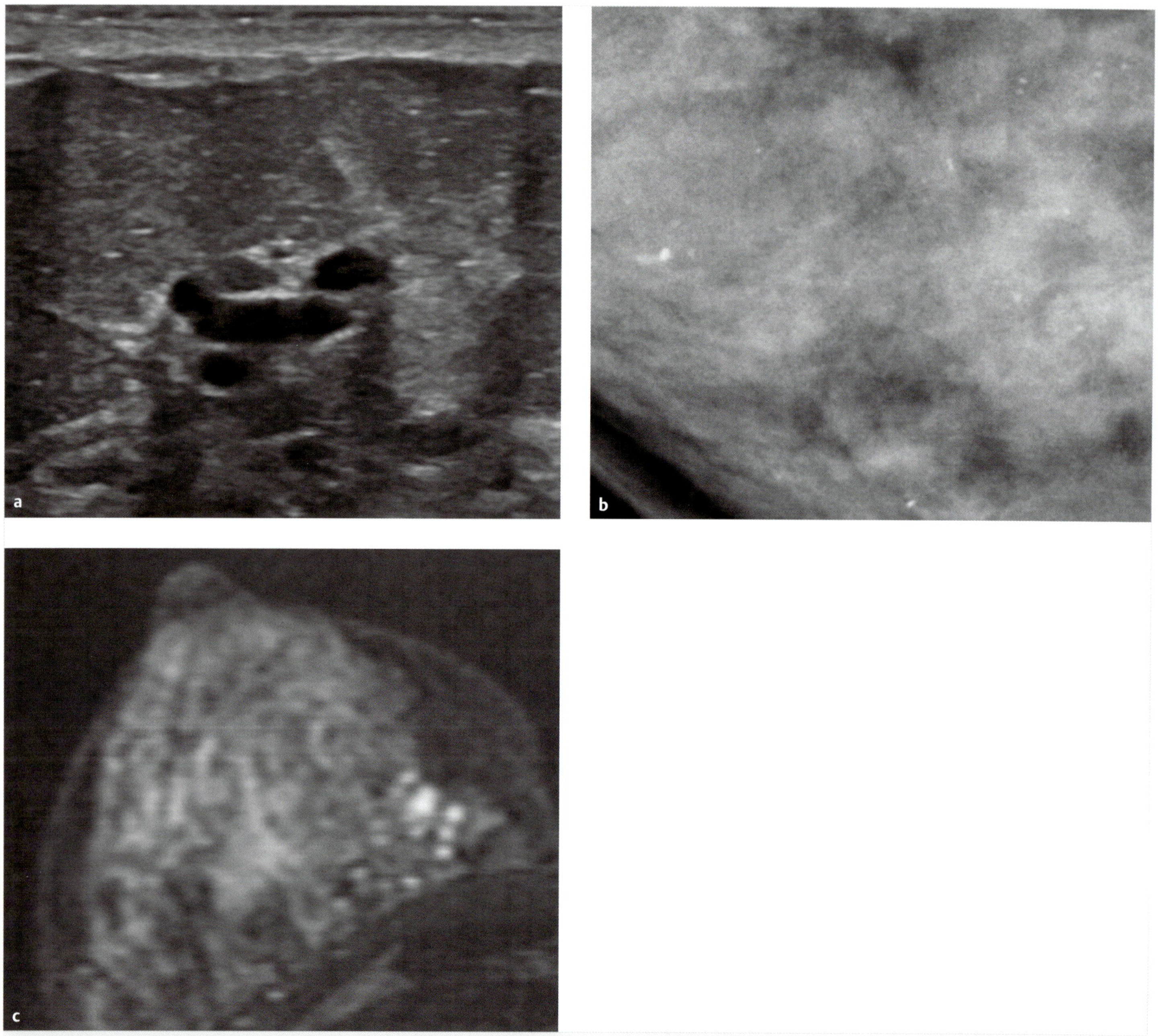

Fig. 8.11 Fibrocystic condition of the breast. (a) Ultrasound image. **(b)** Mammogram. **(c)** T2W breast MR image.

► **Characteristics**

- *Incidence:* very rare.
- *Peak age:* all age groups.
- *Multifocality:* very rare.
- *Bilateral occurrence:* very rare.
- *Complications:* none.
- *Risk of malignancy:* 0%.
- *Prognosis:* excellent.
- *Histological classification after biopsy:* B2.

► **Implications for work-up/treatment.** None. When there is uncertainty in the differential diagnosis, percutaneous biopsy should be undertaken.

► **Shape.** Round, oval.

► **Margins.** Circumscribed or indistinct.

► **Internal composition.** Inhomogeneous proliferative tissue.

► **Clinical findings.** Small lesions normally have no correlative clinical findings. Larger lesions may present as circumscribed, usually firm nodules.

► **Imaging**

- *Sonography* (► Fig. 8.12a): hypoechogenic mass, nonspecific.
- *Mammography* (► Fig. 8.12b): hyperdense mass, nonspecific.
- *Breast MRI* (► Fig. 8.12c): focus, mass:
  - T1W precontrast image: parenchyma-equivalent, nonspecific.
  - T2W image: parenchyma-equivalent, nonspecific.
  - After administration of contrast medium: moderate to strong enhancement; may display postinitial washout.

Note

Adenomyoepitheliomas do not exhibit any specific imaging characteristics. Due to their increased vascularity, they are frequently false-positive findings in MRI.

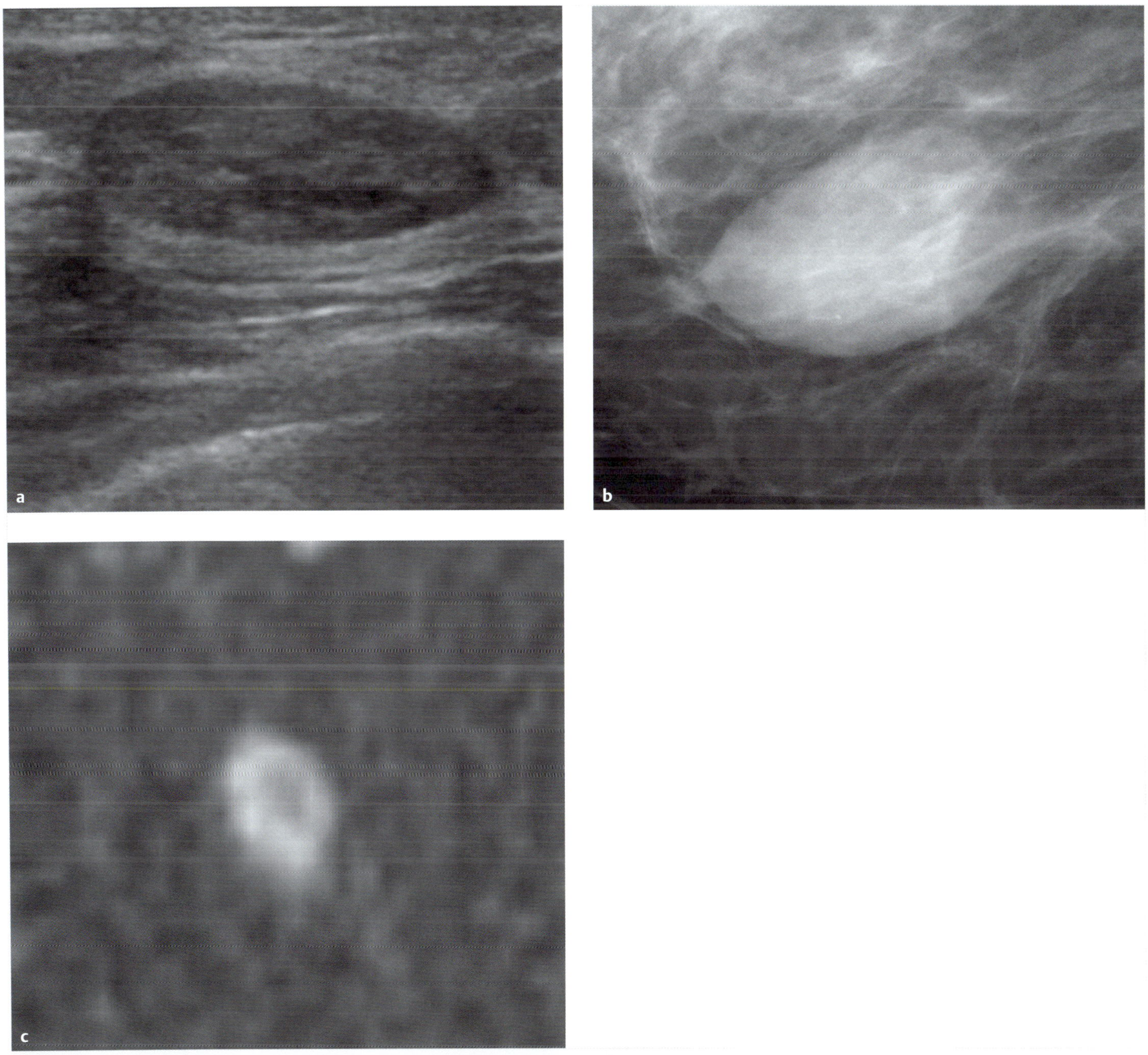

Fig. 8.12 Adenoepithelioma. (a) Ultrasound image. **(b)** Mammogram. **(c)** Contrast-enhanced breast MRI subtraction image.

► **Differential diagnosis.** Fibroadenoma, papilloma, phyllodes tumor, carcinoma of the breast.

8.1.13 Acute Nonpuerperal Mastitis

Acute nonpuerperal mastitis of the breast (synonym: acute inflammation of the breast) is a predominantly interstitial, but occasionally also intraductal, inflammation of the parenchyma outside the lactation period.

► **Characteristics**

- *Incidence:* rare.
- *Peak age:* all age groups.
- *Multifocality:* may extend over the entire breast.
- *Bilateral occurrence:* very rare.
- *Complications:* abscess formation; development of fistulas.
- *Risk of malignancy:* 0%.
- *Prognosis:* good.
- *Histological classification after biopsy:* B2.

► **Implications for work-up/treatment.** Primary treatment of the inflammation: immobilization, cooling, administration of antibiotics. If the symptoms do not completely regress in 7 to 10 days, confirm the diagnosis with a percutaneous biopsy or open representative excision.

► **Shape, distribution.** Irregular, frequently segmental or diffuse (entire breast).

► **Margins.** Indistinct.

► **Internal composition.** Inhomogeneous proliferative, edematous tissue.

► **Clinical findings.** Inflammation triad: warmth to the touch, redness, pain. Possible swelling, skin thickening, lymphadenitis, and fever. May be associated with increased laboratory inflammatory markers.

► **Imaging**

- *Sonography* (► Fig. 8.13a): Hyperechogenicity of the parenchyma and subcutaneous space (edematous imbibition-*compare with the contralateral breast*), skin thickening, locoregional lymphadenitis.

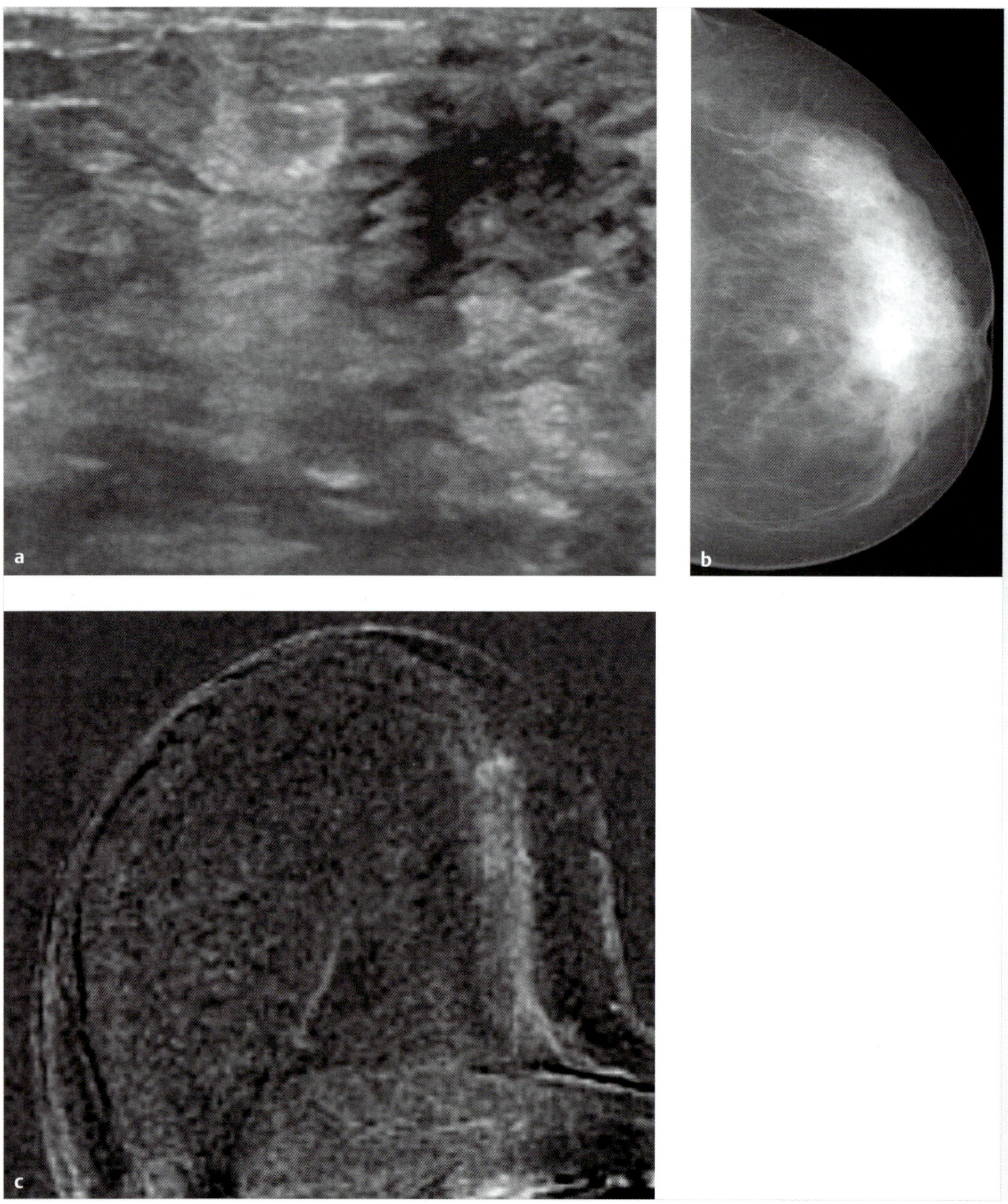

Fig. 8.13 Acute nonpuerperal mastitis. (a) Ultrasound image. **(b)** Mammogram. **(c)** Contrast-enhanced breast MRI subtraction image.

- *Mammography* (▸ Fig. 8.13b): Hyperdensity and decreased transparency (tissue structures with a "washed-out appearance"—compare with the contralateral breast!), skin thickening, locoregional lymphadenitis.
- *Breast MRI* (▸ Fig. 8.13c): non-space-occupying lesion (nonmasslike lesion):
  - T1W precontrast image: no specific intramammary changes; skin thickening; locoregional lymphadenitis.
  - T2W image: hyperintensity of the affected parenchymal areas; skin thickening; locoregional lymphadenitis.
  - After administration of contrast medium: increased enhancement of the thickened skin; often greatly increased perfusion of the inflamed parenchymal areas.

Note

It is frequently difficult or impossible to distinguish between nonpuerperal mastitis and inflammatory breast cancer on a clinical or imaging basis.

▸ **Differential diagnosis.** Inflammatory breast cancer.

8.1.14 Chronic Nonpuerperal Mastitis

Chronic nonpuerperal mastitis (synonym: chronic abacterial inflammation of the breast) is a chronic, abacterial, inflammatory process that occurs in and around the terminal ducts and lobules of the breast. Histopathology differentiates the following 3 forms: galactophoritis, granulomatous mastitis, and lobular mastitis. Plasma cell mastitis is the final stage of circumductal mastitis.

▸ **Characteristics**
- *Incidence:* rare.
- *Peak age:* higher age groups.
- *Multifocality:* very rare.
- *Bilateral occurrence:* very rare.
- *Complications:* none.
- *Risk of malignancy:* 0%.
- *Prognosis:* excellent.
- *Histological classification after biopsy:* B2.

▸ **Implications for work-up/treatment.** None. When there is uncertainty in the differential diagnosis, percutaneous biopsy should be undertaken.

▸ **Shape, distribution.** Irregular, frequently ductal or segmental.

▸ **Margins.** Indistinct .

▸ **Internal composition.** Inhomogeneous proliferative tissue.

▸ **Clinical findings.** Small lesions normally have no correlative clinical findings. Rarely associated with pain, nipple discharge, or retraction of the nipple.

▸ **Imaging**
- *Sonography* (▸ Fig. 8.14a): hypoechogenic tissue; hypoechoic tubular changes (mammary duct ectasia); nonspecific.
- *Mammography* (▸ Fig. 8.14b): hyperdensity; trabecular thickening; retroareolar ductal ectasia; in its final stage radially arranged, lancetlike calcifications are classic (plasma cell mastitis).
- *Breast MRI* (▸ Fig. 8.14c): non-space-occupying lesion (nonmasslike lesion):
  - T1W precontrast image: hypointense, nonspecific.
  - T2W image: hyperintense, nonspecific; occasionally ectatic milk ducts.
  - After administration of contrast medium: mild to strong enhancement, regional or segmental.

Note

The chronic inflammation associated with chronic nonpuerperal mastitis is not typically associated with any symptoms.

▸ **Differential diagnosis.** Adenosis, fibrosis, inflammatory carcinoma of the breast.

8.1.15 Intramammary Lymph Nodes

Intramammary lymph nodes appear morphologically identical to axillary lymph nodes. It is merely the position within the glandular structures of the breast that defines them.

▸ **Characteristics**
- *Incidence:* rare; more common following extensive breast surgery.
- *Peak age:* all age groups.
- *Multifocality:* rare.
- *Bilateral occurrence:* rare.
- *Complications:* none.
- *Risk of malignancy:* 0%.
- *Prognosis:* excellent.
- *Histological classification after biopsy:* B2.

▸ **Implications for work-up/treatment.** None. When there is uncertainty in the differential diagnosis, percutaneous biopsy should be undertaken.

▸ **Shape.** Oval (kidney-shaped) hilum.

▸ **Margins.** Circumscribed.

▸ **Internal composition.** Proliferative tissue with lipomatous vascular pedicle.

▸ **Clinical findings.** Unremarkable.

▸ **Imaging**
- *Sonography* (▸ Fig. 8.15a): hypoechogenic mass, hyperechoic hilum, pathognomonic.
- *Mammography* (▸ Fig. 8.15b): hyperdense mass with hypodense (fat-equivalent) hilum, pathognomonic.
- *Breast MRI* (▸ Fig. 8.15c): mass:
  - T1W precontrast image: hypointense, hyperintense hilum, pathognomonic.

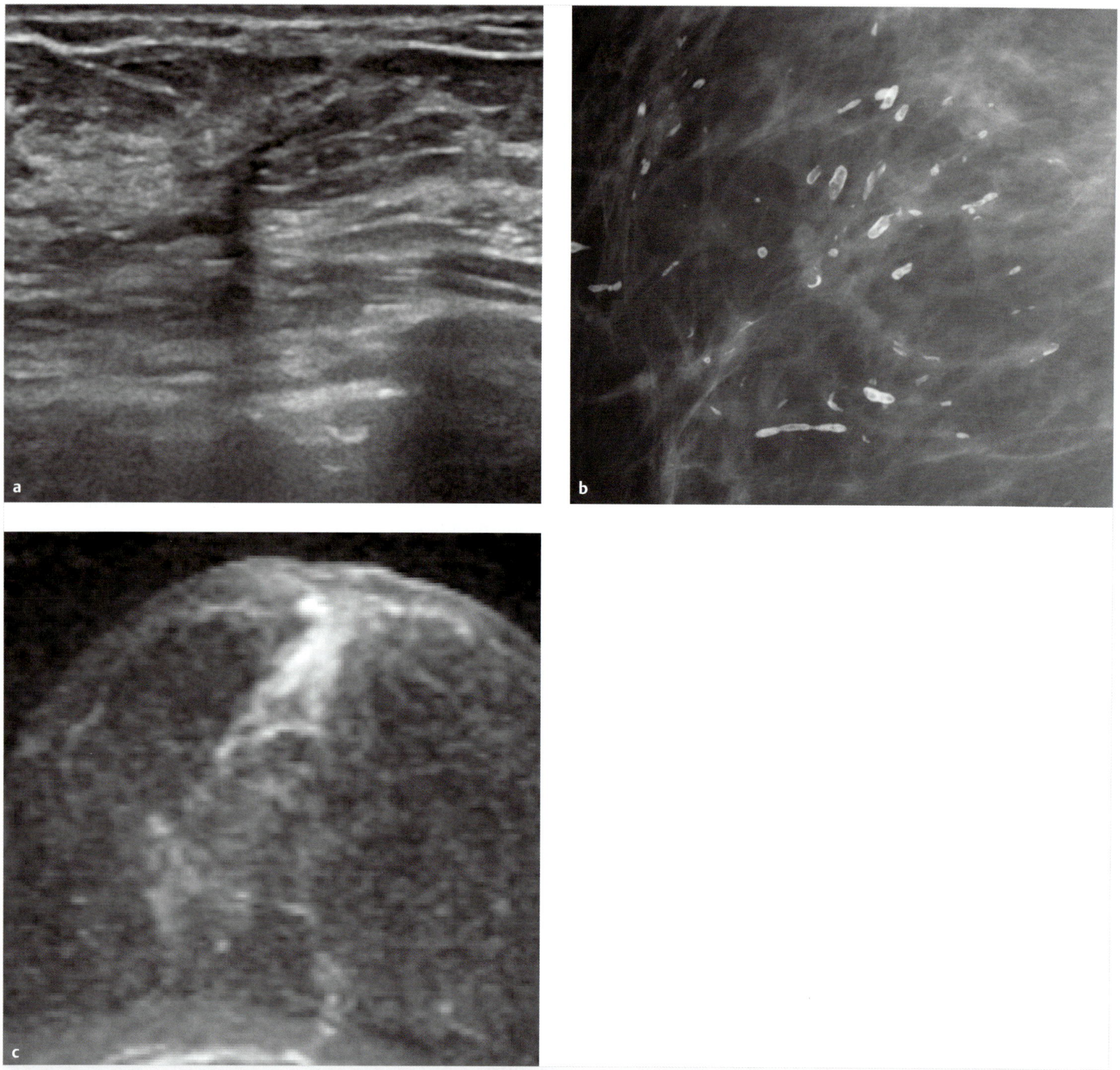

Fig. 8.14 Chronic mastitis. **(a)** Ultrasound image of chronic-segmental inflammation. **(b)** Plasma cell mastitis on mammography. **(c)** Chronic, segmental inflammation in the contrast-enhanced breast MRI subtraction image.

- T2W: hyperintense, hilum hyperintense (T2W) or hypointense (T2W with fat saturation).
- After administration of contrast medium: no enhancement to strong enhancement.

Note

Imaging cannot reliably differentiate between lymphadenitis and the presence of metastatic disease within a lymph node.

► **Differential diagnosis.** In the case of typical lymph node morphology, there is no differential diagnosis.

8.1.16 Pseudoangiomatous Stromal Hyperplasia

Pseudoangiomatous stromal hyperplasia (PASH) is a benign intramammary lesion characterized by concentrically oriented, pseudoangiomatous cavities lined with CD34-positive cells.

► **Characteristics**

- *Incidence:* rare.
- *Peak age:* all age groups.
- *Multifocality:* rare.
- *Bilateral occurrence:* rare.
- *Complications:* none.
- *Risk of malignancy:* 0%.

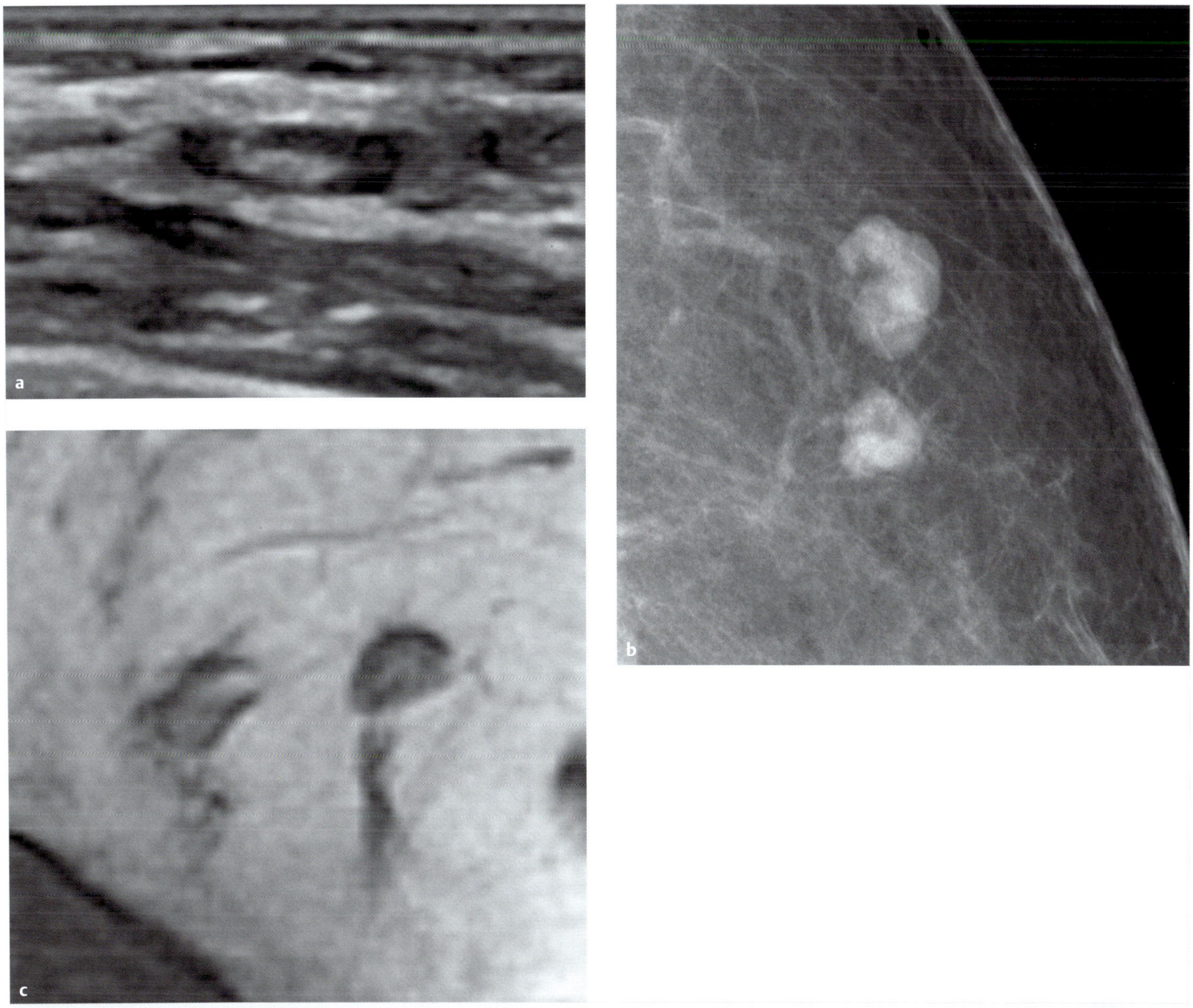

Fig. 8.15 **Intramammary lymph nodes.** Lymph nodes located within the parenchyma. **(a)** Ultrasound image. **(b)** Mammogram. **(c)** T1W precontrast breast MR image.

- *Prognosis:* excellent.
- *Histological classification after biopsy.* B2.

▸ **Implications for work-up/treatment.** None. When there is uncertainty in the differential diagnosis, percutaneous biopsy should be undertaken.

▸ **Shape.** Round to irregular.

▸ **Margins.** Indistinct.

▸ **Internal composition.** Inhomogeneous angiomatous tissue.

▸ **Clinical findings.** Typically, there are no correlative clinical findings, even with larger lesions.

▸ **Imaging**

- *Sonography* (▸ Fig. 8.16a): hypoechogenic area; rarely displays positive color coding in the color-coded Doppler sonography; nonspecific.
- *Mammography* (▸ Fig. 8.16b): hyperdensity, nonspecific.
- *Breast MRI* (▸ Fig. 8.16c): focus, mass, non-space-occupying lesion (nonmasslike lesion):
  - T1W precontrast image: hypointense, nonspecific.
  - T2W: hyperintense, nonspecific.
  - After administration of contrast medium: obligatory moderate to strong enhancement, unspecific time to signal intensity curve.

Note

PASH is associated with increased contrast enhancement on MRI.

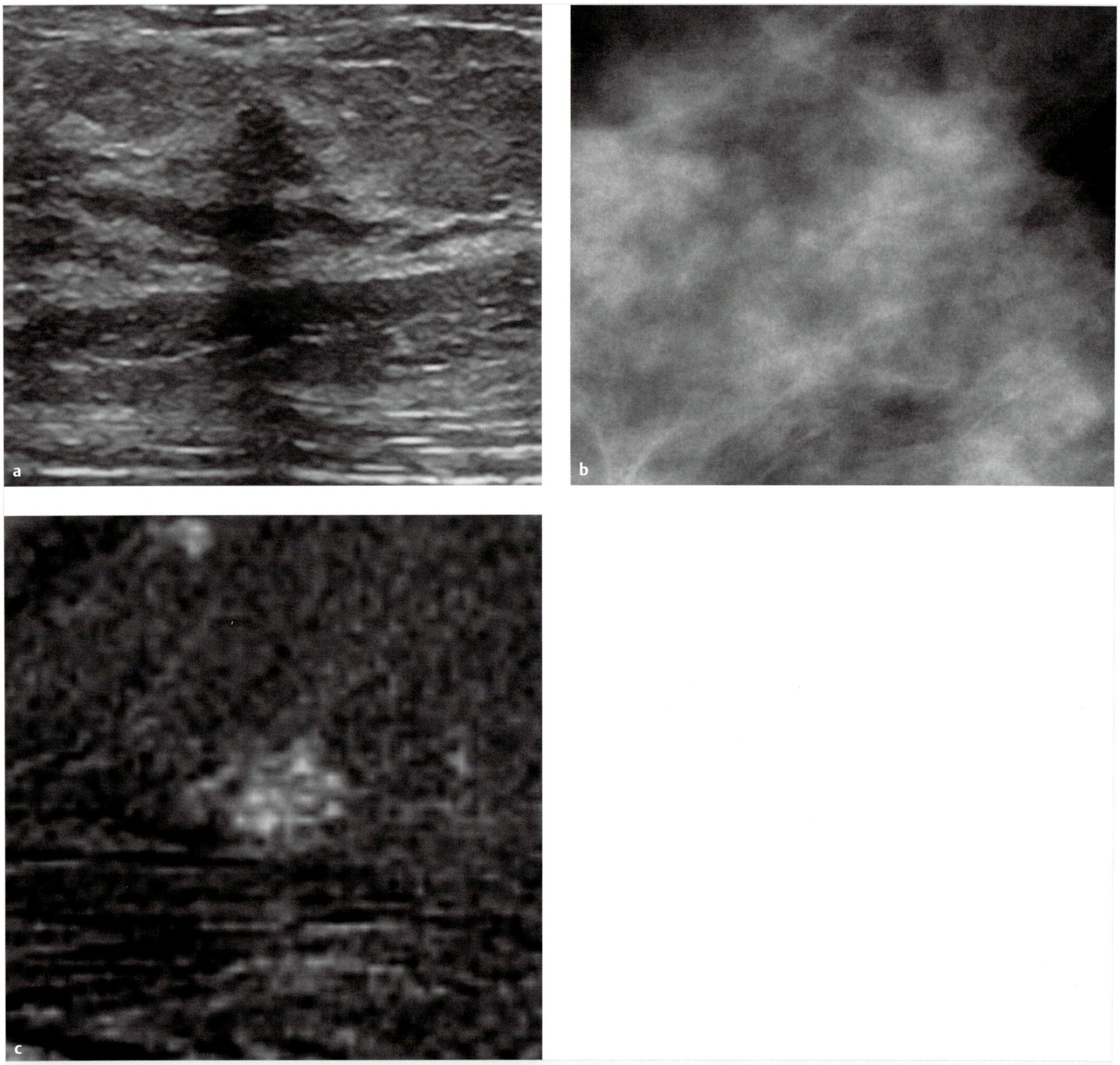

Fig. 8.16 Pseudoangiomatous stromal hyperplasia (PASH). (a) Ultrasound image. **(b)** Mammogram. **(c)** Contrast-enhanced breast MRI subtraction image.

8

► **Differential diagnosis.** Fibrosis, adenosis, hemangioma, mastitis, DCIS, ILC.

8.1.17 Seroma

The designation of "seroma" describes an accumulation of wound secretions (e.g., after a surgical procedure).

► **Characteristics**

- *Incidence:* common following surgical procedures.
- *Peak age:* all age groups.
- *Multifocality:* possible in the case of ipsilateral, multifocal interventional procedures.
- *Bilateral occurrence:* possible in the case of bilateral interventional procedures.
- *Complications:* none.
- *Risk of malignancy:* 0%.
- *Prognosis:* excellent.
- *Histological classification after biopsy:* B2.

► **Implications for work-up/treatment.** None if symptom-free. When symptoms occur, drainage may be indicated.

► **Shape.** Oval.

► **Margins.** Circumscribed.

► **Internal composition.** Homogeneous fluid.

► **Clinical findings.** Small seromas normally have no correlative clinical findings. In the case of larger lesions, a mass may be palpable. Occasionally tenderness or pain is present.

► **Imaging**

- *Sonography* (► Fig. 8.17a): hypoechogenic mass; occasionally displays villiform echo alterations within the fluid.
- *Mammography* (► Fig. 8.17b): hyperdensity, nonspecific.
- *Breast MRI* (► Fig. 8.17c): mass:
  - T1W precontrast image: hypointense.
  - T2W: hyperintense.
  - After administration of contrast medium: usually no enhancement; rare contrast enhancement of the seroma walls.

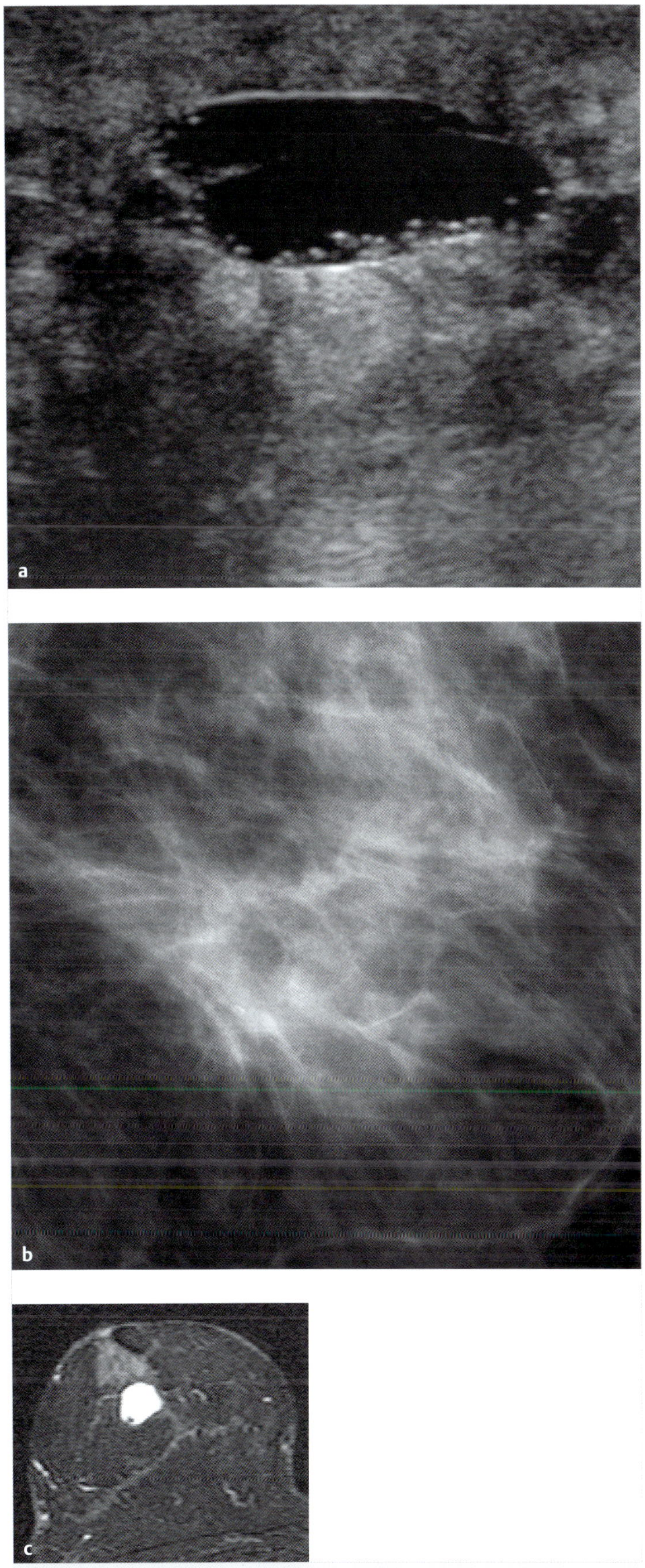

Fig. 8.17 Seroma. Postoperative seroma. **(a)** Ultrasound image. **(b)** Mammogram. **(c)** T2W breast MR image.

Note

On ultrasound, seromas may display solid tissue structures within the fluid interior.

► **Differential diagnosis.** Cyst, complex cyst, cyst with intracystic proliferation, bleeding, abscess.

8.1.18 Hematoma

A hematoma is a mass of, usually clotted, blood resulting from an intramammary hemorrhage, typically after an interventional procedure or surgery, and rarely after trauma.

► **Characteristics**

- *Incidence:* per se, very rare; common after biopsy.
- *Peak age:* all age groups.
- *Multifocality:* following ipsilateral biopsies in various locations.
- *Bilateral occurrence:* following bilateral biopsies.
- *Complications:* infection, abscess.
- *Risk of malignancy:* 0%.
- *Prognosis:* good.
- *Histological classification after biopsy:* B2.

► **Implications for work-up/treatment.** None if symptom-free. When symptoms occur, fluid drainage if indicated (large-caliber needle!).

► **Shape.** Round, oval; irregular in the case of diffuse bleeding.

► **Margins.** Circumscribed; indistinct in the case of diffuse bleeding.

► **Internal composition.** Inhomogeneous blood components.

► **Clinical findings.** Small hematomas normally have no correlative clinical findings. A mass may be palpable in the case of larger hematomas. Occasionally there may be tenderness or pain. Cutaneous discoloration may develop subsequently from black/purple, to red, blue, green, and yellow.

► **Imaging**

- *Sonography* (► Fig. 8.18a): hypoechogenic mass; occasionally displaying inhomogeneous internal echotexture and/or sedimentation effect.
- *Mammography* (► Fig. 8.18b): hyperdense mass; nonspecific.
- *Breast MRI* (► Fig. 8.18c): mass:
  - T1W precontrast image: dependent on the age of the hematoma may be hyperintense (fresh hematoma), hyperintense ring (subacute hematoma), or anechoic due to susceptibility artifacts (old, residual hematoma).
  - T2W: hypointense (fresh hematoma); hypointense ring (old hematoma).

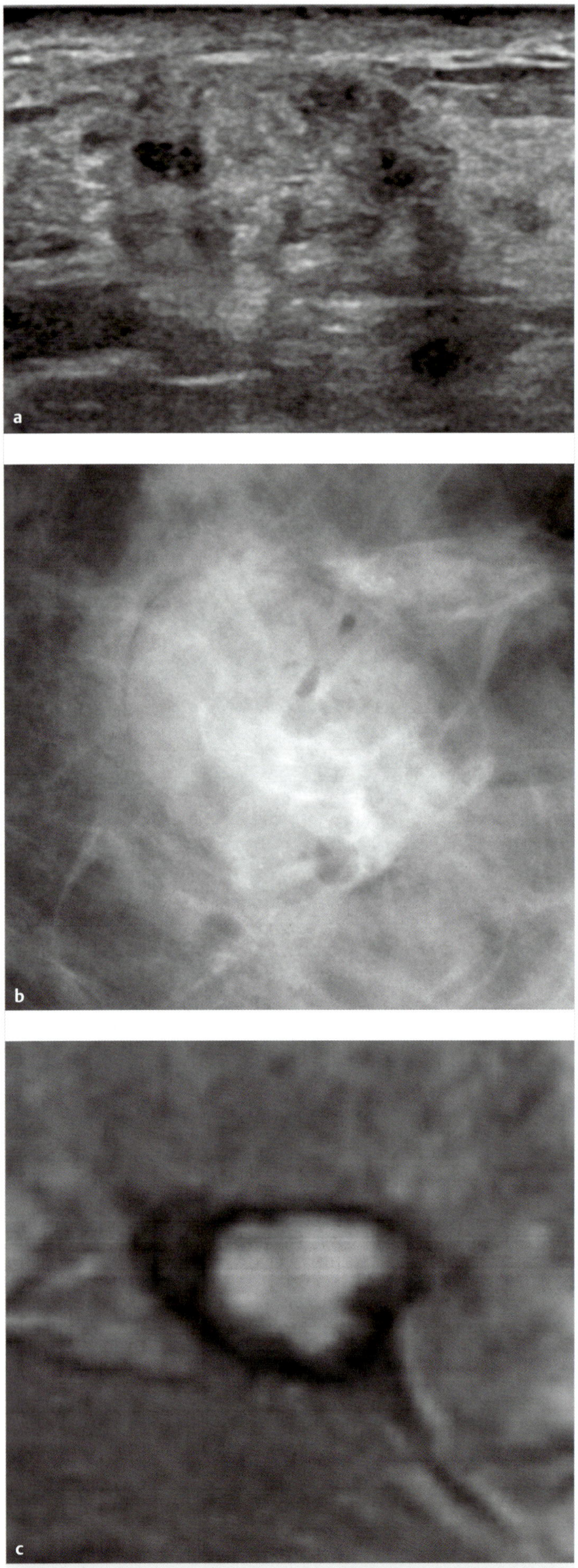

Fig. 8.18 Intramammary hematoma. **(a)** Ultrasound image of a circumscribed hematoma following vacuum-assisted biopsy. **(b)** Mammogram of circumscribed hematoma following vacuum-assisted biopsy. **(c)** Older hematoma in the T1W precontrast breast MR image.

- After administration of contrast medium: mild reactive enhancement in the periphery of the hematoma.

Note

The MRI signal behavior of hematomas changes depending on their age.

► **Differential diagnosis.** Seroma, abscess.

8.1.19 Fat Necrosis (Oil Cyst)

Fat necrosis (synonyms: liponecrosis microcystica calcificata) is a circumscribed area of dead adipose tissue associated with leukocytic and histiocytic infiltrates. The transformation into scar tissue is usually complete in a matter of weeks. Persistence of liquefied adipose tissue results in the formation of so-called "oil cysts."

► **Characteristics**

- *Incidence:* common after surgery, rarely traumatic.
- *Peak age:* all age groups.
- *Multifocality:* following surgery in various ipsilateral locations.
- *Bilateral occurrence:* following bilateral surgery.
- *Complications:* none.
- *Risk of malignancy:* 0%.
- *Prognosis:* excellent.
- *Histological classification after biopsy:* B2.

► **Implications for work-up/treatment.** None. When there is uncertainty in the differential diagnosis, (e.g. fat necrosis versus recurrence after surgery for a carcinoma), a percutaneous biopsy should be undertaken.

► **Shape.** Round to irregular.

► **Margins.** Circumscribed to indistinct.

► **Internal composition.** Adipose tissue.

► **Clinical findings.** Small lesions normally have no correlative clinical findings. In the case of larger lesions, a firm mass may be palpable. Occasionally tenderness or pain is present.

► **Imaging**

- *Sonography* (► Fig. 8.19a): hypoechogenic mass, often nonspecific, often occult in early stages.
- *Mammography* (► Fig. 8.19b): hyperdensity, frequently nonspecific (fresh fat necrosis); mass with fat-equivalent density, macrocalcification of the oil cyst wall is highly specific (old fat necrosis).
- *Breast MRI* (► Fig. 8.19c): focus, mass, non-space-occupying lesion (nonmasslike lesion):
  - T1W precontrast image: isointense, nonspecific (fresh fat necrosis); later fat-equivalent/hyperintense (highly specific for an oil cyst).
  - T2W: hyperintense, nonspecific (fresh fat necrosis); later hypointense in the fat-suppression measurement, highly specific (oil cyst).

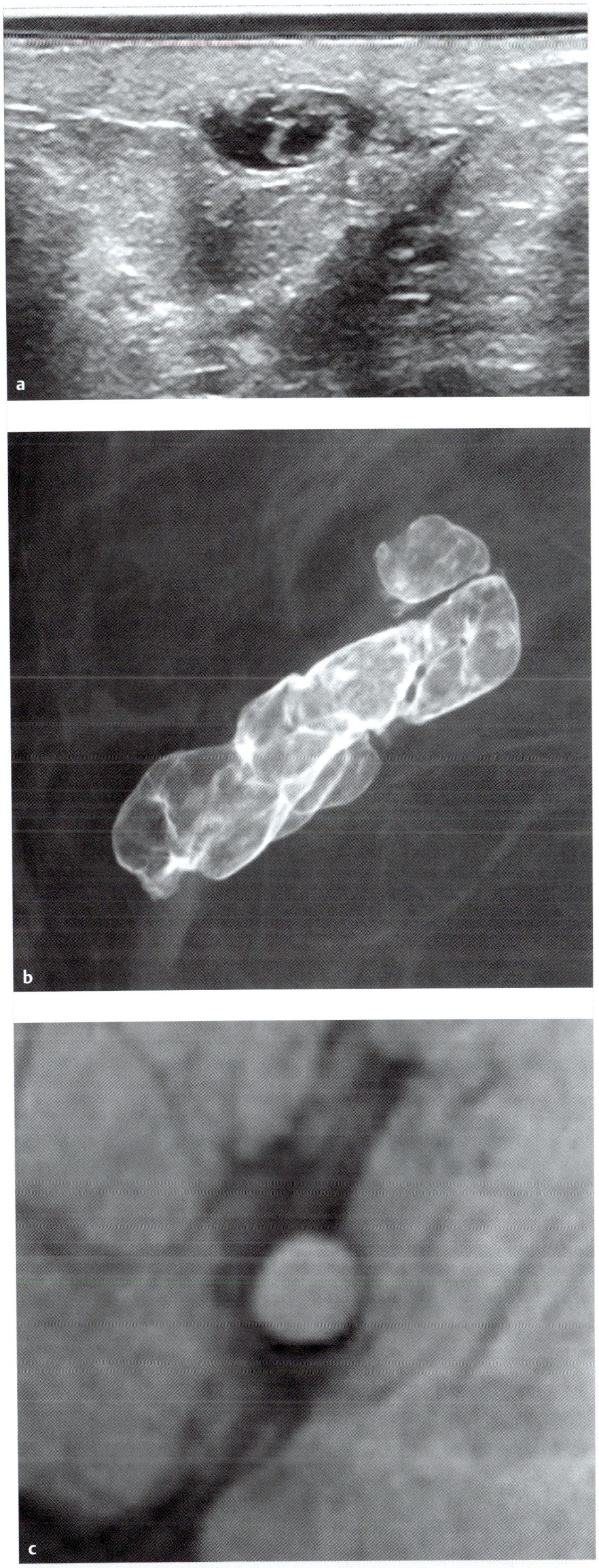

Fig. 8.19 **Old fat necrosis.** Oil cysts. **(a)** Ultrasound image. **(b)** Mammogram. **(c)** T1W precontrast breast MR image.

 - After administration of contrast medium: moderate to strong enhancement (fresh fat necrosis); later, mild reactive enhancement, if any.

Note

Fat necrosis can be regarded as "fresh" within the first 6 months after surgery.

► **Differential diagnosis.** Inflammation, breast carcinoma or tumor recurrence (in the case of fresh fat necrosis).

8.1.20 Abscess

A breast abscess is a localized intramammary collection of pus in a cavity formed by an inflammatory degeneration of tissue. Abscesses may occur after interventional or surgical procedures, or in association with mastitis.

► **Characteristics**

- *Incidence:* very rare.
- *Peak age:* all age groups.
- *Multifocality:* very rare.
- *Bilateral occurrence:* very rare.
- *Complications:* fistula formation, perforation, sepsis.
- *Risk of malignancy:* 0%.
- *Prognosis:* good.
- *Histological classification after biopsy:* B2.

► **Implications for work-up/treatment.** None if symptom-free. When there is uncertainty in the differential diagnosis, percutaneous biopsy should be undertaken.

► **Shape.** Round.

► **Margins.** Indistinct.

► **Internal composition.** Pus, homogeneous.

► **Clinical findings.** Swelling, erythema, pain; fever; increase in inflammatory markers; locoregional lymphadenitis.

► **Imaging**

- *Sonography* (► Fig. 8.20a): hypoechogenic mass with internal echoes; thickened walls.
- *Mammography* (► Fig. 8.20b): hyperdensity, nonspecific.
- *Breast MRI* (► Fig. 8.20c): mass:
  - T1W precontrast image: hypointense, nonspecific.
  - T2W: hyperintense, nonspecific.
  - After administration of contrast medium: obligatory strong enhancement of the thickened abscess wall; reactive hyperemia of the surrounding structures.

Note

The therapeutic management of an abscess of the breast is identical to that of abscesses in other areas of the body.

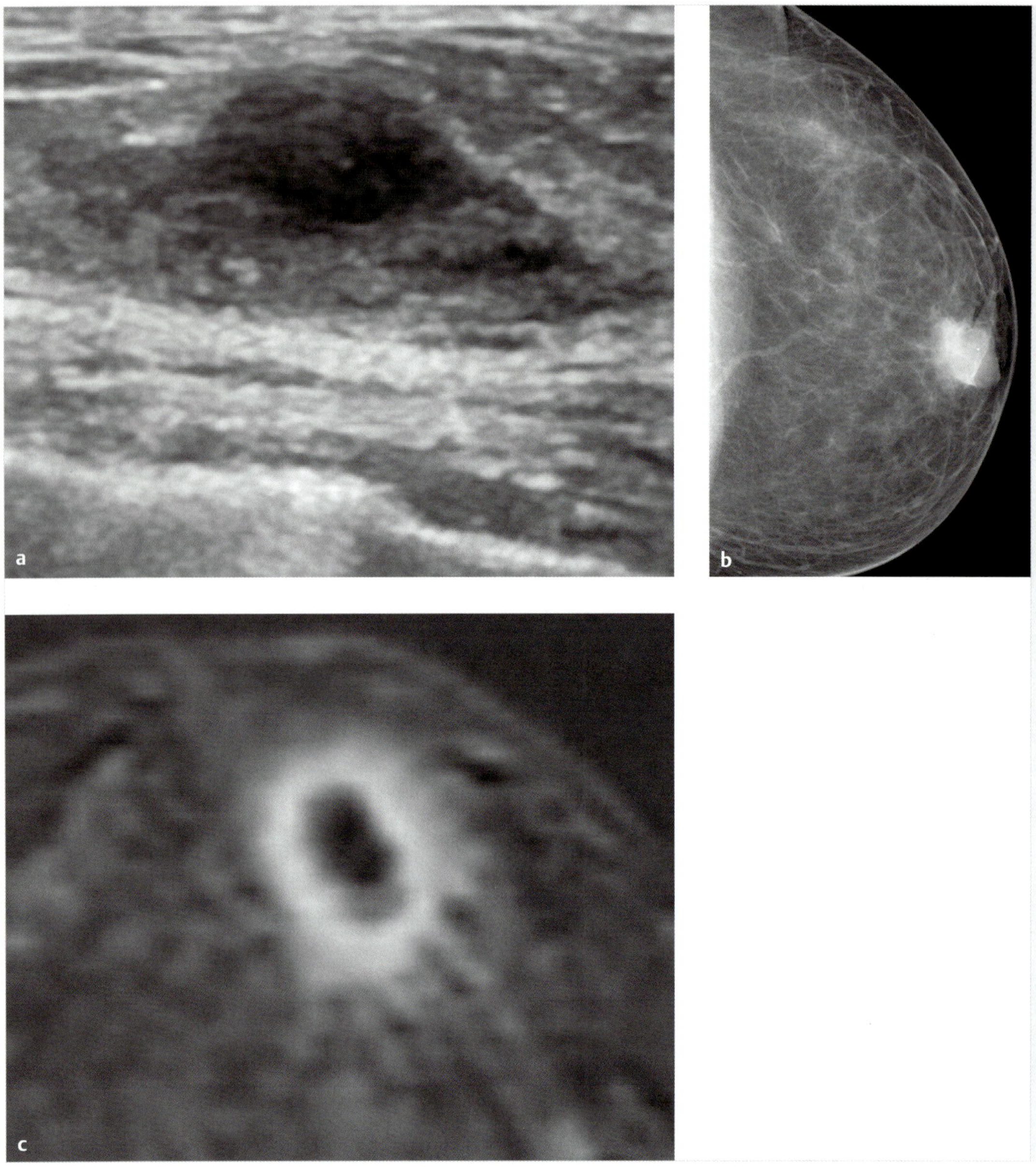

Fig. 8.20 Abscess. (a) Ultrasound image. **(b)** Mammogram. **(c)** Contrast-enhanced breast MRI subtraction image.

▸ **Differential diagnosis.** Carcinoma with central necrosis, triple-negative carcinoma.

8.1.21 Postoperative Scars

A postoperative scar is the final stage of wound healing, with the formation of a tissue complex that is fiber-rich, hypocellular, and hypovascularized. The scar formation is complete approximately 3 to 6 months after surgery.

▸ **Characteristics**

- *Incidence:* inevitable after surgery; occasionally after vacuum-assisted biopsy or trauma.
- *Peak age:* all age groups.
- *Multifocality:* following surgery in multiple ipsilateral locations.
- *Bilateral occurrence:* following bilateral surgery.
- *Complications:* none.
- *Risk of malignancy:* 0%.
- *Prognosis:* excellent.
- *Histological classification after biopsy:* B2.

▸ **Implications for work-up/treatment.** None if symptom free. When there is uncertainty in the differential diagnosis, percutaneous biopsy should be undertaken.

▸ **Shape.** Irregular.

▸ **Margins.** Indistinct.

▸ **Internal composition.** Nonproliferative tissue.

▸ **Clinical findings.** Scar located in the skin. Occasional skin retraction. In the case of larger scars, an induration may be palpable.

▸ **Imaging**

- *Sonography* (▸ Fig. 8.21a): hypoechogenic area, posterior acoustic extinction, nonspecific.
- *Mammography* (▸ Fig. 8.21b): hyperdensity, occasionally associated with micro- and macrocalcifications, nonspecific.
- *Breast MRI* (▸ Fig. 8.21c): non-space-occupying lesion (nonmasslike lesion):

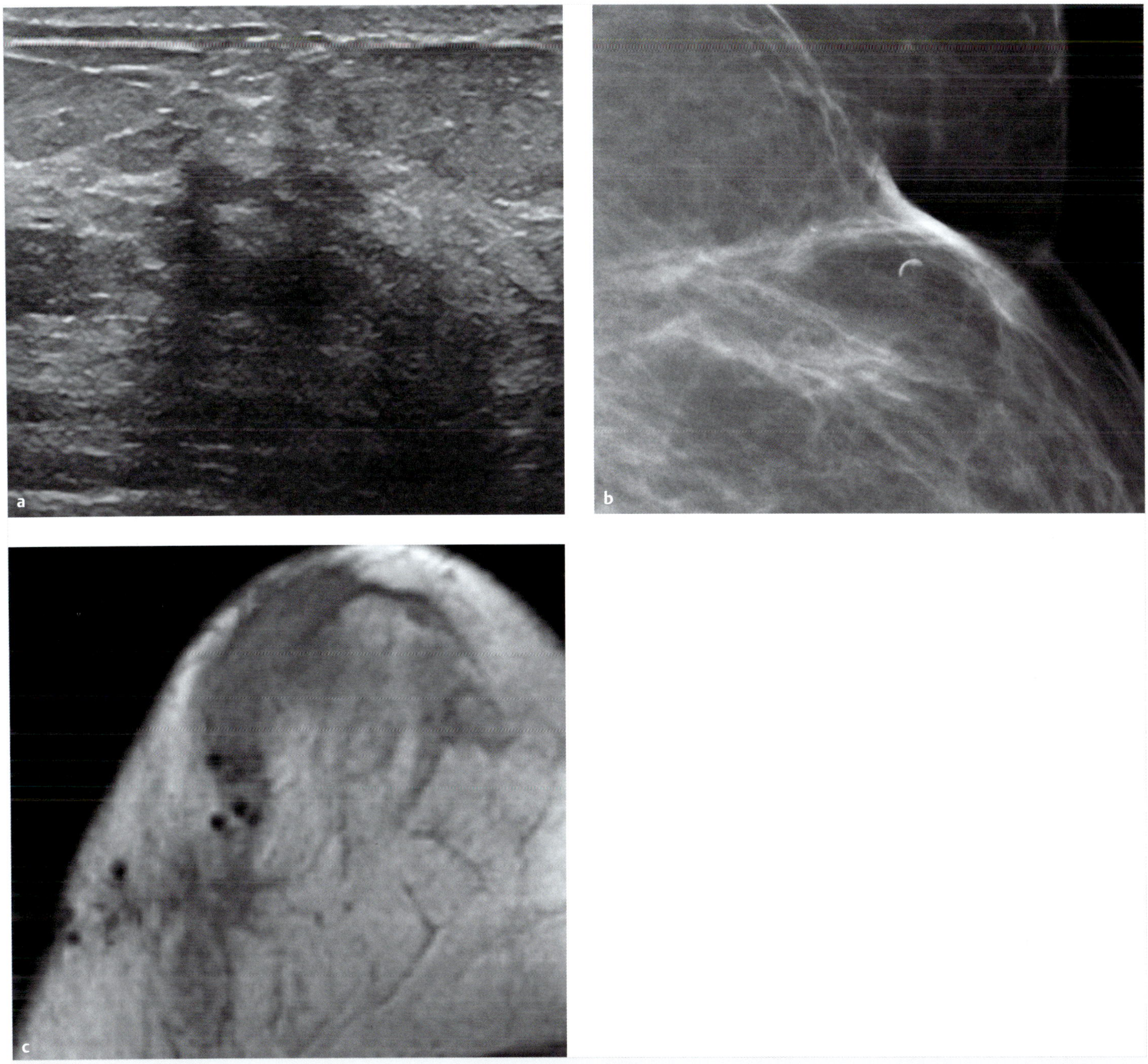

Fig. 8.21 **Postoperative scar.** **(a)** Ultrasound image. **(b)** Mammogram. **(c)** T1W precontrast breast MR image with susceptibility artifacts (black dots).

- ○ T1W precontrast image: hypointense, nonspecific, frequently associated with susceptibility artifacts due to perioperative electrocautery.
- ○ T2W: isointense, occasionally hyperintense, nonspecific.
- ○ After administration of contrast medium: typically, no enhancement; occasional mild enhancement (protracted wound healing or accompanying granulomatous inflammatory changes).

Note

Contrast-enhanced breast MRI enables reliable differentiation between a postoperative scar and a potential tumor recurrence in the vicinity of the scar.

► **Differential diagnosis.** Radial scar, recurrence, breast carcinoma.

8.2 Findings with Ambiguous Biological Potential

8.2.1 Papillomas

Breast papillomas are intraductal proliferations made up of benign epithelial cells covering a central branching, fibrovascular core. The solitary papilloma, typically located in the retromamillary region, is distinguished from the usually multiple peripheral papillomas.

► **Characteristics**

- *Incidence:* rare.
- *Peak age:* 40 to 50 years.
- *Multifocality:* frequent in the case of the peripheral papilloma.
- *Bilateral occurrence:* frequent.
- *Complications:* none.

- *Risk of malignancy:* for solitary papillomas, ca. 2 to 5%; for peripheral papillomas, ca. 10%.
- *Prognosis:* excellent after complete excision of the lesion.
- *Histological classification after biopsy:* B3.

► **Implications for work-up/treatment.** Management should be determined in an interdisciplinary conference. Complete surgical or percutaneous vacuum-assisted excision is commonly recommended due to the increased risk of malignant transformation. For small lesions, follow-up imaging may be recommended as an alternative.

► **Shape, distribution.** Round or oval, solitary or segmental.

► **Margins.** Circumscribed.

► **Internal composition.** Proliferative tissue.

► **Clinical findings.** Pathologic secretion from the corresponding lactiferous duct is occasionally the presenting clinical symptom but usually there are no clinical findings.

► **Imaging**
- *Sonography* (► Fig. 8.22a): hypoechogenic mass, nonspecific; small findings are frequently occult.
- *Mammography:* hyperdensity, occasionally associated with microcalcifications, nonspecific.
- *Galactography* (► Fig. 8.22b): intraductal filling defects and ductal cutoff.
- *Breast MRI* (► Fig. 8.22c): focus, mass:
  - T1W precontrast image: isointense, nonspecific.
  - T2W: isointense, occasionally hyperintense, nonspecific.
  - After administration of contrast medium: mild to strong enhancement; rarely displays ring enhancement.

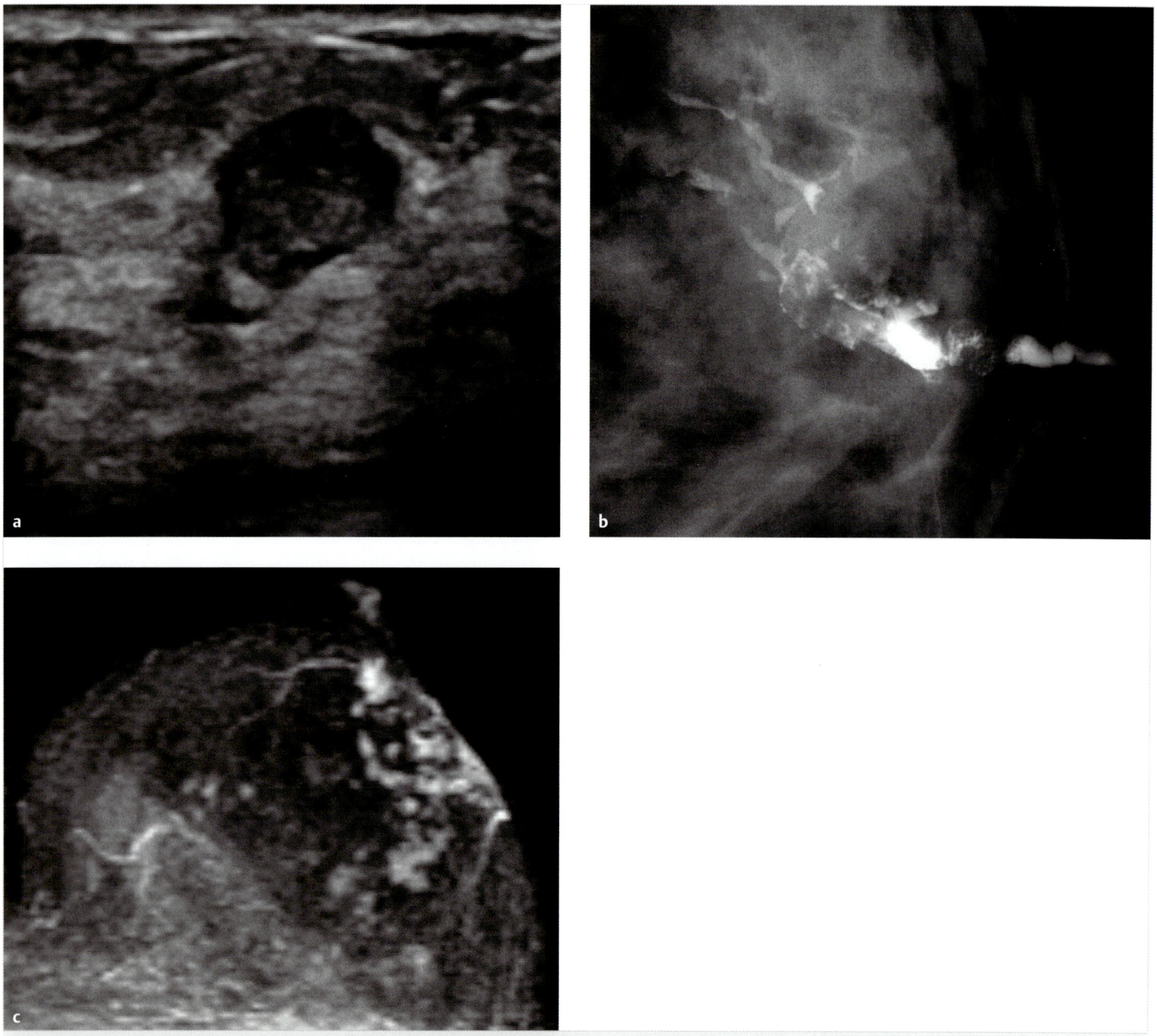

Fig. 8.22 **Papillomas. (a)** Ultrasound image of a solitary papilloma. **(b)** Galactogram showing numerous peripheral papillomas. **(c)** Maximum intensity projection of a contrast-enhanced breast MR image demonstrating peripheral papillomas.

Note

Imaging does not allow differentiation between a pure benign papilloma (or papillomas) and a papilloma in which malignant transformation is incipient or has already taken place.

► **Differential diagnosis.** Fibroadenoma, adenoma, DCIS.

8.2.2 Radial Scars

Histopathologically, a radial scar (synonyms: sclerosing papillary proliferation, indurative mastopathy, sclerosing ductal hyperplasia, infiltrating epithelioma) is a lesion with a defined fibroelastotic core from which ducts with epithelial hyperplasia radiate in a stellate manner.

► **Characteristics**

- *Incidence:* rare.
- *Peak age:* 40 to 60 years.
- *Multifocality:* rare.
- *Bilateral occurrence:* rare.
- *Complications:* none.
- *Risk of malignancy:* high coincidence with carcinoma (20–25%).
- *Prognosis:* excellent after complete excision of the lesion.
- *Histological classification after biopsy:* B3.

► **Implications for work-up/treatment.** Management should be determined in an interdisciplinary conference. Complete surgical excision is usually recommended due to the high coincidence with ADH, LCIS, DCIS, and invasive carcinoma.

► **Shape.** Stellate with a defined center.

► **Margins.** Spiculated.

► **Internal composition.** Proliferative tissue.

► **Clinical findings.** Generally occult without correlative clinical findings.

► **Imaging**

- *Sonography* (► Fig. 8.23a): hypoechogenic area, nonspecific, frequently occult.
- *Mammography* (► Fig. 8.23b): architectural distortion, typically larger than 1 cm; occasionally found in one plane only (two-dimensional finding); occasionally associated with microcalcifications.
- *Breast MRI* (► Fig. 8.23c): non-space-occupying lesion (nonmasslike lesion):
  - T1W precontrast image: hypointense, stellate architectural distortion.
  - T2W: isointense; frequently occult.
  - After administration of contrast medium: no enhancement, mild enhancement or strong enhancement; nonspecific.

Note

Imaging does not allow reliable differentiation between a pure radial scar and a radial scar with an associated malignancy.

► **Differential diagnosis.** Postoperative scars, breast carcinoma (especially tubular carcinoma).

8.2.3 Atypical Ductal Hyperplasia

Atypical ductal hyperplasia (ADH) denotes an intraductal cell proliferation with verified atypia. It is considered a precancerous lesion, i.e., the last benign stage as a precursor to DCIS.

► **Characteristics**

- *Incidence:* rare.
- *Peak age:* 40 to 60 years.
- *Multifocality:* rare.
- *Bilateral occurrence:* rare.
- *Complications:* none.
- *Risk of malignancy:* precancerous.
- *Prognosis:* excellent after complete excision of the lesion.
- *Histological classification after biopsy:* B3.

► **Implications for work-up/treatment.** Management should be determined in an interdisciplinary conference. Complete surgical excision is usually recommended, depending on the size of the initial lesion and the extent of the tissue removed at biopsy. The alternative is follow-up imaging.

► **Shape.** Irregular.

► **Margins.** Indistinct.

► **Internal composition.** Proliferative tissue.

► **Clinical findings.** Generally occult without correlative clinical findings.

► **Imaging**

- *Sonography* (► Fig. 8.24a): hypoechogenic area, nonspecific, frequently occult.
- *Mammography* (► Fig. 8.24b): hyperdensity, occasionally associated with microcalcifications, nonspecific, frequently occult.
- *Breast MRI* (► Fig. 8.24c): non-space-occupying lesion (nonmasslike lesion):
  - T1W precontrast image: hypointense, frequently occult.
  - T2W: isointense or hyperintense; frequently occult.
  - After administration of contrast medium: frequent mild to strong enhancement; nonspecific.

Note

Imaging does not allow a reliable differentiation between ADH and DCIS.

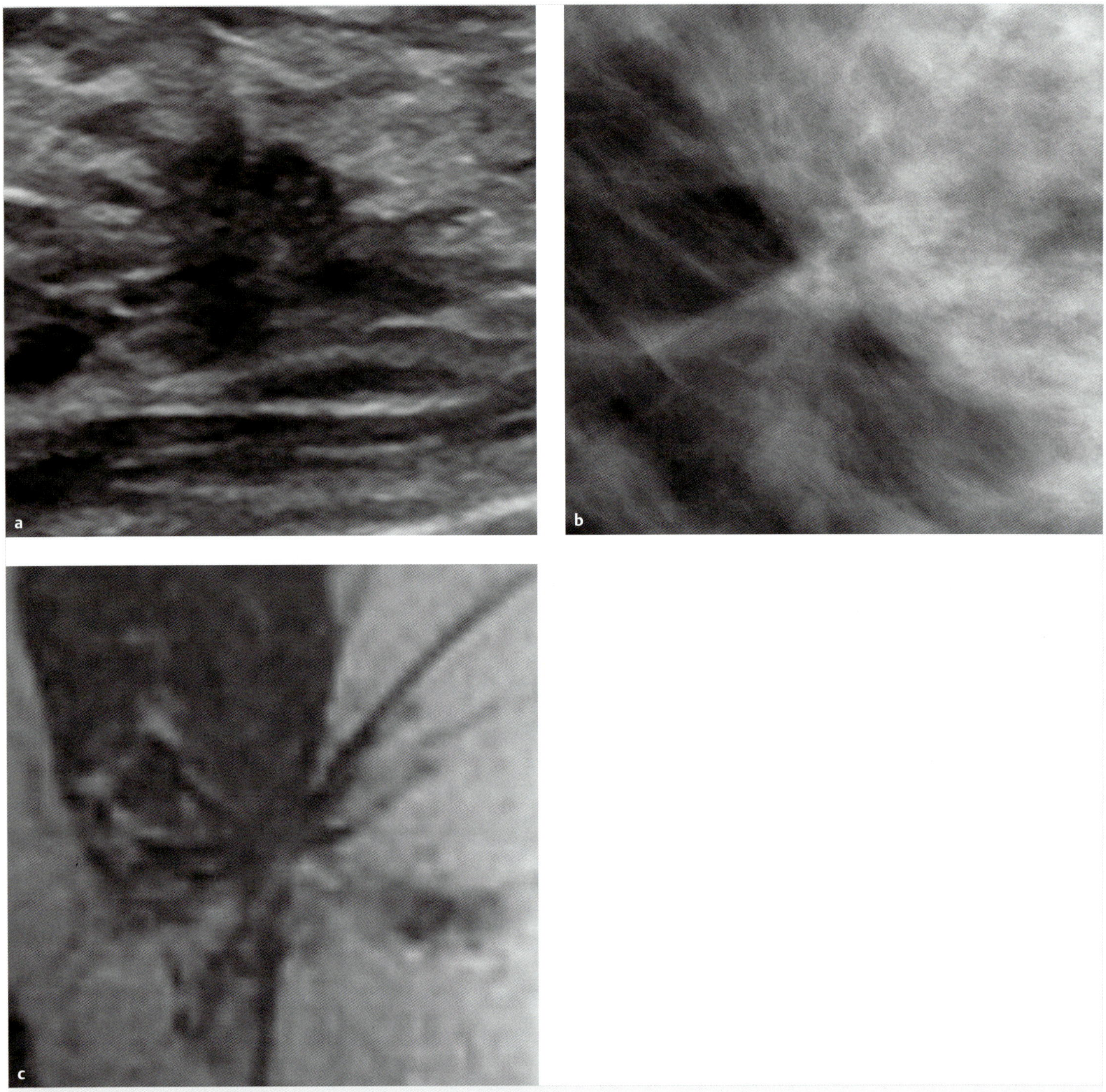

Fig. 8.23 **Radial scar.** **(a)** Ultrasound image. **(b)** Mammogram. **(c)** T1W precontrast breast MR image.

► **Differential diagnosis.** Adenosis, fibrosis, DCIS, breast carcinoma.

8.2.4 Phyllodes Tumors

The benign phyllodes tumor (synonym: fibroadenoma phyllodes, cystosarcoma phyllodes [archaic]) is a fibroepithelial tumor occurring exclusively in the breast.

► **Characteristics**

- *Incidence:* rare.
- *Peak age:* 30 to 50 years.
- *Multifocality:* rare.
- *Bilateral occurrence:* rare.
- *Complications:* increased recurrence rate following surgical excision.
- *Risk of malignancy:* potential for malignant transformation to a malignant phyllodes tumor.
- *Prognosis:* excellent after complete excision of the lesion.
- *Histological classification after biopsy:* B3.

► **Implications for work-up/treatment.** Management should be determined in an interdisciplinary conference. Complete surgical excision is usually recommended because of its frequently very rapid growth and the potential for malignant transformation to a malignant phyllodes tumor.

► **Shape.** Oval.

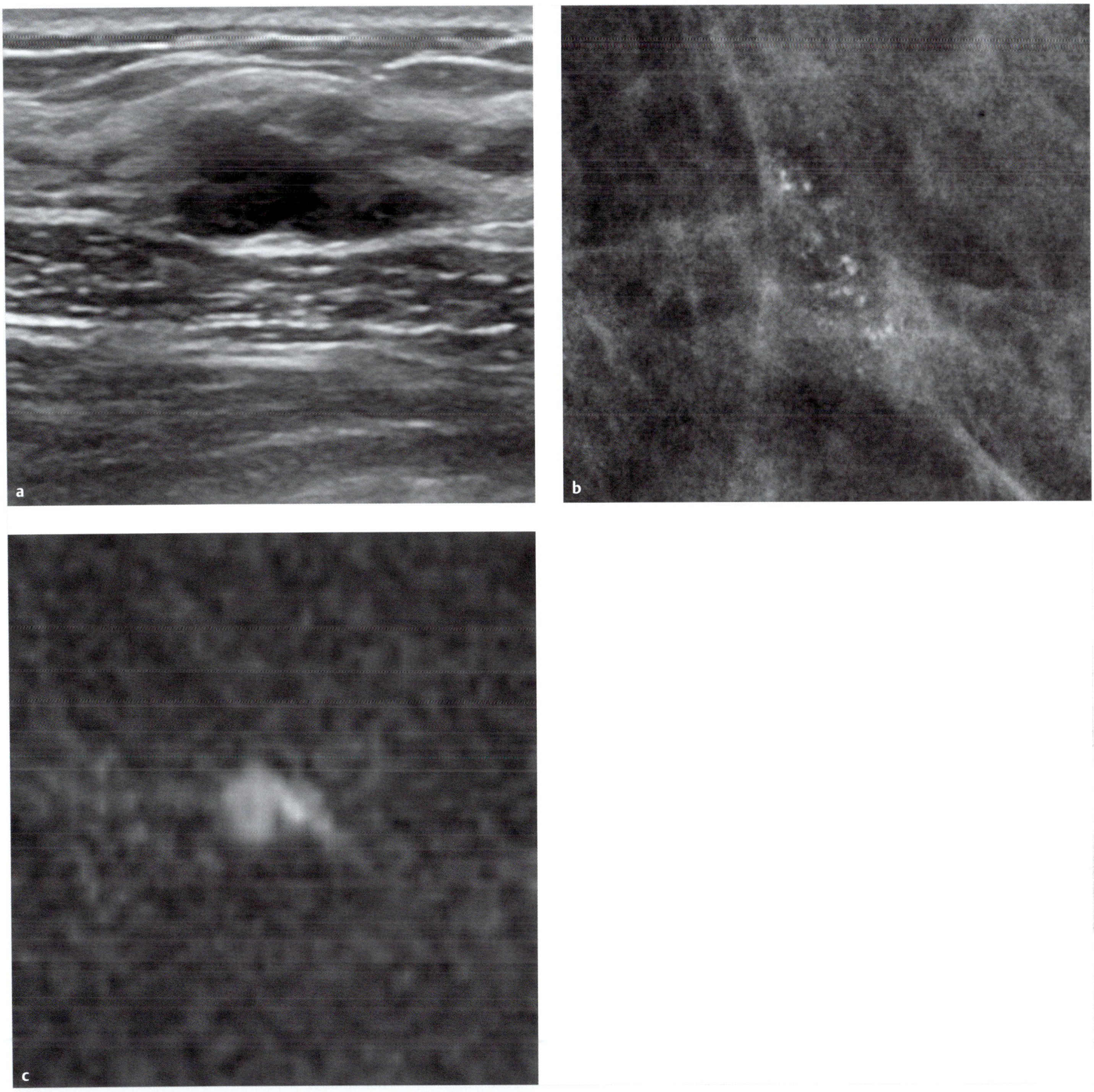

Fig. 8.24 Atypical ductal hyperplasia (ADH). (a) Ultrasound image. **(b)** Mammogram. **(c)** Contrast-enhanced breast MRI subtraction image.

► **Margins.** Circumscribed.

► **Internal composition.** Proliferative tissue.

► **Clinical findings.** Small lesions normally have no correlative clinical findings. Large lesions are often palpable as circumscribed, firm, relatively rapidly developing masses.

► **Imaging**

- *Sonography* (► Fig. 8.25a): inhomogeneous, hypoechogenic mass, occasionally displaying cystic inclusions.
- *Mammography* (► Fig. 8.25b): homogeneous hyperdense mass, occasionally associated with microcalcifications.
- *Breast MRI* (► Fig. 8.25c): mass:
  - T1W precontrast image: hypointense.
  - T2W: isointense or hyperintense; occasionally displays endotumoral cysts.
  - After administration of contrast medium: mild to strong enhancement of the tumor matrix; strong initial signal increase, endotumoral cysts without enhancement.

Note

Imaging does not allow differentiation between benign and malignant phyllodes tumors.

► **Differential diagnosis.** Adenoma, fibroadenoma, papilloma, nodular adenosis, mucinous or medullary carcinoma.

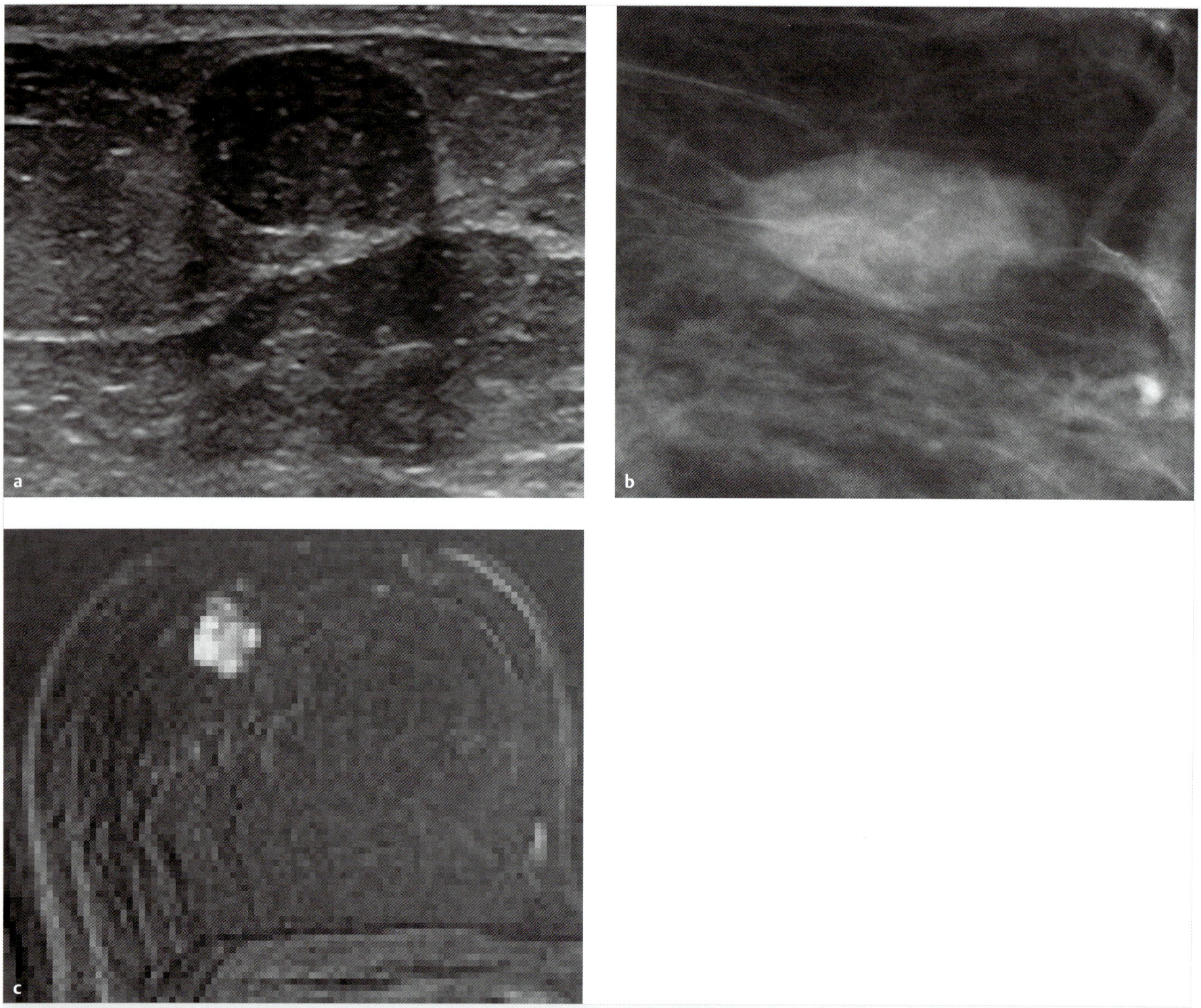

Fig. 8.25 Phyllodes tumor. Benign phyllodes tumor. **(a)** Ultrasound image. **(b)** Mammogram. **(c)** Contrast-enhanced breast MRI subtraction image.

8.2.5 Cysts with Intracystic Proliferation

A cyst with internal nonliquid structures is described as containing an intracystic proliferation (circumscribed wall thickening or polypoid formations).

▸ **Characteristics**
- *Incidence:* rare.
- *Peak age:* all age groups.
- *Multifocality:* very rare.
- *Bilateral occurrence:* very rare.
- *Complications:* none.
- *Risk of malignancy:* dependent on the final histology.
- *Prognosis:* dependent on the final histology.
- *Histological classification after biopsy:* B2 to B5 (dependent on the final histology).

▸ **Implications for work-up/treatment.** Histological diagnostic work-up by means of percutaneous biopsy or primary open excision.

▸ **Shape.** Round, oval.

▸ **Margins.** Circumscribed.

▸ **Internal composition.** Fluid and proliferative tissue.

▸ **Clinical findings.** Small cysts normally have no correlative clinical findings. With larger cysts, a circumscribed, firm-elastic mass may be palpable. The intracystic proliferation itself is not palpable.

▸ **Imaging**
- *Sonography* (▸ Fig. 8.26a): hypoechoic cyst interior with intracystic echogenic structures (polyps or wall-thickening) corresponding to the proliferation; posterior acoustic enhancement.
- *Mammography* (▸ Fig. 8.26b): mass with parenchyma-equivalent, homogeneous density; intracystic proliferation is typically occult; occasionally microcalcifications are located within the proliferation

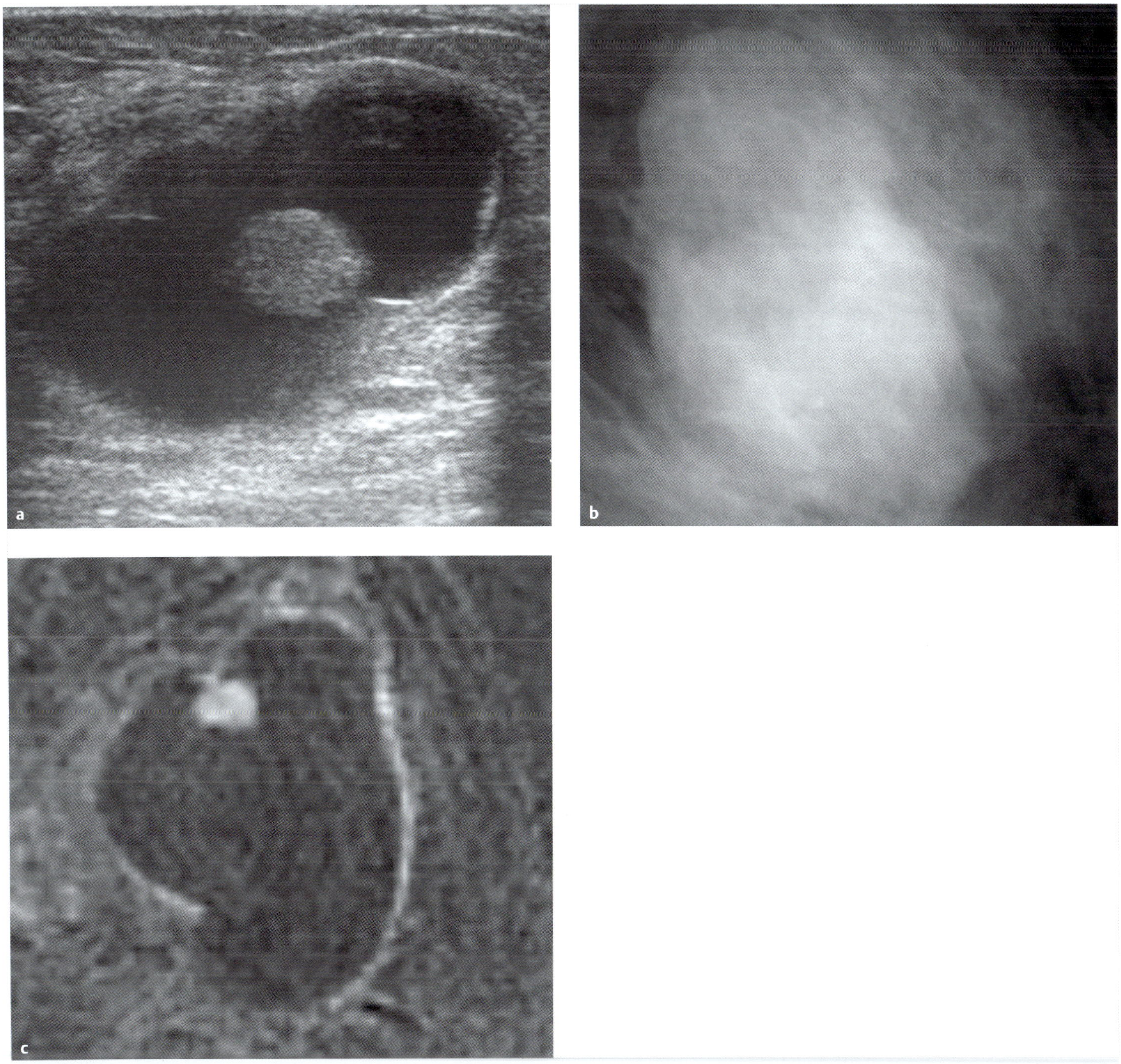

Fig. 8.26 Cyst with intracystic proliferation. (a) Cyst with solid intracystic structure in the ultrasound image. **(b)** Cyst with no identifiable intracystic proliferation in the mammogram. **(c)** Contrast-enhanced breast MRI subtraction image showing a cyst with hypervascularized intracystic growth.

- *Breast MRI* (▸ Fig. 8.26c): mass:
  - T1W precontrast image: isointense (protein) to hypointense (old blood); sedimentation
  - T2W: isointense (protein) to hypointense (old blood); sedimentation.
  - After administration of contrast medium: no to mild cyst wall enhancement; possible enhancement of the intracystic proliferation (depending on histology).

Note

Cysts with intracystic proliferation should be histologically verified.

▸ **Differential diagnosis.** Inflammatory cysts, triple-negative carcinoma, carcinoma with central necrosis.

8.2.6 Lobular Intraepithelial Neoplasia

Lobular intraepithelial neoplasia (LIN), also known as lobular carcinoma in situ (LCIS), is a proliferation of malignant epithelial cells in the lobules and the terminal ducts (see Chapter 3.1.3).

▸ **Characteristics**

- *Incidence:* rare.
- *Peak age:* 40 to 50 years.
- *Multifocality:* frequent.
- *Bilateral occurrence:* frequent.
- *Complications:* none.
- *Risk of malignancy:* increased risk for both breasts, independent of the localization of the LCIS.
- *Prognosis:* good, as long as consistent monitoring is maintained.
- *Histological classification after biopsy:* B3.

▶ **Implications for work-up/treatment.** Management should be determined in an interdisciplinary conference. The alternative courses of action range from surgical excision of the entire lesion to continual monitoring of both breasts with imaging for the purpose of early detection in this high-risk profile group.

▶ **Shape, distribution.** Irregular; focal, regional or diffuse.

▶ **Margins.** Indistinct.

▶ **Internal composition.** Proliferative tissue.

▶ **Clinical findings.** The lesion is usually occult without correlative clinical findings.

▶ **Imaging**

- *Sonography* (▶ Fig. 8.27a): hypoechogenic area, frequently occult
- *Mammography* (▶ Fig. 8.27b): hyperdensity; occasionally associated with microcalcifications; frequently occult.
- *Breast MRI* (▶ Fig. 8.27c): focus, non-space-occupying lesion (nonmasslike lesion):
  - T1W precontrast image: hypointense, frequently occult.
  - T2W: isointense or hyperintense; frequently occult.
  - After administration of contrast medium: mild or strong enhancement; nonspecific.

Note

LCIS typically represents an incidental finding within the context of a percutaneous biopsy performed for the diagnostic work-up for ambiguous findings on imaging. There is no characteristic imaging correlate for LCIS.

▶ **Differential diagnosis.** Fibrosis, adenosis, DCIS.

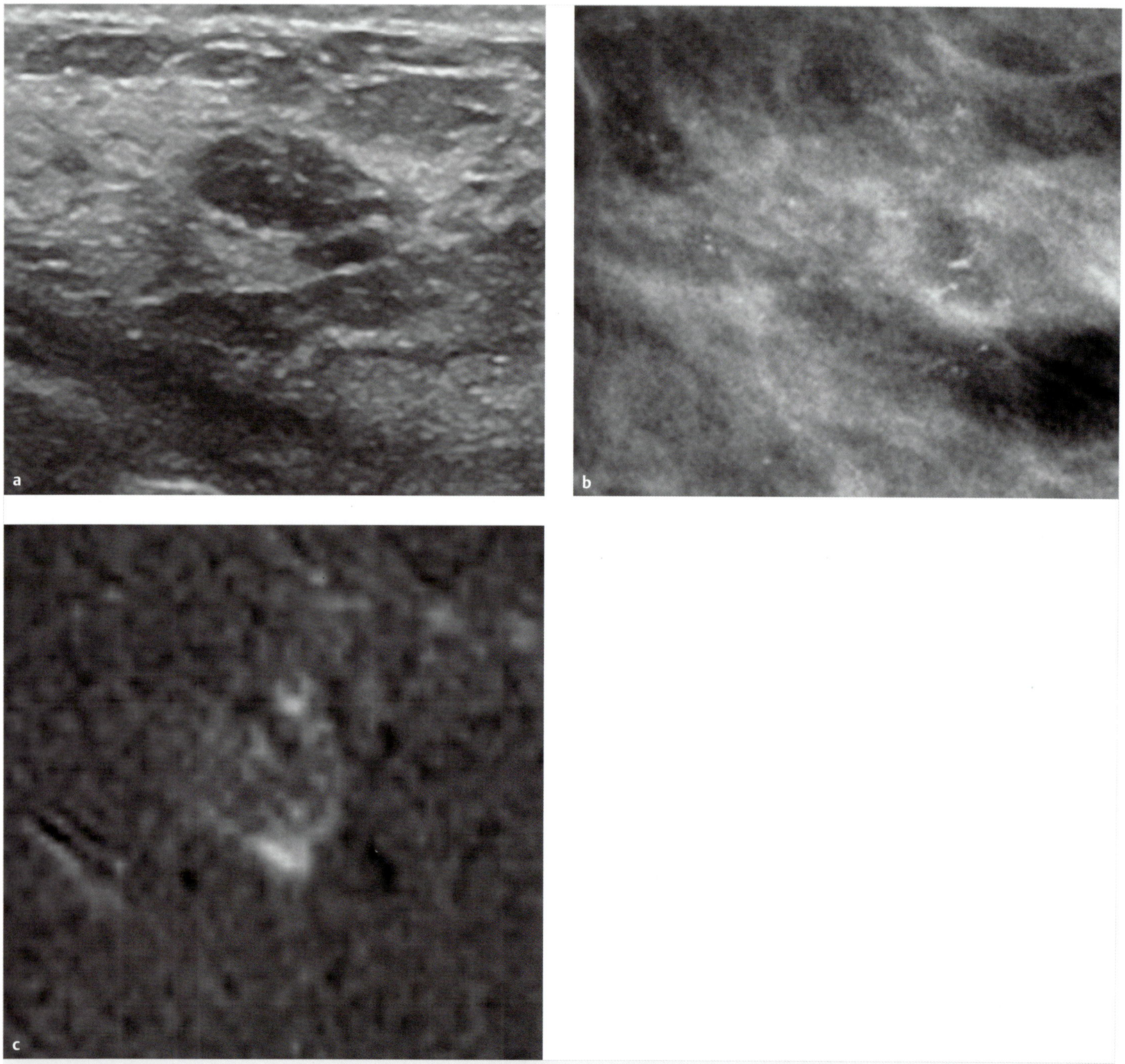

Fig. 8.27 Lobular carcinoma in situ (LCIS). (a) Ultrasound image. **(b)** Mammogram. **(c)** Contrast-enhanced breast MRI subtraction image.

8.3 Intraductal Carcinoma

8.3.1 Ductal Carcinoma In Situ (Low Grade)

Ductal carcinoma in situ (DCIS) is an intraductal proliferation of malignant epithelial cells without cell invasion through the basement membrane. *Low-grade* DCIS (synonym: non-comedo type DCIS) is classified as grade 1.

► **Characteristics**

- *Incidence:* rare.
- *Peak age:* 40 to 60 years.
- *Multifocality:* frequent.
- *Bilateral occurrence:* 5%.
- *Risk of invasion*: the lifetime risk of progression from a low-grade DCIS to an invasive carcinoma is approximately 30%.
- *Prognosis:* good, provided that a complete surgical excision is done.
- *Histological classification after biopsy:* B5a.

► **Implications for work-up/treatment.** Complete surgical excision of the DCIS lesion is usually recommended. Debate continues as to whether this represents overtreatment, since, over a lifetime, approximately 70% of the lesions remain intraductal and do not threaten the patient.

► **Shape, distribution.** Irregular; ductal, segmental, diffuse.

► **Margins.** Indistinct.

► **Internal composition.** Proliferative tissue.

► **Clinical findings.** Usually occult without correlative clinical findings.

► **Imaging**

- *Sonography* (► Fig. 8.28a): hypoechogenic area, frequently occult.
- *Mammography* (► Fig. 8.28b): hyperdensity, often occult; frequently associated with microcalcifications.
- *Breast MRI* (► Fig. 8.28c): non-space-occupying lesion (non-masslike lesion):
  - T1W precontrast image: hypointense, frequently occult.
  - T2W: isointense or hyperintense; frequently occult.
  - After administration of contrast medium: no enhancement or, at most, mild enhancement; nonspecific.

Note

It is still an open question whether G1 DCIS is overdiagnosed.

► **Differential diagnosis.** DCIS G2 or G3, mastitis, adenosis, fibrosis.

8.3.2 Ductal Carcinoma In Situ (Intermediate Type)

Ductal carcinoma in situ (DCIS) is an intraductal proliferation of malignant epithelial cells without cell invasion through the basement membrane. The *intermediate-type* DCIS (synonym: non-comedo type DCIS) is classified as grade 2.

► **Characteristics**

- *Incidence:* rare.
- *Peak age:* 40 to 60 years.
- *Multifocality:* frequent.
- *Bilateral occurrence:* 5%.
- *Risk of invasion*: no exact data are available on the lifetime risk of progression from an intermediate-type DCIS to an invasive carcinoma.
- *Prognosis:* good, provided that a complete surgical excision is done.
- *Histological classification after biopsy:* B5a.

► **Implications for work-up/treatment.** Complete surgical excision of the DCIS lesion is recommended.

► **Shape, distribution.** Irregular; ductal, segmental, diffuse.

► **Margins.** Indistinct.

► **Internal composition.** Proliferative tissue.

► **Clinical findings.** Usually occult without correlative clinical findings.

► **Imaging**

- *Sonography* (► Fig. 8.29a): hypoechogenic area, frequently occult.
- *Mammography* (► Fig. 8.29b): hyperdensity, often occult; frequently associated with microcalcifications.
- *Breast MRI* (► Fig. 8.29c): non-space-occupying lesion (nonmasslike lesion):
  - T1W precontrast image: hypointense, frequently occult.
  - T2W: isointense or hyperintense; frequently occult.
  - After administration of contrast medium: no enhancement, mild enhancement or strong enhancement.

► **Differential diagnosis.** DCIS G1 or G3, mastitis, adenosis, fibrosis.

8.3.3 Ductal Carcinoma In Situ (High Grade)

Ductal carcinoma in situ (DCIS) is an intraductal proliferation of malignant epithelial cells without cell invasion through the basement membrane. *High-grade* DCIS (synonym: comedo type DCIS) is classified as grade 3.

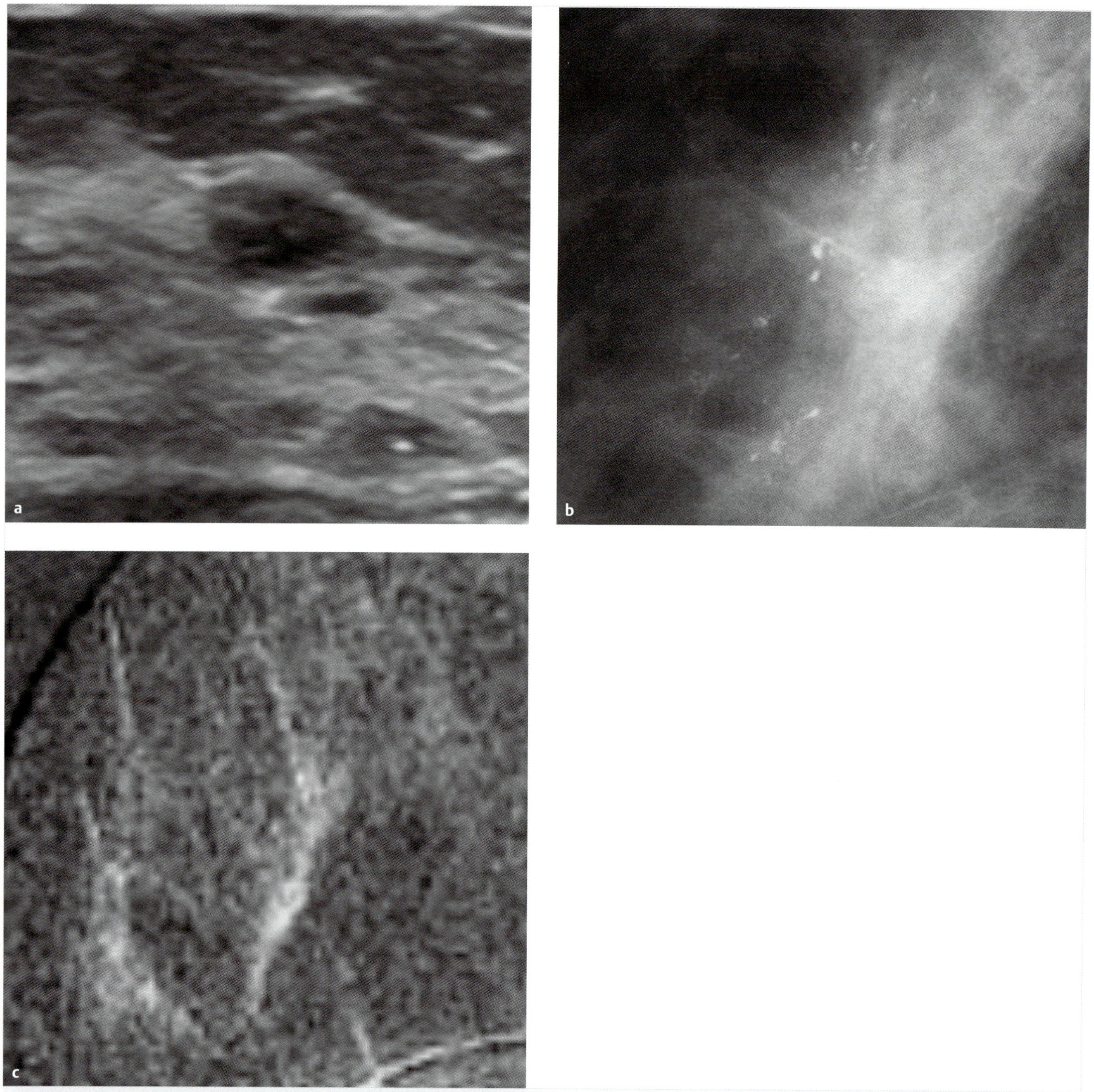

Fig. 8.28 Low-grade ductal carcinoma in situ (DCIS). DCIS G1. **(a)** Ultrasound image. **(b)** Mammogram. **(c)** Contrast-enhanced breast MRI subtraction image.

► **Characteristics**

- *Incidence:* rare.
- *Peak age:* 40 to 60 years.
- *Multifocality:* frequent.
- *Bilateral occurrence:* 5%.
- *Risk of invasion*: Untreated, high-grade DCIS frequently progresses to an invasive carcinoma.
- *Prognosis:* good, provided that a complete surgical excision is done.
- *Histological classification after biopsy:* B5a.

► **Implications for work-up/treatment.** Complete surgical excision of the DCIS lesion is recommended.

► **Shape, distribution.** Irregular; ductal, segmental, diffuse.

► **Margins.** Indistinct.

► **Internal composition.** Proliferative tissue.

► **Clinical findings.** Usually occult without correlative clinical findings.

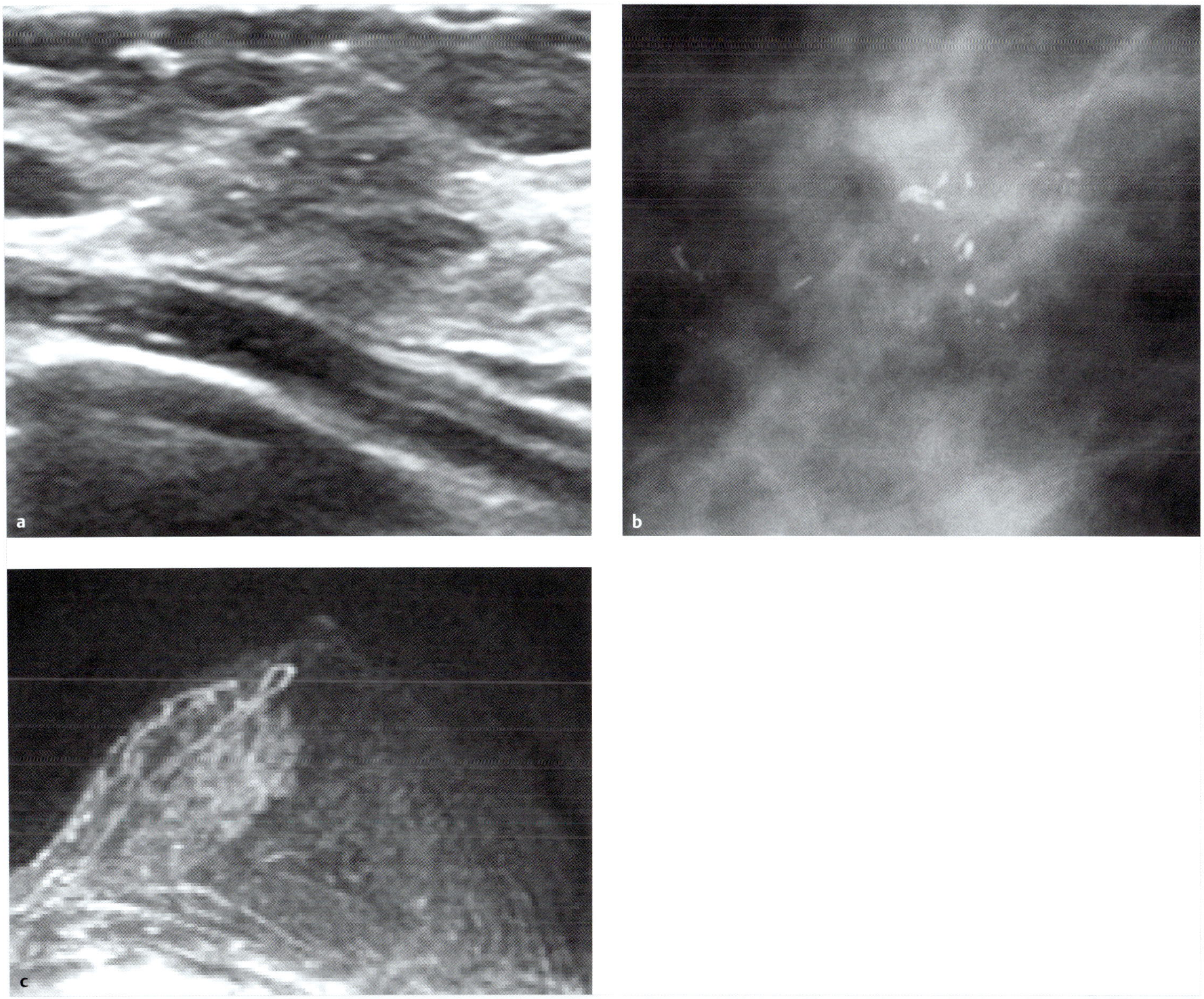

Fig. 8.29 Intermediate-type DCIS. DCIS G2. **(a)** Ultrasound image. **(b)** Mammogram. **(c)** Contrast-enhanced breast MR image, maximum intensity projection. DCIS, ductal carcinoma in situ.

► **Imaging**

- *Sonography* (► Fig. 8.30a): hypoechogenic area, frequently occult.
- *Mammography* (► Fig. 8.30b): hyperdensity, often occult; frequently associated with microcalcifications.
- *Breast MRI* (► Fig. 8.30c): non-space-occupying lesion (nonmasslike lesion):
  - T1W precontrast image: hypointense, frequently occult.
  - T2W: isointense or hyperintense; frequently occult.
  - After administration of contrast medium: mild or strong enhancement.

Note

Current data indicate that high-grade DCIS is detected significantly more frequently with high-resolution breast MRI than with mammography and/or sonography.

► **Differential diagnosis.** DCIS G1 or G2, mastitis, adenosis, fibrosis.

8.4 Invasive Tumors

8.4.1 Invasive Ductal Carcinoma

Invasive ductal carcinoma (IDC) is a malignant breast tumor that initially develops in the terminal lactiferous ducts and grows through the basement membrane. It exhibits no specific histological differentiation pattern and therefore does not fall into any of the other histopathologically defined carcinoma types (invasive carcinoma of "no special type", NST).

► **Characteristics**

- *Incidence:* although rare, it is the most common form of invasive breast cancer.
- *Peak age:* 50 to 60 years.
- *EIC (extensive intraductal component):* 10%.
- *Multifocality:* 10%.
- *Multicentricity:* 5%.
- *Bilateral occurrence:* 5%.
- *Prognosis:* dependent on tumor stage (pT, pN, pM), grade and receptor status, among other factors.

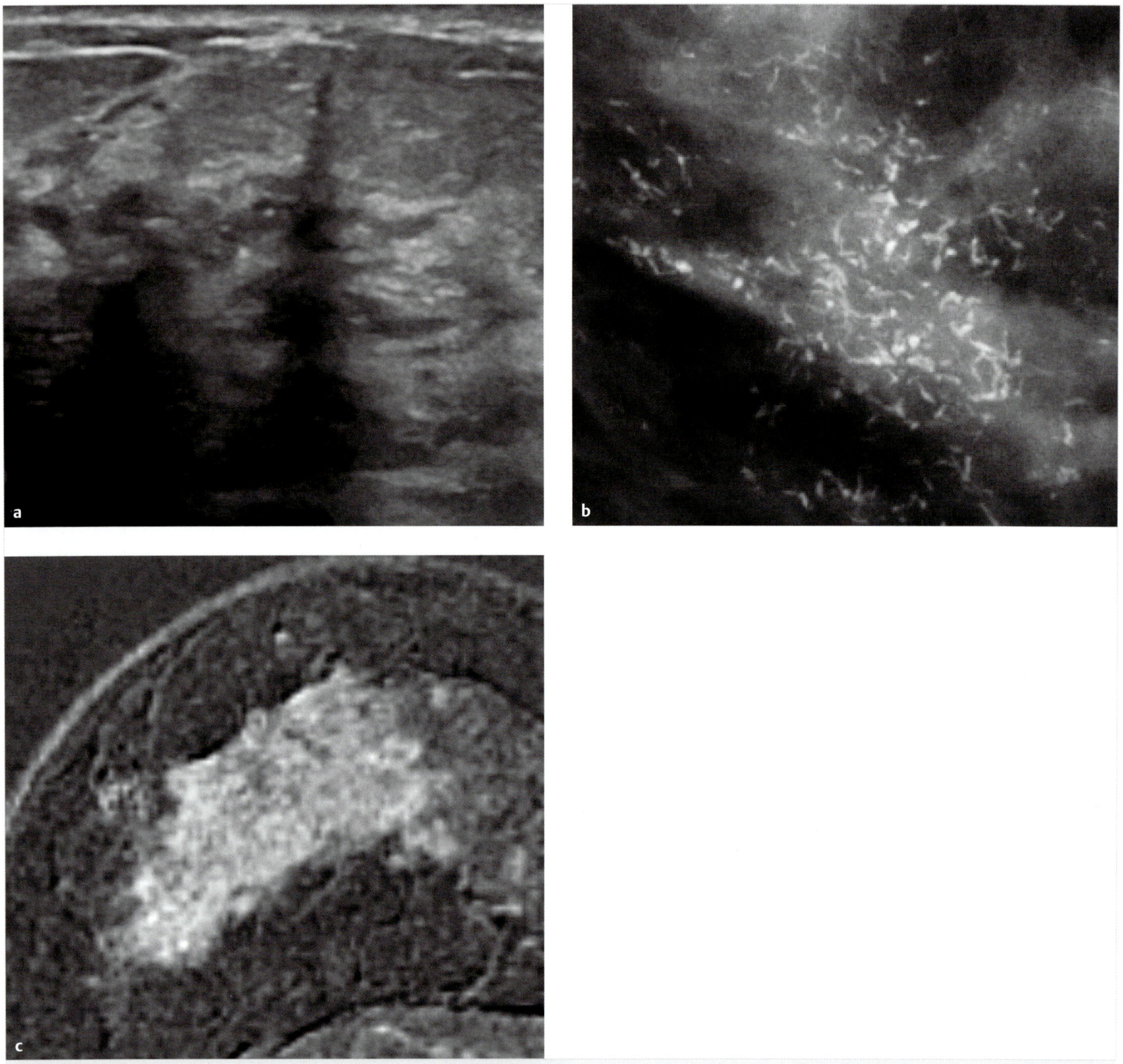

Fig. 8.30 High-grade DCIS. DCIS G3. **(a)** Ultrasound image. **(b)** Mammogram. **(c)** Contrast-enhanced breast MRI subtraction image. DCIS, ductal carcinoma in situ.

- *Histological classification after biopsy:* B5b.

► **Implications for work-up/treatment.** Percutaneous biopsy should be done for histological verification. Once the diagnosis is confirmed, the tumor is usually surgically excised. Occasionally, a form of neoadjuvant therapy (primary chemotherapy) may be recommended.

► **Shape, distribution.** Irregular, rarely round.

► **Margins.** Indistinct, spiculated.

► **Internal composition.** Proliferative tissue.

► **Clinical findings.** Small tumors are usually occult without correlative clinical findings. Palpable masses have a diameter of 2.3 cm on average. Larger tumors may exhibit the classic carcinoma findings of firm mass, limited mobility, occasional skin or nipple retraction.

► **Imaging**

- *Sonography* (► Fig. 8.31a): hypoechogenic mass; may display hyperechogenic rim; disruption of the surrounding ligamentous structures, central or peripheral posterior acoustic shadowing/extinction, longitudinal axis perpendicular to the skin.
- *Mammography* (► Fig. 8.31b): hyperdense; associated with intratumoral microcalcifications in 30% of cases.
- *Breast MRI* (► Fig. 8.31c): focus, mass:
  - T1W precontrast image: hypointense, nonspecific.
  - T2W: 70% hypointense, otherwise isointense, nonspecific.

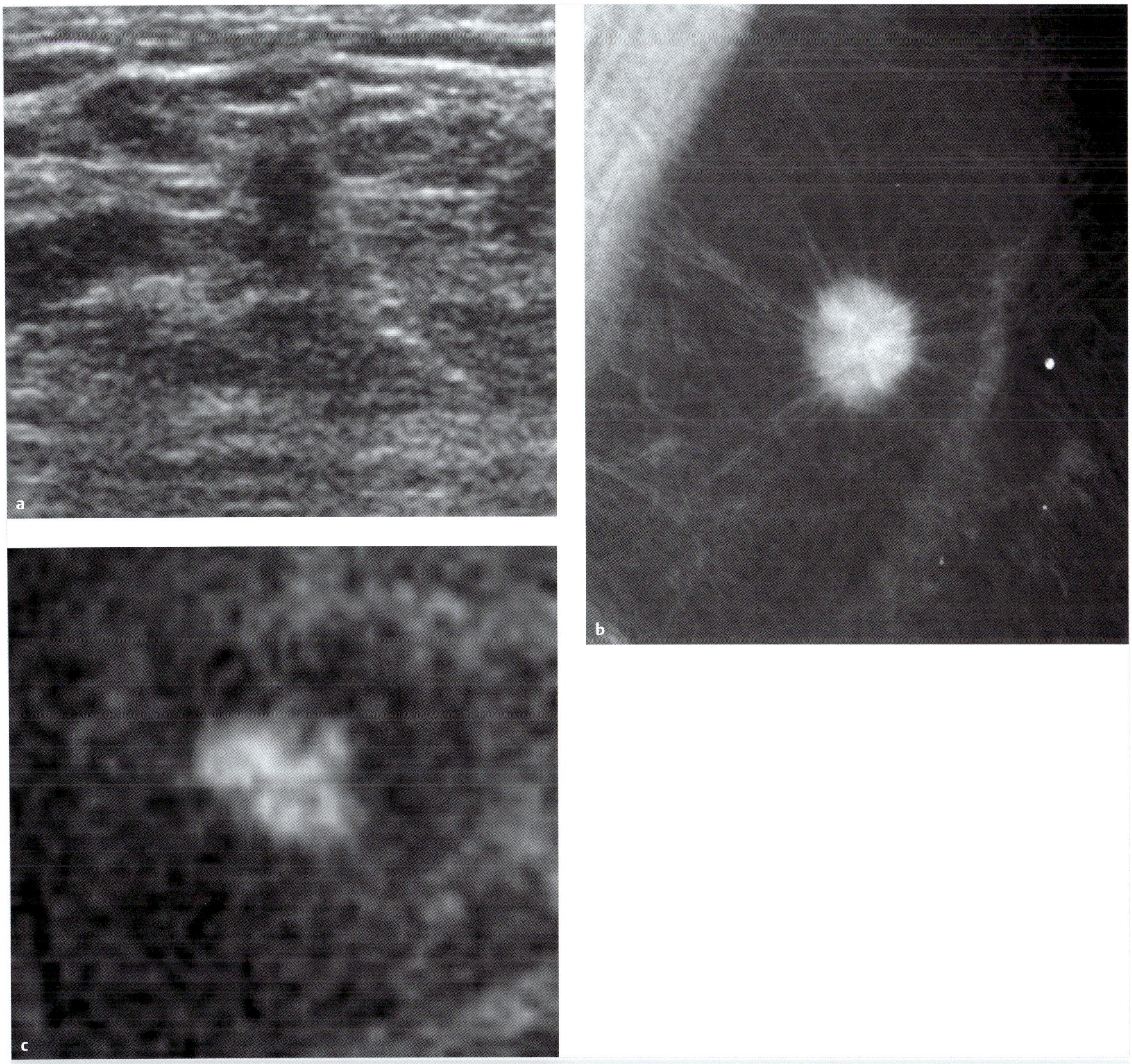

Fig. 8.31 **Invasive ductal carcinoma. (a)** Ultrasound image. **(b)** Mammogram. **(c)** Contrast-enhanced breast MRI subtraction image.

- After administration of contrast medium: typically displays strong initial enhancement; in 30 to 50% of cases there is subsequent washout phenomenon; 50% show ring enhancement.

Note

Early detection—optimally when the IDC is less than 10 mm in size—is significant for survival.

► **Differential diagnosis.** Other malignant tumors, benign mass lesions.

8.4.2 Invasive Lobular Carcinoma

Invasive lobular carcinoma (ILC) represents a malignant breast tumor that is characterized by desmoplastic stroma reactions, small cells, and, in particular, by a linear and "string of pearls" arrangement of the tumor cells (single file). Growth patterns are differentiated into two types-diffuse and nodular.

► **Characteristics**

- *Incidence:* although rare, it is the second most common form of invasive breast cancer.
- *Peak age:* 50 to 60 years.

- *EIC:* no concrete data available.
- *Multifocality:* 10%.
- *Multicentricity:* 5%.
- *Bilateral occurrence:* 5 to 10%.
- *Prognosis:* dependent on tumor stage (pT, pN, pM), grade and receptor status, among other factors.
- *Histological classification after biopsy:* B5b.

▸ **Implications for work-up/treatment.** Percutaneous biopsy for histological verification. Once the diagnosis is confirmed, usually the tumor is surgically excised. Occasionally, a form of neoadjuvant therapy (primary chemotherapy) may be recommended.

▸ **Shape, distribution.** Irregular (diffuse form of ILC, common); round (nodular form of ILC, rare).

▸ **Margins.** Indistinct.

▸ **Internal composition.** Proliferative tissue.

▸ **Clinical findings.** Small tumors are usually occult without correlative clinical findings. Large tumors may be palpated as an area of increased induration. Rarely, a circumscribed mass may be palpated (nodular type of ILC). Occasional skin or nipple retraction.

▸ **Imaging**

- *Sonography* (▸ Fig. 8.32a): hypoechogenic area or mass; isolated banded or streaked posterior acoustic shadowing/extinction; often occult.
- *Mammography* (▸ Fig. 8.32b): hyperdensity; occasional discrete asymmetry in right–left comparison; impression of

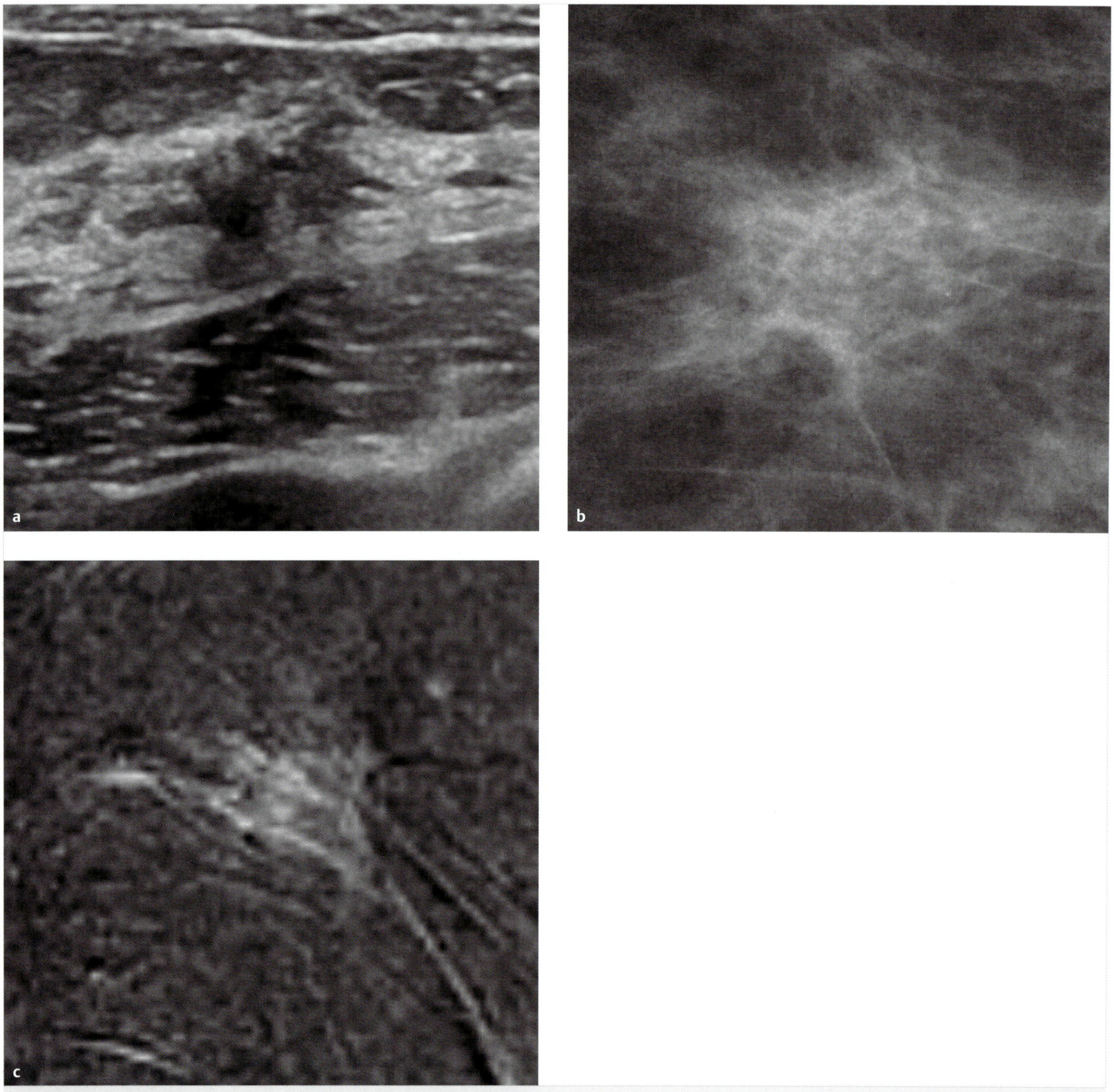

Fig. 8.32 Invasive lobular carcinoma. (a) Ultrasound image. **(b)** Mammogram. **(c)** Contrast-enhanced breast MRI subtraction image.

parenchymal retraction (shrinking sign); rarely associated with microcalcifications.
- *Breast MRI* (► Fig. 8.32c): non-space-occupying lesion (nonmasslike lesion):
  - T1W precontrast image: hypointense, nonspecific.
  - T2W: hypodense or isodense, nonspecific.
  - After administration of contrast medium: mild to strong initial enhancement, sparing lipomatous tissue areas.

Note

ILC is the most difficult type of breast cancer to detect with all imaging modalities. High-resolution breast MRI offers the best results.

► **Differential diagnosis.** Adenosis, DCIS, other invasive tumors, benign diffuse lesions.

8.4.3 Tubular Carcinoma

Tubular carcinomas are characterized by the presence of a very high percentage (over 75%) of highly differentiated tubules that are lined with a single layer of cuboidal epithelial cells. Consequently, they are often grade 1 tumors.

► **Characteristics**
- *Incidence:* rare, less than 2% of invasive breast cancer.
- *Peak age:* 40 to 50 years.
- *EIC:* no concrete data available.
- *Multifocality:* no concrete data available.
- *Multicentricity:* no concrete data available.
- *Bilateral occurrence:* no concrete data available.
- *Prognosis:* rather favorable; dependent on tumor stage (pT, pN, pM) and receptor status.
- *Histological classification after biopsy:* B5b.

► **Implications for work-up/treatment.** Percutaneous biopsy for histological verification. Once the diagnosis is confirmed, the tumor is usually surgically excised.

► **Shape, distribution.** Irregular.

► **Margins.** Spiculated.

► **Internal composition.** Proliferative tissue.

► **Clinical findings.** Small tumors are usually occult without correlative clinical findings. Large tumors may be palpated as a circumscribed mass.

► **Imaging**
- *Sonography* (► Fig. 8.33a): hypoechogenic mass; architectural distortion; posterior acoustic extinction.
- *Mammography* (► Fig. 8.33b): hyperdensity or architectural distortion; in more than half the cases tumors display associated microcalcifications.
- *Breast MRI* (► Fig. 8.33c): non-space-occupying lesion (nonmasslike lesion):
  - T1W precontrast image: hypointense; architectural distortion.
  - T2W: hypointense or isointense, nonspecific.
  - After administration of contrast medium: mild to strong enhancement.

Note

Tubular carcinomas of the breast frequently appear on imaging as an architectural distortion.

► **Differential diagnosis.** Radial scar, postoperative scar.

8.4.4 Medullary Carcinoma

Medullary carcinomas (synonym: carcinoma with medullary features) are breast carcinomas with tumor cells that grow in large, solid cell aggregates without distinct cell borders (syncytial growth pattern).

► **Characteristics**
- *Incidence:* rare, less than 5% of invasive breast cancer.
- *Peak age:* 40 to 50 years; frequent in *BRCA* mutation carriers.
- *EIC:* 5%.
- *Multifocality:* no concrete data available.
- *Multicentricity:* no concrete data available.
- *Bilateral occurrence:* no concrete data available.
- *Prognosis:* rather favorable; dependent on tumor stage (pT, pN, pM), grade and receptor status.
- *Histological classification after biopsy:* B5b.

► **Implications for work-up/treatment.** Percutaneous biopsy for histological verification. Once the diagnosis is confirmed, the tumor is usually surgically excised.

► **Shape, distribution.** Round or oval.

► **Margins.** Microlobulated or indistinct.

► **Internal composition.** Proliferative tissue.

► **Clinical findings.** Small tumors are usually occult without correlative clinical findings. Larger tumors may be palpated as a circumscribed mass.

► **Imaging**
- *Sonography* (► Fig. 8.34a): hypoechogenic mass, intermediate posterior echo pattern.
- *Mammography* (► Fig. 8.34b): hyperdensity; rarely associated with microcalcifications.
- *Breast MRI* (► Fig. 8.34c): mass:
  - T1W precontrast image: hypointense or isointense, nonspecific.
  - T2W: isointense, often hyperintense, nonspecific.
  - After administration of contrast medium: mild, more frequently strong enhancement; rarely ring enhancement; occasional washout phenomenon.

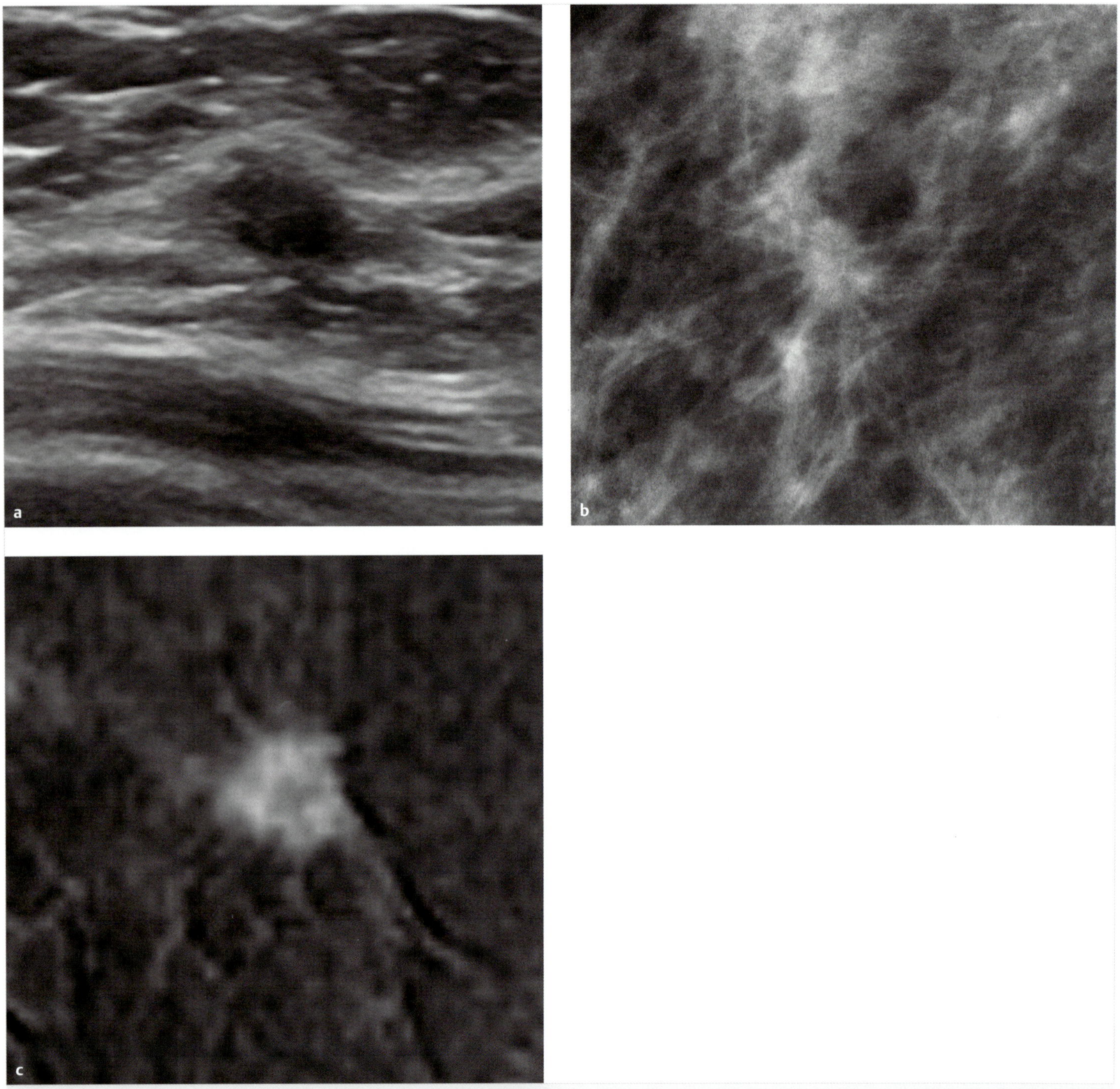

Fig. 8.33 Tubular carcinoma of the breast. (a) Ultrasound image. **(b)** Mammogram. **(c)** Contrast-enhanced breast MRI subtraction image.

Note

In the differential diagnosis, medullary carcinomas of the breast are at times difficult to differentiate from fibroadenomas. In case of doubt, percutaneous biopsy should be undertaken.

▸ **Differential diagnosis.** Fibroadenoma, adenoma, papilloma, other malignant breast tumors.

8.4.5 Mucinous Carcinoma

Mucinous carcinomas (synonym: colloid carcinoma) are breast carcinomas that characteristically exhibit islandlike clusters of tumor cells floating in "lakes" of extracellular mucin.

▸ **Characteristics**

- *Incidence:* rare, less than 5% of invasive breast cancer.
- *Peak age:* 50 to 60 years.
- *EIC:* no concrete data available.
- *Multifocality:* no concrete data available.
- *Multicentricity:* no concrete data available.
- *Bilateral occurrence:* no concrete data available.
- *Prognosis:* rather favorable; dependent on tumor stage (pT, pN, pM), grade and receptor status.
- *Histological classification after biopsy:* B5b.

▸ **Implications for work-up/treatment.** Percutaneous biopsy for histological verification. Once the diagnosis is confirmed, the tumor is usually surgically excised.

▸ **Shape, distribution.** Round or oval.

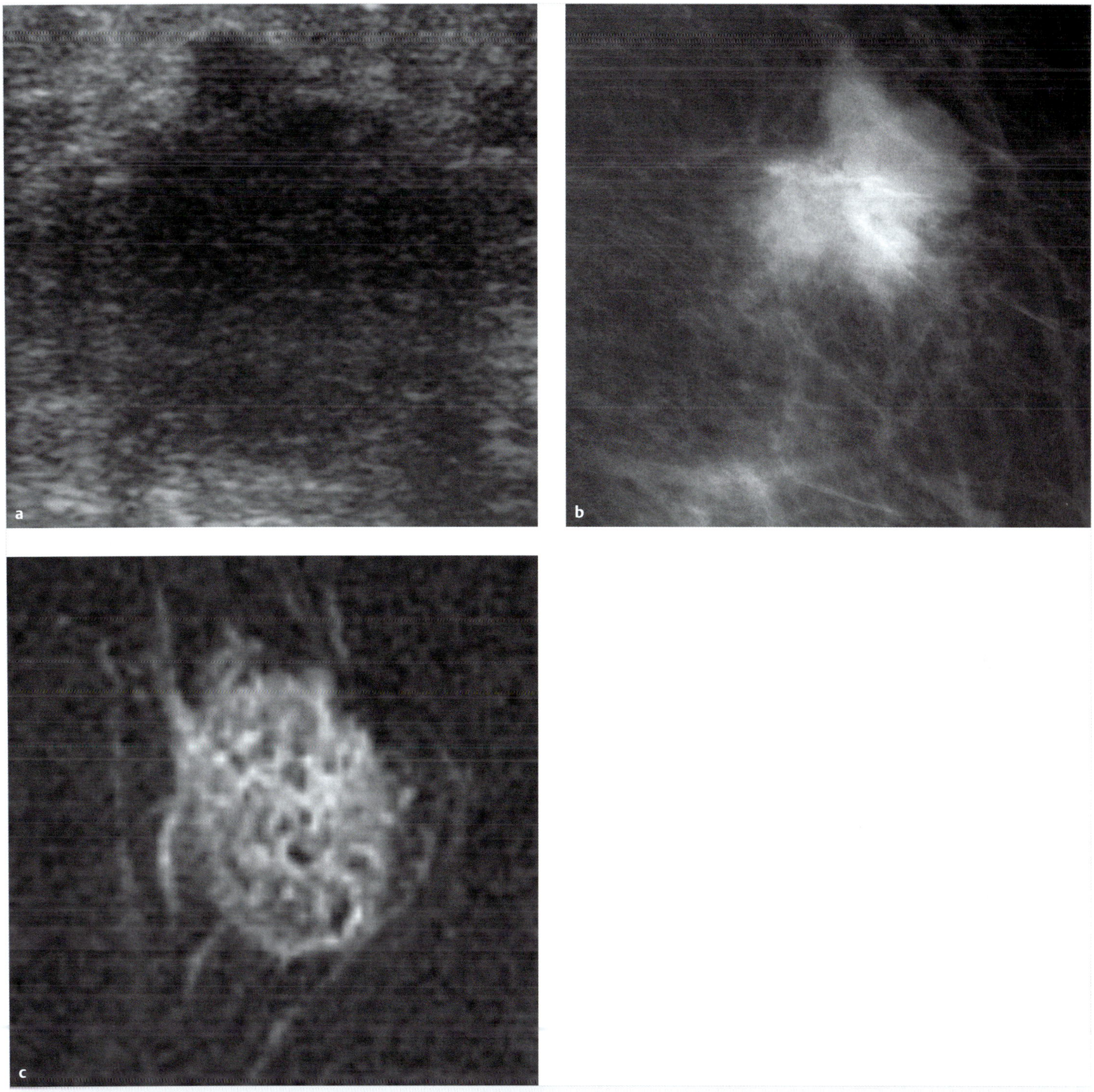

Fig. 8.34 Medullary carcinoma of the breast. (a) Ultrasound image. (b) Mammogram. (c) Contrast-enhanced breast MRI subtraction image.

► **Margins.** Microlobulated or indistinct.

► **Internal composition.** Proliferative tissue.

► **Clinical findings.** Small tumors are usually occult without correlative clinical findings. Larger tumors may be palpated as a circumscribed mass.

► **Imaging**

- *Sonography* (► Fig. 8.35a): hypoechogenic mass; intermediate posterior echo pattern, often displaying a mild acoustic enhancement.
- *Mammography* (► Fig. 8.35b): hyperdensity; very rarely associated with microcalcifications.
- *Breast MRI* (► Fig. 8.35c): mass:
  - T1W precontrast image: hypointense or isointense, nonspecific.
  - T2W: isointense, often hyperintense, nonspecific.
  - After administration of contrast medium: mild, or frequently very strong enhancement; rarely ring enhancement; occasional washout phenomenon.

Note

There are individual case reports in which a mucinous carcinoma showed no enhancement at all on the breast MRI.

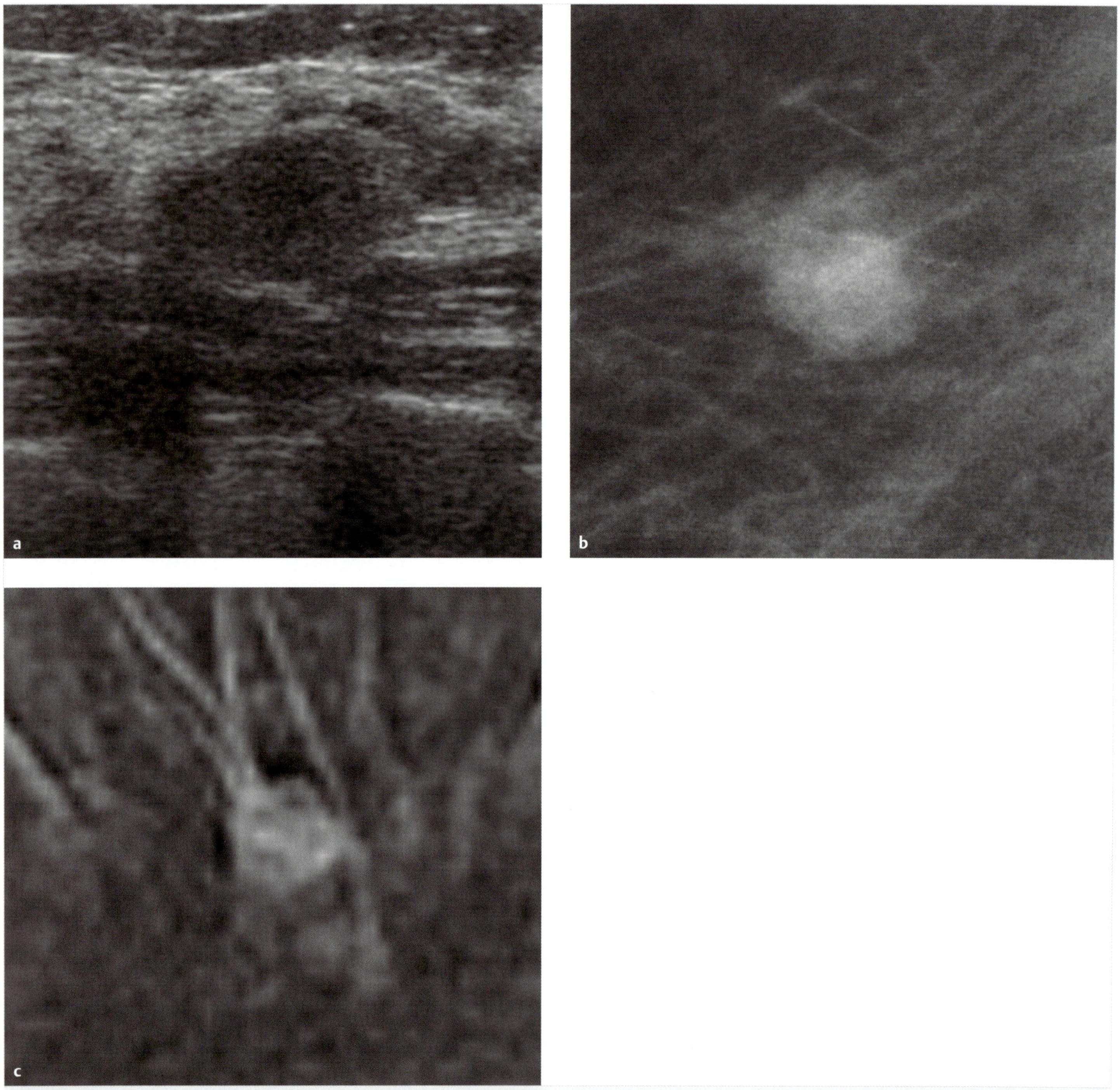

Fig. 8.35 Mucinous carcinoma of the breast. (a) Ultrasound image. **(b)** Mammogram. **(c)** Contrast-enhanced breast MRI subtraction image.

► **Differential diagnosis.** Fibroadenoma, adenoma, papilloma, other malignant breast tumors.

8.4.6 Invasive Papillary Carcinoma

An invasive papillary carcinoma (IPC) can arise from a benign papilloma; the lesions are characterized by a fingerlike (frond-forming) growth pattern.

► **Characteristics**

- *Incidence:* rare, less than 2% of invasive breast cancer.
- *Peak age:* 40 to 50 years.
- *EIC:* no concrete data available.
- *Multifocality:* no concrete data available.
- *Multicentricity:* no concrete data available.
- *Bilateral occurrence:* no concrete data available.
- *Prognosis:* dependent on tumor stage (pT, pN, pM), grade and receptor status.
- *Histological classification after biopsy:* B5b.

► **Implications for work-up/treatment.** Percutaneous biopsy for histological verification. Once the diagnosis is confirmed, the tumor is usually surgically excised.

► **Shape, position.** Lobulated; frequently retroareolar location.

► **Margins.** Indistinct.

▶ **Internal composition.** Proliferative tissue.

▶ **Clinical findings.** Small tumors are usually occult without correlative clinical findings. Larger tumors may be palpated as a circumscribed mass.

▶ **Imaging**

- *Sonography* (▶ Fig. 8.36a): isoechogenic to hypoechogenic mass; hyperechoic rim; nonspecific.
- *Mammography* (▶ Fig. 8.36b): hyperdensity; very rarely associated with microcalcifications; nonspecific.
- *Breast MRI* (▶ Fig. 8.36c): mass:
  - T1W precontrast image: hypointense or isointense, nonspecific.
  - T2W: isointense, often hyperintense, nonspecific.
  - After administration of contrast medium: mild to strong enhancement; rarely ring enhancement; occasional washout phenomenon.

Note

Imaging cannot reliably differentiate between a macropapilloma and an invasive papillary carcinoma.

▶ **Differential diagnosis.** Fibroadenoma, adenoma, papilloma, other malignant breast tumors.

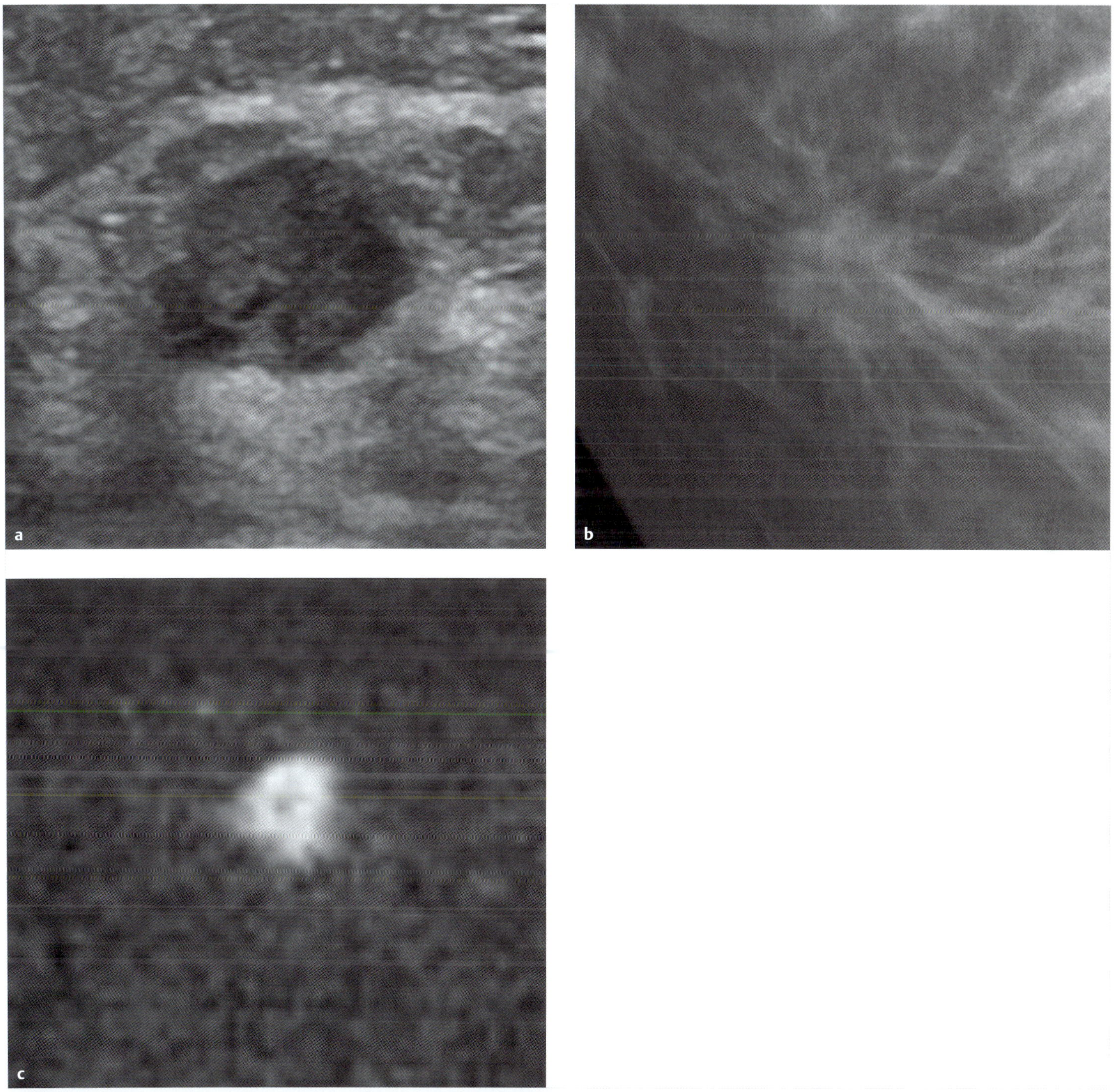

Fig. 8.36 Invasive papillary carcinoma. (a) Ultrasound image. **(b)** Mammogram. **(c)** Contrast-enhanced breast MRI subtraction image.

8.4.7 Sarcomas

Sarcoma of the breast comprises a heterogeneous group of highly malignant tumors originating in periductal or perilobular stroma. Included in this group are fibrous histiocytoma, angiosarcoma, and fibrosarcoma, as well as liposarcoma, leiomyosarcoma, and chondrosarcoma.

► **Characteristics**
- *Incidence:* very rare.
- *Peak age:* 30 to 40 years.
- *Bilateral occurrence:* very rare.
- *Prognosis:* dependent on the tumor entity, but essentially always bad.
- *Histological classification after biopsy:* B5d.

► **Implications for work-up/treatment.** Percutaneous biopsy for histological verification. Once the diagnosis is confirmed, surgical excision of the tumor is usually performed.

► **Shape.** Round or lobular.

► **Margins.** Indistinct.

► **Internal composition.** Proliferative tissue, inhomogeneous.

► **Clinical findings.** Small tumors are usually occult without correlative clinical findings. A larger tumor may be palpated as a circumscribed mass.

► **Imaging**
- *Sonography* (► Fig. 8.37a): isoechogenic to hypoechogenic mass; nonspecific.
- *Mammography* (► Fig. 8.37b): hyperdense; nonspecific.
- *Breast MRI* (► Fig. 8.37c): mass:
  - T1W precontrast image: hypointense or isointense, nonspecific.
  - T2W: isointense or hyperintense, nonspecific.
  - After administration of contrast medium: mild to (more frequently) strong enhancement; rarely ring enhancement; frequent washout phenomenon.

Note

Sarcomas of the breast are extraordinarily rare.

► **Differential diagnosis.** All focal benign or malignant solid breast lesions.

8.4.8 Triple-Negative Carcinoma

Triple-negative carcinoma refers to those breast cancers that are negative for the ER, PgR, and HER2/neu receptors.

► **Characteristics**
- *Incidence:* rare, more frequent in *BRCA1* mutation carriers.
- *Age peak:* 30 to 50 years.
- *EIC:* no concrete data available.
- *Multifocality:* no concrete data available.
- *Multicentricity:* no concrete data available.
- *Bilateral occurrence:* no concrete data available.
- *Prognosis:* rather poor, partly due to the lack of a specific antihormonal therapy option (receptor negative); frequently classified as grade 3.
- *Histological classification after biopsy:* B5b.

► **Implications for work-up/treatment.** Percutaneous biopsy for histological verification. Once the diagnosis is confirmed, the tumor is usually surgically excised.

► **Shape.** Round.

► **Margins.** Indistinct.

► **Internal composition.** Proliferative tissue; often displays central tumor necrosis.

► **Clinical findings.** Small tumors are usually occult without correlative clinical findings. A larger tumor may be palpated as a circumscribed mass.

► **Imaging**
- *Sonography* (► Fig. 8.38a): hypoechogenic mass; anechoic central fluid/necrosis.
- *Mammography* (► Fig. 8.38b): hyperdensity; rarely associated with microcalcifications.
- *Breast MRI* (► Fig. 8.38c): mass:
- T1W precontrast image: isointense; center frequently hypointense.
- T2W: isointense; center frequently hyperintense (fluid-equivalent).
- After administration of contrast medium: strong enhancement; frequent ring enhancement, occasionally with very narrow ring; occasional bright endotumoral septations; frequent washout phenomenon in proliferating tumor fraction.

Note

It is occasionally difficult to differentiate between a triple-negative carcinoma of the breast and a complex cyst. In case of doubt, it is advisable to perform a percutaneous ultrasound-guided core needle biopsy.

► **Differential diagnosis.** Inflammatory or complex cyst, abscess, phyllodes tumor, other types of malignant tumors.

8.4.9 Paget's Disease of the Nipple

Paget's disease is an intraepidermal DCIS with large round or oval tumor cells (Paget cells).

► **Characteristics**
- *Incidence:* rare, less than 2% of invasive breast cancer.
- *Peak age:* 40 to 60 years.
- *Associated intramammary DCIS:* > 60%.
- *Associated invasive carcinoma:* < 30%.
- *Bilateral occurrence:* no concrete data available.
- *Prognosis:* rather favorable; dependent on tumor stage (pT, pN, pM).

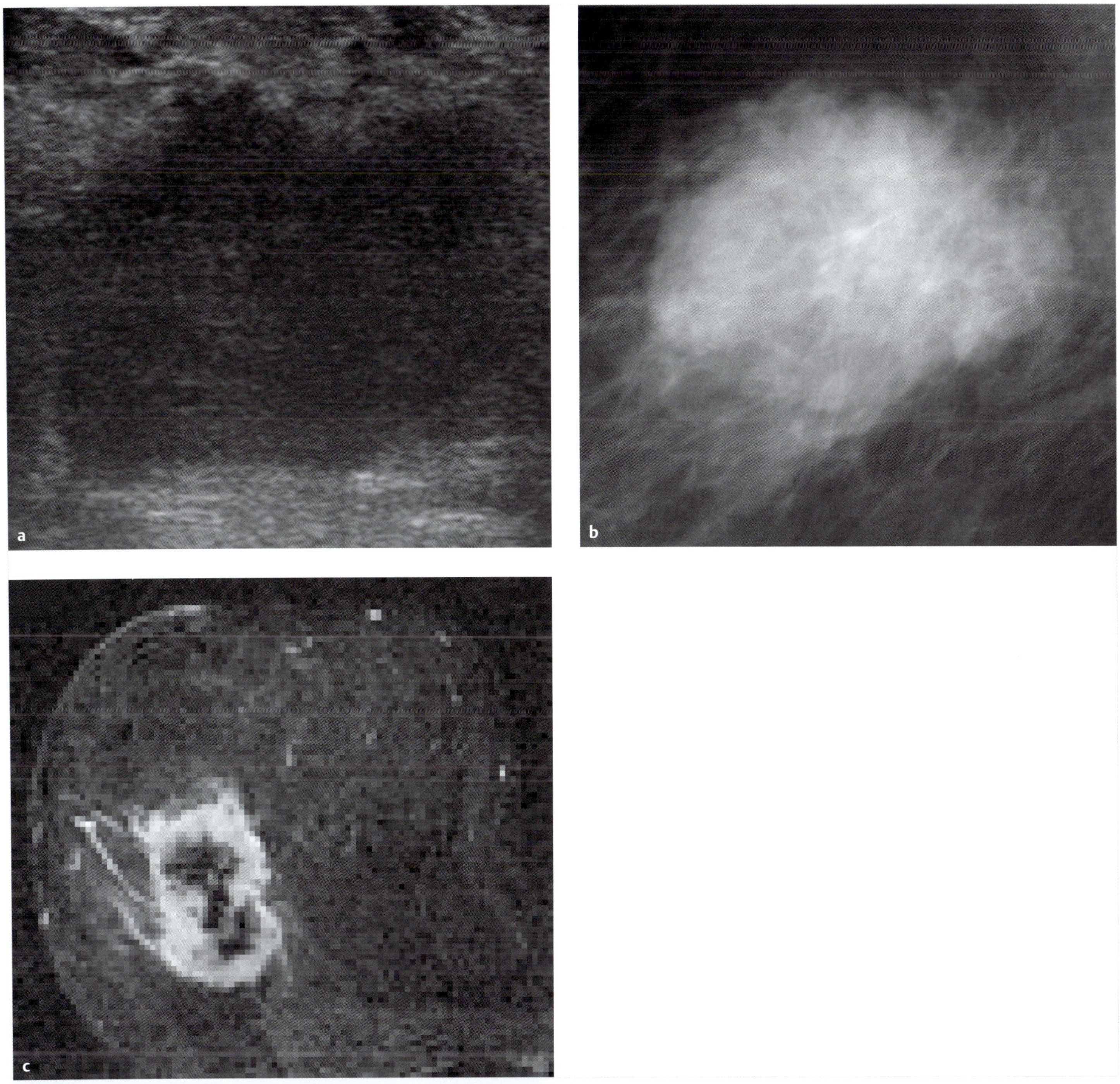

Fig. 8.37 Sarcoma. (a) Ultrasound image. (b) Mammogram. (c) Contrast-enhanced breast MRI subtraction image.

- *Histological classification after biopsy:* B5a.

▸ **Implications for work-up/treatment.** Exfoliative cytology; percutaneous biopsy (punch biopsy) for histological verification. Once the diagnosis is confirmed, the tumor is usually surgically excised.

▸ **Shape, distribution.** Flat (within nipple plateau); mass, segmental or diffuse, when there is breast involvement.

▸ **Margins.** Nipple/areola epidermis; indistinct margins of the intramammary manifestation when the breast is involved.

▸ **Internal composition.** Proliferative tissue.

▸ **Clinical findings.** Eczematous changes of the nipple and, if involved, of the areola. Suspicious palpable findings may be present in the case of an intramammary tumor.

▸ **Imaging**

- *Sonography* (▸ Fig. 8.39a): typically occult, even in the case of intramammary DCIS; typical findings suspicious of malignancy in the case of intramammary invasive carcinoma.
- *Mammography* (▸ Fig. 8.39b): typically occult; intramammary DCIS frequently displays microcalcifications; typical findings suspicious of malignancy in the case of additional invasive carcinoma.
- *Breast MRI* (▸ Fig. 8.39c): usually occult, no features that differentiate from physiological enhancement of the nipple.
  - T1W precontrast image: hypointense or isointense, nonspecific.

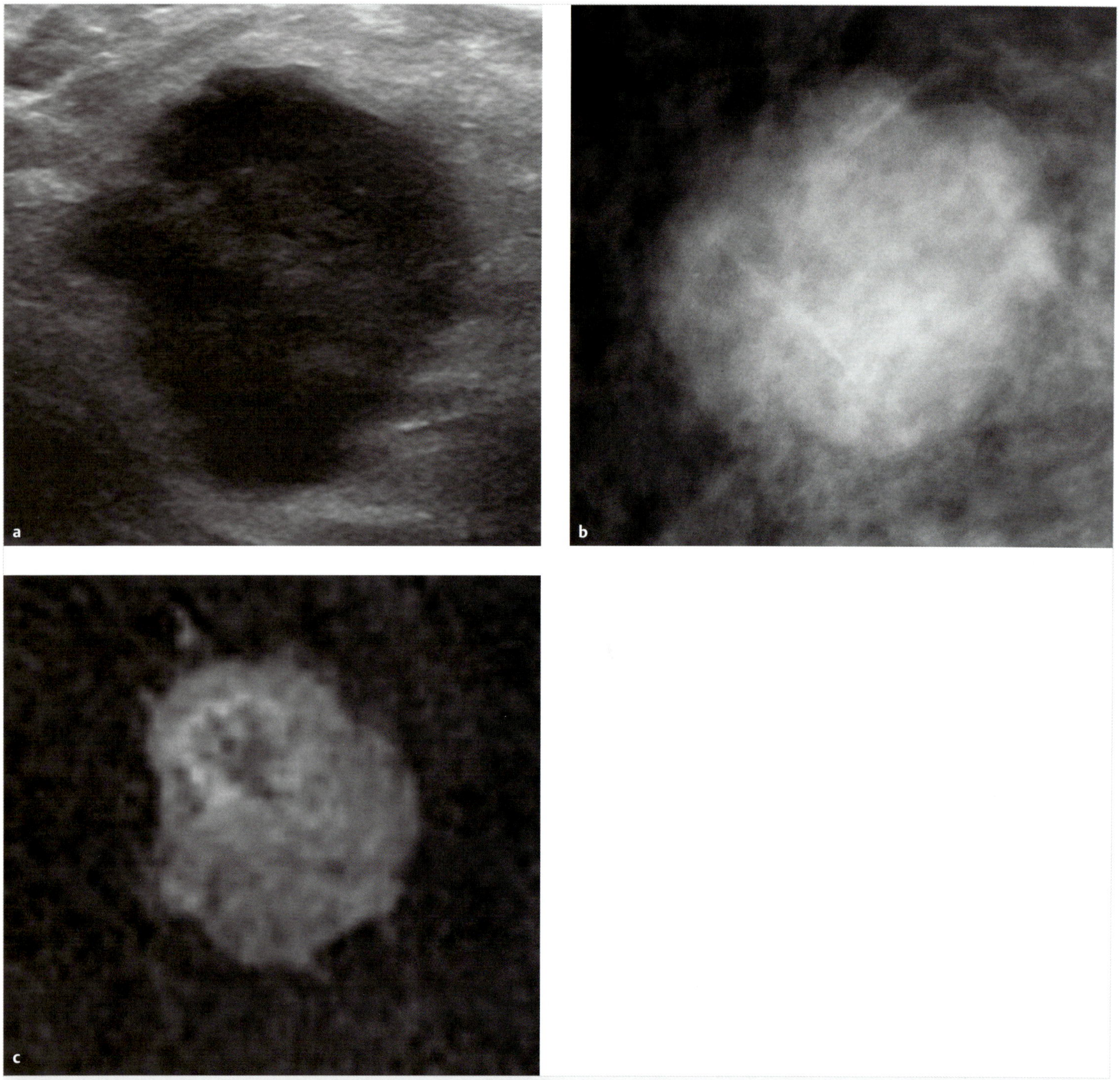

Fig. 8.38 Triple negative carcinoma of the breast. (a) Ultrasound image. **(b)** Mammogram. **(c)** Contrast-enhanced breast MRI subtraction image.

- T2W: isointense or hyperintense, nonspecific.
- After administration of contrast medium: in the presence of DCIS, a DCIS-typical enhancement pattern; in the presence of an intramammary invasive tumor, an enhancement pattern typical of a carcinoma.

Note

Paget's disease is primarily a clinical diagnosis.

► **Differential diagnosis.** Eczema of the nipple, diseases of the skin of other origin.

8.4.10 Inflammatory Carcinoma

Inflammatory breast cancer is characterized by the diffuse spread of a poorly differentiated breast cancer. Clinical signs of inflammation are frequently present.

► **Characteristics**

- *Incidence:* rare, less than 2% of all breast cancer.
- *Peak age:* 40 to 55 years.
- *Bilateral occurrence*: frequent (up to 30%).
- *Prognosis:* very unfavorable.
- *Histological classification after biopsy:* B5b.

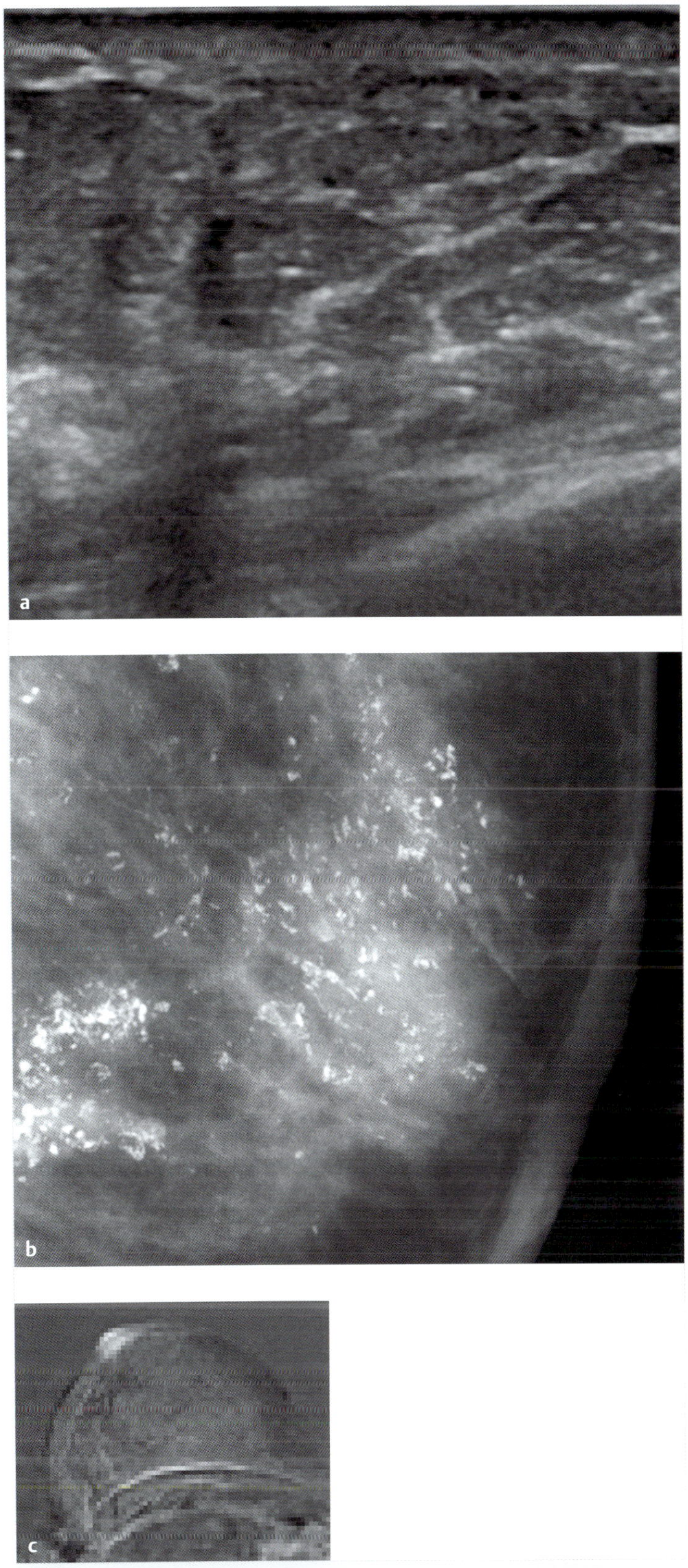

Fig. 8.39 **Paget's disease.** **(a)** Ultrasound image of extensive DCIS associated with Paget's disease. **(b)** Mammogram of extensive DCIS associated with Paget's disease. **(c)** Paget's disease without intramammary involvement in a contrast-enhanced breast MRI subtraction image.

▶ **Implications for work-up/treatment.** Confirmation of the diagnosis by percutaneous biopsy or primary open surgery. Once the diagnosis is established, primary treatment is with chemotherapy (neoadjuvant concept). Secondary mastectomy.

▶ **Shape, extension.** Diffuse, segmental.

▶ **Margins.** None.

▶ **Internal composition.** Proliferative cells, edema, lymph.

▶ **Clinical findings.** Swelling, erythema, hyperthermia, occasional pain. Generalized increased consistency. Coarsely pitted skin (peau d'orange) to cancer en cuirasse. Frequent metastases in locoregional lymph nodes.

▶ **Imaging**

- *Sonography* (▶ Fig. 8.40a): hyperechogenicity of the subcutaneous adipose tissue and parenchyma (interstitial accumulation of fluid); *asymmetry* in side-to-side comparison; frequent locoregional lymphadenopathy.
- *Mammography* (▶ Fig. 8.40b): hyperdensity of the parenchyma (interstitial accumulation of fluid); *asymmetry* in side-to-side comparison; rarely associated with microcalcifications; an intramammary index tumor is present in 30% of cases; frequent locoregional lymphadenopathy.
- *Breast MRI* (▶ Fig. 8.40c): non-space-occupying changes (nonmasslike lesion):
  - T1W precontrast image: skin thickening (compare side-to-side).
  - T2W: hyperintensity of the parenchyma (lymphangiosis); fluid-filled lymph spaces; skin thickening.
  - After administration of contrast medium: enhancement of the entire skin circumference; frequently no or only mild intramammary enhancement; in the case of an index tumor, classical criteria of an invasive carcinoma of the breast.

Note

An inflammatory carcinoma without the clinical signs of "inflammation" is referred to as "diffuse carcinoma of the breast."

▶ **Differential diagnosis.** Nonpuerperal mastitis.

8.4.11 Systemic Diseases Involving the Breast

In addition to primary breast conditions, the breast can also be a target organ of systemic diseases. Here one must consider intramammary metastases (e.g., melanoma), lymphatic diseases (e.g., Hodgkin's disease, non-Hodgkin's lymphoma), or plasmacytoma.

▶ **Characteristics**

- *Incidence:* very rare.
- *Peak age:* dependent on the respective systemic disease.
- *Bilateral occurrence:* very frequent.
- *Prognosis:* dependent on the respective systemic disease.
- *Histological classification after biopsy:* B5d.

▶ **Implications for work-up/treatment.** According to the respective systemic disease, work-up with percutaneous biopsy for histological verification, if appropriate. Once the diagnosis is established, treatment is as recommended for the systemic disease.

8

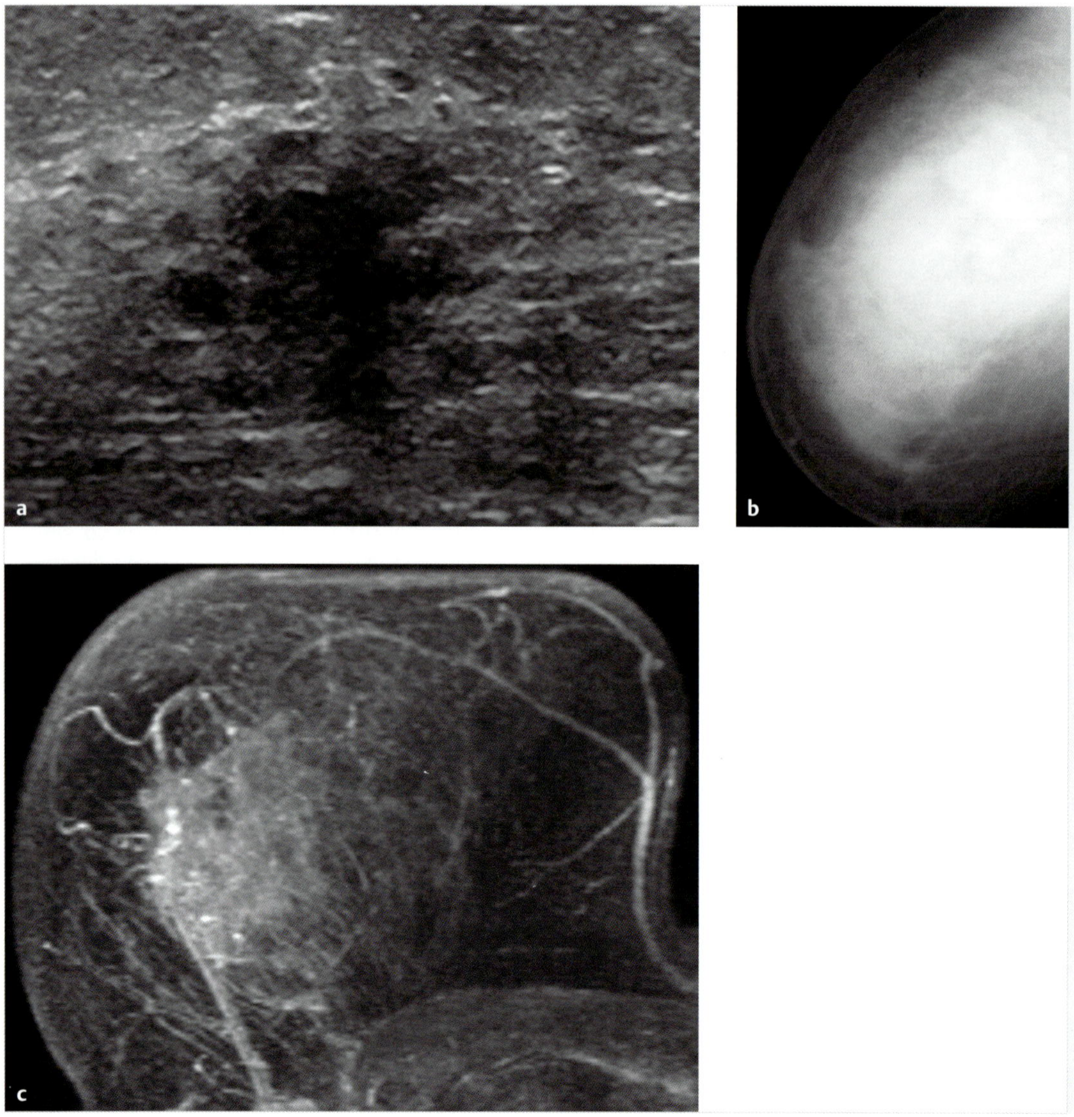

Fig. 8.40 Inflammatory carcinoma of the breast. (a) Ultrasound image. **(b)** Mammogram. **(c)** Contrast-enhanced breast MRI, maximum intensity projection.

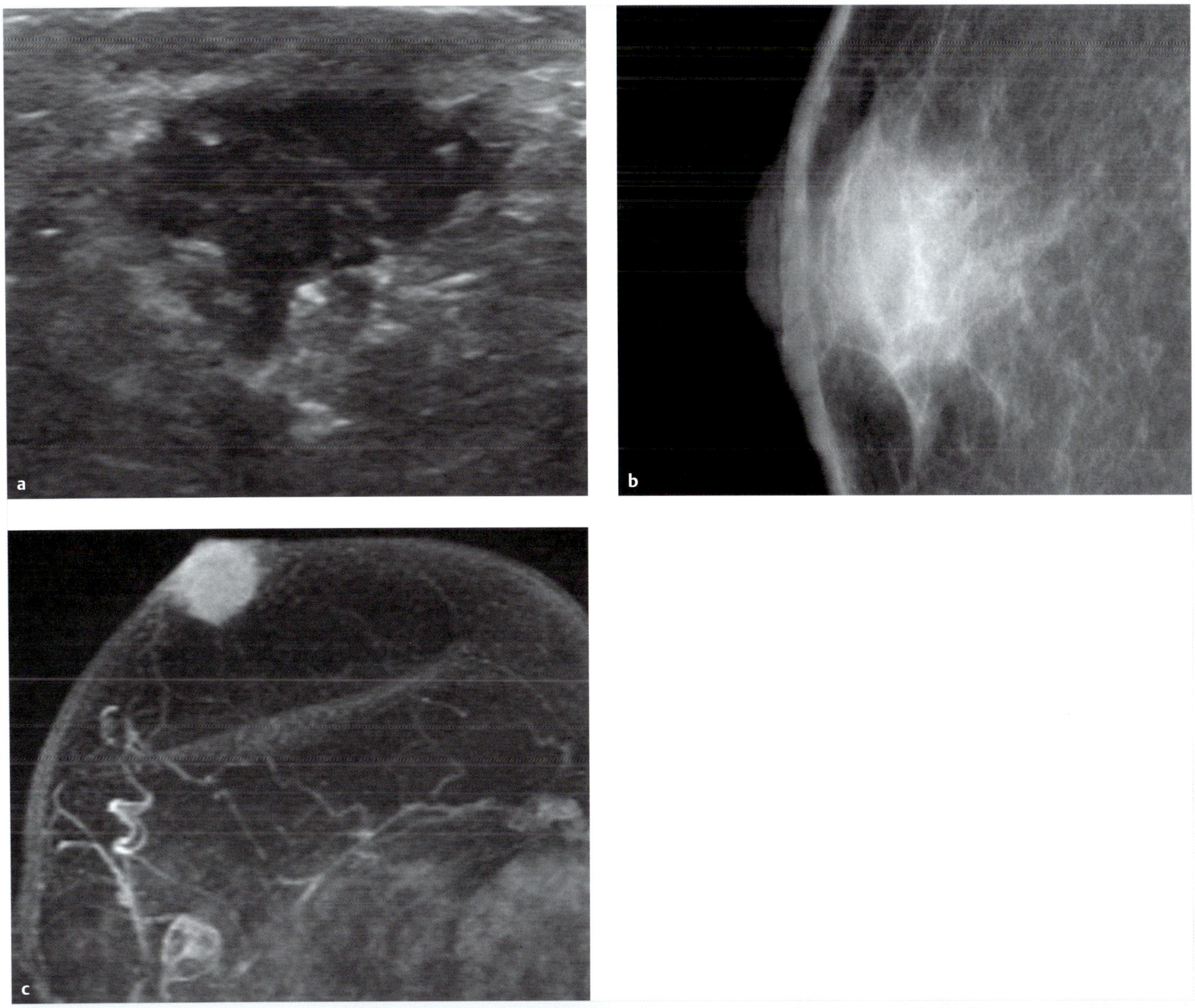

Fig. 8.41 Mammary manifestation of Hodgkin's disease. Manifestation of Hodgkin's disease in the retroareolar region of the breast. **(a)** Ultrasound image. **(b)** Mammogram. **(c)** Contrast-enhanced breast MRI, maximum intensity projection.

► **Imaging**

- *Lymphatic diseases* (► Fig. 8.41): an occurrence of Hodgkin's disease or non-Hodgkin's lymphoma in the breast commonly manifests itself as edematous fluid in the interstitial space, swelling, skin thickening and axillary lymphadenopathy.
- *Plasmacytoma:* intramammary manifestations of a plasmacytoma take the form of round, circumscribed mass lesions with the typical hemodynamic criteria of a malignant tumor.
- *Hematogenous metastasis:* in all the imaging modalities, intramammary metastases originating from extramammary organ tumors typically appear in the form of multiple, often bilateral, circumscribed, round mass lesions.

► **Differential diagnosis.** If the underlying systemic disease is known, there is usually no differential diagnosis to consider.

9 Breast Intervention

F. Baum

9.1 Biopsy

9.1.1 Objective of Percutaneous Tissue Sampling

In the year 2000 in Europe, approximately 80% of open diagnostic excisional breast biopsies yielded benign results on histological examination. Since then, concrete guidelines have been developed that recommend initial percutaneous biopsies in the diagnostic work-up for ambiguous breast findings in order to reduce the number of unnecessary open breast biopsies. As well as avoiding unnecessary scars, women are spared unnecessary hospital stays and potential surgical risks.

Surgical intervention can usually be avoided if there is a history of benign results from the percutaneous tissue sampling. If the results indicate a lesion with histologically ambiguous biological potential, the further action can be determined in an interdisciplinary conference. When the findings are malignant, the subsequent treatment—usually primary surgery or a neoadjuvant treatment plan—has to be decided, and, in the case of surgery, the extent of the surgical intervention should be clearly defined preoperatively and discussed in detail with the patient.

Take Home Points

Image-Guided Minimally Invasive Biopsy

Image-guided tissue sampling for histopathological confirmation of the diagnosis and for treatment planning should be performed in cases with

- Mammographic findings of MX BI-RADS 4/5 and/or
- Sonographic findings of US BI-RADS 4/5 and/or
- MRI findings of MR BI-RADS 4/5.

The sample required to clarify the histological diagnosis should be obtained by core needle biopsy, vacuum-assisted biopsy, or open biopsy. The core needle biopsy and the vacuum-assisted biopsy can be done with mammographic, ultrasound, or MRI guidance.

The image-guided tissue sampling should be done with the imaging method that best visualized the findings and the procedure that is least invasive.

Primary open diagnostic excisional biopsies should be done only in exceptional cases in which an image-guided intervention is not possible or is considered to be high risk.

9.1.2 Percutaneous Tissue Sampling Equipment and Implementation

There are currently several puncture techniques available for carrying out ambulatory percutaneous breast tissue sampling. The main ones are fine needle aspiration, core needle biopsy, and vacuum-assisted biopsy.

Fine Needle Aspiration

In the second half of the twentieth century, fine needle aspiration was at first the only method and later the predominant one used for breast biopsies. Today it has largely been replaced by more complex mechanical biopsy systems.

The goal of fine needle aspiration is to extract fluid or to recover individual cells from a tissue complex for cytologic diagnosis. The lesion is punctured and a vacuum is manually created to aspirate the fluid or diagnostic material. Because diagnostic specimens acquired in this way are cytologically not assessable in up to 50% of cases, this minimally invasive technique has increasingly been abandoned.

However, fine needle puncture still finds use in the diagnostic work-up of ambiguous or painful cysts. The cyst puncture is carried out under sonographic guidance using a special syringe holder (Cameco syringe pistol, Morton Medical). This holder enables the operator to guide the syringe and simultaneously pull back on the plunger to aspirate with one hand while using the other hand to guide the transducer of the ultrasound device (▶ Fig. 9.1). With correct placement of the needle tip, the intracystic fluid is aspirated and the cyst collapses (▶ Fig. 9.2). Thus, for this indication, fine needle aspiration is not only diagnostic but also therapeutic. The aspirate obtained is centrifuged and is then distributed onto a specimen slide, stained, and evaluated microscopically.

Take Home Point

Image-Guided Minimally Invasive Biopsy

Fine needle aspiration should not be used as a standard method to confirm the diagnosis of solid breast tumors.

Another application of the fine needle technique is the diagnostic work-up of ambiguous lymph nodes. Cytology provides very reliable differentiation between normal lymphocytes and metastatic cells from a breast carcinoma. In addition, fine needle puncture carries less risk than core needle biopsy, especially for punctures in the axilla with proximity to arterial blood vessels. To carry out fine needle biopsy, the needle is introduced into the lymph node

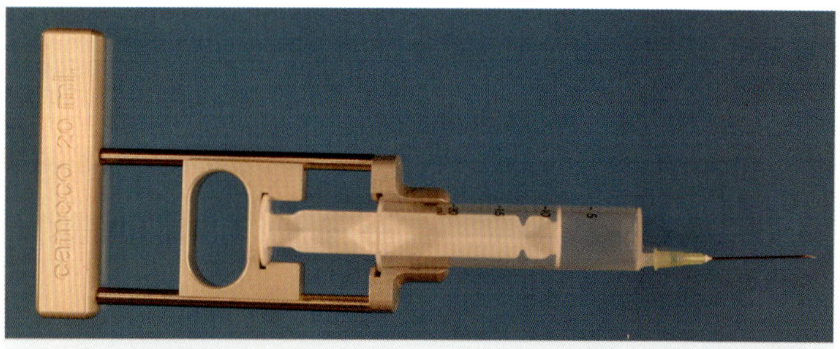

Fig. 9.1 Fine needle puncture. Cameco syringe pistol with syringe and puncture needle (Morton Medical, Cirencester, UK); puncture needle with Ciba beveled tip.

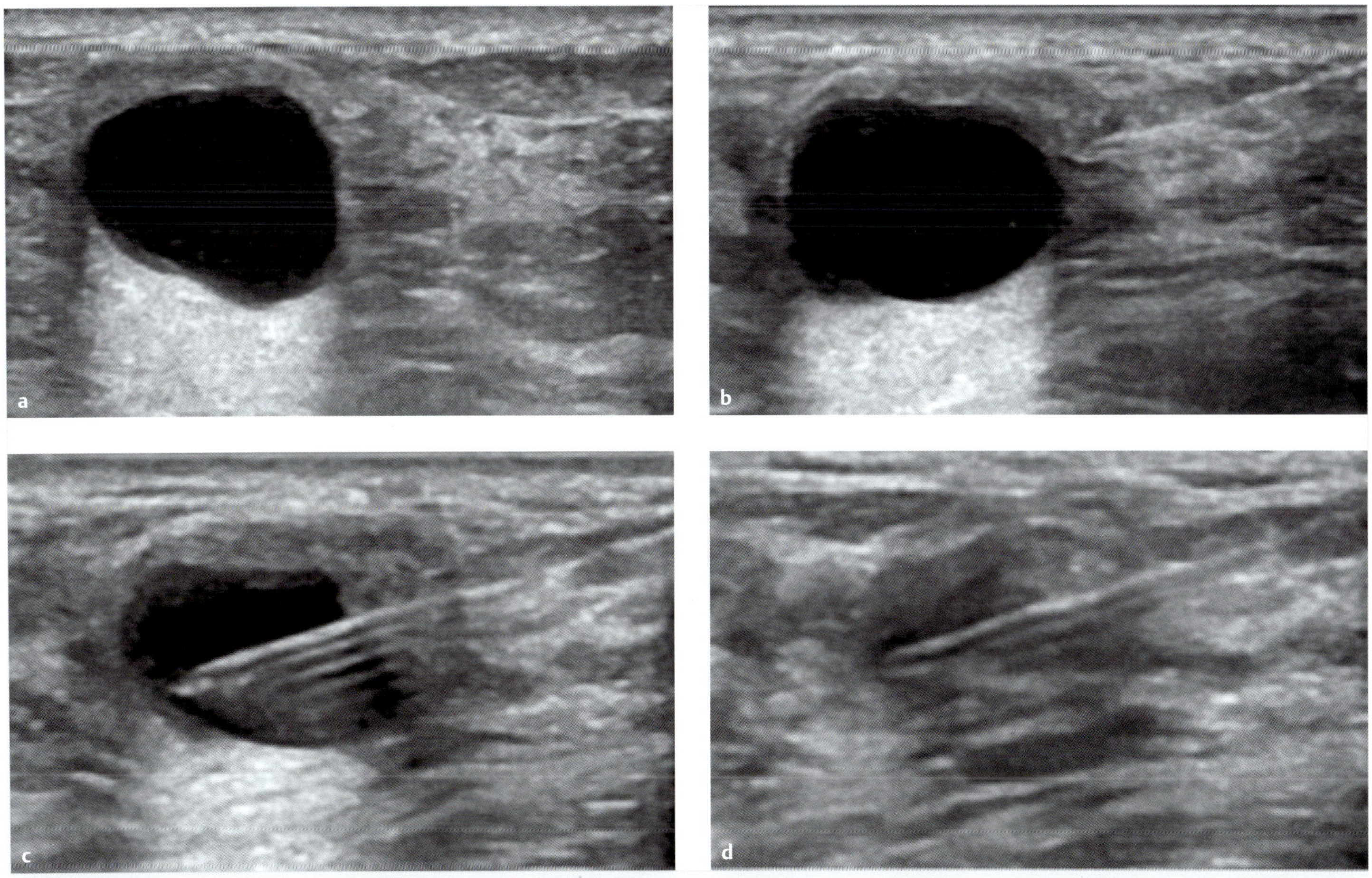

Fig. 9.2 Ultrasound-guided cyst puncture. (a) Anechoic cyst with pronounced posterior acoustic enhancement. **(b)** Acoustic artifact of puncture needle introduced from the right edge of the image with visualization of the needle point before the right cyst wall. **(c)** Puncture needle within the cyst. **(d)** Collapsing cyst during aspiration.

under sonographic guidance. While simultaneously aspirating, the examiner repeatedly withdraws and advances the needle within the lymph node in a fan-like pattern (▶ Fig. 9.3). To end the process, the vacuum is slowly released and the needle is withdrawn. The material in the needle and the Luer adaptor of the syringe is then distributed onto a specimen slide and delivered for cytologic appraisal.

Core Needle Biopsy

In the late 1980s and 1990s, the high-velocity core needle biopsy was increasingly established as the standard technique for extracting tissue samples in breast diagnostics. By this means, entire cell aggregates from an ambiguous lesion can be harvested and thus yield histological diagnoses.

The biopsy needle used in the procedure has two components: an inner puncture needle and an outer hollow-bore needle. The solid puncture needle contains an approximately 2 cm long recess (notch) near the tip. When a tissue sample is being taken, this needle first penetrates into the lesion to be investigated. Then the hollow outer needle advances over the puncture needle and cuts the tissue lying within the recess from the tissue complex. Next, the biopsy needle is withdrawn from the breast to recover the tissue sample. After the spring mechanism of the needle holder has been reset, second, third, and fourth biopsies can be taken. Because it is recommended to take four tissue samples, a coaxial needle is

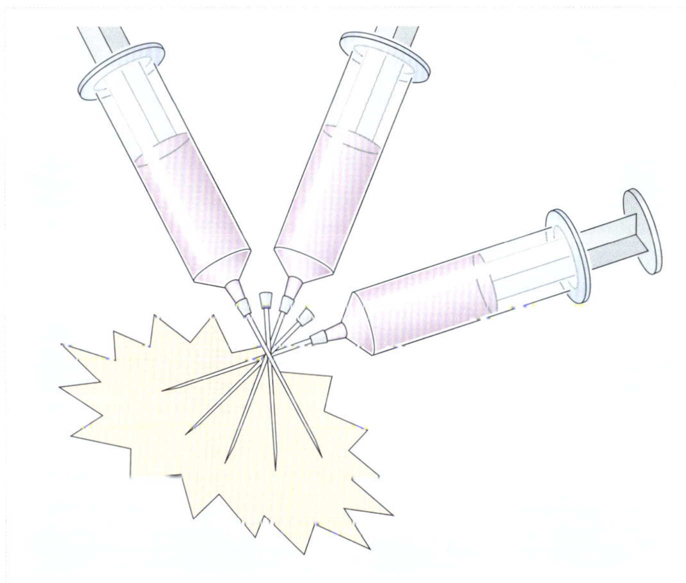

Fig. 9.3 Lymph node puncture. While simultaneously aspirating, the examiner repeatedly advances and withdraws the needle within the ambiguous lymph node in a fanlike pattern.

commonly employed. This is positioned immediately before the lesion to be investigated and left in place in the breast after the first tissue sampling. Repeated piercings of the mammary gland tissue are thus avoided when taking multiple samples. The recommended diameter of the biopsy needle is 14G (ca. 2 mm) (▶ Fig. 9.4).

Take Home Point

Image-Guided Minimally Invasive Biopsy
Percutaneous core needle biopsy is preferably performed under ultrasound guidance using a needle of 14G or larger to obtain three or more representative samples.

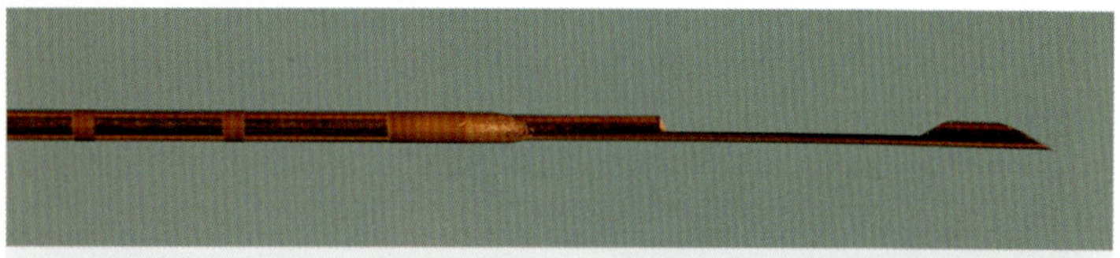

Fig. 9.4 Core biopsy needle. Needle with sampling notch (C.R. Bard GmbH, Karlsruhe, Germany).

Typically, the high-velocity core needle biopsy is performed under sonographic guidance (▶ Fig. 9.5), allowing the procedure to be monitored in real time and the successful needle placement to be documented. In rare cases, this technique is also performed under mammographic imaging guidance with a stereotactic table in the diagnostic work-up of mammographic densities or mass lesions. An ultrasound-guided core needle biopsy can also be done to sample ambiguous MRI findings if, for example, the lesion in question is extensive and diffuse and, correspondingly, a sufficiently high accuracy of the ultrasound-guided biopsy can be assumed.

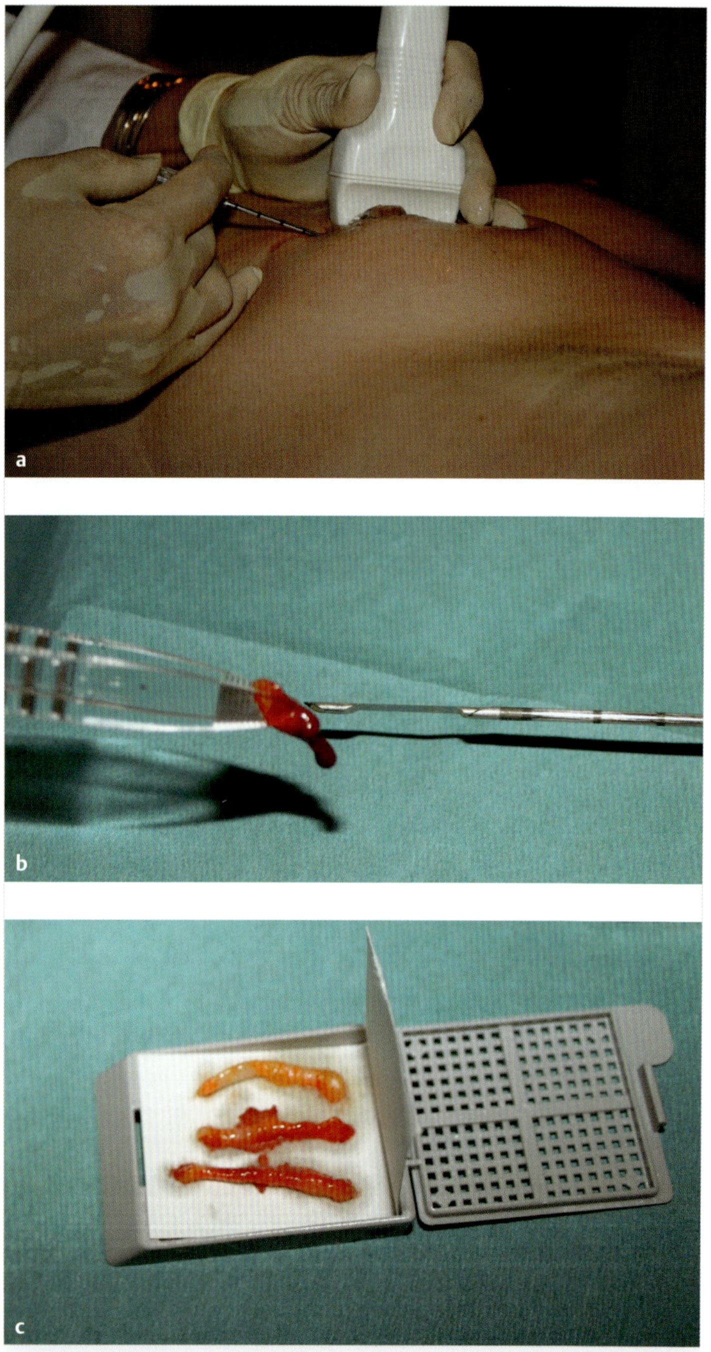

Fig. 9.5 Ultrasound-guided core needle biopsy. (a) Ultrasound transducer with coaxial needle, prepared for puncture. **(b)** Core needle biopsy showing tissue notch and harvested biopsy sample. **(c)** Three tissue samples in a histological specimen cassette (Sanowa, Leimen, Germany).

9

Vacuum-Assisted Biopsy

The vacuum-assisted biopsy technique was developed to improve the limited success rate of the high-velocity core needle biopsy in the diagnostic work-up of clustered microcalcifications. After the biopsy probe has been positioned under imaging guidance, a vacuum is applied to draw tissue from the target lesion into a laterally located "sampling window" near the tip of the needle. In a second step, a rotating cutting device then cuts this tissue from the tissue complex. The sampling window of the biopsy needle can be rotated 360° around the needle axis (▶ Fig. 9.6). In this manner, multiple contiguous cores of tissue can be obtained for histological examination. Depending on the diameter of the biopsy probe, the extraction of about 12 tissue samples (for a 10G probe) is recommended (▶ Fig. 9.7), yielding a volume of approximately 1 cm^2 (roughly equivalent to that of a sugar cube). Because this technique extracts more extensive tissue, it features a higher success rate than core needle biopsy, especially for the diagnostic work-up of microcalcifications, for which the core needle biopsy often yields nonrepresentative tissue samples, termed a "mismatch."

To extract sufficiently representative tissue samples for histological examination and to accommodate the rotating cutting device, vacuum-assisted biopsy probes are larger in diameter than core needle biopsy needles. Vacuum-assisted biopsy probes are usually 8 to 11G (ca. 5 mm to 2.5 mm).

Take Home Point

For a vacuum-assisted biopsy, it is recommended that 12 or more samples be obtained using a 10G probe. When using other calibers (between 8G and 11G), the number of samples taken should be adjusted to obtain an equivalent sample volume of ca. 1 cm^2.

The harvested tissue samples are recovered from the breast in various ways depending on the particular vacuum-assisted biopsy system. Some systems use a collection receptacle, from which all the tissue samples can be extracted at the end of the procedure. Here the biopsy probe remains in the breast during harvesting. In other systems, each tissue sample is removed

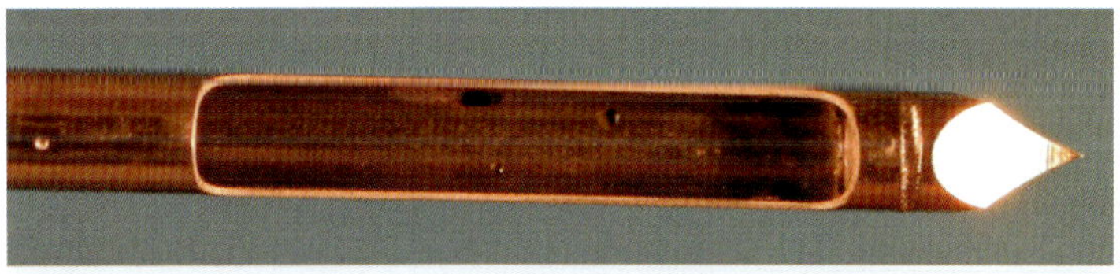

Fig. 9.6 Vacuum-assisted biopsy needle. Needle with lateral sampling window (C.R. Bard GmbH, Karlsruhe, Germany).

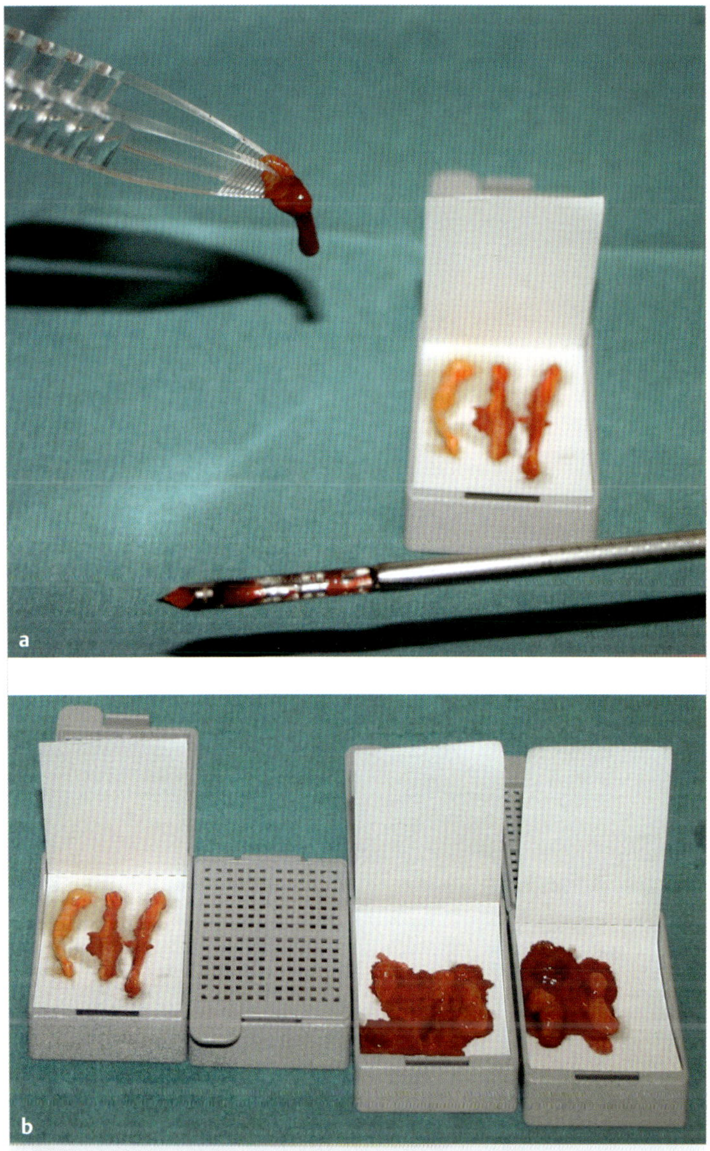

Fig. 9.7 Vacuum-assisted biopsy. (a) Vacuum-assisted biopsy needle with extracted biopsy sample. **(b)** Three tissue samples in each of three histological specimen cassettes (Sanowa, Leimen, Germany).

individually from the biopsy unit, requiring that the probe be completely removed from the breast for retrieval of each biopsy. In such cases, a coaxial guide needle maintains the path to the biopsy target position.

Several different vacuum-assisted biopsy systems are available that are compatible for use with MRI as well as for stereotactic use (▸ Fig. 9.8).

Punch Biopsy

Punch biopsy is used in the diagnostic work-up of ambiguous changes in the skin. For this, a small hollow instrument with extremely sharp metal cutting edges is placed over the lesion in question and pressed down through the skin layers after a local anesthetic has been applied (▸ Fig. 9.9). A cylinder of skin about 3 mm in diameter with about 1 cm of underlying cutaneous and subcutaneous tissue structures is recovered. The tissue cylinder can be removed from the biopsy sheath with forceps. Indications for this procedure in the breast are ambiguous inflammatory or pagetoid changes in the skin (▸ Fig. 9.10).

9.1.3 Interventional Imaging

Ultrasound-Guided Interventions

Over the last two decades, ultrasound-guided fine needle aspiration has been almost completely superseded by ultrasound-guided core needle biopsy. Sonography is the only imaging method that allows a biopsy procedure to be both monitored and documented in real time. Furthermore, ultrasound-guided interventions are simple, uncomplicated procedures that can be promptly implemented without major expense. For these reasons, ultrasound-guided core needle biopsy is by far the most common method used for the minimally invasive, histological diagnostic work-up of ambiguous findings.

For quality assurance purposes, it is advisable to provide imaging documentation of the interventional procedure in the form of an image showing the needle's position before it enters the lesion (pre-fire) and after it has entered the lesion (post-fire; in two planes; ▸ Fig. 9.11).

In rare cases, sonographic imaging is also used in conjunction with therapeutic interventions. These range from the aspiration of painful cysts to the minimally invasive removal of small mass lesions. The latter applies especially to small papillomas, the resection of which is recommended because of an increased cancer risk. Vacuum-assisted biopsy is employed in therapeutic removal of tissue and can be used to completely remove lesions with a diameter up to about 1.5 cm.

Mammography-Guided Interventions (Stereotactic)

Interventional techniques utilizing mammographic imaging for guidance are termed "stereotactic interventions." Because a mammographic image depicts only two planes, using a stereotactic table the exact location of a target lesion in the third plane must be determined by mathematical computation. The depth of a lesion can be calculated using supplemental images acquired at defined projection angles (stereo pairs). Stereotactic breast intervention systems are available for examination of patients in either the sitting or the prone position (▸ Fig. 9.12).

Once the coordinates of the lesion are known, the needle is positioned with computer assistance. During the intervention, the correct needle position is documented before, during, and after the biopsy procedure.

The majority of lesions that are stereotactically evaluated are microcalcifications, since these are reliably depicted only by mammography. Vacuum-assisted biopsy is the biopsy method of choice here, with a success rate between 98 and 100%. Representative sampling can and must be demonstrated by positive evidence of calcifications in the specimen radiograph. Before termination of the procedure, a postbiopsy image should be

9

Fig. 9.8 Vacuum-assisted biopsy systems. (a) Vacora hand-held unit (C.R. Bard GmbH, Karlsruhe, Germany). **(b)** ATEC (Hologic, Bedford, MA, USA). **(c)** Mammotome (Mammotome, Norderstedt, Germany). **(d)** SenorX (C.R. Bard GmbH, Karlsruhe, Germany).

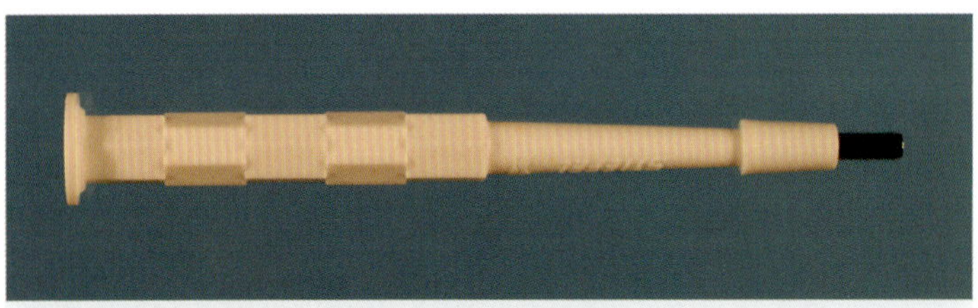

Fig. 9.9 Punch biopsy needle (Stiefel, Wächtersbach, Germany).

obtained to check for microcalcifications remaining in the biopsy area. If none remain, it is recommended that a marker clip is placed in the biopsy cavity. If histology reveals that surgical excision is necessary, the presence of the marker clip enables an accurate and reliable preoperative hookwire localization of the biopsy cavity (▸ Fig. 9.13).

Take Home Point

Image-Guided Minimally Invasive Biopsy

Stereotactic vacuum-assisted biopsy is the standard method of choice for the diagnostic work-up of microcalcifications without an accompanying mass lesion.

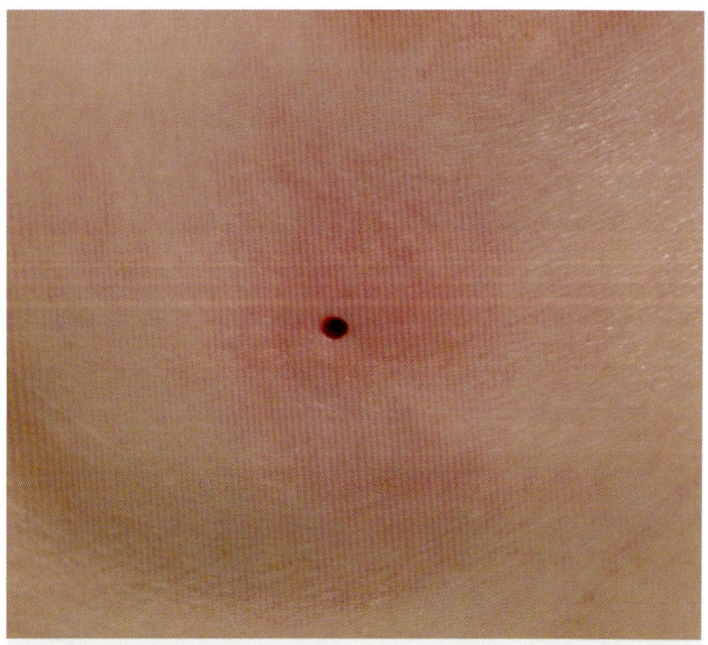

Fig. 9.10 Punch biopsy. Status post tissue sampling.

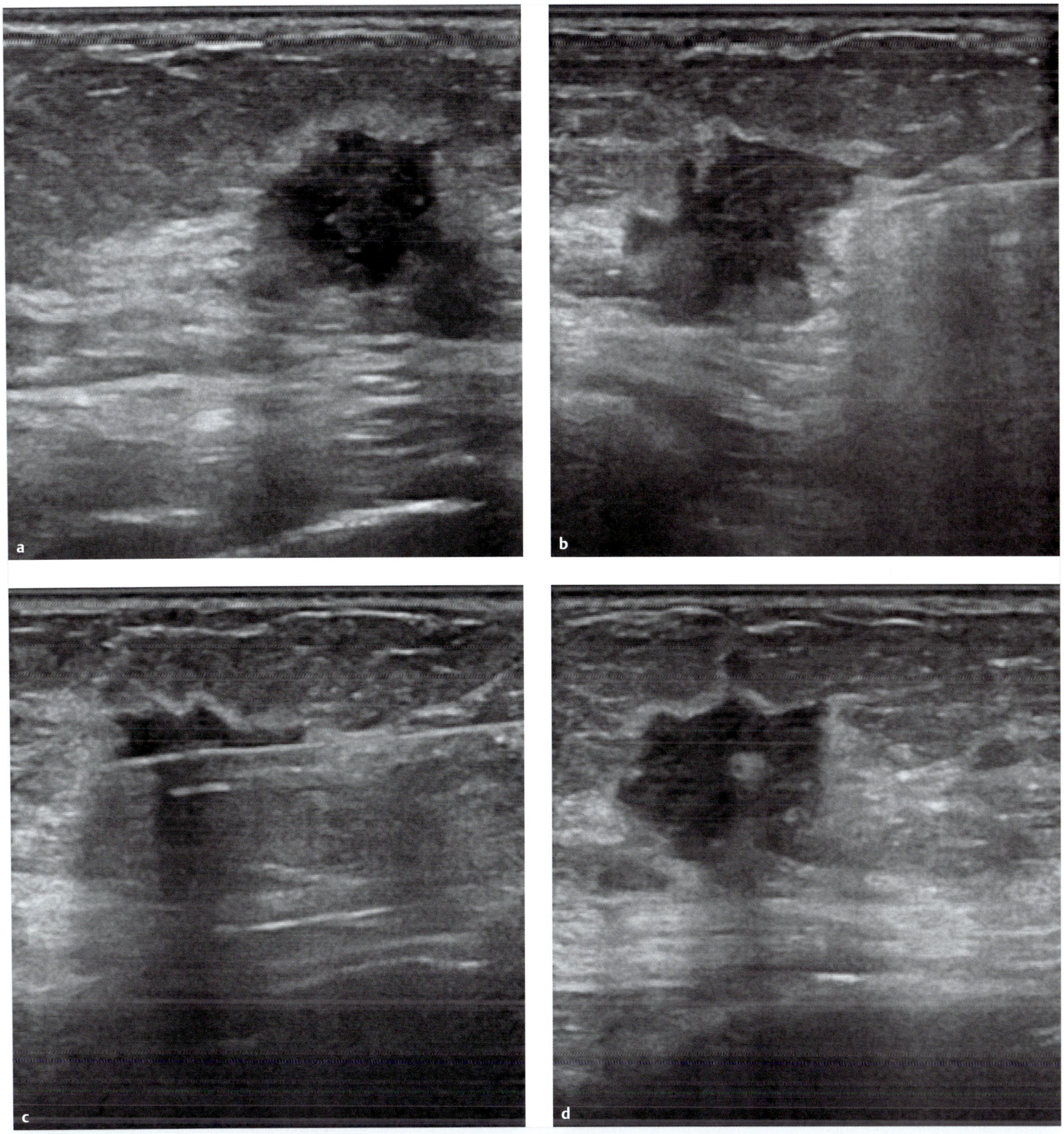

Fig. 9.11 Documentation of the ultrasound-guided core needle biopsy. **(a)** Hypoechoic mass with echogenic rim. **(b)** Acoustic artifact of the puncture needle introduced from the right edge of the image with visualization of the needle point before entering the mass (pre-fire). **(c)** Post-fire documentation of the puncture needle within the mass in the longitudinal axis. **(d)** Post-fire documentation of the puncture needle within the mass in the transverse axis.

Rarely, a mammographic density is also evaluated by stereotactic biopsy if it cannot be localized in the second imaging view. The method is used even more rarely in the diagnostic work-up of mass lesions that are visible in both mammography planes, because such lesions almost always have a sonographic correlate, in which case they should undergo a primary ultrasound-guided core-needle biopsy instead. If there is doubt about the correspondence between the mammographic and sonographic findings, then stereotactic tissue sampling may be undertaken, using high-velocity core needle biopsy or vacuum-assisted biopsy. If there are no associated calcifications, the specimen radiograph can be waived.

MR-Guided Interventions

The sensitivity of breast MRI for carcinoma of the breast is higher than that of mammography or ultrasound and this frequently leads to the detection of suspicious lesions that are visible

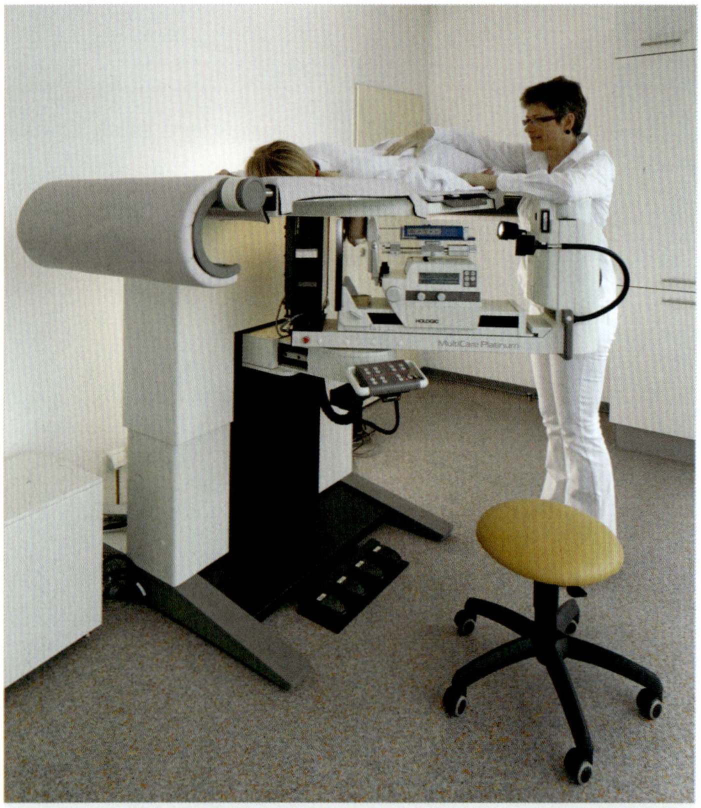

Fig. 9.12 Stereotactic vacuum-assisted biopsy. Positioning of the patient on a Lorad Premium Stereotactic Table (Hologic, Bedford, MA, USA). The breast hangs compressed under the table and the vacuum biopsy needle is positioned according to the target coordinates.

exclusively on MRI. In these cases, the histological diagnostic work-up must be carried out by means of an MRI-guided intervention. Because MRI—in contrast to ultrasound and mammographic interventions—features an extremely strong magnetic field, materials and equipment used in the vicinity of the MRI system must have only very limited or no ferromagnetic properties ("MRI compatible"). This applies to the biopsy grids and systems as well as to the materials inserted into the breast (needles, wires, coils, clips), which are typically made of plastics or titanium- and nickel-based alloys. This insures the clear visualization of the materials without the sort of prominent susceptibility artifacts that would interfere with the process and ruin the MRI image.

Generally, as for the diagnostic breast MRI examination, a dynamic contrast-enhanced MRI examination with image subtraction must be carried out at the beginning of an MRI-guided breast intervention. Only in this way can an ambiguous contrast-enhancing lesion be identified and made accessible for a percutaneous biopsy.

MRI-guided interventions are performed with the patient lying in the prone position and the breast positioned in a special interventional breast compression device. Once the target lesion has been identified, the needle guide is positioned in the appropriate examination plane and aligned with the position of the mass between the nipple and chest wall (▶ Fig. 9.14). After administration of local anesthesia, a nonmagnetic coaxial introduction needle is positioned before or within the target lesion, depending on the biopsy system used. These and all following manual steps entailed during the intervention are performed without changing the position of the breast within the breast coil, but outside the MRI bore. Once the correct coaxial introduction needle position is documented, the biopsy needle can be introduced into the sheath and tissue sampling can begin (▶ Fig. 9.15).

MRI-guided biopsies are generally performed as a vacuum-assisted biopsies. The sampling window of the biopsy needle is step-rotated through 360° around the needle axis and an average of 12 contiguous samples is taken. With small lesions, a unidirectional approach with the sampling window directed toward the lesion, has also demonstrated its value.

Because the tissue sampling is not documented in real time, slight deviations in needle placement must be compensated for. Only after the interventional procedure has been completed can the correct position of the biopsy cavity be checked with reference to the previously determined position of the target lesion. To facilitate preoperative placement of a localization hookwire when surgical excision is required and when MRI-guided localization may not be feasible, then—especially when the lesion is a focus or a mass lesion—it is prudent to place a marker clip in the biopsy cavity at the end of the intervention. In the case of nonmasslike lesions, clip marking is not usually done because, if open biopsy is required, it may be preferable to mark surgically relevant lesion borders.

Take Home Point

Image-Guided Minimally Invasive Biopsy

MRI-guided biopsy should be carried out using the vacuum-assisted technique.

Only in rare cases of diffuse, very extensive lesions should MR guidance be used for high-velocity core needle biopsy. Such use of the cost-effective vacuum-assisted technique is justified when a lesion is of such size that it cannot be missed.

9.1.4 Classification of Findings

C Classification of Cytologic Specimens (Fine Needle Aspiration)

Cytology is performed on a direct smear preparation or a smear of a fluid specimen that has been centrifuged, stained, and placed on a slide for evaluation. The task of the cytologist is to categorize the lesion according the C classification (▶ Table 9.1).

B Classification of Histological Specimens (Core Needle or Vacuum-Assisted Biopsy)

Core needle and vacuum-assisted biopsy specimens are evaluated histologically. The biopsy specimens are placed in special specimen capsules and transferred to the pathology laboratory in formalin. When there are microcalcifications, it may be advisable to send the corresponding specimen radiograph with the specimens. The pathologist classifies the tissue collected by core needle or vacuum biopsies according to the B classification (▶ Table 9.2).

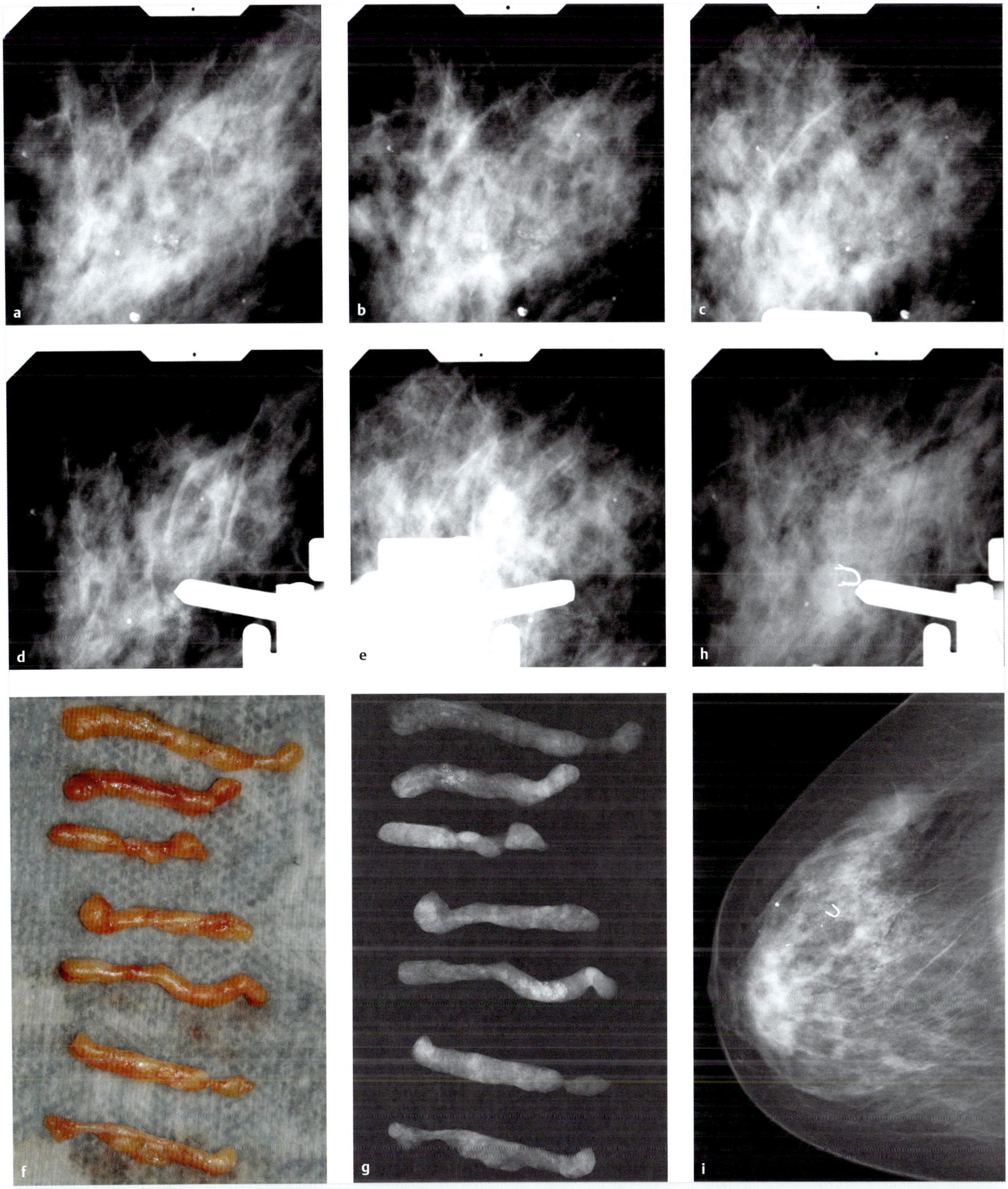

Fig. 9.13 Stereotactic vacuum-assisted biopsy of microcalcifications. Cluster of microcalcifications in the right breast requiring diagnostic work-up. **(a)** Visualization of the microcalcification cluster within the intervention window of the stereotactic examination system (−15° spot image). **(b)** Visualization within the intervention window of the stereotactic examination system (0° spot image). **(c)** Visualization within the intervention window of the stereotactic examination system (+ 15° spot image). **(d)** −15° spot image with visualization of the coaxial needle before it enters the cluster of microcalcifications. **(e)** + 15° spot image with visualization of the coaxial needle before it enters the cluster of microcalcifications. **(f)** Biopsy specimens. **(g)** Specimen radiograph. **(h)** Visualization of the released marker clip in the −15° spot image after biopsy. **(i)** Visualization of the released marker clip in the postinterventional mammogram.

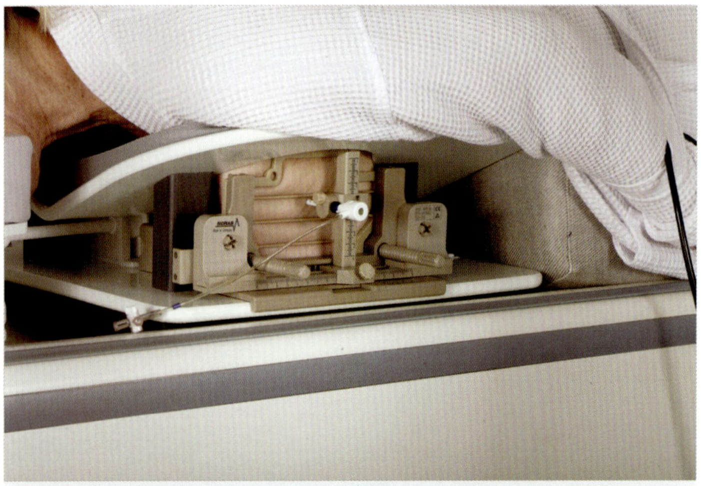

Fig. 9.14 MRI-guided vacuum-assisted biopsy. Positioning of the patient on the MRI table with interventional compression plates. The coaxial needle is positioned according to the target coordinates.

9.1.5 Tumor Seeding and Mechanical Tumor Induction

Meta-analyses of available studies show unequivocally that percutaneous biopsy of the breast does not lead to significant tumor seeding and subsequent tumor growth along the biopsy needle track, provided that the breast cancer is subsequently treated according to the usual recommendations. It is true that seeding of vital tumor cells can be detected in microscopic examination of the needle track in the first few days after a percutaneous biopsy. With increasing time after the intervention, however, the number of these cells decreases and approaches zero, because the seeded tumor cells apparently do not have the potential to grow in the needle track. This may be due in part to the routine administration of radiotherapy after breast-conserving therapy (BCT). There are only a few individual anecdotal case descriptions of tumors growing within the needle track, and in these cases, no adjuvant radiotherapy was given after BCT.

On the basis of the results of these meta-analyses, guidelines no longer call for the surgical excision of the needle track as part of BCT.

The widely held idea that the injury to benign tissue by a biopsy needle might itself cause the tissue to become malignant is likewise unfounded. This mistaken idea is undoubtedly based on the higher puncture failure rate in the past compared with today. Histology after a failed puncture of a lesion would naturally show benign results, but if the lesion were actually malignant this would become clinically apparent over the further course of time. The thought might easily occur that the lesion that was originally benign (with benign histology) at the time of biopsy had had its tissue characteristics changed by the needle manipulation and thus become malignant. From a tumor-biology standpoint, this idea has no basis.

9.1.6 Quality Assurance

Recommendations have been developed to improve the reliability of percutaneous biopsy procedures:

- Equipment used in interventional procedures, such as stereotactic interventional units, should be calibrated and the accuracy of imaging and mechanical parts checked at specified time intervals.
- During the course of tissue sampling, the correct position of the biopsy needle should be documented (in a sonographic intervention in two planes, for instance). In core needle biopsy, at least three tissue samples should be taken.
- The equivalent of 12 biopsies with a 10G needle should be harvested when performing a vacuum-assisted biopsy. In vacuum-assisted biopsy of microcalcifications, the requirement for verification of microcalcifications in a specimen radiograph is standard.
- When the target lesion may have been completely removed by the biopsy, marker clips are inserted in the biopsy cavity to enable localization in case surgery is necessary.

Quality standards have also been defined for correlating the histological examination results with the results expected from the image findings. As an example, a benign histological result after biopsy of a BI-RADS 5 lesion cannot be accepted without misgivings. If there is a potential mismatch, an interdisciplinary conference to decide the further courses of action is mandatory.

Take Home Point

After minimally invasive image-guided biopsy, the histopathological results should be correlated with the results expected from the diagnostic imaging findings.

Note

Guidelines are recommendations. These should generally be followed, but exceptions may be made in justified cases.

9.2 Localization

9.2.1 Objective of Pretherapeutic Localization

As the quality of early detection improves there is accordingly an increase in the number of small cancerous lesions detected that are not palpable. In order to achieve the goal of complete excision it is thus especially important to mark such lesions before surgery is performed. Appropriate marking of lesion borders may also be required if imaging reveals that only part of the lesion is palpable. In addition to wire localizations in the preoperative setting, clips are used for marking biopsied lesions after minimally invasive biopsy. This enables later surgical treatment even if the lesion has been so extensively sampled that it is no longer visible on imaging.

Clip marking is also recommended before neoadjuvant therapy of extensive carcinoma. In this way, the surgical revision of the original tumor borders is assured, especially in the case of partial or complete remission after chemotherapy.

Placement of Wire Markers

Hookwire markers are used for preoperative localization. They are introduced into the breast with the aid of a puncture needle.

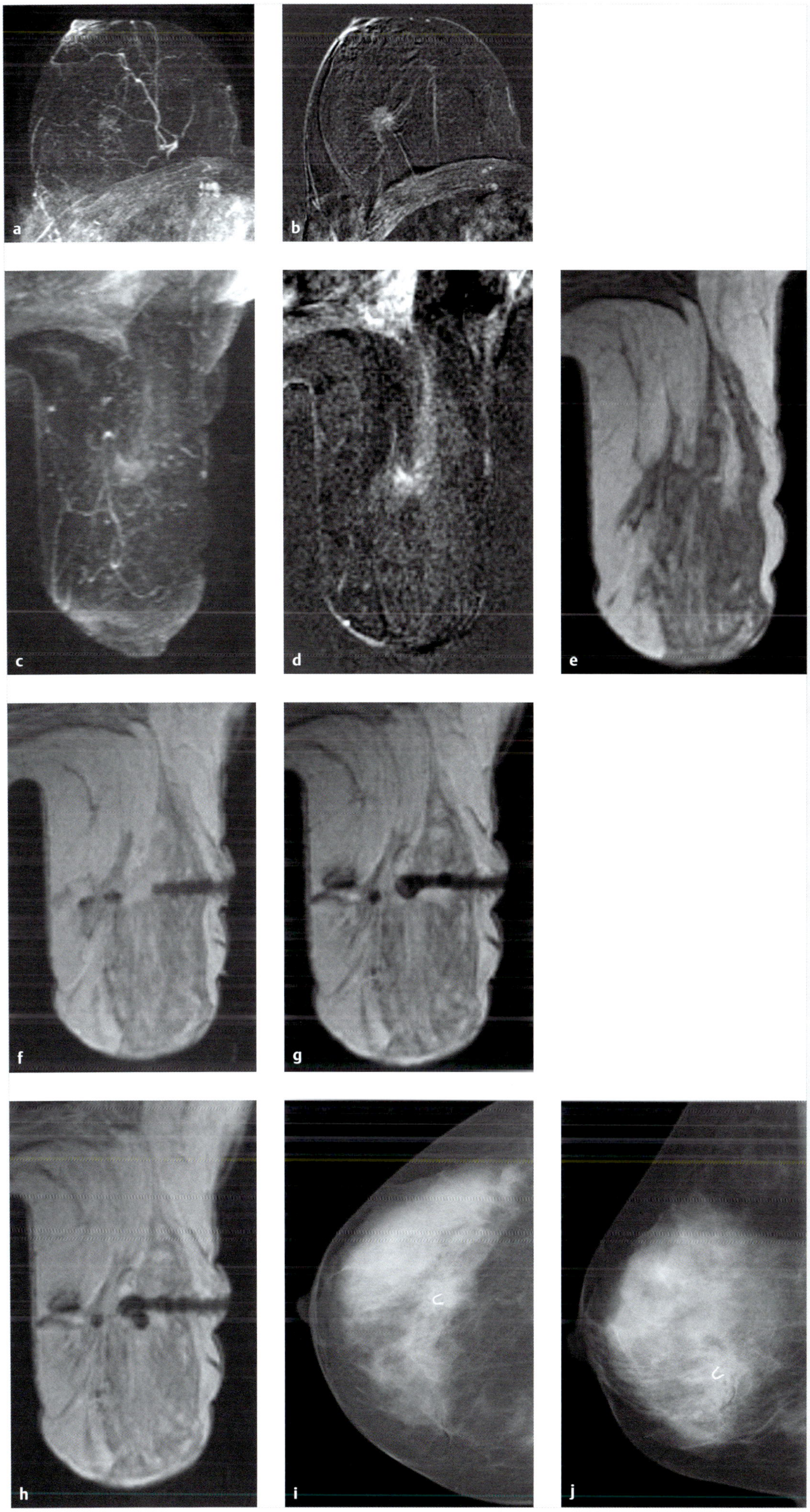

Fig. 9.15 MRI-guided vacuum-assisted biopsy. (a) Hypervascularized mass lesion in diagnostic breast MRI: maximum intensity projection. **(b)** Single-slice subtraction image. **(c)** Hypervascularized mass lesion in interventional breast MRI: maximum intensity projection. **(d)** Corresponding interventional single-slice subtraction image. **(e)** Single-slice precontrast T1W image. **(f)** Visualization of the coaxial needle within the mass lesion. **(g)** T1W image with depiction of a substance defect (leak of tissue) following the harvesting of 12 biopsy samples by means of vacuum-assisted biopsy. **(h)** T1W image with visualization of the released marker clip after biopsy. **(i)** Postinterventional documentation of the marker clip in the mammogram (CC projection). **(j)** Postinterventional documentation of the marker clip in the mammogram (ML projection).

Table 9.1 C classification for evaluation of cytologic specimens

Class	Findings	Explanation
C1	Inadequate/ insufficient	The material or smear is not sufficient for evaluation, or is not assessable because of technical deficiency
C2	Conclusively benign	Cellular material is adequate; cells unequivocally benign
C3	Benign with atypia	Subtle nuclear changes or cellular dissociation within the variation range of benign changes
C4	Suspicious atypia	Diagnostic criteria for malignancy not conclusively fulfilled, e.g., in the case of limited assessability of the specimen
C5	Conclusively malignant	Clearly malignant tumor cell aggregates and isolated malignant cells

Table 9.2 B classification for evaluation of histological specimens

Class	Findings
B1	Normal mammary gland tissue
B2	Conclusively benign changes
B3	Findings with ambiguous biological potential
B4	Suspicious of malignancy but limited assessability of the specimen
B5	Clearly malignant findings
B5a	Ductal carcinoma in situ (DCIS)
B5b	Invasive carcinoma
B5c	Malignant tumor, questionably invasive
B5d	Malignant tumor, but not a primary breast cancer

When the puncture needle is in the desired position, the hookwire is advanced out of the needle. The metal wire is flexible and has a hook configuration at the tip that anchors itself in the lesion like a fish hook. The localization hookwire provides a guide for the surgeon, enabling the correct and complete excision of the localized lesion with a minimal amount of cosmetic disfigurement. When the excised lesion is a mammographic finding (such as microcalcifications), the complete resection can be checked by specimen radiography.

In the localization of mass lesions, the hookwire should ideally penetrate the center of the mass lesion. The maximum distance between the wire and the mass lesion should be 1 cm.

Take Home Point

Open Excisional Biopsy

In preoperative marking of nonpalpable lesions, the wire should penetrate a mass lesion and project less than 1 cm beyond it. If the wire does not penetrate the mass, then the distance between the wire and the border of the mass should not be greater than 1 cm. In the case of non-space-occupying processes, it may be prudent to mark the borders of the surgically relevant target volume.

Large nonmasslike lesions are best marked with multiple hookwires at the surgically relevant boundaries.

The imaging technique used for imaging guided hookwire marking of a lesion should be the one in which the lesion is imaged best. In most cases, this is sonography. The lesion is punctured with the tip of the needle, as for ultrasound-guided biopsy, and the hookwire is then advanced through the needle. Subsequently, the position of the hookwire is documented mammographically in two planes (CC and ML projections) (▶ Fig. 9.16).

When the lesion to be localized shows microcalcifications, the needle is usually placed under mammographic guidance. As with stereotactic vacuum-assisted biopsies, a mammographic stereotactic system is used (▶ Fig. 9.17). Alternatively, a freehand localization may be performed, in which the access path for needle placement is calculated without the aid of a stereotactic unit, and the needle position is checked mammographically and corrected if necessary.

MRI-detected lesions with no sonographic or mammographic correlates are marked using MRI guidance, a procedure that is similar to that for performing an MRI guided biopsy. It is best to use special hookwire markers made from MRI-compatible metals to avoid imaging artifacts (▶ Fig. 9.18).

In rare cases when a lesion is located immediately under the skin, preoperative markers can be drawn directly onto the skin. To locate and document the correct position on the skin to be marked, cutaneous markers are placed at the determined skin position before imaging: lead shot for mammography and oil-containing markers like vitamin or nitroglycerine capsules for MRI.

Perfect communication between the surgeon and diagnostician is a prerequisite for optimal quality of the treatment. In this regard, image documentation of the hookwire position after sonographic and MRI-guided needle placement is helpful, including CC and ML mammography projections. Drawings or photodocumentation of the position of the needle further facilitate the surgeon's topographical perception of the lesions position.

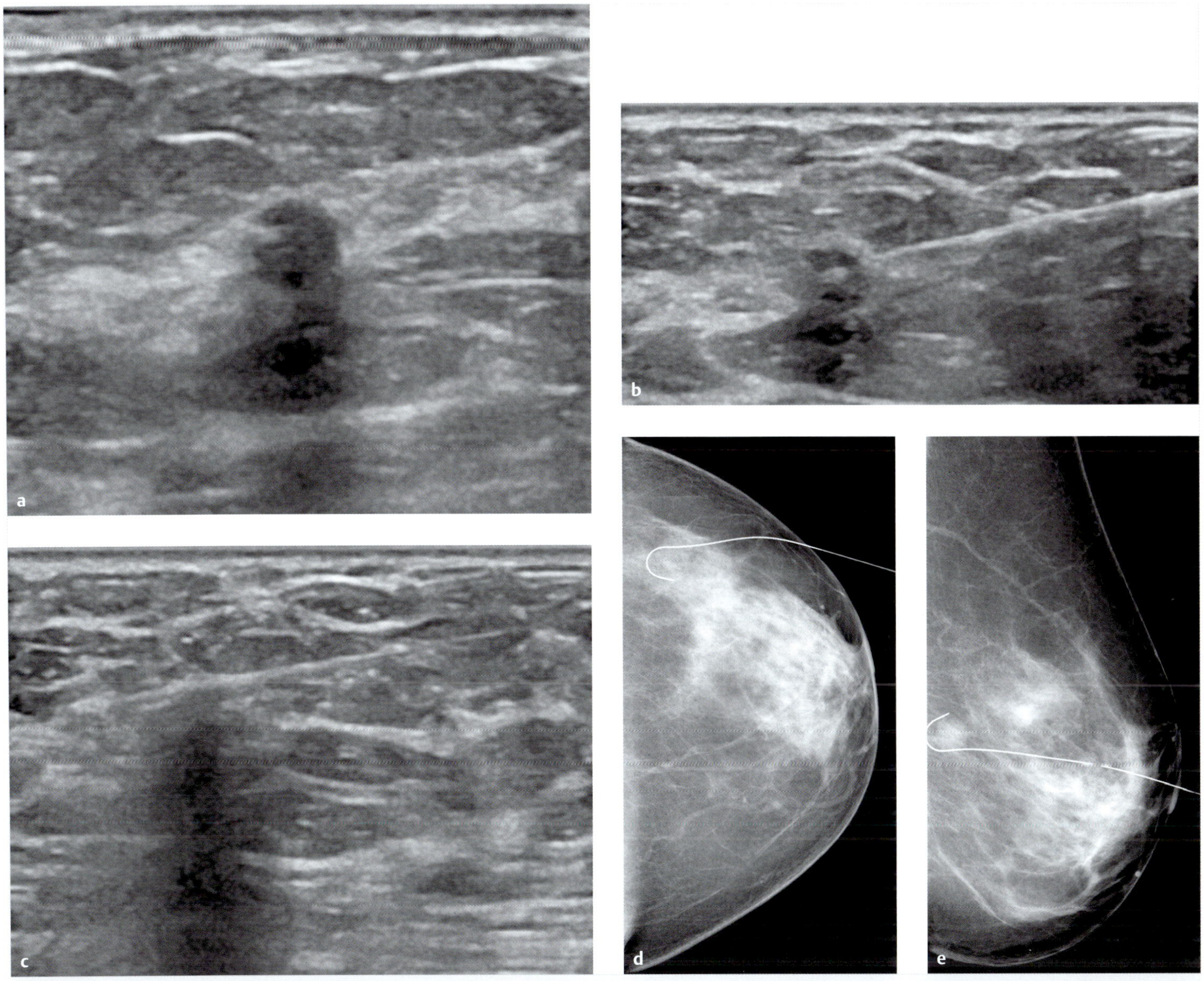

Fig. 9.16 Ultrasound-guided localization with a hookwire. (a) Hypoechoic mass lesion. **(b)** Puncture needle extending from the right edge of the image; the tip lies before the mass. **(c)** The hookwire, now released, penetrates the mass lesion. **(d)** Mammogram of the hookwire in the CC projection. **(e)** Mammogram of the hookwire in the ML projection.

Clip and Coil Marking

Clips and coils are small metal localization hooks or spirals, which can remain in the breast (see ▶ Fig. 9.20). Occasionally, these markers are combined with materials that can also produce an equivalent sonographic image. They enable one to find a biopsied area again even if the lesion has been completely excised by biopsy and, in addition, allow the use of a different image guiding method for preoperative localization. Lesions that are originally imaged only on MRI, for example, can be preoperatively localized using mammographic guidance (localization of the coil/clip).

After minimally invasive harvesting of tissue samples, usually by way of a stereotactic or MRI-guided vacuum-assisted biopsy, an appropriate marker is placed in the biopsy cavity. If surgical treatment is shown to be necessary after histopathological diagnosis, hookwires can be localized accordingly.

Clip and coil marking is also used before the start of neoadjuvant chemotherapy. In case of complete remission with complete

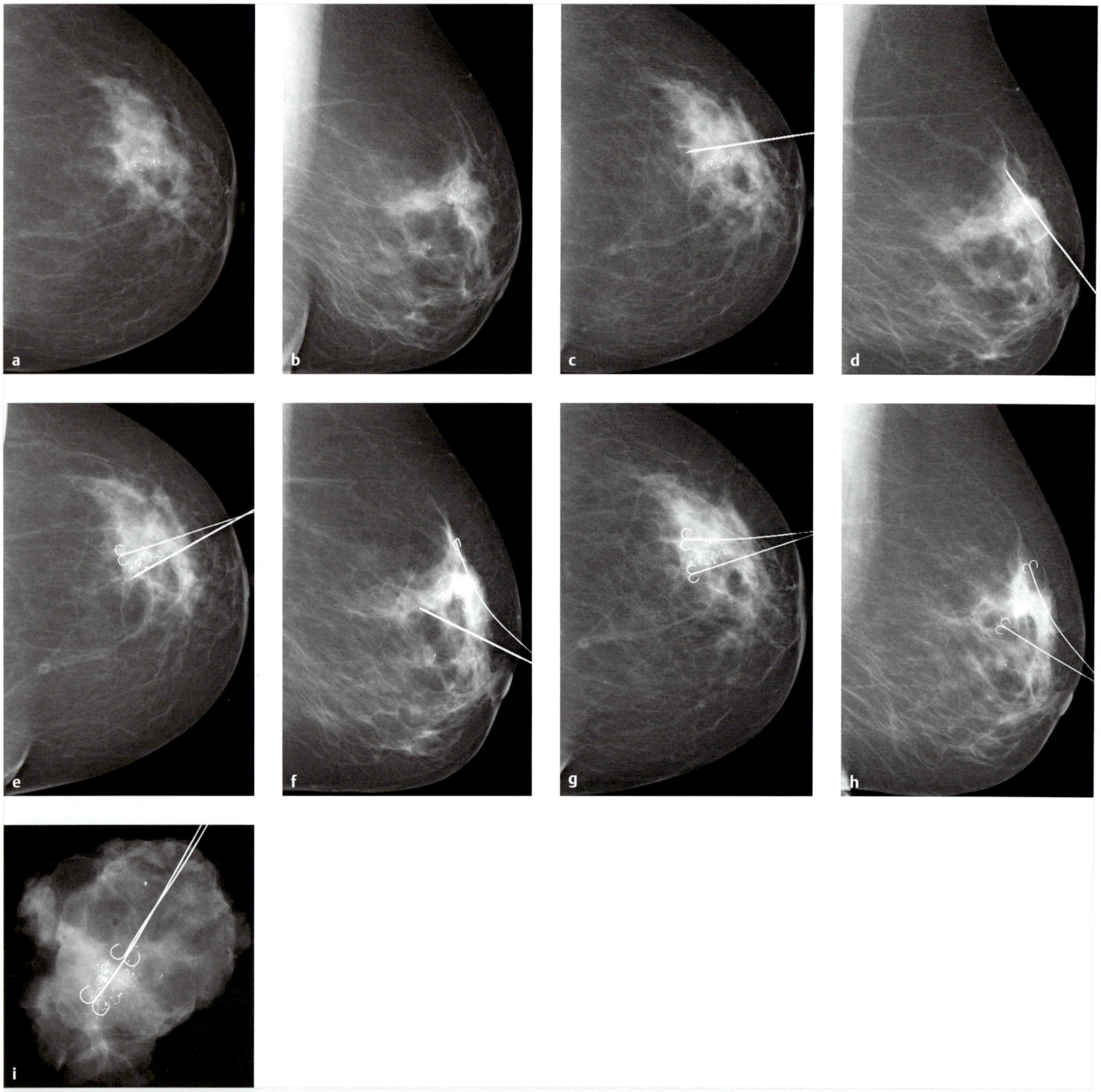

Fig. 9.17 Freehand mammographic hookwire localization. (a) Mammogram in CC projection with a centrally located cluster of pleomorphic microcalcifications. **(b)** Mammogram in ML projection corresponding to **(a)**. **(c)** Puncture needle at the lateral edge of the cluster of calcifications in the CC view. **(d)** Puncture needle at the superior edge of the clustered microcalcifications in the ML view. **(e)** First hookwire advanced; second needle at the medial edge in the CC view. **(f)** First hookwire advanced; second needle at the inferior edge of the clustered microcalcifications in the ML view. **(g)** Localized target volume. The pleomorphic cluster of calcifications lies between the hookwire markers; in the CC view. **(h)** ML image corresponding to **(g)**. **(i)** Specimen radiograph with complete resection of the calcification cluster.

disappearance of the tumor in all imaging modalities, the markers insure the identification of the tumor bed. In this way, histological documentation of complete remission can be achieved. Before pretherapeutic placement of such markers, the attending surgeon should be consulted so that the planned surgical technique can be taken into consideration. In the case of a planned segmental or quadrant resection, for example, the superior and inferior (i.e., the medial and lateral) borders of the target segment must be marked.

9.2.2 Equipment and Implementation

Wire markers are introduced into the tissue using a puncture needle. The tip of the wire has a single or a double hook. As well as nonrepositionable hooks that cannot be drawn back into the puncture needle once advanced, there are also repositionable wires. Once released from the needle, the tips of these wires adopt a curved shape that allows them to be drawn back into the needle again for correction of positioning after suboptimal

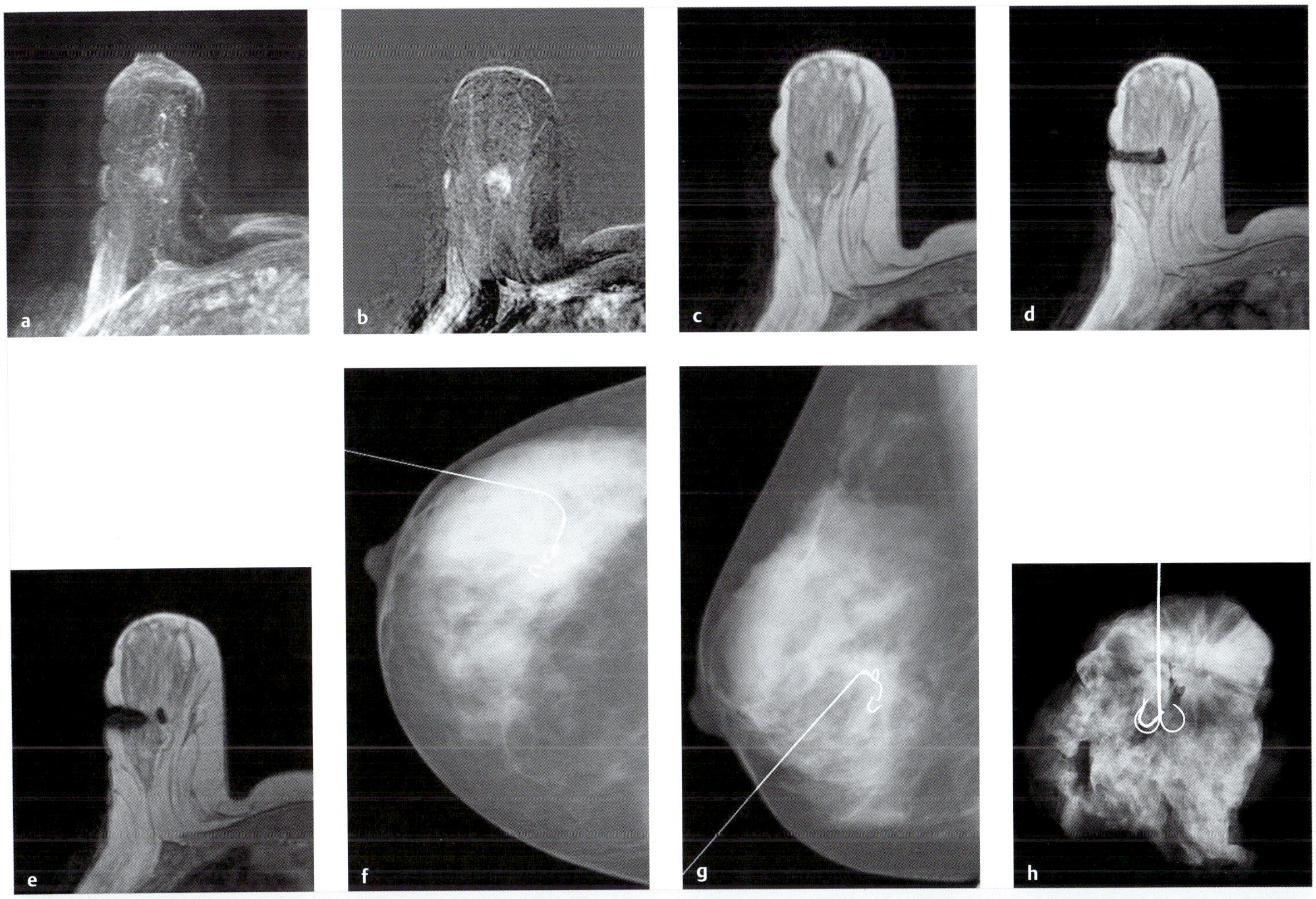

Fig. 9.18 MRI localization with a hookwire. **(a)** Centrally located enhancement in the maximum-intensity projection. **(b)** Single subtraction slice image. **(c)** Precontrast T1W single slice demonstrating a clip after previous vacuum-assisted biopsy. **(d)** Coaxial needle ahead of the mass lesion. **(e)** Hookwire tip in the mass lesion. **(f)** Postlocalization mammogram of the hookwire in the CC projection. **(g)** Postlocalization mammogram of the hookwire in the ML projection. **(h)** Specimen radiograph with visualization of the marker clip.

placement. Some wires consist of a single wire strand; others are composed of two twisted wire strands (▶ Fig. 9.19).

In addition to the usual hookwire markers, coils with a proprietary Ariadne's Thread are occasionally used (▶ Fig. 9.20c) Instead of a continuous marker wire, these coils exhibit a firm thread as an extension of the marker coil. During surgery, the coil can be localized with the help of this thread. Especially when there is a long interval between the localization and the surgery, this type of marker is subjectively more comfortable for the women involved. However, this type of localization is not very popular with surgeons because of the fear of cutting through the thread during the surgical procedure and, as a result being unable to find the coil again.

Within the scope of minimally invasive intervention, when a lesion might not be clearly identified following tissue sampling, marker clips or coils are used to tag the biopsy cavity (▶ Fig. 9.20). This is often the case, for example, after vacuum-assisted biopsy of small mass lesions or small groups of microcalcifications. When necessary, these clips can be mammographically visualized and localized at a later time. They consist of tiny, millimeter sized metal hooks or spiral-shaped metal wires.

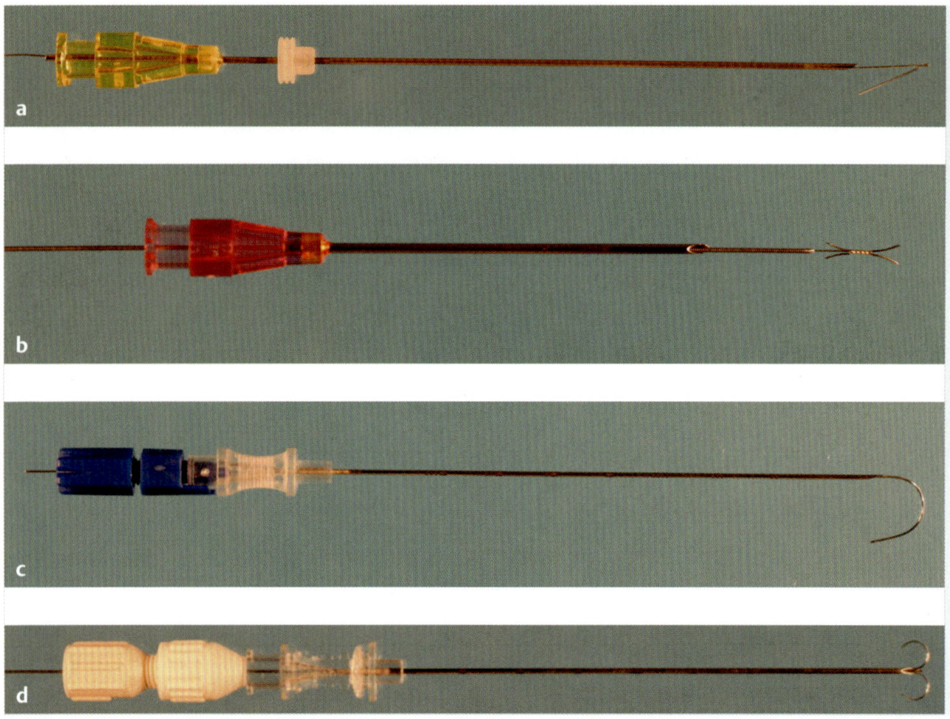

Fig. 9.19 Equipment for wire localizations. Various configurations for preoperative placement of wire markers in the breast: nonrepositionable wires **(a, b)**; repositionable wires **(c, d)**. **(a)** Hook (Cook Medical Inc., Bloomington, USA). **(b)** Double hook (Cook Medical Inc.). **(c)** Homer-Lok (Angiotech, Gainesville, FL, USA). **(d)** DuaLok (Somatex, Teltow, Germany).

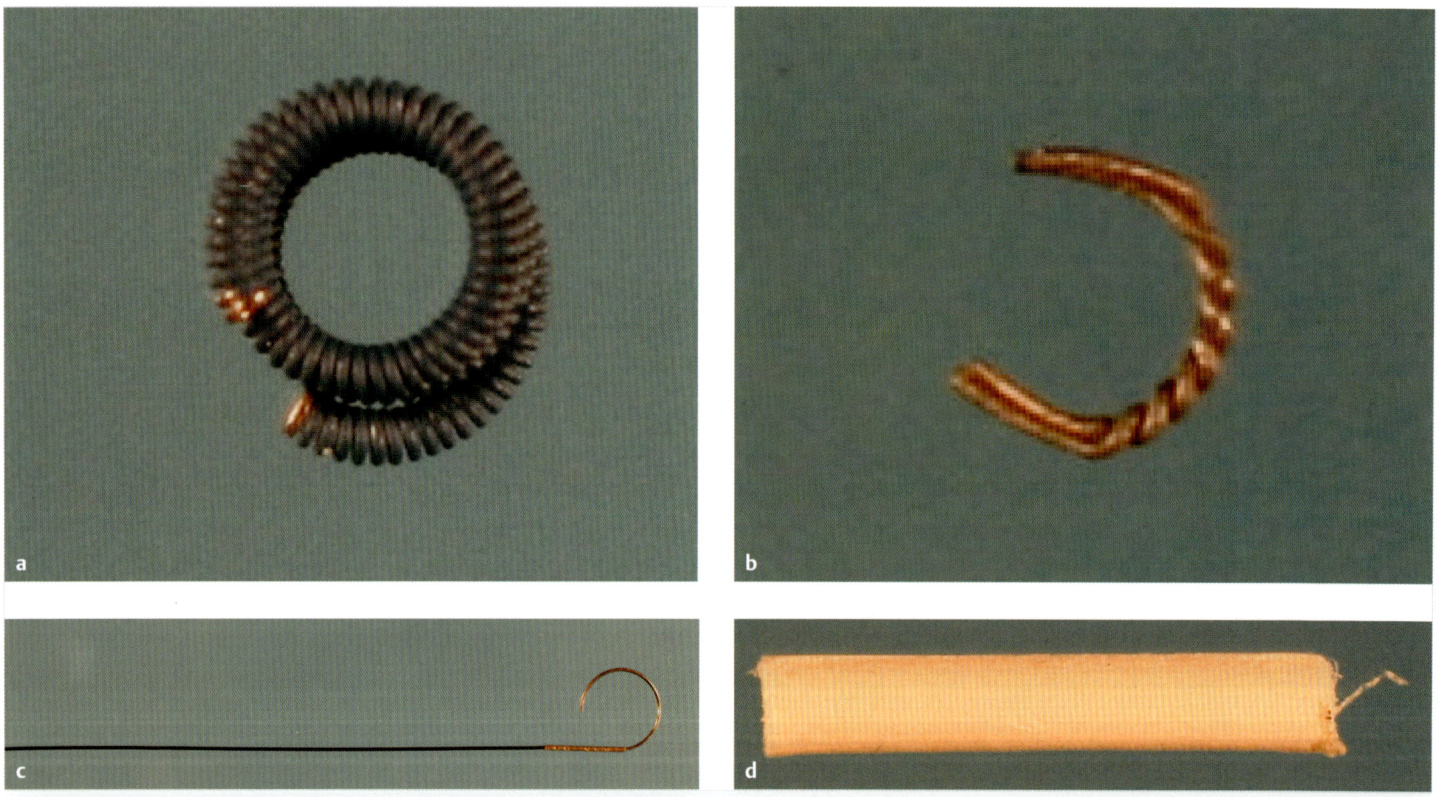

Fig. 9.20 Clips and coils. Various configurations for placement of markers in the breast: **(a)** Echotip (Cook Medical Inc., Bloomington, IN, USA). **(b)** Tumark (Somatex, Teltow, Germany). **(c)** Ariadne (Invivo Germany, Schwerin, Germany). **(d)** MammaMark (Mammatome, Norderstedt, Germany).

To enable the ultrasound-guided localization of a marker, some of these markers are equipped with small sponges that result in a typical finding on ultrasound. Other sonographically reproducible markers consist of collagen, polyglycolic acid, or polylactic acid (▶ Fig. 9.20).

9.2.3 Quality Assurance

Even after ultrasound-guided or MRI-guided placement of a lesion marker, mammograms in two perpendicular projections (CC and ML projections) should be obtained. This makes the

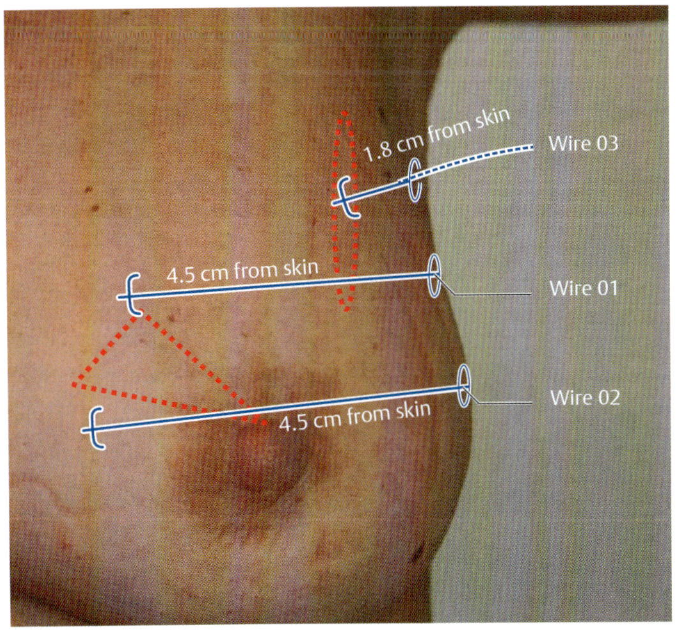

Fig. 9.21 Postinterventional photocollage. Photographic documentation with drawn projection of the marker wire positions and localized lesions. Estimated distances of the wire tips from the skin are given.

orientation easier for the surgeon and sometimes opens up the possibility of a postoperative radiograph of the specimen for documentation of complete excision of possible masses or microcalcifications. Schematic drawings also help the surgeon with orientation, especially if several marker wires were used in the marking (▶ Fig. 9.21).

Bibliography

▶ Biopsy

[1] Abdullah N, Mesurolle B, El-Khoury M, Kao E. Breast imaging reporting and data system lexicon for US: interobserver agreement for assessment of breast masses. Radiology. 2009; 252(3):665–672

[2] American College of Radiology (ACR). Breast Imaging Reporting and Data System Atlas (BI-RADS Atlas). Reston, VA: American College of Radiology; 2003

[3] Berg WA. Image-guided breast biopsy and management of high-risk lesions. Radiol Clin North Am. 2004; 42(5):935–946, vii

[4] Brancato B, Crocetti E, Bianchi S, et al. Accuracy of needle biopsy of breast lesions visible on ultrasound: audit of fine needle versus core needle biopsy in 3233 consecutive samplings with ascertained outcomes. Breast. 2012; 21(4):449–454

[5] Deutsche Gesellschaft für Senologie e. V., Deutsche Krebshilfe e.V. S 3-Leitlinie "Brustkrebs-Früherkennung in Deutschland". 1. Aktualisierung. Munich: W. Zuckschwerdt; 2008

[6] Fischer U, Baum F. Diagnostische Interventionen der Mamma. Stuttgart: Thieme; 2008

[7] Fischer U, Schwethelm L, Baum FT, Luftner-Nagel S, Teubner J. Effort, accuracy and histology of MR-guided vacuum biopsy of suspicious breast lesions—retrospective evaluation after 389 interventions [in German]. Rofo. 2009; 181(8):774–781

[8] Friedrich M. Lehratlas der Mammasonografie. Stuttgart: Wissenschaftliche Verlagsgesellschaft; 1999

[9] He Q, Fan X, Guan Y, Tian J, Fan Z, Zheng L. Percutaneous excisional biopsy of impalpable breast lesions under ultrasound visualization. Breast. 2008; 17(6):666–670

[10] Helbich TH, Matzek W, Fuchsjäger MH. Stereotactic and ultrasound-guided breast biopsy. Eur Radiol. 2004; 14(3):383–393

[11] Heywang-Köbrunner SH, Sinnatamby R, Lebeau A, Lebrecht A, Britton PD, Schreer I, Consensus Group. Interdisciplinary consensus on the uses and technique of MR-guided vacuum-assisted breast biopsy (VAB): results of a European consensus meeting. Eur J Radiol. 2009; 72(2):289–294

[12] Kettritz U, Morack G, Decker T. Stereotactic vacuum-assisted breast biopsies in 500 women with microcalcifications: radiological and pathological correlations. Eur J Radiol. 2005; 55(2):270–276

[13] Leitlinienprogramm Onkologie der AWMF, Deutschen Krebsgesellschaft eV und Deutschen Krebshilfe eV. Leitlinienreport der S3 Leitlinie fur die Diagnostik, Therapie und Nachsorge des Mammakarzinoms. AWMF Register-Nr. 032–045OL. Available at: http://www.awmf.org/leitlinien/detail/ll/032-045OL.html. Accessed February 9, 2017

[14] Liberman L, Gougoutas CA, Zakowski MF, et al. Calcifications highly suggestive of malignancy: comparison of breast biopsy methods. AJR Am J Roentgenol. 2001; 177(1):165–172

[15] Liberman L, Bracero N, Morris E, Thornton C, Dershaw DD. MRI-guided 9-gauge vacuum-assisted breast biopsy: initial clinical experience. AJR Am J Roentgenol. 2005; 185(1):183–193

[16] Madjar H. Kursbuch Mammasonografie. Stuttgart: Thieme: 2012

[17] Perlet C, Heywang-Kobrunner SH, Heinig A, et al. Magnetic resonance-guided, vacuum-assisted breast biopsy: results from a European multicenter study of 538 lesions. Cancer. 2006; 106(5):982–990

[18] Siegmann KC, Wersebe A, Fischmann A, et al. Stereotactic vacuum-assisted breast biopsy—success, histologic accuracy, patient acceptance and optimizing the BI-RADS™-correlated indication [in German]. Rofo. 2003; 175(1):99–104

▶ Marking

[1] Fischer U, Baum F. Diagnostische Interventionen der Mamma. Stuttgart: Thieme; 2008

[2] Friedrich M. Lehratlas der Mammasonografie. Stuttgart: Wissenschaftliche Verlagsgesellschaft; 1999

[3] Leitlinienprogramm Onkologie der AWMF, Deutschen Krebsgesellschaft eV und Deutschen Krebshilfe eV. Leitlinienreport der S3 Leitlinie fur die Diagnostik, Therapie und Nachsorge des Mammakarzinoms. AWMF Register-Nr. 032–045OL. Available at: http://www.awmf.org/leitlinien/detail/ll/032–045OL.html. Accessed February 9, 2017

[4] Madjar H. Kursbuch Mammasonografie. Stuttgart: Thieme; 2005

[5] Obenauer S, Grabbe E, Knollmann F. Value of MR-guided localization and biopsy in breast lesions [in German]. Rofo. 2006; 178(5):477–483

[6] Siegmann KC, Speck S, Baur A, et al. Performance of a newly developed clip (Tumark Professional) for MRI-guided lesion localization after MRI-guided vacuum-assisted biopsy—first results [in German]. Rofo. 2009; 181 (2):147–154

Part 3

Prevention and Therapy of Breast Cancer

10 Examination Concepts

U. Fischer

10.1 Prevention

There are essentially three forms of prevention:

- *Primary prevention:* Primary prevention aims to actually prevent the occurrence of a disease, thus decreasing its incidence. Colonoscopy is a good example of this: polyps, which are (at this stage) benign lesions with a high risk of malignancy, can be removed, so that colon cancer does not develop. Avoiding the use of hormone replacement therapy during menopause is an example of primary prevention of breast carcinoma. However, imaging per se does not qualify as primary prevention.
- *Secondary prevention:* Secondary prevention aims to reduce the mortality of a disease that has already developed. This is most successful when a carcinoma is discovered at an early stage (early detection), which is made possible in practice through imaging examinations.
- *Tertiary prevention:* The aim of tertiary prevention is also to reduce mortality (improving overall survival) and the impact of manifest disease. Follow-up care is a form of tertiary prevention.

10.2 Early Breast Cancer Detection (Secondary Prevention)

In most cases, breast cancer is a slow-growing tumorous process. Depending on the tumor grade, one can assume an average tumor doubling time of between 150 and 300 days. This rather low growth rate is an essential premise for the systematic application of imaging diagnostics with the objective of early tumor detection. It is also the basis for the usually recommended examination interval of 1 to 2 years. Assuming a high-quality examination strategy with a high sensitivity for the detection of breast cancer, a breast carcinoma should not reach an advanced tumor stage within this interval. Exempt from this are the inherently highly aggressive breast carcinomas with extremely high proliferation rates, which fortunately represent the exception to the rule.

The goal of an effective breast cancer early detection strategy must be to detect breast cancer at a prognostically favorable stage. In essence, there are two different tumor constellations to be considered in this regard:

- *Early detection at the DCIS tumor stage:* The diagnosis of intraductal breast carcinoma, in and of itself, implies an early diagnosis even though the tumor occasionally may extend over several centimeters by the time the diagnosis is made. As a localized disease of the breast, an intraductal tumor does not extend into the locoregional lymph nodes or disseminate to generate remote metastases. In considering DCIS, however, it is important to note that—especially in the group of DCIS with a low proliferation rate—a high proportion of tumors will not grow beyond the mammary duct to threaten the woman during her lifetime. For this reason, the issue of potential overdiagnosis with the associated overtreatment of the patients thus affected is an area of ongoing discussion.
- *Early detection at an invasive tumor stage with tumor size less than 10 mm:* An invasive breast carcinoma is considered to be detected early if it has a diameter of less than 1 cm when discovered. Breast cancers at such stages (pT1a, pT1b) exhibit an extremely low rate of metastasis to locoregional lymph nodes or parenchymatous organs, so that the long-term survival rate is usually excellent (over 95%). Here there is no controversy regarding overdiagnosis, because untreated invasive carcinoma usually leads to death.

Note

Although DCIS is considered a very early, noninvasive breast cancer, it may already be several centimeters in size when detected.

In addition to the clinical breast examinations (by inspection and palpation), other means for the early detection of breast cancer generally include the established imaging examination techniques of mammography, breast ultrasonography, and breast MRI.

- *Inspection and palpation:* Study data demonstrate that periodic inspection—and in particular palpation of the female breast—does not offer any survival advantage, because breast cancers discovered by this approach already average 2 to 3 cm in size. However, Women who regularly examine their breasts are also more consistent in their participation in early detection imaging programs.

Take Home Point

Used as the sole method for breast cancer detection, breast self-examination cannot decrease the breast cancer mortality rate.

- *Mammography:* X-ray mammography is the only imaging technique that has been shown to reduce the mortality of breast cancer in a quality-controlled and population-based mass screening program. The relevant data, originating from the United Kingdom in the 1970s, has demonstrated a reduction in mortality of 30% through the implementation of mammographic screening in a group of women who had not previously had mammography.

Take Home Point

General recommendations call for the use of mammography in quality-assured screening programs. Recommendations to attend screening vary across countries and organizations. In general, it is recommended that women begin mammography screening at an age 40 to 50 years, at an interval of once every 1 to 2 years.

- *Breast ultrasonography:* Breast ultrasonography is not appropriate for use as the sole examination method in the early detection of breast cancer. Nevertheless, it represents an important supplementary method to compensate for the limitations of other imaging methods. In younger women, however, it is the primary method of choice.

Take Home Point

Sonography is not suitable for use as the sole diagnostic method in the early detection of breast cancer.

- *Breast MRI:* Data from studies in the last 10 years document the great significance of breast MRI in the detection of early and small carcinomas, in particular. In this regard, MRI is clearly superior to all the other imaging examination methods, provided that high technological and methodological standards are assured and the radiologists who do the evaluation have sufficient expertise.

In European statistics, approximately 25% of all of diagnosed breast carcinomas are detected within the mammography screening program. The remaining 75% are diagnosed either within individualized and risk-adapted early detection frameworks or after the diagnostic work-up of clinical symptoms (▶ Fig. 10.1).

10.2.1 Mammography Screening

"Screening" refers to a systematic testing strategy aiming to identify a particular characteristic within a defined number of specimens or people.

Note

The goal of breast cancer screening programs is to reduce the breast cancer mortality rate in a sustained manner by diagnosing malignant tumors of the breast at the earliest possible and prognostically most favorable stage.

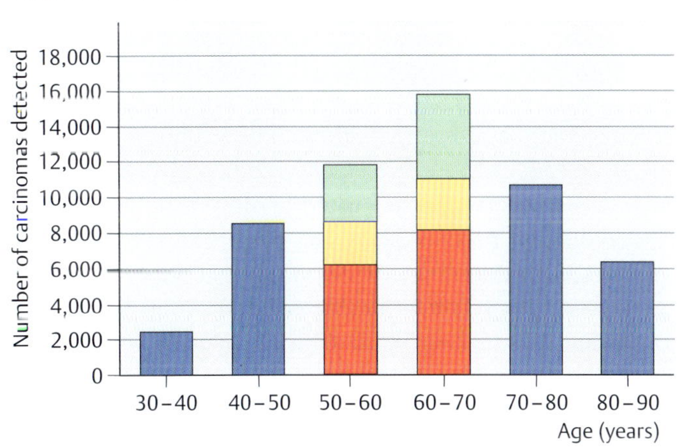

Fig. 10.1 Detection rate of breast cancer within and outside of screening programs. Red bars = carcinomas detected within the screening program in the 50- to 69-year-old group of participants. Yellow bars = rate of interval carcinomas (ca. 30%). Green bars = carcinomas found in women not participating in screening (ca. 30% of women 50–69 years old with an assumed participation rate of 70%). Blue bars = carcinomas detected in women outside the screening age range; i.e. women under 50 years and over 70 years.

History

Comprehensive mammography screening programs have been operating in various countries for over 40 years. The Netherlands, Sweden, Norway, and Finland implemented an organized, mass screening program for women without symptoms in 1974, and the United Kingdom in 1979. Canada has also had an organized program for many years. Studies have demonstrated that a 25–30% decrease in mortality is possible through the institution of a quality-controlled screening program. It should be noted, however, that in many of these countries there had been no previous mammographic examinations in the population before the introduction of mammography screening. This was not the case for the population in Germany when screening was introduced in 2005.

In general, mammography screening programs are population-based, quality-controlled and well-structured. They typically offer women between the ages of 50 and 69 years biennial participation in a program for early detection of breast cancer on the basis of an invitational system. For women under the ages of 40 to 50 years, screening mammography has not been shown to offer any advantages because of the higher percentage of false -positive findings in younger women. This can, at least in part, by the higher breast tissue density in younger women, which limits the overall sensitivity of the X-ray imaging. Data substantiate that the incidence of interval carcinomas in this group (women 40 to 49 years old) in the first screening round is approximately 50%. Mammography screening programs are subject to a comprehensive quality assurance program in terms both of equipment and technical aspects and of the training of the radiologist and the radiologic technologist personnel. The relevant quality-assurance measures are evaluated regularly.

The screening mammography images are assessed by at least two specially trained reviewers, who evaluate the images independently. If at least one of the reviewers determines that the mammography requires clarification, then a final assessment is made by way of a consensus conference under the direction of the physician responsible for the program. ▶ Fig. 10.2 illustrates a typical overall structure of a national mammography screening program, using Germany as an example.

For mammography screening programs, it is assumed that the effect on mortality reduction will be detectable, at the earliest, after a period of at least 10 years. The analysis of surrogate performance parameters enables earlier evaluation of the effectiveness of the program and also allows possible side effects to be registered and minimized. The results are published regularly.

Criteria for a Screening Program

Screening programs designed to examine a large group of people to detect a disease at an early stage make sense only if certain criteria are met:

- The illness must constitute a relevant public health concern.
- The illness must be treatable, that is, have distinctly improved treatment outcome when detected early.
- The screening examination method used should exhibit a high degree of sensitivity so as to detect the disease as reliably as possible.

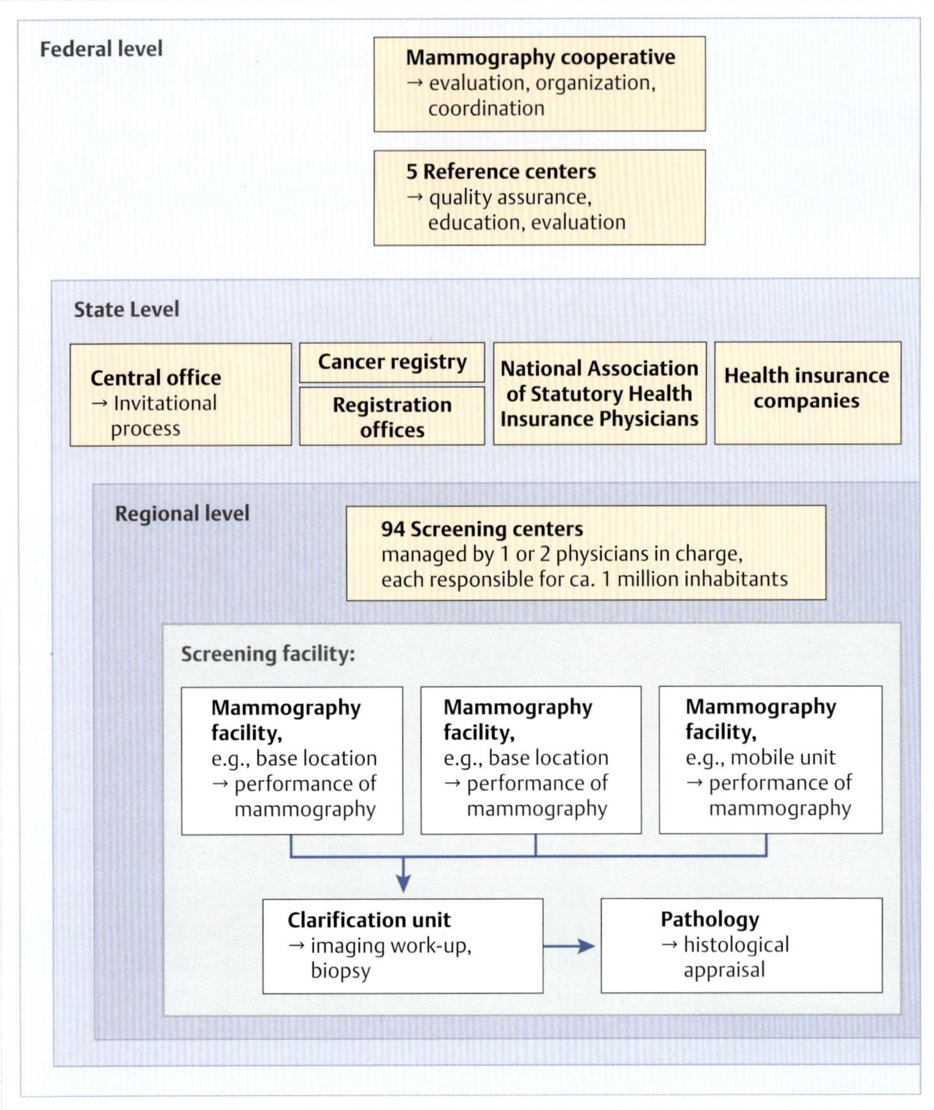

Fig. 10.2 Structure of a mammography screening program, using Germany as an example. (From the 2008–2009 evaluation report of the Kooperationsgemeinschaft Mammographie [Mammography Cooperative, Berlin] 2012, modified.)

- The screening examination method used should exhibit a high degree of specificity so as to recognize healthy people as such and to protect them from overdiagnosis.
- The screening examination method used should be time- and cost-effective.
- The screening examination method used should have as little a negative impact as possible on those examined.

Advantages

The existence of a screening program motivates women who would not otherwise do so to undergo early detection mammography.

The screening quality specifications usually distinctly improve the quality of the examination devices used and also of the participating medical and technical radiology assistants and physicians. The quality requirements and the mandated periodic advanced training courses are instrumental in this regard.

Disadvantages

False-positive screening results can cause unnecessary concern and anxiety to patients. Additional costs may be incurred due to the necessity for further examination procedures to clarify any such ambiguous findings. With false-negative findings, the goal of screening has been completely missed since, in spite of a screening examination having been performed, a malignant tumor that is present has not been detected. In the case of mammography screening, this is particularly true for women with high tissue density, for whom the sensitivity of mammography falls to 30 to 40% when no complementary examination procedure is used.

It is a subject of debate whether some women may be diagnosed at an early breast cancer stage without any resulting extension of their lifespan. Rather there is often a reduction in the quality of life due to the diagnosis of "breast cancer" and the subsequent therapy that in retrospect was unnecessary therapy (for example, low-grade DCIS). Tumors may indeed be detected earlier without this leading to an increase in the life expectancy, resulting in an increase in the patient's time with a diagnosed disease and all the subsequent disadvantages (for example, metastasized breast carcinoma).

Patient Information

Given the potential disadvantages of a screening program, it is essential for individuals considering screening to be appropriately and accurately informed of both its advantages and its disadvantages. Only so informed can they decide for themselves whether they wish to participate in a screening program.

Terminology Used in Screening

- *Invitation rate:* the percentage of eligible women invited to participate in a mammography screening program (including self-invitees).
- *Background incidence:* the incidence of invasive tumors to be expected in the target population in the absence of an active screening program.
- *Interval carcinoma:* breast cancer (invasive or in situ) detected after a negative screening examination (including diagnostic work-up, if performed) and before the next regular screening appointment.
- *Incidence (rate):* the number of new occurrences of a particular disease in a population group of a defined size (usually per 100,000) during a determined time (usually one year).
- *Longitudinal evaluation:* comparison of outcomes in one region over several time intervals.
- *Mobile mammography unit:* semitrailer or van with built-in mammography unit for mobile use.
- *Mammography facility:* a stationary facility or mobile unit that is approved to perform mammography as part of the mammography screening program.
- *Multidisciplinary case conference:* a conference in which all existing study results for a patient are reviewed and correlated, and the future management of the patient is discussed and decided upon. Meeting participants include all the physicians who were involved in the patient's diagnosis and treatment, including nonmedical personnel, as appropriate. This includes, in particular, the physician in charge of screening, pathologist, reviewer, radiology specialists, surgeon, cooperating hospital radiologist or pathologist, and the treating family doctor or gynecologist.
- *Performance indicator:* a parameter used to evaluate the process quality (in the mammography screening program, the performance indicators include all the steps of the examination chain: the invitation process, the findings, and the diagnostic work-up).
- *PPV I* (▶ Table 10.1): the positive predictive value of abnormal findings at screening: the percentage of all positive screening examinations (usually BI-RADS 0, 3, 4, and 5) that result in a tissue diagnosis of cancer within a defined time period (= true-positive examinations/positive screening examinations).
- *PPV II* (▶ Table 10.1): the positive predictive value of examinations resulting in the recommendation for tissue diagnosis: the percentage of all diagnostic and screening examinations (BI-RADS 4, and 5) that result in a tissue diagnosis of cancer within a defined time period (= true-positive examinations/positive examinations recommending tissue diagnosis).
- *Cross-sectional evaluation:* comparison of outcomes from different regions within a specified time interval.
- *Recall:* invitation to a woman who has participated in mammography screening to return for diagnostic work-up after the mammographic reading because of a suspicious finding on mammography; or due to a clinical suspicion, when there is no mammographic suspicion (*clinical recall*); or if it is determined that there is a deficiency in the diagnostic image quality and repeat imaging is recommended, when there is no mammographic or clinical suspicion (*technical recall*).
- *Recall rate (abnormal interpretation rate):* the percentage of all screening examinations interpreted as positive (usually BI-RADS 0, 4, and 5) resulting in an invitation for the woman to return for diagnostic work-up (= positive examinations/all examinations).
- *Surrogate parameter:* a parameter for short and mid-term assessment and monitoring of the effectiveness and quality of a screening program used before the actual efficacy in mortality reduction can be ascertained and analyzed.
- *Certification and recertification of the screening facilities* (using Germany as an example): comprehensive evaluation of the screening facilities including the participating personnel and the organizational structures, with regard to the quality of the structure, process, and results, by means of published protocols on behalf of the German National Association of Statutory

Table 10.1 Screening results. Parameters used to evaluate the quality of the results and the process quality of the German mammography screening program in 2008–2009 as an example, with reference range data from the EU guidelines[8]

Parameter	Reference range	Screening results	
		Initial examination	Follow-up examination
Participation rate (%)	> 70	53.7	
Recall (%)	< 7	6.1	3.0
Breast cancer detection rate (cases per 1,000 participants) (‰)	No data	8.2	5.6
Proportion DCIS (%)	10 / > 15	19.5	19.8
Proportion invasive carcinoma ≤ 10 mm (%)	≥ 25 / ≥ 30	30.0	34.9
Proportion invasive carcinoma < 15 mm (%)	50 / > 50	51.3	57.2
Proportion invasive carcinoma, lymph node negative (%)	> 70 / > 75	74.8	79.0
Proportion carcinoma, UICC stage II + (%)	< 30 / < 25	28.9	24.2
Proportion carcinomas confirmed preoperatively (%)	90 / > 90	91.7	
PPV I (%)	No data	14.8	
PPV II (%)	No data	49.4	

Abbreviations: PPV, positive predictive value; UICC, Union for International Cancer Control.
Source: Kuhl CK, Schrading S, Weigel S, et al. Die "EVA"-Studie: Evaluierung der Leistungsfähigkeit diagnostischer Verfahren (Mammographie, Sonographie, MRT) zur sekundären und tertiären Prävention des familiären Mammakarzinoms—Zwischenergebnisse nach der ersten Hälfte der Förderungsperiode. Rofo 2005;177(6):818–827.

Health Insurance Physicians (the certification takes place after issuing the license and before the onset of the public service mandate; the recertification takes place for the first time after 6 months, and subsequently at regular intervals of 30 months following the public service mandate).

10.2.2 Individualized Examination Concepts

Individualized and risk-adapted examination schemes produce a higher rate of early breast cancer detection, but they also involve higher costs than a comprehensive mammography screening program.

Take Home Point

Breast cancer early detection examinations are beneficial for all women with a significant individual risk profile determined by their age and/or other risk factors, including endogenous and exogenous hormonal factors, mammary gland density, reproductive behavior, lifestyle, and hereditary factors.

The concept of "individualized" examination refers here primarily to the aspects of the woman's age and her breast tissue density as outlined below.

10

Age

In principle, breast cancer early detection is not indicated before the age of 40 years, because the incidence of breast cancer in this age group is extremely low. An exception is made for women with a defined high-risk profile (see Chapter 10.2.3). If there does exists an indication for an early detection examination before the age of 40 years, such as carcinophobia, then sonography should be considered the primary imaging examination method of choice for these women. If cost is not an issue, breast MRI may also be performed. Use of ionizing radiation (i.e., mammography) should be avoided in this age group, however, although it may be indicated if ultrasound or MR images exhibit inconclusive findings.

Even between the ages of 40 and 60 years, the use of X-ray mammography must be considered critically and a legitimate indication for the use of ionizing radiation must be required in each individual case before the procedure is carried out.

Legitimate Indications for Mammography

- Family history of breast cancer:
  - One or more breast cancers in first- or second-degree relatives.
  - Two or more breast cancers in third- or fourth-degree relatives.
  - Ovarian carcinoma in a first-degree relative.
- High breast cancer risk profile.
- Palpable mass.
- Ambiguous findings on palpation.
- Positive sonographic findings.
- Unilateral mastodynia.
- Histologically defined high-risk lesions (e.g., atypical ductal hyperplasia [ADH], radial scar, lobular carcinoma in situ [LCIS]).
- Nipple discharge.
- Status post surgery for breast cancer (noninvasive, invasive).
- Inflammatory changes (mastitis, abscess).
- New occurrence of skin or nipple changes.

These indications accord with guidelines presented by most relevant international organizations, such as the *Guidelines for Imaging Examinations* (Breast Disease) of the German Radiation Protection Commission (▶ Table 10.2). In addition to requiring a legitimate indication before mammography is carried out, individual consideration must be given to any perceived need for obtaining a second mammographic view. Experience to date

Table 10.2 General recommendations for imaging examinations

Clinical findings	Imaging examination	Degree of recommendation	Frequency/indication
Asymptomatic woman from the age of 40 years with increased risk of disease (15–30%), e.g.: • Familial predisposition • Histologically confirmed high risk lesion • S/P radiotherapy to the thorax	Mammography	P	Every 1–2 years
	Sonography	P	Yearly
	Breast MRI	W	When ultrasound/mammography findings are inconclusive
Patients over the age of 30 years with • Ambiguous or suspicious findings on palpation • Local pain or retraction	Sonography	P	
	Mammography	W	Further work-up as necessary
	Breast MRI	W	If ultrasound/mammography is not conclusive
Patients over the age of 40 years with • Ambiguous or suspicious findings on palpation • Local pain or retraction	Mammography	P	
	Sonography	P	
	Breast MRI	W	If ultrasound/mammography is not conclusive
Nonpuerperal inflammation (suspicion of Paget's disease)	Mammography	P	
	Sonography	P	
Pathologic discharge (unilateral from single ducts and/or bloody, brown, watery or cytologically suspicious)	Sonography	P	
	Mammography	P	
	Galactography	W	

Abbreviations: P, primary examination is indicated. W, further work-up indicated.

indicates that a one-view mammogram (MLO projection) is sufficient, for example, when a supplementary breast MRI examination has been performed, sparing approximately one-half of the parenchymal dose. This concept (single view mammography + breast MRI) is known as the "Goettinger Optipack."

When there is a legitimate indication, concerns about radiation protection in mammography no longer apply in women over 60 years of age because the latency between radiation exposure and potentially induced radiation damage (e.g., radiation-induced carcinoma) averages 20 to 25 years. Nevertheless, independently of this, one must always consider whether the health benefits outweigh the radiation risk.

Parenchymal Density

The informative value of X-ray mammography depends crucially on the tissue density of a woman's breast. On this basis, the sensitivity ranges from 32% (for dense breast tissue: ACR types III, IV) to 98% (for transparent breast tissue: ACR type I). For women with a high parenchymal density, mammography represents the weakest of the three available examination methods. Considering early-detection mammograms, where most women are healthy and do not have breast cancer, there is the additional concern of to a threefold to fivefold higher radiation dose being applied to breasts with ACR density types III or IV than that for breasts with ACR density type I. This ultimately raises the question whether it makes sense for women with dense breast tissue to undergo the usual two-view mammography when a one-view mammogram may be completely sufficient, since the significance of mammography in these cases is reduced more or less to the detection of calcified tumors. Moreover, it is obligatory to carry out an examination by a complementary procedure, such as ultrasound or MRI, in these women to increase the overall sensitivity. In contrast, in women with high tissue transparency a two-view mammogram without a complementary examination procedure is usually sufficient.

Note

For women with ACR mammographic tissue density types III or IV, mammography represents the weakest examination method in comparison with breast ultrasonography and breast MRI.

The term "risk-adapted" refers to a woman's individually determined personal lifetime risk for developing breast cancer. The fundamental risk factors here are age, tissue density type, personal medical history (e.g., exposure to radiation as a child, previous surgery for high-risk lesions), and familial predisposition. The occurrence of breast or ovarian cancer in multiple relatives at a very young age should especially raise the question of a potential genetic defect.

For women with a confirmed high-risk constellation, a suitably adapted breast cancer early detection strategy is recommended (see Chapter 10.2.3).

A prospective evaluation of individualized and risk-adapted breast cancer early detection strategies utilizing mammography, sonography, and breast MRI in our own patient collective with 3,750 examinations of asymptomatic women yielded a carcinoma detection rate of 11.5‰. The tumor stage distribution was 23.3% DCIS, 9.3% carcinoma stage pT1a, 32.6% carcinoma stage pT1b, and 25.6% carcinoma stage pT1c. Carcinomas larger than 2 cm were found in 9.3% of those affected. The detection rate was 72% for mammography, 61% for ultrasound, and 98% for MRI; 4.2‰ of all carcinomas were demonstrated only on MRI.

10.2.3 Early Detection in Women with a High-Risk Profile

In approximately 5 to 10% of all women with breast cancer there is a genetic predisposition with a gene mutation displaying an autosomal dominant inheritance pattern. This primarily affects women with germline mutations in one of the known high-risk oncogenes, *BRCA1* and *BRCA2*. With regard to the morbidity rate, the age of onset, and the histopathological typing, the carcinomas in this high-risk group exhibit distinct dissimilarities from sporadic breast cancers. These characteristics are taken into consideration in good, risk-adapted early detection strategies:

- Lifetime breast cancer risk: 70 to 80%.
- Lifetime risk of developing an additional contralateral breast carcinoma: 60%.
- Lifetime ovarian cancer risk: 20 to 40%.
- Increased coincidence of breast cancer and ovarian cancer.
- Increased occurrence of multiple breast cancers.
- Age of onset frequently under 40 years.
- Increased sensitivity of the mammary gland to radiation.
- More frequent occurrence of rapidly growing carcinomas (G3).
- More frequent occurrence of triple-negative breast cancer.
- More frequent occurrence of carcinomas with a medullary growth pattern.
- More frequent occurrence of breast cancer in men.

Note

Women with a > 20% risk of being the carrier of a pathogenic gene variant (i.e., of being heterozygous), or with a lifetime breast cancer risk of over 30% are defined as having a high-risk profile.

These distinct characteristics of breast cancer development and histology in high-risk women with genetic mutations influence the age at which early detection is implemented, the screening interval selected, and the choice of the examination method used (▶ Table 10.3).

Studies to date on intensified early detection strategies consistently show the high significance of breast MRI for women with a high-risk profile. Kriege et al[9] documented a sensitivity of 96% for breast MRI, 29% for mammography, and 30% for ultrasound with a high specificity for all three methods (breast MRI, 94%; mammography, 98%; ultrasound, 91%). A multicenter study (the EVA study) done in 2010[11], also showed the impressive superiority of breast MRI in comparison with clinical examination, mammography, and/or breast ultrasonography (▶ Table 10.4 and ▶ Fig. 10.3). Supplementary clinical examination and supplementary ultrasound did not further improve the results for MRI alone. The addition of mammography revealed two additional carcinomas that were not seen on MRI alone. These were a low-grade DCIS and a 3 mm invasive breast carcinoma. It can be assumed, however, that if both these lesions had been detected one year later on the MRI, the prognosis for both tumors would have remained

Table 10.3 Example of an intensified early detection strategy for women with a high risk profile

Examination	Frequency/Indication
Clinical breast examination by the physician	Every 6 months from age 25 years, or 5 years before the earliest age of onset in the family history
Sonography	Every 6 months from age 25 years, or 5 years before the earliest age of onset in the family history
Mammography	Every 12 months from age 30 years; or from age 35 years in the case of high mammary gland density (ACR IV)[a]
Breast MRI	Every 12 months from age 25 years, or 5 years before the earliest age of onset in the family history, generally only until age 55 years or until involution of the glandular parenchyma reaches ACR I–II; cycle-dependent in premenopausal women

[a]Recommendations since 2014 advise that mammography in women at high risk should not be performed before age 40 years.

Table 10.4 Design of the EVA study

Age (years)	Clinical examination	Sonography	Mammography	Breast MRI
25	X	X		X
25.5	X	X		
26	X	X		X
26.5	X	X		
27	X	X		X
27.5	X	X		
28	X	X		X
28.5	X	X		
29	X	X		X
29.5	X	X		
30	X	X	X	X
30.5	X	X		
31	X	X	X	X

unchanged (very good). In contrast to this, more than half of all carcinomas (8 of 16 invasive carcinomas and 6 of 11 DCIS, all of intermediate type or high grade) were detected exclusively on MRI. Here it must be noted, however, that these excellent MRI results can only be achieved when the quality of the methodology and the technological aspects of breast MRI is assured.

Take Home Point

General Recommendations for Primary Prevention in Women with *BRCA* Gene Mutation

Women with *BRCA1* or *BRCA2* gene mutation may be offered a prophylactic bilateral mastectomy. A prophylactic bilateral salpingo-oophorectomy is also recommended, generally around the age of 40 years.

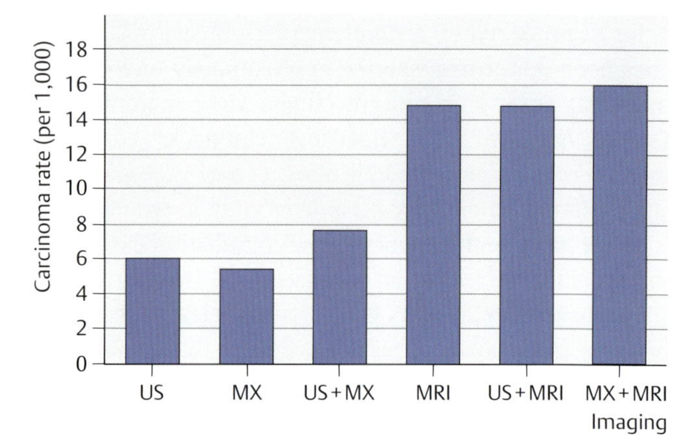

Fig. 10.3 Results of the EVA study. Detection rate of breast cancer in 386 women with a high-risk profile. Mammography (MX) and breast MRI (MRI) were administered yearly, ultrasonography (US) was performed biannually. US + MX; US + MRI; and MX + MRI correspond to combined examination methods.

10.2.4 Future Concepts of Early Breast Cancer Detection

It can be assumed that future diagnostic concepts for early breast cancer detection will increasingly take current data into account. There are three essential aspects that will lead to a paradigm change:

1. Women who are examined within the framework of early detection are, for the most part, healthy, making it all the more important that the examination strategy employed should not cause them any harm or disadvantage.
2. Breast MRI represents the most sensitive diagnostic method in the early detection of breast cancer, with a specificity that is comparable to that of the other methods. On the other hand, in its current form, it is much more expensive than mammography or sonography.
3. X-ray mammography has significant limitations, especially in women with dense breast tissue structures. For these the sensitivity of mammography is limited to 30 to 40%, and the radiation dose required for imaging is three to five times higher than in women with lower parenchymal density ("high dose, low sensitivity").

Taking into consideration the results of the EVA study (▶ Fig. 10.3), future early breast cancer detection concepts will rank the MRI examination as the primary imaging method used for early detection. If a high-quality, artifact-free breast MRI with high tissue transparency does not show any suspicious findings, then any additional complementary clinical examination, breast ultrasonography, or even mammography is unnecessary because these do not offer any additional relevant information. Thus, at least women with known dense breast tissue should be informed that there is an alternative to X-ray mammography and that breast MRI delivers better results while not involving ionizing radiation.

Future changes in the early breast cancer detection strategy must not only address the change in the primary examination technique but also deal with the associated cost issue. In a strategy devised to include breast MRI, the time expenditure and costs of MRI must be brought within a range that is comparable to that of current programs.

Preliminary approaches already address these issues.[5] Studies have shown the feasibility of one such concept—the abbreviated breast MRI, which uses a high spatial resolution MRI and permits the presence of a malignant breast tumor to be excluded with a high degree of certainty within 4 to 5 minutes.[12] If a tumor is excluded, then the examination can be terminated and the next breast MRI examination can be initiated. Using dedicated MRI units for this breast examination scheme, it is possible to examine five or six women per hour. Complementing this concept with real-time computer-aided lesion evaluation would enable evaluation of the acquired MRI examinations within seconds.[4]

Note

Early breast cancer detection without X-radiation: in the future, high-resolution, first-pass breast MRI may be the method of choice.

10.3 Diagnostic Work-up

Clinical symptoms that may indicate the presence of a malignant breast tumor must be investigated. A diagnostic work-up may require the use of clinical and all available imaging techniques to prove definitively that breast cancer is not the cause of the suspicious symptoms. In case of doubt, a percutaneous biopsy should complement the clinical and imaging diagnostics.

Primarily, clinical symptoms potentially indicating the presence of breast cancer are changes in the skin or nipple (retraction, erythema, eczema), breast induration, a palpable lump, and pathological nipple discharge. In women younger than 40 years, primary diagnostic work-up should proceed with ultrasound or MRI. Complementary mammography should only be undertaken when further investigation is still required. In women over 40 years, mammography is an integral component of the diagnostic work-up.

Take Home Points

Basic Diagnostic Procedures for the Work-up of Conspicuous Breast Findings

- Clinical breast examination: inspection and palpation of the breast and its lymphatic drainage areas.
- Mammography.
- Ultrasound.

If the clinical breast examination produces a conspicuous finding, then the diagnostic work-up should include imaging procedures and, if appropriate, a histological examination (biopsy).

For symptomatic findings in women under 40 years of age, sonography should be the imaging method of choice.

10.4 Pretherapeutic Local Staging

In patients with confirmed breast cancer (BI-RADS 6), local staging is recommended before initiation of the appropriate therapy. The following aspects need to be clarified:

- Exact localization and extent of the confirmed index tumor, including potential chest wall infiltration.
- Presence of extensive intraductal component (EIC).
- Multifocality.
- Multicentricity.
- Presence of a synchronous contralateral tumor.
- Presence of locoregional lymph node involvement.

As a rule, the appraisal of these aspects should always take into account all clinical and imaging results. Except for assessment of the locoregional lymph nodes, however, MRI has emerged as by far the most effective method in the evaluation of these aspects, delivering the most reliable results. Studies in the last few years have shown that MRI provides additional information in the affected breast that is relevant to treatment in approximately 20 to 30% of cases. In addition, manifestations of carcinoma in the contralateral breast that could not be visualized with the other imaging methods are found in 5% of cases.

- *Extensive intraductal component (EIC):* This is extensive DCIS in the immediate proximity of invasive breast cancer in which the intraductal component constitutes more than 25% of the surface area of the invasive carcinoma. To be distinguished from this are the PIC (predominant intraductal components, with more than 80% of the tumor being intraductal) and the SIC (small intraductal components, with the intraductal proportion being under 25%). EIC may be apparent on mammography as peritumoral calcifications, or on MRI as linear-dendritic areas of increased enhancement in the periphery of the tumor. The clinical importance of this is that EIC is the main reason for tumor recurrence within the breast if it is not adequately surgically excised (e.g., by segmentectomy instead of a tumorectomy) and the adjuvant treatment does not annihilate it.
- *Multifocality:* This term is not uniformly defined. The specifications in the literature range from the presence of two tumor masses within a distance of 2 to 5 cm, to having two or more tumor manifestations within one quadrant. The most pragmatic approach is to consider a tumor to be multifocal when several tumor foci lie within an area that is still treatable with breast-conserving surgery (segmentectomy, quadrantectomy, extended quadrantectomy).
- *Multicentricity:* Following from the criteria for multifocality, the term "multicentricity" should be used when several intramammary tumor foci are disposed so that a breast-conserving surgical procedure is no longer reasonable or possible. It is critical to note that before mastectomy is performed on the grounds of multicentricity, the relevant lesions that define the tumor as multicentric must be histologically confirmed by means of a percutaneous biopsy. It is not justifiable to perform a mastectomy on the basis of multicentric imaging findings without histological confirmation.

Take Home Points

General Recommendations for Pretherapeutic Diagnostic Work-up in Patients with Suspicious or Suspect Breast Findings:

- For clinically conspicuous findings in women over 40 years of age, mammography should be the imaging method of choice.
- When the mammographic density is high (ACR types III and IV), the sensitivity of mammography is limited, so that a complementary ultrasound examination should be carried out.
- Breast ultrasonography is a complementary examination for the diagnostic work-up of inconclusive clinical or mammographic findings.
- Breast ultrasound should be used in particular to clarify nonpalpable mammographic findings of BI-RADS 0, III, IV, and V.

A complementary contrast-enhanced MRI in the pretherapeutic setting can optimize the visualization of locoregional spread and improve therapeutic decision making. This is especially true when

- Conventional diagnostic work-up cannot delineate locoregional spread.
- Histological examination of the primary tumor reveals a lobular carcinoma.
- Multicentricity is suspected.
- A lymph node metastasis of unknown primary lesion is present.
- The patient has a high genetic risk.
- The patient has breast implants.
- The patient is younger than 40 years of age.
- The response to neoadjuvant chemotherapy has to be evaluated.
- Breast tissue is dense (ACR density types III–IV).
- The patient has had breast cancer previously (local recurrence).

It is not possible to reliably assess local lymph nodes with any imaging modality. The main imaging criteria suggestive of metastatic involvement are lymph node enlargement and deformation, followed by loss or displacement of the intranodal, fatty hilum. If necessary, cytologic diagnostic work-up can be performed by means of an ultrasound-guided fine needle aspiration. If the lymph nodes are morphologically unremarkable, then the sentinel lymph node biopsy technique is recommended.

10.5 Pretherapeutic Peripheral Staging

In patients with confirmed breast cancer (BI-RADS 6), peripheral staging may be recommended before initiation of the appropriate therapy. This depends primarily on intramammary and nodal baseline findings.

Take Home Points

In locally advanced carcinoma or if there is clinical suspicion of metastasis, the following studies should be performed for pretherapeutic staging:

- Chest radiography.
- Ultrasound of the liver.
- Bone scan.

Practical Tip

Preoperative tumor marker determinations are not necessary.

10.6 Follow-up Care

A patient with breast cancer has increased risk for the an ipsilateral tumor recurrence or the formation of a second, metachronous ipsilateral or contralateral breast tumor.

For this reason, regular updating of the medical history, a physical examination, and consultation are recommended: four times a year for the first to third years, biannually for the fourth and fifth years, and annually thereafter.

In addition, regular imaging examination of both breasts is recommended following breast-conserving therapy; or of the contralateral breast following mastectomy. Mammographic and ultrasonographic examinations of a breast that has had breast-conserving surgery are recommended at biannual, or at least annual intervals over a period of 3 years. The contralateral breast should be examined annually. After 3 years, both breasts should be examined at one-year intervals.

Clinical, mammographic, or sonographic differentiation between scar tissue and a recurrence is often difficult or impossible. In such cases, further diagnostic work-up with breast MRI is recommended.

While a clinical examination is of secondary importance in the context of early detection and in the detection of recurrences in the breast, it has an important role in evaluation of the chest wall after mastectomy and of the lymph drainage areas.

Take Home Points

Follow-up Care Interval

Clinical follow-up should take place four times a year for the first 3 years after primary local breast cancer therapy, twice a year in the fourth and fifth years, and annually from the sixth year onward.

Imaging Diagnostics following BCT

In women who are symptom free following breast-conserving therapy, regular diagnostic imaging (mammography, ultrasonography) of the ipsilateral breast is essential.

Mammography of the Contralateral Breast

Mammographic examinations of the contralateral breast should be done at yearly intervals in all breast cancer patients (complemented by breast ultrasound where appropriate).

There is no reliable current assessment data on the value and usefulness of intensive diagnostic imaging and/or blood testing in the follow-up of patients after breast cancer therapy. Existing studies are old and outdated and do not take newer diagnostic options into account.

Take Home Point

General Recommendations for Intensive Imaging and Laboratory Diagnostics in the Follow-up Care Plan

An intensive program of imaging and laboratory diagnostics, including chest radiography, bone scan, CT, PET, or MRI as well as complete blood count, serum biochemistry, and determination of tumor markers is not part of a standard follow-up care plan, and is only indicated when clinical findings are suspicious.

In the era of breast MRI, it is conceivable that there will be a paradigm shift in the follow-up care of breast cancer patients. Because these patients exhibit an increased lifetime risk for the a recurrence or development of a metachronous second tumor and breast MRI displays a greater sensitivity for the detection of these tumors, MRI could become the diagnostic method of choice in this context, making mammography and sonography second-line examinations when MRI demonstrates a suspicious finding. A reasonable strategy would thus seem to be, for example, the aftercare monitoring of patients by breast MRI at intervals of 1 to 2 years, depending upon the estimated biological aggressiveness of the primary tumor. The customary aftercare examination intervals of 6 months (in the first 3 years when performing mammography and ultrasonography) would therefore be superfluous. Another aspect to be reconsidered is whether a recurrence within a breast that has already received radiotherapy after BCT always necessitates a mastectomy, or whether—when transparency and thus sensitivity is high on MRI—a breast-conserving surgical procedure without follow-up radiotherapy may be appropriate.

10.7 Implant Evaluation

Breast imaging in women with implants should preferably use MRI, which is distinctly superior to mammography and sonography. This is true both for the early detection of breast cancer and for the aftercare surveillance after breast cancer therapy with reconstruction.

There are various examination protocols for the MRI examination of prostheses, depending on the point of interest:

- Contrast-enhanced, dynamic breast MRI for carcinoma detection; e.g., transversal.
- Breast MRI (silicone-sensitive, silicone-suppressed) to rule out implant complications; e.g., transversal and sagittal.

The evaluation criteria in MRI examinations undertaken to rule out or confirm breast cancer are those used for women without implants. In the evaluation of examinations to rule out or confirm prosthetic complications, there may be the following findings:

- Implant migration.
- Capsule fibrosis.
- Gel bleeding.
- Intracapsular rupture.
- Extracapsular rupture.
- Siliconoma.

10.8 Evaluation of the Male Breast

Unlike for women, there is no recommendation for regular imaging to detect breast cancer early in men, because of its significantly lower incidence. If symptoms occur that are suggestive of malignancy (changes in the skin or nipples such as retraction, erythema or eczema, induration, masses), then a diagnostic work-up is indicated. This includes inspection and palpation, single-view-mammography (MLO projection), breast ultrasonography, and, if indicated, breast MRI.

On imaging, findings indicative of unilateral or bilateral lipomastia or gynecomastia are common. In such cases it may be prudent to examine for the causes (e.g., medication or primary hormone-producing tumors). In case of doubt, a percutaneous biopsy, usually in the form of an ultrasound-guided core needle biopsy, is performed to rule out malignancy.

Bibliography

[1] Cott Chubiz JE, Lee JM, Gilmore ME, et al. Cost-effectiveness of alternating magnetic resonance imaging and digital mammography screening in BRCA1 and BRCA2 gene mutation carriers. Cancer. 2013; 119

[2] Elmore JG, Nakano CY, Koepsell TD, Desnick LM, D'Orsi CJ, Ransohoff DF. International variation in screening mammography interpretations in community-based programs. J Natl Cancer Inst. 2003; 95(18):1384–1393

[3] Fischer U. Practical MR Mammography. 2nd ed. Stuttgart: Thieme; 2012

[4] Fischer U, Korthauer A, Baum F, Luftner-Nagel S, Heyden D, Marten-Engelke K. Short first-pass MRI of the breast. Acta Radiol. 2012; 53(3):267–269

[5] Kilburn-Toppin F, Barter SJ. New horizons in breast imaging. Clin Oncol (R Coll Radiol). 2013; 25(2):93–100

[6] Kriege M, Brekelmans CT, Obdeijn IM, et al. Factors affecting sensitivity and specificity of screening mammography and MRI in women with an inherited risk for breast cancer. Breast Cancer Res Treat. 2006; 100(1):109–119

[7] Kooperationsgemeinschaft Mammografie. Evaluationsbericht 2008–2009. Ergebnisse des Mammografie-Screening-Programms in Deutschland. Berlin: Kooperationsgemeinschaft Mammografie; 2012

[8] Kriege M, Brekelmans CT, Boetes C, et al. Magnetic Resonance Imaging Screening Study Group. Efficacy of MRI and mammography for breast-cancer screening in women with a familial or genetic predisposition. N Engl J Med. 2004; 351(5):427–437

[9] Kriege M, Brekelmans CT, Obdeijn IM, et al. Kooperationsgemeinschaft Mammografie. Evaluationsbericht 2005–2007. Ergebnisse des Mammografie-Screening-Programms in Deutschland. Cologne: Kooperationsgemeinschaft Mammografie; 2009

[10] Kuhl CK, Schrading S, Weigel S, et al. The "EVA" Trial: Evaluation of the Efficacy of Diagnostic Methods (Mammography, Ultrasound, MRI) in the secondary and tertiary prevention of familial breast cancer. Preliminary results after the first half of the study period [in German]. Rofo. 2005; 177(6):818–827

[11] Kuhl C, Weigel S, Schrading S, et al. Prospective multicenter cohort study to refine management recommendations for women at elevated familial risk of breast cancer: the EVA trial. J Clin Oncol. 2010; 28(9):1450–1457

[12] Kuhl CK, Schrading S, Strobel K, Schild HH, Hilgers RD, Bieling HB. Abbreviated breast magnetic resonance imaging (MRI): first postcontrast subtracted images and maximum-intensity projection-a novel approach to breast cancer screening with MRI. J Clin Oncol. 2014; 32(22):2304–2310

[13] Morrow M, Waters J, Morris E. MRI for breast cancer screening, diagnosis, and treatment. Lancet. 2011; 378(9805):1804–1811

[14] Otten JD, Karssemeijer N, Hendriks JH, et al. Effect of recall rate on earlier screen detection of breast cancers based on the Dutch performance indicators. J Natl Cancer Inst. 2005; 97(10):748–754

[15] Perry N, Broeders M, de Wolf C, et al. European Guidelines for Quality Assurance in Breast Cancer Screening and Diagnosis. 4th ed. Vol. XVI. Luxembourg: Office for Official Publications of the European Communities; 2006

[16] Robert Koch-Institut. Gesellschaft der epidemiologischen Krebsregister in Deutschland e. V. Krebs in Deutschland 2005/2006. Häufigkeiten und Trends. 7th ed. Berlin: Robert Koch-Institut; 2010

[17] Scaranelo A. Breast screening with magnetic resonance imaging. CMAJ. 2012; 184(16):E877

[18] Schell MJ, Yankaskas BC, Ballard-Barbash R, et al. Evidence-based target recall rates for screening mammography. Radiology. 2007; 243(3):681–689

[19] Smith-Bindman R, Chu PW, Miglioretti DL, et al. Comparison of screening mammography in the United States and the United kingdom. JAMA. 2003; 290(16):2129–2137

[20] Tabár L, Vitak B, Chen HH, Yen MF, Duffy SW, Smith RA. Beyond randomized controlled trials: organized mammographic screening substantially reduces breast carcinoma mortality. Cancer. 2001; 91(9):1724–1731

[21] Tozaki M. BI-RADS-MRI terminology and evaluation of intraductal carcinoma and ductal carcinoma in situ. Breast Cancer. 2013; 20(1):13–20

[22] Urbschat I, Kieschke J, Schlanstedt-Jahn U, von Gehlen S, Thiel A, Jensch P. Contributions of the epidemiological cancer registries to the evaluation of mammography screening in Germany [in German]. Gesundheitswesen. 2005; 67(7):448–454

[23] Wittekind C, Meyer HJ, Bootz F. TNM Klassifikation maligner Tumoren. 6. Aufl. Berlin: Springer; 2002

11 Surgical Treatment of Breast Carcinoma

T. Kuehn

11.1 Significance of Surgery in the Context of Multimodal Treatment of Breast Carcinoma

The surgical treatment of breast cancer has changed significantly in the last few decades. The reason for this is an improved understanding of the complexity of the disease, its biological behavior, and the impact on various therapeutic options (surgery, systemic therapy, radiation therapy). Each surgical treatment is integrated into a multidisciplinary adjuvant-therapy scheme (multimodal treatment) that must be individually determined for each patient. With improved early detection, breast cancer is increasingly detected in its preliminary stages, the treatment of which is fundamentally different from that of invasive carcinoma.

Preoperative determination of the histology using core needle or vacuum-assisted biopsies has also led to distinctly improved and better targeted surgical treatment. Preoperative knowledge of important tumor-biological information can simplify surgical planning significantly. To this end, the standardized histological assessment of the core needle biopsy with regard to its histological malignant potential is of great importance.

11.2 Types of Breast Carcinoma

11.2.1 Lesions of Uncertain Biological Potential (B3 Lesions)

Mammography screening is detecting more and more clinically occult lesions, the clinical significance of which still cannot be conclusively clarified. Lesions in this group include

- Flat epithelial hyperplasia.
- Atypical ductal hyperplasia.
- Lobular intraepithelial neoplasia (LIN).
- Papillomas.
- Radial scars.
- Phyllodes tumors.

In addition to their individual risk of progression, these lesions also serve an "indicator" function (giving an indication of malignant events in their immediate vicinity). Moreover, they can indicate a general increased risk of disease (risk marker).

For a B3 lesion verified by core needle or vacuum-assisted biopsy, the intent of surgical intervention is primarily diagnostic. In about 20% of cases, there is an "upgrade" to a preinvasive or invasive lesion on the basis of the final surgical specimen. Due to the very diverse risks involved in B3 lesions, the indication for surgical intervention in each case should be determined in discussion by an interdisciplinary team including radiologist, surgeon, and pathologist.

11.2.2 Preinvasive Carcinoma (Ductal Carcinoma in Situ; B5a)

The morphology and biological risk potential of DCIS differs importantly from that of invasive carcinomas. In contrast to invasive carcinoma, DCIS is a locally circumscribed disease. The tumor cells grow inside the milk ducts without passing through the basement membrane. Thus, metastasis is essentially not possible in the case of pure DCIS. The risk of progression of DCIS to an invasive carcinoma is 14 to 52%. Surgery is the most important treatment modality for DCIS; it is significantly more effective than radiotherapy or systemic medical treatment. Even without adjuvant measures, mastectomy achieves a cure rate approaching 100%. Because of the lack of potential to metastasize, and the corresponding minimal threat to life, breast conservation has been adopted as standard therapy in cases in which the spread of the lesion is limited. This approach involves accepting a small risk of recurrence: this risk is diminished, but not entirely reversed, through the use of adjuvant radiation therapy. It should be noted that 50% of recurrences of DCIS have the appearance of invasive tumor, so that the quality of the primary surgery is extremely important.

11.2.3 Invasive Carcinoma (B5b)

Surgical excision of the primary tumor and regional lymph nodes constitutes an important pillar in the multimodal therapy concept of invasive breast carcinoma treatment. The extent of surgery can be significantly reduced by adjuvant therapy measures, such as systemic and radiation therapy.

In previous times the goal of breast carcinoma surgery was mechanical excision of all accessible tumor cells, both those that are clinically apparent and those that are occult, resulting in very radical surgery. This included resection of the entire breast (including the pectoral muscles, if required) and all accessible lymph nodes. Even for small tumors, a supplementary course of radiotherapy to the chest wall and regional lymph drainage areas was added to ensure local tumor destruction. This radical procedure was associated with high morbidity (▶ Fig. 11.1).

In the second half of the 20th century, large randomized studies showed that surgical radicality did not affect overall survival. Thus the "extended radical mastectomy" (mastectomy including the pectoral muscles) offered no advantage when compared to the "modified radical mastectomy" (mastectomy preserving the pectoral muscles). Later, it was shown that breast-conserving therapy (BCT), with irradiation of the breast tissue that remained after surgery, had the same survival rate as complete excision of the breast (▶ Fig. 11.2).

The findings from these studies have led to an important strategic paradigm shift in the treatment of breast cancer. Invasive carcinoma of the breast was understood to be a systemic disease, the prognosis of which is determined by the biology of the individual tumor type and its potential for distant metastasis. It was much more the improvement of systemic medical treatment (hormonal and chemotherapy) and less the mechanical tumor

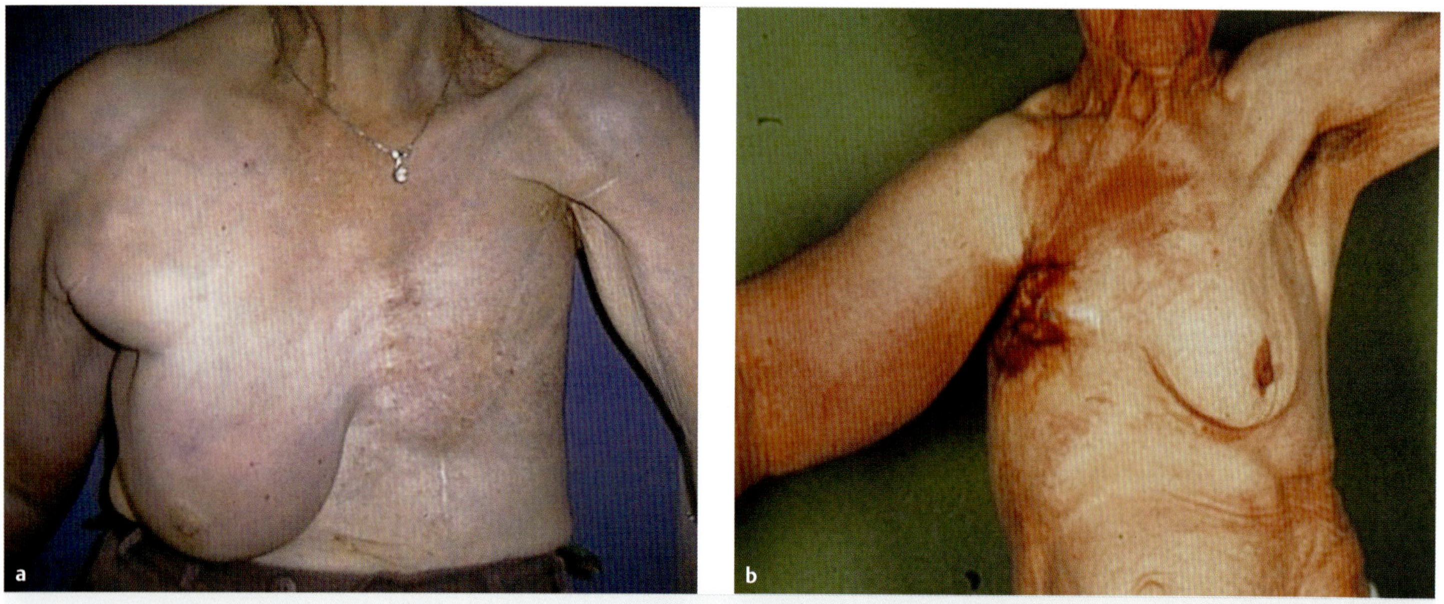

Fig. 11.1 **Radical mastectomy. (a)** Radical mastectomy (Rotter–Halsted) with excision of the breast, the pectoral major muscle, and the axillary lymph nodes. **(b)** Radical mastectomy (Rotter–Halsted) with additional irradiation. Formation of secondary lymphedema of the right arm.

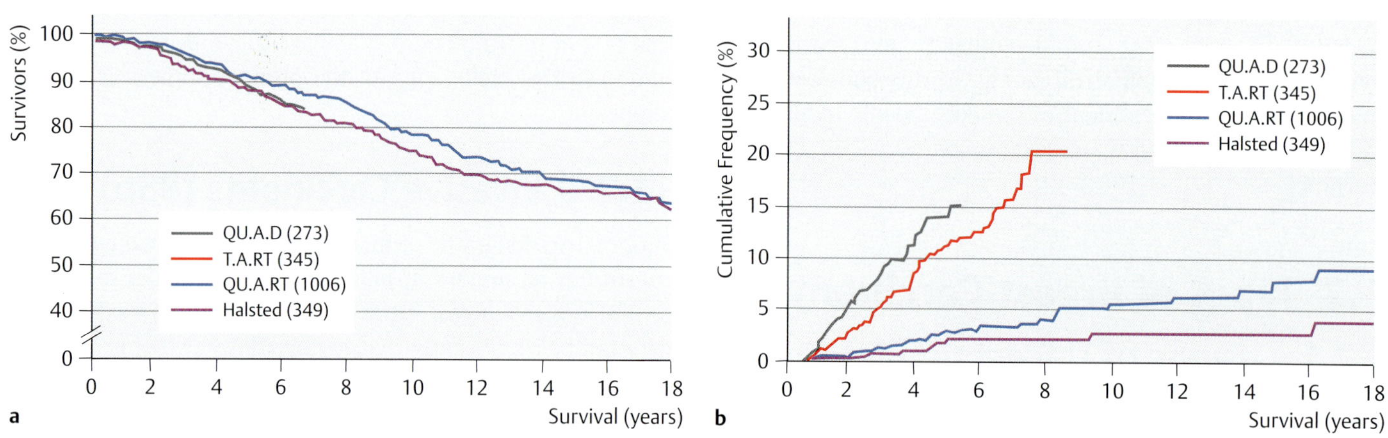

Fig. 11.2 Comparison of surgical techniques and survival curves with the Halsted radical mastectomy. From pooled study data from the Milan Trials (I, II and II). **(a)** Overall survival rate of patients who have undergone various combinations of surgical and radiological treatments. **(b)** Cumulative incidence of local recurrence depending on the treatment used (all patients). Halsted: Halsted radical mastectomy. QU.A.D: quadrantectomy without radiation therapy. QU.A.RT: quadrantectomy with radiation therapy. T.A.RT: tumorectomy with radiation therapy.

destruction (surgery and radiation therapy) that became the focus of interest. This development was the foundation for our current treatment concept that surgery secures local tumor control whereas systemic therapy has the most important influence on prevention of distant metastases and thus on overall survival. Medical treatment also influences locoregional recurrence. Despite of the continuing reduction in extensiveness of surgery, the overall survival in breast carcinoma has been dramatically improved through the use of the multimodal treatment concept.

As well as controlling the tumor locally, surgical excision of the primary tumor and the lymph nodes provides important diagnostic and prognostic information. Both the tumor size (pT staging) as well as the nodal status (pN staging) are determined on the basis of the surgical specimen.

Note

For advanced local tumors or tumors with an unfavorable biology (triple-negative, HER2-positive), the treatment sequence of surgery first followed by systemic therapy is increasingly being reversed—surgery is *preceded* by medication treatment, most often chemotherapy (neoadjuvant therapy).

This procedure allows reduction of radical surgery and the increased use of breast-conserving therapy, and also this “in vivo chemosensitivity test” provides further important prognostic information.

11.3 Surgical Treatment of the Primary Lesion

11.3.1 Oncologic Aspects

Invasive Carcinoma

The fundamental principle for the treatment of primary invasive breast carcinoma is complete surgical resection with free margins ("in sano").[19] This can be carried out either as a mastectomy or as breast-conserving surgery. In appropriate cases, breast-conserving therapy is comparable to mastectomy in terms of survival, and so breast-conserving surgery represents the standard procedure for surgical treatment of invasive carcinoma of the breast.

Take Home Points

General Basic Principle for Tumor Resection of Invasive Carcinoma
The fundamental principle for the treatment of nonadvanced breast carcinoma is resection of the tumor in sano.

Goal of Breast-conserving Therapy of Invasive Carcinoma
The goal of operative treatment is tumor excision. With reference to survival, breast-conserving therapy (BCT) with subsequent radiotherapy of the entire breast is comparable to modified radical mastectomy (MRM) alone.

Breast-conserving procedures are not possible in the following situations:

- Multicentric carcinoma.
- Inflammatory carcinoma.
- Extensive microcalcifications.
- If radiotherapy is not possible.
- If surgical excision of the tumor in sano is not possible while still preserving the breast.

As part of BCT, it is obligatory to treat the remaining breast tissue with radiation therapy. With this, the local recurrence rate has been reduced from over 40% to the current rate of under 5%. Invasive foci of carcinoma are still commonly detectable microscopically at a distance of several centimeters from the primary lesion. A complete and reliable excision of all tumor cells is thus not clinically feasible and, notably, not verifiable. Even a resection "in sano" does not preclude microscopic tumor residues, but these can be controlled by means of the radiation therapy.

The optimal resection margin for invasive carcinoma in BCT has been intensively and controversially debated. Unfortunately, there are no available data from prospective studies concerning this question. A meta-analysis published in 2010, however, unambiguously demonstrated the impact of the free surgical margin on the recurrence rate.[13] Patients with free surgical margins displayed significantly lower recurrence than women with "narrow" or "affected" resection margins, though widening the resection to a healthy tissue margin of 2 to 5 mm did not lead to any further improvement in the results.

Therefore, according to current guidelines, a free surgical margin ("no ink on tumor") is recommended. Widening the resection further can worsen the cosmetic result without reducing the recurrence rate.

Ductal Carcinoma in Situ

As opposed to invasive carcinoma, ductal carcinoma in situ (DCIS) is treated as a local disease rather than a systemic one. The quality of the surgery is therefore of particular importance. The goal of surgical treatment is complete resection of the entire affected tissue area.

DCIS is generally a unicentric lesion that spreads within a mammary duct system, rarely affecting multiple milk ducts. During their spread, the tumor cells can exhibit a discontinuous growth pattern. The "gaps" between the tumor parts can extend up to 1 cm. In contrast to the case with invasive carcinoma, therefore, a wide resection margin is very important here; a free resection margin per se does not guarantee a reliable surgical excision of the entire lesion, and adjuvant treatment modalities are less effective for DCIS than for invasive carcinoma.

As with invasive carcinoma, there are no prospective studies available on which to base the determination of an optimal resection margin. This is especially important because, for the majority of women who receive BCT, adjuvant radiotherapy is indicated. According to a meta-analysis by Dunne et al,[8] a resection margin of 2 mm appears to be adequate for these women. However, a subsequent meta-analysis by Wang et al suggests that even women who receive radiotherapy can benefit from a widened surgical margin of 1 cm.[31]

Take Home Points

Goal of Surgery for Preinvasive Lesions
The resection margin is an important prognostic factor for DCIS. The tumor-free distance to the surgical margin should be at least 2 mm if postoperative irradiation is administered.

As a rule, DCIS is not clinically apparent. It is diagnosed by way of early detection mammography showing suspicious microcalcification formations or by linear-dendritic areas of increased enhancement on MRI. During surgery, the surgeon must make allowance for the fact that the histomorphological extent of the DCIS is usually greater than the area of microcalcification demonstrated on imaging.

Lesions of Uncertain Biological Potential (B3 Lesions)

Surgery for a B3 lesion has both diagnostic and therapeutic functions. After excision, approximately 20% of the changes turn out to be high-grade neoplasia (demonstrating the diagnostic function of the excision). B3 lesions carry their own risk of progression, so resection of them also has therapeutic intent. Adherence to a specific surgical margin is not mandatory for B3 lesions.

11.3.2 Technical Aspects

Resection

After prevention of local recurrence, an important goal in the surgical treatment of breast carcinoma is the preservation of an aesthetic body image. The quality of primary surgery can even influence the long-term surveillance of the breast positively by preventing postoperative seromas, extensive scarring, and fibrosis. To minimize the burden on the patient, the number of procedures required to achieve complete tumor resection should be kept as low as possible. According to currently available data, a re-resection is not mandatory in cases with close surgical margins of 1 mm or less.[19] The rate of second procedures can be minimized by intraoperative imaging (sonography or radiography of the specimen) and by confirmation of the macroscopically determined surgical margin by the pathologist. In doubtful cases, an intraoperative re-resection can be performed.

Note

Pretherapeutic determination of the surgical target volume is of great importance in assuring the complete closure of the tissue defect created by the surgery.

It is important to cover all defects to prevent hollows in the tissue, which could fill with seroma, in turn leading to widespread fibrosis and scar formation. These could hinder further imaging surveillance of the breast and adversely effect the cosmetic result. A preoperative interdisciplinary case conference including radiologist, surgeon, and pathologist, is necessary for optimal breast volume assessment—which in particular must include compilation of associated preoperative tumor elements.

- *Palpable lesions:* For palpable lesions, the surgery can be performed using digital guidance. Intraoperative use of sonography is becoming increasingly established to ensure the reliable excision of the target volume, but also to spare healthy tissue as much as possible.
- *Nonpalpable lesions:* Due to the increased use of early breast cancer detection methods, including mammographic screening, lesions are identified ever more frequently that are only detectable using imaging (sonography, mammography, MRI) and that exhibit no palpable counterpart. Image-guided marking is required for the surgical excision of such a nonpalpable lesion, and should be done using the simplest modality by which the lesion (mass lesion, areas of microcalcification, or non-space-occupying enhancement) is best visualized. The marking is accomplished using sonographic, mammographic, or MRI-guided wire marker placement. The tip of the wire should be placed as close as possible to the target lesion. In the case of extensive areas, such as extensive microcalcifications, it is a good idea to mark several surgically relevant corner points of the target volume to be excised.

An imaging examination of the surgical specimen (specimen sonography or mammography) must be carried out to ensure the reliable removal of the target area when surgically excising nonpalpable lesions.

An intraoperative frozen specimen of the surgical specimen should not be obtained for nonpalpable lesions because the technique of specimen preparation would cause loss of tissue. Intraoperative pathological evaluation is rarely necessary and is only indicated in the case of a palpable tumor. This is performed as a macroscopic evaluation of the surgical margin and merely represents an orienting initial examination to minimize the rate of follow-up resections.

Because the exact spread of a breast carcinoma (especially its accompanying, preinvasive intraductal tumor extensions) is frequently not identifiable on imaging, a secondary reexcision may be determined to be necessary after the complete processing of the surgical specimen. Due to the segmental spread of intraductal tumor parts, it is advisable to perform a resection following the orientation of the mammary duct segments. Using this technique, a more radical resection is pursued in the axis that has the greatest probability of containing occult tumor extensions than in glandular tissues that have a lower probability of residual tumor extension.

Unambiguous topographic marking of the specimen is essential in the evaluation of the surgical margins to enable a targeted follow-up resection in case of narrow or insufficient excisional margins. For this, the specimen is marked with sutures. Ideally, it is also pinned onto a template to document the relationship of the resection margins to the remaining mammary glands in a reproducible way (▶ Fig. 11.3).

Targeted irradiation of the tumor bed (radiation boost), in addition to the adjuvant irradiation of the entire breast, has been shown to significantly reduce the local recurrence rate. The tumor bed or resection margins should be marked intraoperatively with titanium clips to enable the radiation therapists to plan for the radiation boost to the area as precisely as possible, and to minimize the target volume (▶ Fig. 11.4).

Defect Reconstruction–Oncoplastic Surgery

Along with an adequate surgical resection, the preservation or restoration of an aesthetic breast shape is an important target criterion of surgical treatment. To satisfy breast-conserving concepts, every tissue defect must be systematically reconstructed so as to avoid retractions or collections of serum with resultant contractures. Depending on the size of the defect, a variety of different techniques come into play, all of which are included in the term "oncoplastic surgery." This is understood combine an oncologically oriented resection with the immediate reconstruction of the defect. To this end, a multitude of operative techniques are employed.

- *Glandular flap reconstruction:* Smaller tissue defects can be closed by moving local (autologous) glandular tissue. A variety of techniques are used in a procedure that ranges from simple adaptation of the resection margins to a wide-ranging mobilization of the entire mammary gland and the preparation of an extensive glandular rotation flap.
- *Tumor-adapted mastopexy:* In the tumor-adapted mastopexy, the skin envelope surrounding the breast is adapted to fit the smaller breast volume. This method may be required if a defect cannot be covered with autologous tissue. It can also be used if it is necessary excise the overlying skin because of its proximity to the tumor. This usually leads to distortion of the breast and/or the nipple-areola complex. An unfavorable result can be avoided only by reshaping the skin envelope. Depending on the location of the tumor, various incisional patterns can be used. Recentering of the nipple–areola complex is usually required when a mastopexy is performed (▶ Fig. 11.5).

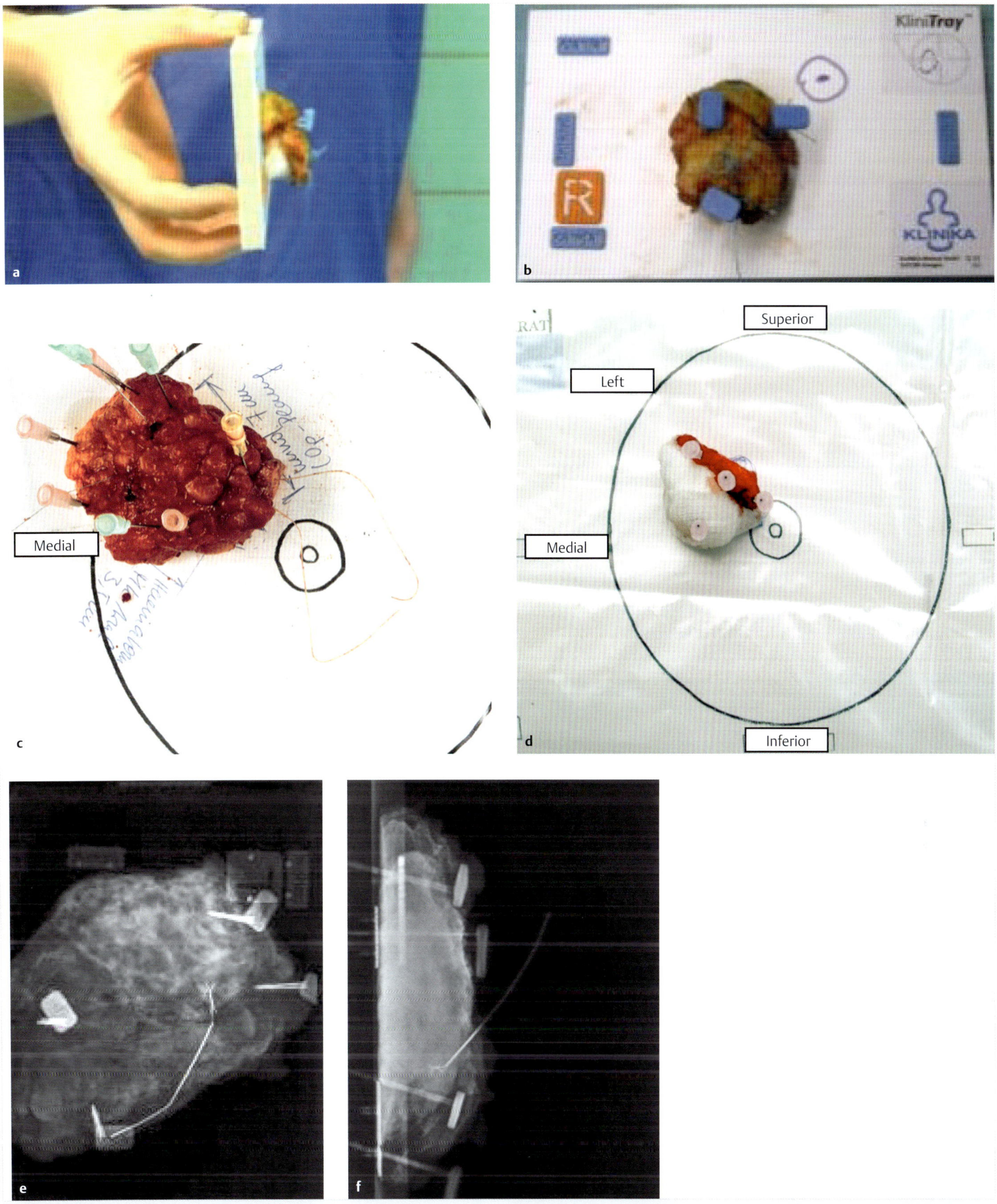

Fig. 11.3 **Perioperative and radiographic processing of the surgical specimen.** **(a)** Lateral view of the template. **(b)** Surgical specimen attached to the template. **(c)** Topographical identification markings for a surgical specimen. **(d)** Topographical identification markings for a follow-up resection. **(e)** Specimen radiograph while maintaining the topographical orientation. **(f)** Specimen radiograph (lateral image).

- *Tumor-adapted reduction mammoplasty:* In this method, further breast tissue is excised in addition to the target volume removed for oncologic reasons. In the case of large tumors with unfavorable locations, a reduction mammoplasty with wide excision margins can result in a good cosmetic result. And the effect of future radiotherapy can be improved in very voluminous breasts. The various techniques used when performing a tumor-adapted reduction mammoplasty differ in the pedicle-flap techniques used to reposition the nipple. As a rule, when a tumor-adapted reduction mammoplasty is done, an equalizing procedure on the other breast will have to be performed as well (▶ Fig. 11.6).
- *Defect coverage using transfer of distant tissue:* When the relationship between the breast and the required resection volume is unfavorable, the resected glandular tissue (and skin, if applicable) can also be covered by the transfer of distant tissue. This comes into play if the skin envelope, or the entire breast itself, is insufficient to allow a tumor-adapted mastopexy or a breast reduction because of the large resection volume or because a breast is primarily very small. In breast-conserving surgery, the latissimus dorsi muscle can be used as a myocutaneous flap. In this procedure, a flap of skin and fat is transferred from the back to the chest wall. To maintain tissue perfusion, the latissimus dorsi muscle, serving as a "guide rail" for the thoracodorsal vascular bundle, is resected along with the overlying skin and adipose tissue. Only the narrow, superior insertion of the muscle remains intact in its original position. On this pedicle, the autologous tissue from the back is rotated to the chest wall (reconstruction using a myocutaneous pedicle flap; ▶ Fig. 11.7)

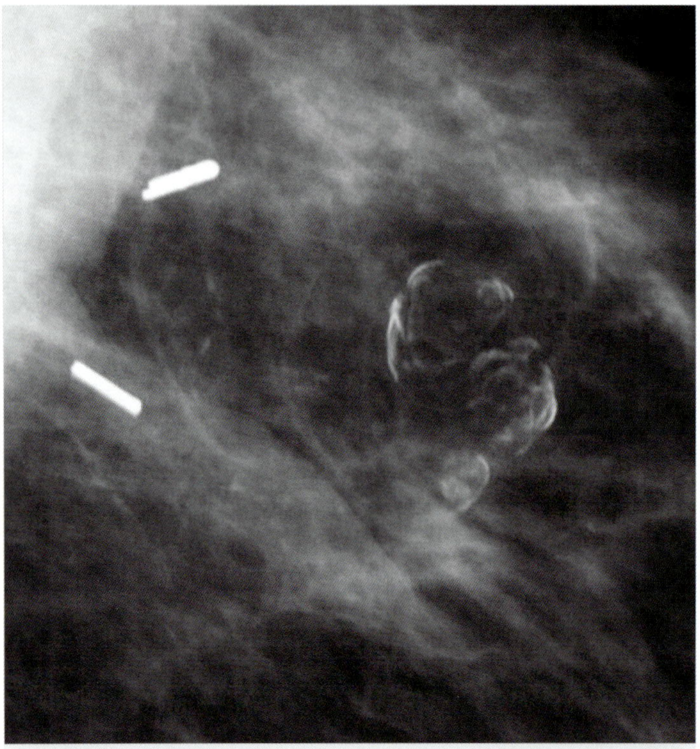

Fig. 11.4 Marking the tumor with clips. Mammographic visualization of two radiopaque clips in the area of the resection.

11.4 Lymph Node Surgery

The surgical excision of lymph nodes from the drainage system of the breast has always been an important component of the surgical treatment of carcinoma of the breast. Both the goal of lymph node excision and its technical procedure have changed considerably in the last few decades. The extent of lymph node surgery has also been markedly reduced.

The first surgical strategy for breast carcinoma was developed over 100 years ago and was maximally radical; its goal, not only in the breast but also in the lymphatic drainage areas, was to surgically excise all accessible, potentially tumor-afflicted tissue. Once invasive carcinoma of the breast was recognized as a systemic disease in which the prognosis is influenced more by systemic therapy (medication) than by local treatment, the goal of surgical excision of lymph nodes changed.

The status of the lymph nodes was identified as the most important prognostic factor. The intensity of medical treatment (chemotherapy vs. hormonal therapy) was increasingly oriented to the prognosis, and therefore to the lymph node status. Thus, axillary lymph node dissection (ALD), which for many decades had been considered the gold standard for the surgical treatment of invasive breast carcinoma, became a diagnostic intervention

11

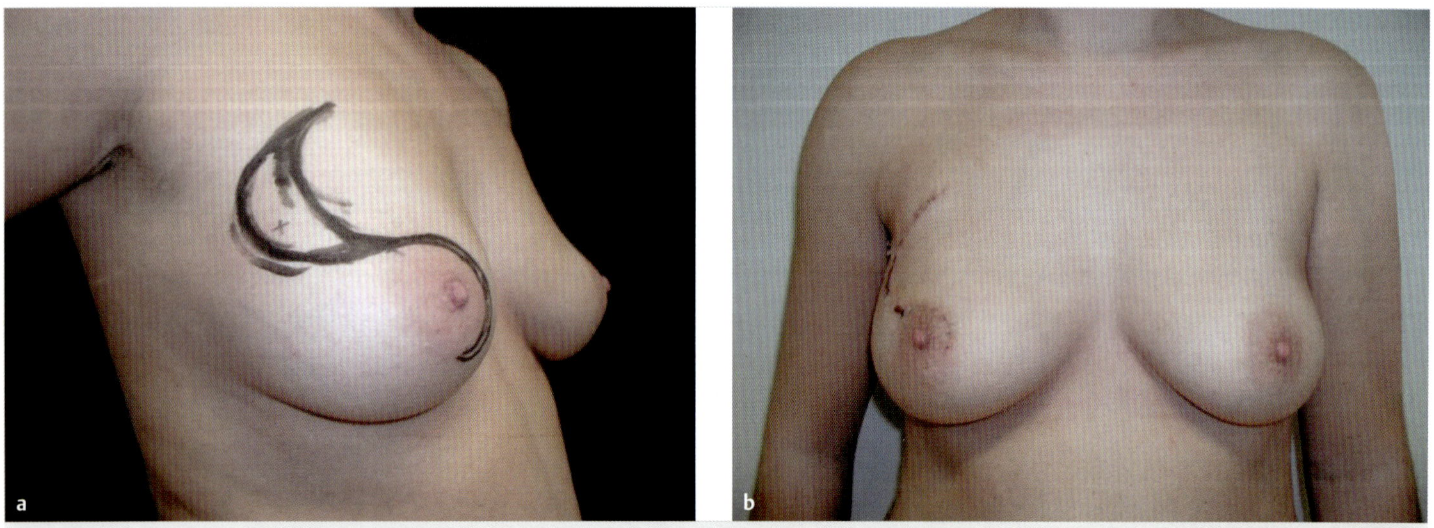

Fig. 11.5 Mastopexy. Tumor-adapted mastopexy. Reduction and restructuring of the skin envelope to preserve an aesthetic result (e.g., when the tumor is located close to the skin). **(a)** Reference lines for the surgical incision. **(b)** Results of the surgery.

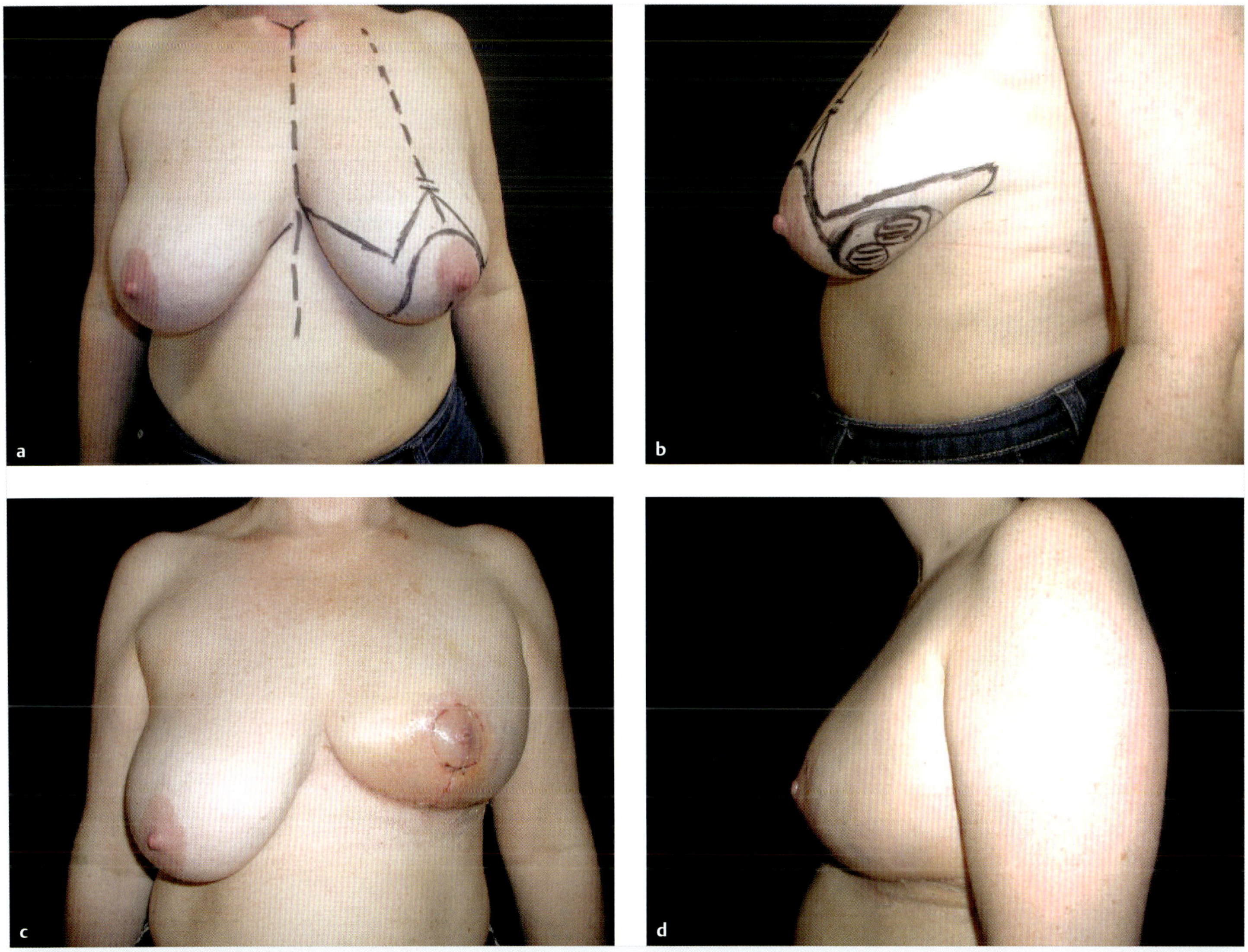

Fig. 11.6 Breast reduction surgery for unfavorable results of tumor resection or unfavorable tumor site. The surgical technique used (e.g., pedicle flap for the nipple) depends on the tumor location (tumor-adapted breast reduction). The equalizing procedure for the other breast can be carried out either at the same time as primary surgery, or later. **(a)** Reference lines for the surgical incision: frontal view. **(b)** Reference lines for the surgical incision: lateral view. **(c)** Results of the surgery: frontal view. **(d)** Result of the surgery: lateral view.

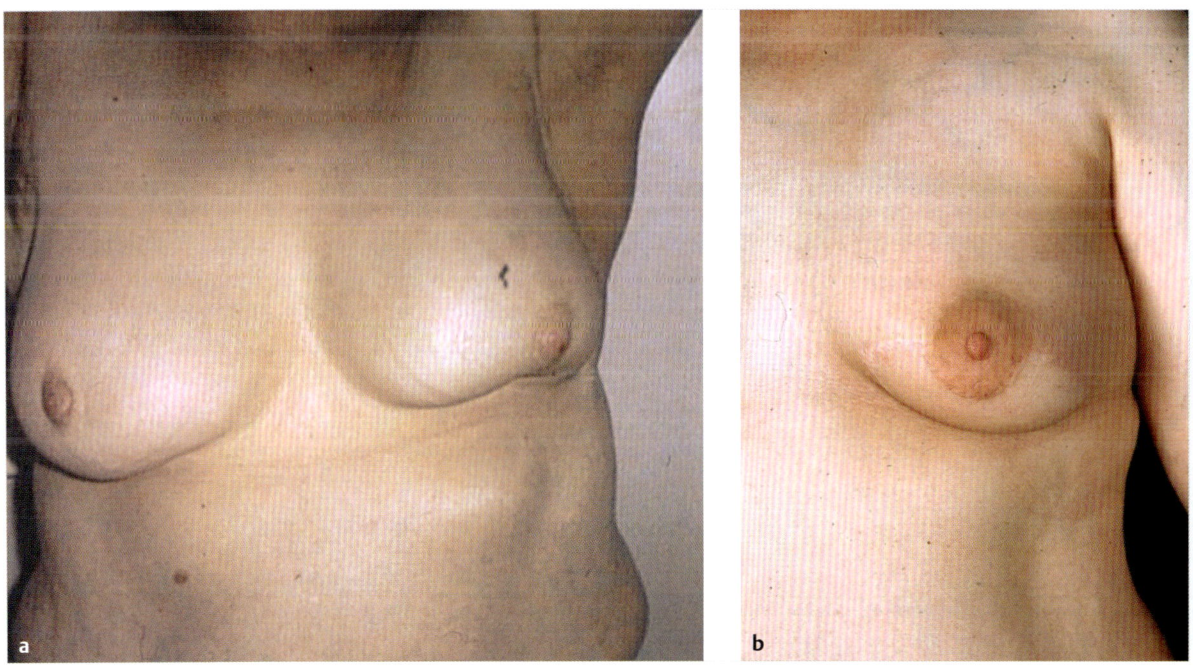

Fig. 11.7 Latissimus dorsi flap: plastic surgery using a myocutaneous flap. (a) Simple resection at the lower pole of the breast without tissue replacement. **(b)** Reconstruction of the lower pole of the breast using autologous tissue (here, a latissimus dorsi flap).

tool in patients with clinically negative lymph nodes. Simply confirming lymph node involvement histologically meant that the ALD would be carried out for therapeutic reasons.

Note

We are at present experiencing a paradigm shift: increasingly, medical treatment of breast carcinoma is based more on the biology of a tumor and less on its prognosis.

This means that the medical treatment (chemotherapy vs. hormone therapy) depends more on the expected therapeutic effect on the individual tumor type than is on prognostic parameters. The classification of the tumor into its genetic profile (luminal A, luminal B, HER2-positive, triple-negative) contributes more to the decisions regarding medical treatment than do prognostic factors such as the nodal status. This paradigm shift forms the basis for the present discussions whether the extent of regional treatment should be reduced still further. According to current guidelines, however, determination of the nodal status remains an integral component of the surgical treatment of carcinoma of the breast.

Until only a few years ago, complete ALD with removal of at least 10 lymph nodes from axillary levels I and II was the only option for verifying or excluding lymph node involvement and thereby determining the pN status. No imaging methods—including sonography, CT, MRI, and even positron emission tomography (PET)—demonstrate sufficient sensitivity and/or specificity to determine lymph node involvement with the necessary precision.

However, ALD is associated with high postoperative morbidity (pain, lymphedema, and limitation of movement and strength). In fact it has been shown that in patients with long-term survival the morbidity after ALD represents the most important negative factor affecting the quality of life (▶ Fig. 11.8).

In the late 1990s, the concept of sentinel lymph node biopsy (SLNB) was introduced as a minimally invasive method of determining the nodal status in cases of breast carcinoma. According to this principle, solid tumors have an orderly lymphatic drainage pattern. This means that tumor cells spread by way of a reproducible lymph node that drains the lymph first—the "sentinel node" (SLN)—before affecting other lymph nodes. This principle enables one to mimic the path of lymphogenic metastasis: a tracer (dye or radioactive substance) is injected into the breast and then accumulates in the SLN. Using an image of the tracer, the SLN can be identified and resected while sparing additional axillary tissue. The best results are obtained through the use of radioactive colloid; the SLN can then be visualized and localized by the lymph node scintigraphy.

Intraoperatively, the SLN is located by with the help of a gamma probe. Because SLNB is a minimally invasive procedure, the morbidity in the shoulder–arm region can be dramatically reduced (▶ Fig. 11.9).

According to evaluation of SLNB in large, randomized studies, it was comparable with the ALD not only in terms of the locoregional recurrence rate but also with regard to overall survival. The reduction in shoulder–arm morbidity was highly significant, however. A detailed examination (palpation and sonography) of the axilla is required for all women who undergo SLNB; further surgical procedures are based on the findings of that examination.

SLNB is only required for invasive carcinomas; axillary staging is not necessary in cases of DCIS. Only women who undergo a mastectomy (for DCIS) should undergo SLNB, because, if

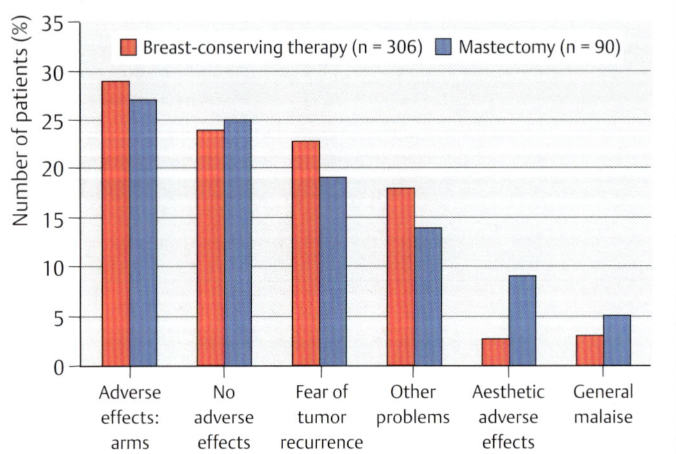

Fig. 11.8 Long-term morbidity secondary to axillary lymph node dissection. Problems in the shoulder–arm area represented the biggest limitation to long-term quality of life for the patients in the study who had undergone ALD[36]. For the patients, the encumbrance from the arm problems was more severe than the loss of the breast.

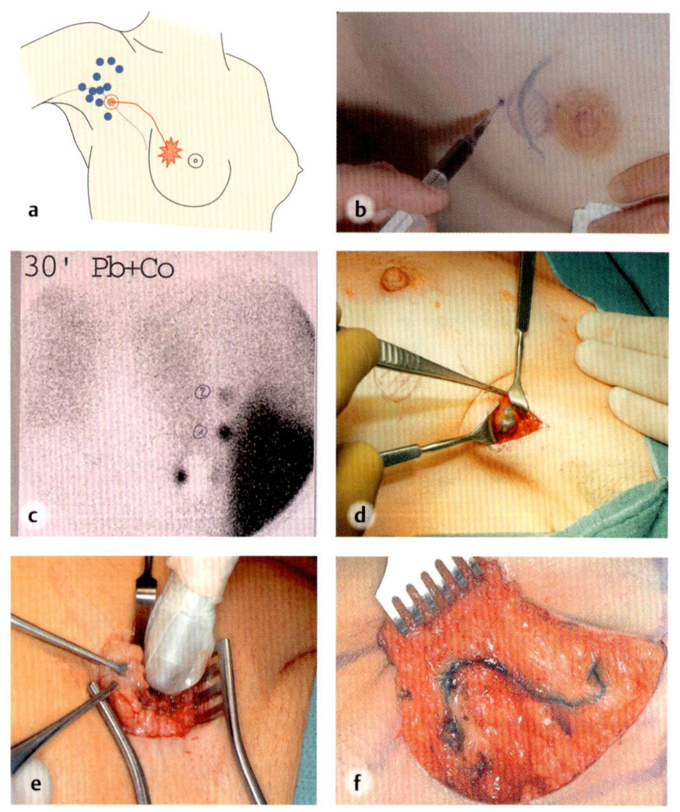

Fig. 11.9 Sentinel lymph node biopsy (SLNB). (a) Marking the SLN (schematic representation). **(b)** Peritumoral injection of a dye solution. **(c)** Lymphoscintigraphy: robust hot spot in the breast (primary injection) and visualization of one strong and one weak area of accumulation in the SLN. **(d)** Small skin incision to localize the SLN. The SLN is dyed blue. **(e)** Searching for the SLN with the gamma probe. **(f)** Blue-colored lymph channel with dyed SLN after the dye injection.

unexpected invasion is confirmed later a secondary SLNB would no longer be possible because of the destruction of the primary lymph channels.

11.4.1 Procedure for Clinically Negative Node Status

SLNB is indicated in all patients who have surgery for an invasive carcinoma of the breast and who exhibit unremarkable node status on clinical and/or sonographic examination. SLNB can also be carried out in the case of multicentric carcinomas and large tumors.

Take Home Points

General Recommendations and Insights for the Surgical Treatment of the Axilla for Invasive Carcinoma

- Determination of the histological nodal status (pN status) is a component of the surgical treatment of invasive breast carcinoma. This should be done with the help of sentinel lymph node resection (SLNB).
- SLNB is comparable to axillary dissection in terms of local control in patients with a negative SLN.

Injection of a radioactive tracer leads to the detection of an axillary SLN in approximately 99% of patients. In the rare cases in which no SLN can be identified, primary ALD may be considered because, in these cases, there is frequently a clinically occult, highly positive nodal status.

11.4.2 Procedure for Clinically Positive Nodal Status

The lymph nodes in the axilla are frequently enlarged and palpable in patients who have undergone a core needle or vacuum-assisted biopsy. A thorough clinical examination is necessary to spare these women an unnecessary ALD; an ultrasound examination of the axilla provides the most valuable information in these cases. If the structure of the hilum and cortex of the lymph nodes is not pathologically altered, even an enlarged and/or palpable lymph node can be classified as normal. In contrast, if the hilar/cortical structures are displaced, the lymph node must be classified as suspect, even if this lesion has not been found to have positive nodal status. Increasingly, interventional lymph node diagnostic work-up is carried out in these cases.

Sonographically suspicious lymph nodes are investigated by fine needle aspiration or core needle biopsy. If tumor cells are verified in the lymph node, then a primary ALD is carried out. If proof of a tumor is still lacking, then SLNB is done as the next step (▶ Fig. 11.10, ▶ Fig. 11.11).

11.4.3 Procedure for Clinically Negative Node Status and Positive Sentinel Node

About 30% of the women who (with a clinically negative nodal status) undergo a SLNB, exhibit a SLN that is histologically affected by tumor (pN1 sn). Until a few years ago, axillary lymph node dissection was the gold standard for all women with a positive sentinel lymph node, irrespective of whether it was a micrometastasis or a macrometastasis.

The approach in women with a positive sentinel lymph node is being reappraised. Multiple randomized studies indicate that the surgical removal of lymph nodes in women with a positive SLN does not have a high therapeutic effect.

Even in earlier studies (before the introduction of SLN), a randomized approach was used to compare the therapeutic effect of axillary lymph node dissection with the approach of withholding any kind of lymph node removal. Even though each of the available studies was carried out in selected low-risk collectives, no advantage for ALD was demonstrated in these studies.

A recently published randomized study (ACOSOG Z0011)[35], included patients with small tumors (T1 and T2) who had undergone breast-conserving surgery with subsequent tangential field radiation to the breast and who exhibited at most two affected SNLs. They were divided into two groups: one

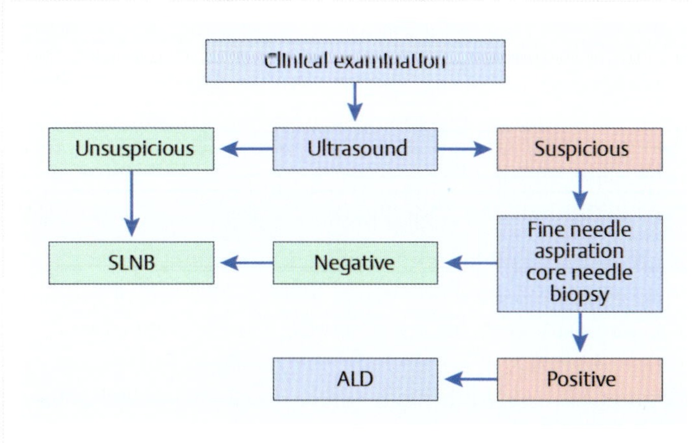

Fig. 11.10 Algorithm for clarification of the axillary lymph node status. ALD, axillary lymph node dissection; SLNB, sentinel lymph node biopsy.

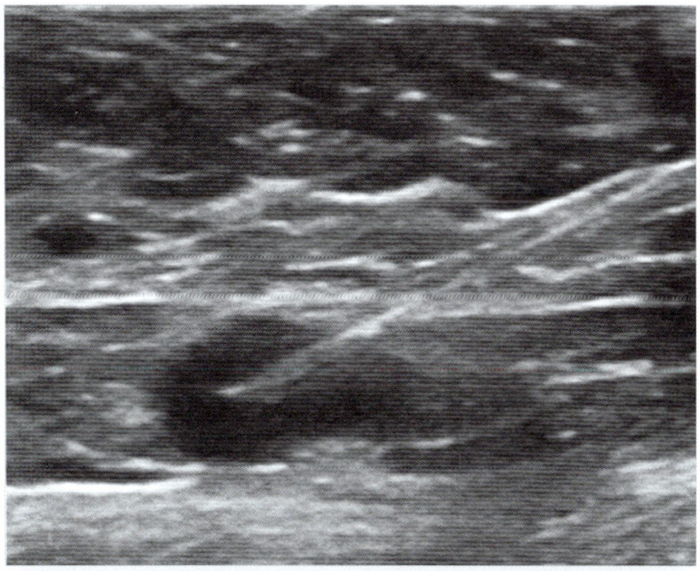

Fig. 11.11 Ultrasound-guided core needle biopsy of a sonographically suspect axillary lymph node.

group was merely monitored, with no further lymph node removal; the other group was treated with ALD. After more than 6 years of follow-up, the two groups did not differ in the rate of locoregional relapse or in overall survival. This study strengthens the concept that the extent of axillary surgery is not associated with any great therapeutic value.

The ACOSOG study is still hotly debated. Because the study was able to recruit just under half the required patients and exhibited several methodological ambiguities, the optimal patient selection was not conclusively clarified for the group with positive SLN in which ALD was withheld. Furthermore, it is ambiguous regarding with which effect the accompanying radiotherapy was to be associated because extensive deviations from the designated radiation protocol were observed within the study.[33]

The ACOSOG study did, however, confirm earlier studies that demonstrated no benefit from therapeutic lymph node excision. The general consensus remained that withholding therapy to the lymphatic drainage areas was not indicated in cases with positive SLN in patients undergoing mastectomy without subsequent radiation. In the Amaros study published in 2014, postoperative radiation of the lymphatic drainage areas was compared with axillary lymph node dissection in patients with a positive sentinel lymph node. Both methods demonstrated a very low regional recurrence rate, but radiotherapy was associated with a more favorable morbidity. Consequently, especially for patients with positive sentinel lymph nodes undergoing mastectomy, there is the additional option of radiotherapy of the lymphatic drainage areas.[34]

11

Note

The current evidence does not allow a clear recommendation for the optimal regional management of patients with low tumor burden in the axillary lmph nodes.

However, patients who undergo mastectomy should not forgo axillary lymph node dissection. This is particularly true if no radiotherapy to the chest wall is given.

Take Home Points

General Recommendations for the Surgical Treatment of the Axilla for Invasive Carcinoma

- In patients with a positive SLN (macrometastasis), an axillary dissection is indicated, with the removal of at least 10 lymph nodes from levels I and II.
- For patients with pT1 or pT2/cN0 tumors who undergo breast-conserving therapy with subsequent percutaneous irradiation over a tangential field and who exhibit one or two positive sentinel lymph nodes, there is the option of forgoing the axillary dissection. An axillary dissection is no longer indicated for patients who exhibit only micrometastasis or isolated tumor cells in the SLN. This is true both for women who undergo breast-conserving surgery and for those who are treated with a breast amputation.
- Axillary dissection can be withheld in cases with micrometastasis exclusively in the SLN.

11.5 Secondary Breast Reconstruction

Even today, some 10 to 20% of all breast carcinoma patients require a breast amputation (see Chapter 11.2 for the spectrum of indications). A variety of possibilities are available to these women for primary or secondary breast reconstruction. The reconstruction itself does not affect the oncologic course of the disease (rate of locoregional recurrence or distant metastases). Breast reconstruction is possible with autologous tissue as well as with an implant. Each technique is associated with its own set of risks, and the quality of the reconstructed breast is limited by the shape of the normal contralateral breast, the patient's body habitus, the accompanying therapy (e.g., radiation therapy) ,and other risk factors (smoking, for example). Therefore it is necessary to have an individual consultation with the patient, with explanation and disclosure of the possibilities and technical limitations of the reconstruction method under consideration, the potential complications, and the possibility that a second or third operation may be required.

The tissue that has been removed—not only the missing skin envelope but also the lost breast volume—must be regenerated during the postmastectomy breast reconstruction. When autologous tissue is used for reconstruction, the missing skin envelope is transferred from another body region. When the reconstruction uses implants, the skin envelope is gained through tissue expansion. And whereas autologous reconstruction creates the volume through the use of skin, adipose tissue, and muscle tissue, silicone implants are used in the alloplastic methods.

11.5.1 Timing the Reconstruction: Primary vs. Secondary Reconstruction

In principle, all the reconstruction methods can be carried out either at the time of the amputation (primary reconstruction) or at a later time (secondary reconstruction). One advantage of primary reconstruction is that parts of the patient's own skin envelope and important anatomical structures, such as the inframammary fold, can be maintained. Furthermore, the patients can be spared the traumatic experience of losing a breast. Disadvantages of primary reconstruction include the short time available in which to make a decision, and the missed opportunity to weigh the advantages and disadvantages of the various techniques. Additionally, there may be a delay in the oncologically important adjuvant therapy (e.g., chemotherapy) in those methods that can be associated with impaired wound-healing for a prolonged time. In some techniques, the addition of radiotherapy to the chest wall can affect the outcome (especially with autologous methods), so that a secondary procedure may provide a better cosmetic result.

Note

Against the background of these considerations, the more complicated types of surgery are usually carried out as secondary procedures. These include the microsurgical flap and pedicled TRAM (transverse rectus abdominis musculocutaneous) flap reconstructions. On the other hand, simple procedures and those with few complications, such as the insertion of a skin expander or a simple implant reconstruction without significant risks, may be carried out as primary surgery.

11.5.2 Alloplastic Reconstruction (Implant Reconstruction)

Included in alloplastic methods are breast reconstruction through use of tissue expanders and permanent silicone implants. A tissue expander consists of a silicone covering and an expandable cavity. It is provided with a valve, through which the expander can be filled with saline solution to produce an expansion of the tissue. When performing implant reconstruction, one must take care to provide sufficient soft tissue cover for the implant. Accordingly, breast implants are usually inserted under the pectoral muscle. A disadvantage of implant reconstruction is that the body demarcates the foreign material from the surrounding tissue by forming a connective tissue shell around it. This implant capsule can be subject to shrinkage (capsule fibrosis), leading to hardening that can be both painful and cosmetically disturbing. Another disadvantage is that there is often a considerable difference in consistency between the two breasts. Perfect symmetry is frequently not attained. The symmetry actually worsens with time, because the reconstructed breast is not subject to age-related changes in shape that the other breast experiences (▶ Fig. 11.12).

Secondary implant reconstruction after a complete mastectomy always requires a two-stage procedure:

- In the first step, the muscles and skin are stretched using an expander to form the pocket needed for the implant.
- In the second step, the breast volume is reconstructed using a permanent silicone implant.

Because of the limited variation in shape of the available implants, this method frequently requires an equalization procedure on the opposite breast to attain a symmetrical body image.

Primary breast reconstruction can be carried out as either a one-stage or a two-stage procedure. The two-stage procedure takes place analogously to the secondary reconstruction. In the one-stage method (immediate reconstruction), the tissue expansion is waived, because the primary skin envelope is preserved. This assumes that the mastectomy was carried out either as a skin-sparing mastectomy (minimal skin resection including the nipple-areolar complex) or as a subcutaneous mastectomy, with preservation of the entire skin of the breast, including the nipple–areolar complex. Because the pectoral muscle is not expanded in the one-time immediate reconstruction, however, this technique would result in either an insufficient projection of the lower pole of the breast or a deficient soft tissue cover for the lower part of the implant (by detachment of the muscle). Therefore, the inferior pole of the implant is covered either with a titanium mesh (synthetic foreign material) or with an acellular matrix (biological foreign material). Pig skin that has been treated by a method to extract the cells and prevent a rejection reaction is used for the acellular matrix. The remaining skin can be used as tissue replacement to replace the patient's own inadequate tissue to cover an implant.

11.5.3 Autologous Reconstruction (Reconstruction Using Endogenous Tissue)

In autologous reconstruction, skin, fat, and muscle tissue from other body regions on the chest wall are transferred and used to fashion a new breast. Advantages of this method lie in the improved plasticity of the breast and the natural consistency and aging of the patient's own tissue. Disadvantages involve, in particular, the distinctly higher surgical costs, the additional morbidity in the donor region, and the potentially increased possibilities of serious complications when compared with alloplastic methods.

Note

When weighing the options between alloplastic and autologous breast reconstruction techniques, it is important to keep in mind that an allopathic procedure can usually be converted to an autologous reconstruction, whereas, after an autologous reconstruction, it is no longer possible to perform an allopathic procedure.

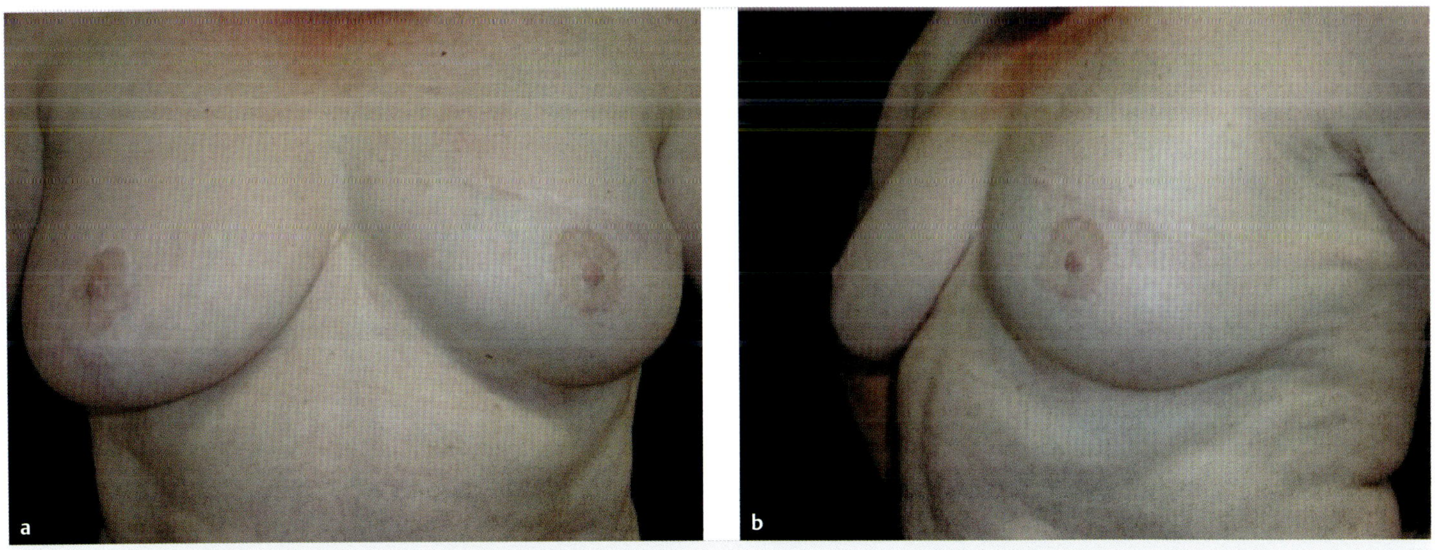

Fig. 11.12 Alloplastic breast reconstruction with an implant. Status post secondary left mastectomy (post amputation): equalization procedure on the opposite breast and left nipple reconstruction. **(a)** Frontal view. **(b)** View from obliquely anterior.

Pedicle Flap Reconstruction

Autologous reconstruction includes pedicle and free flap reconstructions. When transferring autologous tissue, maintenance of perfusion must be guaranteed. In the pedicle flap reconstruction, the muscle serves as "guide rail" for the vascular supply. The flap of skin, fat, and muscle is rotated from its extraction site to the desired new location while maintaining its vascular supply through the original system of blood vessels.

Latissimus dorsi flaps (see ▶ Fig. 11.7) and TRAM flaps (▶ Fig. 11.13) are used for pedicle flap reconstruction. In the latissimus dorsi muscle flap, a tissue block from the back is removed and transferred to the chest wall. The tissue is supplied by the strong thoracodorsal vascular bundle. Because of this excellent supply, this surgical technique is very reliable in terms of the flap perfusion, and necrosis is decidedly rare. A disadvantage of the latissimus dorsi flap is the relatively small amount of tissue that can be extracted (a maximum of about 100–400 grams), which means that large-volume breasts cannot be reconstructed using this technique. Some patients can experience symptoms at the donor site on the back, which can also impair function in the shoulder–arm region.

In essence, it is possible to combine a latissimus dorsi flap with an implant. Due to potential capsular fibrosis and scar formation, however, the combination of autologous and alloplastic material

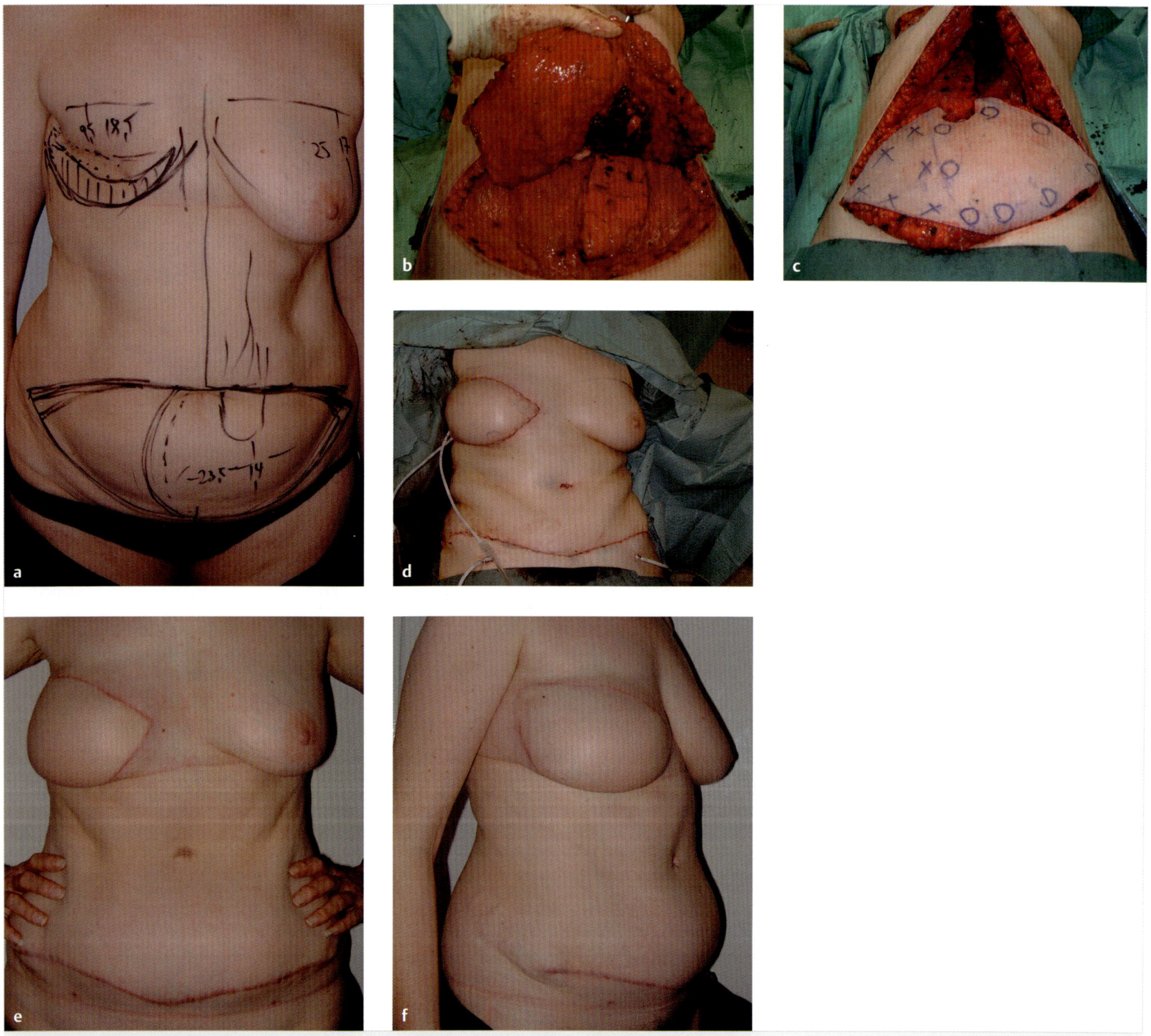

Fig. 11.13 Pedicle TRAM Flap reconstruction. (a) Preoperative planning of a pedicled contralateral TRAM as secondary breast reconstruction post mastectomy. **(b)** Intraoperative flap elevation; the posterior aspect of the left rectus muscle is visible. **(c)** Sufficient flap perfusion is present in the flap zone near the pedicle to allow approximately 60% of the entire circumcised area to be transplanted, while the less-perfused areas of the right side of the flap, located further from the pedicle, are resected. **(d)** Fitting the flap into the mastectomy defect after resection of the less-perfused areas; closure of the abdomen and suturing the navel into the tightened abdominal wall. **(e)** Postoperative result after healing of the TRAM flap reconstruction: frontal view. **(f)** View from obliquely anterior. Image corresponds to **(e)**. (Photographs courtesy of Dr. Nestle-Krämling, Gerresheim, Germany.)

is associated with an increased rate of complications such as capsular fibrosis and movement dependent dimpling and distortion of the breast. In the pedicled TRAM flap reconstruction, tissue from the lower abdomen is used to reconstruct the skin envelope and breast volume. This method enables the transplantation of a significantly larger amount of tissue volume than is possible with the latissimus dorsi flap. The rectus abdominis muscle serves as the muscular pedicle and the blood supply is provided by the epigastric vessels. However, perfusion problems occur frequently when using this method, with partial necrosis of the skin and adipose tissue of the breast. In addition, the removal of abdominal muscle and adipose tissue can lead to hernias in the donor area of the abdominal wall.

Free Flap Reconstruction

In the free flap reconstruction, the perfusing vessels from the donor region are clipped and cut and then reattached to the available vascular system in the receiving region by means of microsurgical anastomosis. These notably elaborate techniques enable the use of more donor sites. In addition, the morbidity in the donor region is reduced as much as possible.

Free flap reconstructions used are muscle-sparing TRAM flap; DIEP (deep inferior epigastric perforator) flap; and the SGAP or IGAP (superior or inferior gluteal artery perforator) flap. The advantage of these methods is the potentially lower morbidity in the donor region, because here muscle to serve as a "guide rail" for the blood vessel supply does not need to be transposed. Furthermore, a large proportion of the volume of the harvested tissue block can be utilized. In addition to the high cost of the surgery, disadvantages include the danger of a total flap loss due to problems in the area of vessel anastomosis, such as thrombosis.

- *Free TRAM flap reconstruction:* In this flap reconstruction, the inferior epigastric vessels are reattached to recipient vessels on the chest wall, e.g., in the perfusion area of the internal mammary vessels. The advantage here is that a large part of the rectus abdominis muscle can be maintained. Because of improved perfusion, larger tissue volumes can be used than in the pedicled TRAM flap reconstruction.
- *DIEP flap reconstruction:* This procedure also uses fat from the abdomen. Here the smallest blood vessels, which exit from the epigastric vessels beneath the rectus abdominis muscle and then perforate the muscle and fascia to perfuse the skin and subcutaneous fat regions in the abdominal wall area (perforating vessels), are isolated and reattached in the area of the chest wall (▶ Fig. 11.14). Using this technique, musculofascial defects can be avoided almost completely.
- *SGAP and IGAP flap reconstructions:* The development of microsurgery and the use of perforator vessels for the vessel anastomoses has also enabled the extraction of autologous tissue from other body regions, such as the gluteal area. Especially for very thin women with little autologous tissue in the back or abdominal areas, the SGAP or IGAP flap reconstructions represent a promising solution.

Practical Tip

For every breast reconstruction, it is a prerequisite to determine the exact wishes and objectives of each individual patient. It is also necessary to assess the surgical conditions (previous treatment, e.g., radiation therapy; the condition of the soft tissue; physiognomy; and the body fat distribution) and to investigate the technical feasibility of various surgical options. Finally, it is vital to review all the advantages and disadvantages with the patient, including the surgical costs, the risks of surgery and the necessity for further procedures.

11.5.4 Nipple Reconstruction

There is a choice from numerous methods for the reconstruction of the nipple–areolar complex. As well as tattooing, there is the possibility of reconstructing the areola through skin transplantation. The papilla can be rebuilt by nipple sharing from the opposite breast or by using small tissue flaps from autochthonous tissue. The reconstruction of a nipple does not require a large expenditure of time and effort and usually provides an excellent outcome.

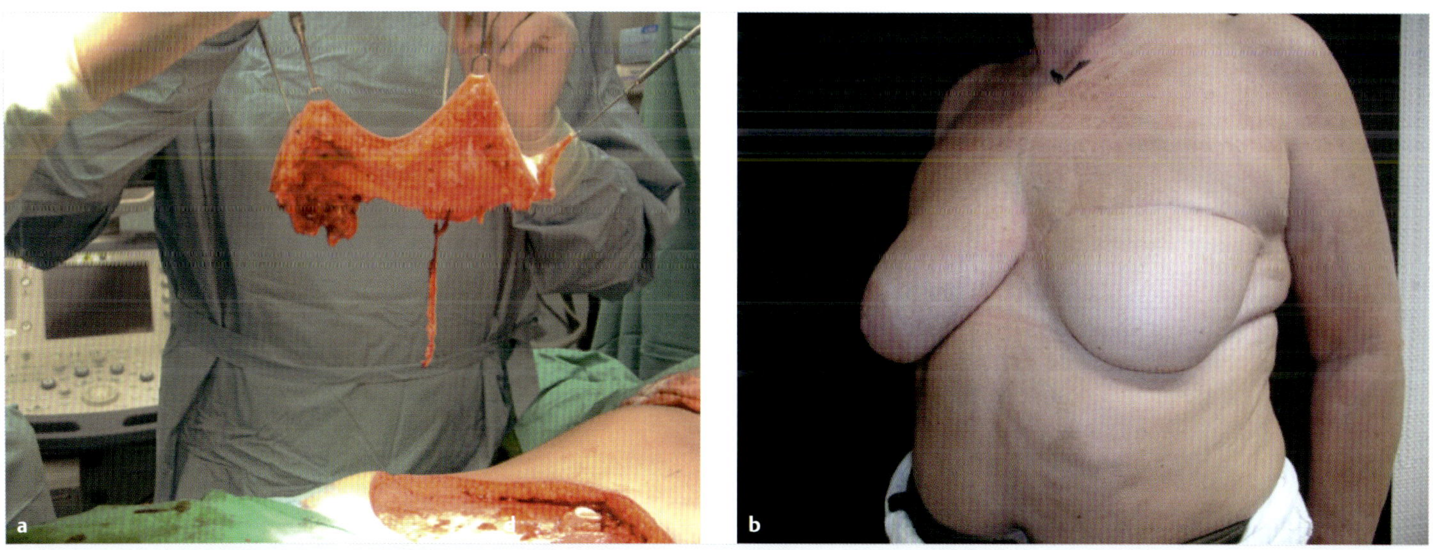

Fig. 11.14 Free deep inferior epigastric perforator (DIEP) flap reconstruction. The skin and adipose tissue are extracted along with two small afferent blood vessels (perforators). The reconnection to the blood vessel system is carried out on the chest wall through microanastomosis with vessels in the axilla **(a)** DIEP flap reconstruction. **(b)** Postoperative results. (Photographs courtesy of Dr. Mario Marx, Radebeul, Germany.)

Bibliography

[1] Bartelink H, Horiot JC, Poortmans PM, et al. Impact of a higher radiation dose on local control and survival in breast-conserving therapy of early breast cancer: 10-year results of the randomized boost versus no boost EORTC 22881–10882 trial. J Clin Oncol. 2007; 25(22):3259–3265

[2] Bijker N, Peterse JL, Duchateau L, et al. Risk factors for recurrence and metastasis after breast-conserving therapy for ductal carcinoma-in-situ: analysis of European Organization for Research and Treatment of Cancer Trial 10853. J Clin Oncol. 2001; 19(8):2263–2271

[3] Bijker N, Peterse JL, Fentiman IS, et al. Effects of patient selection on the applicability of results from a randomised clinical trial (EORTC 10853) investigating breast-conserving therapy for DCIS. Br J Cancer. 2002; 87(6):615–620

[4] Blichert-Toft M, Smola MG, Cataliotti L, O'Higgins N. Principles and guidelines for surgeons—management of symptomatic breast cancer. On behalf of the European Society of Surgical Oncology. Ann Chir Gynaecol. 1998; 87(1):101–109

[5] Boyages J, Delaney G, Taylor R. Predictors of local recurrence after treatment of ductal carcinoma in situ: a meta-analysis. Cancer. 1999; 85(3):616–628

[6] Burstein HJ, Polyak K, Wong JS, Lester SC, Kaelin CM. Ductal carcinoma in situ of the breast. N Engl J Med. 2004; 350(14):1430–1441

[7] Chan KC, Knox WF, Sinha G, et al. Extent of excision margin width required in breast conserving surgery for ductal carcinoma in situ. Cancer. 2001; 91(1):9–16

[8] Dunne C, Burke JP, Morrow M, Kell MR. Effect of margin status on local recurrence after breast conservation and radiation therapy for ductal carcinoma in situ. J Clin Oncol. 2009; 27(10):1615–1620

[9] Correa C, McGale P, Taylor C, et al. Early Breast Cancer Trialists' Collaborative Group (EBCTCG). Overview of the randomized trials of radiotherapy in ductal carcinoma in situ of the breast. J Natl Cancer Inst Monogr. 2010; 2010 (41):162–177

[10] EC Working Group on Breast Screening Pathology. Quality assurance guidelines for pathology in mammography screening—nonoperative diagnosis. In: Perry N, Broders M, de Wolf C, Törnberg S, Schouten J, eds. European Guidelines for Quality Assurance in Mammography Screening. 3rd ed. Ch 6. Luxembourg: Office for Official Publications of the European Communities; 2001:159–172 [DKG-N]

[11] Fisher B, Anderson S, Tan-Chiu E, et al. Tamoxifen and chemotherapy for axillary node-negative, estrogen receptor-negative breast cancer: findings from National Surgical Adjuvant Breast and Bowel Project B-23. J Clin Oncol. 2001; 19(4):931–942

[12] Hoffmann J, Wallwiener D. Classifying breast cancer surgery: a novel, complexity-based system for oncological, oncoplastic and reconstructive procedures, and proof of principle by analysis of 1225 operations in 1166 patients. BMC Cancer. 2009; 9:108–117

[13] Houssami N, Macaskill P, Marinovich ML, et al. Meta-analysis of the impact of surgical margins on local recurrence in women with early-stage invasive breast cancer treated with breast-conserving therapy. Eur J Cancer. 2010; 46 (18):3219–3232

[14] Jones HA, Antonini N, Hart AA, et al. Impact of pathological characteristics on local relapse after breast-conserving therapy: a subgroup analysis of the EORTC boost versus no boost trial. J Clin Oncol. 2009; 27(30):4939–4947

[15] Kuehn T, Klauss W, Darsow M, et al. Long-term morbidity following axillary dissection in breast cancer patients—clinical assessment, significance for life quality and the impact of demographic, oncologic and therapeutic factors. Breast Cancer Res Treat. 2000; 64(3):275–286

[16] Kuehn T, Bembenek A, Decker T, et al. Consensus Committee of the German Society of Senology. A concept for the clinical implementation of sentinel lymph node biopsy in patients with breast carcinoma with special regard to quality assurance. Cancer. 2005; 103(3):451–461

[17] Lagios MD. Practical pathology of duct carcinoma in situ: how to derive optimal data from the pathologic examination. In: Silverstein MJ, ed. Ductal Carcinoma In Situ of the Breast. 2nd ed. Philadelphia: Lippincott, Williams & Wilkins; 2002:207–221

[18] Lebeau A. Prognostic factors in ductal carcinoma in situ [in German]. Pathologe. 2006; 27(5):326–336

[19] Leitlinienprogramm Onkologie der AWMF, Deutschen Krebsgesellschaft eV und Deutschen Krebshilfe eV. Leitlinienreport der S 3 Leitlinie fur die Diagnostik, Therapie und Nachsorge des Mammakarzinoms. AWMF Register-Nr. 032–045OL. Available at: http://www.awmf.org/leitlinien/detail/ll/032–045OL.html. Accessed February 9, 2017

[20] Morrow M, Harris JR, Schnitt SJ. Local control following breast-conserving surgery for invasive cancer: results of clinical trials. J Natl Cancer Inst. 1995; 87(22):1669–1673

[21] National Institute for Clinical Excellence (NICE). Early and Locally Advanced Breast Cancer: Diagnosis and Treatment. 2009. Available at: https://www.nice.org.uk/guidance/cg80/evidence/full-guideline-242201629. Accessed February 02, 2017

[22] New Zealand Guidelines Group (NZGG). Management of early breast cancer. Wellington: NZGG; 2009

[23] Silverstein MJ, Lagios MD, Recht A, et al. Image-detected breast cancer: state of the art diagnosis and treatment. J Am Coll Surg. 2005; 201(4):586–597

[24] Schnitt SJ, Abner A, Gelman R, et al. The relationship between microscopic margins of resection and the risk of local recurrence in patients with breast cancer treated with breast-conserving surgery and radiation therapy. Cancer. 1994; 74(6):1746–1751

[25] Silverstein MJ, Poller DN, Waisman JR, et al. Prognostic classification of breast ductal carcinoma-in-situ. Lancet. 1995; 345(8958):1154–1157

[26] Renton SC, Gazet JC, Ford HT, Corbishley C, Sutcliffe R. The importance of the resection margin in conservative surgery for breast cancer. Eur J Surg Oncol. 1996; 22(1):17–22

[27] Vargas C, Kestin L, Go N, et al. Factors associated with local recurrence and cause-specific survival in patients with ductal carcinoma in situ of the breast treated with breast-conserving therapy or mastectomy. Int J Radiat Oncol Biol Phys. 2005; 63(5):1514–1521

[28] Veronesi U, Salvadori B, Luini A, et al. Breast conservation is a safe method in patients with small cancer of the breast. Long-term results of three randomised trials on 1,973 patients. Eur J Cancer. 1995; 31A(10):1574–1579

[29] Veronesi U, Cascinelli N, Mariani L, et al. Twenty-year follow-up of a randomized study comparing breast-conserving surgery with radical mastectomy for early breast cancer. N Engl J Med. 2002; 347(16):1227–1232

[30] Wald NJ, Murphy P, Major P, Parkes C, Townsend J, Frost C. UKCCCR multicentre randomised controlled trial of one and two view mammography in breast cancer screening. BMJ. 1995; 311(7014):1189–1193

[31] Wang S-Y, Chu H, Shamliyan T, et al. Network meta-analysis of margin threshold for women with ductal carcinoma in situ. J Natl Cancer Inst. 2012; 104 (7):507–516

[32] Weaver DL, Krag DN, Ashikaga T, Harlow SP, O'Connell M. Pathologic analysis of sentinel and nonsentinel lymph nodes in breast carcinoma: a multicenter study. Cancer. 2000; 88(5):1099–1107

[33] Jagsi R, Chadha M, Moni J, et al. Radiation field design in the ACOSOG Z0011 (Alliance) Trial. J Clin Oncol. 2014; 32(32):3600–3606

[34] Donker M, van Tienhoven G, Straver ME, et al. Radiotherapy or surgery of the axilla after a positive sentinel node in breast cancer (EORTC 10981–22023 AMAROS): a randomised, multicentre, open-label, phase 3 non-inferiority trial. Lancet Oncol. 2014; 15(12):1303–1310

[35] Giuliano AE, McCall L, Beitsch P, Withworth PW, Leitch AM, Saha S, Hunt KK, Morrow M, Ballman K. Locoregional recurrence after sentinel lymph node dissection with or without axillary dissection in patients with sentinel lymph node metastases: the American Collegue of Surgeons Oncology Group Z 0011 randomized trial. Ann Surg. 2010; 252(3):426–32

[36] Kühn T, Klauss W, Darsow M, Regele R, Flock F, Maiterth C, Dahlbender R, Wendt I, Kreienberg R. Long term morbidity following axillary dissection in breast cancer patients, clinical assessment, significance for life quality and the impact of demographic, oncologic and therapeutic factors. Breast Cancer Res Treat. 2000; 64(3):275–286

12 Medical Treatment of Breast Cancer

M. Hellriegel and G. Emons

12.1 Basic Principles and Objectives

Primary breast carcinoma is a systemic, heterogeneous disease. It is considered a systemic disease because it is now known that even the smallest carcinoma can trigger metastases. Systemic medical treatment should therefore come after localized treatment of the breast carcinoma. Classes of substances available for systemic medical treatment are:

- Cytostatics.
- Hormones.
- Antibodies.
- New anticancer agents, such as angiogenesis inhibitors and receptor tyrosine kinase inhibitors.

Breast carcinoma is a heterogeneous disease in terms of its emergence, its individual characteristics, its treatment, and its course. The development of a breast carcinoma involves numerous cellular changes, on which the cancer cell depends completely for its survival and growth. Using molecular genetic identification of these changes and understanding how they interact, it is possible to define the breast carcinoma's particular characteristics and from this to develop targeted points of attack for the therapy. The daily clinical routine includes, in particular, defining the cells' immunohistological properties (hormone receptor status, HER2 status) and typing the carcinoma at the molecular biological level, which is increasingly important. Some of these points of attack will be dealt with in greater detail later in this chapter.

Cytostatic therapy follows a completely different approach. Empirically researched combinations of cytostatic substances damage the carcinoma tissue by interfering with various metabolic pathways that influence cell growth and cell division at different stages. In essence, the relatively selective effect of the cytostatics is achieved through the increased rate of cell division of the carcinoma cells.

In the drug-based treatment of breast cancer we can essentially differentiate four distinct disease categories with different therapeutic approaches and goals:

- *Adjuvant therapy of breast cancer:* In adjuvant disease treatment, the primary breast carcinoma is first operated on and then treated by radiation therapy. The aim is cure. Both localized irradiation and the subsequent systemic therapy are preventive measures designed to reduce the risk of recurrence in a clinically tumor-free patient.
- *Neoadjuvant therapy of breast cancer:* In neoadjuvant therapy, the systemic treatment precedes localized treatment. This form of therapy is also known as "primary systemic" or "preoperative" therapy. Here too, the therapeutic goal is cure of the patient. A certain histopathological diagnosis is always a prerequisite. Here too, the therapeutic goal is cure of the patient. Inflammatory and inoperable, locally far-advanced cases of carcinoma of the breast have always been indications for neoadjuvant chemotherapy. A further goal is to increase the rate of breast-conserving surgery[9]: studies by the NSABP (National Surgical Adjuvant Breast and Bowel Project) group using primary systemic therapy have demonstrated efficacy equal to that of adjuvant therapy.[21,22] The better the tumor's response to the primary systemic therapy, the better is the patient's prognosis.
- *Therapy of locoregional and thoracic wall relapses*
  - *Locoregional relapse* is defined as the recurrence of the breast carcinoma in the remaining breast or axillary tissue on the same side. The therapeutic goal here is surgical cure. The 10-year survival rate is 40 to 58%.[66] With an isolated ipsilateral diagnosis the medical prognosis is considerably better. It is therefore vital to establish the absence of local and axillary tumors during aftercare examinations.
  - *Thoracic wall relapse* refers to recurrence of the breast carcinoma at the surgical site or in the irradiated area after a mastectomy. With isolated occurrence and successful removal of the recurrent lesion, there is a good chance of cure and a 5-year survival rate is about 65%.[52] The critical prognostic indicator here is the occurrence of distant metastases—approximately one-third of patients already have distant metastases when diagnosed.
- *Treatment of distant metastases:* Cure is not possible in this advanced stage of the disease with the current state of knowledge. The aim is to maintain the best possible quality of life for the patient for as long as possible. The decision on the most suitable form of therapy is made for each individual situation. The average life expectancy, depending on the affected organ, is about 2 years.

Take Home Point

General Recommendations in the Medical Treatment of Primary Disease

Pharmacologic treatment of the primary disease is by chemotherapy, endocrine therapy, and/or anti-HER2 antibody therapy, given in various sequences and combinations, either before or after surgery.

12.2 Adjuvant Drug Therapy

12.2.1 Adjuvant Chemotherapy

Adjuvant chemotherapy is the standard treatment for the following indications:

- Hormone receptor-negative tumors (endocrine nonsensitive: ER- and PgR-negative).
- Equivocally hormone receptor-negative tumors (equivocally endocrine-sensitive).
- HER-2-positive tumors treated with simultaneous antibody therapy using trastuzumab, over a period of 1 year, combined with a guideline-appropriate polychemotherapy.
- Nodal-positive tumors.
- Grade: G3.
- Young age of onset (below 35 years).
- Luminal B tumors with a "high proliferation rate."

12

- Triple-negative tumors (ER-/PgR-negative, HER2-negative = basal type).

The current evaluations by the EBCTCG (Early Breast Cancer Trialists' Collaborative Group) under the direction of the statistician Sir Richard Peto showed that women below the age of 50 years, in particular, benefit from adjuvant chemotherapy in terms of the risk of death and recurrence. However, even in postmenopausal and older patients, the benefits of standard chemotherapy outweigh the higher rate of side effects when the corresponding risk constellation is taken into account.

Based on current understanding, chemotherapy should be started within the first 4 to 6 weeks after surgery, applied over 6 to 8 cycles, and repeated every 3 weeks. Standard drugs are anthracycline and taxane in combination or in sequence. The EBCTCG 2011 meta-analysis[16] evaluated the effectiveness of taxane for treating nodal-positive and nodal-negative breast carcinoma with high-risk constellations (e.g., G2/3; ER-/PgR-negative; pT stage above 2; age less than 35 years).[41] Taxane and anthracycline-based regimes show a significant advantage in their effect on survival free from relapse and distant metastases and on overall survival compared with purely anthracycline-based regimes. Current studies corroborate these results.

More recent studies establish that dose-dense chemotherapy is beneficial in nodal-positive high-risk patients (four or more affected lymph nodes).[10,43] Additional substances such as capecitabine and gemcitabine, and high-dose chemotherapy currently play no role in the adjuvant therapy of early breast cancer.

The conditions of therapy are being continuously improved through analysis of patients in clinical studies. As well as ensuring optimized care for the individual patient through participation in the study protocols per se, it also often provides the only means by which further data can be collected, new substances introduced, biomarkers determined, and unanswered questions resolved.

The meta-analyses from the EBCTCG have repeatedly shown that, for primary breast carcinoma, treatment with both modern polychemotherapy and endocrine therapy can significantly reduce the cumulative 15-year mortality risk by about one-third.[18,53]

Take Home Point

Systemic Adjuvant Therapy—What We Know For Sure

The use of systemic therapy reduces the relapse and mortality rates. This applies to polychemotherapy, in particular with anthracycline and taxane, the use of drugs to eliminate ovarian function, tamoxifen, aromatase inhibitors, and trastuzumab.

The indication for adjuvant systematic therapy is determined by considering the risk profile of the individual breast cancer patient and the likely benefits of this therapy. Even with current established methods it is not really possible to determine exactly which patient groups would not benefit from this therapy (due to overtreatment or lack of response).[46] The following classical clinical pathological parameters are considered in assessing the patient's individual risk rating:

- Tumor size.
- Lymph node status.
- Tumor grade.
- Hormone receptor status.
- HER2-status.
- Menopausal status.
- Age.

These are used to decide whether medication is necessary and the type of drug to be used. In every case, the need for the adjuvant therapy must be described to the patient in a detailed explanatory consultation; the side effects and the possibility of long-term health issues must be explained as well as the benefits.

Following the recommendations of the 2011 St. Gallen Consensus Conference,[28] it is for the first time possible to accommodate the types of breast carcinoma as defined by immunohistochemical properties in the decision making process. The immunohistological markers used for this are:

- ER.
- PgR.
- HER2.
- Ki-67 (as a proliferation index).

Along with the classic categorization, certain constellations of markers are increasingly being integrated into the nomenclature. These new classification subtypes are also referred to as "intrinsic types" of breast carcinoma (▶ Table 12.1).

It is important to realize, however, that there is currently no validated threshold level for the Ki-67 labeling index, and that indices between 10 and 25% have no clear predictive value as to the proliferation rate.

12.2.2 Neoadjuvant Therapy

Systemic neoadjuvant therapy includes all the types of medical treatment that are administered after the confirmed histological diagnosis of a breast carcinoma but before any surgical measures. Here too, the therapeutic goal is cure.

Systemic neoadjuvant therapy has for decades been the standard treatment in patients with a locally advanced, primary inoperable or inflamed breast carcinoma. In these patients, the therapeutic goal was, through pharmacologic tumor reduction, to achieve an operable state that enables the complete excision of the carcinoma. Numerous studies on this concept have shown

Table 12.1 Classification of intrinsic types of breast cancer

Breast carcinoma type	ER	PgR	HER2	Grade	Ki-67 (%)
Luminal A	+	+/−	−	1–2	< 14
Luminal B–HER2-negative	+	+/−	−	1–3	≥ 14
Luminal B–HER2-positive	+	+/−	−	1–3	Any value
HER2-positive	−	−	+	1–3	Any value
Triple negative or basal type	−	−	−	1–3	Any value

Abbreviations: ER, estrogen receptor; PgR, progesterone receptor; HER2, human epidermal growth factor receptor 2; Ki-67 (%), proliferation index (a nuclear antigen).

not only that operability can be improved but that neoadjuvant therapy offers additional advantages:

- Improved options for breast-conserving surgery.
- Individualization of the therapy through regular in vivo testing and monitoring of response.
- Faster validation of new pharmaceuticals in the course of clinical trials.

In terms of the long-term survival rate, there are no differences between the adjuvant and neoadjuvant therapy experiences. Accordingly, neoadjuvant chemotherapy is now regarded as being equivalent in standing to adjuvant therapy.

Take Home Points

General Recommendations and Findings on Systemic Neoadjuvant Therapy

- Neoadjuvant (primary, preoperative) systemic therapy is considered the standard treatment for patients with locally advanced, inoperable primary, or inflammatory breast carcinoma in the course of a multimodal therapy plan.
- If chemotherapy is the indicated treatment, it can be given either before (neoadjuvant) or after surgery (adjuvant). The two methods are of equal value in terms of overall survival rates, but neoadjuvant therapy can result in a higher rate of breast-conserving treatment.
- The effect is greatest with hormone receptor-negative carcinoma.
- A resection within the new tumor boundaries is possible if a R0 resection can be achieved with a sufficient safety margin.

According to the Breast Committee of the AGO (Arbeitsgemeinschaft für gynäkologische Onkologie [Working Group for Gynecological Oncology]), according to international guidelines, suitable chemotherapeutic regimens are combinations containing anthracycline and taxane. Participation in clinical studies is strongly recommended. The neoadjuvant therapy should take place for a minimum of six cycles prior to surgery. Patients with a hormone receptor-positive carcinoma who respond well after two cycles of TAC (a combination of docetaxel 75 mg/m^2 BSA + doxorubicin 50 mg/m^2 BSA + cyclophosphamide 500 mg/m^2 BSA every 3 weeks; BSA, body surface area) will benefit even more from eight cycles, according to data from the neoadjuvant GePAR-TRIO Study.[64] The therapeutic effect (therapeutic response) should be checked after two cycles of therapy and regularly thereafter. Clinical examination (including palpation), breast ultrasound, and mammography are well suited for this. If there is no response, action must be swift, with a corresponding adjustment to the neoadjuvant therapy or, depending on clinical findings, cessation of the treatment. This is where the advantages of neoadjuvant therapy come to the fore: the sensitivity of the tumor to the chemotherapy can be assessed and corresponding action can be taken. During a course of adjuvant therapy, a lack of response cannot be detected. However, the response of the tumor stem cells, which are responsible for the metastatic spread, is not evaluated in primary systemic therapy.

When there is lack of clinical response following two cycles of neoadjuvant chemotherapy, one can consider switching to a non-cross-resistant scheme with reduced side effects. In the GePAR-TRIO Study,[32,62] it was shown that switching the chemotherapy from two cycles of TAC to four cycles of vinorelbine/capecitabine (days 1 and 8: 25 mg/m^2 BSA; days 1 to 14: 2000 mg/m^2 BSA, every 3 weeks) resulted in significantly prolonged relapse-free and overall survival rates. Patients with hormone receptor-positive breast carcinoma benefited particularly well from the changeover. However, the prognosis was not improved by the change in treatment for patients with triple-negative breast carcinoma, who showed no corresponding effects after two cycles of TAC therapy. A large proportion of patients suffered recurrence and metastases.

Pathological complete response (pCR) is defined as the absence of invasive tumor cell in the breast and the axilla following the completion of a course of neoadjuvant chemotherapy.[65] The pCR rate is a validated surrogate marker for the response to neoadjuvant chemotherapy and the long-term survival rate. Patients who respond after only two to four cycles of neoadjuvant therapy, have an approximately 4 times higher chance of a pCR and can be expected to have a very favorable long-term prognosis. This is especially true for patients with a HER2-positive/hormone receptor-negative or triple-negative breast carcinoma (basal type) and, with newly available data, also for those with Luminal B-/HER2-negative tumors.[65]

Note

All patients whose breast carcinoma shows no change after the first cycles of neoadjuvant therapy have a distinctly worse prognosis.

The pCR rate serves as a surrogate marker in patients with the following tumor constellations:

- Hormone receptor negative/HER2-positive.
- Triple negative.
- Luminal B/HER2-negative.

The pCR rate is not useful as a surrogate marker in Luminal A- and Luminal B/HER2-positive tumors.

Other predictive factors for a high likelihood of pCR have been collected by the Breast Committee of the AGO: in addition to younger age, smaller tumor size, higher grading, nonlobular histology, triple negativity, and positive HER2-status, the most important predictive factor for the effectiveness of neoadjuvant chemotherapy containing taxane–anthracycline is a negative hormone receptor status. A pCR rate of up to 40% can be achieved in patients who have a negative hormone receptor status.

If there is progression, or after completion of the neoadjuvant therapy, plans should be made for surgery. Before surgery, tumor localization should be precisely documented, for example, using pretherapeutic clip marking. The extent of the surgery should be determined by the effectiveness of the neoadjuvant therapy and should be carried out within the new tumor boundaries. The indications for postoperative systemic and local therapeutic measures are the same as for adjuvant therapy.[37]

Neoadjuvant Therapy for HER2-Positive Breast Carcinoma

Standard treatment for patients with breast carcinoma overexpressing HER2 includes additional therapy with the antibody trastuzumab. This should be given preoperatively, simultaneously with chemotherapy; it significantly improves the pCR and survival rates.[65] These conclusions are from a meta-analysis data of 6,377 patients from seven German breast cancer studies involving neoadjuvant therapy. As with the adjuvant therapy, the antibody therapy should be continued for a full year.[1]

Neoadjuvant Endocrine Therapy of Hormone Receptor-positive Breast Carcinoma

Endocrine therapy is a further neoadjuvant therapeutic option. Postmenopausal, highly hormone receptor-positive patients with locally advanced, primary inoperable or inflammatory breast carcinoma, where surgery or chemotherapy are contraindicated or have been declined, will benefit from endocrine therapy. With this indication, therapy with third-generation aromatase inhibitors for longer than 3 months is recommended.[55]

12.2.3 Adjuvant Endocrine Therapy

According to a meta-analysis of the EBCTCG, adjuvant hormonal therapy with tamoxifen can significantly reduce the cumulative risk of death over a 15-year period in patients with a hormone receptor-positive breast carcinoma by about one-third or by an absolute 9.2%. This effect is independent of the age of the patient, the nodal status, the menopausal status, the size of the tumor, and the administration of adjuvant chemotherapy.

Take Home Point

General Recommendation for Endocrine Treatment of Receptor-Positive Tumors

Endocrine treatment is indicated in patients with estrogen receptor-and/or progestogen receptor-positive tumors.

12

As part of the primary diagnosis process, the hormone receptor status is determined by way of a quality-controlled evaluation of the paraffin section or of tissue previously obtained from a core needle biopsy. The content of ER(α) and PgR in the cells is determined immunohistochemically. If no receptor staining is observed (ER content 0%), then this form of therapy is ineffective.

The progesterone receptor, PgR, has particular prognostic significance in breast carcinoma patients—a less favorable prognosis may be assumed in the absence of PgR expression.

Note

A response to endocrine therapy is assumed if the cell ER content is > 1%, with endocrine sensitivity correlating with the strength of the receptor expression.[60]

At the time of diagnosis, approximately one-third of all breast cancer patients are pre- or perimenopausal. About 60% of the tumors in these patients overexpress the ER. The goal of systemic endocrine therapy is blockade of the intracellular ER of the tumor cells and/or reduction of the circulating estrogens to a postmenopausal level of 20 pg/mL or less. Before endocrine therapy with aromatase inhibitors is proposed to a patient, her menopausal status should be clarified and an exact menstrual history should be documented. It should be noted that, in women younger than 50 years, chemotherapy can induce amenorrhea with high levels of follicle-stimulating hormone (FSH) and luteinizing hormone (LH). In these women, even years later, follicles can continue to ripen, producing estrogen. This process is supported by the administration of aromatase inhibitors. Thus, high estrogen levels in association with aromatase inhibitor therapy can be fatal because they stimulate tumor growth.[56] Only from the age of 55 years can a patient reliably be considered to be postmenopausal.

Endocrine Therapeutic Agents

Tamoxifen

The standard therapeutic option for patients who are premenopausal is selective ER-blockade with tamoxifen, using a dosage of 20 mg/day over a period of 5 years.

Take Home Point

General Recommendations for Endocrine Therapy in Premenopausal Patients

In premenopausal patients, tamoxifen is the endocrine therapy of choice. Antiestrogen therapy with tamoxifen at 20 mg/day should be administered over a period of 5 years or until a relapse.

As a SERM (selective ER modulator), tamoxifen competitively inhibits the binding of estrogens to hormone receptors, so with its use there is a decrease in cell division in estrogen-dependent tissues, leading to apoptosis of the corresponding tumor cells. Tamoxifen is a prodrug; in an initial reaction, it is converted by the cytochrome P450 enzyme CYP2D6 into its active metabolites, endoxifen and 4-hydroxytamoxifen. In addition to its antiestrogenic effect, tamoxifen also exhibits a slight organ-specific agonistic, estrogenic effect. Because of this, on the one hand tamoxifen exerts osteoprotective properties, while on the other it leads to an increase in endometrial changes and thromboembolic events (in ca. 1 to 3% of cases). The relative risk for the occurrence of endometrial carcinoma after a 5-year course of tamoxifen therapy is 2.2 compared with the average for the population. This risk is low in contrast with its benefits (reducing the risk of recurrence of breast carcinoma), and is only relevant from the age of 55 years and above. Furthermore, the mortality rate is not increased by these side effects.[16]

The conclusions of the Consensus Conferences in St. Gallen in 2005 and 2007 are helpful in developing treatment recommendations (► Table 12.2). The current AGO recommendations[1] extend the indication spectrum for the sole use of an endocrine therapy with tamoxifen to patients with a moderate risk of recurrence:

- *Patients with high risk:* According to the current AGO treatment recommendations[1], the standard treatment for premenopausal patients with an endocrine-sensitive breast carcinoma and a high-risk constellation is adjuvant sequential chemoendocrine therapy with tamoxifen over a period of 5 years. However, the current EBCTCG meta-analysis[16] does not demonstrate any disadvantage in using different therapy types simultaneously.
- *Patients with moderate risk:* These patients can be treated either with a sequential chemoendocrine therapy or with

Table 12.2 Risk categories (consensus of 2007 St. Gallen Conference[26])

Status	Low risk	Moderate risk	High risk
Nodal status	0 and all other criteria	1–3 and HER2-negative or 0 with at least one other criterion	≥ 4, or 1–3 and HER2-positive
Tumor size (cm)	1 (maximum 2)	2–4	
Degree of differentiation	1	2 and 3	
Vascular collapse	0	1	
Hormone status (ER/PgR)	+		
HER2/neu status	–	+	+
Age (years)	≥ 35	> 35	< 35

endocrine therapy alone, depending on the individual risk assessment, the level of endocrine sensitivity, and the safety considerations.[1]

- *Patients with low risk:* Endocrine therapy with tamoxifen alone or combined with GnRH analogs may be recommended to these patients (see the following section). In this group all of the following conditions must be met:
  - Age at least 35 years.
  - Tumor size no greater than 2 cm (T1).
  - A good degree of differentiation (G1).
  - ER and/or PgR positive.
  - HER2/neu status negative.
  - No positive axillary lymph nodes (N0).

Practical Tip

Patients who have a moderate risk profile, for whom there are contraindications to chemotherapy, who decline chemotherapy, or who belong to a lower risk group can be offered tamoxifen monotherapy as an alternative.

Gonadotropin-Releasing Hormone (GnRH) Analogs

Ovarian ablation is another proven therapeutic principle for hormone suppression, but it has not yet been definitively established when GnRH analogs should be used and for which indication.

GnRH analogs are synthetically produced derivatives of the neurohormone GnRH, which suppress the gonadotropic function of the pituitary gland and thereby ovarian function. These analogs bind to the pituitary receptor with a 100 times greater affinity than natural GnRH and exhibit a longer half-life. After the initial binding there is a temporary increase in release of FSH and luteinizing hormone ("flare-up effect"), but in the course of therapy there is a reduction in the number of receptors. The secretion of the pituitary hormones, LH and FSH, is thereby permanently inhibited, preventing the maturation of ovarian follicles and the associated estrogen production. A hypogonadotropic, hypoestrogenic state is achieved within 2 weeks that is similar to that of a postmenopausal woman. The effects of a GnRH analog therapy are reversible and there are relatively few side effects; radiotherapeutic ovarian ablation or oophorectomy provides irreversible suppression.

A course of treatment with a GnRH analog over a 2- to 5-year period in addition to tamoxifen therapy to eliminate ovarian function would appear to benefit only premenopausal, hormone receptor-positive, and nodal-positive breast cancer patients who are younger than 40 years of age.[8,14,15]

Note

GnRH monotherapy for the small population of premenopausal, endocrine-sensitive breast cancer patients with low or medium risk should only be considered for those with contraindications to tamoxifen, because it has not to date been shown to have any advantage over standard chemotherapy.

Aromatase Inhibitors

Aromatase inhibitors are contraindicated as a primary therapy in premenopausal and perimenopausal patients. Prescription of aromatase inhibitors as monotherapy intended to inhibit estrogen biosynthesis by the ovary is not appropriate. In fact, if there is residual ovarian function in a premenopausal woman, adjuvant chemotherapy can lead to stimulation of the ovaries and an increase in the body's estrogen production.

Premenopausal Patients

The utility and ideal duration of chemotherapeutically induced amenorrhea before menopause and the significance of ovarian suppression are still uncertain. When and in what combination the GnRH analogs should best be used, and whether there is an advantage from combination of aromatase inhibitors with a GnRH analog are also questions requiring clarification. Evaluation of the randomized, multicenter Phase III SOFT and TEXT studies of the IBCSG (International Breast Cancer Study Group) and the BIG (Breast International Group), help to answer these questions.

Further questions arise from the classification of breast cancer patients into different risk groups (▶ Table 12.2), which is not uncontroversial. For example, it is unclear which particular form of therapy is optimal for a patient with moderate risk, and how large the benefit is for patients with a low hormone-sensitivity (1–9% positive cells). Even within the risk groups, it is not possible to identify patients who remain recurrence free without adjuvant system therapy, especially because approximately two-thirds of patients would not benefit from adjuvant endocrine therapy. In addition, we cannot identify the subgroups of premenopausal patients who could benefit from continuous therapy with tamoxifen for longer than 5 years with a low side effect profile and without developing resistance.

Molecular biology approaches are required for better characterization of the heterogeneous disease termed "breast cancer." We need to be able to identify the relevant signal pathways and their interactions and to predict the effectiveness of therapy. A start in identifying individually decisive tumor properties may be made through gene expression profiles. For example, at the time of writing the application of the Oncotype DX test is being monitored and the results are being evaluated for hormone receptor-positive patients in the Plan B study. By means of this 21-gene test, the risk of a relapse in the course of their disease can be determined for the individual patient. A low risk value of 1 to 10 predicts few benefits from adjuvant chemotherapy and should thus help the patient in the decision whether to forgo chemotherapy.[57] The initial data for such prognosis assessment is available for breast carcinoma patients who have been treated with tamoxifen. Patients with a high-risk rating, in particular, have benefited from additional chemotherapy.[51]

Lastly, a premenopausal patient's question whether she can bear a child after completing therapy and again demonstrating premenopausal ovarian function can only be answered case by case.

Postmenopausal Patients

The use of tamoxifen as part of an adjuvant endocrine therapy in postmenopausal women with breast cancer was considered the standard therapy for more than 20 years. This changed upon the introduction of aromatase inhibitors, with the increased odds of extending disease-free and metastasis-free survival for the patients. However, survival benefits from the use of aromatase inhibitors could only be demonstrated in a few studies.

Take Home Point

Endocrine Therapy in Postmenopausal Women—What We Know for Sure

For women confirmed as postmenopausal, third-generation aromatase inhibitors are superior to tamoxifen in terms of disease-free survival.

Orally administered third-generation aromatase inhibitors currently approved for endocrine therapy of postmenopausal, hormone receptor-positive breast cancer patients are of steroidal and nonsteroidal types. These prevent the creation of estrogens (estrone and estradiol) from the androgenic precursors androstenedione and testosterone, through the inhibition of the enzyme aromatase, which occurs in numerous tissues such as ovaries, adipose tissue, skin, liver, muscles, and adrenal glands, as well as in cancerous breast tissue. In premenopausal women, aromatase has highest activity in the ovaries and this cannot be inhibited by the use of conventional aromatase inhibitors. In these women, the administration of aromatase inhibitors leads to stimulation of the ovaries and to an increased secretion of estrogen. In postmenopausal women, ovarian function is quiescent and the remaining estrogen biosynthesis occurs primarily due to aromatase activity in the peripheral tissues.

Types of Aromatase Inhibitors in Use

- Exemestane (25 mg/day): Exemestane is an irreversible, steroidal aromatase inhibitor that is similar in its structure to the natural substrate androstenedione. Irreversible covalent bonding at the active center of the aromatase permanently inactivates it. However, as the enzyme is reproduced rapidly, there is no advantage compared to the reversible inhibitors.
- Letrozole (25 mg/day): Letrozole is a selective, nonsteroidal aromatase inhibitor that binds competitively to the heme of the cytochrome P450 of the aromatase. In this way, the specific enzyme function and therefore estrogen biosynthesis are inhibited in the peripheral tissues and the tumor tissue.
- Anastrozole (1 mg/day): Anastrozole is a selective, nonsteroidal aromatase inhibitor that binds competitively to the heme group of aromatase and so blocks it.

The described various pharmacokinetic properties of the aromatase inhibitors do not show any differences in relation to their clinical effects.[6]

Potential clinically relevant effects of lowering endogenous estrogen levels are the resultant decrease of bone mass and increased risk of osteoporosis, which further results in bone, muscle, and joint pain and even pathologic fractures. Also, vaginal atrophy is intensified. In addition, there is a higher risk of cardiovascular disease than with tamoxifen. The majority of adverse events occur within the first year of treatment and can be associated with the normal pharmacologically induced effects of estrogen withdrawal.

Practical Tip

To prevent loss of bone mass, concomitant treatment with a combination of calcium (1.5 g/day) and vitamin D (400–800 IU/day) supplements together with physical activity is recommended.

Therapeutic Options for Aromatase Inhibitors

The use of aromatase inhibitors has given rise to new therapeutic options that have been studied in large randomized studies involving the examination of nearly 40,000 patients:

- *Option 1: The use of aromatase inhibitors as primary adjuvant therapy (Upfront).* There is one clear indication for monotherapy with an aromatase inhibitor—a contraindication to tamoxifen. When used alone, the recommendation is for aromatase inhibitor therapy to be continued for a total of 5 years. Approval exists for anastrozole (ATAC study: Anastrozole, Tamoxifen Alone or in Combination) and letrozole (BIG 1–98 study). The ATAC study, with the longest period of subsequent observation of 68 months, showed a significantly better disease-free survival rate[30] in anastrozole monotherapy and a significant reduction in the risk of relapse, but no effect on overall survival.
- *Option 2: The use of aromatase inhibitors as "switch therapy."* Switch therapy includes, for example, starting endocrine therapy with tamoxifen over a 2- to 3-year period and then, if there is no relapse, switching to an aromatase inhibitor over a 2- to 3- year period, with a total therapy duration of 5 years. The BIG 1–98 study showed that, compared with the use of tamoxifen alone, there was a significant reduction in relapses.[58] The individual evaluation of the ARNO-95 study had already demonstrated a significant improvement in the overall survival rate in favor of the switch group (with a switch to anastrozole) compared with the group continuing on tamoxifen. For sequential therapy in the adjuvant therapy setting, tamoxifen should be used first for 2 to 3 years, then a switch to an aromatase inhibitor should be made to achieve a total therapy duration of 5 years. Anastrozole and letrozole are approved for this purpose. According to AGO recommendations, all three aromatase

inhibitors are of equal value. Another alternative, according to AGO recommendations, is sequential therapy, starting with letrozole for 2 years, followed by tamoxifen for another 3 years; this model is used particularly in the case of nodal-positive patients.

- *Option 3: The use of aromatase inhibitors as extended adjuvant therapy.* The continuation of endocrine therapy after 5 years of tamoxifen treatment with an aromatase inhibitor for another 3 to 5 years can significantly reduce the risk of a recurrence. An improvement in overall survival is demonstrated in women with a nodal-positive breast carcinoma. It appears to be different at the beginning of the therapy in the case of premenopausal patients, who will be classed as postmenopausal after the 5-year primary treatment with tamoxifen. These patients in particular benefit from subsequent aromatase inhibitor therapy with regard to the disease-free and distant metastasis–free survival rates, compared with women who are already in an initial postmenopausal stage (regardless of the nodal status). This effect was also demonstrated after a break in therapy for an average of 2.8 years.[29]
- *Option 4: Administration of tamoxifen 20 mg/day over 5 years.* Although a 5-year treatment with tamoxifen remains a good option, it should only be used as monotherapy in the postmenopausal stage where there are contraindications to and intolerance of aromatase inhibitors.

12.2.4 Antibody Therapy

Targeted therapy with the monoclonal humanized antibody trastuzumab (Herceptin) is considered as standard in all patients with a HER2-overexpressing primary breast carcinoma larger than 10 mm.[40] The recurrence rate can be reduced to 45 to 50% using simultaneous or sequential adjuvant therapy with trastuzumab combined with chemotherapy and the mortality rate decreases by ca. 30% according to results from five independent studies. However, several retrospective case series and studies show that even patients with smaller, nodal-negative tumors, in which chemotherapy is indicated, could benefit from trastuzumab therapy. In these cases the risk/benefit profile discussed above should be considered.

The treatment may be administered as either adjuvant or neoadjuvant therapy and should last in total for 1 year. If the antibody was not initially given simultaneously, results of one study show that patients benefit from a 1-year treatment with trastuzumab even a year after conclusion of the primary therapy.[68]

Take Home Points

Approved Indications for Antibody Therapy
Patients with HER2-overexpressing tumors of diameter ≥ 1 cm (immunohistochemical score 3 + and/or ISH-positive) should receive a (neo)adjuvant treatment with trastuzumab over a 1 year course.

General Recommendations for Antibody Therapy
Adjuvant treatment with trastuzumab should preferably be started simultaneously with the taxane phase of the adjuvant chemotherapy. If there is an indication for chemotherapy for tumors < 10 mm, then trastuzumab should be given in addition.

Quality-assured determination of the HER2-status is a prerequisite for trastuzumab therapy, and its administration should meet the following conditions:

- Sequential start of treatment with chemotherapy regimens containing anthracyclines.
- Simultaneous initiation of therapy with taxanes.
- Optimum start: up to 3 months after chemotherapy.
- Dosage (every 3 weeks): initial start-up dose 8 mg/kg BW, then 6 mg/kg BW.
- Dosage (weekly): initial start-up dose 6 mg/kg BW, then 4 mg/kg BW.
- After two pauses in the therapy, it is restarted with the start-up dose.
- Assessment of adequate cardiac function before starting treatment and regularly throughout treatment; echocardiography with analysis of LVEF (left ventricular ejection fraction) before the onset of therapy and then every 3 months (as an approximate value for possible cardiac dysfunction, be alert for a 15% decrease in LVEF; watch out for heart failure!).

Practical Tip

Radiotherapy can also be implemented simultaneously with trastuzumab treatment.

Take Home Point

General Recommendation for the Therapeutic Sequence for Antibody Therapy and Radiotherapy
Simultaneous administration of trastuzumab and radiotherapy is permissible as long as no irradiation of the internal mammary lymph drainage is planned.

Other anti-HER2 substances, such as pertuzumab, are not currently recommended as adjuvant therapy and are the subject of numerous clinical studies.

12.3 Medical Treatment in Locoregional Recurrence

A locoregional recurrence (intramammary, chest wall, axillary) is surgically excised to the maximum extent possible with curative intent after further diagnostic investigation of the surrounding area. The best local control of an intramammary recurrence is achieved by mastectomy.[33] Should BCT be desired, the increased risk of further intramammary recurrence must be explained to the patient. For hormone receptor-positive patients, additional endocrine therapy may be considered.[39] It has not been verified whether this actually improves overall survival.[67]

Note

Because no present data support an increase in survival rates, the indication for “adjuvant” chemotherapy for locoregional breast carcinoma recurrence must be very strictly defined.

12

12.4 Medical Treatment of Distant Metastases

Patients with metastatic breast carcinoma are not curable with the current state of knowledge; here, preservation of quality of life is the primary goal of treatment. That sounds easy, but in practice it places the highest demands on the treating physician. The therapy has to be individualized and adapted to the disease and the symptoms: a concept also termed "palliative" drug therapy. However, since patients with metastasized breast carcinoma may achieve a good or satisfactory quality of life over many years, the term should not be confused with palliative medicine involving care of patients in the terminal phase of life.

If it is possible and appropriate, a tissue sample of the metastatic tissue should be obtained to re-identify the current immunohistological properties of the tumor. For positive hormone receptor status, endocrine therapy—which has relatively few side effects—represents a very good therapeutic option as long as there is no acute threat. The remission phase persists longer than with chemotherapy. The response rate of endocrine therapy, depending on the study, is between 20 and 35%, but its effectiveness can only be assessed after about 2 months. If the patient responds to the endocrine therapy, it is continued until there is proven tumor progression.

In a HER2-positive breast carcinoma, the trastuzumab treatment should, if possible, be combined with mild monochemotherapy. A particularly effective regimen is weekly administration of paclitaxel or administration of vinorelbine every 3 weeks (days 1 and 8). This provides substantially greater efficacy within the palliative therapy concept. Worldwide, combination with endocrine therapy may also be considered as another option. Only the aromatase inhibitors (letrozole) are approved in this context. If endocrine therapy is not possible, as in the case of a hormone receptor-negative metastatic breast carcinoma or because of contraindications, then a combination of trastuzumab with monochemotherapy should be administered here as well. Usually the antibody therapy is continued until disease progression occurs, but it is often extended beyond that point.

12

Take Home Point

Systemic Endocrine Therapy—What We Know for Sure

Endocrine therapy is the treatment of choice in the case of a positive hormone receptor status.

12.4.1 Endocrine Therapy in Premenopausal Patients with Distant Metastases

The treatment of choice in the palliative treatment of the premenopausal woman with a hormone receptor-positive breast carcinoma is suppression of ovarian function (see Chapter 12.2) combined with tamoxifen.[13,38] If there is recurrent progression of disease after a response to the endocrine therapy, then an aromatase inhibitor with simultaneous ovarian suppression should be used.[23] If there is further progression or intolerance, high-dose progestogens are a therapeutic option (▶ Fig. 12.1). If primary progression occurs during endocrine therapy, then a course of (mono)chemotherapy is indicated.[1]

Take Home Point

General Recommendations for Ovarian Suppression and Tamoxifen in Premenopausal Patients

In premenopausal patients, suppression of ovarian function (with GnRH analogs, oophorectomy, or radiotherapy) combined with tamoxifen is the first line of therapy.

12.4.2 Endocrine Therapy in Postmenopausal Patients with Distant Metastases

In a postmenopausal patient with a metastatic, hormone receptor-positive breast carcinoma, the appropriate choice of treatment in the first instance depends on the previous therapy and

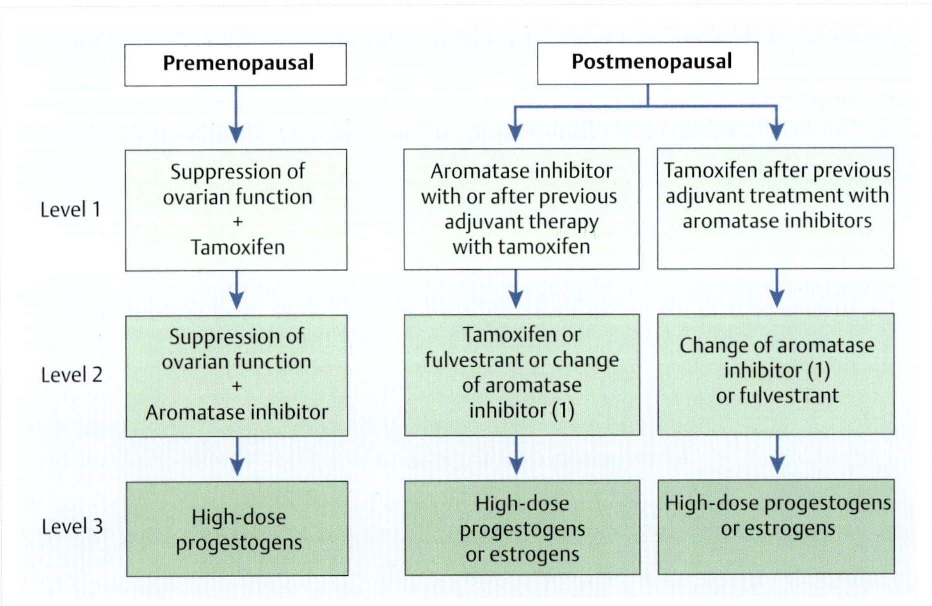

Fig. 12.1 Algorithm for palliative endocrine therapy according to the recommendations of AGO,[1] with everolimus as appropriate.

also on the side-effect profile. If the patient was treated with tamoxifen as adjuvant therapy, then aromatase inhibitor therapy is the treatment of choice in the palliative setting.[19,25] Tamoxifen is one alternative. If the first line fails, a second-line or even a third-line treatment should be tried, using the equally effective drug fulvestrant.[31,59]

If the patient has already had an aromatase inhibitor as adjuvant therapy, then, in case of metastasis, initial treatment with tamoxifen is indicated. If there is no response or if the disease progresses, another aromatase inhibitor or fulvestrant may be used (▶ Fig. 12.1).

Other forms of endocrine therapy are the administration of high-dose gestagens and estrogens, or alternation between steroidal and nonsteroidal aromatase inhibitors (▶ Fig. 12.1). All drug classes mentioned in this text are approved for this purpose.

Everolimus (Afinitor [Novartis]/RAD001) is a new drug for this purpose. It has been approved since July 2012 in combination with exemestane for use in postmenopausal patients with advanced, hormone receptor-positive and HER2-negative breast carcinoma who have been pretreated with a nonsteroidal aromatase inhibitor. The pivotal phase III, Bolero-2 study demonstrated a median progression-free survival of 11 months (versus 4.1 months in the control group). Everolimus is a selective inhibitor of mTOR kinase (mammalian target of rapamycin kinase). Through binding of everolimus with this serine/threonine kinase and subsequent formation of an intracellular complex, the kinase is deactivated, thereby inhibiting important signal transduction pathways of the tumor cell. Everolimus is a strong growth and proliferation inhibitor of tumor cells, among other cell types. In addition, everolimus reduces the levels of VEGF (vascular endothelial growth factor), which promotes the processes of tumor angiogenesis (see technical information). The recommended daily dose of everolimus is 10 mg. The side-effect spectrum of this new combination is noteworthy, however: in the Bolero-2 study, the rate of side effects resulting in a discontinuation of therapy was 10%, compared with that in the placebo-exemestane arm of 3%. Noninfectious pneumonitis occurred very frequently (12%); some cases were severe and some individual cases even fatal. Additional common systemic and local infections occurred due to the immunosuppressive effects. Other common to very common side effects (see the prescribing information for Afinitor) were:

- Stomatitis.
- Kidney failure.
- Pancytopenia.
- Metabolic and nutritional disorders.
- Diarrhea.
- Hypertension.
- Bleeding.
- Pulmonary embolism.
- Headache.
- Hypersensitivity reactions.
- Increased transaminase levels.
- Skin rash.
- Hand–foot syndrome.
- Peripheral edema.
- Arthralgia.
- Fatigue.
- Weight loss.

12.5 Endocrine Maintenance Therapy after Completing Chemotherapy

The benefits of endocrine maintenance therapy after completion of chemotherapy have not yet been proven, but the regimen may be considered in individual cases because it prolongs progression-free survival (AGO recommendation[1]).

12.6 Chemotherapy of Metastatic Breast Cancer Combined with New Agents

If the endocrine therapy approach has been exhausted or is not possible in the face of metastatic disease, a well-tolerated monochemotherapy should be attempted. It is true that polychemotherapy gives faster results and more often leads to remission, possibly conferring a small survival benefit. However, these advantages are paid for through numerous serious side effects, and it should always be critically questioned as to whether they meet the requirements of an overall palliative approach. They are nevertheless certainly indicated for use with patients with rapid tumor progression, acutely life-threatening visceral metastasis, or the urgent need for rapid control of the disease or reduction of symptoms.[7] The duration of therapy should be individually adjusted and the response regularly monitored so as to avoid unnecessary toxicities.

If possible, patients with a HER2-positive breast carcinoma should be treated from the start of palliative therapy with trastuzumab regardless of whether they have already been treated with that in the adjuvant stage.

The combination of taxanes with trastuzumab remains the preferred choice of treatment.[42,54] With the appropriate indication, trastuzumab or even lapatinib (see below) can also be combined with an aromatase inhibitor,[34,35] providing an interesting approach in which, with a single blow, cancer cells are targeted using two different methods. Studies show a prolonged disease-free interval compared with monotherapy. Finally, even if there is progression during antibody therapy, treatment with trastuzumab combined with monochemotherapy should be pursued if possible.

On the basis of study results on the first-line treatment of metastatic HER2-positive breast carcinoma, the combination of docetaxel with trastuzumab and pertuzumab (a monoclonal antibody that inhibits the dimerization of the HER receptors[2]), pertuzumab has been approved, as of March 4, 2013, by the European Medicines Agency for the following indications:

Pertuzumab is indicated for use combined with trastuzumab and docetaxel in the case of adult patients with HER2-positive metastatic or locally recurrent, inoperable breast cancer, who have previously received no anti-HER2 therapy or chemotherapy treatment for their metastatic disease.

Because of promising results in the palliative therapy setting, pertuzumab is already being intensively investigated in adjuvant studies.

12

As a second-line therapy, capecitabine, for example, can be combined with lapatinib (a dual tyrosine kinase inhibitor) if trastuzumab therapy has already been given.[24]

An option in the first-line treatment of HER2-negative breast carcinoma is the angiogenesis inhibitor bevacizumab combined with paclitaxel (weekly) or with capecitabine for individual cases. Bevacizumab is a monoclonal humanized immunoglobulin G antibody that inhibits angiogenesis via blockade of the VEGF receptors. However, studies have so far shown at best moderate success in terms of disease-free survival.[50]

Eribulin represents the first time a cytostatic has been available in the treatment of a HER2-negative breast carcinoma. Even after pretreatment with anthracycline and taxane, a significant increase in life expectancy with good tolerability has been achieved using eribulin.[12]

Bibliography

[1] Arbeitsgemeinschaft Gynäkologische Onkologie E.V. (AGO), Kommission Mamma. Aktuelle Empfehlungen zur Diagnostik und Therapie primärer und metastasierter Mammakarzinome. 2012. Available at: http://www.ago-online.de/de/fuer-mediziner/leitlinien/mamma/. Accessed February 28, 2013

[2] Baselga J, Cortés J, Kim SB, et al. CLEOPATRA Study Group. Pertuzumab plus trastuzumab plus docetaxel for metastatic breast cancer. N Engl J Med. 2012; 366(2):109–119

[3] Bear HD, Anderson S, Smith RE, et al. Sequential preoperative or postoperative docetaxel added to preoperative doxorubicin plus cyclophosphamide for operable breast cancer: National Surgical Adjuvant Breast and Bowel Project Protocol B-27. J Clin Oncol. 2006; 24(13):2019–2027

[4] Brito RA, Valero V, Buzdar AU, et al. Long-term results of combined-modality therapy for locally advanced breast cancer with ipsilateral supraclavicular metastases: The University of Texas M.D. Anderson Cancer Center experience. J Clin Oncol. 2001; 19(3):628–633

[5] Burstein HJ, Prestrud AA, Seidenfeld J, et al. American Society of Clinical Oncology. American Society of Clinical Oncology clinical practice guideline: update on adjuvant endocrine therapy for women with hormone receptor-positive breast cancer. J Clin Oncol. 2010; 28(23):3784–3796

[6] Buzdar AU, Robertson JF, Eiermann W, Nabholtz JM. An overview of the pharmacology and pharmacokinetics of the newer generation aromatase inhibitors anastrozole, letrozole, and exemestane. Cancer. 2002; 95(9):2006–2016

[7] Cardoso F, Bedard PL, Winer EP, et al. ESO-MBC Task Force. International guidelines for management of metastatic breast cancer: combination vs sequential single-agent chemotherapy. J Natl Cancer Inst. 2009; 101(17): 1174–1181

[8] Castiglione-Gertsch M, O'Neill A, Price KN, et al. International Breast Cancer Study Group (IBCSG). Adjuvant chemotherapy followed by goserelin versus either modality alone for premenopausal lymph node-negative breast cancer: a randomized trial. J Natl Cancer Inst. 2003; 95(24):1833–1846

[9] Chen AM, Meric-Bernstam F, Hunt KK, et al. Breast conservation after neoadjuvant chemotherapy: the MD Anderson Cancer Center experience. J Clin Oncol. 2004; 22(12):2303–2312

[10] Citron ML, Berry DA, Cirrincione C, et al. Randomized trial of dose-dense versus conventionally scheduled and sequential versus concurrent combination chemotherapy as postoperative adjuvant treatment of node-positive primary breast cancer: first report of Intergroup Trial C9741/Cancer and Leukemia Group B Trial 9741. J Clin Oncol. 2003; 21(8):1431–1439

[11] Ferguson T, Wilcken N, Vagg R, Ghersi D, Nowak AK. Taxanes for adjuvant treatment of early breast cancer. Cochrane Database Syst Rev. 2007; 4(4):CD004421

[12] Cortes J, O'Shaughnessy J, Loesch D, et al. EMBRACE (Eisai Metastatic Breast Cancer Study Assessing Physician's Choice Versus E7389) investigators. Eribulin monotherapy versus treatment of physician's choice in patients with metastatic breast cancer (EMBRACE): a phase 3 open-label randomised study. Lancet. 2011; 377(9769):914–923

[13] Crump M, Sawka CA, DeBoer G, et al. An individual patient-based meta-analysis of tamoxifen versus ovarian ablation as first line endocrine therapy for premenopausal women with metastatic breast cancer. Breast Cancer Res Treat. 1997; 44(3):201–210

[14] Cuzick J, Ambroisine L, Davidson N, et al. LHRH-agonists in Early Breast Cancer Overview group. Use of luteinising-hormone-releasing hormone agonists as adjuvant treatment in premenopausal patients with hormone-receptor-positive breast cancer: a meta-analysis of individual patient data from randomised adjuvant trials. Lancet. 2007; 369(9574):1711–1723

[15] Davidson NE, O'Neill AM, Vukov AM, et al. Chemoendocrine therapy for premenopausal women with axillary lymph node-positive, steroid hormone receptor-positive breast cancer: results from INT 0101 (E5188). J Clin Oncol. 2005; 23(25):5973–5982

[16] Davies C, Godwin J, Gray R, et al. Early Breast Cancer Trialists' Collaborative Group (EBCTCG). Relevance of breast cancer hormone receptors and other factors to the efficacy of adjuvant tamoxifen: patient-level meta-analysis of randomised trials. Lancet. 2011; 378(9793):771–784

[17] Early Breast Cancer Trialists' Collaborative Group (EBCTCG). Polychemotherapy for early breast cancer: an overview of the randomised trials. Lancet. 1998; 352(9132):930–942

[18] Early Breast Cancer Trialists' Collaborative Group (EBCTCG). Effects of chemotherapy and hormonal therapy for early breast cancer on recurrence and 15-year survival: an overview of the randomised trials. Lancet. 2005; 365(9472):1687–1717

[19] Ferretti G, Bria E, Giannarelli D, et al. Second- and third-generation aromatase inhibitors as first-line endocrine therapy in postmenopausal metastatic breast cancer patients: a pooled analysis of the randomised trials. Br J Cancer. 2006; 94(12):1789–1796

[20] Fisher B, Brown A, Mamounas E, et al. Effect of preoperative chemotherapy on local-regional disease in women with operable breast cancer: findings from National Surgical Adjuvant Breast and Bowel Project B-18. J Clin Oncol. 1997; 15(7):2483–2493

[21] Fisher B, Dignam J, Wolmark N, et al. Tamoxifen and chemotherapy for lymph node-negative, estrogen receptor-positive breast cancer. J Natl Cancer Inst. 1997; 89(22):1673–1682

[22] Fisher ER, Wang J, Bryant J, Fisher B, Mamounas E, Wolmark N. Pathobiology of preoperative chemotherapy: findings from the National Surgical Adjuvant Breast and Bowel (NSABP) protocol B-18. Cancer. 2002; 95(4):681–695

[23] Forward DP, Cheung KL, Jackson L, Robertson JF. Clinical and endocrine data for goserelin plus anastrozole as second-line endocrine therapy for premenopausal advanced breast cancer. Br J Cancer. 2004; 90(3):590–594

[24] Geyer CE, Forster J, Lindquist D, et al. Lapatinib plus capecitabine for HER2-positive advanced breast cancer. N Engl J Med. 2006; 355(26):2733–2743

[25] Gibson LJ, Dawson C, Lawrence DH, Bliss JM. Aromatase inhibitors for treatment of advanced breast cancer in postmenopausal women. Cochrane Database Syst Rev. 2007; 1(1):CD003370

[26] Goldhirsch A, Wood WC, Gelber RD, Coates AS, Thürlimann B, Senn HJ, Panel members. Progress and promise: highlights of the international expert consensus on the primary therapy of early breast cancer 2007. Ann Oncol. 2007; 18(7):1133–1144

[27] Goldhirsch A, Ingle JN, Gelber RD, Coates AS, Thürlimann B, Senn HJ, Panel members. Thresholds for therapies: highlights of the St Gallen International Expert Consensus on the primary therapy of early breast cancer 2009. Ann Oncol. 2009; 20(8):1319–1329

[28] Goldhirsch A, Wood WC, Coates AS, Gelber RD, Thürlimann B, Senn HJ, Panel members. Strategies for subtypes—dealing with the diversity of breast cancer: highlights of the St. Gallen International Expert Consensus on the Primary Therapy of Early Breast Cancer 2011. Ann Oncol. 2011; 22(8):1736–1747

[29] Goss PE, Ingle JN, Martino S, et al. Randomized trial of letrozole following tamoxifen as extended adjuvant therapy in receptor-positive breast cancer: updated findings from NCIC CTG MA.17. J Natl Cancer Inst. 2005; 97 (17):1262–1271

[30] Howell A, Cuzick J, Baum M, et al. ATAC Trialists' Group. Results of the ATAC (Arimidex, Tamoxifen, Alone or in Combination) trial after completion of 5 years' adjuvant treatment for breast cancer. Lancet. 2005; 365(9453):60–62

[31] Howell A. Fulvestrant ('Faslodex'): current and future role in breast cancer management. Crit Rev Oncol Hematol. 2006; 57(3):265–273

[32] Huober J, von Minckwitz G, Denkert C, et al. Effect of neoadjuvant anthracycline-taxane-based chemotherapy in different biological breast cancer phenotypes: overall results from the GeparTrio study. Breast Cancer Res Treat. 2010; 124(1):133–140

[33] Jobsen JJ, van der Palen J, Meerwaldt JH. The impact of age on local control in women with pT1 breast cancer treated with conservative surgery and radiation therapy. Eur J Cancer. 2001; 37(15):1820–1827

[34] Johnston S, Pippen J , Jr, Pivot X, et al. Lapatinib combined with letrozole versus letrozole and placebo as first-line therapy for postmenopausal hormone receptor-positive metastatic breast cancer. J Clin Oncol. 2009; 27(33): 5538–5546

[35] Kaufman B, Mackey JR, Clemens MR, et al. Trastuzumab plus anastrozole versus anastrozole alone for the treatment of postmenopausal women with human epidermal growth factor receptor 2-positive, hormone receptor-positive metastatic breast cancer: results from the randomized phase III TAnDEM study. J Clin Oncol. 2009; 27(33):5529–5537

[36] Kaufmann M, von Minckwitz G, Smith R, et al. International expert panel on the use of primary (preoperative) systemic treatment of operable breast cancer: review and recommendations. J Clin Oncol. 2003; 21(13):2600–2608

[37] Kaufmann M, Hortobagyi GN, Goldhirsch A, et al. Recommendations from an international expert panel on the use of neoadjuvant (primary) systemic treatment of operable breast cancer: an update. J Clin Oncol. 2006; 24(12):1940–1949

[38] Klijn JG, Blamey RW, Boccardo F, Tominaga T, Duchateau L, Sylvester R, Combined Hormone Agents Trialists' Group and the European Organization for Research and Treatment of Cancer. Combined tamoxifen and luteinizing hormone-releasing hormone (LHRH) agonist versus LHRH agonist alone in premenopausal advanced breast cancer: a meta-analysis of four randomized trials. J Clin Oncol. 2001; 19(2):343–353

[39] Kuo SH, Huang CS, Kuo WH, Cheng AL, Chang KJ, Chia-Hsien Cheng J. Comprehensive locoregional treatment and systemic therapy for postmastectomy isolated locoregional recurrence. Int J Radiat Oncol Biol Phys. 2008; 72(5):1456–1464

[40] Leitlinienprogramm Onkologie der AWMF, Deutschen Krebsgesellschaft eV und Deutschen Krebshilfe eV. Leitlinienreport der S 3 Leitlinie fur die Diagnostik, Therapie und Nachsorge des Mammakarzinoms. AWMF Register-Nr. 032–045OL. Available at: http://www.awmf.org/leitlinien/detail/ll/032–045OL.html. Accessed June 6, 2017

[41] Martín M, Seguí MA, Antón A, et al. GEICAM 9805 Investigators. Adjuvant docetaxel for high-risk, node-negative breast cancer. N Engl J Med. 2010; 363(23):2200–2210

[42] Marty M, Cognetti F, Maraninchi D, et al. Randomized phase II trial of the efficacy and safety of trastuzumab combined with docetaxel in patients with human epidermal growth factor receptor 2-positive metastatic breast cancer administered as first-line treatment: the M77001 study group. J Clin Oncol. 2005; 23(19):4265–4274

[43] Moebus V, Jackisch C, Lueck HJ, et al. Intense dose-dense sequential chemotherapy with epirubicin, paclitaxel, and cyclophosphamide compared with conventionally scheduled chemotherapy in high-risk primary breast cancer: mature results of an AGO phase III study. J Clin Oncol. 2010; 28(17):2874–2880

[44] Muss HB, Berry DA, Cirrincione CT, et al. CALGB Investigators. Adjuvant chemotherapy in older women with early-stage breast cancer. N Engl J Med. 2009; 360(20):2055–2065

[45] National Comprehensive Cancer Network (NCCN). Practice Guidelines in Oncology. Version 2. 2006. Available at: https://www.nccn.org/professionals/physician_gls/f_guidelines.asp. Accessed XXXX XXXX

[46] National Comprehensive Cancer Network (NCCN). Breast Cancer. Version 2. 2011. Available at: http://www.nccn.org/professionals/physician_gls/f_guidelines.asp. Accessed February 28, 2013

[47] National Institute for Clinical Excellence (NICE). Clinical Guideline CG80, Early and locally advanced breast cancer: diagnosis and treatment. 2009. Available at: https://www.nice.org.uk/guidance/conditions-and-diseases/cancer/breast-cancer. Accessed December 27, 2016

[48] National Institutes of Health Consensus Development Panel (NIH). National Institutes of Health Consensus Development Conference statement: Adjuvant therapy for breast cancer. November 1–3, 2000. Bethesda (MD). Available at: http://consensus.nih.gov/2000/2000AdjuvantTherapyBreastCancer114html.htm. Accessed February 28, 2013

[49] New Zealand Guidelines Group (NZGG). Management of Early Breast Cancer. Wellington: NZGG; 2009

[50] O'Shaughnessy J, Miles D, Gray RJ. et, al. A meta-analysis of overall survival data from three randomized trials of bevacizumab (BV) and first-line chemotherapy as treatment for patients with metastatic breast cancer (MBC). J Clin Oncol. 2010; 28:(Suppl; Abstr 1005):15s

[51] Albain KS, Barlow WE, Shak S, et al. Breast Cancer Intergroup of North America. Prognostic and predictive value of the 21-gene recurrence score assay in postmenopausal women with node-positive, oestrogen-receptor-positive breast cancer on chemotherapy: a retrospective analysis of a randomised trial. Lancet Oncol. 2010; 11(1):55–65

[52] Pameijer CR, Smith D, McCahill LE, Bimston DN, Wagman LD, Ellenhorn JD. Full-thickness chest wall resection for recurrent breast carcinoma: an institutional review and meta-analysis. Am Surg. 2005; 71(9):711–715

[53] Peto R, Davies C, Godwin J, et al. Early Breast Cancer Trialists' Collaborative Group (EBCTCG). Comparisons between different polychemotherapy regimens for early breast cancer: meta-analyses of long-term outcome among 100,000 women in 123 randomised trials. Lancet. 2012; 379(9814):432–444

[54] Slamon DJ, Leyland-Jones B, Shak S, et al. Use of chemotherapy plus a monoclonal antibody against HER2 for metastatic breast cancer that overexpresses HER2. N Engl J Med. 2001; 344(11):783–792

[55] Smith IE, Dowsett M, Ebbs SR, et al. IMPACT Trialists Group. Neoadjuvant treatment of postmenopausal breast cancer with anastrozole, tamoxifen, or both in combination: the Immediate Preoperative Anastrozole, Tamoxifen, or Combined with Tamoxifen (IMPACT) multicenter double blind randomized trial. J Clin Oncol. 2005; 23(22):5108–5116

[56] Smith IE, Dowsett M, Yap YS, et al. Adjuvant aromatase inhibitors for early breast cancer after chemotherapy-induced amenorrhoea: caution and suggested guidelines. J Clin Oncol. 2006; 24(16):2444–2447

[57] Sotiriou C, Pusztai L. Gene-expression signatures in breast cancer. N Engl J Med. 2009; 360(8):790–800

[58] Breast International Group (BIG) 1–98 Collaborative Group, Thürlimann B, Keshaviah A, Coates AS, et al. A comparison of letrozole and tamoxifen in postmenopausal women with early breast cancer. N Engl J Med. 2005; 353(26):2747–2757

[59] Valachis A, Mauri D, Polyzos NP, Mavroudis D, Georgoulias V, Casazza G. Fulvestrant in the treatment of advanced breast cancer: a systematic review and meta-analysis of randomized controlled trials. Crit Rev Oncol Hematol. 2010; 73(3):220–227

[60] Viale G, Regan MM, Maiorano E, et al. Prognostic and predictive value of centrally reviewed expression of estrogen and progesterone receptors in a randomized trial comparing letrozole and tamoxifen adjuvant therapy for postmenopausal early breast cancer: BIG 1–98. J Clin Oncol. 2007; 25(25):3846–3852

[61] von Minckwitz G, Blohmer JU, Raab G, et al. German Breast Group. In vivo chemosensitivity-adapted preoperative chemotherapy in patients with early-stage breast cancer: the GEPARTRIO pilot study. Ann Oncol. 2005; 16(1):56–63

[62] von Minckwitz G, Kümmel S, Vogel P, et al. German Breast Group. Neoadjuvant vinorelbine-capecitabine versus docetaxel-doxorubicin-cyclophosphamide in early nonresponsive breast cancer: phase III randomized GeparTrio trial. J Natl Cancer Inst. 2008; 100(8):542–551

[63] von Minckwitz G, Kümmel S, Vogel P, et al. German Breast Group. Intensified neoadjuvant chemotherapy in early-responding breast cancer: phase III randomized GeparTrio study. J Natl Cancer Inst. 2008; 100(8):552–562

[64] von Minckwitz G, Untch M, Nüesch E, et al. Impact of treatment characteristics on response of different breast cancer phenotypes: pooled analysis of the German neo-adjuvant chemotherapy trials. Breast Cancer Res Treat. 2011; 125(1):145–156

[65] von Minckwitz G, Untch M, Blohmer JU, et al. Definition and impact of pathologic complete response on prognosis after neoadjuvant chemotherapy in various intrinsic breast cancer subtypes. J Clin Oncol. 2012; 30(15):1796–1804

[66] Voogd AC, van Oost FJ, Rutgers EJ, et al. Dutch Study Group on Local Recurrence after Breast Conservation (BORST Group). Long-term prognosis of patients with local recurrence after conservative surgery and radiotherapy for early breast cancer. Eur J Cancer. 2005; 41(17):2637–2644

[67] Waeber M, Castiglione-Gertsch M, Dietrich D, et al. Swiss Group for Clinical Cancer Research (SAKK). Adjuvant therapy after excision and radiation of isolated postmastectomy locoregional breast cancer recurrence: definitive results of a phase III randomized trial (SAKK 23/82) comparing tamoxifen with observation. Ann Oncol. 2003; 14(8):1215–1221

[68] Cameron D, Piccart-Gebhart MJ, Gelber RD, et al. 11 years' follow-up of trastuzumab after adjuvant chemotherapy in HER2-positive early breast cancer: final analysis of the HERceptin Adjuvant (HERA) trial. Lancet. 2017; 389 (10075):1195–1205

13 Radiotherapy of Breast Cancer

C. F. Hess

For many decades now, radiotherapy has occupied a firm place in the multimodal treatment of carcinoma of the breast. It is an essential part of breast-conserving therapy, but it is also needed after mastectomy for advanced tumors. The use of radiotherapy leads to a substantial reduction of locoregional risk and to significant improvement in the long-term survival rate.

These positive effects of radiotherapy cannot be achieved by the use of systemic therapeutic measures. In addition, modern radiotherapy methods have evolved dynamically such that significant acute reactions and late complications have become much rarer. As a result, the anticipated benefits of radiotherapy have, as a rule, clearly outstripped its potential risks. This is equally true for the primary treatment of very advanced and inoperable tumors[2] and the palliative treatment of symptomatic metastasis, especially in the skeletal system. As an integral component of what is always a multimodal treatment, the use of radiotherapy requires close interdisciplinary cooperation, especially with the gynecologist, the breast surgeon, and the internist/oncologist. Its organization for the benefit of the patients has significantly improved in the more recent years with the establishment of breast cancer centers.

The essential aspects of radiotherapy of breast cancer are summarized in the respective current international guidelines for breast cancer[9,15], including the most important references. Here the most important considerations are described, and additional current results are discussed and their practical uses clarified. Accordingly, the references presented here are limited to those aspects that go beyond the current guidelines.

13.1 Adjuvant Radiotherapy after Breast-Conserving Surgery

Breast-conserving surgery today is by far the predominant modality in the primary treatment of breast cancer—used in over 80% of the cases in designated breast cancer centers. This is always true for patients with early breast cancer. Even large primary tumors or involvement of an axillary lymph node do not constitute contraindications to breast-conserving surgery. In all these cases, adjuvant radiotherapy is an integral component of multimodal treatment. This also applies to most cases of ductal carcinoma in situ (DCIS).

Earlier studies had already shown that, for most affected patients, a total mastectomy usually offers no advantage in comparison to breast-conserving surgery with subsequent radiation therapy. It is also applicable for younger patients with early breast cancer.[13]

After its contribution to breast-conserving surgery, the most important effect of adjuvant radiotherapy is that the local recurrence rate is reduced by approximately one-third to one-fourth. Additionally, the tumor-related mortality also declines significantly in absolute terms by 3.3 to 8.5%, depending on risk factors with postoperative radiation therapy.[4] Similar improvements are also seen in the overall survival of affected patients, but manifested only after a period of at least 5 years (in contrast to the reduction of local recurrence rate) (▶ Table 13.1).[21] Thus, adjuvant radiotherapy after breast-conserving surgery has a similarly large long-term effect on tumor-specific and long-term survival as adjuvant systemic therapeutic measures.[20]

Take Home Point

Radiotherapy after Breast-Conserving Surgery

In invasive carcinoma, the affected breast should be irradiated following breast-conserving surgery.

13.2 Adjuvant Radiotherapy after Mastectomy

The performance of a mastectomy in patients with breast cancer has substantially declined over the last few decades. In large centers today, it is carried out at most on only 20% of all affected patients, predominantly those in locoregionally advanced stages.

Even in this situation, adjuvant percutaneous radiotherapy not only significantly improves the locoregional tumor control but improves the long-term survival rate as well (▶ Table 13.1). This also holds if additional effective systemic treatment is given, such

Table 13.1 Substantial improvement of mid and long-term tumor control from adjuvant radiation therapy. All patients received additional medical treatment as indicated[3]

Treatment	Time after treatment (years)	Locoregional recurrence (%)		Mortality from breast cancer (%)		All-cause mortality (%)	
		Without RT	With RT	Without RT	With RT	Without RT	With RT
After breast-conserving surgery (7,311 patients, of which 17% were node-positive)	5	25.9	7.3	11.4	10.1	12.9	11.9
	10	32.0	10.3	24.8	20.7	27.7	24.5
	15			35.9	30.5	40.5	35.2
After mastectomy (8,505 node-positive patients)	5	22.8	5.8	34.0	32.1	35.5	33.6
	10	27.6	7.5	50.9	46.7	53.7	49.9
	15	29.2	7.8	60.1	54.7	64.2	59.8

as hormonal therapy and chemotherapy.[4,21] The positive effect of adjuvant radiotherapy has been strengthened even further to the extent that, as part of an effective systemic therapy program, it can prevent systemic metastasis.[15] There have been further positive effects resulting from the continued improvement of radiotherapeutic modalities and techniques.

Take Home Points

Radiotherapy to the Chest Wall after Mastectomy

- Postsurgical radiotherapy to the chest wall after mastectomy decreases the risk of locoregional recurrence. In patients with high risk of local recurrence, the overall survival is also improved.
- Therefore, subsequent radiotherapy to the chest wall after mastectomy is indicated in the following situations:
  - Tumor > 5 cm.
  - Tumor < 5 cm only with the existence of other risk factors.
  - R1/R2 resection with no possibility of a reexcision to clean up the margins.
  - Axillary lymph node involvement.

Overall, it may be assumed that, in patients with advanced locoregional tumors, the locoregional risk of relapse decreases to about one-fourth and that the long-term survival rate (after 5–10 years) shows an absolute increase of at least 5%. The more advanced the disease is at the time of diagnosis, the greater are these effects. Today it can be expected that the risk of locoregional recurrence is less than 8% (see ▶ Table 13.1).

One indication for adjuvant radiotherapy following mastectomy is in patients with T3 tumors and/or positive axillary lymph nodes—especially if other unfavorable factors, such as invasion of lymph or blood vessels, are present (L1, V1), or if the primary tumor could not be completely surgically excised (R1, R2). In patients with T1/T2 tumors without axillary lymph node involvement in whom a mastectomy is performed instead of breast-conserving surgery, there is no sufficient advantage to be gained from adjuvant radiation therapy. However, all patients who undergo neoadjuvant chemotherapy and in whom a locoregional advanced carcinoma was diagnosed before this treatment, should undergo radiation therapy. In these patients, the indication is correspondingly oriented to the constellation of findings at the time the diagnosis is made.[15]

Note

If possible, the nodal status of a breast cancer should already be determined by sentinel lymph node biopsy before neoadjuvant chemotherapy is given.

13.3 Effectiveness of Adjuvant Radiation Therapy: Prognostic Factors

The positive effect of adjuvant radiotherapy is recognized in all age groups. However, a younger age, especially an age of less than 40 years, is the most important risk factor for local recurrence. Consequently, in this age group, the absolute benefit of adjuvant radiotherapy is of the greatest importance. Only in older patients with favorable prognostic risk factors and a limited life expectancy for other reasons, can adjuvant radiotherapy possibly be waived, and then only after extensive consultation with the patient. One must be explicitly warned of a therapeutic nihilism in this age group. Such behavior in the past has actually led to unjustifiably unfavorable results.[5]

An important risk factor for local recurrence is a positive surgical margin after surgery. In the few cases with positive margins without the opportunity for a follow-up resection, the radiation therapist should consider a local increase in the radiation dose. Moreover, there is no doubt that systemic hormonal therapy, chemotherapy, and immunotherapy likewise contribute to a reduction of local recurrence. However, the positive effect of adjuvant radiotherapy on the local risk of recurrence and on long-term survival is sustained even after systemic therapy.[4,21]

There are other prognostic factors that have a confirmed influence on the tumor-specific mortality to some extent but have no importance in the indication for adjuvant radiation therapy. This is especially true for various molecular-genetic factors or gene-expression analyses, which are considered in making the choice of adjuvant medical treatment. Overall, current data indicate that—possibly with the exception of a subset of older patients with good prognostic factors and limited life expectancy—the withholding of adjuvant radiotherapy is associated with an approximately threefold higher risk of local recurrence in the affected breast and a significantly worsened survival rate. Correspondingly, all national and international guidelines include adjuvant radiotherapy as a standard part of the multimodal treatment scheme after breast-conserving surgery.[15]

13.4 Integration of Adjuvant Radiotherapy into the Multimodal Treatment Concept

As a rule, when there is an indication for systemic chemotherapy, it is administered first, with the adjuvant radiotherapy being initiated about 3 to 4 weeks later. If there is merely an indication for adjuvant antiestrogen therapy, on the other hand, the adjuvant radiotherapy should be performed first and not later than 4 to 6 weeks after surgery. Because the use of tamoxifen before the course of radiotherapy could possibly increase the risk of side effects, it should preferably be started only after completion of the radiation therapy.[23] On the other hand, treatment with trastuzumab can safely be continued, because this treatment represents an essential building block in the overall treatment of the high-risk patient (with positive HER2/neu status).[15]

13.5 Target Volume and Dose Concept

A reasonable definition of the target volumes, the overall dose, and the respective individual doses to be applied daily are of vital importance for the effectiveness and safety of the course of radiation therapy. Thus the *clinical target volume* (CTV) designates the regions in which residual tumor cells may be expected. The *planning target volume* (PTV), on the other hand, includes additional possible sources of spatial variability, such as movements of individual organs (some caused by breathing), and must take into account the technical aspects of optimal radiation therapy.

13.5.1 Clinical Target Volume: Former Tumor Region, Mammary Gland, Chest Wall, and Regional Lymph Channels

For many years, adjuvant radiotherapy consisted of application to the entire chest wall on the affected side of 50 Gy, preferably homogeneously, with individual doses of 1.8 to 2.0 Gy over 5 to 6 weeks. After breast-conserving therapy, the entire affected mammary gland is included; the tumor bed receives a dose that is about 10 to 16 Gy higher (▶ Fig. 13.1, ▶ Fig. 13.2). As a rule, such a boost can be omitted after breast-conserving surgery in sano for DCIS, or after a mastectomy for locoregional advanced breast cancer.

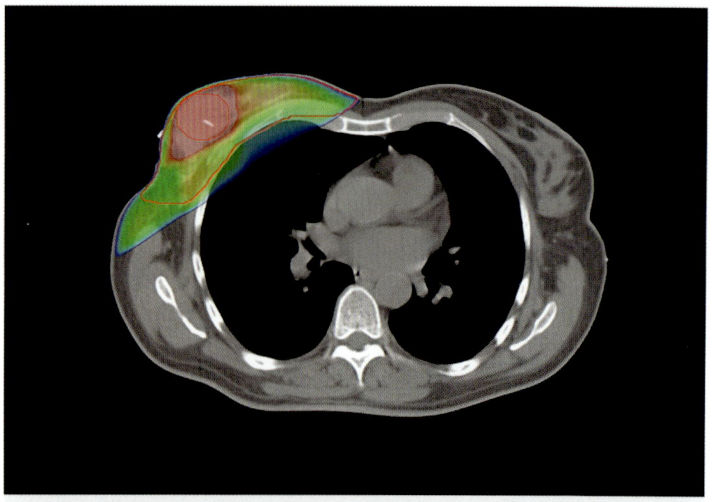

Fig. 13.2 Computer-aided planning (selected cross section) of radiotherapy to the tumor bed (red) and the remaining right mammary gland (green). Various colors signify different individual and cumulative doses for the various risk regions.

Take Home Points

Radiotherapy after Breast-conserving Surgery

- The target volume for the percutaneous radiotherapy should include the entire remaining breast and the adjacent chest wall.
- The dose should total approximately 50 Gy using conventional fractionation (5 × [1.8–2.0] Gy/week).
  - A local radiation boost to the tumor bed decreases the local recurrence rate in the breast but does not confer a survival advantage. The boost radiation is usually indicated. The recommended boost dose amounts to 10 to 16 Gy with conventional fractionation.

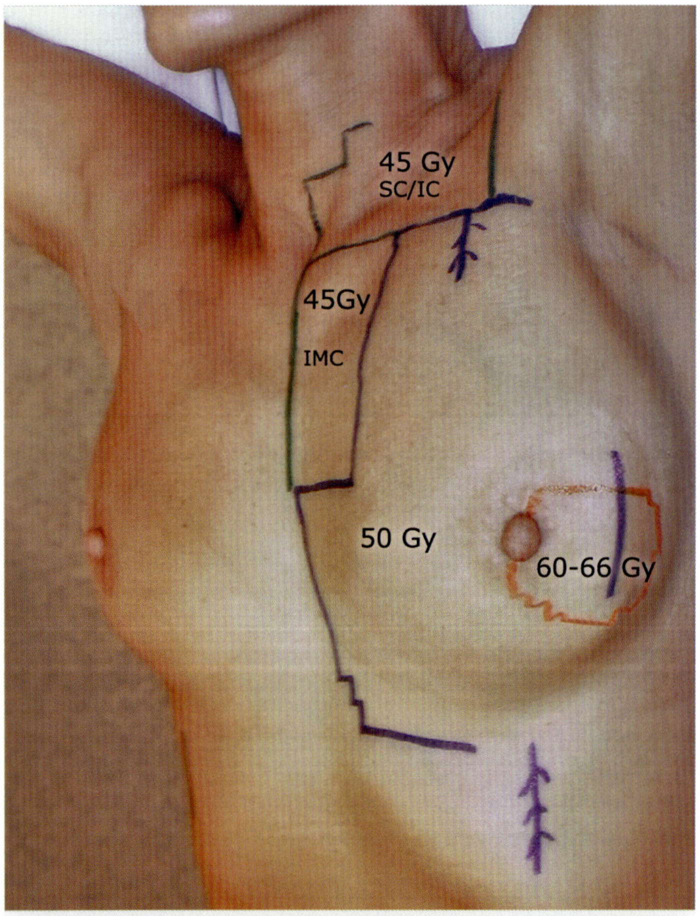

Fig. 13.1 Radiotherapy after breast-conserving surgery. The entire mammary gland is treated with 50 Gy and the tumor bed with 60 to 66 Gy. When positive axillary lymph nodes are found, the possibility of additional irradiation of the supraclavicular (SC), infraclavicular (IC), and lymph nodes of the internal mammary chain (IMC) of up to 45 Gy is discussed with the patient. In planning the radiation therapy, additional markings (arrows) are used that reliably ensure the correct positioning of the patient with the use of lasers.

Whether and in what manner the regional lymphatic drainage should be part of the clinical target volumes has been a matter of debate for many years. Recent results of randomized trials (MA.20 and EORTC 22922/10925) have demonstrated that the addition of regional nodal irradiation to the internal mammary nodes and the supraclavicular region significantly improves outcome in patients, in whom the axillary lymph node was positive at initial diagnosis.[14,29] In the MA.20 and EORTC trial, regional recurrences were reduced to 2.7% and 4.2% respectively with adjuvant breast irradiation only, to 0.7% and 2.7% respectively with the addition of nodal irradiation. Most importantly, the distant recurrences were significantly reduced from 17.3% and 19.6% respectively to 13.4% and 15.9% respectively at a median follow-up of more than 10 years. In addition, there was a corresponding improvement of disease-free survival. Interestingly, these positive aspects of additional nodal irradiation have been shown not only for patients with a least four positive lymph nodes but similarly for those with one to three affected lymph nodes.[31,29]

Furthermore, fears that radiotherapy of the retrosternal lymph nodes could lead to a significant increase in cardiac mortality (especially when combined with anthracycline chemotherapy and treatment with trastuzumab) have not been confirmed to date.[19,21]

All of this advocates discussion with the patient, in whom at least one axillary lymph node has been found to be positive, about also giving radiotherapy for the ipsilateral supraclavicular and retrosternal lymph nodes also (see ▶ Fig. 13.1). As a result, the National Comprehensive Cancer Network

(NCCN) panel strongly recommends additional irradiation of infraclavicular and supraclavicular areas and the internal mammary nodes in patients with positive axillary lymph nodes. In contrast, irradiation of the regional nodal area is generally not indicated for those with negative axillary nodes.[14] It is unknown at present whether special diagnostic measures, such as positron emission tomography (PET) combined with CT, or a representation of the lymph channels with scintigraphy, can contribute to a more targeted choice for radiotherapy of the retrosternal lymph nodes.[8,27]

Take Home Points

Radiotherapy of the Regional Lymphatic Drainage Areas

- Radiotherapy should not be administered to the regional lymphatic drainage areas for stage pN0.
- Radiotherapy of the supraclavicular and infraclavicular lymph drainage channels and the internal mammary nodes is recommended in the case of positive lymph nodes.
- Radiotherapy should be considered for any part of the axilla at risk:
  - Postoperative residual tumor in the axilla.
  - Extensive tumor load in the axilla.
  - Positive axillary nodes if only sentinel node has been removed (no lymph node dissection).[30]
- The utility of radiotherapy of the regional lymphatic drainage channels in cases of verified isolated tumor cells or of micrometastases in the regional lymph nodes has not been demonstrated, so that this should not be routinely carried out.

13.5.2 Partial Breast Irradiation

There have been numerous attempts after breast-conserving surgery to replace percutaneous radiotherapy of the entire residual breast with partial breast irradiation. The main objective here was to further reduce the acute side effects and late sequelae and to shorten the treatment time through accelerated treatments, or even a single treatment.[14,17,32] Several methods were employed for this, including percutaneous irradiation with electron beams and a three-dimensional conformal or intensity-modulated technique (both with and without proton beams); most frequently, interstitial forms of therapy with multicatheter or balloon catheter techniques, or intraoperative techniques with electron beams or X-rays have been used.[25,32] After the first limited experiences, the debate over partial breast irradiation was reignited by the release of the first results of the Targit A trial.[25] In this study, a total of 2,232 patients were randomized into a standard arm involving treatment exclusively with percutaneous radiotherapy of the breast, and an arm involving treatment with intraoperative radiation therapy. In the later treatment arm, 854 patients received exclusively intraoperative radiation therapy; only 142 received an additional percutaneous irradiation of the entire mammary gland. The intraoperative radiotherapy was carried out with conventional X-irradiation, which was delivered to the surgical area by means of a specially adapted applicator after tumor resection. The published 3-year follow-up data showed no significant difference with regard to local risk of recurrence to that point (approximately 1% in both arms of the study). The incidence of observed side effects of the therapy also did not differ significantly.[25]

However, because the follow-up time for the exclusively intraoperative radiotherapy group is still very short, this method has to date only been approved as boost therapy in Germany, except for its use in medical studies. Further follow-up of the Targit trials and the performance of further studies (e.g., for patients over 70 years of age) will reveal more clearly which form of partial irradiation is particularly suitable for which indication area and whether this can replace the current standard methods. As this unfolds, questions about the local recurrence rate, the long-term overall survival rate, and possible late sequelae (in particular the cosmetic results) will be in the foreground. Previous experience shows that follow-up periods of at least 10 years need to pass before significant effects on the survival rate can be expected (see ▶ Table 13.1).

Take Home Point

Partial Breast Irradiation as the Exclusive Form of Irradiation, Including Exclusive Intraoperative Radiotherapy

Partial breast irradiation as the exclusive intraoperative or postoperative radiation treatment does not represent standard therapy.

13.5.3 Shortened Treatment Time: Alternative Fractionation Schemes

Radiotherapy of the entire mammary gland usually lasts approximately 5 to 6 weeks and the additional treatment of the tumor bed (boost) another 1 to 2 weeks, resulting in a total treatment time of about 7 weeks. In countries with great distances between or limited numbers of radiation treatment centers, this has led to the testing of treatment schemes with shortened treatment times, attempting to decrease the number of radiation units and to shorten the treatment time (e.g., a total dose of 42.5 Gy given over 16 treatment days). The published reports do, indeed, show similar treatment results compared to the classical, longer-lasting schemes, but to date studies have only been carried out in a small number of cases and with limited follow-up time,[28] and for these reasons the methods have not yet gained widespread acceptance.

In the interim, with modern radiation techniques it has been demonstrated to be possible to coordinate the treatment of the tumor bed (which requires a higher dose) with the radiotherapy of the remaining affected mammary gland. With these irradiation methods, the individual dose is increased only in the most important areas of risk in the mammary gland. Thus, an optimally adapted dose is achieved in all parts of the target volume, and the treatment time can be shortened to about 5½ weeks (see ▶ Fig. 13.2).[1,26] Similar results can be achieved when the boost dose is applied as intraoperative radiation therapy.

13.6 Acute Side Effects and Complications of Adjuvant Radiation Therapy

The use of modern multimodal treatment plans and radiation therapeutic techniques has led not only to a marked improvement in tumor control for the affected patients but also to improved tolerability. In these terms, adjuvant radiotherapy is now considered to be a very safe treatment method that, even in the long term, involves no disproportionate risks. Nevertheless, there are several, mostly temporary, acute reactions that need to be kept in mind. The treating radiotherapist pays particular attention to minimizing the risk of functionally important late sequelae. In doing this, he or she must also consider interaction with other treatment methods, especially with systemic medication therapy. Non-treatment-related factors can also affect potential risks of radiation therapy, however.

13.6.1 Acute Side Effects

- *Erythema:* During the course of radiotherapy, it is normal to develop redness of the skin to varying degrees. The skin in the inframammary fold (beneath the mammary gland) and that in the axillary fold is especially sensitive. Skin reactions caused by irradiation usually start in the third or fourth week of radiotherapy and can lead to some degree of peeling of the superficial skin layer. This may be associated with an increase in the volume of the breast that is usually mild and a feeling of tightness of the breast. This is abetted by obstruction of lymphatic drainage, especially following extensive lymph node surgery, and by large breast size. It frequently occurs after previous chemotherapy and in smokers. Concurrent administration of tamoxifen or immunomodulation as an alternative medication (e.g., mistletoe application) can possibly increase the risk for such acute reactions, which is why these measures should be stopped during radiation therapy.
- *Paresthesias:* Another possible acute reaction is paresthesia in the treated breast, particularly in the surgical area, or in the ipsilateral arm, which is often associated with extensive treatment of the lymphatic drainage areas.
- *Moderate temporary pain with swallowing:* This is to be expected only if radiotherapy is delivered to the lymphatic channels, and it occurs most often when treating carcinoma of the left breast.
- *Radiation sickness:* Many patients also experience fatigue and, more rarely, therapy-induced nausea as well, but such a "radiation hangover" has become quite rare.
- *Changes in the blood count:* Checking the blood count regularly is particularly important if the patient has had previous systemic chemotherapy. In these cases, a decrease of the white blood cells and/or platelets occurs, but this is only moderate and almost never threatening. Infections or bleeding caused by this are exceedingly rare; the administration of hematologic growth factors or transfusion of platelets or erythrocyte concentrates is practically never necessary. Significant changes in other laboratory parameters caused by radiotherapy are rarely seen.

All the acute reactions mentioned can be treated symptomatically and usually regress within the first week after cessation of radiation therapy. Sometimes they may persist in a milder form but they very rarely affect the patient's quality of life.

Note

After cessation of radiotherapy following breast-conserving surgery, rare cases (0.2–1.3%) of lung tissue inflammation, termed radiation pneumonitis, can develop.[29] In these cases, cough, fever, and/or dyspnea develop, typically 1 to 2 months after the end of radiation therapy, accompanied by exhaustion.

In situations of radiation pneumonitis, prompt action is required by the treating radiation therapist, first to determine the cause by CT scanning and then to initiate appropriate therapy. Ordinarily, antibiotics and corticosteroids are given for several weeks. With prompt and adequate treatment, permanent functional pulmonary limitation occurs only in very rare cases. The risk of this delayed acute reaction is possibly increased by the concurrent administration of tamoxifen, by nicotine abuse, and by the administration of immunomodulating substances.[23] One must accordingly pay particular attention to the circumstances that accompany this rare but very important acute reaction.

13.6.2 Late Complications of Radiation Therapy

Many of the consequences of earlier breast cancer treatment, which were common and quite burdensome for the patient, have become rarer and less severe with contemporary treatment.[11,12] Nevertheless, even with the use of modern strategies and radiation therapy techniques, there are still many aspects that must be considered, even before the initiation of treatment, not only to increase the chance of recovery but also to optimize the posttherapeutic quality of life:

- *Lymphedema:* The risk of lymphedema following the treatment of breast cancer has decreased substantially over the years; functional limitations from damage to the brachial plexus within the axilla have become extremely rare. This development is due to the introduction of less invasive surgical methods, especially the of the sentinel lymph node technique, and also to more stringent indications and lower radiation loads for radiotherapy of the axilla and the remaining regional lymphatic channels.[24] However, even today, lymphedema (usually only moderate) or paresthesias in the area of the affected breast and the ipsilateral arm can develop. With adequate lymph drainage, these symptoms only rarely lead to a significant disturbance for the affected patients. The patient herself should pay particular attention to making sure that the affected breast is well supported and the affected arm is not overburdened or even injured.
- *Blood vessel changes:* In planning treatment, the radiation therapist pays particular attention to exposing the adjacent heart and lungs to as little radiation as possible. This is particularly important for the heart because patients sometimes take medications as well, such as chemotherapy with anthracycline or treatment with trastuzumab, that also involve a certain cardiac risk. Current long-term analyses show that with modern

methods of radiation therapy, the potential additional risk from radiotherapy is considerably less than the expected positive effect on relapse and survival rates.[19,21] This applies also to the treatment of left-sided breast cancer, including the regional lymph nodes. The available data on survival probabilities do not indicate that the benefits gained from radiotherapy are significantly offset by other possible causes of death, such as by radiogenic changes to the blood vessels (see ▶ Table 13.1).[21]

- *Other malignant tumors:* Similarly, small additional risks of later developing another type of tumor do not significantly offset the positive aspects of radiotherapy. The radiation therapist must pay particular attention that the other breast does not receive any appreciable scattered radiation. With this precaution, it is unlikely that the use of radiotherapy would significantly endanger the patient through the development of a tumor in the other breast. In this context, it is self-evident that the patient herself should consistently avoid all possible carcinogens—especially from smoking tobacco.[6,16]

There are a very few tumor-independent diseases that preclude the use of radiation therapy or in which extreme caution must be taken in determining the indication for radiotherapy and in its execution. Included in these is, in particular, previous irradiation of the same body region or an immediately adjacent region (for malignant lymphoma, for example). Also included are certain rheumatologic and immunologic diseases, such as scleroderma and lupus erythematosis. In addition, very rare genetic predispositions (e.g., ataxia telangiectasia) can rule out the use of adjuvant radiotherapy or allow its use only in a reduced form, such as partial breast irradiation.[15]

13.7 Planning and Implementing Radiation Therapy

Planning of the most precise, most effective, and yet most conservative radiotherapy possible represents an especially difficult challenge for the treating radiation therapist. The very complex anatomy involved, different in each individual, makes this task particularly difficult. Another complicating factor is the need to target varying individual and cumulative doses to the various areas of risk in the breast and lymphatic channels (see ▶ Fig. 13.1).

This planning is based on an initial CT scan carried out using a special stable positioning system. All the necessary anatomical details are identified and input electronically into dedicated systems for computer-assisted treatment planning. In some individual cases, it may be useful to coordinate the delivery of the radiotherapy with the respiratory cycle. With this acquired data, the radiotherapist prescribes the various target volumes and risk organs in a virtual form stored in the computer. With the input and support of specially trained physicists, engineers, and radiation oncology technologists, the optimal choice of beam projection and its modulation is calculated in an extremely complex process. Three-dimensional segmentation can be carried out into further partial volumes, or it may be advantageous to plan intensity-modulated radiotherapy on a clinical, anatomical, or physical basis.[7,10,14] Several days may be required to accomplish the stepwise formulation of the best solution for the individual patient.

After the choice of the individually optimized scheme (by "virtual" simulation), it is then applied to the patient. This can be immediately implemented on the radiation unit or at an additional appointment on a simulator ("real" simulation). During this visit, relevant markings are made on the patient's skin to ensure both secure positioning of the patient (using a 3D laser arrangement) and reliable positioning of the radiation beam for the following radiotherapy series, usually consisting of 25 to 35 sessions.

Practical Tip

It is important that these markings remain reliably intact and legible throughout the entire duration of therapy; the patient needs to be extremely careful not to disturb them during personal hygiene sessions (see ▶ Fig. 13.1).

The therapy plans might have to be changed depending on individual circumstances, which means that the necessary markings must also be changed. In such a case the initial procedure has to be repeated. In addition, care must be consistently taken that the daily application of ionizing radiation follows the planning guidelines exactly. First, the treating technician and doctor implement a visual check; they also use a special image, which is acquired using either the therapeutic beam ("portal vision") or a separate radiographic system ("on-board imaging"). In some cases, a repeat CT may be necessary.

The patient and her family are to be kept informed of all this and all the other relevant aspects of the radiotherapy in a thorough discussion with the treating physician. In this dialogue, the doctor explains the goals of treatment and any potential acute reactions and late sequelae of the radiotherapy in an understandable form, usually with the help of a leaflet specially developed for the purpose. The timing and sequences for the entire treatment period should be known beforehand, and the doctor reviews with the patient what is expected of her during the very few minutes spent in the daily treatment sessions. With adequate information and effective explanation, few patients refuse the radiotherapy that is considered to be necessary for them.[22]

Practical Tip

It is also important to emphasize that the patient should avoid possible risk factors, such as smoking and use of alternative-therapy "immunomodulatory" substances. For skin care, application of a thin oil-based cream usually suffices.

In cases of more severe acute reactions during the treatment series, the treating physician will modify the order if necessary and prescribe other measures.

The initial patient consultation should take place in a quiet environment without time pressures. Only in this way is it possible to reduce unnecessary anxiety about a treatment method that is usually very safe. The treating physician should also always be available to address any questions the patient might have at any time during her course of treatment. After completion of radiation therapy, the radiation oncologist discusses the further course of action thoroughly with the patient. The radiation oncologist will address in depth the necessary measures from a broad radiotherapeutic point of view; but it is also very important that he or she communicate with the referring and aftercare physicians concerning the patient's future care, and then convey this information to the patient. Usually this will be the practicing gynecologist, in whom the patient will have placed her trust. However, it should not be overlooked to include the patient's primary physicians and to send them, and all the other doctors involved with the patient's care, a written report on the measures that were carried out and the planed future management.

If the patient is treated in a certified breast cancer center, these measures are all centrally documented and monitored. This not only provides for the optimal security of the patient but also allows all the treating partners to quickly recognize and correct potential deficits, especially in organization and coordination. Unusual situations or courses of treatment, which in the case of treatment according to the guidelines are very rare, are analyzed in detail and a solution is found. It is the job of the radiation therapist to monitor possible unexpected effects of the therapy; such occurrences require complete documentation. After the final examination, which usually takes place 2 to 6 weeks after the end of the course of radiation therapy, the patient will normally be seen at most yearly by the treating radiotherapist. Both the patient and the treating gynecologist must arrange for an appointment with the treating radiotherapist as soon as possible if they suspect an unusual radiation-related effect. The treating radiation therapist must thoroughly investigate such suspicions and institute appropriate treatment. Complications of therapy are unusual, but particularly involve the skin and mammary gland; rarely, a pulmonary reaction (pneumonitis) can occur (see Chapter 13.6.1).

13.8 Radiotherapy in Primarily Inoperable Tumors, Recurrences, and Metastatic Disease

13

In these situations, all appropriate surgical, medical, and radiotherapeutic measures must be considered and coordinated, and then individually adapted to meet the problems of the individual patient. Inclusion of the patient and her family in the process is especially important, as is providing open and sensitive support. All necessary supportive measures must be offered, including intensive psychological care and support. This care also should be provided in an accredited breast cancer center if possible.

The objectives of the radiation therapy here are to bring locoregionally circumscribed tumor manifestations under control as much as possible, hopefully for the long term; to resolve or palliate pain and malfunction as much as possible; and to prevent threatening complications, such as bone fracture or paraplegia. In addition, most patients require accompanying medications, especially analgesics, but frequently also chemotherapy, immunotherapy, or hormone therapy.

In the case of a very large, primary inoperable tumor or a locoregional recurrence without distant metastasis, a treatment duration of 5 to 7 weeks with a total dose of at least 60 Gy is frequently required.[2] Often, high-precision radiation methods are necessary, either as percutaneous radiotherapy or as interstitial brachytherapy. In addition, treatment with hyperthermia may be worthwhile. This is particularly true if locoregional recurrence occurs in a region that has previously been irradiated.

In contrast, for symptomatic distant metastases, especially of the skeletal system, a short course of radiotherapy (30 Gy in 2 weeks, for example) provides symptom control and reliable prevention of fracture of the threatened bone in over 80% of patients. In a few cases, effective pain reduction is possible with only one dose or a few single doses (1 × 8 Gy or 4 × 5 Gy). Even regions that have previously been irradiated can be re-treated if modern radiation methods are used. Depending on the known risk factors and the previous course of the disease, long-term survival (sometimes many years) can be achieved with minimal symptoms.

13.9 Summary

Radiotherapeutic measures have an established place in the primary adjuvant treatment of patients with breast cancer, but also in the treatment of advanced, recurrent, and metastatic disease. It is important that the treating radiotherapy clinic offers all the modern methods of radiation therapy, including brachytherapy and possibly hyperthermia, as well as an experienced, highly qualified staff. It is just as important, however, that multimodal therapy be carried out with close coordination between the participating associates, who must communicate coherently with the patients and contribute to a trusting environment.

Bibliography

[1] Bantema-Joppe EJ, Schilstra C, de Bock GH, et al. Simultaneous integrated boost irradiation after breast-conserving surgery: physician-rated toxicity and cosmetic outcome at 30 months' follow-up. Int J Radiat Oncol Biol Phys. 2012; 83(4):e471–e477

[2] Bourgier C, Khodari W, Vataire AL, et al. Breast radiotherapy as part of locoregional treatments in stage IV breast cancer patients with oligometastatic disease. Radiother Oncol. 2010; 96(2):199–203

[3] Clarke M, Collins R, Darby S, et al. Early Breast Cancer Trialists' Collaborative Group (EBCTCG). Effects of radiotherapy and of differences in the extent of surgery for early breast cancer on local recurrence and 15-year survival: an overview of the randomised trials. Lancet. 2005; 366(9503):2087–2106

[4] Darby S, McGale P, Correa C, et al. Early Breast Cancer Trialists' Collaborative Group (EBCTCG). Effect of radiotherapy after breast-conserving surgery on 10-year recurrence and 15-year breast cancer death: meta-analysis of individual patient data for 10,801 women in 17 randomised trials. Lancet. 2011; 378(9804):1707–1716

[5] Giordano SH. Radiotherapy in older women with low-risk breast cancer: why did practice not change? J Clin Oncol. 2012; 30(14):1577–1578

[6] Gujral DM, Sumo G, Owen JR, et al. Ipsilateral breast tumor relapse: local recurrence versus new primary tumor and the effect of whole-breast radiotherapy on the rate of new primaries. Int J Radiat Oncol Biol Phys. 2011; 79(1):19–25

[7] Harsolia A, Kestin L, Grills I, et al. Intensity-modulated radiotherapy results in significant decrease in clinical toxicities compared with conventional wedge-based breast radiotherapy. Int J Radiat Oncol Biol Phys. 2007; 68(5):1375–1380

[8] Hindié E, Groheux D, Hennequin C, et al. Lymphoscintigraphy can select breast cancer patients for internal mammary chain radiotherapy. Int J Radiat Oncol Biol Phys. 2012; 83(4):1081–1088

[9] Leitlinienprogramm Onkologie der AWMF, Deutschen Krebsgesellschaft eV und Deutschen Krebshilfe eV. Leitlinienreport der S 3 Leitlinie fur die Diagnostik, Therapie und Nachsorge des Mammakarzinoms. AWMF Register-Nr. 032–045OL. Available at: "https://www.nccn.org/professionals/physician_gls/pdf/breast.pdf", accessed June 22, 2017

[10] Ludwig V, Schwab F, Guckenberger M, Krieger T, Flentje M. Comparison of wedge versus segmented techniques in whole breast irradiation: effects on dose exposure outside the treatment volume. Strahlenther Onkol. 2008; 184(6):307–312

[11] Lundstedt D, Gustafsson M, Malmström P, et al. Symptoms 10–17 years after breast cancer radiotherapy data from the randomised SWEBCG91-RT trial. Radiother Oncol. 2010; 97(2):281–287

[12] Lundstedt D, Gustafsson M, Steineck G, et al. Long-term symptoms after radiotherapy of supraclavicular lymph nodes in breast cancer patients. Radiother Oncol. 2012; 103(2):155–160

[13] Mahmood U, Morris C, Neuner G, et al. Similar survival with breast conservation therapy or mastectomy in the management of young women with early-stage breast cancer. Int J Radiat Oncol Biol Phys. 2012; 83(5):1387–1393

[14] Mukesh M, Harris E, Jena R, Evans P, Coles C. Relationship between irradiated breast volume and late normal tissue complications: a systematic review. Radiother Oncol. 2012; 104(1):1–10

[15] National Comprehensive Cancer Network. (NCCN). Clinical Practice Guidelines in Oncology. Breast Cancer. Version 2. 2016. Available at: http://www.nccn.com/cancer-guidelines.html. Accessed February 28, 2017

[16] Ng J, Shuryak I, Xu Y, Clifford Chao KS, Brenner DJ, Burri RJ. Predicting the risk of secondary lung malignancies associated with whole-breast radiation therapy. Int J Radiat Oncol Biol Phys. 2012; 83(4):1101–1106

[17] Offersen BV, Overgaard M, Kroman N, Overgaard J. Accelerated partial breast irradiation as part of breast conserving therapy of early breast carcinoma: a systematic review. Radiother Oncol. 2009; 90(1):1–13

[18] Offersen BV, Brodersen H-J, Nielsen MM, Overgaard J, Overgaard M. Should postmastectomy radiotherapy to the chest wall and regional lymph nodes be standard for patients with 1–3 positive lymph nodes? Breast Care (Basel). 2011; 6(5):347–351

[19] Offersen B, Højris I, Overgaard M. Radiation-induced heart morbidity after adjuvant radiotherapy of early breast cancer - Is it still an issue? Radiother Oncol. 2011; 100(2):157–159

[20] Peto R, Davies C, Godwin J, et al. Early Breast Cancer Trialists' Collaborative Group (EBCTCG). Comparisons between different polychemotherapy regimens for early breast cancer: meta-analyses of long-term outcome among 100,000 women in 123 randomised trials. Lancet. 2012, 379(9814).432–444

[21] Ragaz J, Olivotto IA, Spinelli JJ, et al. Locoregional radiation therapy in patients with high-risk breast cancer receiving adjuvant chemotherapy: 20-year results of the British Columbia randomized trial. J Natl Cancer Inst. 2005; 97(2):116–126

[22] Ravdin PM, Siminoff IA, Harvey JA. Survey of breast cancer patients concerning their knowledge and expectations of adjuvant therapy. J Clin Oncol. 1998; 16(2):515–521

[23] Schmidberger H, Hermann RM, Hess CF, Emons G. Combination of antiestrogenic therapy with radiation in breast cancer: simultaneous or sequential treatment? Onkologie. 2005; 28(5):275–280

[24] Shah C, Wilkinson JB, Baschnagel A, et al. Factors associated with the development of breast cancer-related lymphedema after whole-breast irradiation. Int J Radiat Oncol Biol Phys. 2012; 83(4):1095–1100

[25] Vaidya JS, Joseph DJ, Tobias JS, et al. Targeted intraoperative radiotherapy versus whole breast radiotherapy for breast cancer (TARGIT-A trial): an international, prospective, randomised, non-inferiority phase 3 trial. Lancet. 2010; 376(9735):91–102

[26] van der Laan HP, Dolsma WV, Maduro JH, Korevaar EW, Hollander M, Langendijk JA. Three-dimensional conformal simultaneously integrated boost technique for breast-conserving radiotherapy. Int J Radiat Oncol Biol Phys. 2007; 68(4):1018–1023

[27] Walker GV, Niikura N, Yang W, et al. Pretreatment staging positron emission tomography/computed tomography in patients with inflammatory breast cancer influences radiation treatment field designs. Int J Radiat Oncol Biol Phys. 2012; 83(5):1381–1386

[28] Whelan TJ, Pignol J-P, Levine MN, et al. Long-term results of hypofractionated radiation therapy for breast cancer. N Engl J Med. 2010; 362(6):513–520

[29] Whelan TJ, Olivotto IA, Parulekar WR, et al. MA.20 Study Investigators. Regional nodal irradiation in early-stage breast cancer. N Engl J Med. 2015; 373(4):307–316

[30] Donker M, van Tienhoven G, Straver ME, et al. Radiotherapy or surgery of the axilla after a positive sentinel node in breast cancer (EORTC 10981-22023 AMAROS): a randomised, multicentre, open-label, phase 3 non-inferiority trial. Lancet Oncol. 2014; 15(12):1303–1310

[31] Poortmans PM, Collette S, Kirkove C, et al. EORTC Radiation Oncology and Breast Cancer Groups. Internal mammary and medial supraclavicular irradiation in breast cancer. N Engl J Med. 2015; 373(4):317–327

[32] Strnad V, Ott OJ, Hildebrandt G, et al. Groupe Européen de Curiethérapie of European Society for Radiotherapy and Oncology (GEC-ESTRO). 5-year results of accelerated partial breast irradiation using sole interstitial multicatheter brachytherapy versus whole-breast irradiation with boost after breast-conserving surgery for low-risk invasive and in-situ carcinoma of the female breast: a randomised, phase 3, non-inferiority trial. Lancet. 2016; 387(10015):229–238

14 Management of a Diagnostic Breast Center

F. Baum

Cancer is a topic that provokes anxiety. In our industrialized society, breast cancer is the most common cancer in women, and represents a particular threat. It is not surprising, therefore, that not only the actual revelation of ambiguous breast findings but even the early diagnostic setting itself arouses feelings of apprehension in many women. Thus, the evident quality and friendliness of the professional medical personnel, the quality and range of the devices and procedures offered, and provision of a pleasant ambience help create an environment that is as anxiety-free as possible and in which care can be provided without adding to the patient's insecurities.

High-quality breast diagnostics is founded and supported on three pillars (▶ Fig. 14.1). It is important to note that deficiencies in one pillar cannot be compensated for by making a special effort in one of the others. It follows that each individual pillar must provide maximum quality so that they jointly ensure the highest level of breast diagnostic quality.

14.1 Expertise

In addition to the technical-methodological aspects, breast diagnostics requires a special expertise on the part of the personnel involved in the entire process. This affects the radiologic technologists on the one hand and the physicians on the other.

Given the natural anatomical differences between the women to be examined, generating the mammograms makes high demands of the diagnostic radiographers. For this reason, specific courses have been established in American and European countries to certify that the radiologic technologists performing mammograms meet specific qualification requirements. The quality of all completed mammograms should be reviewed and coded according to the PGMI classification (perfect, good, moderate, or inadequate). The proportion of images classified as "perfect" or "good" should be at least 75%.

A physician's expertise correlates first and foremost with the extent of their experience in evaluating mammographic images. It is well established that the learning curve for physicians who interpret mammograms rises over a time period of 8 to 10 years, assuming that 3,000 to 5,000 mammograms are read per year. These physicians then have stored mental image patterns that allow them to recognize pathologic processes in the mammogram within a few seconds.

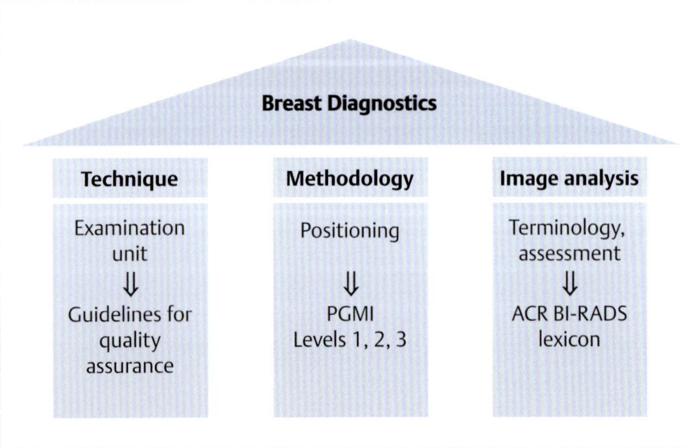

Fig. 14.1 Quality aspects of high-quality breast diagnostics. ACR, American College of Radiology. PGMI (classification) perfect, good, moderate, or inadequate.

The American College of Radiology (ACR) Mammography Accreditation Program for interpreting physicians calls initially for board certification in radiology, interpretation, and documentation of 240 mammograms within 6 months, and 15 units of category I continuing medical education (CME) in mammography. For continuing experience, 960 mammographic examinations must be interpreted over a period of 2 years. Continuing education requires 15 units of category I CME within 3 years.

The ACR also gives specifications for breast ultrasound accreditation. Interpreting physicians are required, among other things, to perform or oversee, interpret, and document 300 breast ultrasounds within 36 months. To maintain the certification, for example, 200 breast ultrasound examinations can be performed and a total of 15 hours CME completed within 3 years. Minimum criteria to be met by a physician who performs or supervises ultrasound-guided biopsies are the qualification to interpret mammograms under MQSA (Mammography Quality Standards Act) and the performance of 36 image-guided biopsies within 36 months.

Minimum criteria required by the ACR for breast MRI accreditation are, for example, board certification in radiology or diagnostic radiology, interpretation, and documentation of 150 breast MRI examinations within 3 years and completion of 15 hours of category I CME in MRI. To renew the certification, 75 independent examinations must be interpreted within the prior 36 months. In order to meet ACR general quality standards, certain requirements, such as a dedicated breast coil, bilateral examinations, and the means for performing MRI-guided interventions are specified.

In European countries there are similar certification requirements and course programs for all breast imaging modalities. As an example, in Germany, the AIM e.V. (Arbeitsgemeinschaft für individuelle Mammadiagnostik [Working Group for Individualized Breast Diagnostics]) has defined two quality levels: "Level 1" (high level) and "Level 2" (professional level) to ensure the quality of high-resolution breast MRI. In addition to device and technical-methodological requirements, certified radiologists must independently interpret 250 breast MRI examinations per year for Level 1, and 500 per year for Level 2 (▶ Table 14.1). In addition, certification requirements for the performance of percutaneous biopsy examinations prescribe country-specific numbers of ultrasound-, mammography-, and MRI-guided interventions within a certain period for accreditation.

14.2 Equipment

As well as providing all the devices and means for breast imaging, a diagnostic breast center should also have available all the equipment needed to perform percutaneous biopsies in the diagnostic work-up of ambiguous findings. Because of its

Table 14.1 Certification for breast MRI. Requirement specifications for high level (Level 1) and professional level (Level 2) according to AIM e.V.

Requirement	Level 1	Level 2
System	≥ 1.5 T	≥ 1.5 T
Spatial resolution	512 × 512	512 × 512
Temporal resolution	< 2 min/sequence	< 2 min/sequence
Coil	Open, bilateral	Open, bilateral
Breast fixation	Yes	Yes
Postprocessing	Subtraction, maximum intensity projection.	Subtraction, maximum intensity projection.
Number of patients per examiner per year	> 250	> 500
Interventions per examiner per year	Cooperation	> 50 biopsies > 50 localizations
CME in breast MRI	> 8 h/year	> 16 h/year

Abbreviation; AIM, Arbeitsgemeinschaft individuelle Mammadiagnostik (Working Group for Individualized Breast Diagnostics). CME, continuing medical education.

Source: http://www.aim-mamma.de/tl_files/downloads/open/aim-info-aerzte.pdf (accessed March 28, 2013).

many advantages in terms of radiation exposure, data storage, and image postprocessing and evaluation, digital mammography has now almost completely superseded analog imaging. In addition to the digital mammography unit itself, digital imaging necessitates the provision of a diagnostic viewing station with screen monitors that must meet the high demands of breast diagnostics. Both the mammography unit and the review workstation must meet performance requirements in terms of resolution, contrast, and diagnostic quality. In the United States, these units are tested by a medical physicist and must meet the requirements of the Mammography Quality Standards Act to be certified.

Note

A significant advantage of digital mammography is the rapid availability of the images: They are available for interpretation within about 1 minute after image generation and can, if desired, be reviewed with the woman examined.

In addition to mammography, breast ultrasonography represents an important element of imaging that is primarily undertaken in younger women and as a complementary examination in patients with dense tissue structures. It also serves to visualize the needle in ultrasound-guided percutaneous biopsy procedures. Breast ultrasound examinations are performed using a linear transducer with a variable frequency that usually ranges from 7 to 15 MHz. Occasionally, the quality of the ultrasound images may be improved by employing sophisticated imaging techniques such as "tissue harmonic imaging" and "cross beam." Additional options, such as color-coded Doppler sonography or ultrasound elastography, can also be used to increase the informative value of the procedure.

As the third breast imaging method alongside mammography and sonography, breast MRI is an essential component of modern breast diagnostics. The hardware required includes a 1.5-T (or higher) whole-body magnet and appropriate equipment such as an open, high-resolution breast coil (matrix 512 × 512 or higher) with integrated compression plates.

For standardized imaging procedures, such as mammography and breast MRI, the sensitivity and accuracy of the image analysis can be improved by the practice of double reading. Additionally, computer-aided detection (CAD) systems can be used as a "third reading," but preferably only after each of the individual reviewers has made their evaluation.

Ideally, the provision of a full range of imaging devices is complemented by equipment necessary for the histological work-up of suspicious mammographic, sonographic or MRI findings. In this regard, the installation of a second ultrasound unit facilitates the performance of ultrasound-guided interventional procedures without tying up the first unit used for diagnostic examinations. A stereotactic breast biopsy table is useful for the histological work-up of suspicious mammographic findings, especially microcalcifications. A specimen radiograph of the samples retrieved during a stereotactic intervention can then be obtained with the diagnostic mammography unit. MRI-guided interventions require an additional MRI-compatible targeting device and special MRI-compatible biopsy and localization materials.

14.3 Facility Design

14.3.1 Doctor's Consultation Room

Before any clinical or diagnostic examinations in the context of early breast cancer detection or diagnostic work-up, there should be a consultation with the patient in a separate doctor's office. During this consultation, the physician provides the patient with information and explains the capabilities and possible negative effects of the individual examination methods. It also provides the physician with an opportunity to assess the individual risk of disease, develop an appropriate individualized diagnostic strategy for the examination of the patient, and obtain informed consent. After the individualized examination strategy has been determined, this is conveyed to the diagnostic radiographer working on the case, who then takes responsibility for the patient and accompanies her through the diagnostic imaging process.

14.3.2 Mammography and Sonography Rooms

The arrangement of the examination units in a diagnostic facility should be chosen to optimize patient comfort and workflow (▸ Fig. 14.2). Today, the combination of mammography and ultrasound represents a standard strategy in breast diagnostics. A complementary ultrasound examination is recommended, particularly when mammographic assessability is limited due to dense glandular parenchyma (ACR density types III, IV). If the ultrasound unit is located directly next to the mammography unit, it is not necessary for the patient to re-dress between the examinations. In the age of digital mammography, the information concerning mammographic parenchymal density is quickly available, and the doctor's interpretation of the mammogram can generally be accomplished before the ultrasound examination is done. Ideally, the entrance to the dressing room should open to both the mammography and ultrasound rooms, so that the patient need only disrobe once (▶ Fig. 14.3). Optimally, the mammography unit can be used for the next examination while the ultrasound study is taking place in the adjoining room.

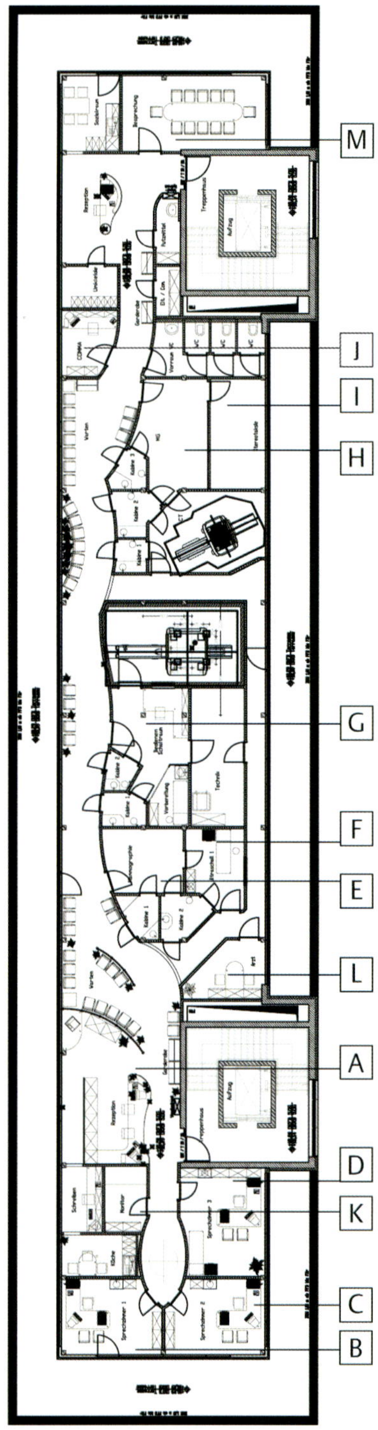

Fig. 14.2 Architectural layout of a diagnostic breast center. Spatial layout of a comprehensive diagnostic breast center, using the Diagnostisches Brustzentrum Göttingen [Women's Health Care Center in Göttingen] as an example. Entry area with the reception desk directly across from the entrance (A). Area for the doctors' offices, with three consultation rooms (B, C, D). Examination area 1 with mammography (E) and diagnostic ultrasound (F). Examination area 2 with breast MRI and two adjacent dressing rooms (G). Examination area 3 (interventional) with the second ultrasound system for ultrasound-guided biopsy (H), a stereotactic biopsy table (I), and a recovery room for a cooling and rest period after the intervention (J). Consultation rooms 1 and 2 to review the results with women after their examinations (K, L). Conference room for interdisciplinary conferences and interdepartmental meetings (M).

14.3.3 Breast MRI Room

Breast MRI is firmly established alongside mammography and sonography, and is often part of an effective and comprehensive breast cancer diagnostic strategy. It is offered primarily to women with mammographic and/or breast ultrasound examinations that are difficult to assess, and to women with high-risk profiles. Optimally, the MRI unit is also located in close proximity to the ultrasound and mammography units. This allows an MRI to be carried out in a timely manner, thus saving the woman from the inconvenience and worry of having to make another appointment. To assure an optimal workflow, the MRI room should have two adjacent changing rooms available, so that while one examination is being performed the next patient can be prepared for the subsequent examination.

The quality of a breast MRI examination depends to a large extent on the patient's comfort during the examination as well as on suitable equipment and qualified personnel. Dressing the patient in a suitable bath robe with a belt into which the arms can be tucked and held alongside the body during the examination, providing an anterior head support with an integrated angled mirror to view a chosen picture motif, and supplying protective head phones transmitting recorded music all contribute to a feeling of wellbeing and inner peace during the examination. Disruptive motion artifacts on MRI images can be dramatically reduced or completely eliminated in this way.

14.3.4 Rooms for a Second Ultrasound Unit and for Interventional Procedures

Because ultrasound-guided core needle biopsies are regularly done in a diagnostic breast center, a second ultrasound unit

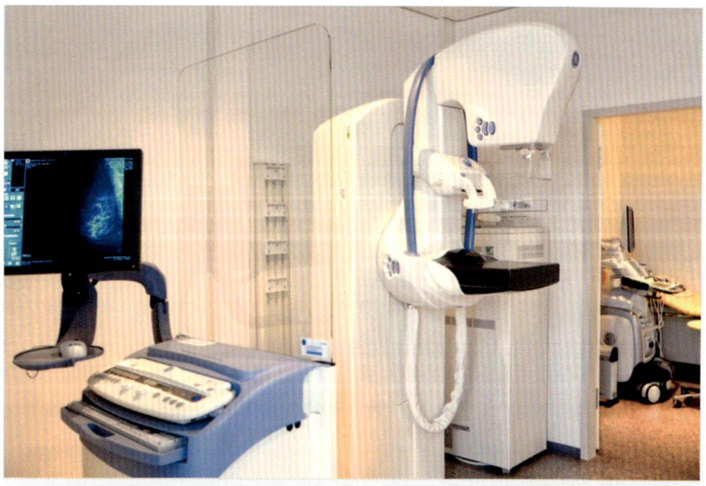

Fig. 14.3 Mammography and ultrasound room. The arrangement of adjoining mammography and ultrasound rooms (E and F in ▶ Fig. 14.2) allows the examinations to take place sequentially without interruption for dressing and undressing. The patient enters the mammography room from the dressing room, and goes from there directly into the neighboring ultrasound room. After completion of the sonographic study, the patient returns directly to the dressing room through a connecting door.

should be available for this minimally invasive procedure, leaving the first unit free for diagnostic examinations. This allows a timely diagnostic work-up of ambiguous ultrasound findings without a disconcerting delay for the patient or interference in the day's work flow.

A comprehensive diagnostic breast center should also have a stereotactic biopsy table for the histological work-up of suspicious microcalcifications. Stereotactic interventions can also be carried out using an add-on device for diagnostic mammography units with which the patient is usually examined in the sitting position. Optimally, however, these interventions are performed on a separate, dedicated examination unit that allows the patient to lie down during the examination, thus avoiding orthostatic complications. Another advantage is that a specimen radiograph can be taken on the diagnostic mammography unit before the patient is removed from the biopsy table. In this way, if no microcalcifications are seen on the radiograph, then it is still possible to extract further tissue samples before terminating the procedure. However, alternative systems are available specifically for performing specimen radiography.

Because about 50% of suspicious breast MRI findings cannot be reproduced in either ultrasound or mammography, it is also crucial for a comprehensive diagnostic breast center to provide the prerequisites for performing MRI-guided biopsies and lesion markings.

Practical Tip

MRI-guided interventions cannot be carried out on the same day as a diagnostic MRI examination. Because the contrast media must be completely excreted in order to reliably identify the ambiguous lesion on the interventional MRI, a time of 10 to 12 hours should elapse between performing these MRI procedures.

14.3.5 Recovery Room

It is helpful to have a separate recovery room for the postprocedure care of patients after core needle or vacuum-assisted biopsies. To prevent significant hematomas, the biopsy area should be cooled and compressed for about 30 minutes after such interventions. In addition to the calming effect of the cozy, comfortable ambiance in a special recovery area, the use of such a room also frees the relevant examination units for the examination of subsequent patients and optimizes the workflow (▶ Fig. 14.4).

14.4 Ambiance

It is common for women to find both an early breast cancer detection examination and the diagnostic work-up of an ambiguous finding to be quite threatening. In addition to attentive care, an open and friendly atmosphere helps prevent, or at least lessen, the anxiety.

As patients often remark, space constricts as one makes one's way toward a medical examination. From the street one enters the building, from the waiting room one goes into the dressing cubicle, then enters an often windowless room for the examination. Thus the threatening aspect of the already anxiety-provoking topic of "breast cancer" is further increased. An appropriate lighting scheme with the maximum use of daylight is the main focus for setting the mood. Thus, a skylight in the dressing room, for example, can significantly counteract a feeling of constriction.

The colors used should also harmonize and the selection should preferably draw on warm tones. Open seating with the avoidance of a closed-in waiting room likewise helps prevent anxiety. The availability of appealing magazines and newspapers helps to shorten the perception of possible waiting times (▶ Fig. 14.5).

14.5 Communication

Even in a diagnostic breast center, various levels of communication are encountered. In addition to communication between the patient and the practice staff, good internal communication between staff members is essential. Furthermore, regular exchange between pathologists, surgeons, and other medical colleagues is a necessary part of the daily office routine.

The initial communication with the patient creates the basis for the necessary level of trust and conveys important information concerning patient history, risk constellation, and current

14

Fig. 14.4 Recovery room. A separate comfortable room for the postinterventional rest and cooling phase.

Fig. 14.5 Open waiting area. The waiting area is generously flooded with light with warm color tones.

symptoms. The necessary examinations should not be initiated until after this consultation.

Practical Tip

After the examinations have been completed, it is desirable to share the results of the study with the patient as soon as possible (▶ Fig. 14.6). Thus, imaging should be interpreted directly after completion of the examination to avoid unnecessary uncertainty and anxiety.

Timely interpretation of the initial images usually makes it possible to initiate any additional studies that may be required on the same day. In this way it is often possible to avoid having the patient return for complementary studies on following days.

Fig. 14.6 Image viewing room. A friendly atmosphere is created in the viewing room through the use of small, colorful accessories.

To encourage communication beyond the personal interaction, offering a forum on a facilities homepage gives patients—especially reticent patient—the opportunity to voice their comments and to critique. The effectiveness of the diagnostic facility is enhanced by offering patients further opportunities for engaging in dialog.

When tissue samples must be taken in the diagnostic work-up of an ambiguous finding, the histological results should be made available as soon as possible. Experience has shown that when results can be expected on the following workday it is much more stressful for the patient to have a needle biopsy performed on a Friday than on any other workday. The earliest possible opportunity to get the study results for biopsies performed on a Friday is after the weekend. Having the biopsy done on another workday, with the histological study results available the next day, is usually associated with significantly less tension.

The internal communication among team members is also very important. When appropriate, the staff involved in the patient's care should be aware of more than just her past medical history. In particular, the staff should be informed about special anxieties, such as claustrophobia or a fear of needles, that the patient may have voiced in the preliminary consultation. Relevant clinical or imaging findings should be communicated between the radiologic technologists and the examining physicians when this might benefit the patient.

In addition to the regular exchange between the radiologist and the pathologist concerning histological study results, special interdisciplinary discussions and/or joint meetings are required in cases of potential mismatch between imaging and histopathological findings. When this occurs, a joint decision should be made regarding the further course of action.

Communication with the operating surgeon is also a necessary part of the daily office routine in a diagnostic breast center. Before surgery, the positions of relevant lesions must be known and the surgical procedure should be jointly planned. When preoperative hookwire markers are placed, the position of the wires must be communicated and documented, and explained, when necessary. If the surgical lesion was visible on mammography, then it is prudent to obtain a specimen radiograph. In such cases, whether or not the lesion is seen on the specimen radiograph must be conveyed to the surgeon as quickly as possible, especially because the operating team is usually waiting for these results before finishing the surgery.

15 Logistics in an Interdisciplinary Breast Center

G. Emons

15.1 Background

In the last third of the 20th century it became clear that efficient diagnosis and treatment of breast cancer requires close cooperation between various areas of expertise. The establishment of interdisciplinary professional societies was the logical consequence. In Germany, this was the German Society of Senology (Deutsche Gesellschaft für Senologie, DGS), and for Europe, the European Society of Mastology (EUSOMA). In these organizations, it quickly became clear that it was not sufficient to organize interdisciplinary conventions. Much more important for optimal breast diagnosis and treatment was close cooperation between the various disciplines involved in the daily care of the affected women.

In 1998, a consensus statement—the "Florence Statement"—was adopted by a group of European organizations, including EUSOMA, the Breast Cancer Cooperative Group of the European Organization for the Research and Treatment of Cancer (EORTC), and the European self-help group Europa Donna (the European Breast Cancer Coalition). This consensus statement called for multidisciplinary breast clinics with mandatory quality assurance programs to be available for all women living in the European Union. In it, a set of specifications—the "Requirements of a Specialist Breast Unit"—was developed to be used throughout Europe. With the support of the European parliaments, the voluntary certification program was instituted, which placed high demands on the breast units in terms of content and organization (see http://www.eusoma.org; accessed February 28, 2017).

Parallel to this, the DGS, in cooperation with the German Cancer Society (Deutsche Krebsgesellschaft, DKG), initiated the national breast center certification program. In most respects, this is similar to the EUSOMA program but it requires a defined quality management system according to industry norms set up by the German Institute for Standardization (Deutsches Institut für Normung, DIN) and based on the appropriate current German guidelines regarding the diagnosis, treatment and aftercare of breast cancer.[1] Both certification systems agree that a minimum number of 150 new patients with primary breast cancer per year are required to justify the large structural and organizational expenditure required by a certified breast center and to assure its efficient use. Both systems carry out a recertification every 3 years and an annual review audit, in which the structural and process quality and the quality of the results are reviewed. During this process, not only the individual partners (e.g., diagnostic radiology, pathology, gynecology, and radiation therapy) are evaluated but also, in particular, the interdisciplinary cooperation between the members. Particular value is placed on the psycho-oncologic support and qualified oncologic care. For example, EUSOMA mandates additional breast care nurses, who, after a specialized training, accompany the affected women from her first entry into the breast center through the entire process of diagnosis, treatment, and aftercare (see the EUSOMA website, http://www.eusoma.org).

Today there are 26 EUSOMA and 204 DGS/DKG certified breast centers in Germany, in which approximately 50,000 primary cases are treated annually, representing about 85% of the German primary cases (http://www.krebsgesellschaft.de; accessed February 28, 2017). The success of the certified breast centers has led the German Cancer Society to establish further cancer facilities, the "organ cancer centers," with corresponding certification programs—centers for gynecologic cancer, bowel cancer, skin cancer, head and neck tumors, lung cancer, and pancreatic and prostate cancer. Clinics that have a minimum of three such certified organ cancer centers are known as certified oncologic centers. As oncologic centers of excellence, comprehensive cancer centers (CCCs), which pursue additional oncologic research, constitute the highest echelon in this three-level model.

15.2 Structure of a Certified Breast Center

The breast cancer center of the University Medical Center Göttingen serves as an example of a breast cancer center certified by the German Society of Senology/German Cancer Society and the European Society of Breast Cancer Specialists (▶ Fig. 15.1). For all participating external partners, the cooperation and commitment are contractually regulated according to the stipulated quality criteria. All central partners are extensively evaluated annually by external evaluators, and the quality criteria and quality management system are reviewed according to the standards of the German Institute for Standardization (DIN) and the International Organization for Standardization (ISO).

Key to the breast center is the interdisciplinary case conference, in which the findings of each individual patient are discussed at least once by the central partners and a treatment recommendation for adjuvant therapy is made. Most patients are also discussed in the preoperative setting to establish the type of preoperative tumor marking and the surgical strategy to be employed.

The members of the breast center meet regularly to form a consensus on how the guidelines and new scientific findings are to be implemented. The findings of the external and internal audits are analyzed regularly.

15.3 Treatment Pathways in a Breast Center

As an example, ▶ Fig. 15.2 shows the treatment pathway for patients with primary breast cancer who can be treated for cure. ▶ Fig. 15.3 portrays the procedure for recurrent, metastasized, and advanced primary carcinoma.

15.4 Outlook

For all those involved, the establishment of a certified breast center initially requires considerable time and effort. However, this is quickly compensated through the simplification of interdisciplinary cooperation and, in an ideal situation, through the improvement of structural and process quality, attested to by

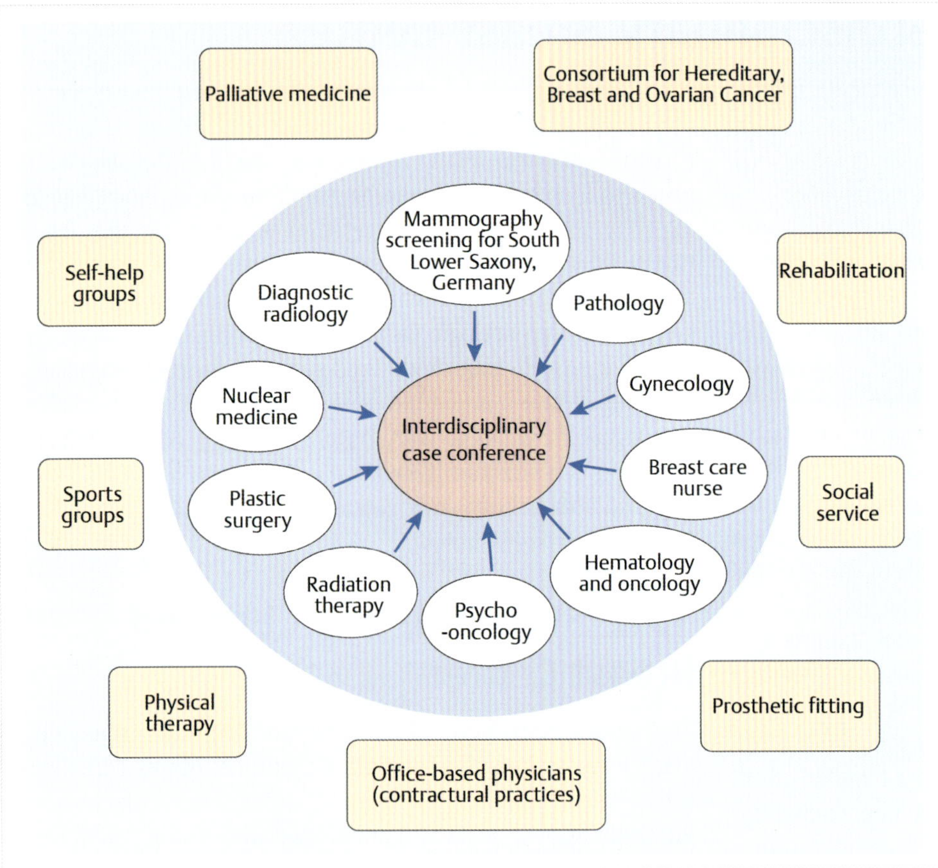

Fig. 15.1 Structure of Certified Breast Center of the University of Medicine, Göttingen.

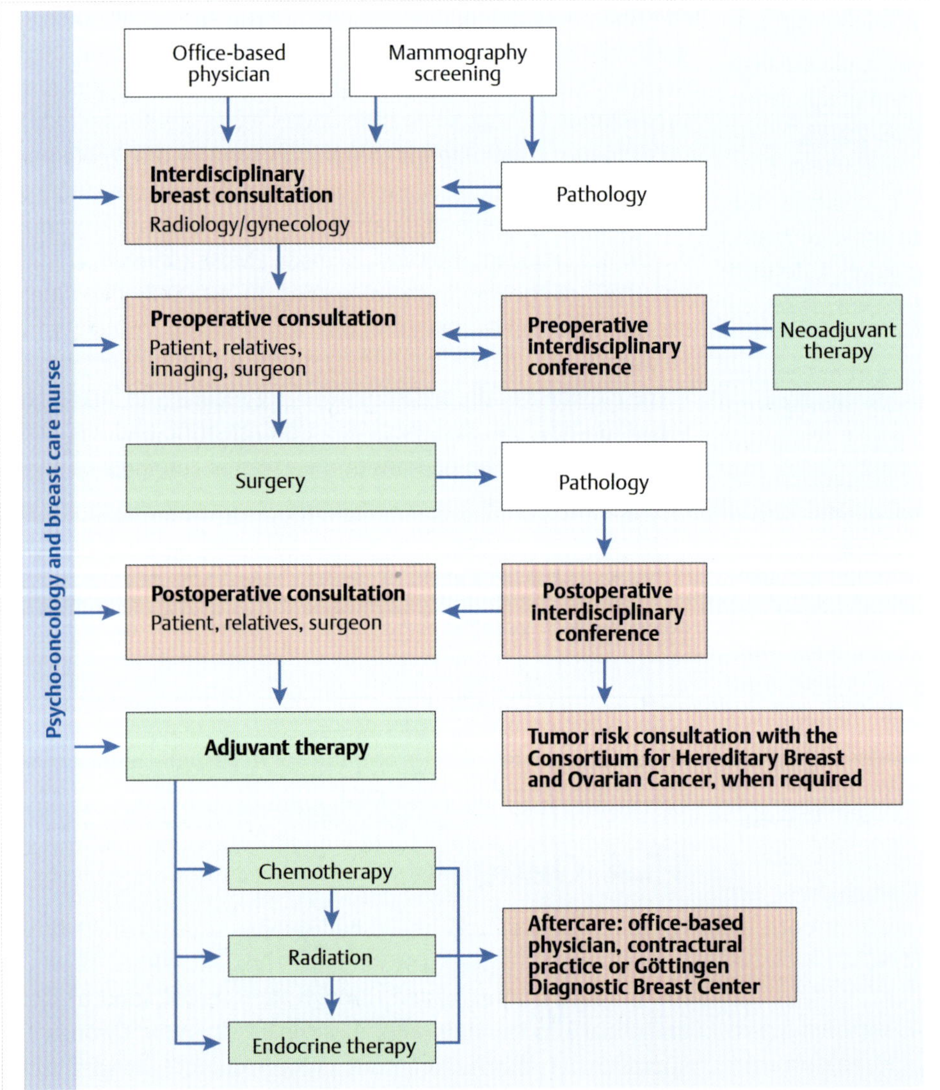

Fig. 15.2 Treatment pathway for patients with primary breast cancer in the Certified Breast Center of the University of Medicine, Göttingen.

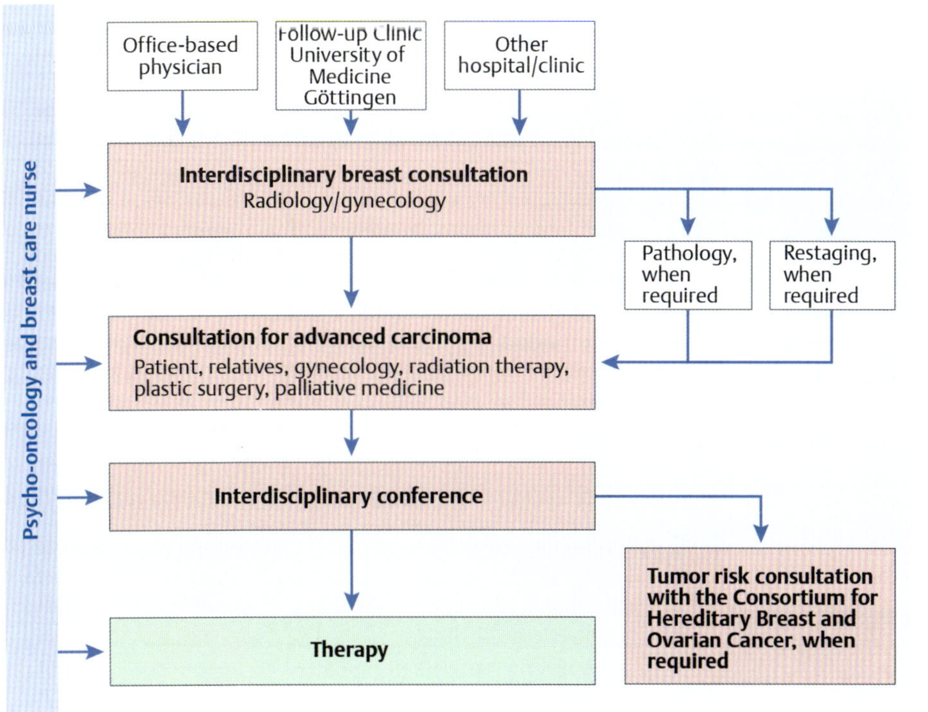

Fig. 15.3 Treatment pathway for patients with recurrent, metastasized, and advanced primary breast cancer in the Certified Breast Center of the University of Medicine, Göttingen.

external auditors. The satisfaction of patients and referring physicians, which is monitored on an ongoing basis, also increases significantly. In the next few years, the quality of the results (recurrence-free survival, overall survival) will be monitored for a sufficient length of time to document the superiority of the certified breast centers. The facts that over 80% of newly diagnosed breast cancer patients in Germany are now treated in certified centers and that this model has been adopted for treating other types of cancer speak for its basic coherence and consistency.

Bibliography

[1] Leitlinienprogramm Onkologie der AWMF, Deutschen Krebsgesellschaft eV und Deutschen Krebshilfe eV. Leitlinienreport der S3 Leitlinie für die Diagnostik, Therapie und Nachsorge des Mammakarzinoms. AWMF Register-Nr. 032–045OL. Available at: http://www.awmf.org/leitlinien/detail/ll/032–045OL.html. Accessed February 9, 2017

16 Counseling Techniques and Psychosocial Support

H. Lorch, A. Kuechemann, and J. Rueschoff

16.1 Compliance

Carcinoma of the breast is the most common tumor affecting women in the western world. Many people have friends or relatives who have had breast cancer.

Modern breast diagnosis and treatment includes a multiplicity of well-differentiated and effective diagnostic and therapeutic procedures. These procedures can only be fully effective if, after receiving a comprehensive explanation based on an honest, cooperative, and trusting relationship with her doctor, the patient understands, endorses, and actively supports them (compliance). The following factors contribute to patient satisfaction:

- Quality of the medical services.
- General and personal requirements.
- Structural, organizational, and procedural components.
- Interactive and communicative competence.

16.1.1 Quality of the Medical Services

High-quality medical services form the basis for a trusting and long-lasting doctor–patient relationship and are presumed and expected by the patient as a matter of course. Lack of quality care leads to dissatisfaction and has negative impacts.

16.1.2 General and Personal Requirements

Breast diagnostics are characterized by particular features over which the examiner has little or no control:

- The patient population is (almost) exclusively female.
- Patients are frequently well-informed beforehand as a result of personal experience and reports from friends and relatives, the internet, and other media. The quality of this information varies considerably.
- Many patients find the examination to be anxiety provoking or uncomfortable; women's personal attitudes toward mammography are strongly influenced by their overall views of their own personal environment.[21]
- The examination requires close physical contact between the radiologic technologist or doctor and the patient.
- Over the course of time, a personal relationship develops between the patient and the mammography team.

16.1.3 Structural, Organizational, and Procedural Components

Patient satisfaction is also influenced by nonmedical factors, such as:

- Telephone accessibility.
- Promptness of appointment scheduling.
- Parking facilities.
- Waiting times.
- The arrangement of the waiting areas, examination rooms, and sanitary facilities.
- Protection of patient privacy and personal space.
- Response to individual needs when performing the examination.

Patients can discern poor cooperation and communications between team members and inadequate organization within the examination procedure. This leads to dissatisfaction and can diminish compliance. Anxious patients become more insecure.[16]

16.1.4 Interactive and Communicative Competence

Interactive and communicative competence is demonstrated by:

- Friendliness of the staff.
- Addressing the patients personally.
- Making explanations understandable to the patient.
- Responding to individual needs when performing the examination.
- The timeliness and manner with which results of tests are shared with the patient.
- Making the patient aware of further diagnostic and therapeutic options.

In most diagnostic breast centers, the patient is given the results immediately after the examination. Sharing the results with the patient, especially if there are abnormal findings, places a high demand on the doctor's communication ability, social skills, and empathy.

16.2 Communication

Although the examination team members will make every effort to communicate successfully with the patient, multiple factors can interfere with the daily clinical routine. These include both external factors, such as time pressures, the obligation to be cost-effective, and focusing on technical details, as well as internal factors, including the working environment, preexisting communication culture, routines, and organizational tunnel vision.

In what follows, our aim is to sensitize both medical and nonmedical staff members to important aspects of communication with the patients, to update and enlarge their knowledge base, and to facilitate the application of this knowledge in everyday practice through the use of practical tips and checklists.

After a short excursion into the fundamentals of communication, we will turn to specific situations during the examination and offer tips to help make this experience a positive one and to improve the relationship with the patient. Perhaps readers will recognize here one or two behavior patterns from their everyday work and will take the opportunity to reflect on their own behavior and to see it from the patient's perspective. Small changes can have a large impact!

16.2.1 General Principles of Communication

Communication occurs as soon as two people become aware of each other, thus forming a relationship. Communication includes every behavior of one individual that is perceived by another. According to Watzlawick, communication consequently goes beyond verbal expression and incorporates conscious and unconscious behavior.[28] This includes body posture, gestures, and facial expression, but also vocal characteristics such as volume and pitch.

Note

"One cannot not communicate." (Paul Watzlawick)[28]

The act of communication has two aspects:

1. *The content aspect* (the verbal, factual information content of a message).
2. *The relationship aspect* (messages concerning the relationship between the sender and receiver, primarily nonverbal, both conscious and unconscious).

Both conscious and unconscious relationship messages are transmitted from the sender and interpreted by the receiver. The result can be that the message from the sender, meant to be on the relationship level, is understood quite differently by the receiver, or that unconscious signals transmitted from the sender on the relationship level lead to reactions by the receiver that the sender does not understand.[27]

Approximately 10% of the effect of a message is determined from its content and 90% from the nonverbal behavior of the partner in the conversation. To a significant degree, the relationship level thus determines the subjective impression of speech, the course of conversation, and the interaction between the speech partners.[17]

Note

The receiver defines the message. Nonverbal behavior is crucial to communication.

16.2.2 Communication: Dealing with the Patient

The communication relationship between the doctor or radiologic technologist and the patient is characterized by a primary asymmetry. The technical staff has a knowledge advantage compared with the patient. For her part, the patient is psychologically and physically "at the mercy" of the departmental procedures and finds herself in an exceptionally emotional situation, frequently characterized by fear or anxiety.

Note

For the patient, the examination environment in breast diagnostics is often characterized by fear or anxiety—because the results of the testing can be of vital significance to the patient!

Many patients are already well informed, or feel that they are. During history-taking or in obtaining informed consent, it often becomes apparent that there is misinformation. The medical team then has the task of gently correcting this misinformation or false information and informing the patient without being patronizing.

Medical jargon is understood only partially or not at all by most laypersons, and it should be avoided in discussions with the patient, to keep the asymmetry of communication levels as slight as possible. In stress situations, factual information is heard and understood even less effectively than when a person is in a balanced, neutral frame of mind. This explains, at least in part, why patients retain such a small proportion of the content of medical information given to them.

The patient's "antennae" are highly sensitized in the examination environment. The nonverbal behavior of the doctor or technologist, in particular, is closely observed and interpreted:

- "Did the doctor wrinkle his forehead during the ultrasound examination?"
- "Did the technologist look at me with pity?"
- "Did the doctor appear cheerful or downcast on entering the consultation room?"

The following structural conditions and behavior patterns on the part of staff members influence the communication with the patient in a positive way:

- A friendly atmosphere in the department and an appealing ambiance have a positive effect on the general mood of the patient and inspire confidence.
- Adherence to schedules and short waiting times demonstrates organizational competence, shows appreciation for the patient, and avoids irritation, brooding, and the development of anxiety in the waiting room.
- Impartial acceptance of every patient creates the basis for successful communication.
- Empathetic understanding of the patient's frame of mind creates trust. It should be understood that "empathy" here means empathetic understanding, not pity.
- Medical issues should be expressed in generally understandable language that is appropriate for the patient.
- The patient must have the opportunity to express her concerns and questions.
- The physician should avoid delivering monologues.
- "Active listening" conveys receptiveness and attentiveness to the patient. In this context, appropriate eye contact is particularly important—preoccupation with the computer should be reduced to a minimum during consultation with the patient.
- The patient's privacy and personal space must be respected and protected.

Note

The relationship between the doctor or technologist and the patient should be shaped by organizational competence, acceptance, respect, empathy, eye contact, appropriate language, and active listening.

Take Home Points

Basic Principles of Patient-centered Communication in the Diagnosis, Treatment, and Aftercare of Carcinoma of the Breast

The manner in which information is conveyed to the patient and informed consent is obtained should ideally be consistent with the following basic principles of patient-centered communication, in which the patient is enabled to participate in decision-making:

- Express empathy and show active listening.
- Address difficult issues in a direct and sensitive manner.
- Whenever possible, avoid using medical terms, or explain the technical terms.
- Use strategies to improve understanding (repetition, summarizing important points, use of graphs, etc.).
- Encourage the patient to ask questions.
- Allow and encourage the patient to express emotions.
- Offer further assistance.

16.2.3 Communicating Results to the Patient

It has now been customary for some time in breast diagnostics for the radiologist to give the results to the patient.[8] This is remarkable, in that previously radiologists traditionally were denied the right to share results directly with patients.[22]

In the 1980s, women increasingly sought out radiologists directly for breast diagnostics on their own initiative. As a result the role of the radiologist was newly redefined—the radiologist now found herself or himself in the role of primary care and as the physician directly responsible to the patient, thus relinquishing the role as the "doctor's doctor."[24]

In addition, since 1998 in the United States, the Mammography Quality Standards Reauthorization Act (MQSA) provided legal grounds mandating that the patient personally receives written information regarding her test results. Studies had shown that a significant percentage of the patients whose mammogram demonstrated a lesion that required clarification were not notified and that the findings were not properly investigated, when only the referring physician, but not the patient herself, was informed of the results.[5,23]

This development is supported by the changing understanding of patients' rights and autonomy. Whereas the paternalistic view prevailed until the 1980s, and physicians decided at their own discretion what information they would want their patients to have, today the autonomy of the patient has the highest priority.

But what do patients want from breast diagnostics? And what positions do the referring physicians and radiologists hold concerning this topic?

Since the 1990s, studies have been carried out in the United States investigating the expectations of mammography patients and referring physicians regarding radiologists communicating the diagnosis to the patient. One study, carried out in 1988 by Lind et al, showed that patients still preferred to hear the results of a mammogram from their treating physician, not from the radiologist (in the clinic where the study took place, the radiologist did not communicate the results of the mammogram).[13] In contrast, studies in 1994 and 1995 showed that approximately 90% of the women wanted to receive the mammography results directly from the radiologist, even if the findings required further work-up or were clearly malignant.[11,14,25]

A much more recent study in 2011 demonstrated the increasing importance of rapid communication of test results in radiology. The majority of patients now expected the results to be determined and conveyed to them within a few hours. Who communicates the results is less important ("whoever is faster"), although the preference tended toward the referring physician.[4]

Referring physicians' acceptance of the radiologist giving the results was dependent on the degree of severity of the diagnosis—the more severe the diagnoses, the less acceptable it was for the radiologist to give the results.[3,12] This attitude is also shared by the radiologists.[12]

A study by Schreiber in 1996 showed that obstetricians and gynecologists strongly supported the radiologists' giving the diagnosis, even more strongly than the group of radiologists themselves.[26] A study in 2007 showed that 100% of the gynecologists questioned supported the idea of radiologists giving the results.[15]

Note

Patients' satisfaction with how they receive the diagnosis depends more strongly on "how" the diagnosis is given and less on the diagnosis itself.[7]

16.3 The Patient's Flow through the Department

We will now accompany a patient on her way through the department and observe the different stations she passes through. Each of these stations poses specific challenges in dealing with the patient. A checklist with useful tips and information is provided for each station.

16.3.1 Station 1: Registration

Note

You will never get a second chance for a first impression.

As the first point of contact, registration (▶ Fig. 16.1) is the showpiece of every department or practice. It gives the critical first impression of the department that the patient receives. The activity of the receptionists often goes unnoticed or barely noticed by the doctors and the personnel working with patient

Fig. 16.1 Reception.

examination and care. Mistakes that occur at the first contact can be compensated only with great effort or not at all.

The receptionists greet the patient, take care of the registration formalities, check the completeness of the documents, and prepare for the examination by inputting the data and explaining the procedure to the patient. The receptionists also commonly have the task of receiving incoming calls for appointments or regarding questions about the examination, and of then answering the question directly or redirecting the calls.

The receptionists must know the general examination procedure and the important indications and contraindications. Furthermore, they should be able to recognize emergency situations and competently answer commonly recurring questions. On the grounds of data protection and to safeguard patient privacy, medical matters should be discussed in such a manner that they cannot be overheard by everyone in the registration area.

In case of complicated technical questions or if a question is beyond the receptionist's ability to answer, then he or she should pass the issue on to the doctor or technologist. Good training for the front-desk personnel is therefore very important. This applies particularly if the staff member at the registration desk or switchboard has not had any formal medical training.

The workers at the registration desk are often insufficiently integrated into the practice procedures and information flow in the department because of their spatial, and at times professional, distance from patient examination. Respect and appreciation for the team members at the registration desk and their work should be self-evident. When staffing allows, it can be a good idea to have the registration personnel sit in on patient examinations. Conversely, "registration days" can be planned for the personnel involved exclusively with diagnostics. This increases mutual appreciation and the understanding of the work of others in the department.

The work at the registration desk is important and challenging, and it is not free of conflict. The receptionist cannot meet the expectations of every patient. For instance, the receptionist is at times called on to reschedule patients for an appointment, which can lead to angry or aggressive reactions. Telephone complaints from patients and referring physicians are first received by the registration workers. Targeted training is available to train the staff to handle such difficult situations professionally. This in turn provides the team members with "status," increases their professionalism, and supports job satisfaction, which ultimately benefits the entire department, but the patient in particular.

Department management should be involved with registration and the receptionists on an ongoing basis—giving feedback, asking about difficulties, recognizing the need for further training, and observing departmental procedures. Patient surveys can provide valuable feedback.

Practical Tips

Registration checklist:

- Maintain eye contact (if engaged on the telephone to someone else, nod and have at least short eye contact).
- Give a personal greeting.
- Maintain a polite conversational tone.
- When problems arise, attempt to find an amicable solution.
- Do not reproach or accuse (verbally or nonverbally).
- Get help in difficult situations.
- Maintain a businesslike manner.
- Protect privacy (data protection!)
- Give a clear explanation for further procedure.

16.3.2 Station 2: History and Physical Examination

Note

It is important to use a respectful, appreciative communication style.

History-taking (▶ Fig. 16.2) is usually carried out by the doctor on the basis of a history form that the patient fills out beforehand and other relevant documents.

In designing the history questionnaire, it is important to use clear, understandable language and an adequate type size. Avoid overloading the form; remember "Less is more!" Depending on the composition of the patient population, it may be worthwhile to have history forms available in foreign languages or with pictures to aid understanding.

Fig. 16.2 History-taking.

Not every patient can fill out a history questionnaire completely without mistakes. Patients frequently do not remember when or where their last examination took place. Many patients forget their reading glasses. (Tip: keep a selection of reading glasses on hand!). The back of the information sheet is often not noticed by the patient, and therefore not filled out. Previous studies from outside sources are forgotten or are "in the car." The practice personnel should treat such situations in a friendly and good-humored manner, without patronizing or reproachful reactions; avoid comments such as "But you know you have to fill the form out completely!", "You must know where your last examination was!", "Isn't it clear that we need the previous X-rays!".

Note

Gaps in the history should be filled out on the form jointly with the patient in conditions of privacy (not in the waiting room, where other patients could hear).

Patients who are already known to the practice do not need to fill out the complete history form. The patients wish to experience the sense that the team knows them and is prepared for them. The physician should inform himself or herself of the previous findings and course of treatment before the patient consultation. The physician can then immediately turn his or her total attention to the patient and her current situation.

Taking the history is commonly associated with the physical examination by the physician. One must take care that the time between the patient disrobing the upper body and the carrying out of the examination is kept as short as possible. During the history taking, even if this occurs when the examination is taking place, the upper body of the patient should be covered. Nonverbal signals during the examination, such as wrinkling the forehead, a possibly worried facial expression, or examining a particular point for a prolonged time, will be sensed and interpreted by the patients. The examiner should be aware of this. Describing and explaining as one proceeds can be helpful in relieving anxiety. When a man is conducting the examination, it is advisable to have a female staff member in attendance in the examination room for legal reasons.

Practical Tips

Checklist for history-taking and physical examination:

- Have a clearly structured history form, avoiding unnecessary details.
- Make foreign-language history forms available when needed.
- When problems develop, offer friendly support without reproof.
- Prepare for the history-taking interview ahead of time.
- Avoid duplicate questions.
- Give the patient sufficient opportunity to bring up her concerns.
- Protect patient dignity and personal space (keep the upper body covered as much as possible).
- Explain any deviations from the usual examination procedure.

16.3.3 Station 3: The Diagnostic Procedure

Note

The way the technologist communicates with the patient has a large impact on the quality of the examination.

Care of the Patient during the Diagnostic Procedure

After specifications from the doctor, the diagnostic examinations of the breast (X-ray mammography and MRI of the breast, ▶ Fig. 16.3) are carried out independently by the radiologic technologist. The technologist introduces herself to the patient by name. Changing the examiner during the examination should be avoided, as should disturbances, such as people entering the examination room.

Giving the patient a comprehensive and truthful explanation is a prerequisite for the test to be performed as smoothly as

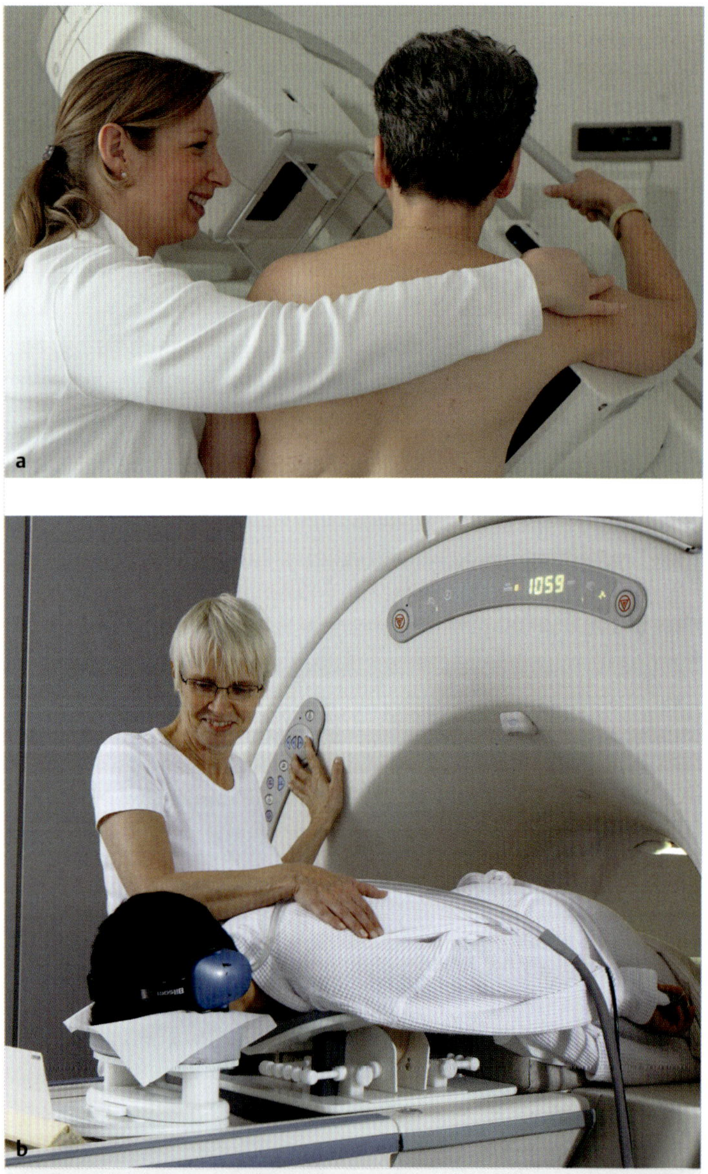

Fig. 16.3 Instrumental diagnostic procedures. (a) Mammography. (b) Breast MRI.

possible and for it to be experienced by all participants with as little stress as possible. The course of the examination is explained to the patient in clear, understandable language.[16,19] The reason for any onerous measures or situations, such as breast compression, positioning, and injection of contrast agent (for an MRI examination), should be explained so as to elicit the patient's maximal cooperation. The course of the examination should be realistically represented, avoiding downplaying or exaggeration.

Patients' anxieties and concerns must be taken seriously. Everyone involved is aware that the patient finds herself in an exceptional situation and may possibly not be in the position to understand and implement the informational content of a request. Therefore, if the patient does not comply with the technologist's requests immediately or in the desired manner, the technologist should try to help the patient by repeating the instructions without rancor or expressing them in other words.

The same principles apply for the diagnostic examination as were given above for the physical examination: The time the patient waits for the examination with the upper body disrobed should be kept as short as possible. It is important to maintain a comfortable temperature and ventilation in the examination rooms to guarantee good test conditions, both for the patient and for the personnel[19].The technologist's responsibility does not end with the completion of the diagnostic examination but only after postprocedural patient care and preparation of the image material for the reviewing radiologist, including all relevant ancillary information. Patients should never be "forgotten" in the changing room.

Practical Tips

Checklist for the instrumental diagnostic examination:

- Provide a comfortable conversational atmosphere and room ambience.
- Avoid disturbances and changes of examiners.
- Speak *with* the patient, not *about* the patient.
- Use understandable language.
- Observe and protect the patient's privacy, dignity, and personal space.
- Provide postprocedural care for the patient.
- Provide information about what happens next.

Fig. 16.4 Informed-consent discussion for the breast MRI.

Fig. 16.5 Patient transfer.

Care of the Patient during a Breast MRI Examination

Because of the organizational, personnel, and time commitments required, the course of events involved in the breast MRI examination has considerable significance for the department, but also for the patient. The MRI unit has an imposing effect because of its size, and the positioning of the patient within the unit is uncomfortable for many patients. With this in mind, the course of events during a breast MRI examination should therefore be described in detail and its special features elucidated.

In the informed-consent discussion (▶ Fig. 16.4), the physician informs the patient of the indications for the examination and of the procedure that takes place during the examination. The doctor answers the patient's questions and treats anxieties empathetically (claustrophobia, for example). Contraindications are excluded.

The physician reports important information concerning the examination to the technologist, so that the technologist is well informed about the patient and can adequately address special situations and take the appropriate measures; such situations might include, for example, anxious patients, claustrophobia, special considerations in positioning. and venepuncture.

After the patient talks with the doctor in the reception area, the technologist introduces herself or himself to the patient by name and accompanies the patient to the changing room (▶ Fig. 16.5). Ideally, the technologist accompanies and takes care of the patient during all subsequent examinations. It is particularly comforting for the patient if a bath robe is made available to her for the duration of the examination.

During the preparatory discussion (▶ Fig. 16.6), the technologist acquaints the patient with the details of the examination procedure (positioning, the duration and course of events of the examination, sound intensity of the machine, possible sensation of coolness or feeling of fluid flowing during the injection of contrast material). Previous information from the doctor's consultation flows into this conversation so that the technologist can introduce or implement appropriate measures, such as sedation.

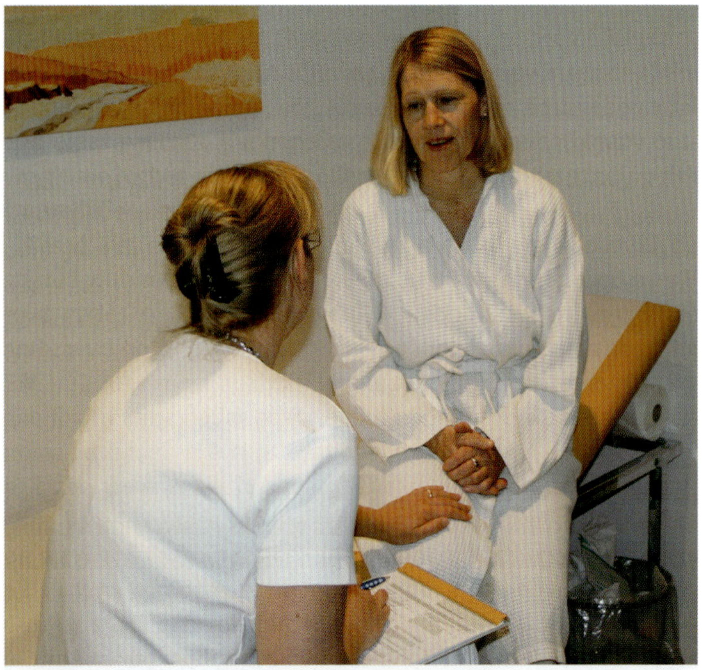

Fig. 16.6 Preparatory discussion.

The breast imaging is performed in the supine position. Relaxed and comfortable positioning is crucial to obtaining optimal image quality (▶ Fig. 16.7a). The alarm bulb gives the patient the option of interrupting the examination in case of emergency.

The technologist announces the sequences, the remaining exposure time, and the administration of contrast material over the headphones. During the procedure, the patient can listen to a personal choice of music (▶ Fig. 16.7b). The headrest offers the possibility to view a postcard or photograph by way of a mirror (▶ Fig. 16.7c, d). The personal care given supports the patient, provides security, reduces stress and anxiety, and engenders trust. Sedated patients are accompanied back to the waiting area after the examination. After the study, the technologist advises the patient of the further procedure in the department.

Practical Tips

Breast MRI checklist:

- Provide understandable explanation and preparation for the examination procedure.

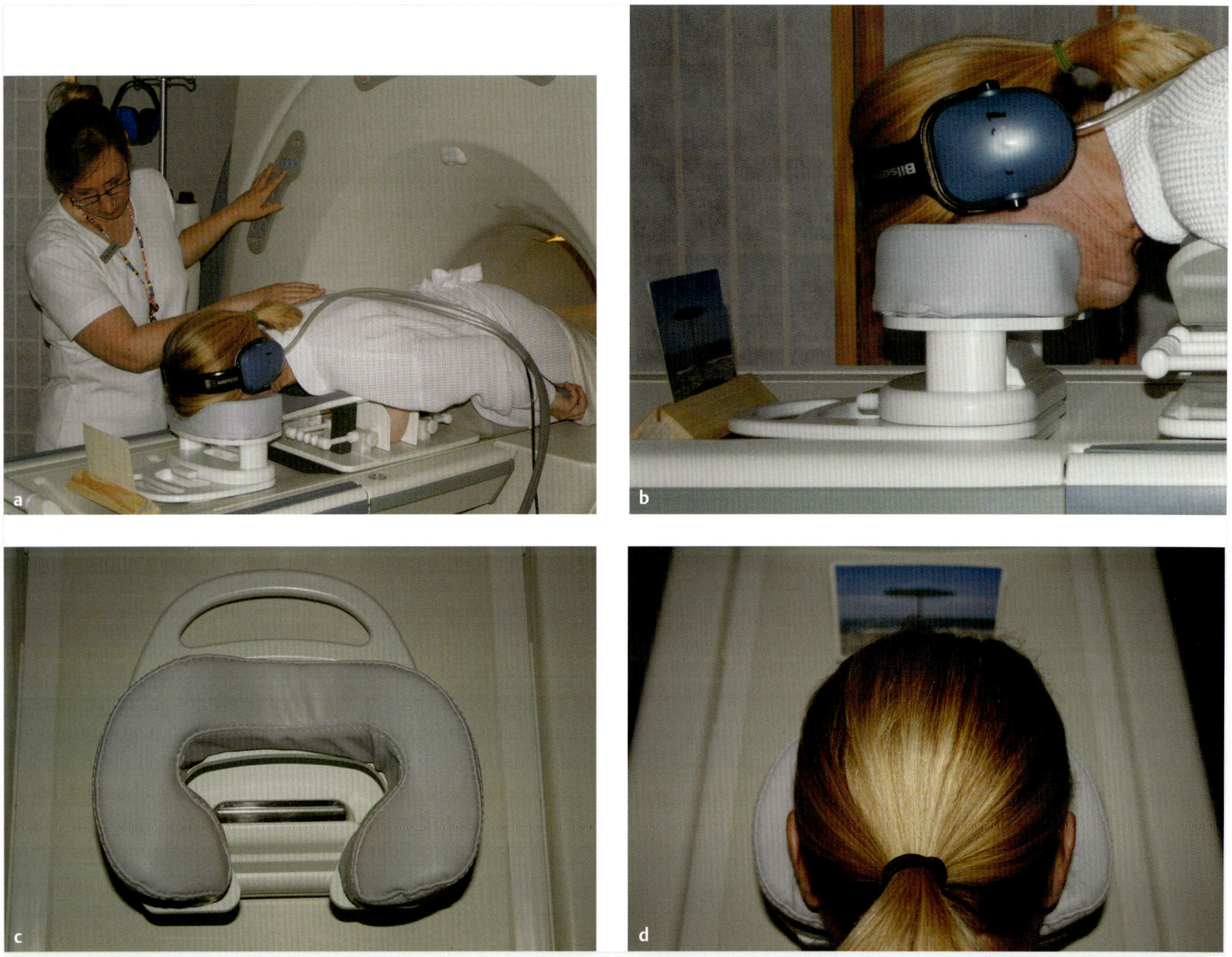

Fig. 16.7 Positioning for breast MRI. (a) Positioning and assistance. (b) Relaxing with "good music." (c) Head support with mirror. (d) Viewing a motif card.

- Handle patients' anxiety with empathy.
- Provide the patient with time checks during the test.
- Take particular care of sedated patients; accompany them to the waiting room.

16.3.4 Station 4: Communication of Results and Concluding Consultation

Note

Use of a multilevel approach when disclosing difficult diagnoses enables an individual response to the patient's needs.

In breast diagnostics, communicating the test results is usually done directly after the examination.[8,9] To prevent patient anxiety, any deviations from the usual departmental procedure that the patient is accustomed to must be previously announced and substantiated (e.g., missing previous images, a complex examination with extensive findings).

The results should be given in an environment that is as free from disturbance as possible. This provides a better opportunity to respond to emotional reactions (▶ Fig. 16.8). It is important to use a clear choice of words that are understandable by the patient.[7,16] The patient needs to have sufficient opportunity to ask questions. When meeting to discuss complex findings, it is therefore often a good idea to arrange a separate consultation appointment. One must schedule adequate time when the patient has to be informed of a serious diagnosis.[18]

If the patient would rather discuss the results with the treating gynecologist or family doctor, this wish must be respected. In such cases, however, the radiologist should inform the responsible doctor personally if there are pathologic findings that must be further evaluated, to ensure a prompt diagnostic work-up.

Note

Informing the patient of a serious diagnosis, especially a malignant disease, represents a particularly stressful situation for both the doctor and the patient.

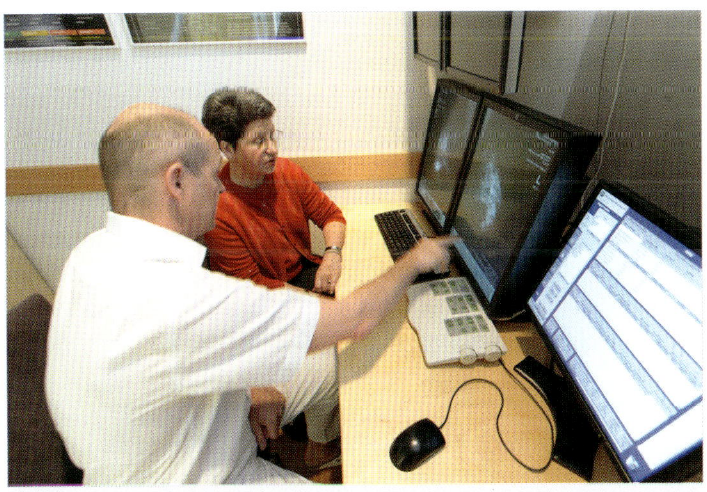

Fig. 16.8 Communicating the results.

Baile et al recommend a six-step protocol when informing the patient of a malignant diagnosis.[2] This multilevel approach enables the doctor to listen to and respond to the patient's specific previous knowledge, her personal experiences, and her need for information.[21]

Recommendations for disclosing a malignant disease:

1. Create an appropriate setting:
   - Undisturbed conversational atmosphere.
   - Adequate timeframe.
2. Assess the patient's previous knowledge (What does she already know? What does she think about her illness?).
3. Ascertain the patient's need for information (What does the patient want to know?).
4. Give the diagnostic information to the patient.
5. Show an empathetic response to the patient's emotions.
6. Summarize the situation and determine further strategy.

Radiologists are not properly prepared during their specialty training to deal with disclosing difficult diagnoses to patients, although this is a scenario that must be dealt with not infrequently in breast diagnostics. Many authors therefore advocate a special training program for radiologists to prepare them for such situations in a focused manner.[1,2,6,10,20,25]

16.4 Summary

Empathetic interaction with the patient and adequate psychosocial support before, during, and after the examination increase patient satisfaction. In this regard, structural and procedural features and the verbal and nonverbal conduct of the medical and nonmedical staff members have to be considered. Communicating results is a particularly sensitive part of the doctor–patient relationship and requires special care.

Every member of the diagnostic unit team contributes to the development of a successful relationship with the patient and to a good working atmosphere.

The quality of the communication and the manner of the relationship with the patient and within the team are an expression of the department or practice culture, which can be sensed by every patient. Competent and empathetic conduct, honesty, reliability, openness to new ideas and a respectful feedback culture promote a departmental climate in which the patients, with all their cares and difficulties, feel that they are in good hands.

Bibliography

[1] Alderson PO. Customer service and satisfaction in radiology. AJR Am J Roentgenol. 2000; 175(2):319–323

[2] Baile WF, Buckman R, Lenzi R, Glober G, Beale EA, Kudelka AP. SPIKES-A six-step protocol for delivering bad news: application to the patient with cancer. Oncologist. 2000; 5(4):302–311

[3] Bassett LW, Bomyea K, Liu S, Sayre J. Communication of mammography results to women by radiologists: attitudes of referring health care providers. Radiology. 1995; 195(1):235–238

[4] Basu PA, Ruiz-Wibbelsmann JA, Spielman SB, Van Dalsem VF , III, Rosenberg JK, Glazer GM. Creating a patient-centered imaging service: determining what patients want. AJR Am J Roentgenol. 2011; 196(3):605–610

[5] Cardenosa G, Eklund GW. Rate of compliance with recommendations for additional mammographic views and biopsies. Radiology. 1991; 181(2):359–361

[6] Carney PA, Kettler M, Cook AJ, et al. An assessment of the likelihood, frequency, and content of verbal communication between radiologists and

women receiving screening and diagnostic mammography. Acad Radiol. 2009; 16(9):1056–1063
[7] Dolan NC, Feinglass J, Priyanath A, Haviley C, Sorensen AV, Venta LA. Measuring satisfaction with mammography results reporting. J Gen Intern Med. 2001; 16(3):157–162
[8] Hammond I, Franche RL, Black DM, Gaudette S. The radiologist and the patient: breaking bad news. Can Assoc Radiol J. 1999; 50(4):233–234
[9] Hoffman NY, Janus J, Destounis S, Logan-Young W. When the patient asks for the results of her mammogram, how should the radiologist reply? AJR Am J Roentgenol. 1994; 162(3):597–599
[10] Leclère J, Ollivier L, Dolbeault S, Neuenschwander S. [Communication between radiologist and cancer patients]. J Radiol. 2006; 87(2 Pt 1):99–104
[11] Levin KS, Braeuning MP, O'Malley MS, Pisano ED, Barrett ED, Earp JA. Communicating results of diagnostic mammography: what do patients think? Acad Radiol. 2000; 7(12):1069–1076
[12] Levitsky DB, Frank MS, Richardson ML, Shneidman RJ. How should radiologists reply when patients ask about their diagnoses? A survey of radiologists' and clinicians' preferences. AJR Am J Roentgenol. 1993; 161(2):433–436
[13] Lind SE, Kopans D, Good MJ. Patients' preferences for learning the results of mammographic examinations. Breast Cancer Res Treat. 1992; 23(3):223–232
[14] Liu S, Bassett LW, Sayre J. Women's attitudes about receiving mammographic results directly from radiologists. Radiology. 1994; 193(3):783–786
[15] Lorch H, Scherer P. Disclosure of diagnosis in ambulatory radiology practice: expectations of patients and referring physicians [in German]. Rofo. 2007; 179(10):1043–1047
[16] Lown BA, Roy E, Gorman P, Sasson JP. Women's and residents' experiences of communication in the diagnostic mammography suite. Patient Educ Couns. 2009; 77(3):328–337
[17] Mehrabian A, Ferris SR. Inference of attitudes from nonverbal communication in two channels. J Consult Psychol. 1967; 31(3):248–252
[18] Miller RG. Breast cancer screening: can we talk? J Gen Intern Med. 2001; 16 (3):206–207
[19] Morris KJ, Tarico VS, Smith WL, Altmaier EM, Franken EA , Jr. Critical analysis of radiologist-patient interaction. Radiology. 1987; 163(2):565–567
[20] Ollivier L. Communication between radiologists and patients with cancer. Cancer Imaging. 2005; 5:114–115
[21] Pearlman DN, Rakowski W, Clark MA, et al. Why do women's attitudes toward mammography change over time? Implications for physician-patient communication. Cancer Epidemiol Biomarkers Prev. 1997; 6(6):451–457
[22] Reiser SJ. Words as scalpels: transmitting evidence in the clinical dialogue. Ann Intern Med. 1980; 92(6):837–842
[23] Robertson CL, Kopans DB. Communication problems after mammographic screening. Radiology. 1989; 172(2):443–444
[24] Schreiber MH, Winslade WJ. Rights, roles, and relationships in radiology. Radiology. 1987; 163(1):269–270
[25] Schreiber MH, Leonard M , Jr, Rieniets CY. Disclosure of imaging findings to patients directly by radiologists: survey of patients' preferences. AJR Am J Roentgenol. 1995; 165(2):467–469
[26] Schreiber MH. Direct disclosure by radiologists of imaging findings to patients: a survey of radiologists and medical staff members. AJR Am J Roentgenol. 1996; 167(5):1091–1093
[27] Schulz von Thun F. Miteinander reden. Störungen und Klärungen. Band 1. Reinbek bei Hamburg: Rowohlt; 1981: 44ff
[28] Watzlawik P, Beavin JH, Jackson DD. Menschliche Kommunikation. Bern: Huber; 1969

Index

E

F

G

H

I

L

M